Indeterminate Structural Analysis

Indeterminate Structural Analysis

K.U. Muthu
Dean
BRCE Bangalore

H. Narendra
Associate Professor
MSRIT, Bangalore

Maganti Janardhana
Associate Professor
JNTU Hyderabad

M. Vijayanand
Assistant Professor
MSRIT, Bangalore

I.K. International Publishing House Pvt. Ltd.

NEW DELHI

Published by
I.K. International Publishing House Pvt. Ltd.
S-25, Green Park Extension
Uphaar Cinema Market
New Delhi–110 016 (India)
E-mail: info@ikinternational.com
Website: www.ikbooks.com

ISBN 978-93-82332-60-2

Reprint 2017

Published by Krishan Makhijani for I.K. International Publishing House Pvt. Ltd., S-25, Green Park Extension, Uphaar Cinema Market, New Delhi–110 016 and Printed by Rekha Printers Pvt. Ltd., Okhla Industrial Area, Phase II, New Delhi–110 020.

Preface

Structural analysis of elements including indeterminate structures is an in depth study of the response of structures to the specified applied loads. The response is reflected in the computation of stress resultants and estimation of deformations.

Ever since Castigliano published a book on the above theme in 1879, major developments of structural analysis took place and several milestones have been reached. Axel Bendixson (1914), Maney (1913) developed the slope deflection method for the analysis of continuous beams and frames. Hardy Cross (1930) has presented the moment distribution method— an important contribution of all times. Southwell (1935) proposed the relaxation method. Gasper Kani (1947) developed a self-corrective, iteration method for the analysis of beams and frames. There are numerous contributors for the development of structural analysis, notable among them are Clapeyron, Muller Breslau and Klouczeck.

The computer oriented structural analysis has made a revolution in the analysis of structures. Though the computer oriented method of structural analysis gives rapid results, it is very important to validate the output using the results of classical methods, viz., moment distribution method or Kani's method. These manuals help to arrive at the preliminary designs. Incessant efforts and dedication of pioneers resulted in several software packages which are in great use.

In the present book, an attempt has been made to explain basics of indeterminate structural analysis. The book has been designed to cater to the needs of the undergraduate students and design engineers. The classical methods, viz., slope deflection, moment distribution and Kani's method has been explained in the beginning which forms the basis of analysis, followed by the flexibility method and stiffness method using system approach. The introduction of system stiffness approach in a systematic manner makes the student to develop an interest in analyzing the complex structures using direct stiffness approach later. The approximate method of analysis of multistorey frames subjected to lateral loads and vertical loads is presented herewith. The approximate analysis includes the Portal method, Cantilever method and Factor method for lateral loads. Also, the multistorey frames subjected to vertical loads have also been analyzed using point of inflextion method and substitute frame method. Analysis of two hinged arches and introduction to structural dynamics have been added.

A number of illustrative examples are presented with various degrees of difficulty which would make the student to understand and appreciate the subject in depth. We hope that this book will lay firm foundation for the design subjects, viz., design of reinforced concrete, prestressed concrete and steel structures.

Authors

Preface

The analysis of elements including indeterminate structures is an in-depth study of response of structures to the specified applied loads. The response is achieved by the computation of stress resultants and estimation of deformations.

Ever since Castigliano published a book on the above theme in 1879, major developments of structural analysis took place and several milestones have been reached. Axel Bendixson (1914), Maney (1915) developed the slope-deflection method for the analysis of continuous beams and frames. Hardy Cross (1930) has presented the moment distribution method— an important contribution of all times. Southwell (1935) proposed the relaxation method. Gasper Kani (1947) developed a self-corrective, iteration method for the analysis of beams and frames. There are numerous contributors for the development of structural analysis notable among them are Clapeyron, Muller Breslau, Macaulay etc.

The computer oriented structural analysis has made a revolution in the analysis of structures. Though the computer oriented method of structural analysis gives rapid results, it is very important to validate the output using the results of classical methods, viz., moment distribution method or Kani's method. These manuals help to arrive at the preliminary designs. Incessant efforts and dedication of pioneers resulted in several software packages which are in great use.

In the present book, an attempt has been made to explain basics of indeterminate structural analysis. The book has been designed to cater to the needs of the undergraduate students and design engineers. The classical methods, viz., slope deflection, moment distribution and Kani's method has been explained in the beginning which forms the basis of analysis, followed by the flexibility method and stiffness method using system approach. The introduction of system stiffness approach in a systematic manner makes the student to develop an interest in analyzing the complex structures using direct stiffness approach later. The approximate method of analysis of multistorey frames subjected to lateral loads and vertical loads is presented herewith. The approximate analysis includes the Portal method, Cantilever method and Factor method for lateral loads. Also, the multistorey frames subjected to vertical loads have also been analysed using point of inflextion method and substitute frame method. Analysis of two hinged arches and introduction to structural dynamics have been added.

A number of illustrative examples are presented with various degrees of difficulty which would make the student to understand and appreciate the subject in depth. We hope that this book will lay firm foundation for the design subjects, viz., design of reinforced concrete, prestressed concrete and steel structures.

Authors

Acknowledgements

The authors respectfully express their homage to the founder chairman Late Sri Dr M S Ramaiah. The authors gratefully acknowledge the encouragement extended by chairman of GEF Dr M R Jayaram, Honourable Director MSRIT and Vice Chairman Sri M R Seetharam, Member of Legislative Council, Karnataka, Secretary GEF Sri M R Ramaiah and Chief executive (Technical) Sri S M Acharya.

The authors express their gratitude to chairman of BRCE Dr Majid A A Sabha, Vice Chairman Dr Mustaq Ahmed and Dr R Noor Ahmed, Principal of BRCE, Mr Ismail, Financial Controller and Mr Anantharamaiah DGM of BCRE for their encouragement.

The authors express their respect and gratitude in a special way to Prof. Prakash Desayi Dsc of Indian Institute of Science, Bangalore and Dr A Meher Prasad, Dr Devdas Menon, Dr A R Santhakumar (former Emeritus Professor), Dr B Nageswara Rao of IIT Madras and Dr Alapati Prasada Rao of JNT University, Hyderabad.

The authors express their indebtness to their mentors Dr A Krishna Sarma, former Principal MSRIT and BRCE, Dr V Ramarajan, former principal MVJCE, Dr K Rajanikanth, former Principal, MSRIT, Dr S Y Kulkarni, Principal, MSRIT, Dr Sureshkumar, Registrar Academic, MSRIT, Dr Hashem Mattarneh, Jerash University, Jordan, Dr Azmi Ibrahim and Dr Hadariah Bahron, UiTM, Malaysia, Dr Nicolae Angelescu, Vallahia University, Romania, Dr Gorgy Balazs, BME, Budapest, Hungary, Dr Veeraragavan, Dr Ravindra Gettu, Dr V Kalyanaraman, Dr K Rajagopal, Dr C V R Murty, Dr Amlan K Sengupta, Dr S R Satishkumar, Dr C Lakshmana Rao and Dr S T G Ragukanth IIT Madras, Dr R Sundara Rajan, former Principal GCT Coimbatore, Dr J V Ramasamy, Dean PSG Tech, Coimbatore, Dr M G Rajendran, Dr Mercy Shanthi of Karunya University, ER AT Samuel, Director, STUP Consultants Pvt Ltd, Bangalore, Er R Jeevanandam, Design Consultant, Chennai. The authors thank the colleagues at JNTU Hyderabad, JNTU Kakinada, JNTU Anantapur, Andhra University, Osmania University, NIT - Warangal, SV University - Tirupati, Anna University - Chennai and friends Dr P Robin Davis, Dr Pradip Sarkar of NIT Rourkela, Dr R Shiv Shanker, Dr P Dinakar of IIT Bhubaneshwar, Dr Vasant Matsagar of IIT Delhi, Sri Perla Rama Mohana Rao and Sri Vurugouda Raju of IIT, Madras for their moral support and encouragement.

Thanks are due to senior Professors and authorities in structural engineering, Dr N Krishna Raju, former Professor, MSRIT, Dr D L Prabhakara, Director, Sayadri College of Engineering, Mangalore, Dr Shakeebur Rehman, Dean of VTU, SJCE, Mysore, and Dr Karisidappa, Principal, Government Engineering College, Hassan, Dr Premkumar, RITM, Bangalore. Special thanks are due to Dr M U Aswath, Professor,

BIT, Bangalore, Dr Avanti, Dr P J Sasturkar, PDAC Gulbarga and Dr Sunil Kumar Tengli, JSSCE, Mauritius, Dr Arunachalam, Dr Arul Mary, Sri Ravichandran of TCE and Smt Jayaprabha, VCE, Madurai.

The authors are indebted to many colleagues in various universities and institutions. It gives immense pleasure to record the encouragement received from Dr Harimurthy, Prof Jagadeesha Chandra of MSRIT, Dr K Amarnath, Dean, Oxford College of Engineering, Bangalore, Dr C G Puttappa, Dr R Prabhakara, Dr E T Arasu, Dr S M Naik, Dr K P Nagaraj, Dr Umadevi, Dr Jagadeesha Kumar of MSRIT, Dr Udaya Kumar, Dr Radha Krishna, Dr H R Manjunath from RVCE, Dr N S Kumar, GCE, Bangalore, Sri Raghupathi and Sri S Ravikumar, Sir MVIT, Dr Ramalinga Reddy, Professor, RITM, Bangalore and Sri B Ravishankar Gowda, Structural Consultant, Bangalore.

Special thanks are due to Dr Raghavendra V Kulkarni, Professor ECE, Dr Prem Swaroop Reddy, RITM, Bangalore, Sri Harish and Sri R Mourougane of MSRIT for their help, suggestions, constructive ideas and support during the preparation of the manuscript. Sincere thanks are due to Prof. Venkat Raju, Prof. Venkatesh, Prof. Sashikumar, Prof. Dinesh, Dr Ravikumar, Prof. Sumitra Devi, Dr Abrar Ahmed of BRCE, Bangalore who had helped in many ways.

Special thanks to Prof. S Subramanya, Prof. Ramesh S Manoli, Prof. M N Shiva Kumar, Prof. C S Vijay Kumar, Prof. R S Chikkana Gowder and Prof. C P Anila Kumar of Bapuji Institute of Engineering and Technology, Davangere, Prof. B D V Chandra Mohan Rao and Prof. K Ramujee of VNRVJIET, Hyderabad, for their moral support during the preparation of the manuscript.

A special thanks to our beloved ones. Sincere thanks to Dr K U Muthu's wife Smt M Santhanalakshmi and daughter Mrs Anitha Muthu for constant support. A special note of thanks to Mrs Narendra and his son. A special word of thanks to Dr Maganti Janardhana's parents Sri Maganti Narasimha Rao (former Principal, Andhra Jateeya, Kalasala, Machilipattnam) and Smt Maganti Sree Rangakumari and his cousins Sri Maganti Naga Raju, Dr Maganti Ravindra Krishna for extending moral support during the preparation of the script.

Many more thanks to Mr Mahesh, Miss Khamruz Zama and Miss Namita Amadalli for their assistance.

We thank the authors of the textbooks and research papers which were liberally referred in preparing the text. The authors thank all those who assisted in preparation of this book.

Authors

Course Document

Department of Civil Engineering

Course Name: Indeterminate Structural Analysis	Code: CV52
Total Contact Hours: 56	Semester: 05

Course Objectives

The main objectives of this course are to provide

- Analysis of indeterminate beams and portal frames upto three degree of freedom to determine the end moments.
- An insight to the matrix analysis of structures.
- Analysis for axial forces in indeterminate trusses.
- Analysis for radial thrust, normal force and shear force in two hinged arches.
- An introduction to study the characteristics of structural system under dynamic loads with single degree freedom.

Course Learning Objectives (CLOs)

CL01	At the end of the course the student should be able to determine the moment in indeterminate beams and frames having variable moment of inertia and subsidence using slope deflection method.
CL02	Determine the moment in indeterminate beams and frames of no sway and sway using moment distribution. Determine the end moments of symmetrical structures by cantilever moment distribution method.
CL03	Construct the bending moment diagram for beams and frames using Kani's method.
CL04	Construct the bending moment diagram for beams and frames by system flexibility method.
CL05	Analyze the beams and indeterminate frames by system stiffness method.
CL06	Estimate the bending moments and axial forces in multistorey frames by approximate methods.
CL07	Calculate the internal forces of indeterminate trusses by energy approach.
CL08	Determine the internal forces in two hinged arches.
CL09	Describe the characteristics of vibration of single degree freedom system.

06CV52 Indeterminate Structural Analysis

Topic No.	Topic Name	Hours
1.	Slope Deflection Method	08
2.	Moment Distribution Method	08
3.	Kani's Method	06
4.	Flexibility Method	07
5.	Stiffness Method	07
6.	Approximate Methods of Analysis	06
7.	Indeterminate Trusses	04
8.	Two Hinged Arches	04
9.	Basic Principles of Structural Dynamics	06
	Total Hours	**56**

Course Code & Title:	CV52 Indeterminate Structural Analysis	
Topic Name: Slope Deflection Method	Topic No. 1	Hours: 8

Topic Level Objectives (TLOs)

1. Identification of kinematic redundant to be determined.
2. Introduce the concept of force-displacement relation.
3. Formulation of equilibrium equation for each kinematic redundant using free body diagram and solving for the same.
4. Evaluation of member end moments using the kinematic redundants.
5. Construction of member bending moment, shear force and axial force diagrams.

Intended Learning Outcome (ILO)

After successful completion of the topic, the learner will be able to

- Demonstrate the conversion of given loading on the span to the end moments.
- Recognize the kinematic redundants.
- Formulate the equilibrium equation for each redundant.
- Explain the joint equilibrium and shear equations.
- Construct the shear force and bending moment diagram.

Course Code & Title:	CV52 Indeterminate Structural Analysis	
Topic Name: Moment Distribution Method	Topic No. 2	Hours: 8

Topic Level Objectives (TLOs)

After successful completion of the topic, the learner will be able to

1. Introduce the concept of member stiffness, joint distribution factor and computation of fixed end moments.
2. Introduce the concept of out of balance moments and balancing moments.
3. Introduce carry over moments and complete the first cycle of moment distribution.
4. Introduce the releasing, balancing and coarse over moments and then balancing which completes the second cycle of moment distribution.
5. Provide a method of analysis of sway and nonsway frames including the settlement of supports.

Intended Learning Outcome (ILO)

After successful completion of the topic, the learner will be able to

- Demonstrate the ability to convert the loading into fixed end moments.
- Explain the member stiffness, distribution factor and carry over factor.
- Analyze the continuous beams and indeterminate frames using Hardy Cross moment distribution method.
- Explain the joint ratios and the sway angles.
  (a) Analyze the symmetrical frames with Naylor's cantilever moment distribution method.
  (b) Construct the bending moment diagram and shear force diagram for indeterminate beams and frames.

Course Code & Title:	CV52 Indeterminate Structural Analysis	
Topic Name: Kani's Method	Topic No. 3	Hours: 6

Topic Level Objectives (TLOs)

1. Introduce the concept of rotation contribution and rotation factor.
2. Introduce the concept of displacements contribution and displacement factor.
3. Introduce the treatment for hinged ends.
4. Provide a method of analysis of nonsway and sway frames.

Intended Learning Outcome (ILO)

After successful completion of the topic, the learner will be able to

- Demonstrate the ability to convert the given span and loading to fixed end moments.

- Demonstrate the ability to compute the rotation and displacement factors.
- Determine the storey shears and moments.
- Treatment of frames and columns with heights and unequal heights.
- Construct the bending moment diagram and shear force diagram of indeterminate beams and frame.

Course Code & Title:	CV52 Indeterminate Structural Analysis	
Topic Name: Flexibility Method	Topic No. 4	Hours: 7

Topic Level Objectives (TLOs)

1. Introduce the flexibility method and establish the relationship between flexibility and stiffness matrices.
2. Introduce the concept of flexibility coefficients.
3. Introduce the flexibility method of analysis of structures.
4. Provide a method of analysis of continuous beam indeterminate frames and trusses.

Intended Learning Outcome (ILO)

After successful completion of topic the learner will be able to

- Recognize the importance of analysis of complex structures using matrix analysis of structures.
- Demonstrate the ability to differentiate flexibility and stiffness system approach.
- Formulate the system flexibility matrix.
- Analyse the continuous beam with and without support settlement.
- Analyse the continuous beam with more releases by flexibility system approach.
- Analysis of symmetrical and unsymmetrical frames using flexibility system approach.
- Determine the forces in trusses.
- Construct the bending moment diagram, shear force diagram for continuous beams, rectilinear frames and portal frames.

Course Code & Title:	CV52 Indeterminate Structural Analysis	
Topic Name: Stiffness Method: System Approach	Topic No. 5	Hours: 7

Topic Level Objectives (TLOs)

1. Introduce the stiffness method of analysis of structures.
2. Introduce the concept of stiffness coefficients.
3. Provide a method of analysis of continuous beams, indeterminate frames and trusses.

Intended Learning Outcome (ILO)

After successful completion of the topic, the learner will be able to

- Explain the superposition of displacement and the formulation of equilibrium equations.
- Formulate the system stiffness matrix.
- Analyse the continuous beam with and without support settlement.
- Analysis of continuous beam with more releases by stiffness system approach.
- Analysis of rectilinear frames by stiffness system approach.
- Analysis of symmetrical and unsymmetrical frames by stiffness approach.
- Determine the axial force in the truss by stiffness approach.
- Construct the bending moment diagram and shear force diagram for continuous beams, rectilinear frames and portal frames.

Course Code & Title:		CV52 Indeterminate Structural Analysis
Topic Name: Approximate Method of Analysis	Topic No. 6	Hours: 6

Topic Level Objectives (TLOs)

1. Introduce the concept of approximate methods and their importance.
2. Provide a method of analysis of statically indeterminate beams using inflexion points.
3. Provide a method of analysis of rectilinear frames and portal frames subjected to horizontal and vertical loads.
4. Provide substitute frame analysis multistorey frames subjected for vertical loads and other approximate analysis for multistorey frames subjected to lateral loads.
5. Provide a method of analysis of multistorey frames of trussed portals and mill bents.
6. Provide a method of analysis of indeterminate trusses.

Intended Learning Outcome (ILO)

After successful completion of the topic, the learner will be able to

- Distinguish between approximate methods and classical method of analysis of structures.
- Explain the location of inflexion points to make statically indeterminate structures to determinate structures.
- Analyse the continous beams and indeterminate frames.
- Analyse the multistorey frames subjected to lateral loads.
- Demonstrate the ability to analyse the frame by substitute frame method.
- Analyse the trussed portals and mill bents.

Course Code & Title:	CV52 Indeterminate Structural Analysis	
Topic Name: Indeterminate Trusses	Topic No. 7	Hours: 04

Topic Level Objectives (TLOs)

1. Introduce the concept of strain energy and application of Castigliano's theorem.
2. Provide a method of analysis of indeterminate truss of one degree of redundancy/ two degrees of redundancy.
3. Introduce the concept of lack of fit, stresses due to change of temperature and effect of variation in cross-sectional area.

Intended Learning Outcome (ILO)

After successful completion of the topic, the learner will be able to

- Determine the axial forces in the trusses strain energy method.
- Demonstrate the ability to analyse the indeterminate truss including the secondary effects, viz. lack of fit and change in temperature.
- Demonstrate the ability to analyse the truss with two degrees of freedom by different approximate methods.
- Calculate the internal forces in trusses with sectional variations.

Course Code & Title:	CV52 Indeterminate Structural Analysis	
Topic Name: Two Hinged Arches	Topic No. 8	Hours: 04

Top Level Objectives (TLOs)

1. To introduce the arch action and its significance.
2. Develop a general equation to determine the horizontal thrust in a two hinged arch.
3. Provide an analysis for two hinged segmental arch by integration method and graphical integration method.
4. Provide an analysis of two hinged parabolic arches.
5. Introduce the analysis of linear arch.
6. Provide an analysis for secondary stresses in two hinged arch.
7. Introduce the concept of ILD for two hinged parabolic arches.

Intended Learning Outcome (ILO)

After successful completion of the topic, the learner will be able to

- Explain the differences between beam action and arch action.
- Calculate the horizontal thrust in a two hinged arch by integration method.
- Calculate the horizontal thrust in a two hinged arch by graphical summation method.

- Determine the horizontal thrust in a linear arch.
- Calculate the secondary stresses for temperature effects, elastic shortening and yielding of supports.
- Explain the application of ILD for the two hinged arches.

Course Code & Title:	CV52 Indeterminate Structural Analysis	
Topic Name: Basic Principles of Structural Dynamics	Topic No. 9	Hours : 6

Top Level Objectives (TLOs)

1. Introduce the types of analysis, loading, solution procedures for dynamic problems, spring element and degrees of freedom.
2. Provide tan analysis to formulate the motion of the vibrating systems.
3. Provide an analysis to predict the dynamic response of linear single degree of freedom systems subjected to initial excitations.
4. Introduction to forced vibrations.

Intended Learning Outcome (ILO)

After successful completion of the topic, the learner will be able to

- Explain the basic principles of vibrations, periodic and aperiodic motion, harmonic and nonharmonic motion, period and frequency.
- Explain the differences between free vibration and forced vibration.
- Analyse the response ofsingle degree of freedom un-damped system subjected to initial excitations.
- Analyse the response of single degree of freedom viscously damped system subjected to initial excitations.

Course Articulation Matrix

Course Learning	Activities	Programme Outcomes										
		a	b	c	d	e	f	g	h	i	j	k
CL01	2, 5, 6, 7, 8, 11	H	H		H		M		H		L	H
CL02	2, 5, 6, 7, 8, 11	H	H		H		M		H		M	H
CL03	2, 5, 6, 7, 8, 11	H	H		M		M		H			H
CL04	2, 5, 6, 7, 8, 11	H	H				H		H		L	H
CL05	2, 5, 6, 7, 8, 11	M	M				H		H		M	H
CL06	1, 3, 5, 6, 7, 8, 10, 11	M	M				H		H		M	H
CL07	2, 5, 6, 7, 8, 11	H	H				H		H		M	H
CL08	2, 5, 6, 7, 8, 11, 13	H	H				H		H		M	H
CL09	1, 2, 5, 6, 7, 8, 10, 11	H	H	M		M	H		H		M	H

Activities to be selected

1. Course Seminar
2. Course Project
3. Case Studies
4. Technical Events
5. Assignments
6. Quizzes
7. Internal Assessment Tests
8. Viva Voce
9. Subject Proficiency/Skill Assessment
10. Demonstration/Presentation
11. Tutorial Classes

Programme Outcomes

POS	Students graduating from Civil Engineering discipline will be expected and prepared to exercise the skills and abilities as listed below.
a	Ability to apply knowledge of mathematics, science and engineering.
b	Ability to identify, formulate and solve engineering problems in civil enginering.
c	Ability to design and conduct experiments, analyse and interpret data.
d	Ability to design systems to meet the given specifications.
e	Ability to demonstrate skills in multidisciplinary environment.
f	Ability to use modern engineering tools to analyse engineering problem.
g	Ability to demonstrate knowledge of professionalism and ethics.
h	Ability to communicate effectively in both and verbal written form.
i	Ability to develop the acumen and awareness of the impact.
j	Recognition of the need for and an ability to engage in lifelong learning.
k	Develop ability to attempt and succeed in competitive exams.

Assessment Pattern

Sl. No.	Bloom's Category	Test 1	Test 2	Test 3/Sem End Exam
1.	Remember	10	10	10
2.	Understand	10	10	10
3.	Apply	80	80	80

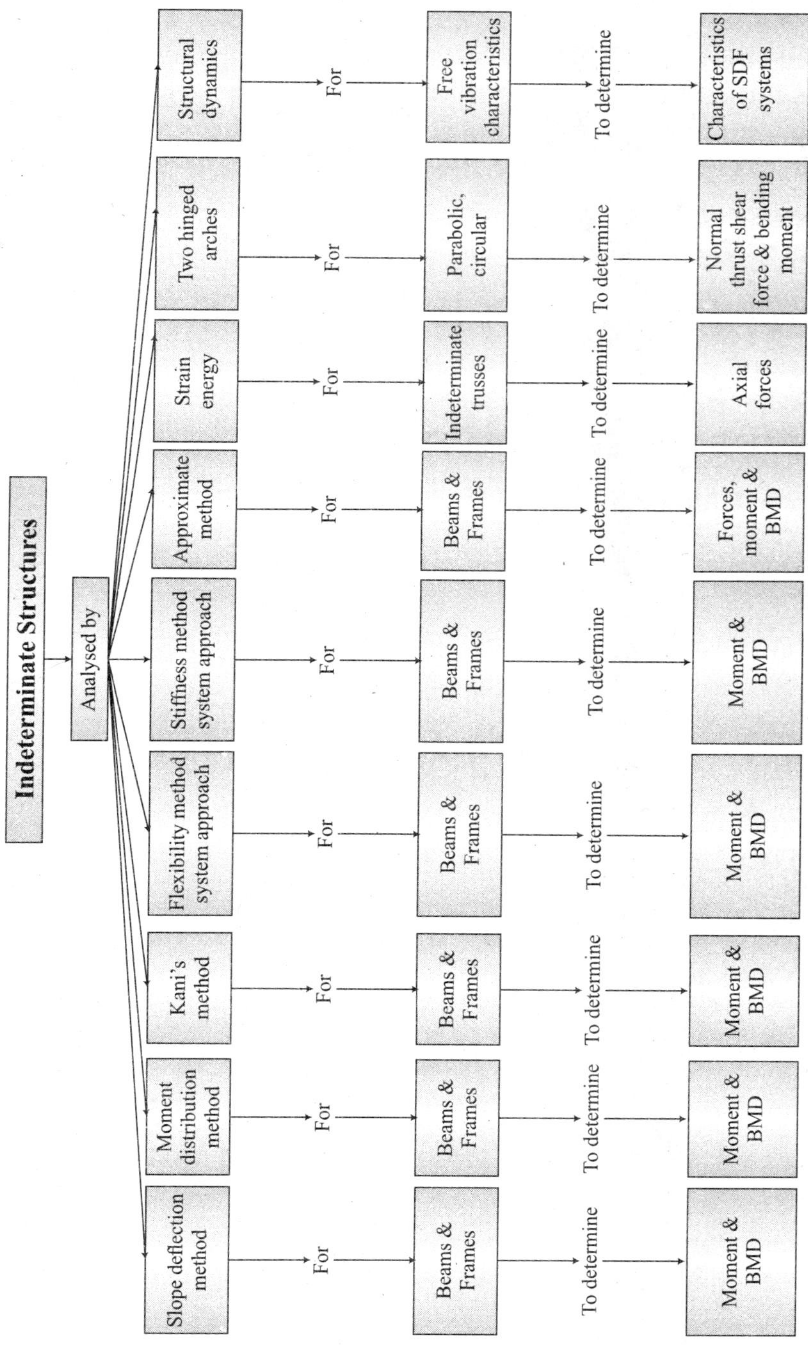
Indeterminate Structures
Analysed by
Slope deflection method
Moment distribution method
Kani's method
Flexibility method system approach
Stiffness method system approach
Approximate method
Strain energy
Two hinged arches
Structural dynamics
For
Beams & Frames
Beams & Frames
Beams & Frames
Beams & Frames
Beams & Frames
Beams & Frames
Indeterminate trusses
Parabolic, circular
Free vibration characteristics
To determine
Moment & BMD
Moment & BMD
Moment & BMD
Moment & BMD
Moment & BMD
Forces, moment & BMD
Axial forces
Normal thrust shear force & bending moment
Characteristics of SDF systems

Contents

1 The Slope Deflection Method

Objectives:

Derivation of slope deflection equations. Analysis of continuous beams without and with support settlement. Analysis of portal frames and Gable frames.

1.1 INTRODUCTION

The slope deflection method was developed by Axel Bendixson in Germany in 1914. Subsequently, Wilson and Maney in the year 1915 reported the analysis for the effect of lateral loads on tall buildings. This method forms the fundamental basis for the development of analysis of indeterminate structures. The other methods include Hardy Cross Moment distribution method, Gasper Kani's method and Klouck's distribution of deformations. As this method is a basic one, the readers are expected to understand this method thoroughly.

In the analysis of statically determinate structures, the static equilibrium conditions are adequate to determine the stress resultants. In case of statically indeterminate structures, we have to satisfy static equilibrium conditions and compatibility conditions of deformations. It is to be noted that in the slope deflection method, the compatibility conditions are used first which will be followed by equilibrium conditions.

The compatibility condition is satisfied by assigning a single value of rotation to all members meeting at that joint. The joints in the structure are assumed to be rigid. It implies that the angles between the members at the joint do not change, when the structure is subjected to loads. It is also assumed that distortions due to axial and shear stresses are negligible.

The equilibrium condition is satisfied by considering the moment equilibrium at each joint. In case of frames, horizontal equilibrium equations are derived by considering the column shear equation. In case of inclined leg frame and Gable frames, the equilibrium of each member is analysed individually as well as the equilibrium of the whole structure.

This method is useful for rigid joint structures such as continuous beams or rigid frames. The slope deflection method is found to be advantageous when the number of joints that can move is small in comparison with the number of members.

The slope deflection method is preferable when the kinematic redundants (slopes and displacements) are less in number. Hence, the number of simultaneous equations (required) to determine the above kinematic redundants is less. Three simultaneous equations can be easily solved with the help of a calculator. The slope deflection method requires more arithmetic if the number of unknown kinematic redundants is more than three. For example, consider an unsymmetrical single bay portal frame with hinged supports. In this case, the number of kinematic redundants is five. They are two rotations at the hinged supports, two rotations at the rigid joints and a sway displacement of the frame. As such it is difficult to solve five simultaneous equations and using arithmetic they can be reduced to three simultaneous equations as follows. At the hinged support, the moment is zero. This condition is used to express the rotations at the hinged support in terms of the slopes at the rigid joints and a sway displacement. Thus, five equations can be transformed into three equations. The slope deflection method is tedious when the number of simultaneous equations to be solved is more. In such cases, other methods of analysis like the moment distribution method and Kani's method would be resorted to.

1.2 DERIVATION OF SLOPE DEFLECTION EQUATION

Consider a beam element of uniform section of length l subjected to set of loads. Let V_{AB}, V_{BA} are the reactions and M_{AB}, M_{BA} are the end moments due to the effect of loads and neighbouring members. θ_A and θ_B are the corresponding rotations at A and B respectively. Let Δ be the settlement at the far end B. We derive the relationship between the changes of slope and the restraining moments applied at the ends of members by the joints.

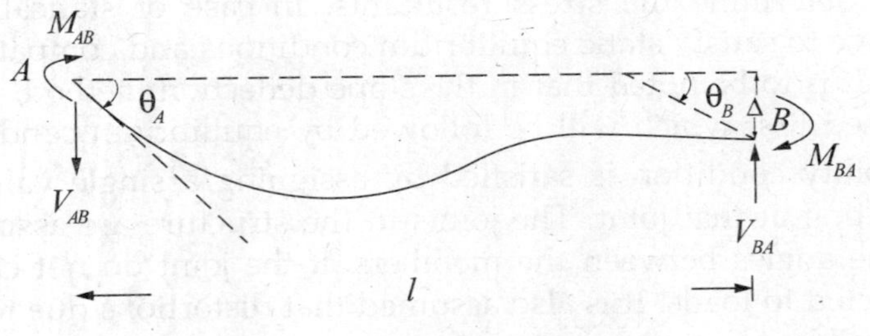

FIG. 1.1

Assumptions

(i) All the joints of the frame are rigid/stiff. It means the angles between the members at the joints do not change before and after loading. For example, the angle at interior support of a continuous beam is 180° before and after the applied loading. Similarly, the angle between column and beam members in a portal frame is 90° after the application of the load too. It is to be noted that when indeterminate beams and frames get deflected due to loading, the joints rotate as a whole. In simple words, the angles between tangents to the various elements of the elastic curve meeting at a joint remain the same as those in original structure. Figure 1.2 illustrates the above assumption.

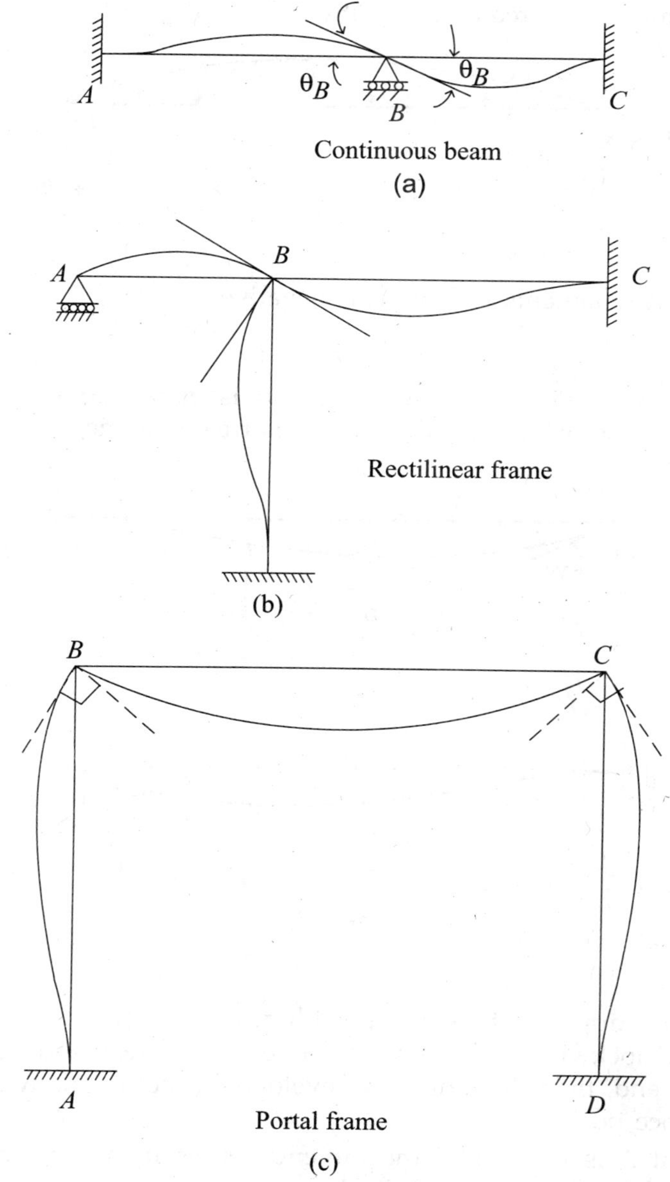

FIG. 1.2

(ii) The deformations due to shear and axial thrust are negligible.

1.3 SIGN CONVENTION

Moments

The clockwise moments are treated as positive.

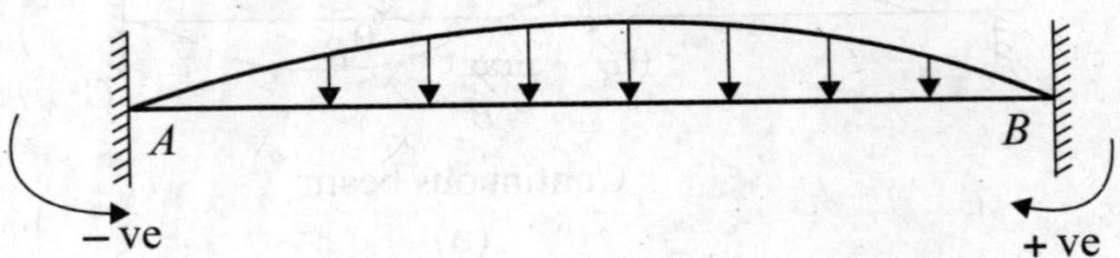

FIG. 1.3

The anticlockwise moments are taken as negative.

Slopes

When the tangent on elastic curve rotates in clockwise is treated as positive. When the tangent rotates in the anticlockwise direction it is treated as negative.

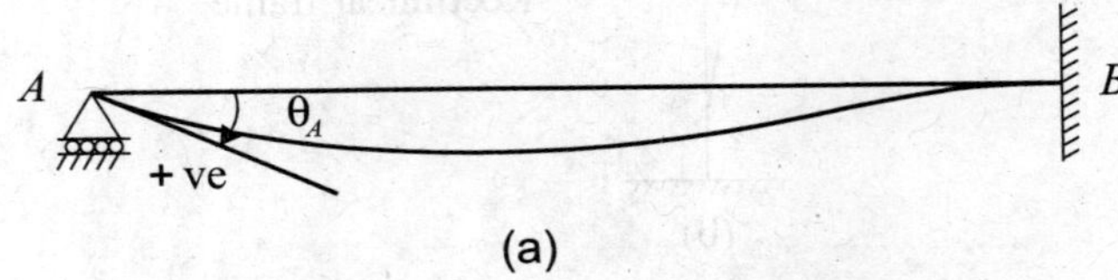

(a)

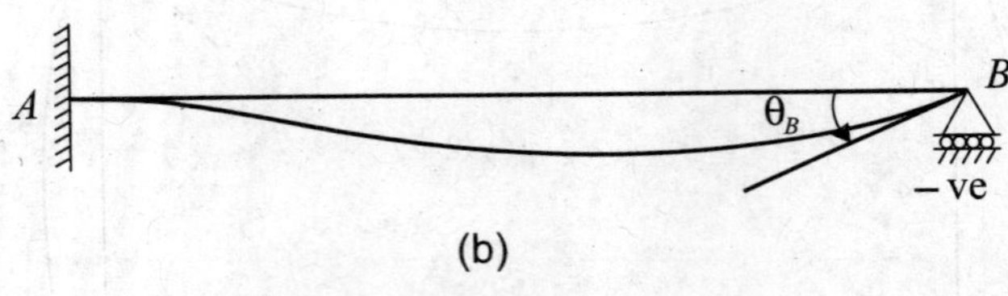

(b)

FIG. 1.4

Settlement

When the far end support settles down with respect to the near end, the induced moments are anticlockwise and hence negative. If the near end settles down with respect to the far end, then the moments developed at both ends will be in clockwise direction and hence positive.

The settlement Δ is positive if the far end settles in downward direction. The settlement Δ is negative if the far end displaces upwards. In otherwords, the clockwise displacements of end relative to the other is considered positive, otherwise it is considered negative. Figure 1.5 illustrates the same.

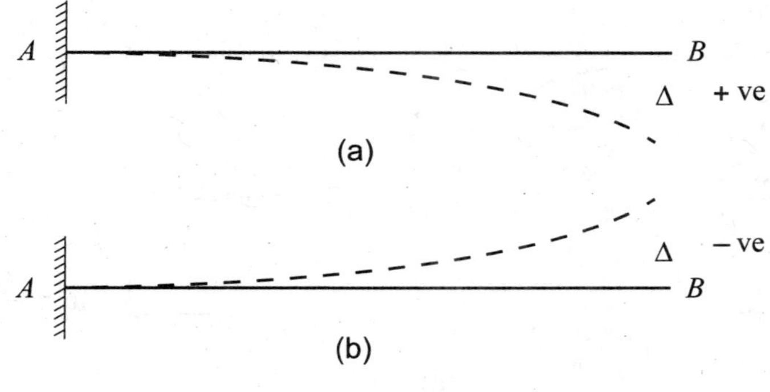

FIG. 1.5

The settlement is also called sinking or subsidence.

The slope deflection equation is derived using principle of superposition. Initially, the ends of the member are fixed. Due to this fixity the restraining moments develop. But in reality, the ends are not fully restrained. Hence, these moments are released for each degree of freedom and the superposition of the effects gives the end moments of the member. This is illustrated in the following figures.

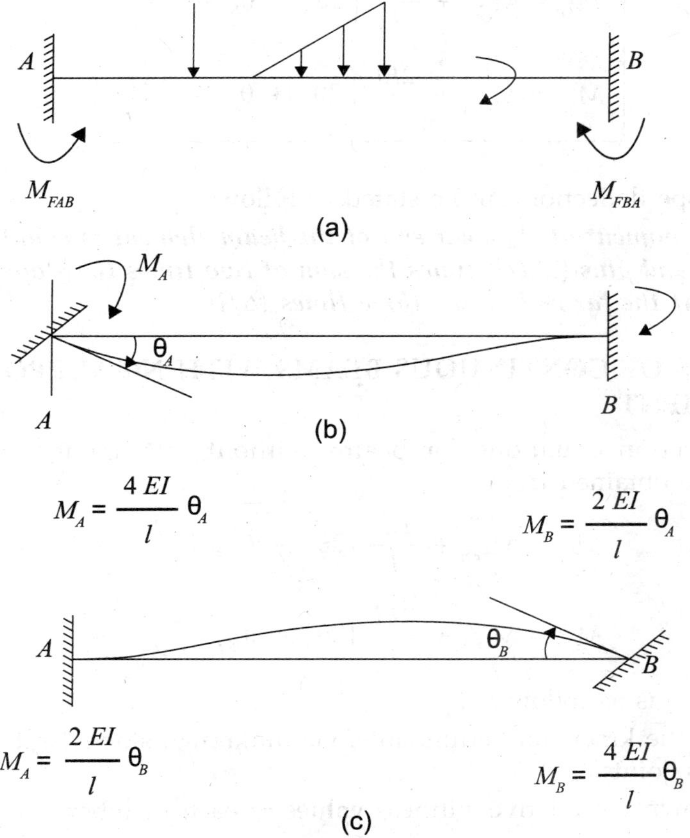

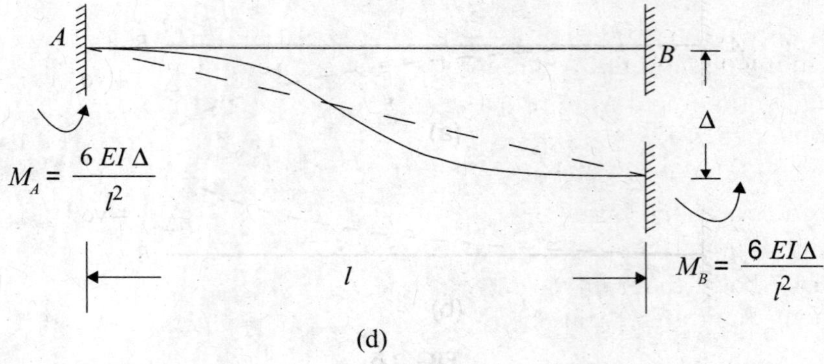

(d)

FIG. 1.6

Referring to the above figures and using the principle of superposition, we obtain

$$M_{AB} = M_{FAB} + \frac{4EI}{l}\theta_A + \frac{2EI}{l}\theta_B - \frac{6EI\Delta}{l^2}$$

$$M_{AB} = M_{FAB} + \frac{2EI}{l}\left(2\theta_A + \theta_B - \frac{3\Delta}{l}\right)$$

$$M_{BA} = M_{FBA} + \frac{2EI}{l}\left(2\theta_B + \theta_A - \frac{3\Delta}{l}\right)$$

Thus, the slope deflection can be stated as follows.

The internal moment at the near end of the beam element is equal to the fixed end moment at that end plus (2*EI*/*l*) ***times the sum of two times the slope at the near end, one time slope at the far end minus three times*** (Δ/*l*)

1.4 ANALYSIS OF CONTINUOUS BEAMS WITH NO SUPPORT SETTLEMENT

The slope deflection equations for beams without settlement/subsidence/lateral displacement are obtained from

$$M_{AB} = M_{FAB} + \frac{2EI}{l}\left(2\theta_A + \theta_B\right)$$

$$M_{BA} = M_{FBA} + \frac{2EI}{l}\left(2\theta_B + \theta_A\right)$$

The procedure is as follows:

Step 1. Identify the kinematic redundants, viz. unknown slopes and displacements at supports/joints.

Step 2. Write down the relative stiffness values of each member.

Step 3. Compute the fixed end moments in each span.

Step 4. Write down the expression for end moments in each span in terms of the fixed end moments and end rotations using slope deflection equations.

Step 5. Using joint moment equilibrium equations, a set of linear simultaneous equations are obtained.

Step 6. Determine the slopes by solving the simultaneous equations.

Step 7. The end moments are obtained by back substitution of slopes in the slope deflection equations.

Step 8. The free body diagrams are drawn for each span separately and the bending moment diagrams/shear force diagrams are drawn.

Example 1.1 Analyse the continuous beam shown in figure by slope deflection method. Draw the shear force diagram and bending moment diagram. *EI* is constant

(Anna Univ, Dec. 2006).

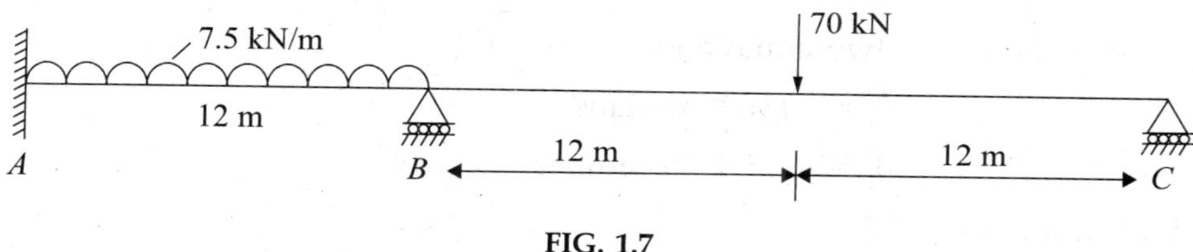

FIG. 1.7

Solution

1. Kinematic Redundants

   As the end *A* is fixed, $\theta_A = 0$. The support *B* is an intermediate support and therefore $\theta_B \neq 0$. At the roller support $\theta_C \neq 0$. Thus, we have to compute the two kinematic redundants θ_B and θ_C.

2. Relative Stiffness Values

Member	(I/l)	k
AB	$I/12$	2
BC	$I/24$	1

   The values of *k* were obtained by multiplying (I/l) with $(24/I)$.

3. Fixed End Moments

$$M_{AB} = -7.5 \times 122/12 = -90 \text{ kNm}$$

$$M_{BA} = +7.5 \times 122/12 = +90 \text{ kNm}$$

$$M_{BC} = -70 \times 24/8 = -210 \text{ kNm}$$

$$M_{CB} = +70 \times 24/8 = +210 \text{ kNm}$$

4. Slope Deflection Equations

$$M_{AB} = -90 + 2(2\theta_A + \theta_B) = 2\theta_B - 90$$

$$M_{BA} = +90 + 2(2\theta_B + \theta_A) = 4\theta_B + 90$$

$$M_{BC} = -210 + 1(2\theta_B + \theta_C) = 2\theta_B + \theta_C - 210$$

$$M_{CB} = +210 + 1(2\theta_C + \theta_B) = \theta_B + 2\theta_C + 210$$

5. Equilibrium Equations

Joint $B \Rightarrow$ $\quad M_{BA} + M_{BC} = 0$

$$6\theta_B + \theta_C = 120 \tag{1}$$

Joint $C \Rightarrow$ $\quad M_{CB} = 0$

$$\theta_B + 2\theta_C = -210 \tag{2}$$

Solving the above two equations;

$$\theta_B = 40.9091 \text{ radians}$$

$$\theta_C = -125.4545 \text{ radians}$$

6. End Moments

$$M_{AB} = 2\,(40.9091) - 90 = -8.18 \text{ kNm}$$

$$M_{BA} = 4\,(40.9091) + 90 = +253.64 \text{ kNm}$$

$$M_{BC} = 2\,(40.9091) - 125.4545 - 210 = -253.64 \text{ kNm}$$

$$M_{CB} = 40.9091 + 2\,(-125.4545) + 210 = 0$$

7. Shear Forces

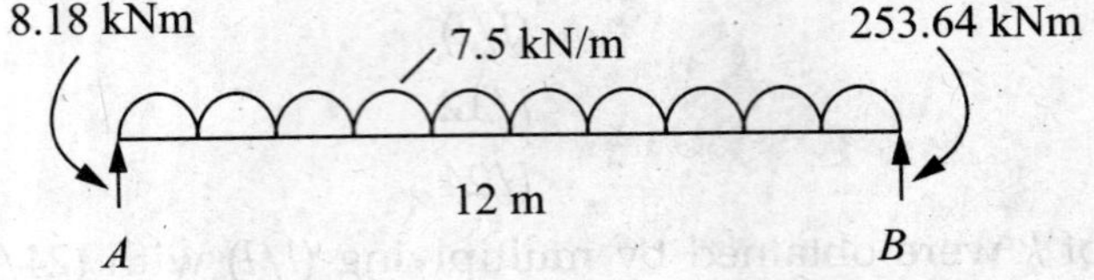

FIG. 1.8

$$\Sigma V = 0; \qquad V_{AB} + V_{BA} = 7.5(12) = 90 \tag{1}$$

$$\Sigma M_A = 0; \qquad -8.18 + 253.64 - 12V_{BA} + 7.5 \times \frac{12^2}{2} = 0 \tag{2}$$

$$V_{BA} = 65.46 \text{ kN}$$

$$V_{AB} = 24.54 \text{ kN}$$

253.64 kNm 12 70 kN 12 m

B C

FIG. 1.9

$$\Sigma V = 0; \qquad V_{BC} + V_{CB} = 70 \tag{3}$$

$$\Sigma M_B = 0; \qquad -253.64 + 70(12) - 24V_{CB} = 0$$

$$V_{CB} = 24.43 \text{ kN}$$

$$V_{BC} = 45.57 \text{ kN}$$

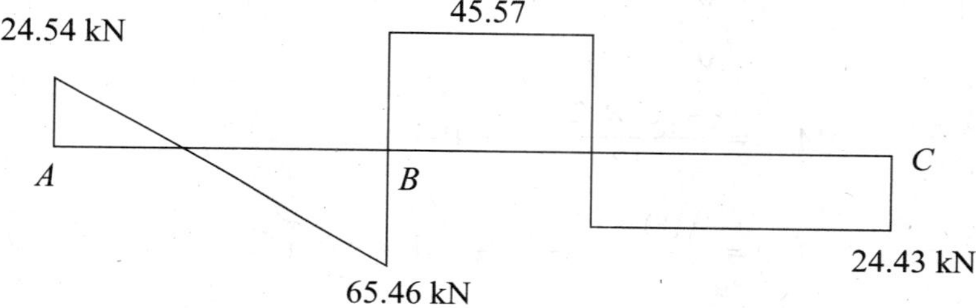

FIG. 1.10 Shear force diagram

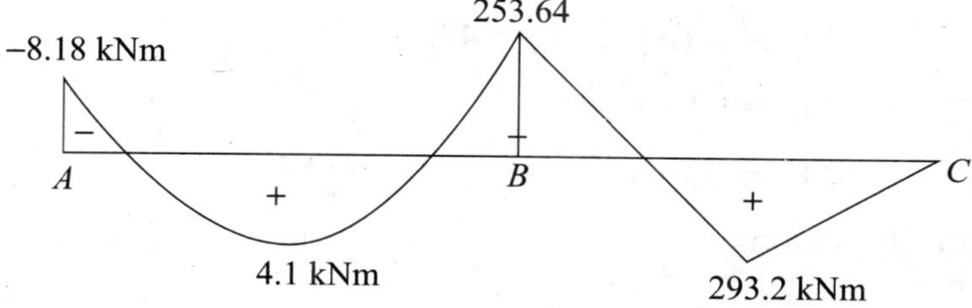

FIG. 1.11 Bending moment diagram

Example 1.2 Analyse the beam by the slope deflection method. Sketch the bending moment diagram.

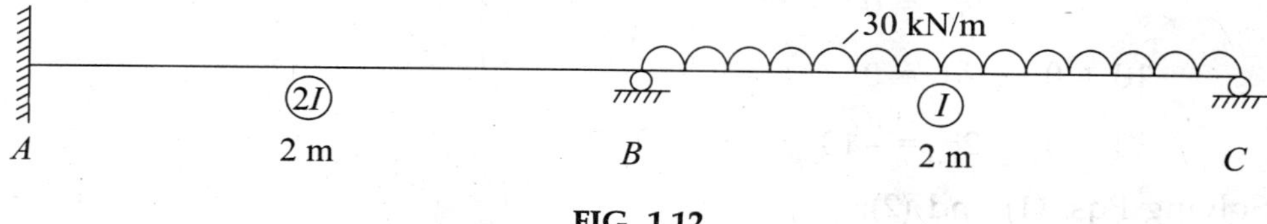

FIG. 1.12

Solution

1. Kinematic Redundants

$$\theta_A = 0, \quad \theta_B \neq 0, \quad \theta_C \neq 0$$

2. Relative Stiffness Values ($k = I/l$)

Member	(I/l)	k
AB	$2I/2$	2
BC	$I/2$	1

The values of k were obtained by multiplying the values of (I/l) with $(2/I)$.

3. Fixed End Moments

$$M_{FAB} = 0$$

$$M_{FBA} = 0$$

$$M_{FBC} = \frac{-30 \times 2^2}{12} = -10\,\text{kNm}$$

$$M_{FCB} = \frac{+30 \times 2^2}{12} = +10\,\text{kNm}$$

4. Slope Deflection Equations

$$M_{AB} = 2\,(2\theta_A + \theta_B) = 2\theta_B$$

$$M_{BA} = 2\,(2\theta_B + \theta_A) = 4\theta_B$$

$$M_{BC} = -10 + 2\theta_B + \theta_C$$

$$M_{CB} = 10 + \theta_B + 2\theta_C$$

5. Equilibrium Equations

At Joint B

$$M_{BA} + M_{BC} = 0$$

$$6\theta_B + \theta_C = 10 \qquad (1)$$

At Joint C

$$M_{CB} = 0$$

$$10 + \theta_B + 2\theta_C = 0$$

$$\theta_B + 2\theta_C = -10 \qquad (2)$$

Solving Eqs. (1) and (2);

$$\theta_B = 2.7273 \text{ radians}$$

$$\theta_C = -6.3636 \text{ radians}$$

6. End Moments

$$M_{AB} = 2(2.7273) = 5.46 \text{ kNm}$$

$$M_{BA} = 4(2.7273) = 10.91 \text{ kNm}$$

$$M_{BC} = 2(2.7273) - 6.3636 - 10 = -10.91 \text{ kNm}$$

$$M_{CB} = 10 + (2.7273) + 2(-6.3636) = 0$$

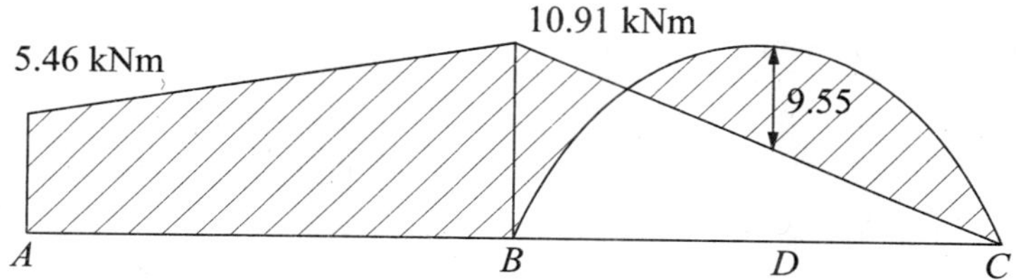

FIG. 1.13 Bending moment diagram

Example 1.3 Analyse the continuous beam by the slope deflection method. Determine the end moments and the reactions. *EI* is constant.

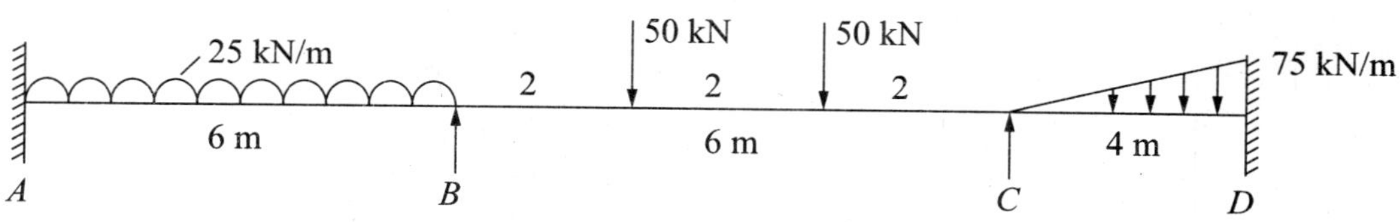

FIG. 1.14

Solution

1. Kinematic Redundants

$$\theta_A = 0, \quad \theta_B \neq 0, \quad \theta_C \neq 0 \quad \text{and} \quad \theta_D = 0$$

2. Relative Stiffness Values

Member	(I/l)	k
AB	$I/6$	4
BC	$I/6$	4
CD	$I/4$	6

The values of k were obtained by multiplying the values of (I/l) with $(24/I)$.

3. Fixed Moments

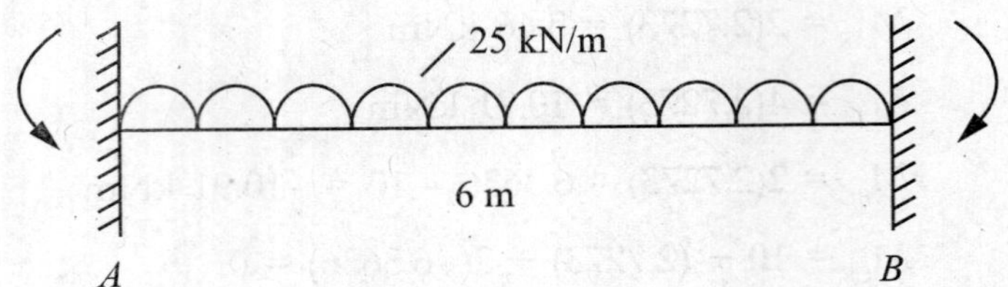

$$\frac{-25 \times 6^2}{12} = -75 \text{ kNm} \qquad \frac{+25 \times 6^2}{12} = +75 \text{ kNm}$$

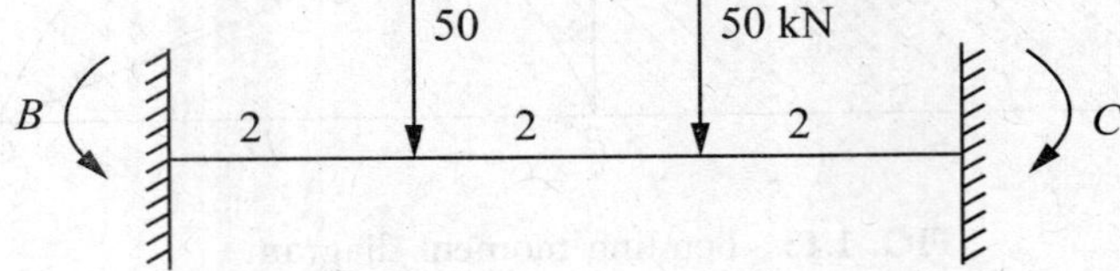

$$\frac{-2 \times 50 \times 4}{6} = -66.67 \text{ kNm} \qquad \frac{+2 \times 50 \times 4}{6} = +66.67 \text{ kNm}$$

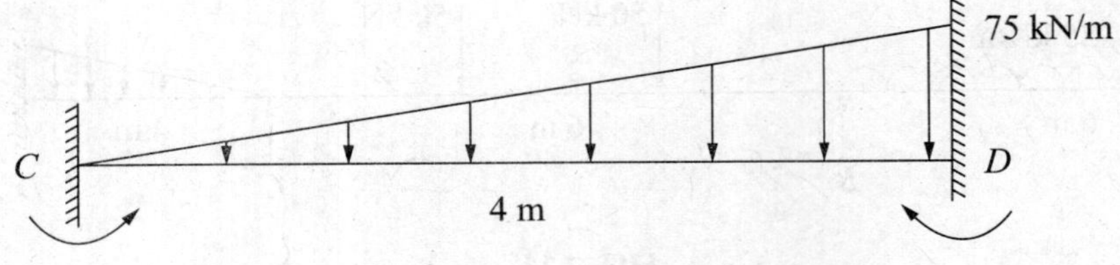

FIG. 1.15

$$\frac{-75 \times 4^2}{30} = -40 \text{ kNm} \qquad \frac{+75 \times 4^2}{20} = +60 \text{ kNm}$$

4. Slope Deflection Equations

$$M_{AB} = -75 + 4\,(2\cancel{\theta_A} + \theta_B) = 4\theta_B - 75$$

$$M_{BA} = +75 + 4\,(2\theta_B + \cancel{\theta_A}) = 8\theta_B + 75$$

$$M_{BC} = -66.67 + 4\,(2\theta_B + \theta_C) = 8\theta_B + 4\theta_C - 66.67$$

$$M_{CB} = +66.67 + 4\,(2\theta_C + \theta_B) = 4\theta_B + 8\theta_C + 66.67$$

$$M_{CD} = -40 + 6\,(2\theta_C + 2\cancel{\theta_D}) = 12\theta_C - 40$$

$$M_{DC} = +60 + 6\,(2\cancel{\theta_D} + \theta_C) = 6\theta_C + 60$$

5. Equilibrium Equations

Joint B $\Rightarrow$ $M_{BA} + M_{BC} = 0$

$$16\theta_B + 4\theta_C + 8.33 = 0$$

$$16\theta_B + 4\theta_C = -8.33 \quad (1)$$

Joint C $\Rightarrow$ $M_{CB} + M_{CD} = 0$

$$4\theta_B + 20\theta_C + 26.67 = 0$$

$$4\theta_B + 20\theta_C = -26.67 \quad (2)$$

Solving Eqs. (1) and (2);

$$\theta_B = -0.1971,\ \theta_C = -1.2941 \text{ radians}$$

6. End Moments

$$M_{AB} = 4\,(-0.1971) - 75 = -75.8 \text{ kNm}$$

$$M_{BA} = 8\,(-0.1971) + 75 = +73.4 \text{ kNm}$$

$$M_{BC} = 8\,(-0.1971) + 4\,(-1.2941) - 66.67 = -73.4 \text{ kNm}$$

$$M_{CB} = 4\,(-0.1971) + 8\,(-1.2941) \ +66.67 = 55.53 \text{ kNm}$$

$$M_{CD} = 12\,(-1.2941) - 40 = -55.53 \text{ kNm}$$

$$M_{DC} = 6\,(-1.2941) + 60 = +52.23 \text{ kNm}$$

7. Shear Forces

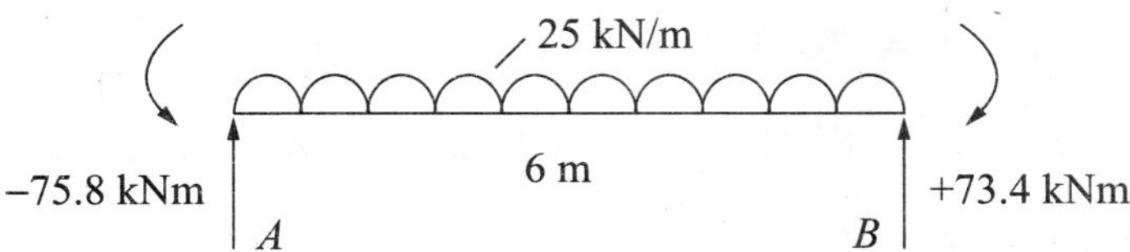

FIG. 1.16

$$\Sigma V = 0; \quad V_{AB} + V_{BA} = 25 \times 6 = 150 \text{ kN} \quad (3)$$

$$\Sigma M_A = 0; \quad -75.8 + 73.4 + 25 \times \frac{6^2}{2} - 6V_{BA} = 0 \quad (4)$$

$$V_{BA} = 74.6 \text{ kN}$$

$$V_{AB} = 75.4 \text{ kN}$$

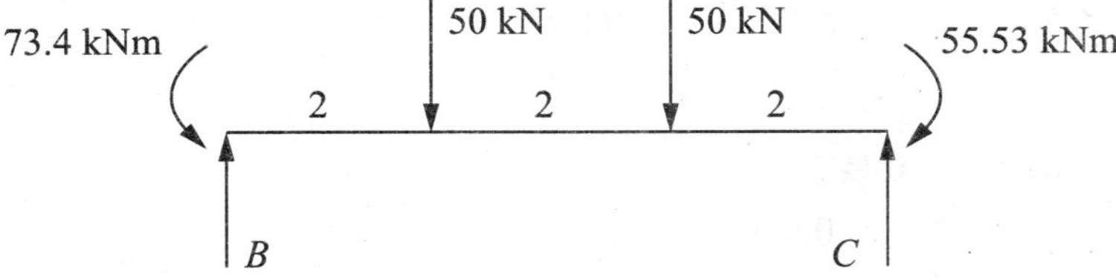

FIG. 1.17

$$\Sigma V = 0; \qquad V_{BC} + V_{CB} = 100 \tag{5}$$

$$\Sigma M_B = 0; \qquad -73.4 + 55.53 + 50(2) + 50(4) - 6V_{CB} = 0 \tag{6}$$

$$V_{CB} = 47.02 \text{ kN}$$

$$V_{BC} = 52.98 \text{ kN}$$

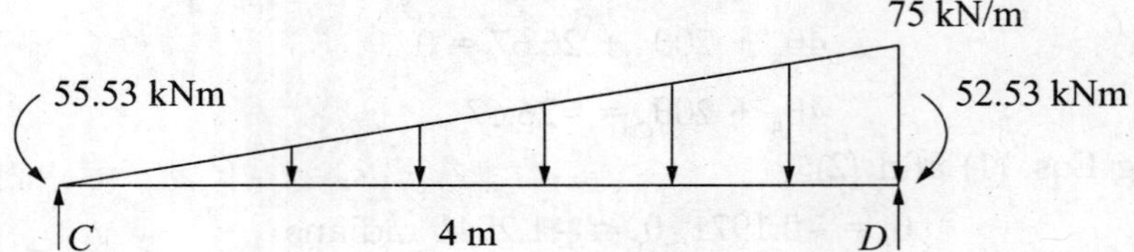

FIG. 1.18

$$\Sigma V = 0; \qquad V_{CD} + V_{DC} = \left(\frac{1}{2}\right)4 \times 75 = 150 \tag{7}$$

$$\Sigma M_C = 0; \qquad +52.53 - 55.33 - 4V_{DC} + \frac{1}{2} \times 4 \times 75 \times \left(\frac{2}{3} \times 4\right) = 0$$

$$V_{DC} = 97 \text{ kN}$$

$$V_{CD} = 53 \text{ kN}$$

The reactions are

$$V_A = 75.4 \text{ kN}, V_B = V_{BA} + V_{BC} = 127.58,$$

$$V_C = 47.02 + 53 = 100.02, \; V_D = 97.00 \text{ kN}$$

Check

$$V_A + V_B + V_C + V_D \cong 400 \text{ kN}$$

Example 1.4 Analyse the continuous beam by the slope deflection method. Draw the bending moment diagram and shear force diagram.

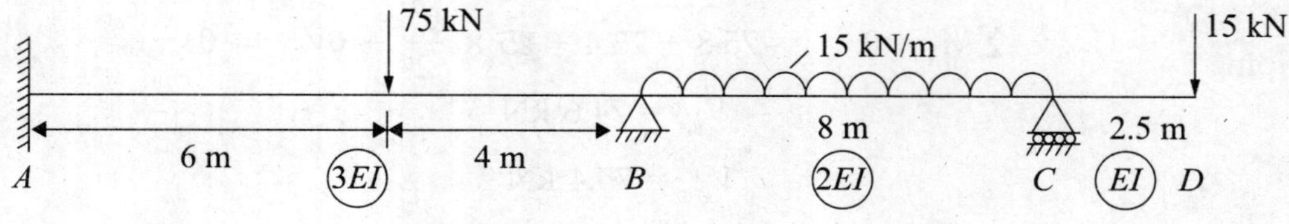

FIG. 1.19

Solution

1. Kinematic Redundants

$$\theta_A = 0, \quad \theta_B \neq \theta_C \neq 0$$

2. Relative Stiffness Values

Member	I/l	k
AB	$3I/10$	1.2
BC	$2I/8$	1.0

3. Fixed Ends Moments

$$M_{FAB} = \frac{-6 \times 75 \times 4^2}{10^2} = -72\ \text{kNm}$$

$$M_{FBA} = \frac{+4 \times 75 \times 6^2}{10^2} = +108\ \text{kNm}$$

$$M_{FBC} = -15 \times \frac{8^2}{12} = -80\ \text{kNm}$$

$$M_{FCB} = +15 \times \frac{8^2}{12} = +80\ \text{kNm}$$

$$M_{FCD} = -15 \times 2.5 = -37.5\ \text{kNm}$$

4. Slope Deflection Equations

$$M_{AB} = -72 + 1.2\ (2\theta_A + \theta_B) = 1.2\theta_B - 72$$

$$M_{BA} = +108 + 1.2\ (2\theta_B + 2\theta_A) = 2.4\theta_B + 108$$

$$M_{BC} = -80 + 1\ (2\theta_B + \theta_C) = 2\theta_B + \theta_C - 80$$

$$M_{CB} = +80 + 1\ (2\theta_C + \theta_B) = \theta_B + 2\theta_C + 80$$

$$M_{CD} = -15(2.5) = -37.5$$

5. Equilibrium Equations

Joint B $\Rightarrow$

$$M_{BA} + M_{BC} = 0$$

$$4.4\theta_B + \theta_C = -28 \qquad (1)$$

Joint C $\Rightarrow$

$$M_{CB} + M_{CD} = 0$$

$$\theta_B + 2\theta_C + 80 - 37.5 = 0$$

$$\theta_B + 2\theta_C = -42.5 \qquad (2)$$

Solving Eqs. (1) and (2);

$$\theta_B = -1.7308 \quad \theta_C = -20.3846\ \text{radians}$$

6. End Moments

$$M_{AB} = 1.2\ (-1.7308) - 72 = -74.08\ \text{kNm}$$

$$M_{BA} = 2.4\ (-1.7308) + 108 = +103.85\ \text{kNm}$$

$$M_{BC} = 2\,(-1.7308) - 20.3846 - 80 = -103.85 \text{ kNm}$$

$$M_{CB} = -1.7308 + 2\,(-20.3846) + 80 = 37.5 \text{ kNm}$$

7. Shear Forces

Equilibrium of span AB

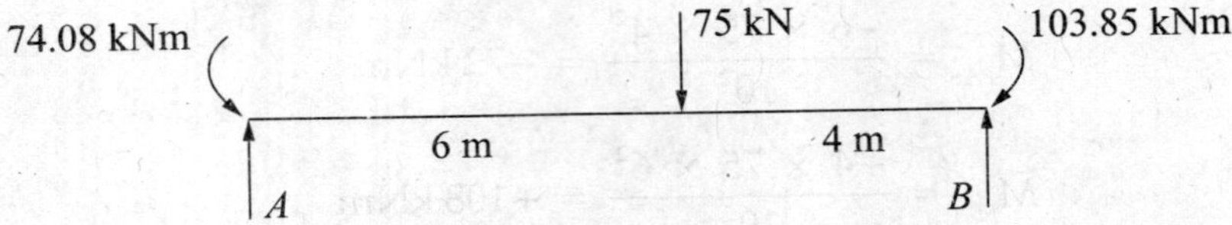

FIG. 1.20

$$\Sigma V = 0; \qquad V_{AB} + V_{BA} = 75 \tag{3}$$

$$\Sigma M_A = 0; \qquad -74.08 + 103.85 + 75 \times 6 - 10V_{BA} = 0 \tag{4}$$

$$V_{BA} = 47.98 \text{ kN}$$

$$V_{AB} = 27.02 \text{ kN}$$

Equilibrium of span BC

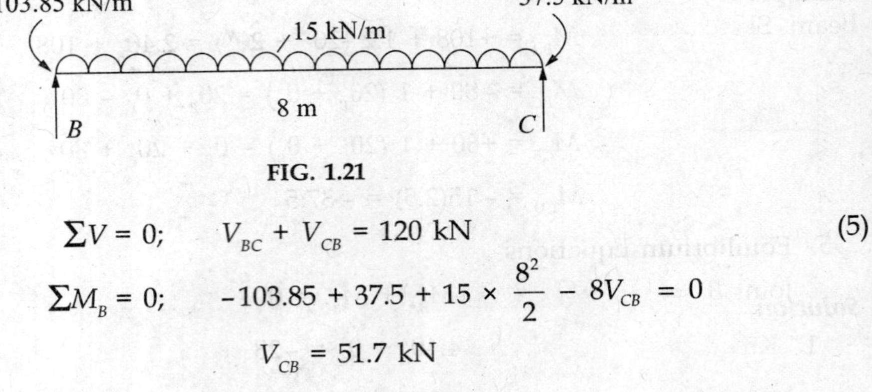

FIG. 1.21

$$\Sigma V = 0; \qquad V_{BC} + V_{CB} = 120 \text{ kN} \tag{5}$$

$$\Sigma M_B = 0; \qquad -103.85 + 37.5 + 15 \times \frac{8^2}{2} - 8V_{CB} = 0$$

$$V_{CB} = 51.7 \text{ kN}$$

$$V_{BC} = 68.3 \text{ kN}$$

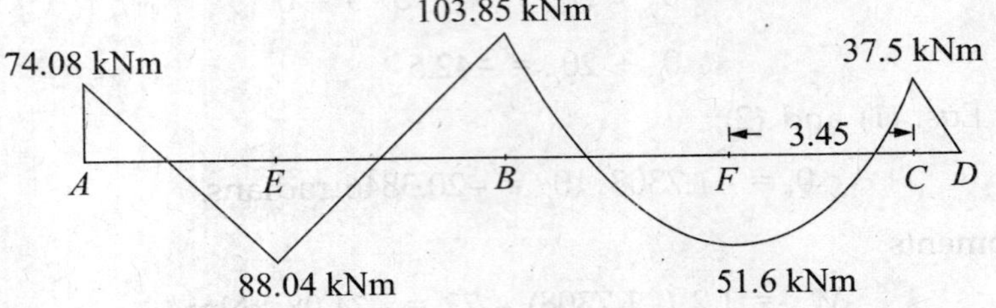

FIG. 1.22 Bending moment diagram

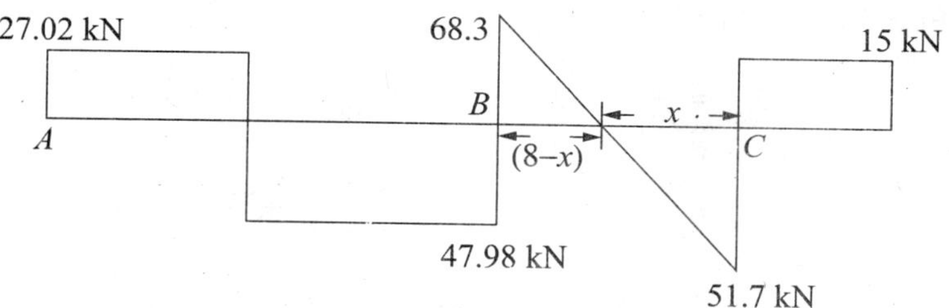

FIG. 1.23 Shear force diagram

The zero shear occurs in BC and from similar triangles

$$\frac{x}{(8-x)} = \frac{51.7}{68.3}$$

$$68.3x = 413.6 - 51.7x$$

$$x = 3.45 \text{ m}$$

$$\therefore \quad \text{Absolute max } BM = 51.7\,(3.45) - 37.5 - 15 \times \frac{3.45^2}{2}$$

$$= 51.6 \text{ kNm}$$

Example 1.5 Determine the reactions at the right supports A, B and C of the continuous beam. Sketch the shear force and bending moment diagram.

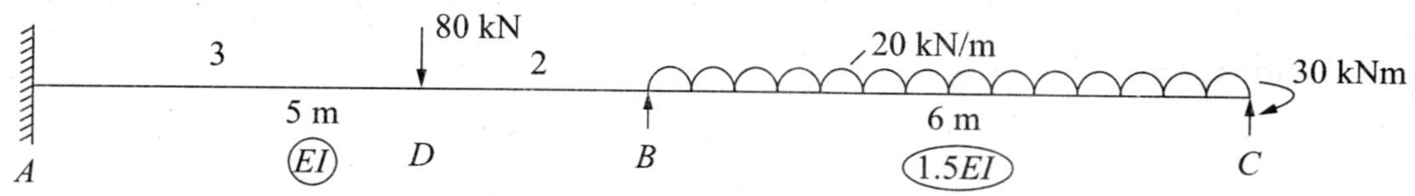

FIG. 1.24

Solution

1. Kinematic Redundants

$$\theta_A = 0, \quad \theta_B \neq \theta_C \neq 0$$

2. Relative Stiffness Values

Members	(I/l)	k
AB	$I/5$	6
BC	$1.5/6$	7.5

The values of k was obtained by multiplying a common factor of $(30/I)$.

3. Fixed End Moments

$$M_{FAB} = \frac{-Wa^2b}{l^2} = -80(2)^2\frac{3^2}{5^2} = -38.4 \text{ kNm}$$

$$M_{FBA} = \frac{+Wab^2}{l^2} = +80(2)\frac{3^2}{5^2} = +57.6 \text{ kNm}$$

$$M_{FBC} = \frac{-wl^2}{12} = -20 \times \frac{6^2}{12} = -60.0 \text{ kNm}$$

$$M_{FCB} = \frac{wl^2}{12} = +20 \times \frac{6^2}{12} = +60.0 \text{ kNm}$$

4. Slope Deflection Equations

$$M_{AB} = -38.4 + 6\,(2\theta_A + \theta_B) = 6\theta_B - 38.4$$

$$M_{BA} = +57.6 + 6\,(2\theta_B + \theta_A) = 12\theta_B + 57.6$$

$$M_{BC} = -60.0 + 7.5\,(2\theta_B + \theta_C) = 15\theta_B + 7.5\theta_C - 60.0$$

$$M_{CB} = +60.0 + 7.5\,(2\theta_C + \theta_B) = 7.5\theta_B + 15\theta_C + 60.0$$

5. Equilibrium Conditions

Joint B ⇒ $M_{BA} + M_{BC} = 0$

$$27\theta_B + 7.5\theta_C = 2.4 \qquad (1)$$

Joint C ⇒ $M_{CB} = 30$

$$7.5\theta_B + 15\theta_C + 60 - 30 = 0$$

$$7.5\theta_B + 15\theta_C = -30 \qquad (2)$$

Note:

At C, the applied moment is 30 kNm in the clockwise direction. The resisting moment is in the anticlockwise direction of 30 kNm. Hence, the resisting moment is substituted in the above equilibrium equation.

Solving the above equations;

$$\theta_B = 0.7484$$

$$\theta_C = -2.3742$$

6. End Moments

The slopes θ_B and θ_C obtained are back substituted in the equilibrium equation as

$$M_{AB} = -38.4 + 6\,(0.7484) = -33.91 \text{ kNm}$$

$$M_{BA} = 12\,(0.7484) + 57.6 = +66.58 \text{ kNm}$$

$$M_{BC} = 15\ (0.7484) + 7.5\ (-2.3742) - 60.0 = -66.58 \text{ kNm}$$

$$M_{CB} = 7.5\ (0.7484) + 15\ (-2.3742) + 60.0 = +30 \text{ kNm}$$

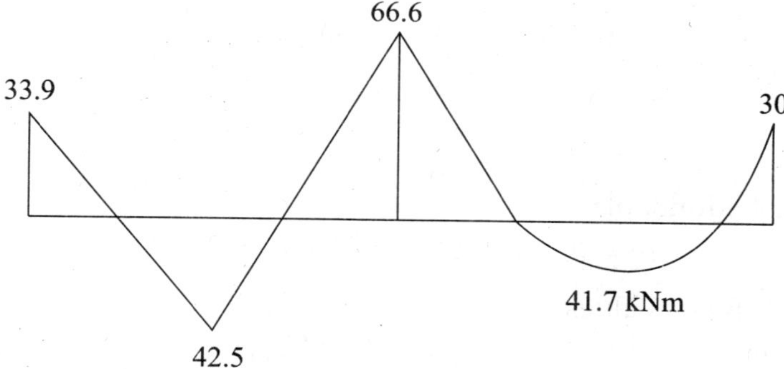

FIG. 1.25 BMD

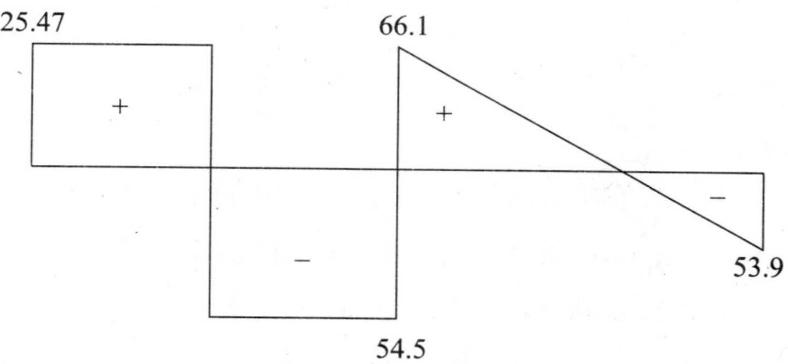

FIG. 1.26 SFD

1.5 ANALYSIS OF CONTINUOUS BEAMS WITH SUPPORT SETTLEMENT

The slope deflection equation for this case has been derived in the earlier section. The relevant equations are

$$M_{AB} = M_{FAB} + \frac{2EI}{l}\left(2\theta_A + \theta_B - \frac{3\Delta}{l}\right)$$

$$M_{BA} = M_{FBA} + \frac{2EI}{l}\left(2\theta_B + \theta_A - \frac{3\Delta}{l}\right)$$

Example 1.6 Figure shows a two-span continuous beam with uniform section. The support *B* sinks by 10 mm below *B* and the support *C* sinks a further 20 mm below *B*. Analyse the beam by the slope deflection method. Draw the bending moment diagram and the elastic curve.

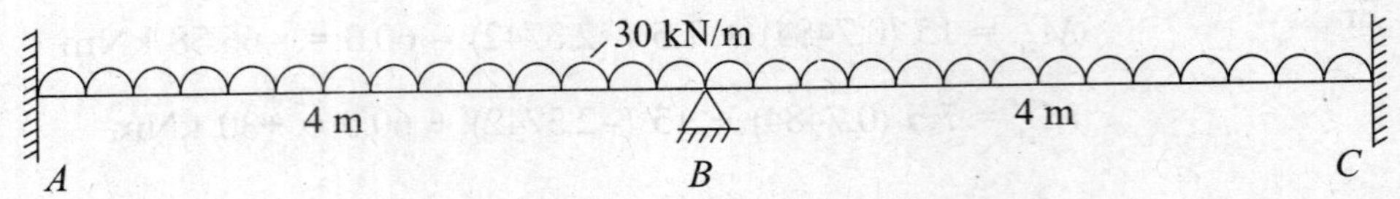

$EI = 10000\ \text{kNm}^2$

FIG. 1.27

Solution

1. Kinematic Redundants

$$\theta_A = 0, \quad \theta_B \neq 0, \quad \theta_C \neq 0$$

2. Relative Stiffness Values

Member	$(2EI/l)$	k
AB	$\dfrac{2 \times 10000}{4}$	5000
BC	$\dfrac{2 \times 10000}{4}$	5000

3. Fixed End Moments

$$M_{FAB} = -40\ \text{kNm}, \quad M_{FBA} = +40\ \text{kNm},$$

$$M_{FBC} = -40\ \text{kNm}, \quad M_{FCB} = +40\ \text{kNm}$$

4. Slope Deflection Equations

$$M_{AB} = -40 + 5000\left(2\cancel{\theta_A} + \theta_B - \frac{3 \times 10}{4000}\right) = 5000\theta_B - 77.5$$

$$M_{BA} = +40 + 5000\left(2\theta_B + \cancel{\theta_A} - \frac{3 \times 10}{4000}\right) = 10000\theta_B + 2$$

$$M_{BC} = -40 + 5000\left(2\theta_B + \cancel{\theta_C} - \frac{3 \times 20}{4000}\right) = 10000\theta_B - 115$$

$$M_{CB} = +40 + 5000\left(2\cancel{\theta_C} + \theta_B - \frac{3 \times 20}{4000}\right) = 5000\theta_B - 35$$

The value of Δ is positive for both spans AB and BC as in both spans the beam rotates in a clockwise direction as a consequence of the sinking of the support.

5. Equilibrium Conditions

Joint $B \Rightarrow$
$$M_{BA} + M_{BC} = 0$$
$$20000\theta_B - 112.5 = 0$$
$$\theta_B = 5.625 \times 10^{-3}\ \text{radians}$$

6. End Moments

$$M_{AB} = 5000\ (5.625 \times 10^{-3}) - 77.5 = -49.4 \text{ kNm}$$

$$M_{BA} = 10000\ (5.625 \times 10^{-3}) + 2.5 = +58.75 \text{ kNm}$$

$$M_{BC} = 10000\ (5.625 \times 10^{-3}) - 115 = -58.75 \text{ kNm}$$

$$M_{CB} = 5000\ (5.625 \times 10^{-3}) - 35 = -6.88 \text{ kNm}$$

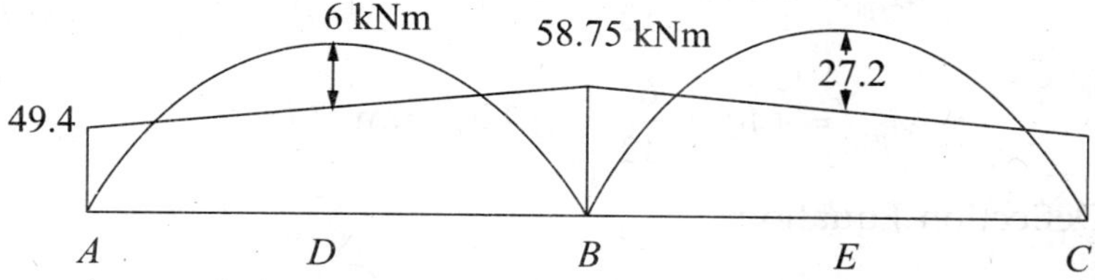

FIG. 1.28 Bending moment diagram

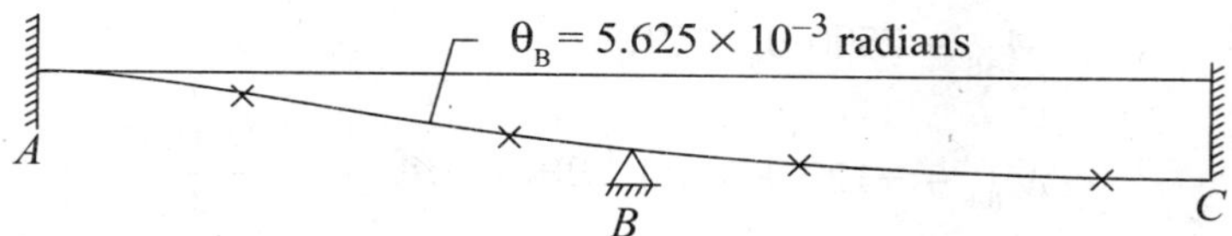

FIG. 1.29 Elastic curve when B and C settles

Example 1.7 Analyse the continuous beam shown in figure by the slope deflection method. Draw the shear force diagram and bending moment diagram. During loading, the supports B and C settle by $(240/EI)$ and $(120/EI)$ respectively. $EI = 15000$ kNm^2.

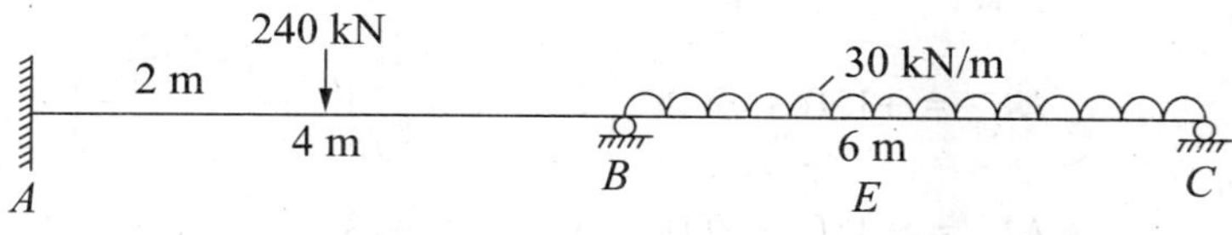

FIG. 1.30

Solution

1. Kinematic Redundants

$$\theta_A = 0, \quad \theta_B \neq \theta_C \neq 0$$

2. Relative Stiffness Values

Member	$(2EI/l)$	k
AB	$2 \times 15000/4$	7500
BC	$2 \times 15000/6$	5000

3. Fixed End Moments

$$M_{FAB} = -240 \times \frac{4}{8} = -120\text{ kNm} \qquad \delta_B = \frac{240}{15000} = 0.016$$

$$M_{FBA} = +240 \times \frac{4}{8} = +120\text{ kNm}$$

$$M_{FBC} = -40 \times \frac{6^2}{12} = -120\text{ kNm} \qquad \delta_C = \frac{120}{15000} = 0.008$$

$$M_{FCB} = +40 \times \frac{6^2}{12} = +120\text{ kNm}$$

4. Slope Deflection Equations

$$M_{AB} = -120 + 7500\left(2\cancel{\theta_A} + \theta_B - \frac{3 \times 0.016}{4}\right)$$

$$M_{AB} = 7500\ \theta_B - 210$$

$$M_{BA} = +120 + 7500\left(2\theta_B + \cancel{\theta_A} - \frac{3 \times 0.016}{4}\right)$$

$$M_{BA} = 15000\ \theta_B + 30$$

$$M_{BC} = -120 + 5000\left(2\theta_B + \theta_C + \frac{3 \times 0.008}{6}\right)$$

$$M_{BC} = 10000\ \theta_B + 5000\ \theta_C - 100$$

$$M_{CB} = +120 + 5000\left(2\theta_C + \theta_B + \frac{3 \times 0.008}{6}\right)$$

$$M_{CB} = +120 + 5000\ (2\theta_C + \theta_B + 3)$$

$$M_{CB} = 5000\ \theta_B + 10000\ \theta_C + 140$$

5. Equilibrium Conditions

Joint $B \Rightarrow$ $M_{BA} + M_{BC} = 0$

$$25000\ \theta_B + 5000\ \theta_C - 70 = 0$$

$$25000\ \theta_B + 5000\ \theta_C = 70 \qquad (1)$$

Joint $C \Rightarrow$ $M_{CB} = 0$

$$5000\ \theta_B + 10000\ \theta_C = -140 \qquad (2)$$

Solving Eqs. (1) and (2);

$$\theta_B = 6.2222 \times 10^{-3} \text{ radians}$$

$$\theta_C = -0.0171 \text{ radians}$$

6. End Moments

$$M_{AB} = 7500\,(6.2222 \times 10^{-3}) - 210 = -163.33 \text{ kNm}$$

$$M_{BA} = 15000\,(6.2222 \times 10^{-3}) + 30 = +123.33 \text{ kNm}$$

$$M_{BC} = 10000\,(6.2222 \times 10^{-3}) + 5000 \times -0.0171 - 100 = -123.33 \text{ kNm}$$

$$M_{CB} = 5000 \times 6.2222 \times 10^{-3} + 10000 \times -0.0171 + 140 = 0 \text{ kNm}$$

7. Shear Force Diagrams
   Equilibrium of span *AB*

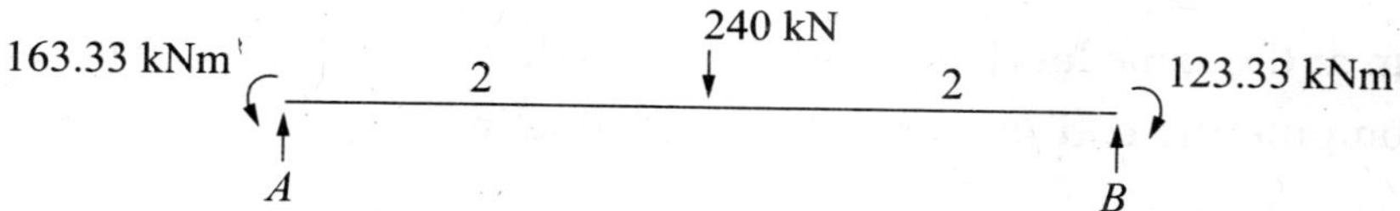

FIG. 1.31

$$\Sigma V = 0; \quad V_{AB} + V_{BA} = 240 \text{ kN} \tag{1}$$

$$\Sigma M_A = 0; \quad -163.33 + 123.33 + 240(2) - 4V_{BA} = 0 \tag{2}$$

$$V_{BA} = 110 \text{ kN}$$

$$V_{AB} = 130 \text{ kN}$$

Equilibrium of Span *BC*

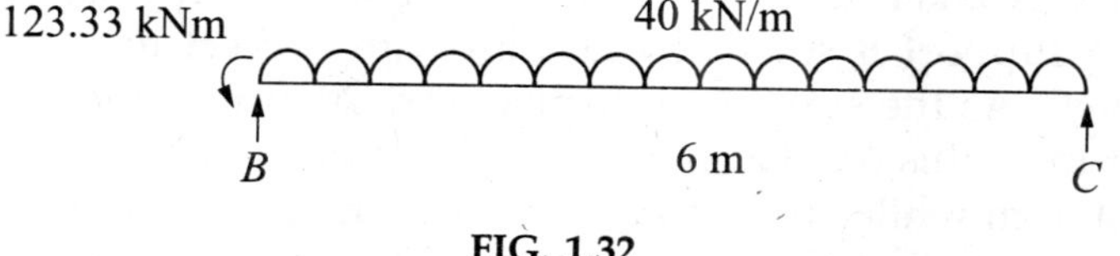

FIG. 1.32

$$\Sigma V = 0; \quad V_{BC} + V_{CB} = 240 \text{ kN} \tag{3}$$

$$\Sigma M_B = 0; \quad -123.33 + 0 + 40 \times \frac{6^2}{2} - 6V_{CB} = 0$$

$$V_{CB} = 99.45 \text{ kN}$$

$$V_{BC} = 140.55 \text{ kN}$$

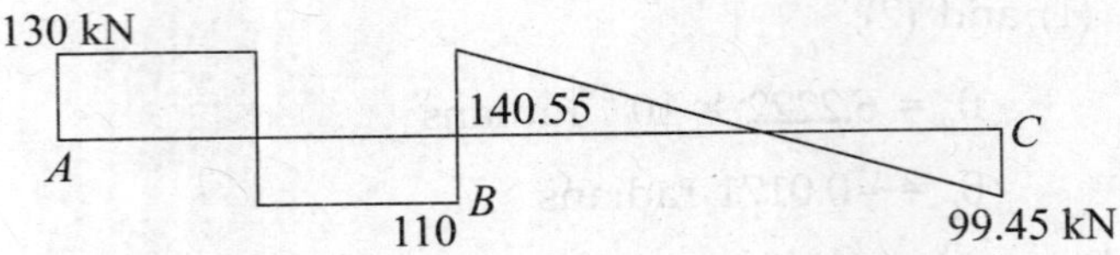

FIG. 1.33 Shear force diagram

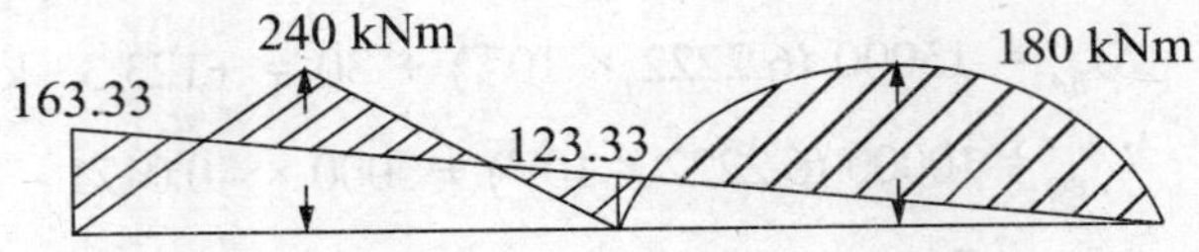

FIG. 1.34 Bending moment diagram

Example 1.8 Figure shows a continuous beam *ABCD* with uniform section. When the beamis loaded, the supports settle. *B* settles by 20 mm and *C* subsides by 30 mm. *A* and *D* remain at the same level. Support *A* rotates through $\left(\frac{1}{200}\right)$ radians in clockwise direction. Compute the end moments. EI = 15000 kNm^2.

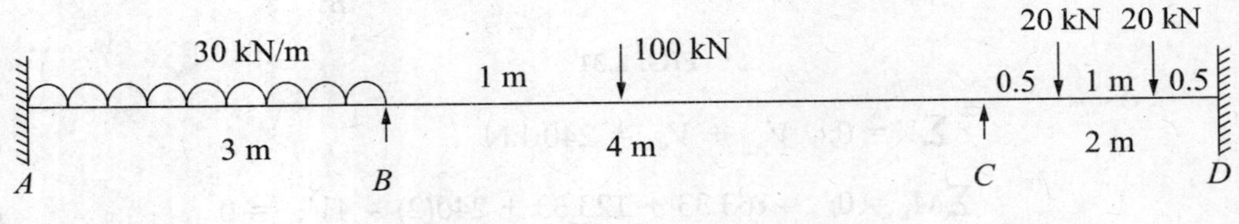

FIG. 1.35

Solution

1. Kinematic Redundants

   It is to be mentioned here that support *A* rotates in the clockwise direction. Hence, the value of θ is taken as positive as per the sign convention. Considering the span *AB*, the support *B* settles by 20 mm with respect to *A* (which remains at the same level). As the support *B* displaces by 20 mm in the clockwise direction with respect to *A*, this displacement is taken as positive. In span *BC*, the support *C* sinks by 30 mm while support *B* sinks by 20 mm. Hence, the net displacement of *C* with respect to *B*, is (30 – 20) = 10 mm. This displacement is in the clockwise direction with respect to *B*. Hence, the value of Δ for the span *BC* is +10 mm. The value of Δ for the span *CD* is as follows. The support *C* sinks by 30 mm with respect to *D* in the anticlockwise direction. Hence, Δ is taken as –30 mm. As the supports *A* and *D* are at the same level and as such they are fixed supports, the displacements are zeros.

   $$\theta_A = + \;;\quad \theta_B \neq 0,\ \Delta_B = 20 \text{ mm},\ \theta_C \neq 0,\ \Delta_C = 10 \text{ mm},$$

   $$\Delta_D = 0$$

2. Relative Stiffness Values

Member	$(2EI/l)$
AB	$2 \times 15000/3 = 10000$
BC	$2 \times 15000/4 = 7500$
CD	$2 \times 15000/2 = 15000$

3. Fixed End Moments

$$M_{FAB} = -30 \times \frac{3^2}{12} = -22.5\text{ kNm}, M_{FBA} = +22.5\text{ kNm}$$

$$M_{FBC} = -1 \times 100 \times \frac{3^2}{4^2} = -56.25\text{ kNm}, M_{FCB} = +\frac{3 \times 100 \times 1^2}{4^2} = 18.75\text{ kNm}$$

$$M_{FCD} = \frac{-20 \times 0.5 \times 1.5}{2} = -7.5\text{ kNm}, M_{FDC} = +7.5\text{ kNm}$$

4. Slope Deflection Equations

$$M_{AB} = -22.5 + 10000\left(2 \times \frac{1}{200} + \theta_B - \frac{3 \times 20}{3000}\right)$$

$$M_{AB} = -22.5 + 10000\,\theta_B - 100 = 10000\,\theta_B - 122.5$$

$$M_{BA} = +22.5 + 10000\left(2\theta_B + \frac{1}{200} - \frac{3 \times 20}{3000}\right)$$

$$M_{BA} = 20000\,\theta_B - 127.5$$

$$M_{BC} = -56.25 + 7500\left(2\theta_B + \theta_C - \frac{3 \times 10}{4000}\right)$$

$$M_{BC} = 15000\,\theta_B + 7500\,\theta_C - 112.5$$

$$M_{CB} = 18.75 + 7500\left(2\theta_C + \theta_B - \frac{3 \times 10}{4000}\right)$$

$$M_{CB} = 7500\,\theta_B + 15000\,\theta_C - 37.5$$

$$M_{CD} = -7.5 + 15000\left(2\theta_C + \theta_D + \frac{3 \times 10}{2000}\right)$$

$$M_{CD} = 30000\,\theta_C + 667.5$$

$$M_{DC} = +7.5 + 15000\left(2\theta_D + \theta_C + \frac{3 \times 30}{2000}\right)$$

$$M_{DC} = 15000\,\theta_C + 682.5$$

5. Equilibrium Equations

Joint $B \Rightarrow$ $M_{BA} + M_{BC} = 0$

$$35000\ \theta_B + 7500\ \theta_C = 240 \quad (1)$$

Joint $C \Rightarrow$ $M_{CB} + M_{CD} = 0$

$$7500\ \theta_B + 45000\ \theta_C = -630 \quad (2)$$

Solving Eqs. (1) and (2);

$$\theta_B = 0.01022;\ \theta_C = -0.0157 \text{ radians}$$

6. End Moments

$$M_{AB} = 10000(0.01022) - 122.5 = -20.3 \text{ kNm}$$
$$M_{BA} = 20000(0.01022) - 127.5 = +76.9 \text{ kNm}$$
$$M_{BC} = 15000(0.01022) + 7500(-0.0157) - 112.5 = -76.25 \text{ kNm}$$
$$M_{CB} = 7500 \times 0.01022 + 15000 \times (-0.0157) - 37.5 = -196.5 \text{ kNm}$$
$$M_{CD} = 30000 \times (-0.0157) + 667.5 = +196.5 \text{ kNm}$$
$$M_{DC} = 15000 \times (-0.0157) + 682.5 = +447 \text{ kNm}$$

Example 1.9 Analyse the continuous beam shown in figure by the slope deflection method. Sketch the shear force and bending moment diagrams. Relative to support A, support B sinks by 1 mm and support C rises by 0.5 mm EI = 30000 kNm^2.

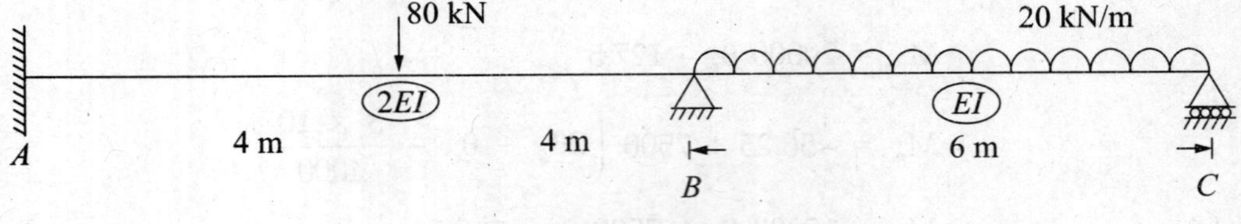

FIG. 1.36

Solution

1. Kinematic Redundants

At fixed support A, slope is zero. Hence, θ_A is zero. The intermediate support B rotates and hence θ_B is not equal to zero. The support at B sinks by 1 mm with respect to A. As this displacement is clockwise with respect to A, Δ_B is positive. Support C rises by 3.5 mm. Thus, C support is displaced by 1.5 mm relative to B. As the displacement is in the anticlockwise direction with respect to B; Δ_C is taken as negative. The roller support at C rotates and hence θ_C is not equal to zero.

$$\theta_A = 0,\quad \theta_B \neq 0,\quad \Delta_B = 1 \text{ mm } (\downarrow),\ \Delta_C = 1.5 \text{ mm } (\uparrow)$$
$$\theta_C \neq 0$$

2. Relative Stiffness Values

Member	$(2EI/l)$	k
AB	$\dfrac{2 \times (2 \times 30000)}{8}$	15000
BC	$2 \times 30000/6$	10000

3. Fixed End Moments

$$M_{FAB} = -80 \times \frac{8}{8} = -80 \text{ kNm}$$

$$M_{FBA} = +80 \text{ kNm}$$

$$M_{FBC} = -\frac{20 \times 6^2}{12} = -60 \text{ kNm}$$

$$M_{FCB} = +20 \times \frac{6^2}{12} = +60 \text{ kNm}$$

4. Slope Deflection Equations

$$M_{AB} = -80 + 15000\left(2\theta_A + \theta_B - \frac{3 \times 1}{8000}\right) = 15000\,\theta_B - 85.625$$

$$M_{BA} = +80 + 15000\left(2\theta_B + \theta_A - \frac{3 \times 1}{8000}\right) = 30000\,\theta_B + 74.375$$

$$M_{BC} = -60 + 10000\left(2\theta_B + \theta_C + \frac{3 \times 1.5}{6000}\right) = 20000\,\theta_B + 10000\,\theta_C - 52.5$$

$$M_{CB} = +60 + 10000\left(2\theta_C + \theta_B + \frac{3 \times 1.5}{6000}\right) = 10000\,\theta_B + 20000\,\theta_C + 67.5$$

5. Equilibrium Equations

Joint B ⇒ $M_{BA} + M_{BC} = 0$

$$50000\,\theta_B + 10000\,\theta_C = -21.875 \qquad (1)$$

Joint C ⇒ $M_{CB} = 0$

$$10000\,\theta_B + 20000\,\theta_C = -67.5 \qquad (2)$$

Solving Eqs. (1) and (2);

$$\theta_B = 2.6389 \times 10^{-4} \text{ radians}$$

$$\theta_C = -3.5070 \times 10^{-3} \text{ radians}$$

6. End Moments

$$M_{AB} = 15000\ (2.6389 \times 10^{-4}) - 85.625 = -81.67 \text{ kNm}$$

$$M_{BA} = 30000\ (2.6389 \times 10^{-4}) + 74.375 = 82.29 \text{ kNm}$$

$$M_{BC} = 20000\ (2.6389 \times 10^{-4}) + 10000\ (-3.5070 \times 10^{-3}) - 52.5 = -82.29 \text{ kNm}$$

$$M_{CB} = 10000\ (2.6389 \times 10^{-4}) + 20000\ (-3.5070 \times 10^{-3}) + 67.5 = 0$$

7. Free Body Diagrams

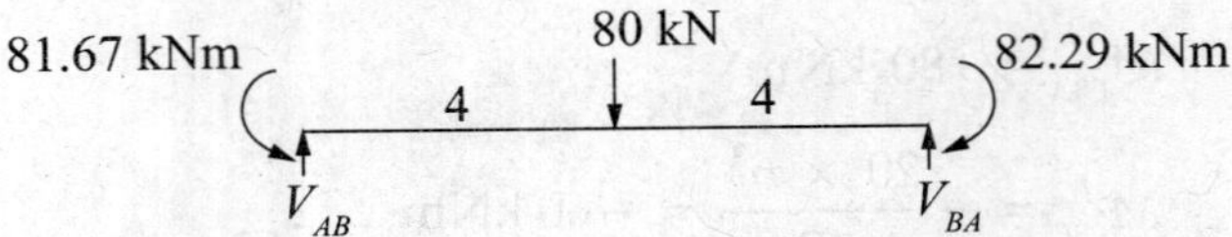

FIG. 1.37

$$\Sigma V = 0; \qquad V_{AB} + V_{BA} = 80 \tag{1}$$

$$\Sigma M = 0; \qquad -81.67 + 82.29 + 320 - 8V_{BA} = 0 \tag{2}$$

$$V_{BA} = 40.1 \text{ kN}$$

$$V_{AB} = 39.9 \text{ kN}$$

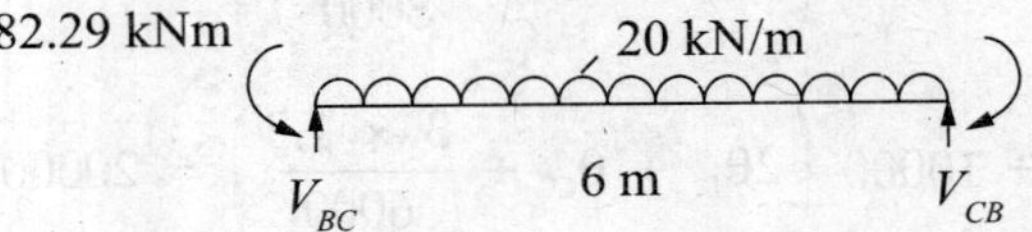

FIG. 1.38

$$\Sigma V = 0; \qquad V_{BC} + V_{CB} = 120 \tag{3}$$

$$\Sigma M = 0; \qquad -82.29 + 20 \times \frac{6^2}{2} - 6\,V_{CB} = 0$$

$$V_{CB} = 46.3 \text{ kN}$$

$$V_{BC} = 73.7 \text{ kN}$$

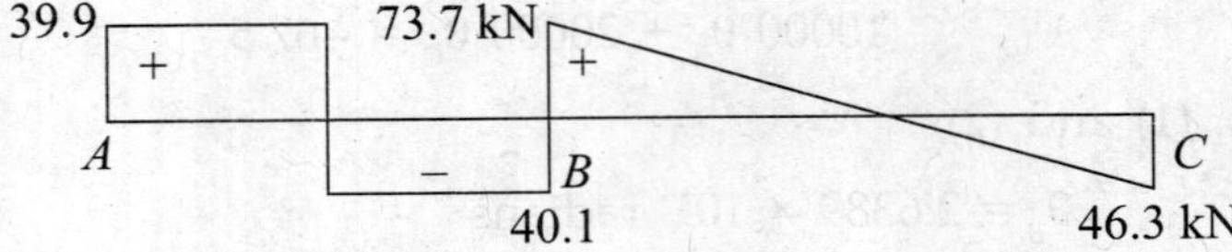

FIG. 1.39 Shear force diagram

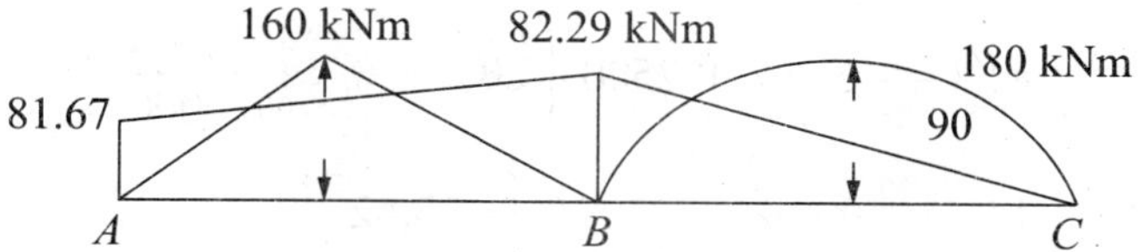

FIG. 1.40 Bending moment diagram

Example 1.10 Analyse the beam shown below by slope deflection method. The support A rotates or yield by (1/250) radian clockwise and B sinks by 6 mm.

$$EI = 5000 \text{ kNm}^2.$$

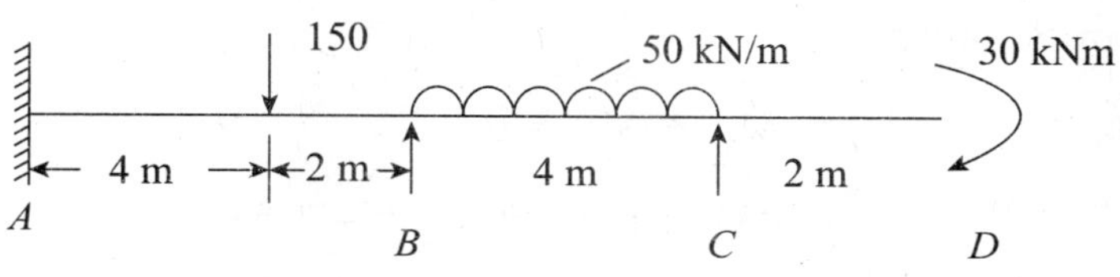

FIG. 1.41

1. Kinematic Redundants

$$\theta_A = +(1/250), \quad \theta_B \neq 0, \quad \theta_C \neq 0, \quad \Delta_B = 6 \text{ mm}$$

2. Relative Stiffness Values

Member	$(2EI/l)$	k
AB	$2 \times 5000/6$	1666.7
BC	$2 \times 5000/4$	2500.0

3. Fixed End Moments

$$M_{FAB} = -Wa^2b/l^2 = -150(4)2^2/6^2 = -66.67 \text{ kNm}$$

$$M_{FBA} = +Wab^2/l^2 = +150(2)4^2/6^2 = +133.33 \text{ kNm}$$

$$M_{FBC} = -wl^2/12 = -50(4)^2/12 = -66.67 \text{ kNm}$$

$$M_{FCB} = +wl^2/12 = +50(4)^2/12 = +66.67 \text{ kNm}$$

4. Slope Deflection Equations

$$M_{AB} = -66.67 + 1666.7\left((2/250) + \theta_B - \frac{3 \times 6}{6000}\right)$$

$$M_{AB} = 1666.67\,\theta_B - 58.34 \quad (1)$$

$$M_{BA} = 133.33 + 1666.7\left(2\theta_B + (1/250) - \frac{3 \times 6}{6000}\right)$$

$$M_{BA} = 3333.4\,\theta_B + 135.00 \quad (2)$$

$$M_{BC} = -66.67 + 2500\left(2\theta_B + \theta_C + \frac{3 \times 6}{6000}\right)$$

$$M_{BC} = 5000\theta_B + 2500\theta_C - 55.42 \quad (3)$$

$$M_{CB} = +66.67 + 2500\left(2\theta_C + \theta_B + \frac{3 \times 6}{6000}\right)$$

$$M_{CB} = 2500\theta_B + 5000\theta_C + 77.92 \quad (4)$$

$$M_{CD} = -30 \quad (5)$$

5. Equilibrium Equations

Joint $B \Rightarrow$

$$M_{BA} + M_{BC} = 0$$

$$8333.4\theta_B + 2500\theta_C = -79.58 \quad (6)$$

Joint $C \Rightarrow$

$$M_{CB} = -30$$

$$2500\theta_B + 5000\theta_C + 77.92 - 30 = 0$$

$$2500\theta_B + 5000\theta_C = -47.92 \quad (7)$$

Solving the above equations;

$$\theta_B = -7.852(10)^{-3}$$

$$\theta_C = -5.658(10)^{-3}$$

6. End Moments

Substituting the values of θ_B and θ_C in equilibrium equations, we obtain

$$M_{AB} = 58.34 + 1666.7(-7.852 \times 10^{-3}) = -71.43 \text{ kNm}$$

$$M_{BA} = 3333.4(-7.852 \times 10^{-3}) + 135.0 = +108.82 \text{ kNm}$$

$$M_{BC} = 5000(-7.852 \times 10^{-3}) + 2500(-5.658 \times 10^{-3}) - 55.42 = -108.82$$

$$M_{CB} = 2500(-7.852 \times 10^{-3}) + 5000(-5.658 \times 10^{-3}) + 77.92 = +30.00$$

$$M_{CD} = -30.00 \text{ kNm}$$

The shear force and bending moment diagrams can be drawn using the above end moments and free body diagram.

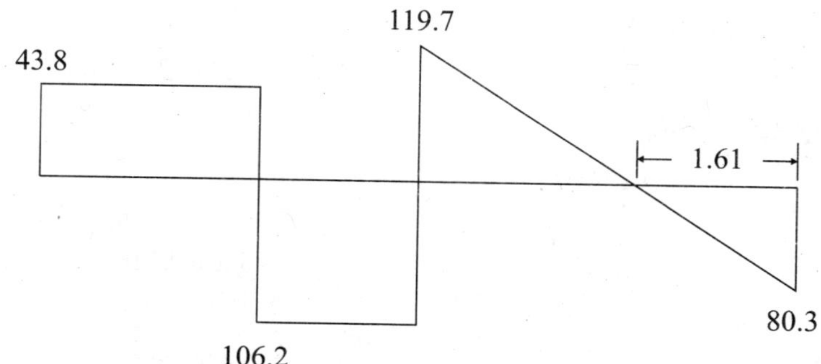

FIG. 1.42 Shear force diagram

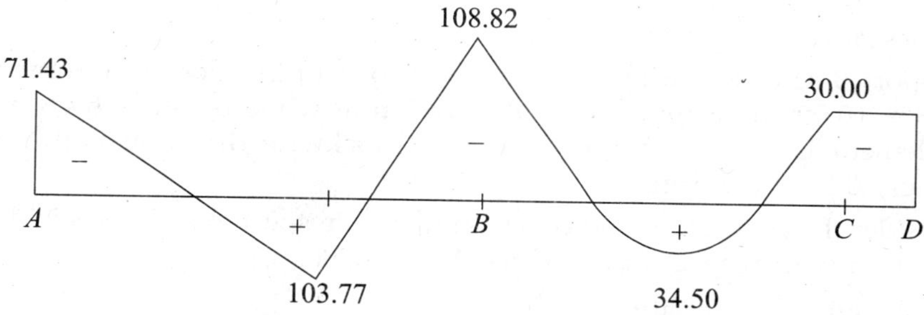

FIG. 1.43 Bending moment diagram

Example 1.11 Analyse the beam slope deflection method. Draw the bending moment diagram and shear force diagram. Support B and C settle by 8 mm and 3 mm respectively. Take $EI = 2(10)^4$ kNm^2

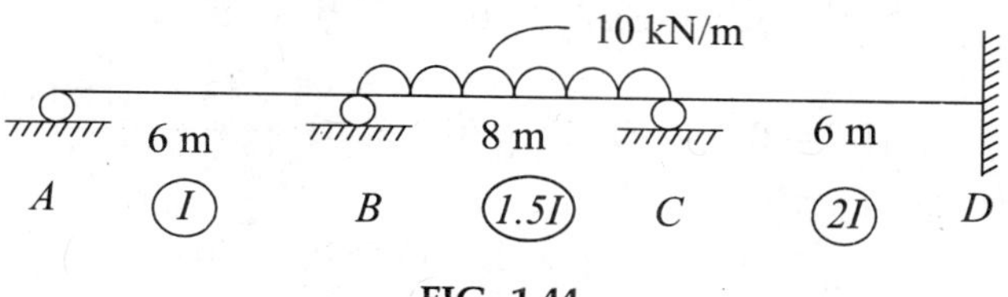

FIG. 1.44

1. Kinematic Redundants

   As the end D is fixed $\theta_D = 0$. As the near end A is on simple support $\theta_A \neq 0$. At the intermediate support B and C, $\theta_B \neq 0$, $\theta_C \neq 0$.

2. Relative Stiffness Values

   The support B and C settles and hence the induced moments depend on $(2EI/l)$. Their values are given below.

Member	$(2EI/l)$
AB	$2 \times 2 \times 10^4/6 = 6666.67$

BC	$2 \times 1.5 \times 2 \times 10^4/8 = 7500.00$
CD	$2 \times 2 \times 2 \times 10^4/6 = 13333.33$

3. Fixed End Moments

$$M_{FAB} = M_{FBA} = 0$$

$$M_{FBC} = \frac{-wl^2}{12} - -10 \times \frac{8^2}{12} = -53.33 \text{ kNm}$$

$$M_{FCB} = \frac{+wl^2}{12} - +10 \times \frac{8^2}{12} = +53.33 \text{ kNm}$$

$$M_{FCD} = M_{FDC} = 0$$

4. Values of Δ

   Support B settles by 8 mm with respect to A in the clockwise direction. Hence, ΔAB = +0.008 m. Support C settles by 3 mm while B settles by 8 mm. The net settlement is 8 – 3 = 5 mm in the anticlockwise direction with respect to C. Hence, Δ_{BC} = –0.005 m.

   Consider the span CD, C settles by 3 mm with respect to D. This is in anticlockwise direction with respect to D. Hence Δ_{CD} = –0.003 m.

5. Slope Deflection Equations

$$M_{AB} = M_{FAB} + \frac{2EI}{l}\left(2\theta_A + \theta_B - \frac{3\Delta}{l}\right)$$

$$M_{AB} = 0 + 6666.67(2\theta_A + \theta_B - 3 \times 0.008/6)$$

$$M_{AB} = 13333.34\,\theta_A + 6666.67\,\theta_B - 26.67 \tag{1}$$

$$M_{BA} = 0 + 6666.67\left(2\theta_B + \theta_A - \frac{3 \times 0.008}{6}\right)$$

$$M_{BA} = 6666.67\,\theta_A + 13333.34\,\theta_B - 26.67 \tag{2}$$

$$M_{BC} = -53.33 + 7500\left(2\theta_B + \theta_C - \frac{3 \times 0.005}{8}\right)$$

$$M_{BC} = 15000\,\theta_B + 7500\,\theta_C - 39.27 \tag{3}$$

$$M_{CB} = +53.33 + 7500\left(2\theta_C + \theta_B - \frac{3 \times 0.005}{8}\right)$$

$$M_{CB} = 7500\,\theta_B + 15000\,\theta_C + 67.39 \tag{4}$$

$$M_{CD} = 0 + 13333.33\left(2\theta_C + \theta_D + \frac{3 \times 0.003}{6}\right)$$

$$M_{CD} = 26666.66\,\theta_C + 20 \quad (5)$$

$$M_{DC} = 0 + 13333.33\left(2\theta_D + \theta_C + \frac{3 \times 0.003}{6}\right)$$

$$M_{DC} = 13333.33\,\theta_C + 20 \quad (6)$$

6. Equilibrium Equations

Joint $B \Rightarrow$ $M_{BA} + M_{BC} = 0$

$$6666.67\theta_A + 28333.34\theta_B + 7500\theta_C - 65.94 = 0$$

$$6666.67\theta_A + 28333.34\theta_B + 7500\theta_C = 65.94 \quad (7)$$

Joint $C \Rightarrow$ $M_{CB} + M_{CD} = 0$

$$7500\theta_B + 41666.66\theta_C + 87.39 = 0$$

$$7500\theta_B + 41666.66\theta_C = -87.39 \quad (8)$$

Joint $A \Rightarrow$ $M_{AB} = 0$

$$13333.34\theta_A + 6666.67\theta_B = +26.67 \quad (9)$$

Solving Eqs. 7 - 9

$$\theta_A = 5.555(10)^{-4}$$

$$\theta_B = 2.889(10)^{-3}$$

$$\theta_C = -2.617(10)^{-3}$$

7. End Moments

$$M_{AB} = 13333.34(5.555)\,10^{-4} + 6666.67(2.889) \times 10^{-3} - 26.67 = 0$$

$$M_{BA} = 6666.7(5.555)\,10^{-4} + 13333.34(2.889) \times 10^{-3} - 26.67 = 15.55 \text{ kNm}$$

$$M_{BC} = 15000(2.889)\,10^{-3} + 7500(-2.617 \times 10^{-3}) - 39.27 = -15.56 \text{ kNm}$$

$$M_{CB} = 7500(2.889)\,10^{-3} + 15000(-2.617)\,10^{-3} + 67.39 = 49.80 \text{ kNm}$$

$$M_{CD} = 26666.6(-2.617)\,10^{-3} + 20 = -49.80 \text{ kNm}$$

$$M_{DC} = 13333.3(-2.617)\,10^{-3} + 20 = -14.89 \text{ kNm}$$

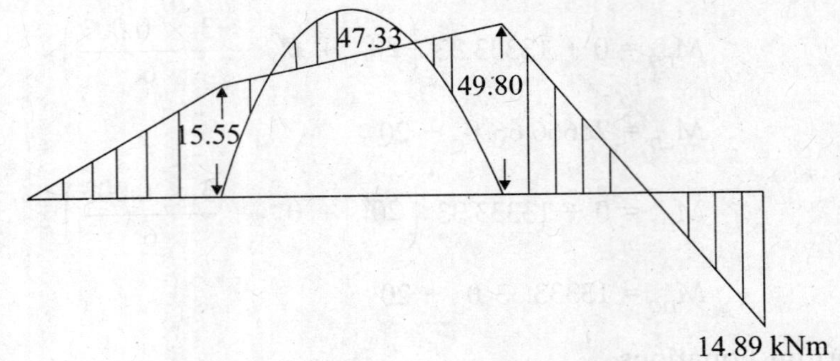

FIG. 1.45 Bending moment diagram

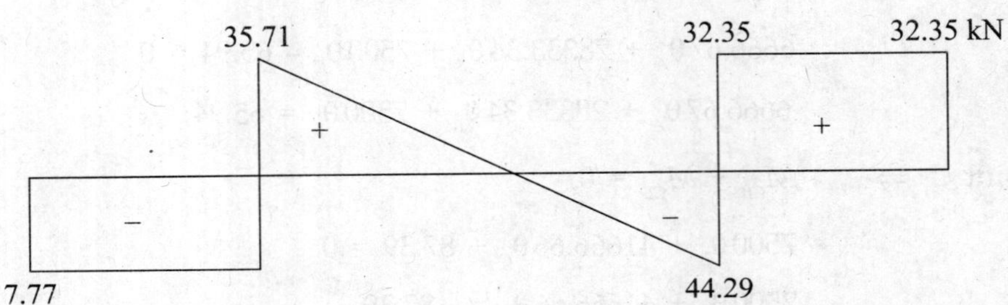

FIG. 1.46 Shear force diagram

Example 1.12 A continuous beam *ABCD* is 12 metre long is fixed at *A* and *D* and is loaded as shown in Fig. 1.47. Analyse the beam completely by slope deflection method. The following deformation takes place simultaneously.

(1) The end *A* yields twinning through (1/250) radians in a clockwise direction

(2) The end *B* sinks 30 mm in downward direction

(3) The end *C* sinks 20 mm in downward direction

The beam has a constant $I = 38.20(10)^5$ mm^4 and $E = 2(10)^5$ N/mm^2. Draw BMD

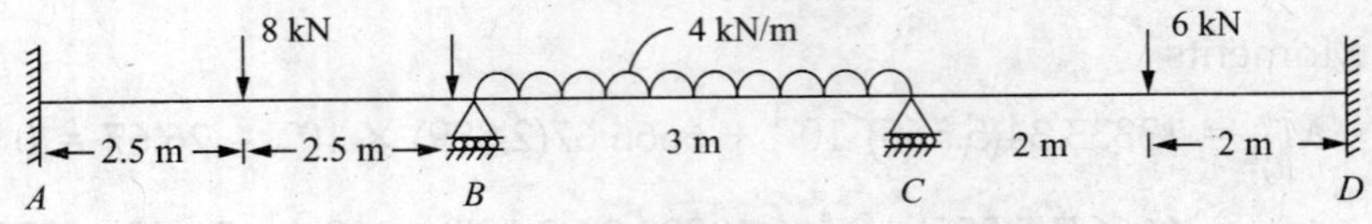

FIG. 1.47

Solution

1. Kinematic Redundants

As the supports *A* and *D* are fixed $\theta_A = \theta_D = 0$. Thus, the redundants are θ_B and θ_C as they are slopes at intermediate supports.

2. Relative Stiffness Values

Member	$(2EI/l)$
AB	$\dfrac{2 \times 2 \times 10^5 \times 38.2 \times 10^5}{5000 \times 1000 \times 1000} = 305.6$
BC	$\dfrac{2 \times 2 \times 10^5 \times 38.2 \times 10^5}{3000 \times 1000 \times 1000} = 509.33$
CD	$\dfrac{2 \times 2 \times 10^5 \times 38.2 \times 10^5}{4000 \times 1000 \times 1000} = 382.00$

3. Fixed End Moments

$$M_{FAB} = \frac{-Wl}{8} = -8 \times \frac{5}{8} = -5\,\text{kNm}$$

$$M_{FBA} = \frac{+Wl}{8} = +8 \times \frac{5}{8} = +5\,\text{kNm}$$

$$M_{FBC} = \frac{-Wl^2}{12} = -4 \times \frac{3^2}{12} = -3\,\text{kNm}$$

$$M_{FCB} = \frac{+Wl^2}{12} = +4 \times \frac{3^2}{12} = +3\,\text{kNm}$$

$$M_{FCD} = \frac{-Wl}{8} = -6 \times \frac{4}{8} = -3\,\text{kNm}$$

$$M_{FDC} = \frac{+Wl}{8} = +6 \times \frac{4}{8} = +3\,\text{kNm}$$

4. Slope Deflection Equations

$$M_{AB} = -5 + 305.6\left(2 \times \frac{1}{250} + \theta_B - \frac{3 \times 30}{5000}\right)$$

$$M_{AB} = 305.6\,\theta_B - 8.06 \tag{1}$$

$$M_{BA} = +5 + 305.6\left(2\theta_B + \frac{1}{250} + - \frac{3 \times 30}{5000}\right)$$

$$M_{BA} = 611.2\,\theta_B + 0.72 \tag{2}$$

$$M_{BC} = -3 + 509.33\left(2\theta_B + \theta_C + \frac{3 \times 10}{3000}\right)$$

$$M_{BC} = 1018.66\,\theta_B + 509.33\,\theta_C + 2.09 \tag{3}$$

$$M_{CB} = +3 + 509.33\left(2\theta_C + \theta_B + \frac{3 \times 10}{3000}\right)$$

$$M_{CB} = 509.33\,\theta_B + 1018.66\,\theta_C + 8.09 \tag{4}$$

$$M_{CD} = -3 + 382\left(2\theta_C + \cancel{\theta_D} + \frac{3 \times 20}{4000}\right)$$

$$M_{CD} = 2.73 + 764\,\theta_C$$

$$M_{DC} = +3 + 382\left(2\cancel{\theta_D} + \theta_C + \frac{3 \times 20}{4000}\right) = 382\,\theta_C + 8.73 \tag{5}$$

5. Equilibrium Equations

Joint B ⇒ $M_{BA} + M_{BC} = 0$

$$1629.86\,\theta_B + 509.33\,\theta_C = -2.81 \tag{6}$$

Joint C ⇒ $M_{CB} + M_{CD} = 0$

$$509.33\,\theta_B + 1782.66\,\theta_C = -10.82 \tag{7}$$

Solving Eqs. (5) and (6);

$$\theta_B = 1.8959(10)^{-4}$$

$$\theta_C = -6.1238(10)^{-3}$$

6. End Moments

$$M_{AB} = -8.06 + 305.6(1.8959)10^{-4} = -8.00 \text{ kNm}$$

$$M_{BA} = 611.2(1.8959)10^{-4} + 0.72 = +0.84 \text{ kNm}$$

$$M_{BC} = 1018.66(1.8959)10^{-4} + 509.33(-6.1238)10^{-3} + 2.09 = -0.84 \text{ kNm}$$

$$M_{CB} = 509.33(1.8959)10^{-4} + 1018.66(-6.1238)10^{-3} + 8.09 = +1.95 \text{ kNm}$$

$$M_{CD} = 2.73 + 764(-6.1238)10^{-3} = -1.95 \text{ kNm}$$

$$M_{DC} = 382(-6.1238)10^{-3} + 8.73 = +6.39 \text{ kNm}$$

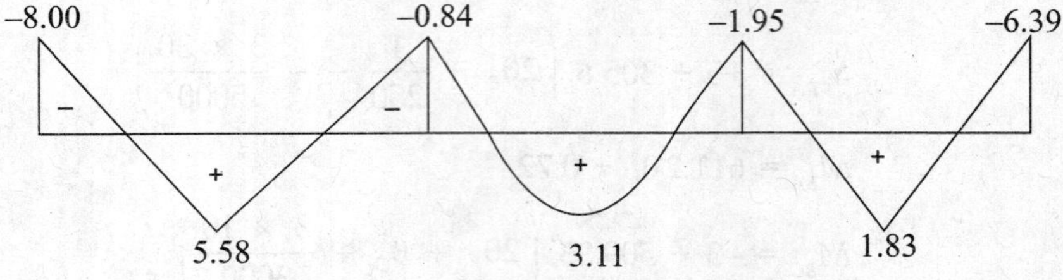

FIG. 1.48 Bending moment diagram

Example 1.13 Evaluate the kinematic redundants for the beam shown below. The support B sinks by 5 mm. Take E = 210 GPa, I = 0.1 G mm^4

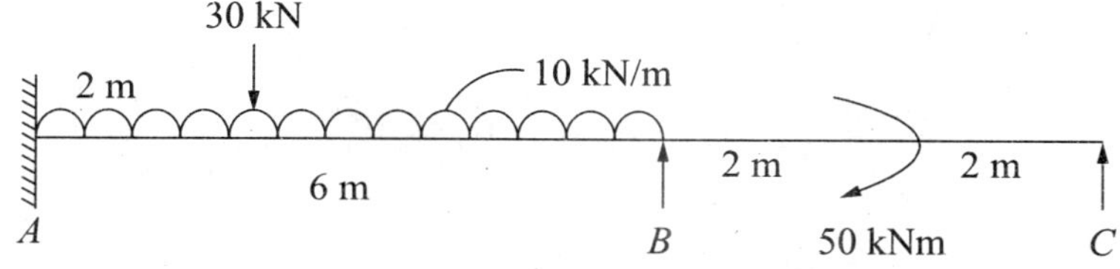

FIG. 1.49

Solution

1. Kinematic Redundants

   As the end A is fixed, $\theta_A = 0$. The support B is an intermediate support and therefore $\theta_B \neq 0$. At the roller support $\theta_C \neq 0$. Thus, we have to compute the two kinematic redundants θ_B and θ_C.

2. Relative Stiffness Values

Member	$(2EI/l)$	k
AB	$2 \times 21000/6$	7000
BC	$2 \times 21000/4$	10500

Note: $E = 210 \text{ GPa} = 210(10)^9 \times 10^{-6} = 2.1 \times 10^5 \text{ N/mm}^2$

$I = 0.1 \text{ G mm}^4 = 0.1 \times 10^9 \text{ mm}^4$

$$EI = \frac{2.1(10^5) \times 0.1 \times 10^9}{10^3 \times (10^3)^2} = 21000 \text{ kNm}^2$$

3. Fixed End Moments

$$M_{FAB} = -\left[\frac{wl^2}{12} + \frac{Wab^2}{l^2}\right] = -\left[\frac{10 \times 6^2}{12} + \frac{30 \times 2 \times 4^2}{6^2}\right] = -56.67 \text{ kNm}$$

$$M_{FBA} = +\left[\frac{wl^2}{12} + \frac{Wa^2b}{l^2}\right] = -\left[\frac{10 \times 6^2}{12} + \frac{30 \times 4 \times 2^2}{6^2}\right] = +43.33 \text{ kNm}$$

$$M_{FBC} = \frac{M}{4} = \frac{50}{4} = M_{FCB}$$

4. Slope Deflection Equations

$$M_{AB} = -56.67 + 7000\left(2\theta_A + \theta_B - \frac{3 \times 5}{1000 \times 6}\right)$$

$$M_{AB} = -56.67 + 7000(\theta_B - 2.5 \times 10^{-3})$$

$$M_{AB} = 7000\,\theta_B - 74.17 \quad (1)$$

$$M_{BA} = +43.33 + 7000\left(2\theta_B + \theta_A + \frac{3 \times 5}{1000 \times 6}\right)$$

$$M_{BA} = 14000\,\theta_B + 25.83 \quad (2)$$

$$MBC = 12.5 + 10500\left(2\theta_B + \theta_C + \frac{3 \times 5}{1000 \times 4}\right)$$

$$M_{BC} = 21000\,\theta_B + 10500\,\theta_C + 51.88 \quad (3)$$

$$M_{CB} = 12.5 + 10500\left(2\theta_C + \theta_B + \frac{3 \times 5}{1000 \times 4}\right)$$

$$M_{CB} = 10500\,\theta_B + 21000\,\theta_C + 51.88 \quad (4)$$

5. Equilibrium Conditions

   Joint B,

$$M_{BA} + M_{BC} = 0$$

$$35000\,\theta_B + 10500\,\theta_C = -77.71 \quad (5)$$

   Joint C,

$$M_{CB} = 0$$

$$10500\,\theta_B + 21000\,\theta_C = -51.88 \quad (6)$$

   Solving the above;

$$\theta_B = -1.74(10)^{-3}$$

$$\theta_C = -1.60(10)^{-3}$$

Example 1.14 Analyse the continuous beam using slope deflection method. If the support B sinks by 5 mm. Draw bending moment diagram and shear force diagram. Assume EI = 4000 kNm² (VTU, July 2011)

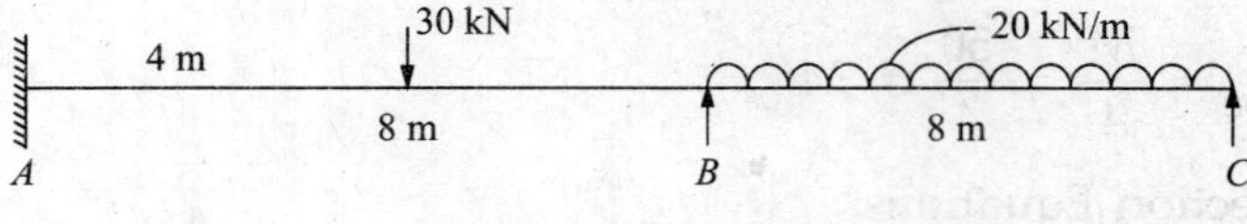

FIG. 1.50

Solution

1. Kinematic Redundants

$$\theta_A = 0, \quad \theta_B \neq 0 \text{ and } \theta_C \neq 0$$

2. Relative Stiffness Values

The value of $2EI/l$ for member AB and BC are $2(4000/8) = 1000$ each.

Member	$(2EI/l)$	k
AB	1000	1000
BC	1000	1000

3. Fixed End Moments

$$M_{FAB} = \frac{-Wl}{8} = -30 \times \frac{8}{8} = -30 \text{ kNm}$$

$$M_{FBA} = \frac{+Wl}{8} = +30 \times \frac{8}{8} = +30 \text{ kNm}$$

$$M_{FBC} = \frac{-wl^2}{12} = -\frac{20 \times 8^2}{12} = -106.67 \text{ kNm}$$

$$M_{FCB} = \frac{+wl^2}{12} = +\frac{20 \times 8^2}{12} = +106.67 \text{ kNm}$$

4. Slope Deflection Equation

$$M_{AB} = M_{FAB} + \frac{2El}{l}\left(2\theta_A + \theta_B - \frac{3\Delta}{l}\right)$$

$$M_{AB} = -30 + 1000\left(0 + \theta_B - \frac{3 \times 5 \times 10^{-3}}{8}\right)$$

$$M_{AB} = -30 + 1000\,\theta_B - 1.875$$

$$M_{AB} = 1000\,\theta_B - 31.875 \qquad (1)$$

$$M_{BA} = 30 + 1000\left(2\theta_B - \frac{3 \times 5 \times 10^{-3}}{8}\right)$$

$$M_{BA} = 30 + 2000\,\theta_B - 1.875$$

$$M_{BA} = 2000\,\theta_B + 28.125 \qquad (2)$$

$$M_{BC} = -106.67 + 1000\left(2\theta_B + \theta_C + \frac{3 \times 5 \times 10^{-3}}{8}\right)$$

$$M_{BC} = -106.67 + 2000\,\theta_B + 1000\,\theta_C + 1.875$$

$$M_{BC} = 2000\,\theta_B + 1000\,\theta_C - 104.795 \qquad (3)$$

$$M_{CB} = +106.67 + 1000\left(2\theta_C + \theta_B + \frac{3 \times 5 \times 10^{-3}}{8}\right)$$

$$M_{CB} = 1000\theta_B + 2000\theta_C + 108.545 \quad (4)$$

Note:
The value of Δ is positive for span *AB* as *B* rotates in clockwise direction with respect to *A* and Δ is negative for span *CB* as *B* rotates in anticlockwise direction with respect to *C*.

5. Joint Equilibrium Equations

$$M_{BA} + M_{BC} = 0 \quad (5)$$

$$4000\theta_B + 1000\theta_C = 76.67 \quad (6)$$

$$M_{CB} = 0 \quad (7)$$

$$1000\theta_B + 2000\theta_C = -108.55 \quad (8)$$

Solving Eqs. (6) and (8)

$$\theta_B = 0.0374$$

$$\theta_C = -0.0730$$

6. End Moments

$$M_{AB} = -30 + 1000(0.0374) - 1.875 = 5.53 \text{ kNm}$$

$$M_{BA} = 2000(0.0374) + 28.125 = 102.93 \text{ kNm}$$

$$M_{BC} = 2000(0.0374) + 1000(-0.073) - 104.795 = -102.9 \text{ kNm}$$

$$M_{CB} = 0$$

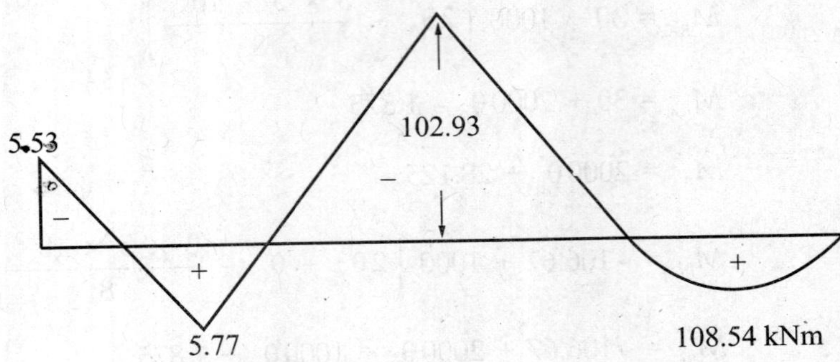

FIG. 1.51 Bending moment diagram

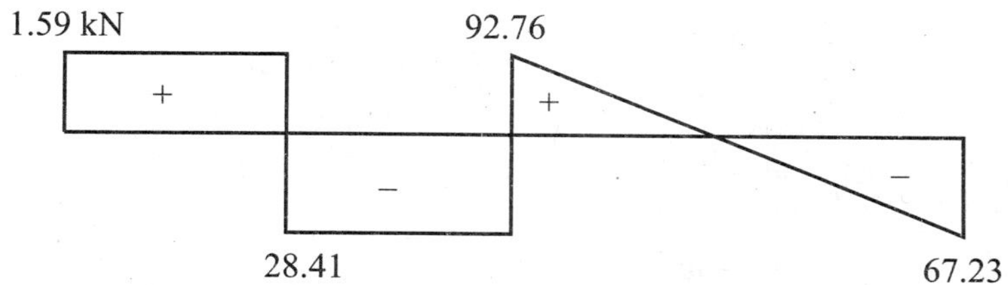

FIG. 1.52 Shear force diagram

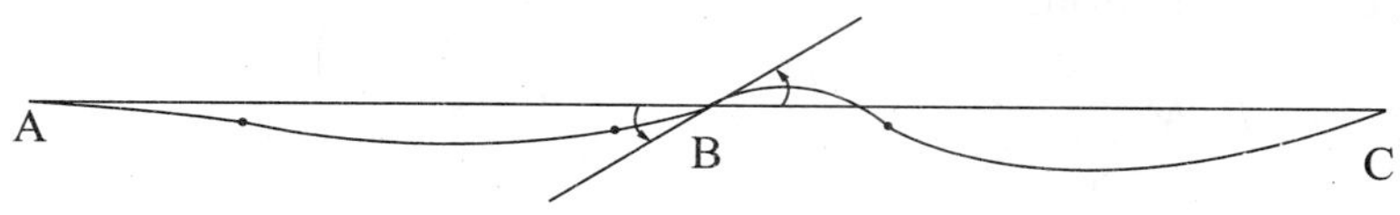

FIG. 1.53 Elastic curve

Example 1.15 Analyse the continuous beam shown in Fig. 1.54 by slope deflection method. The support B sinks by 10 mm. Take $EI = 3000 \text{ kNm}^2$ (VTU, Dec. 2010)

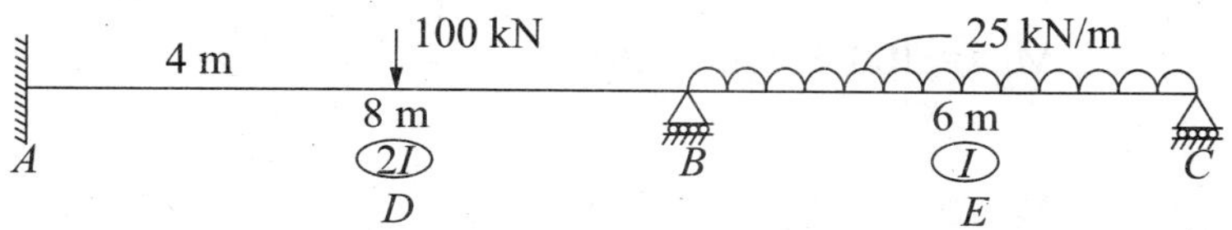

FIG. 1.54

Solution

1. Kinematic Redundants

$$\theta_A = 0, \quad \theta_B \neq 0 \quad \text{and} \quad \theta_C \neq 0$$

2. Relative Stiffness Values

   As the support B sinks, the induced moment depends on the flexural rigidity and the span. Hence, their contributing values are calculated as follows.

Member	$(2EI/l)$	k
AB	$\frac{2E(2I)}{8} = 0.5EI$	1500
BC	$\frac{2EI}{6} = 0.33EI$	1000

3. Fixed End Moments

$$M_{FAB} = \frac{-W \times l}{8} = -100 \times \frac{8}{8} = -100 \text{ kNm}$$

$$M_{FBA} = \frac{+Wl}{8} = 100 \times \frac{8}{8} = +100 \text{ kNm}$$

$$M_{FBC} = \frac{-wl^2}{12} = -25 \times \frac{6^2}{12} = -75 \text{ kNm}$$

$$M_{FCB} = \frac{+wl^2}{12} = +25 \times \frac{6^2}{12} = +75 \text{ kNm}$$

4. Slope Deflection Equations

$$M_{AB} = M_{FAB} + \frac{2EI}{l}\left(2\theta_A + \theta_B + \frac{3\Delta}{l}\right)$$

$$M_{AB} = -100 + 1500\left(\theta_B - \frac{3 \times 10 \times 10^{-3}}{8}\right)$$

$$M_{AB} = 1500\theta_B - 105.625 \tag{1}$$

$$M_{BA} = 100 + 1500\left(2\theta_B - \frac{3 \times 10 \times 10^{-3}}{8}\right)$$

$$M_{BA} = 3000\theta_B + 94.375 \tag{2}$$

$$M_{BC} = -75 + 1000\left(2\theta_B + \theta_C + \frac{3 \times 10 \times 10^{-3}}{6}\right)$$

$$M_{BC} = 2000\theta_B + 1000\theta_C - 70 \tag{3}$$

$$M_{CB} = 75 + 1000\left(2\theta_C + \theta_B + \frac{3 \times 10 \times 10^{-3}}{6}\right)$$

$$M_{CB} = 1000\theta_B + 2000\theta_C + 80 \tag{4}$$

5. Equilibrium Equations

At joint B;

$$M_{BA} + M_{BC} = 0$$

$$5000\theta_B + 1000\theta_C = -24.375 \tag{5}$$

At joint C;

$$M_{CB} = 0$$

$$1000\theta_B + 2000\theta_C = -80.000 \tag{6}$$

On solving Eqs. (5) and (6);

$$\theta_B = 3.475(10)^{-3}$$

$$\theta_C = -0.0417$$

6. End Moments

On substitution;

$$M_{AB} = 1500\,(3.472(10)^{-3}) - 105.625 = -100.42 \text{ kNm}$$

$$M_{BA} = 3000\,(3.472(10)^{-3}) + 94.375 = +104.79 \text{ kNm}$$

$$M_{BC} = 2000\,(3.472(10)^{-3}) + 1000(-0.0417) - 70 = -104.76 \text{ kNm}$$

$$M_{CB} = 1000\,(3.472(10)^{-3}) + 2000(-0.0417) + 80 = 0.0$$

7. Bending Moment Values

$$M_A = -100.42 \text{ kNm;}$$

$$M_D = 100 \times \frac{8}{4} - \left(\frac{100.42 + 104.79}{2}\right) = 97.4 \text{ kNm}$$

$$M_B = -104.79 \text{ kNm}$$

$$M_E = 25 \times \frac{6^2}{8} - \left(\frac{104.79 + 0}{2}\right) = +60.1 \text{ kNm}$$

$$M_C = 0$$

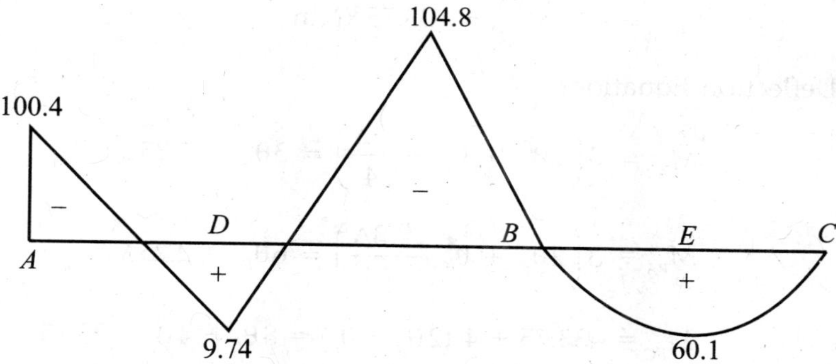

FIG. 1.55 Bending moment diagram

1.6 ANALYSIS OF RECTILINEAR FRAMES

Example 1.16 Analyse the frame by the slope deflection method and draw the bending moment diagram.

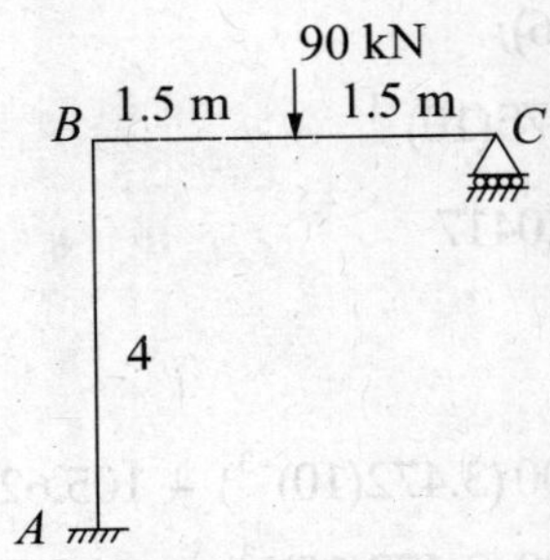

FIG. 1.56

Solution

1. Kinematic Redundants

$$\theta_B, \theta_C \text{ and } \Delta$$

2. Relative Stiffness Values

Member	(I/l)	k
AB	$I/4$	3
BC	$I/3$	4

3. Fixed End Moments

$$M_{FBC} = -90 \times \frac{3}{8} = -33.75 \text{ kNm}$$

$$M_{FCB} = +90 \times \frac{3}{8} = +33.75 \text{ kNm}$$

4. Slope Deflection Equations

$$M_{AB} = 3\left(2\theta_A + \theta_B - \frac{3\Delta}{4}\right) = 3\theta_B - 2.25\Delta$$

$$M_{BA} = 3\left(2\theta_B + \theta_A - \frac{3\Delta}{4}\right) = 6\theta_B - 2.25\Delta$$

$$M_{BC} = -33.75 + 4\,(2\theta_B + \theta_C) = 8\theta_B + 4\theta_C - 33.75$$

$$M_{CB} = +33.75 + 4\,(2\theta_C + \theta_B) = 4\theta_B + 8\theta_C + 33.75$$

5. Equilibrium Conditions

At Joint B ⇒ $M_{BA} + M_{BC} = 0$

$$14\theta_B + 4\theta_C - 2.25\Delta = 33.75 \qquad (1)$$

At Joint $C \Rightarrow$ $M_{CB} = 0$

$$4\theta_B + 8\theta_C = -33.75 \quad (2)$$

Column shear condition

$$\frac{M_{AB} + M_{BA}}{4} = 0$$

$$9\theta_B - 4.5\Delta = 0 \quad (3)$$

Solving the above equations,

$$\theta_B = 6.75 \qquad \theta_C = -7.5938 \qquad \Delta = 13.5$$

6. End Moments

$$M_{AB} = 3\ (6.75) - 2.25\ (13.5) = -10.125 \text{ kNm}$$

$$M_{BA} = 6\ (6.75) - 2.25\ (13.5) = +10.125 \text{ kNm}$$

$$M_{BC} = 8\ (6.75) + 4\ (-7.5938) - 33.75 = -10.125 \text{ kNm}$$

$$M_{CB} = 4\ (6.75) + 8\ (-7.5938) + 33.75 = 0$$

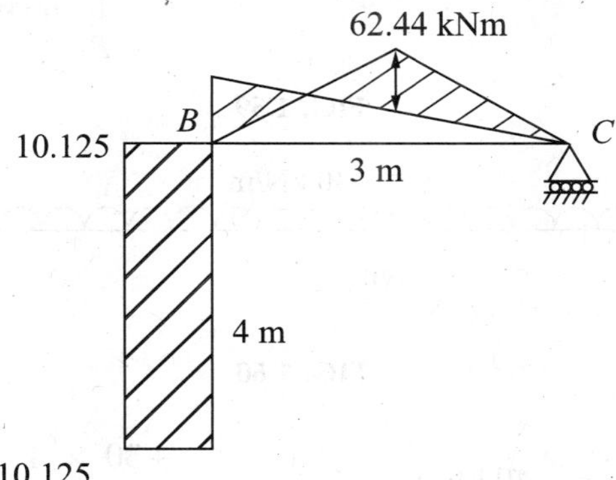

FIG. 1.57 Bending moment diagram

Example 1.17 Analyse the rigid frame by the slope deflection method. Draw bending moment diagram and elastic curve.

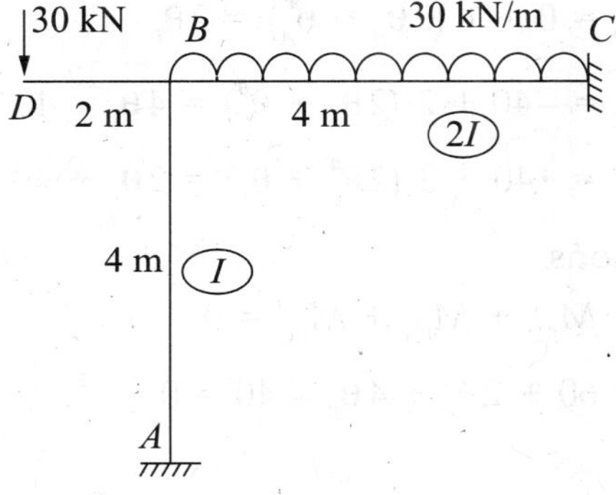

FIG. 1.58

Solution

1. Kinematic Redundants

   In the above frame, supports A and C are fixed. Hence, $\theta_A = \theta_C = 0$. The joint B rotates. Hence, the kinematic redundants to be determined is θ_B only.

2. Relative Stiffness Values

Member	(I/l)	k
AB	$I/4$	1
BC	$2I/4$	2

   The values of k are obtained by multiplying the individual members (I/l) with $(4/I)$.

3. Fixed End Moments

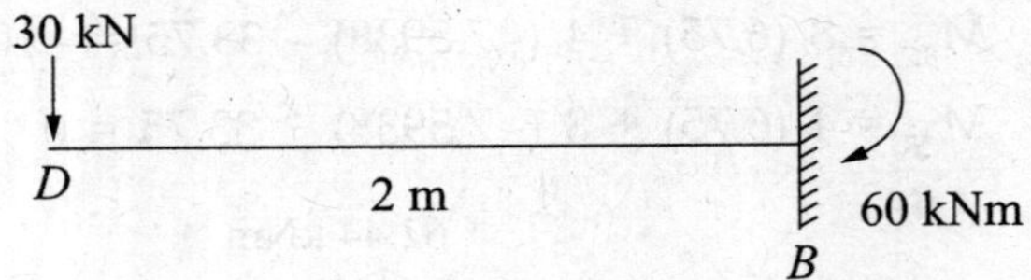

FIG. 1.59

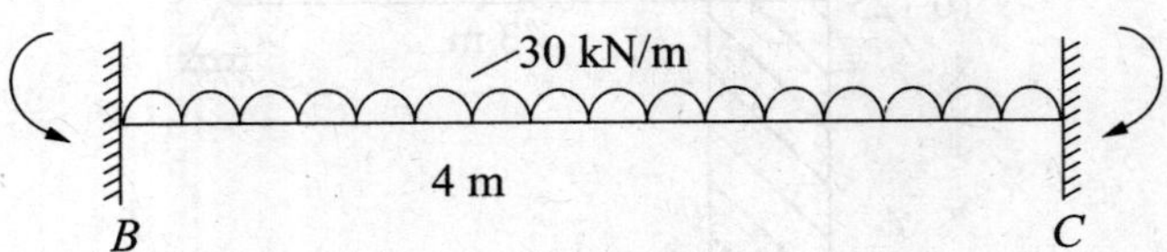

FIG. 1.60

$$\frac{-30 \times 4^2}{12} = -40 \text{ kNm} \qquad \frac{+30 \times 4^2}{12} = +40 \text{ kNm}$$

4. Slope Deflection Equations

$$M_{AB} = 0 + 1\,(2\theta_A + \theta_B) = \theta_B$$

$$M_{BA} = 0 + 1\,(2\theta_B + \theta_A) = 2\theta_B$$

$$M_{BC} = -40 + 2\,(2\theta_B + \theta_C) = 4\theta_B - 40$$

$$M_{CB} = +40 + 2\,(2\theta_C + \theta_B) = 2\theta_B + 40$$

5. Equilibrium Conditions

$$M_{BD} + M_{BA} + M_{BC} = 0$$

$$60 + 2\theta_B + 4\theta_B - 40 = 0$$

$$6\theta_B = -20$$

$$\boxed{\theta_B = -3.333 \text{ radians}}$$

Back substituting in the slope deflection equations.

6. End Moments

$$M_{AB} = -3.333 \text{ kNm}$$

$$M_{BA} = -6.667 \text{ kNm}$$

$$M_{BC} = 4(-3.333) - 40 = -53.332 \text{ kNm}$$

$$M_{CB} = 2(-3.333) + 40 = +33.334 \text{ kNm}$$

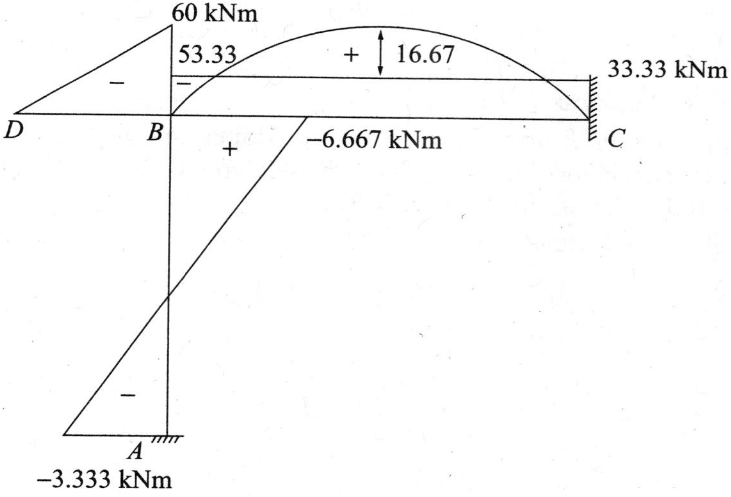

FIG. 1.61 Bending moment diagram

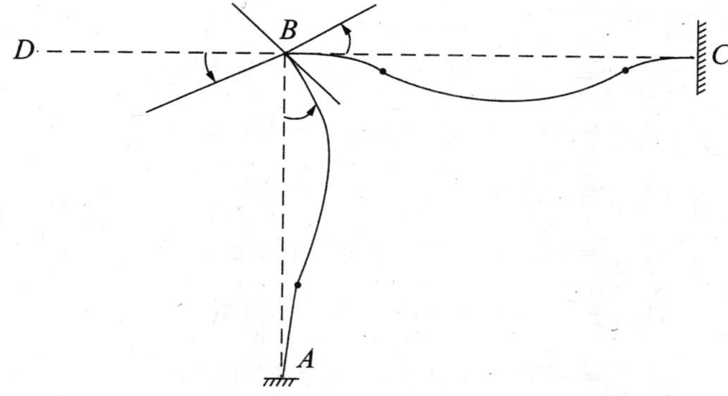

FIG. 1.62 Elastic curve

Example 1.18 Analyse the given frame by the slope deflection method. Draw the bending moment diagram.

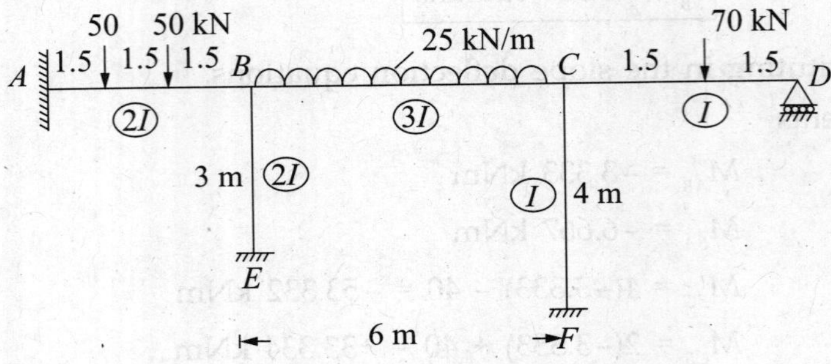

FIG. 1.63

Solution

1. Kinematic Redundants

   As the supports A, E and F are fixed, the slopes are zero. At the hinged end D, it can rotate. Hence, $\theta_D \neq 0$. The joints B and C are rigid joints. Thus, the kinematic redundants are θ_B, θ_C and θ_D.

2. Relative Stiffness Values

Member	(I/l)	k
AB	$(2I/4.5)$	12.0
BC	$(3I/6)$	13.5
CD	$(I/3)$	9.0
BE	$(2I/3)$	18.0
CF	$(I/4)$	6.75

3. Fixed End Moments

$$M_{FAB} = -1.5 \times 50 \times 3/4.5 = -50 \text{ kNm}$$

$$M_{FBA} = +1.5 \times 50 \times 3/4.5 = +50 \text{ kNm}$$

$$M_{FBC} = -25 \times 6^2/12 = -75 \text{ kNm}$$

$$M_{FCB} = +25 \times 6^2/12 = +75 \text{ kNm}$$

$$M_{FCD} = -70 \times 3/8 = -26.25 \text{ kNm}$$

$$M_{FDC} = +70 \times 3/8 = +26.25 \text{ kNm}$$

4. Slope Deflection Equations

$$M_{AB} = -50 + 12\,(2\theta_A + \theta_B) = 12\theta_B - 50$$

$$M_{BA} = +50 + 12\,(2\theta_B + \theta_A) = 24\theta_B + 50$$

$$M_{BC} = -75 + 13.5\,(2\theta_B + \theta_C) = 27\theta_B + 13.5\theta_C - 75$$

$$M_{CB} = +75 + 13.5\,(2\theta_C + \theta_B) = 13.5\theta_B + 27\theta_C + 75$$

$$M_{CD} = -26.25 + 9.0\,(2\theta_C + \theta_D) = 18\theta_C + 9\theta_D - 26.25$$

$$M_{DC} = +26.25 + 9.0\,(2\theta_D + \theta_C) = 9\theta_C + 18\theta_D + 26.25$$

$$M_{BE} = 18\,(2\theta_B + \theta_E) = 36\theta_B$$

$$M_{EB} = 18\,(2\theta_E + \theta_B) = 18\theta_B$$

$$M_{CF} = 6.75\,(2\theta_C + \theta_F) = 13.5\theta_C$$

$$M_{FC} = 6.75\,(2\theta_B + \theta_E) = 6.75\theta_C$$

5. Equilibrium Equations

Joint $B \Rightarrow$ $M_{BA} + M_{BC} + M_{BE} = 0$

$$87\theta_B + 13.5\theta_C = 25 \qquad (1)$$

Joint $C \Rightarrow$ $M_{CB} + M_{CD} + M_{CF} = 0$

$$13.5\theta_B + 58.5\theta_C + 9\theta_D = -48.75 \qquad (2)$$

Joint $\Rightarrow$ $M_{DC} = 0$

$$9\theta_C + 18\theta_D = -26.25 \qquad (3)$$

Solving the above equations

$$\theta_B = 0.4055 \quad \theta_C = -0.7611 \quad \theta_D = -1.0778 \text{ radians}$$

6. End Moments

$$M_{AB} = 12\,(0.4055) - 50 = -45.13 \text{ kNm}$$

$$M_{BA} = 24\,(0.4055) + 50 = +59.73 \text{ kNm}$$

$$M_{BC} = 27\,(0.4055) + 13.5\,(-0.7611) - 75 = -74.33 \text{ kNm}$$

$$M_{CB} = 13.5\,(0.4055) + 27\,(-0.7611) + 75 = +59.92 \text{ kNm}$$

$$M_{CD} = 18.0\,(-0.7611) + 9\,(-1.0778) - 26.25 = -49.65 \text{ kNm}$$

$$M_{DC} = 9\,(-0.7611) + 18\,(-1.0778) + 26.25 = 0 \text{ kNm}$$

$$M_{BE} = 36\,(0.4055) = 14.60 \text{ kNm}$$

$$M_{EB} = 18\,(0.4055) = 7.30 \text{ kNm}$$

$$M_{CF} = 13.5\ (-0.7611) = -10.28 \text{ kNm}$$

$$M_{FC} = 6.75\ (-0.7611) = -5.14 \text{ kNm}$$

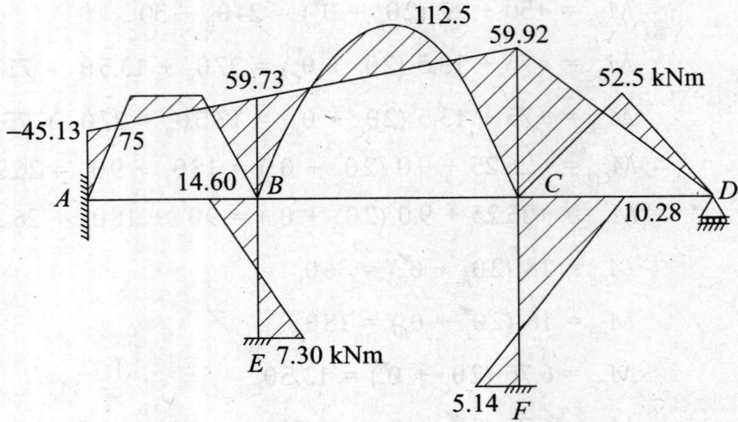

FIG. 1.64 Bending moment diagram

Example 1.19 A frame is loaded and supported as shown in figure. During loading, the hinge support D slips towards right by 2 mm. All the members have the same moment of inertia $4(10)^8$ mm^4 and $E = 2 \times 10^5$ N/mm^2. Determine the end moments and plot the bending moment diagram.

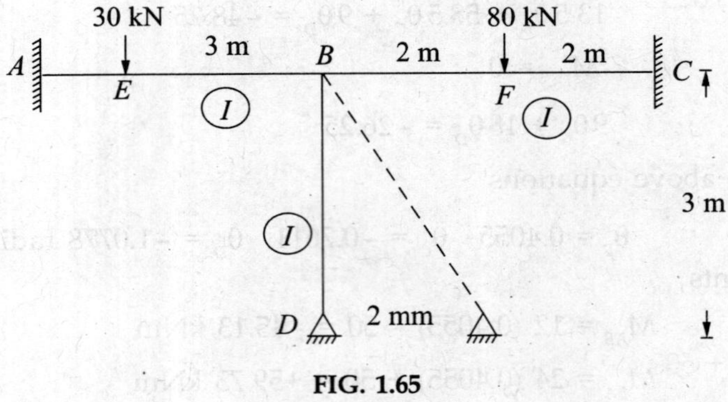

FIG. 1.65

Solution

1. Values of $(2EI/l = k)$

Member	$2EI/l$
AB	$\dfrac{2 \times 2 \times 10^5 \times 4 \times 10^8}{4000 \times 10^6} = 40000$

BC $\dfrac{2\times 2\times 10^5\times 4\times 10^8}{4000\times 10^6} = 40000$

BD $\dfrac{2\times 2\times 10^5\times 4\times 10^8}{3000\times 10^6} = 53333.3$

2. Fixed End Moments

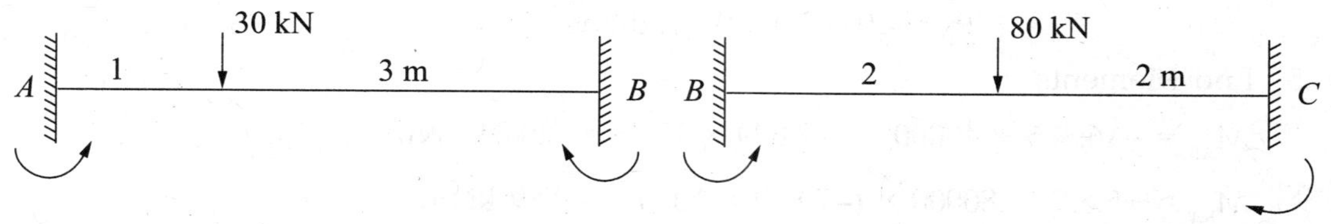

FIG. 1.66

$\dfrac{-1\times 30\times 3^2}{4^2}$ $\dfrac{+3\times 30\times 1^2}{4^2}$ $\dfrac{-80\times 4}{8}$ $\dfrac{+80\times 4}{8}$

$= -16.875$ kNm $= 5.625$ $= -40$ kNm $+40$ kNm

3. Slope Deflection Equations

$$M_{AB} = -16.875 + 40000\,(2\theta_A + \theta_B) = -16.875 + 40000\,\theta_B$$

$$M_{BA} = +5.625 + 40000\,(2\theta_B + \theta_A) = +5.625 + 80000\,\theta_B$$

$$M_{BC} = -40.000 + 40000\,(2\theta_B + \theta_C) = -40.000 + 80000\,\theta_B$$

$$M_{CB} = +40.000 + 40000\,(2\theta_C + \theta_B) = +40.000 + 40000\,\theta_B$$

$$M_{BD} = 0 + 53333.3\left(2\theta_B + \theta_D - \frac{-3(-2/1000)}{3}\right)$$

$$= -106.667 + 106666.6\,\theta_B + 53333.3\,\theta_D$$

$$M_{DB} = 0 + 53333.3\left(2\theta_D + \theta_B - \frac{-3(-2/1000)}{3}\right)$$

$$= 106.667 + 53333.3\,\theta_B + 106666.6\,\theta_D$$

4. Equilibrium Equations

$$M_{BA} + M_{BC} + M_{BD} = 0$$

$$266666.6\,\theta_B + 53333.3\,\theta_D = -72.292 \qquad (1)$$

At the hinge *D*;

$$M_{DB} = 0$$

$$53333.3\,\theta_B + 106666.6\,\theta_D = -106.667 \quad (2)$$

Solving the above

$$\theta_B = -7.899 \times 10^{-5} \text{ radians}$$

$$\theta_D = -9.605 \times 10^{-4} \text{ radians}$$

5. End Moments

$$M_{AB} = -16.875 + 40000 \times (-7.899 \times 10^{-5}) = -20.04 \text{ kNm}$$

$$M_{BA} = +5.625 + 80000 \times (-7.899 \times 10^{-5}) = -0.69 \text{ kNm}$$

$$M_{BC} = -40.000 + 80000 \times (-7.899 \times 10^{-5}) = -46.32 \text{ kNm}$$

$$M_{CB} = +40.000 + 40000 \times (-7.899 \times 10^{-5}) = +36.84 \text{ kNm}$$

$$M_{BD} = 106.667 + 106666.6 \times (-7.899 \times 10^{-5}) + 53333.3 \times (-9.605 \times 10^{-4}) = +47.02 \text{ kNm}$$

$$M_{DB} = 106.667 + 53333.3 \times (-7.899 \times 10^{-5}) + 106666.6 \times (-9.605 \times 10^{-4}) = 0.00 \text{ kNm}$$

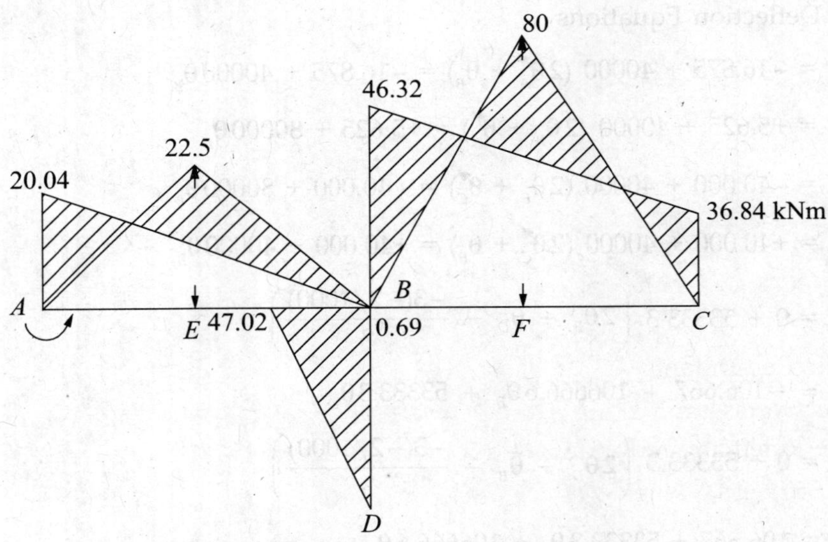

FIG. 1.67 Bending moment diagram

1.7 ANALYSIS OF FRAMES WITHOUT SIDESWAY

The joint translation is called sway. It is nothing but in plane lateral movement of the frame. In symmetric frame, sidesways do not occur. A frame is said to be symmetric when the right half of the frame is the mirror image of the left half. The symmetry is to be maintained identically with respect to geometry (length and moment of inertia), loading and boundary conditions. The following are the examples of nonsway frames.

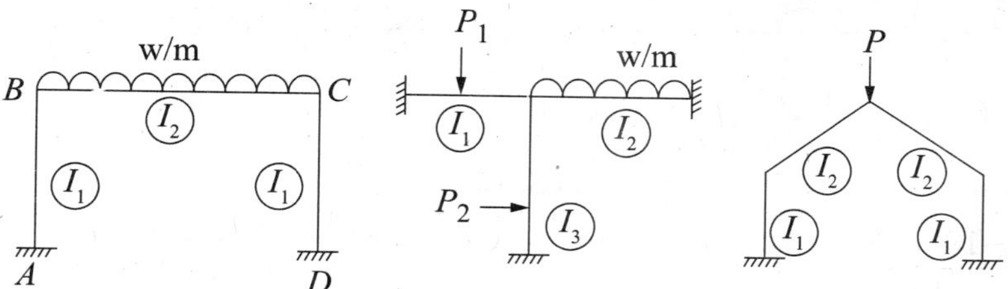

FIG. 1.68

Example 1.20 Analyse the portal fràme as shown in figure by the slope deflection method. Sketch the bending moment and shear force diagram.

(Anna Univ, Coimbatore, June 2010)

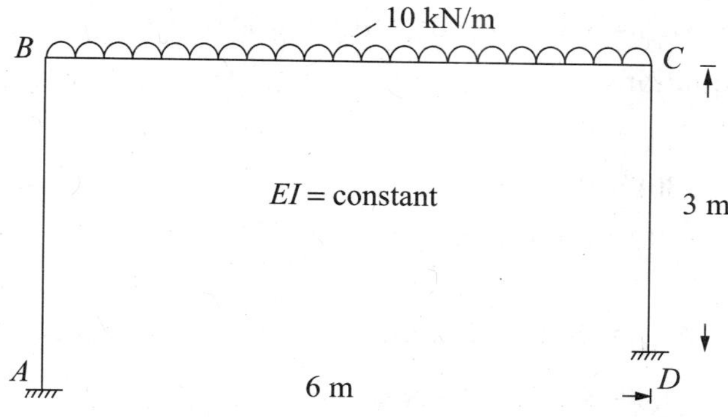

FIG. 1.69

Solution

1. Kinematic Redundants

   As the supports A and D are fixed, θ_A and θ_D are zero. Due to symmetry $\theta_B = -\theta_C$. Thus, the kinematic redundant reduces to one only, i.e. θ_B.

2. Relative Stiffness Values

Member	(I/l)	k
AB	$I/3$	2
BC	$I/6$	1

The values of k are obtained by multiplying (I/l) with $(6/I)$.

3. Fixed End Moments

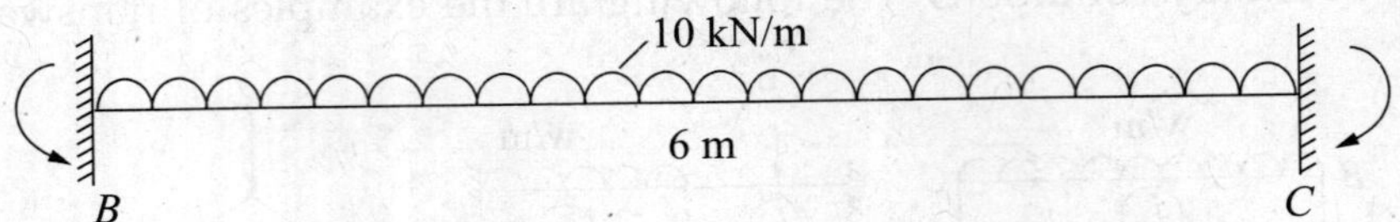

FIG. 1.70

$$\frac{-10 \times 6^2}{12} = -30 \text{ kNm} \qquad \frac{+10 \times 6^2}{12} = +30 \text{ kNm}$$

4. Slope Deflection Equations

$$M_{AB} = 0 + 2\,(2\theta_A + \theta_B) = 2\theta_B$$

$$M_{BA} = 0 + 2\,(2\theta_B + \theta_A) = 4\theta_B$$

$$M_{BC} = -30 + 1(2\theta_B + \theta_C) = -30 + \theta_B$$

5. Equilibrium Equations

$$M_{BA} + M_{BC} = 0$$

$$4\theta_B + (\theta_B - 30) = 0$$

$$\theta_B = 6 \text{ radians}$$

6. End Moments

$$M_{AB} = 2(6) = 12 \text{ kNm}$$

$$M_{BA} = 4(6) = 24 \text{ kNm}$$

$$M_{BC} = -30 + 6 = -24 \text{ kNm}$$

$$M_E = \frac{10 \times 6^2}{8} - 24 = +21 \text{ kNm}$$

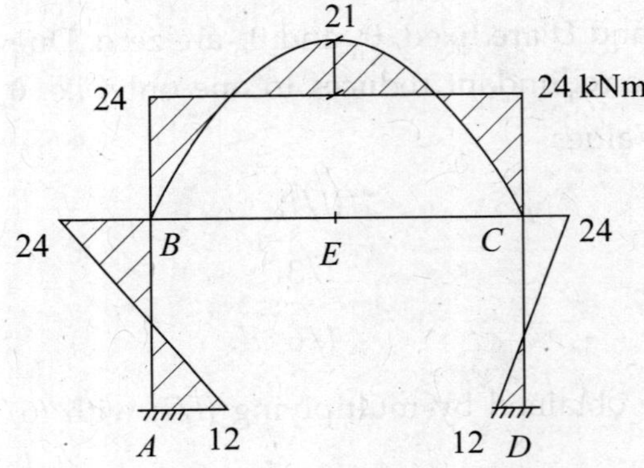

FIG. 1.71 Bending moment diagram

Example 1.21 Analyse the frame shown in figure by the slope deflection method. Draw the bending moment diagram.

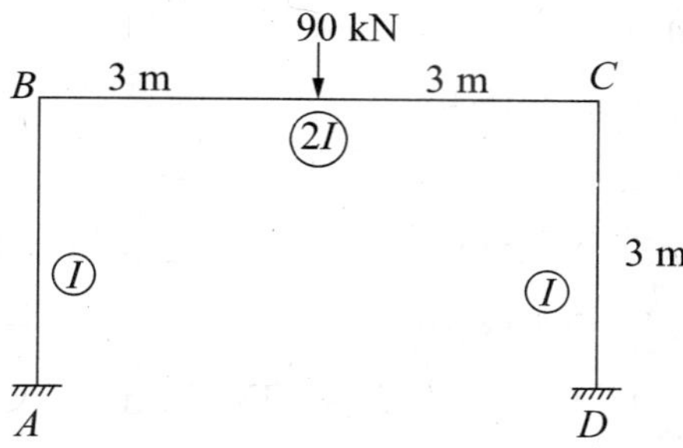

FIG. 1.72

Solution

1. Kinetic Redundants $\theta_A = \theta_D = 0$; $\theta_B \neq \theta_C \neq 0$; $\theta_C = -\theta_B$
2. Relative Stiffness Values

Member	(I/l)	k
AB	$I/3$	1
BC	$2I/6$	1

3. Fixed End Moments

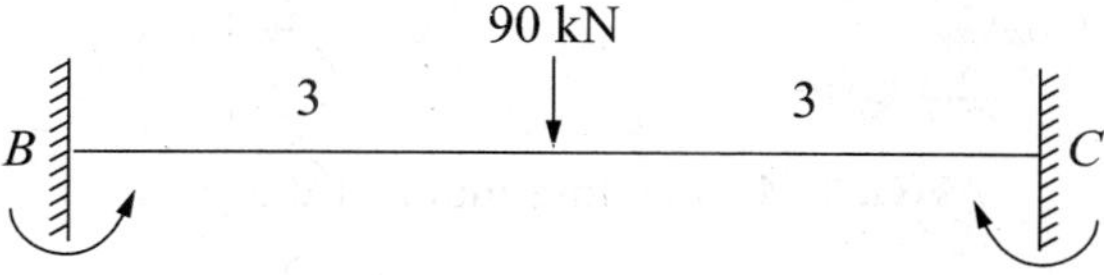

FIG. 1.73

$$\frac{-90 \times 6}{8} = -67.5 \text{ kNm} \qquad \frac{+90 \times 6}{8} = +67.5 \text{ kNm}$$

4. Slope Deflection Equations

$$M_{AB} = 0 + 1\,(2\cancel{\theta_A} + \theta_B) = \theta_B$$

$$M_{BA} = 0 + 1\,(2\theta_B + \cancel{\theta_A}) = 2\theta_B$$

$$M_{BC} = -67.5 + 1\,(2\theta_B - \theta_B) = -67.5 + \theta_B \qquad [\because \theta_C = -\theta_B]$$

$$M_{CB} = +67.5 + 1\,(-2\theta_B + \theta_B) = 67.5 - \theta_B \qquad [\because \theta_C = -\theta_B]$$

$$M_{CD} = 0 + 1\,(2\theta_C + \cancel{\theta_D}) = 2\theta_C$$

5. Equilibrium Equations

$$M_{BA} + M_{BC} = 0$$

$$2\theta_B + (\theta_B - 67.5) = 0$$

$$\theta_B = 22.5 \text{ radian}$$

6. End Moments

$$M_{AB} = 22.5 \text{ kNm}$$

$$M_{BA} = 45.0 \text{ kNm}$$

$$M_{BC} = -67.5 + 22.5 = -45 \text{ kNm}$$

$$M_{CB} = 67.5 - 22.5 = +45 \text{ kNm}$$

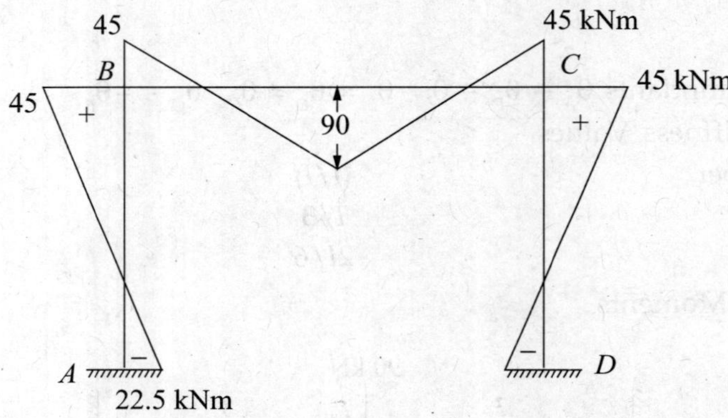

FIG. 1.74 Bending moment diagram

$$M_{CD} = 2(-22.5) = -45 \text{ kNm}$$

Example 1.22 Analyse the frame by the slope deflection method and draw the bending moment diagram.

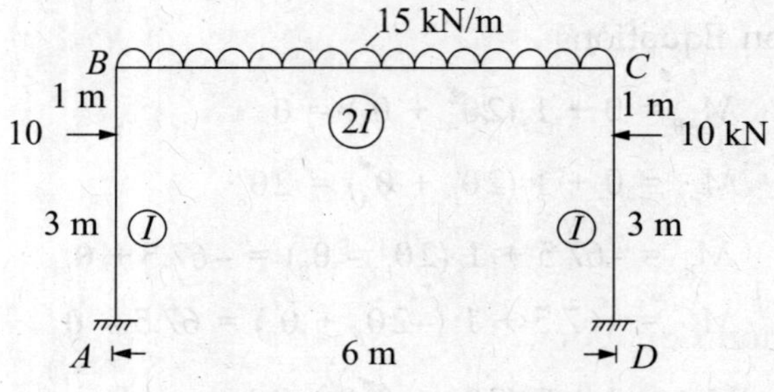

FIG. 1.75

Solution

1. Kinematic Redundants

$$\theta_A = 0, \quad \theta_B \neq 0, \quad \theta_C = -\theta_B, \quad \theta_D = 0$$

2. Relative Stiffness Values

Member	(I/l)	k
AB	$I/4$	6
BC	$2I/6$	8
CD	$I/4$	6

The k values were obtained by multiplying the value of individual (I/l) values by $(24/I)$.

3. Fixed End Moments

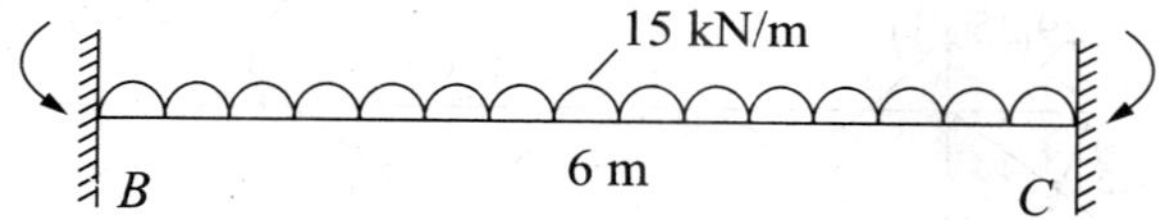

FIG. 1.76

$$\frac{-15 \times 6^2}{12} = -45 \text{ kNm} \qquad \frac{+15 \times 6^2}{12} = +45 \text{ kNm}$$

$$\frac{+3^2 \times 1 \times 10}{4^2} = +5.625 \text{ kNm}$$

FIG. 1.77

$$\frac{-3 \times 10 \times 1^2}{4^2} = -1.875 \text{ kNm}$$

4. Slope Deflection Equations

$$M_{AB} = -1.875 + 6\,(2\cancel{\theta_A} + \theta_B) = 6\theta_B - 1.875$$

$$M_{BA} = 5.625 + 6\,(2\theta_B + \cancel{\theta_A}) = 12\theta_B + 5.625$$

$$M_{BC} = -45 + 8\,(2\theta_B - \theta_B) = 8\theta_B - 45 \qquad [\because \theta_C = -\theta_B]$$

$$M_{CB} = 45 + 8\,(-2\theta_B + \theta_B) = 45 - 8\theta_B \qquad [\because \theta_C = -\theta_B]$$

5. Equilibrium Equations

Joint B $\Rightarrow$

$$M_{BA} + M_{BC} = 0$$

$$20\theta_B - 39.375 = 0$$

$$\boxed{\theta_B = 1.9688 \text{ radian}}$$

6. End Moments

$$M_{AB} = 6\,(1.9688) - 1.875 = 9.94 \text{ kNm}$$

$$M_{BA} = 12\,(1.9688) + 5.625 = 29.25 \text{ kNm}$$

$$M_{BC} = 8\,(1.9688) - 45.000 = -29.25 \text{ kNm}$$

$$M_{CB} = 45.00 - 8\,(1.9688) = 29.25 \text{ kNm}$$

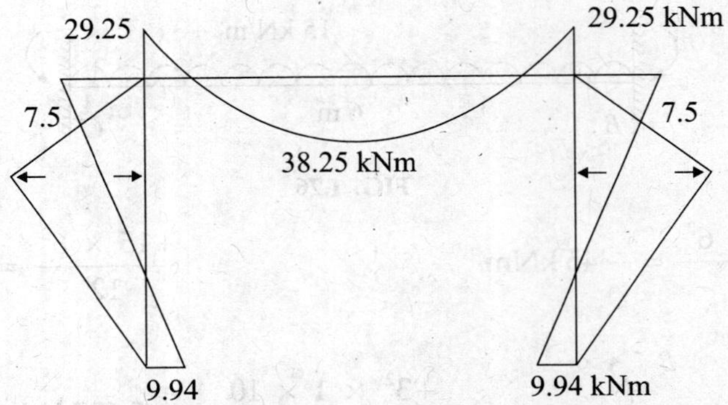

FIG. 1.78 Bending moment diagram

Example 1.23 Analyse the frame by the slope deflection method. Draw the bending moment diagram and the elastic curve.

100 | B | C | 100 kN

E 2 m | 4 m | 2 m F

4 m | 4 m

A | D

EI = Constant

FIG. 1.79

Solution: Due to symmetry analyse one half of the frame.

1. Kinematic Redundants

$$\theta_A \neq 0, \quad \theta_B \neq 0, \quad \theta_C = -\theta_B, \quad \theta_D = -\theta_A$$

2. Relative Stiffness Values

Member	(I/l)	k
AB	$I/4$	1
BC	$I/4$	1

3. Slope Deflection Equations

$$M_{AB} = 1(2\theta_A + \theta_B) = 2\theta_A + \theta_B$$

$$M_{BA} = 1(2\theta_B + \theta_A) = \theta_A + 2\theta_B$$

$$M_{BC} = 1(2\theta_B + \theta_C) = \theta_B \qquad [\because \theta_C = -\theta_B]$$

4. Equilibrium Equations

$$\text{Joint } B \Rightarrow M_{BE} + M_{BC} + M_{BA} = 0$$

$$200 + \theta_B + (\theta_A + 2\theta_B) = 0$$

$$\theta_A + 3\theta_B = -200 \qquad (1)$$

$$\text{Joint } A \Rightarrow M_{AB} = 0$$

$$2\theta_A + \theta_B = 0 \qquad (2)$$

Solving Eqs. (1) & (2),

$$\theta_A = +40$$

$$\theta_B = -80$$

5. End Moments

$$M_{BA} = 40 - 2(80) = -120 \text{ kNm}$$

$$M_{BC} = -80 \text{ kNm}$$

$$M_{AB} = 0$$

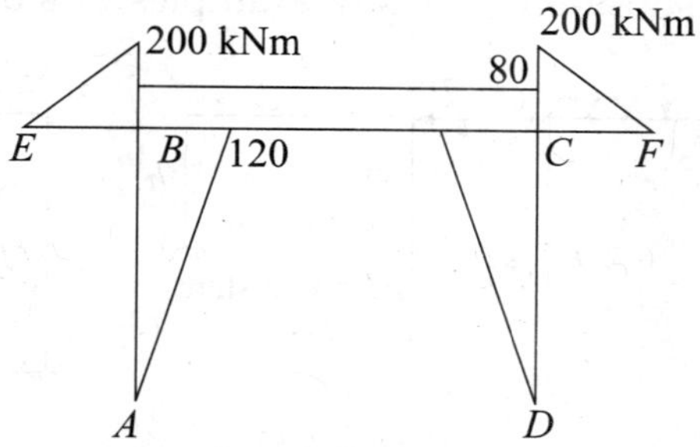

FIG. 1.80 Bending moment diagram

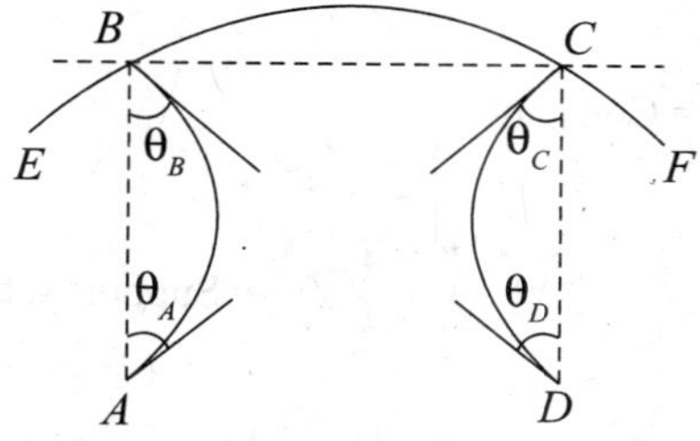

FIG. 1.81 Elastic curve

1.8 ANALYSIS OF FRAMES WITH SIDESWAY

Consider the following portal frame with eccentric loading.

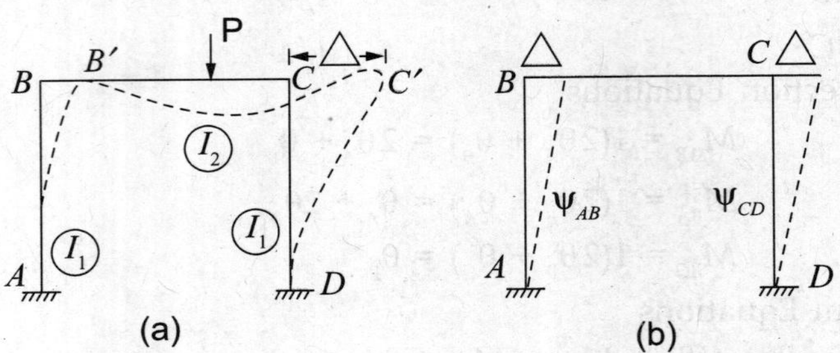

FIG. 1.82 Single bay single storey frame

The frame is symmetrical, i.e. the columns *AB* and *CD* are of same height with moment of inertia I_1. The boundary conditions of columns *AB* and *CD* are same with fixed ends. Though the frame is symmetrical, the frame sways to the right because of eccentric loading (∵ the load is not acting through the line of symmetry). The amount of sway is Δ as shown in the above figure, i.e. the translation of joint *B* as well as at *C*. It should be known that how much the right column sways the left column also sways to the same amount. The sway can occur in the portal frames due to the following: (i) column can be of unequal heights (ii) the moment of inertia of columns are not equal (iii) a symmetrical frame subjected to antisymmetric loading (iv) a symmetrical frame subjected to eccentric loading (v) the columns are not supported identically (one column could be fixed and the other is hinged or vice versa.) (vi) The columns are subjected to settlements/subsidence. A few examples of sidesway are given below.

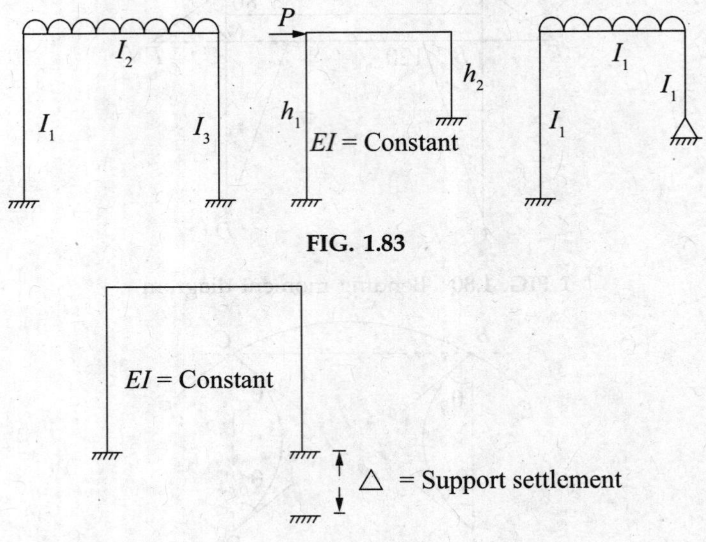

FIG. 1.84

Column Shear Equation

In the above frame, the joints A and D are held in position and restrained in direction. It means A and D do not translate and do not rotate. On the other hand, joints B and C are not restrained and hence they translate laterally by an amount Δ. It is persumed that there is no change in the length of the members as the deformations are considered small. Referring to Fig. 1.82(b) from geometry – the relative displacement to the length of the member can be written as

$$\psi_{AB} = \frac{\Delta}{L_{AB}}$$

and
$$\psi_{CD} = \frac{\Delta}{L_{CD}}$$

Thus, the general governing slope deflection can be written as

$$M_{AB} = M_{FAB} + \frac{2EI}{L_{AB}}(2\theta_A + \theta_B - 3\psi_{AB})$$

and
$$M_{BA} = M_{FBA} + \frac{2EI}{L_{BA}}(2\theta_B + \theta_A - 3\psi_{AB})$$

Thus, in the above problem, the kinematic redundants are θ_B, θ_C and Δ. Two-moment equilibrium can be written at joint B and C respectively.

At joint B $\Rightarrow$ $M_{BA} + M_{BC} = 0$

At joint C $\Rightarrow$ $M_{CB} + M_{CD} = 0$

The above equations are to be supplemented by an extra equation which is termed column shear equation. This column shear equation is obtained by applying the horizontal equilibrium condition for the entire frame as a whole. In the above case

$$H_A + H_D = 0$$

i.e.,
$$\left(\frac{M_{AB} + M_{BA}}{L_{AB}}\right) + \left(\frac{M_{CD} + M_{DC}}{L_{DC}}\right) = 0$$

The column shear equation can be illustrated by considering the frame shown below.

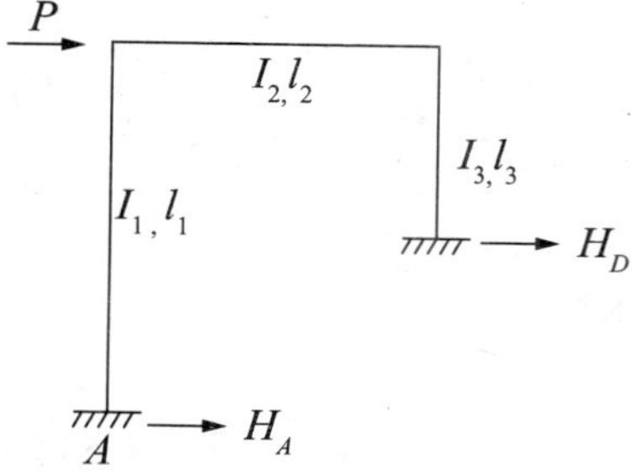

FIG. 1.85

Considering the horizontal equilibrium for the whole frame as

$$H_A + H_D + P = 0$$

$$\left(\frac{M_{AB} + M_{BA}}{l_1}\right) + \left(\frac{M_{CD} + M_{DC}}{l_3}\right) + P = 0$$

The column shear equation along with equilibrium equations at joints are adequate to compute the slopes at joints, viz. θ_B and θ_C as well as the sidesway Δ. On back substitution, the end moments of the beam element can be computed.

Example 1.24 Analyse the frame shown in figure by the slope deflection method. Draw the bending moment diagram.

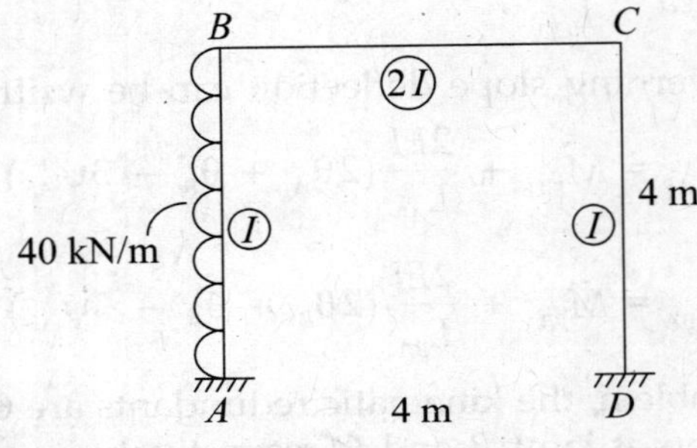

FIG. 1.86

Solution

1. Kinematic Redundants

$$\Delta, \quad \theta_A = 0, \quad \theta_B \neq 0, \quad \theta_C \neq 0, \quad \theta_D = 0.$$

2. Relative Stiffness Values

Member	(I/l)	k
AB	$I/4$	1
BC	$2I/4$	2
CD	$I/4$	1

The stiffness values of k were obtained by multiplying by $(4/I)$.

3. Fixed End Moments

B $\quad \dfrac{+\,40 \times 4^2}{12} = 53.33$ kNm

4 m

40 kN/m

A $\quad \dfrac{-\,40 \times 4^2}{12} = -\,53.33$ kNm

FIG. 1.87

4. Slope Deflection Equations

$$M_{AB} = -53.33 + 1\left(2\cancel{\theta_A} + \theta_B - \frac{3\Delta}{4}\right) = -53.33 + \theta_B - 0.75\Delta$$

$$M_{BA} = +53.33 + 1\left(2\theta_B + \cancel{\theta_A} - \frac{3\Delta}{4}\right) = +53.33 + 2\theta_B - 0.75\Delta$$

$$M_{BC} = 0 + 2(2\theta_B + \theta_C) = 4\theta_B + 2\theta_C$$

$$M_{CB} = 0 + 2(2\theta_C + \theta_B) = 2\theta_B + 4\theta_C$$

$$M_{CD} = 0 + 1\left(2\theta_C + \cancel{\theta_D} - \frac{3\Delta}{4}\right) = 2\theta_C - 0.75\Delta$$

$$M_{DC} = 0 + 1\left(2\theta_D + \theta_C - \frac{3\Delta}{4}\right) = \theta_C - 0.75\Delta$$

5. Equilibrium Equations

Joint B $\Rightarrow$ $M_{BA} + M_{BC} = 0$

$$6\theta_B + 2\theta_C - 0.75\Delta = -53.33 \quad (1)$$

Joint C $\Rightarrow$ $M_{CB} + M_{CD} = 0$

$$2\theta_B + 6\theta_C - 0.75\Delta = 0 \quad (2)$$

Column shear equation

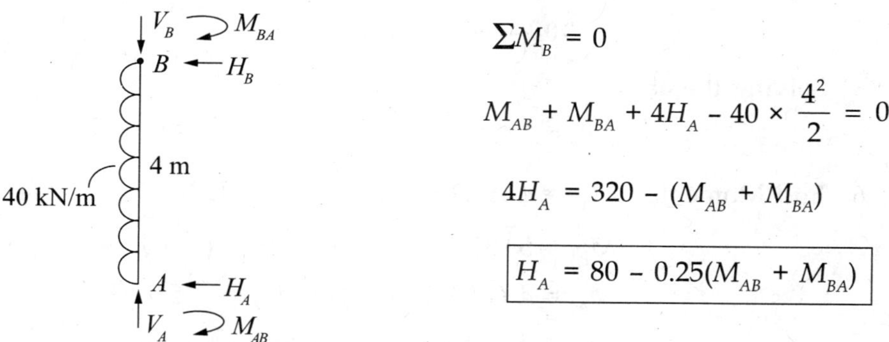

FIG. 1.88

$$\Sigma M_B = 0$$

$$M_{AB} + M_{BA} + 4H_A - 40 \times \frac{4^2}{2} = 0$$

$$4H_A = 320 - (M_{AB} + M_{BA})$$

$$\boxed{H_A = 80 - 0.25(M_{AB} + M_{BA})}$$

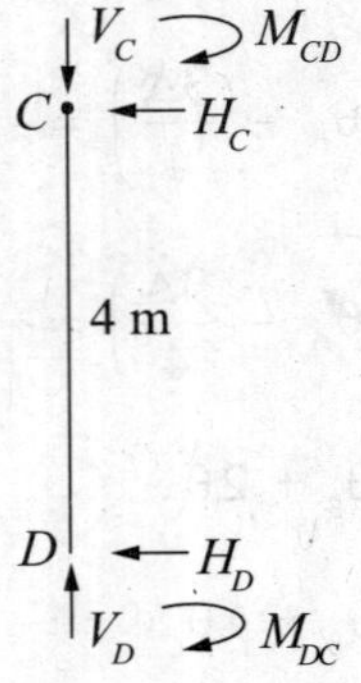

FIG. 1.89

$$\Sigma M_C = 0$$

$$M_{CD} + M_{DC} + 4H_D = 0$$

$$H_D = -\frac{(M_{CD} + M_{DC})}{4}$$

$$\boxed{H_D = -0.25(M_{CD} + M_{DC})}$$

$$H_A + H_D = 160$$

$$80 - 0.25(M_{AB} + M_{BA} + M_{CD} + M_{DC}) = 160$$

$$-0.25(M_{AB} + M_{BA} + M_{CD} + M_{DC}) = 80$$

$$M_{AB} + M_{BA} + M_{CD} + M_{DC} = -320$$

$$3\theta_B - 1.5\Delta + 3\theta_C - 1.5\Delta = -320$$

$$\boxed{3\theta_B + 3\theta_C - 3\Delta = -320}$$

Hence,

$$6\theta_B + 2\theta_C - 0.75\Delta = -53.33$$

$$2\theta_B + 6\theta_C - 0.75\Delta = 0$$

$$3\theta_B + 3\theta_C - 3\Delta = -320$$

Solving the above

$$\theta_B = 1.5391, \quad \theta_C = 14.872, \quad \Delta = 123.0774$$

6. End Moments $M_{AB} = -53.33 + 1.5391 - 0.75\,(123.0774) = -144.1$ kNm

$$M_{BA} = 53.33 + 2\,(1.5391) - 0.75\,(123.0774) = -35.9 \text{ kNm}$$

$$M_{BC} = 4\,(1.5391) + 2\,(14.872) = +35.9 \text{ kNm}$$

$$M_{CB} = 2\,(1.5391) + 4\,(14.872) = 62.6 \text{ kNm}$$

$$M_{CD} = 2\,(14.872) - 0.75\,(123.0774) = -62.6 \text{ kNm}$$

$$M_{DC} = 14.872 - 0.75(123.0774) = -77.44 \text{ kNm}$$

Example 1.25 Analyse the given frame by the slope deflection method. Draw the bending moment diagram.

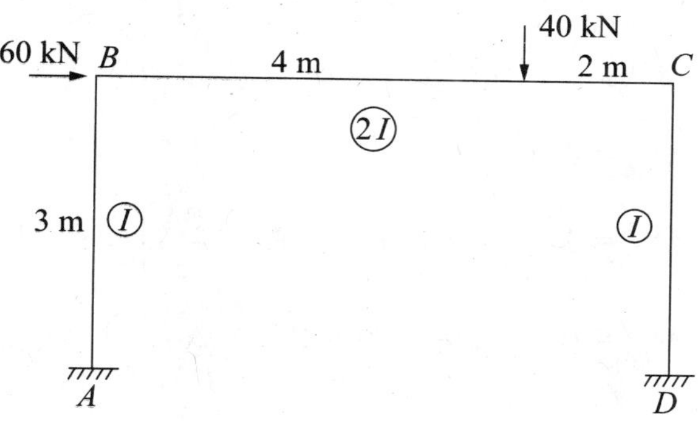

FIG. 1.90

Solution

1. Kinematic Redundants

   The kinematic indeterminants are θ_B, θ_C and Δ. The displacements and slopes at *A* and *D* are zeros. Thus, the kinematic redundants are the slopes at *B* and *C* and the lateral sway Δ. The slope deflections are to be formulated to obtain the same.

2. Relative Stiffness Values (I/l)

Member	(I/l)	k
AB	$I/3$	1
BC	$2I/6$	1
CD	$I/3$	1

   The values of k were obtained by multiplying the relative values by a common factor. In this case, the common factor chosen was $(3/I)$. The above modification was done for simplicity in formulating the slope deflection equations.

B — 4 m — 40 kN — 2 m — C

FIG. 1.91

3. Fixed End Moments of the Members

$$M_F^{BC} = -\frac{4(40)2^2}{6^2} = -17.8 \text{ kNm}$$

$$M_F^{CB} = +\frac{2(40)4^2}{6^2} = +35.6 \text{ kNm}$$

4. Slope Deflection Equations

$$M_{AB} = M_{FAB} + k\left(2\cancel{\theta_A} + \theta_B - \frac{3\Delta}{3}\right)$$

i.e.

$$M_{AB} = 0 + 1(\theta_B - \Delta) = \theta_B - \Delta$$

$\therefore$

$$M_{AB} = \theta_B - \Delta \tag{1}$$

$$M_{BA} = M_{FBA} + k\left(2\theta_B + \cancel{\theta_A} - \frac{3\Delta}{3}\right)$$

$$M_{BA} = 0 + 1(2\theta_B - \Delta) = 2\theta_B - \Delta \tag{2}$$

Similarly,

$$M_{BC} = -17.8 + 1(2\theta_B + \theta_C) = -17.8 + 2\theta_B + \theta_C \tag{3}$$

$$M_{CB} = +35.6 + 1(2\theta_C + \theta_B) = +35.6 + 2\theta_C + \theta_B \tag{4}$$

$$M_{CD} = 0 + 1\left(2\theta_C + \cancel{\theta_D} - \frac{3\Delta}{3}\right) = 2\theta_C - \Delta \tag{5}$$

$$M_{DC} = 0 + 1\left(2\cancel{\theta_D} + \theta_C - \frac{3\Delta}{3}\right) = \theta_C - \Delta \tag{6}$$

5. Equilibrium Equations and Shear Condition

Joint B

$$M_{BA} + M_{BC} = 0 \tag{7}$$

Hence, from Eqs. (2) and (3)

$$4\theta_B + \theta_C - \Delta = 17.8 \tag{8}$$

Joint C

$$M_{CB} + M_{CD} = 0 \tag{9}$$

Hence, from Eqs. (4) and (5)

$$\theta_B + 4\theta_C - \Delta = -35.6 \tag{10}$$

Horizontal shear condition

The horizontal shear condition is formulated by considering the equilibrium of column AB and column CD separately.

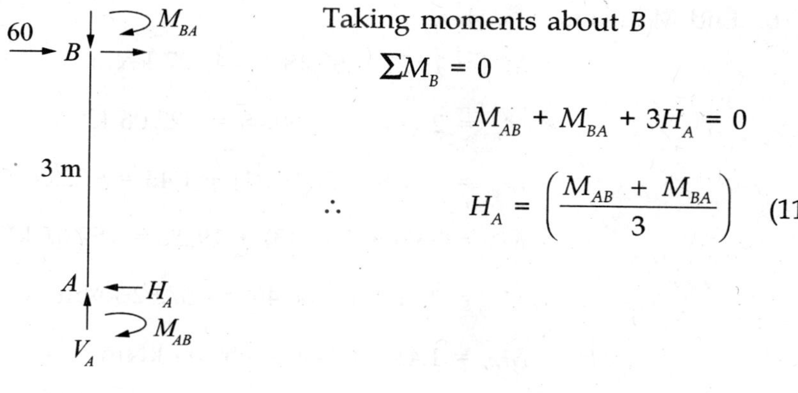

Taking moments about B

$$\Sigma M_B = 0$$

$$M_{AB} + M_{BA} + 3H_A = 0$$

$$\therefore \qquad H_A = \left(\frac{M_{AB} + M_{BA}}{3}\right) \qquad (11)$$

FIG. 1.92

For body diagram AB

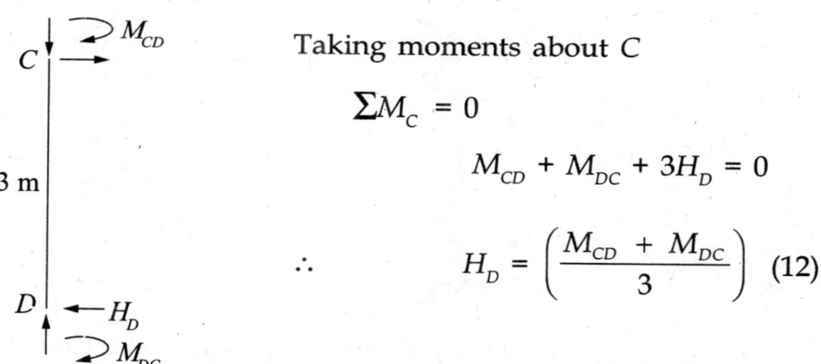

Taking moments about C

$$\Sigma M_C = 0$$

$$M_{CD} + M_{DC} + 3H_D = 0$$

$$\therefore \qquad H_D = \left(\frac{M_{CD} + M_{DC}}{3}\right) \qquad (12)$$

FIG. 1.93

For body diagram CD

Using the horizontal equilibrium condition

$$H_A + H_D = 60 \qquad (13)$$

Substituting Eqs. (11) and (12) in (13)

$$M_{AB} + M_{BA} + M_{CD} + M_{DC} = -180 \qquad (14)$$

Substituting

$$(\theta_B - \Delta) + (2\theta_B - \Delta) + (2\theta_C - \Delta) + (\theta_C - \Delta) = -180 \qquad (15)$$

Which reduces to

$$3\theta_B - 3\theta_C - 4\Delta = -180 \qquad (16)$$

On solving Eqs. (8), (10) and (16)

$$\therefore \qquad \theta_B = 19.21, \quad \theta_C = 1.43 \text{ and } \Delta = 60.48$$

6. End Moments

$$M_{AB} = 19.21 - 60.48 = -41.27 \text{ kNm}$$

$$M_{BA} = 2\,(19.21) - 60.48 = -22.06 \text{ kNm}$$

$$M_{BC} = -17.8 + 2\,(19.21) + 1.43 = +22.06 \text{ kNm}$$

$$M_{CB} = 35.6 + 2\,(1.43) + 19.21 = +57.67 \text{ kNm}$$

$$M_{CD} = 2\,(1.43) - 60.48 = -57.62 \text{ kNm}$$

$$M_{DC} = 1.43 - 60.48 = -59.05 \text{ kNm}$$

The above moments are marked and shown in the following free body diagrams.

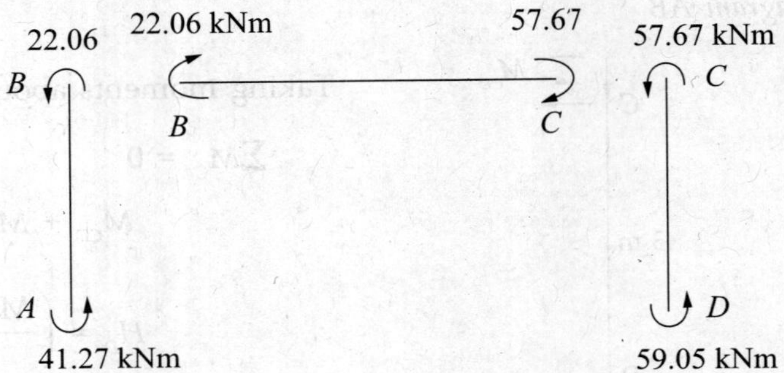

FIG. 1.94 Free body diagram of the elements

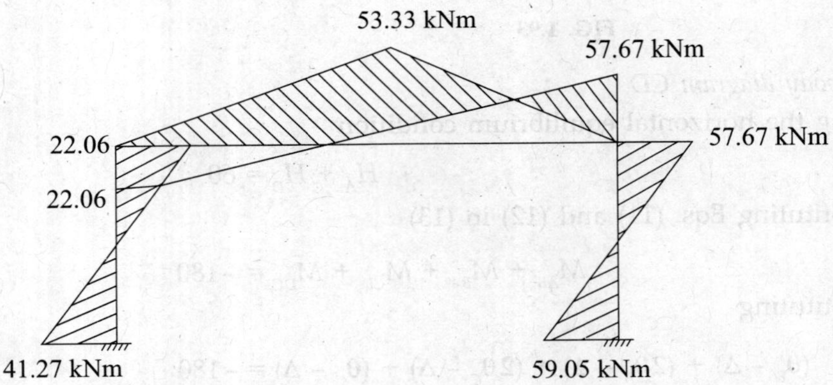

FIG. 1.95 Bending moment diagram

Example 1.26 Analyse the frame shown in figure by the slope deflection method. Draw the bending moment diagram.

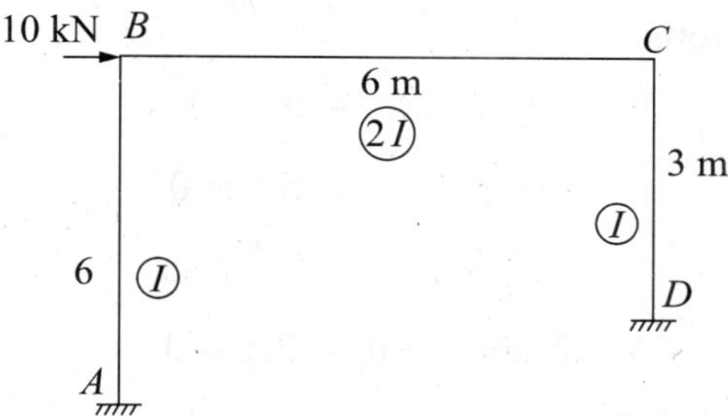

FIG. 1.96

Solution

1. Kinematic Redundants

   The supports at A and D are fixed. Hence, $\theta_A = 0$, $\theta_D = 0$. Joints B and C rotate. As the frame is not symmetrical, it sways. Hence, the kinematic redundants are θ_B, θ_C and Δ.

2. Relative Stiffness Values

Member	(I/l)	k
AB	$I/6$	1
BC	$2I/6$	2
CD	$I/3$	2

   The values of k were obtained by multiplying the values of (I/l) with $(6/I)$.

3. Fixed End Moments

   The fixed end moments are zero as there are no loads acting on the individual member.

4. Slope Deflection Equations

$$M_{AB} = 1\left(2\cancel{\theta_A} + \theta_B - \frac{3\Delta}{6}\right) = \theta_B - 0.5\Delta.$$

$$M_{BA} = 1\left(2\theta_B + \cancel{\theta_A} - \frac{3\Delta}{6}\right) = 2\theta_B - 0.5\Delta$$

$$M_{BC} = 2(2\theta_B + \theta_C) = 4\theta_B + 2\theta_C$$

$$M_{CB} = 2(2\theta_C + \theta_B) = 2\theta_B + 4\theta_C$$

$$M_{CD} = 2\left(2\theta_C + \cancel{\theta_D} - \frac{3\Delta}{3}\right) = 4\theta_C - 2\Delta$$

$$M_{DC} = 2\left(2\cancel{\theta_D} + \theta_C - \frac{3\Delta}{3}\right) = 2\theta_C - 2\Delta$$

5. Equilibrium Equations

At Joint $B \Rightarrow$ $M_{BA} + M_{BC} = 0$

$$6\theta_B + 2\theta_C - 0.5\Delta = 0 \quad (1)$$

At Joint $C \Rightarrow$ $M_{CB} + M_{CD} = 0$

$$2ss\theta_B + 8\theta_C - 2\Delta = 0 \quad (2)$$

Column shear condition

$$\frac{M_{AB} + M_{BA}}{6} + \frac{M_{CD} + M_{DC}}{3} = -10$$

$$\frac{(3\theta_B - \Delta)}{6} + \frac{(6\theta_C - 4\Delta)}{3} = -60$$

$$3\theta_B - \Delta + 12\theta_C - 8\Delta = -60$$

$$3\theta_B + 12\theta_C - 9\Delta = -60 \quad (3)$$

On solving Eqs. (1) - (3), we get

$$\theta_B = 0 \;\; \theta_C = +2.5 \;\; \Delta = +10$$

6. End Moments

$$M_{AB} = 0 - 0.5\,(+10) = -5 \text{ kNm}$$

$$M_{BA} = 2\,(0) - 0.5\,(+10) = -5 \text{ kNm}$$

$$M_{BC} = 2\,(0 + 2.5) = +5 \text{ kNm}$$

$$M_{CB} = 2\,(0) + 4\,(+2.5) = +10 \text{ kNm}$$

$$M_{CD} = 4\,(+2.5) - 2\,(+10) = -10 \text{ kNm}$$

$$M_{DC} = 2\,(+2.5) - 2\,(+10) = -15 \text{ kNm}$$

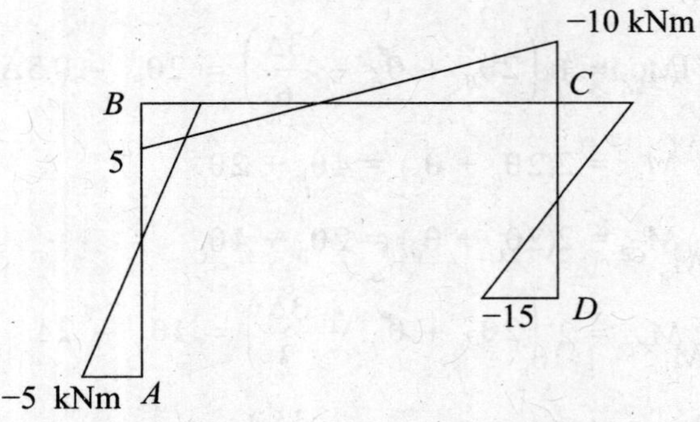

FIG. 1.97 Bending moment diagram

Example 1.27 Analyse the given frame by the slope deflection method. Evaluate the end moments.

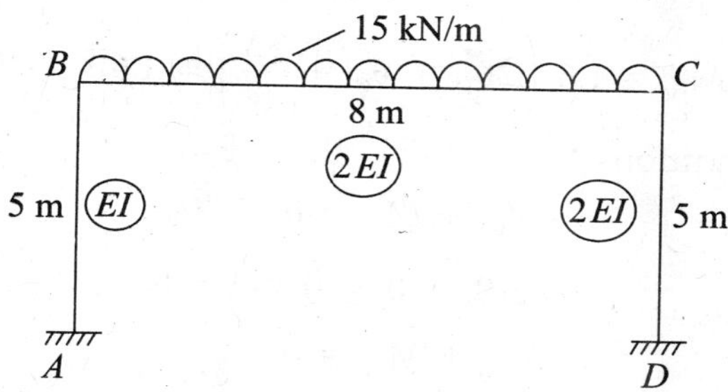

FIG. 1.98

Solution

1. Kinematic Redundants

$$\theta_A = 0, \quad \theta_B \neq 0, \quad \theta_C \neq 0, \quad \Delta \neq 0, \quad \theta_D = 0$$

2. Relative Stiffness Values

Member	(I/l)	k
AB	$I/5$	0.8
BC	$2I/8$	1.0
CD	$2I/5$	1.6

The values of k were obtained by multiplying individual values of (I/l) with $(40/I)$.

3. Fixed End Moments

$$M_{FBC} = -\frac{15 \times 8^2}{12} = -80 \text{ kNm}$$

$$M_{FCB} = +\frac{15 \times 8^2}{12} = +80 \text{ kNm}$$

4. Slope Deflection Equations

$$M_{AB} = 0.8\left(2\theta_A + \theta_B - \frac{3\Delta}{5}\right) = 0.8\theta_B - 0.48\Delta$$

$$M_{BA} = 0.8\left(2\theta_B + \theta_A - \frac{3\Delta}{5}\right) = 1.6\theta_B - 0.48\Delta$$

$$M_{BC} = 1(2\theta_B + \theta_C) - 80 = 2\theta_B + \theta_C - 80$$

$$M_{CB} = 1(2\theta_C + \theta_B) + 80 = \theta_B + 2\theta_C + 80$$

$$M_{CD} = 1.6\left(2\theta_C + \theta_D - \frac{3\Delta}{5}\right) = 3.2\theta_C - 0.96\Delta$$

$$M_{DC} = 1.6\left(2\theta_D + \theta_C - \frac{3\Delta}{5}\right) = 1.6\theta_C - 0.96\Delta$$

5. Equilibrium Conditions

Joint B $\Rightarrow$ $M_{BA} + M_{BC} = 0$

$$3.6\theta_B + \theta_C - 0.48\Delta = 80 \qquad (1)$$

Joint C $\Rightarrow$ $M_{CB} + M_{CD} = 0$

$$\theta_B + 5.2\theta_C - 0.96\Delta = -80 \qquad (2)$$

Column shear condition

$$\frac{M_{AB} + M_{BA}}{5} + \frac{M_{CD} + M_{DC}}{5} = 0$$

$$(M_{AB} + M_{BA}) + (M_{CD} + M_{DC}) = 0$$

$$2.4\theta_B - 0.96\Delta + 4.8\theta_C - 1.92\Delta = 0$$

$$2.4\theta_B + 4.8\theta_C - 2.88\Delta = 0 \qquad (3)$$

Solving Eqs. (1) - (3), we get

$$\theta_B = 26.4808\,\theta_C = -23.6933\,\Delta = -17.4216$$

6. End Moments

$$M_{AB} = 0.8\,(26.4808) - 0.48\,(-17.4216) = 29.55 \text{ kNm}$$

$$M_{BA} = 1.6\,(26.4808) - 0.48\,(-17.4216) = 50.73 \text{ kNm}$$

$$M_{BC} = 2\,(26.4808) - 23.6933 - 80 = -50.73 \text{ kNm}$$

$$M_{CB} = 26.4808 + 2\,(-23.6933) + 80 = 59.09 \text{ kNm}$$

$$M_{CD} = 3.2\,(-23.6933) - 0.96\,(-17.4216) = -59.09 \text{ kNm}$$

$$M_{DC} = 1.6\,(-23.6933) - 0.96\,(-17.4216) = -21.18 \text{ kNm}$$

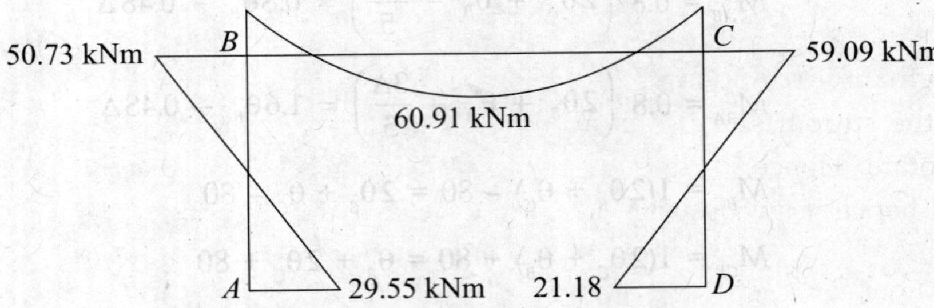

FIG. 1.99 Bending moment diagram

In the above problem find the force P to be applied at C to prevent sway, i.e. $\Delta = 0$

$$3.6\theta_B + \theta_C = 80 \tag{4}$$

$$\theta_B + 5.2\theta_C = -80 \tag{5}$$

$$2.4\theta_B + 4.8\theta_C + 5P = 0 \tag{6}$$

Solving Eqs. (4) and (5);

$$\theta_B = 27.9909; \quad \theta_C = -20.7675 \text{ radian}$$

From Eq. (6);

$$5P = 32.51 \text{ kN}$$

$$\boxed{P = 6.5 \text{ kN}}$$

Example 1.28 The figure below shows a fixed portal frame. Analyse the frame by the slope deflection method. Draw the bending moment diagram, shear force diagram and elastic curve.

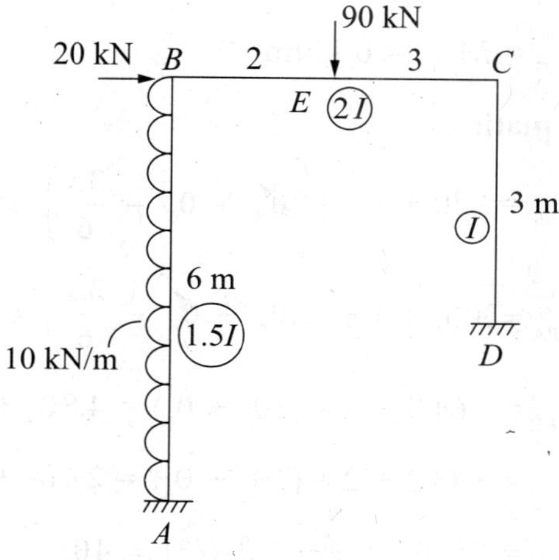

FIG. 1.100

Solution

1. Kinematic Redundants

   As the supports A and D are fixed, the slopes θ_A and θ_D are zero. Joints B and C rotate. Hence, θ_B and θ_C are to be determined. The frame is unsymmetrical and hence it sways by Δ. Thus, the kinematic redundants are θ_B, θ_C and Δ.

2. Relative Stiffness Values

Member	(I/l)	k
AB	$1.5I/6$	1.5
BC	$2I/5$	2.4
CD	$I/3$	2.0

The values of k were obtained by multiplying the values of (I/l) by $(6/I)$.

3. Fixed End Moments

$$M_{FAB} = \frac{-10 \times 6^2}{12} = -30 \text{ kNm}$$

$$M_{FBA} = +10 \times \frac{6^2}{12} = +30 \text{ kNm}$$

$$M_{FBC} = -2 \times 90 \times \frac{3^2}{5^2} = -64.8 \text{ kNm}$$

$$M_{FCB} = +\frac{3 \times 90 \times 2^2}{5^2} = +43.2 \text{ kNm}$$

$$M_{FCD} = M_{FDC} = 0 \text{ kNm}$$

4. Slope Deflection Equations

$$M_{AB} = -30 + 1.5\left(2\theta_A + \theta_B - \frac{3\Delta}{6}\right) = 1.5\theta_B - 0.75\Delta - 30$$

$$M_{BA} = +30 + 1.5\left(2\theta_B + \theta_A - \frac{3\Delta}{6}\right) = 3\theta_B - 0.75\Delta + 30$$

$$M_{BC} = -64.8 + 2.4\,(2\theta_B + \theta_C) = 4.8\theta_B + 2.4\theta_C - 64.8$$

$$M_{CB} = +43.2 + 2.4\,(2\theta_C + \theta_B) = 2.4\theta_B + 4.8\theta_C + 43.2$$

$$M_{CD} = 2(2\theta_C + \theta_D - 3\Delta/3) = 4\theta_C - 2\Delta$$

$$M_{DC} = 2\,(2\theta_D + \theta_C - 3\Delta/3) = 2\theta_C - 2\Delta$$

5. (i) Equilibrium Equations

Joint $B \Rightarrow$ $M_{BA} + M_{BC} = 0$

$$7.8\theta_B + 2.4\theta_C - 0.75\Delta = 34.8 \qquad (1)$$

Joint $C \Rightarrow$ $M_{CB} + M_{CD} = 0$

$$2.4\theta_B + 8.8\theta_C - 2\Delta = -43.2 \qquad (2)$$

(ii) Column Shear Condition

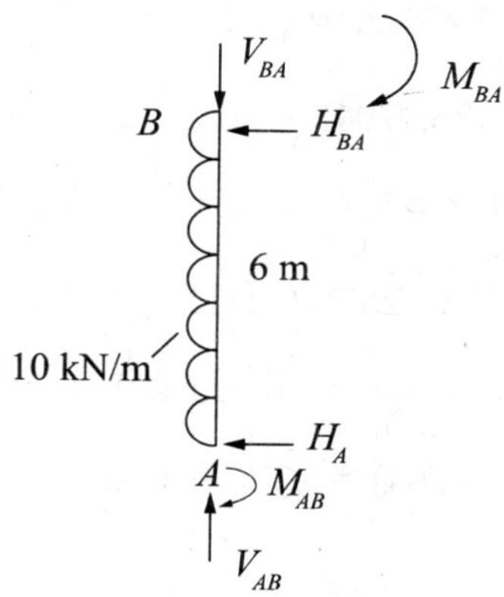

FIG. 1.101

Taking moment about B

$$6H_A + M_{AB} + M_{BA} - 10 \times \frac{6^2}{2} =$$

$$6H_A = 180 - (M_{AB} + M_{BA})$$

$$\boxed{H_A = 30 - (M_{AB} + M_{BA})/6}$$

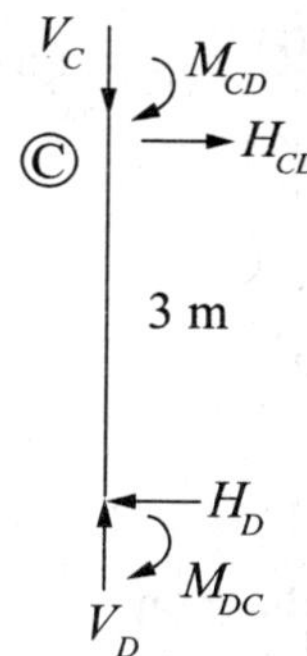

FIG. 1.102

Taking moment about C

$$3H_D + M_{DC} + M_{CD} = 0$$

$$\boxed{H_D = -(M_{CD} + M_{DC})/3}$$

$$H_A + H_D = 10(6) = 80$$

$$+20$$

$$\therefore \quad 30 - \frac{(M_{AB} + M_{BA})}{6} - \frac{(M_{CD} + M_{DC})}{3} = 80$$

$$\frac{M_{AB} + M_{BA}}{6} + \frac{(M_{CD} + M_{DC})}{3} = -50$$

$$(M_{AB} + M_{BA}) + 2(M_{CD} + M_{DC}) = -300$$

$$4.5\theta_B - 1.5\Delta + 2(6\theta_C - 4\Delta) = -300$$

$$4.5\theta_B + 12\theta_C - 8\Delta - 1.5\Delta = -300$$

$$4.5\theta_B + 12\theta_C - 9.5\Delta = -300 \qquad (3)$$

Solving Eqs. (1), (2) and (3);

$$\theta_B = 7.5774 \quad \theta_C = 1.4267 \quad \Delta = 36.9704$$

6. End Moments

$$M_{AB} = 1.5\ (7.5774) - 0.75\ (36.9704) - 30 = -46.36 \text{ kNm}$$

$$M_{BA} = 3\ (7.5774) - 0.75\ (36.9704) + 30 = +25.00 \text{ kNm}$$

$$M_{BC} = 4.8\ (7.5774) + 2.4\ (1.4267) - 64.8 = -25.00 \text{ kNm}$$

$$M_{CB} = 2.4\ (7.5774) + 4.8\ (1.4267) + 43.2 = +68.23 \text{ kNm}$$

$$M_{CD} = 4\ (1.4267) - 2\ (36.9704) = -68.23 \text{ kNm}$$

$$M_{DC} = 3\ (1.4267) - 3\ (36.9704) = -71.08 \text{ kNm}$$

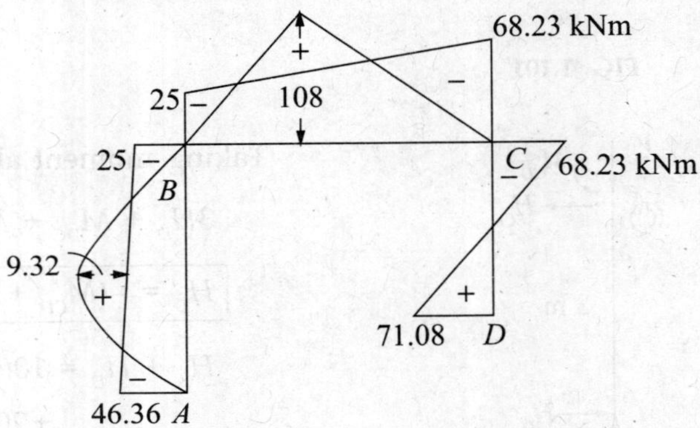

FIG. 1.103 Bending moment diagram

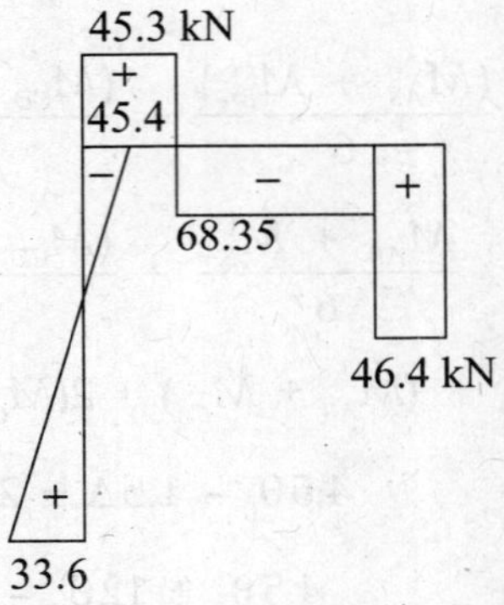

FIG. 1.104 Shear force diagram

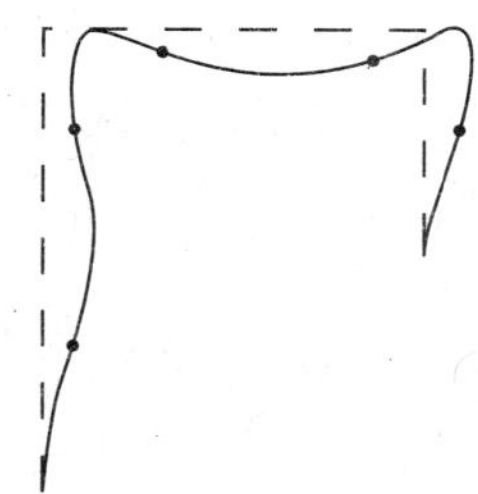

FIG. 1.105 Elastic curve

Example 1.29 Using the slope deflection method, analyse the portal frame with a loaded cantilever as shown. Draw the bending moment diagram and the elastic curve.

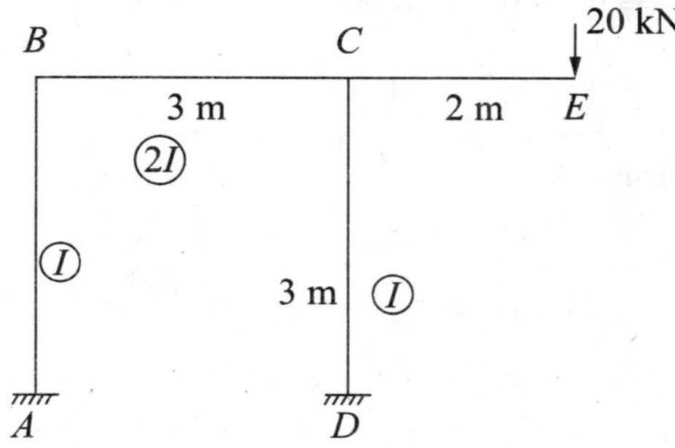

FIG. 1.106

Solution

1. Kinematic Redundants

$$\theta_A = 0, \quad \theta_B \neq 0, \quad \theta_C \neq 0, \quad \theta_D = 0, \quad \Delta \neq 0$$

2. Relative Stiffness Values

Member	(I/l)	k
AB	$(I/3)$	1
BC	$(2I/3)$	2
CD	$(I/3)$	1

3. Fixed End Moments

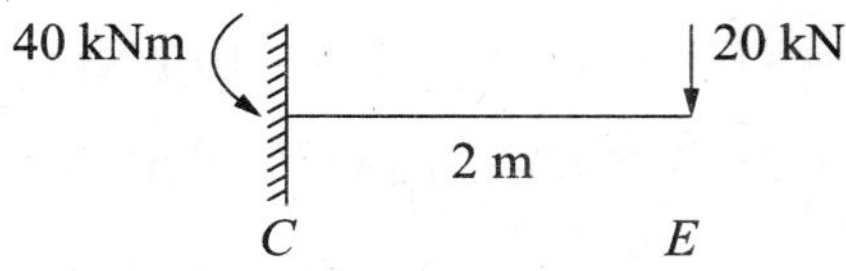

FIG. 1.107

$$M_{CE} = -40 \text{ kNm}$$

4. Slope Deflection Equations

$$M_{AB} = 1\left(2\theta_A + \theta_B - \frac{3\Delta}{3}\right) = \theta_B - \Delta$$

$$M_{BA} = 1\left(2\theta_B + \theta_A - \frac{3\Delta}{3}\right) = 2\theta_B - \Delta$$

$$M_{BC} = 2(2\theta_B + \theta_C) = 4\theta_B + 2\theta_C$$

$$M_{CB} = 2(2\theta_C + \theta_B) = 2\theta_B + 4\theta_C$$

$$M_{CD} = 1\left(2\theta_C + \theta_D\right) = 2\theta_C$$

$$M_{DC} = 1\left(2\theta_D + \theta_C\right) = \theta_C$$

5. Equilibrium Conditions

Joint B ⇒ $M_{BA} + M_{BC} = 0$

$$6\theta_B + 2\theta_C - \Delta = 0 \quad (1)$$

Joint C ⇒ $M_{CB} + M_{CD} + M_{CE} = 0$

$$2\theta_B + 4\theta_C + 2\theta_C - 40 = 0$$

$$2\theta_B + 6\theta_C = 40 \quad (2)$$

Column shear condition

$$\frac{M_{AB} + M_{BA}}{3} + \frac{M_{CD} + M_{DC}}{3} = 0$$

$$M_{AB} + M_{BA} + M_{CD} + M_{DC} = 0$$

$$3\theta_B - 2\Delta + 3\theta_C = 0 \quad (3)$$

Solving the above $\theta_B = -0.7692$, $\theta_C = 6.9231$, $\Delta = 9.2307$

6. End Moments

$$M_{AB} = -0.7692 - 9.2307 = -10 \text{ kNm}$$

$$M_{BA} = 2(-0.7692) - 9.2307 = -10.77 \text{ kNm}$$

$$M_{BC} = 4(-0.7692) + 2(6.9231) = +10.77 \text{ kNm}$$

$$M_{CB} = 2(-0.7692) + 4(6.9231) = +26.15 \text{ kNm}$$

$$M_{CD} = 2(0.69231) = 13.85 \text{ kNm}$$

$$M_{DC} = 6.92 \text{ kNm}$$

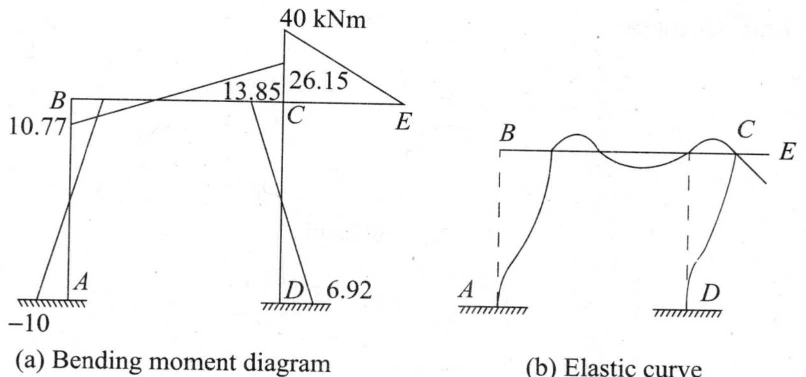

(a) Bending moment diagram (b) Elastic curve

FIG. 1.108

Example 1.30 Analyse the frame by the slope deflection method and draw the bending moment diagram.

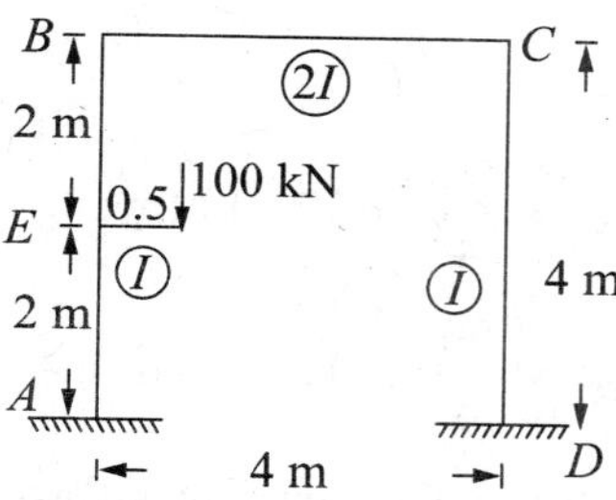

FIG. 1.109

Solution

1. Kinematic Redundants

   The kinematic redundants are θ_B, θ_C and Δ. The supports A and D are fixed. Hence, the deflections and slopes are zero at A and D.

2. Relative Stiffness Values ($k = I/l$)

Member	(I/l)	k
AB	$I/4$	1
BC	$2I/4$	2
CD	$I/4$	1

The value of (I/l) are normalised by multiplying the above by $(4/I)$.

3. Fixed End Moments

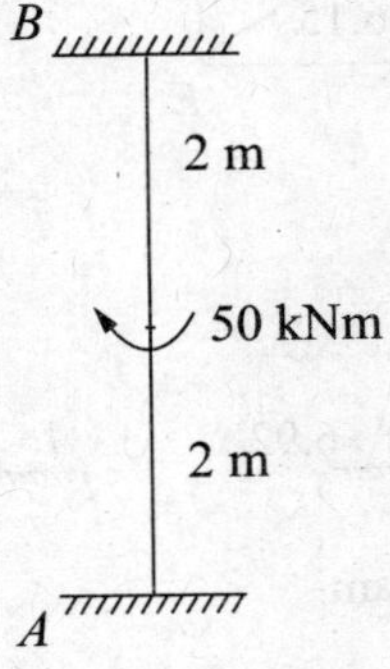

FIG. 1.110

It is to be noted here that when a moment M is applied at midspan, it induces the fixed end moments at each end by one-fourth of the applied moment of the same nature.

Hence,

$$M_{AB}^{F} = \frac{50}{4} = 12.5 \text{ kNm}$$

$$M_{BA}^{F} = \frac{50}{4} = 12.5 \text{ kNm}$$

The beam BC and the column CD are not subjected to any loading and hence

$$M_{BC}^{F} = M_{CB}^{F} = M_{CD}^{F} = M_{DC}^{F} = 0$$

4. Slope Deflection Equations

$$M_{AB} = 12.5 + 1\left(2\cancel{\theta_A} + \theta_B - \frac{3\Delta}{4}\right) = 12.5 + \theta_B - 0.75\Delta$$

$$M_{BA} = 12.5 + 1\left(2\theta_B + \cancel{\theta_A} - \frac{3\Delta}{4}\right) = 12.5 + 2\theta_B - 0.75\Delta$$

$$M_{BC} = 0 + 2(2\theta_B + \theta_C) = 4\theta_B + 2\theta_C$$

$$M_{CB} = 0 + 2(2\theta_C + \theta_B) = 4\theta_C + 2\theta_B$$

$$M_{CD} = 0 + 1\left(2\theta_C + \cancel{\theta_D} - \frac{3\Delta}{4}\right) = 2\theta_C - 0.75\Delta$$

$$M_{DC} = 0 + 1\left(2\cancel{\theta_D} + \theta_C - \frac{3\Delta}{4}\right) = \theta_C - 0.75\Delta$$

5. Equilibrium Equations

Joint $B \Rightarrow$ $M_{BA} + M_{BC} = 0$

$$6\theta_B + 2\theta_C - 0.75\Delta = -12.5 \quad (1)$$

Joint $C \Rightarrow$ $M_{CB} + M_{CD} = 0$

$$2\theta_B + 6\theta_C - 0.75\Delta = 0 \quad (2)$$

The horizontal shear condition is formulated as follows.

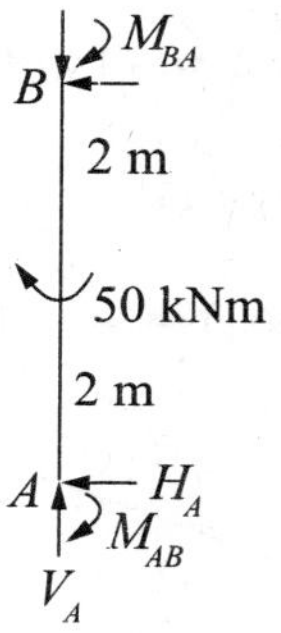

FIG. 1.111

Taking moment about the point B;

$$M_{AB} + M_{BA} + 50 + 4H_A = 0$$

$$\boxed{H_A = -\frac{1}{4}(M_{AB} + M_{BA} + 50)}$$

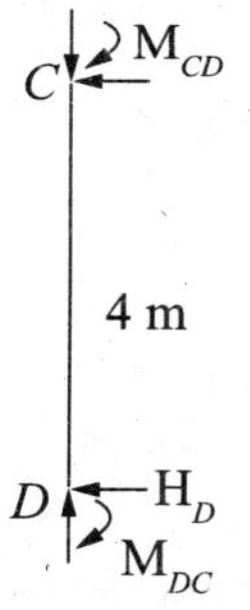

FIG. 1.112

Taking moment about the point C;

$$M_{CD} + M_{DC} + 4H_D = 0$$

$$\boxed{HD = -\frac{1}{4}(M_{CD} + M_{DC})}$$

$$H_A + H_D = 0$$

i.e. $$M_{AB} + M_{BA} + M_{CD} + M_{DC} + 50 = 0$$

$$12.5 + \theta_B - 0.75\Delta + 12.5 + 2\theta_B - 0.75\Delta$$

$$+2\theta_C - 0.75\Delta + \theta_C - 0.75\Delta + 50 = 0$$

$$3\theta_B + 3\theta_C - 3\Delta = -75 \quad (3)$$

Solving the above equations;

$$\theta_B = 0.3606 \text{ radian}$$

$$\theta_C = 3.4856 \text{ radian}$$

$$\Delta = 28.8461$$

6. End Moments

$$M_{AB} = 12.5 + 0.3606 - 0.75 \times 28.8461 = -8.774 \text{ kNm}$$

$$M_{BA} = 12.5 + 2\ (0.3606) - 0.75 \times 28.8461 = -8.413 \text{ kNm}$$

$$M_{BC} = 4(0.3606) + 2(3.4856) = +8.413 \text{ kNm}$$

$$M_{CB} = 4(3.4856) + 2\ (0.3606) = +14.664 \text{ kNm}$$

$$M_{CD} = 2(3.4856) - 0.75(28.8461) = -14.664 \text{ kNm}$$

$$M_{DC} = 3.4856 - 0.75\ (28.8461) = -18.15 \text{ kNm}$$

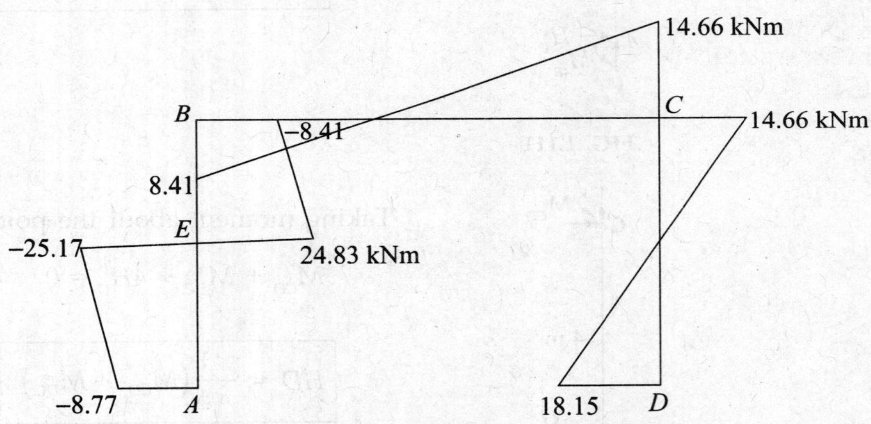

FIG. 1.113 Bending moment diagram

$$H_A = -\frac{1}{4}(-8.774 - 8.413 + 50) = -8.20 \text{ kN}$$

The negative sign indicates that the horizontal reaction direction is to be changed. Thus,

$$M_E = -8.774 - 8.2(2) = -25.17 \text{ kNm}$$

$$H_D = -\frac{1}{4}(-14.664 - 18.15) = +8.20 \text{ kN}$$

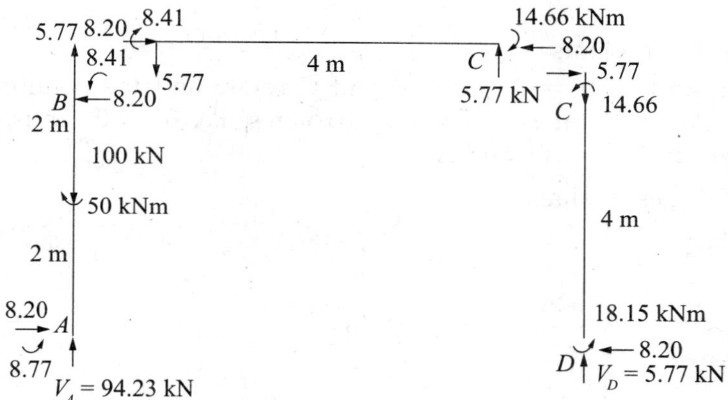

FIG. 1.114 Free body diagram

The vertical reactions V_A and V_D are obtained using the free body diagrams. Considering the span B_C, and taking moments about B, the value of V_C is obtained as $V_C = (8.41 + 14.66)/4 = 5.77$ kN. For vertical equilibrium, there should be a downward force of 5.77 kN at B. For equilibrium, an upward force of 5.77 kN acts at B. Now, consider the equilibrium of column A_B, $V_A + 5.77 = 100$ and hence $V_A = 94.23$ kN. The shear force diagram is drawn by using the horizontal and vertical reactions at A and D.

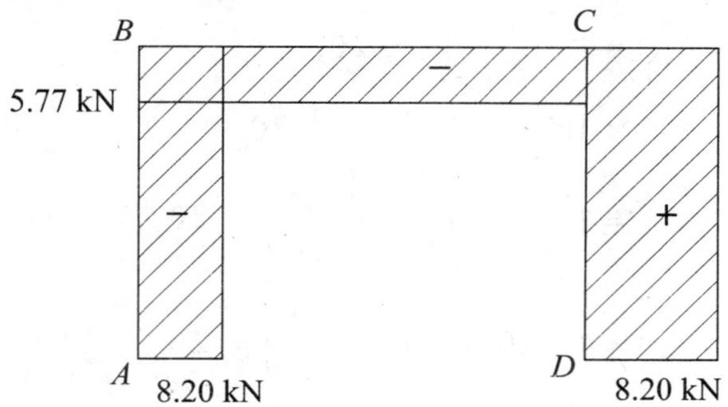

FIG. 1.115 Shear force diagram

Example 1.31 Analyse the frame by the slope deflection method. Draw the bending moment diagram.

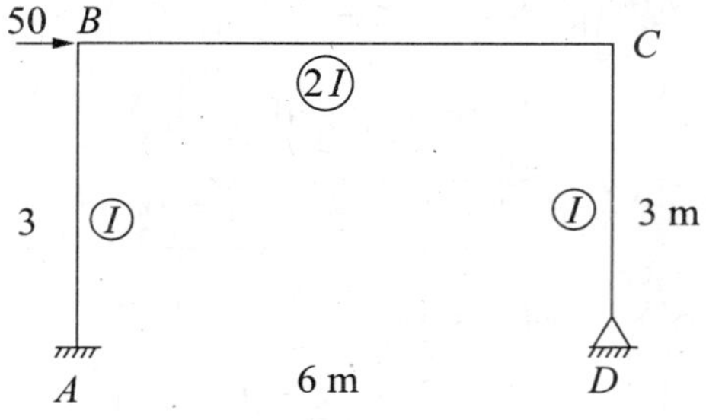

FIG. 1.116

Solution

1. Kinematic Redundants

   At the support A, $\theta_A = 0$. Joints B and C rotate and the column AB sways to the right. As the joint D is hinged, it rotates, i.e. $\theta_D \neq 0$. Thus, the kinematic redundants are θ_B, θ_C, θ_D and Δ.

2. Relative Stiffness Values

Member	I/l	k
AB	$I/3$	1
BC	$2I/6$	1
CD	$I/3$	1

3. Fixed End Moments

   As the frame members are not loaded; the fixed end moments are zero.

4. Slope Deflection Equations

$$M_{AB} = 1\left(2\cancel{\theta_A} + \theta_B - \frac{3\Delta}{3}\right) = \theta_B - \Delta$$

$$M_{BA} = 1\left(2\theta_B + \cancel{\theta_A} - \frac{3\Delta}{3}\right) = 2\theta_B - \Delta$$

$$M_{BC} = 1(2\theta_B + \theta_C) = 2\theta_B + \theta_C$$

$$M_{CB} = 1(2\theta_C + \theta_B) = \theta_B + 2\theta_C$$

$$M_{CD} = 1\left(2\theta_C + \theta_D - \frac{3\Delta}{3}\right) = 2\theta_C + \theta_D - \Delta$$

$$M_{DC} = 1\left(2\theta_D + \theta_C - \frac{3\Delta}{3}\right) = 2\theta_D + \theta_C - \Delta$$

5. Equilibrium Equations

Joint $B \Rightarrow$ $\quad M_{BA} + M_{BC} = 0$

$$\boxed{4\theta_B + \theta_C - \Delta = 0} \tag{1}$$

Joint $C \Rightarrow$ $\quad M_{CB} + M_{CD} = 0$

$$\boxed{\theta_B + 4\theta_C + \theta_D - \Delta = 0} \tag{2}$$

Joint $D \Rightarrow$ $\quad M_{DC} = 0$

$$2\theta_D + \theta_C - \Delta = 0$$

$$2\theta_D = \Delta - \theta_C$$

$$\theta_D = 0.5\Delta - 0.5\theta_C \qquad (3)$$

Substitute Eq. (3) in Eq. (2);

$$\theta_B + 4\theta_C + 0.5\Delta - 0.5\theta_C - \Delta = 0$$

$$\theta_B + 4\theta_C - 0.5\theta_C - 0.5\Delta = 0$$

$$\theta_B + 3.5\theta_C - 0.5\Delta = 0 \qquad (4)$$

6. Column Shear Equation

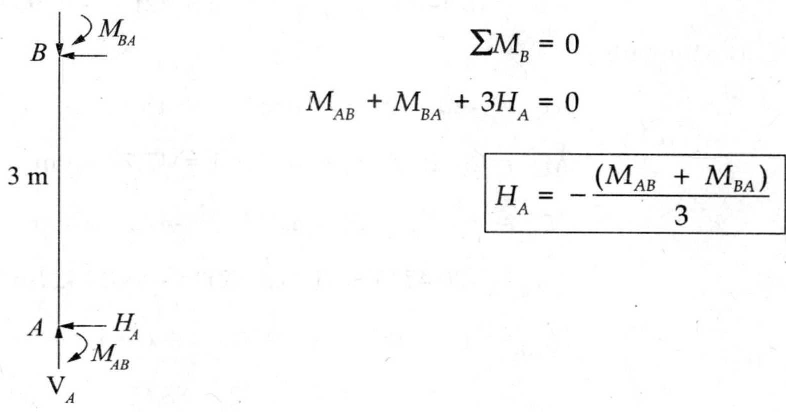

$$\Sigma M_B = 0$$

$$M_{AB} + M_{BA} + 3H_A = 0$$

$$\boxed{H_A = -\frac{(M_{AB} + M_{BA})}{3}}$$

FIG. 1.117 Free body diagram of *AB*

$$\Sigma M_C = 0$$

$$M_{CD} + M_{DC} + 3H_D = 0$$

$$\boxed{H_D = -\frac{(M_{CD} + M_{DC})}{3}}$$

FIG. 1.118 Free body diagram of *CD*

$$50 - \frac{(M_{AB} + M_{BA})}{3} - \frac{(M_{CD} + M_{DC})}{3} = 0$$

$$(M_{AB} + M_{BA}) + (M_{CD} + M_{DC}) = 150$$

$$3\theta_B - 2\Delta + 3\theta_C + 3\theta_D - 2\Delta = 150$$

$$3\theta_B + 3\theta_C - 4\Delta + 3(0.5\Delta - 0.5\theta_C) = 150$$

$$3\theta_B + 3\theta_C - 4\Delta - 1.5\theta_C + 1.5\Delta = 150$$

$$\boxed{3\theta_B + 1.5\theta_C - 2.5\Delta = 150} \quad (5)$$

Solving the above equations;

$$\theta_B = -20.4545,\ \theta_C = -6.8182 \text{ radian},\ \Delta = -88.6364 \text{ mm}$$

$$\theta_D = 0.5(-88.6364) - 0.5(-6.8182) = -40.9091 \text{ radians}$$

7. End Moments

$$M_{AB} = -20.4545 + 88.6364 = 68.18 \text{ kNm}$$

$$M_{BA} = 2(-20.4545) + 88.6364 = 47.72 \text{ kNm}$$

$$M_{BC} = 2(-20.4545) - 6.8182 = -47.72 \text{ kNm}$$

$$M_{CB} = -20.4545 + 2(-6.8182) = -34.09 \text{ kNm}$$

$$M_{CD} = 2(-6.8182) - 40.9091 + 88.6364 = +34.09 \text{ kNm}$$

$$M_{DC} = 2(-40.9091) - 6.8182 + 88.6364 = 0.00$$

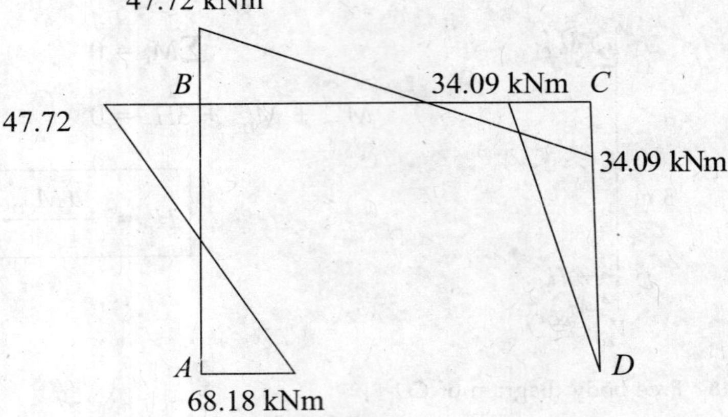

FIG. 1.119 Bending moment diagram

Example 1.32 Analyse the rigid frame shown in figure by slope deflection method. The support D settles down by 10 mm and rotates 0.001 radians in the clockwise direction.

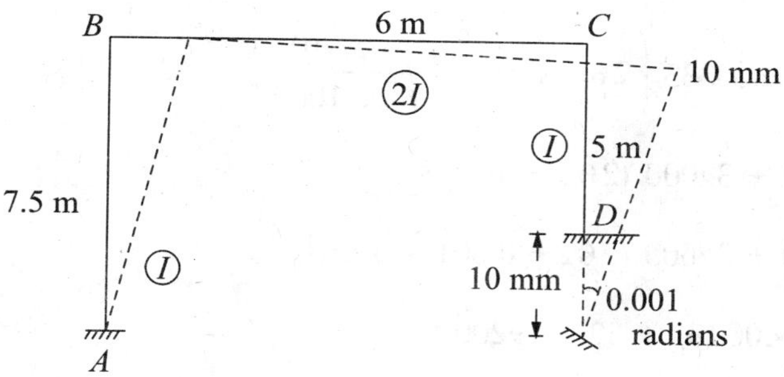

FIG. 1.120

The modulus of elasticity can be taken as 200×10^6 kN/m^3 and the second moment of the area is 400×10^{-6} m^4.

1. Kinematic Redundants

   The kinematic redundants at the fixed support at A is zero. At support D; $\theta_D = 0.001$. The joints B and C rotate and the column CD sways to the right by an amount Δ. Hence, the kinematic redundants to be found out are θ_B, θ_C and Δ.

2. Values of 'k'

   As no loads are applied on the entire frame, the fixed end moments are zeros. The appropriate values of k are calculated as follows.

Member	$(k = 2EI/l)$
AB	$\dfrac{2 \times 200 \times 10^6 \times 400 \times 10^{-6}}{7.5} = 21333$ kNm
BC	$\dfrac{2 \times 200 \times 10^6 \times 2 \times 400 \times 10^{-6}}{6.0} = 53333$ kNm
CD	$\dfrac{2 \times 200 \times 10^6 \times 400 \times 10^{-6}}{5.0} = 32000$ kNm

3. Slope Deflection Equations

$$M_{AB} = 0 + 21333\left(2\theta_A + \theta_B - \frac{3\Delta}{7.5}\right) = 21333\theta_B - 8533\Delta$$

$$M_{BA} = 0 + 21333\left(2\theta_B + \theta_A - \frac{3\Delta}{7.5}\right) = 42666\theta_B - 8533\Delta$$

$$M_{BC} = 0 + 53333\left(2\theta_B + \theta_C - \frac{3 \times 10}{6 \times 1000}\right) = 106666\theta_B + 53333\theta_C - 266.67$$

$$M_{CB} = 0 + 53333\left(2\theta_C + \theta_B - \frac{3 \times 10}{6 \times 1000}\right) = 106666\theta_C + 53333\theta_B - 266.67$$

$$M_{CD} = 0 + 32000\ (2\theta_C + \theta_D - 3\Delta/5)$$

$$= 0 + 32000\ (2\theta_C + 0.001 - 3\Delta/5)$$

$$= 64000\theta_C + 32 - 19200\Delta$$

$$M_{DC} = 0 + 32000\ (2\theta_D + \theta_C - 3\Delta/5)$$

$$= 32000\ (2 \times 0.001 + \theta_C - 3\Delta/5)$$

$$= 32000\theta_C + 64 - 19200\Delta$$

4. Equilibrium Conditions

At joint $B \Rightarrow$ $\quad M_{BA} + M_{BC} = 0$

$$149332\theta_B + 53333\theta_C - 8533\Delta = 266.67$$

$$\theta_B + 0.3571\theta_C - 0.0571\Delta = 1.7857 \times 10^{-3}$$

At joint $C \Rightarrow$ $\quad M_{CB} + M_{CD} = 0$

$$53333\theta_B + 1.70.666\theta_C - 19200\Delta = 234.67$$

$$\theta_B + 3.2\theta_C - 0.3600\Delta = 4.4000 \times 10^{-3}$$

Horizontal shear condition

By taking moment about the top hinge at B;

$$M_{AB} + M_{BA} + 7.5\ H_A = 0$$

$$H_A = \frac{-(M_{AB} + M_{BA})}{7.5}$$

FIG. 1.121

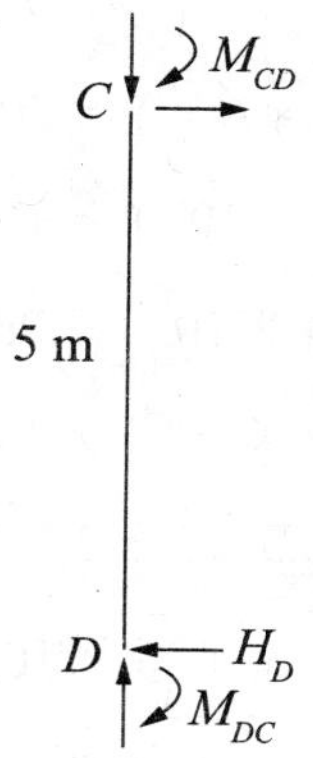

Taking moment about the top hinge at C;

$$M_{CD} + M_{DC} + 5\,H_D = 0$$

$$H_D = \frac{-(M_{CD} + M_{DC})}{5}$$

FIG. 1.122

$$H_A + H_D = 0$$

$$\frac{M_{AB} + M_{BA}}{7.5} + \frac{M_{CD} + M_{DC}}{5} = 0$$

$$2\,(M_{AB} + M_{BA}) + 3\,(M_{CD} + M_{DC}) = 0$$

$$2\,(21333\,\theta_B - 8533\,\Delta + 42666\,\theta_B - 8533\,\Delta) + 3\,(64000\,\theta_C + 32$$

$$-19200\,\Delta + 32000\,\theta_C + 64 - 19200\,\Delta) = 0$$

$$2\,(63999\,\theta_B - 17066\,\Delta) + 3\,(96000\,\theta_C - 38400\,\Delta + 96) = 0$$

$$127998\,\theta_B - 34132\,\Delta + 288000\,\theta_C - 115200\,\Delta + 288 = 0$$

$$127998\,\theta_B + 288000\,\theta_C - 149332\,\Delta = -288$$

$$\theta_B + 2.250\,\theta_C - 1.1667\,\Delta = 2.25 \times 10^{-3}$$

Hence, the equations to be solved are

$$\begin{bmatrix} 1 & 0.3571 & -\ 0.0571 \\ 1 & 3.2 & -\ 0.3600 \\ 1 & 2.250 & -\ 1.1667 \end{bmatrix} \begin{bmatrix} \theta_B \\ \theta_C \\ \Delta \end{bmatrix} = \begin{bmatrix} 1.7857 \\ 4.4000 \\ 2.25 \end{bmatrix} \times 10^{-3}$$

On solving

$$\theta_B = 1.5788 \times 10^{-3} \text{ radians}$$

$$\theta_C = 1.5974 \times 10^{-3} \text{ radians}$$

$$\Delta = 6.3625 \times 10^{-3} \text{ radians}$$

5. End Moments

$$M_{AB} = (21333\ (1.5788 \times 10^{-3}) - 8533\ (6.3625 \times 10^{-3}) = 20.616 \text{ kNm}$$

$$M_{BA} = (42666\ (1.5788 \times 10^{-3}) - 8533\ (6.3625 \times 10^{-3}) = +13.07 \text{ kNm}$$

$$M_{BC} = 106666\ (1.5788 \times 10^{-3}) + 53333\ (1.5974 \times 10^{-3}) - 266.67 = -13.07 \text{ kNm}$$

$$M_{CB} = 106666\ (1.5974 \times 10^{-3}) + 53333\ (1.5788 \times 10^{-3}) - 266.67 = -12.08 \text{ kNm}$$

$$M_{CD} = 64000\ (1.5974 \times 10^{-3}) + 32 - 19200\ (6.3625 \times 10^{-3}) = +12.07 \text{ kNm}$$

$$M_{DC} = 32000\ (1.5974 \times 10^{-3}) + 64 - 19200 \times 6.3625 \times 10^{-3} = -7.04 \text{ kNm}$$

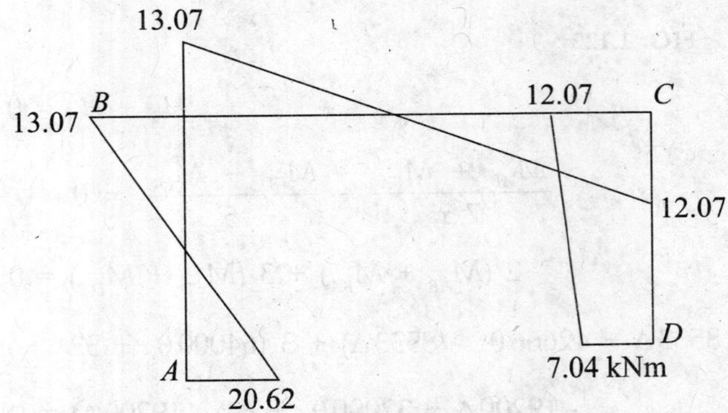

FIG. 1.123 Bending moment diagram

1.9 ANALYSIS OF BOX CULVERT

Example 1.33 Analyse the box type frame by the slope deflection method. Compute the end moments. There is no vertical settlements of the ends *A* and *D* of the frame.

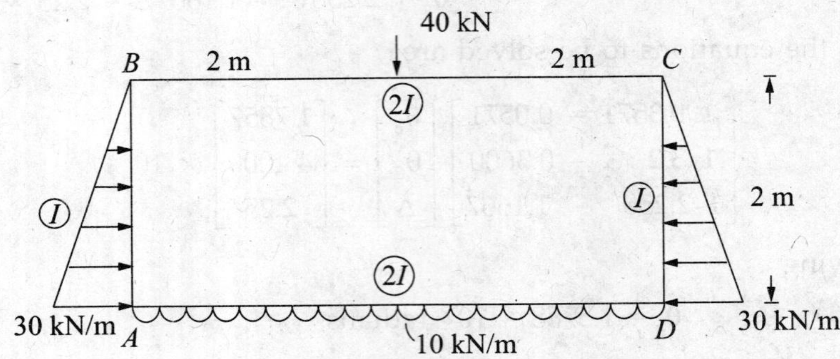

FIG. 1.124

1. Kinematic Redundants

   In the above box culvert, the unknowns to be determined kinematically are θ_A, θ_B, θ_C and θ_D. However, due to symmetry $\theta_B = -\theta_C$ and $\theta_A = -\theta_D$

2. Relative Stiffness Values

Member	I/l	k
AB	$I/2$	1
AD	$2I/4$	1
BC	$2I/4$	1

The values of k were obtained by multiplying the values of (I/l) with $(2/I)$.

3. Fixed End Moments

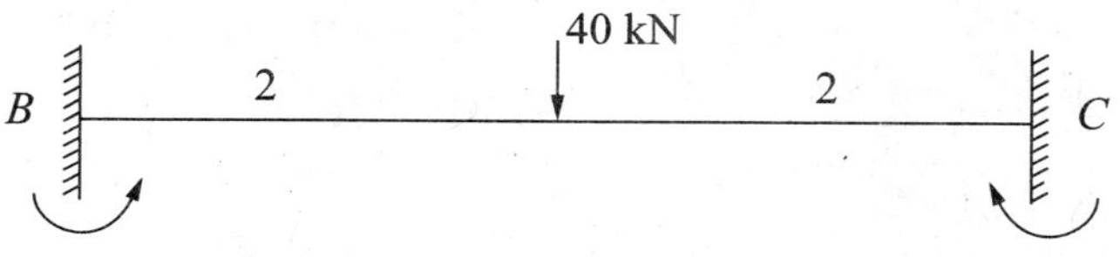

FIG. 1.125

$$\frac{-40 \times 4}{8} = -20 \text{ kNm} \qquad + 40 \times 4/8 = +20 \text{ kNm}$$

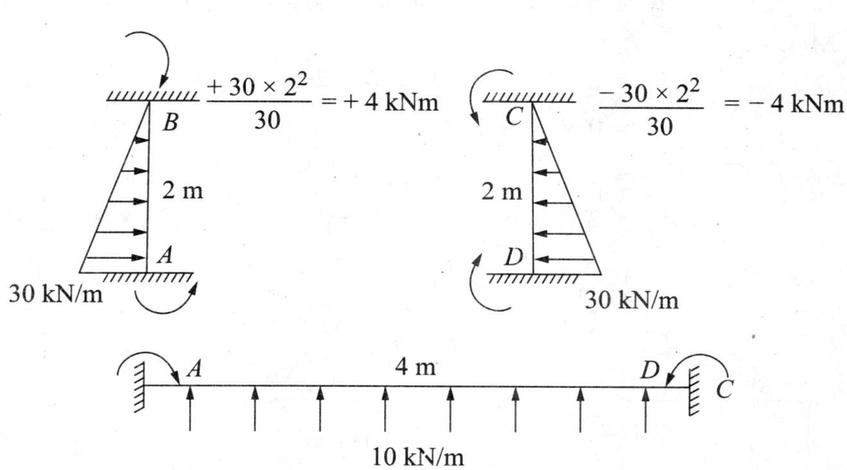

FIG. 1.126

$$\frac{+10 \times 4^2}{12} = +13.33 \text{ kNm} \qquad \frac{-10 \times 4^2}{12} = -13.33 \text{ kNm}$$

4. Slope Deflection Equations

$$M_{AB} = -20 + 1\ (2\theta_A + \theta_B) = 2\theta_A + \theta_B - 20$$

$$M_{BA} = +4 + 1\ (2\theta_B + \theta_A) = \theta_A + 2\theta_B + 4$$

$$M_{BC} = -20 + 1\ (2\theta_B + \theta_C) = 2\theta_B + \theta_C - 20$$

$$M_{AD} = 13.33 + 1\ (2\theta_A + \theta_D) = 2\theta_A + \theta_D + 13.33$$

5. Equilibrium Equations

$$M_{BA} + M_{BC} = 0$$

$$\theta_A + 2\theta_B + 4 + 2\theta_B + \theta_C - 20 = 0$$

$$\theta_A + 4\theta_B + \theta_C = 16$$

But $\theta_C = -\theta_B$ and hence

$$\theta_A + 3\theta_B = 16 \qquad (1)$$

$$M_{AB} + M_{AD} = 0$$

$$2\theta_A + \theta_B - 20 + 2\theta_A + \theta_D + 13.33 = 0$$

But $\theta_D = -\theta_A$ and hence (2)

$$3\theta_A + \theta_B = 6.67 \qquad (3)$$

Solving Eqs. (1) and (2);

$$\theta_A = 0.5013, \quad \theta_B = 5.1663 \text{ radians}$$

6. End Moments

$$M_{AB} = 2\ (0.5013) + 5.1663 - 20 = -13.83 \text{ kNm}$$

$$M_{BA} = 0.5013 + 2\ (5.1663) + 4 = +14.83 \text{ kNm}$$

$$M_{BC} = 2\ (5.1663) - 5.1663 - 20 = -14.83 \text{ kNm}$$

$$M_{AD} = 2\ (0.5103) - 0.5103 + 13.33 = +13.83 \text{ kNm}$$

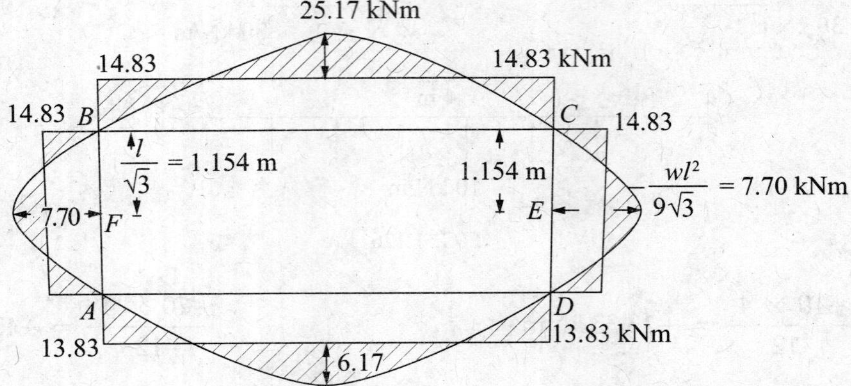

FIG. 1.127 Bending moment diagram

The free bending moments at E and F are calculated as follows. When a simply supported beam is subjected to uniformly varying load, the maximum bending moment occurs at a distance $(l/\sqrt{3})$ from where the linearly varying load is of zero magnitude. Also, the corresponding maximum bending moment is $(wl^2/9\sqrt{3})$.

1.10 ANALYSIS OF BENT FRAMES

Example 1.34 Analyse the given frame by the slope deflection method. Draw the bending moment diagram.

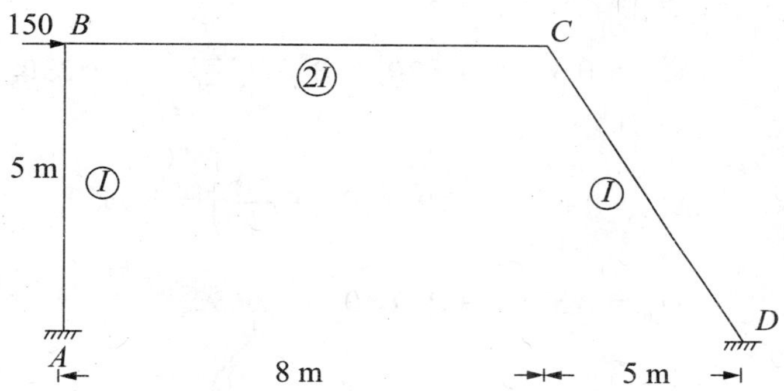

FIG. 1.128

1. Kinematic Redundants

$$\theta_A = 0; \quad \theta_B \neq 0 \quad \theta_C \neq 0, \quad \theta_D = 0$$

2. Relative Stiffness Values

Member	(I/l)	k
AB	$I/5$	1.414
BC	$I/4$	1.768
CD	$I/5\sqrt{2}$	1.000

3. Displacement Relationship

$$\sin\theta_1 = \sin 90 = 1 \qquad \Delta_1 = +\text{ve}$$

$$\cos\theta_1 = \cos 90 = 0 \qquad \Delta = -\text{ve}$$

$$\sin\theta_2 = \sin 45 = 0.707 \qquad \Delta_2 = +\text{ve}$$

$$\cos\theta_2 = \cos 45 = 0.707$$

$$\Delta_2 = \Delta_1 \frac{\sin\theta_1}{\sin\theta_2} = 1.414\,\Delta_1$$

$$\Delta = \Delta_1 \cos\theta_1 + \Delta_2 \cos\theta_2$$

$$\Delta = 0 + 1.414\,\Delta_1(0.707) = \Delta_1$$

4. Slope Deflection Equations

Assume the sway to the right

$$M_{AB} = 0 + 1.414\left(2\theta_A + \theta_B - \frac{3\Delta_1}{5}\right) = 1.414\theta_B - 0.8484\Delta$$

$$M_{BA} = 0 + 1.414\left(2\theta_B + \theta_A - \frac{3\Delta_1}{5}\right) = 2.828\theta_B - 0.8484\Delta_1$$

$$M_{BC} = 0 + 1.768\left(2\theta_B + \theta_C + \frac{3\Delta}{8}\right) = 3.536\theta_B + 1.768\theta_C + 0.663\Delta$$

$$M_{BC} = 3.536\theta_B + 1.768\theta_C + 0.663\Delta_1$$

$$M_{CB} = 0 + 1.768\left(2\theta_C + \theta_B + \frac{3\Delta}{8}\right) = 1.768\theta_B + 3.536\theta_C + 0.663\Delta$$

$$M_{CB} = 1.768\theta_B + 3.536\theta_C + 0.663\Delta_1$$

$$M_{CD} = 0 + 1(2\theta_C + \theta_D - 3\Delta_2/5\sqrt{2})$$

$$M_{CD} = 2\theta_C - \frac{3}{5\sqrt{2}} \times \sqrt{2}\,\Delta_1 = 2\theta_C - 0.6\Delta_1$$

$$M_{DC} = 0 + 1(2\theta_D + \theta_C - 0.6\Delta_1) = \theta_C - 0.6\Delta_1$$

5. Equilibrium Equations

Joint $B \Rightarrow$ $M_{BA} + M_{BC} = 0$

$$6.364\theta_B + 1.768\theta_C - 0.1854\Delta_1 = 0 \qquad (1)$$

Joint $C \Rightarrow$ $M_{CB} + M_{CD} = 0$

$$1.768\theta_B + 5.536\theta_C + 0.063\Delta_1 = 0 \qquad (2)$$

Horizontal shear equation

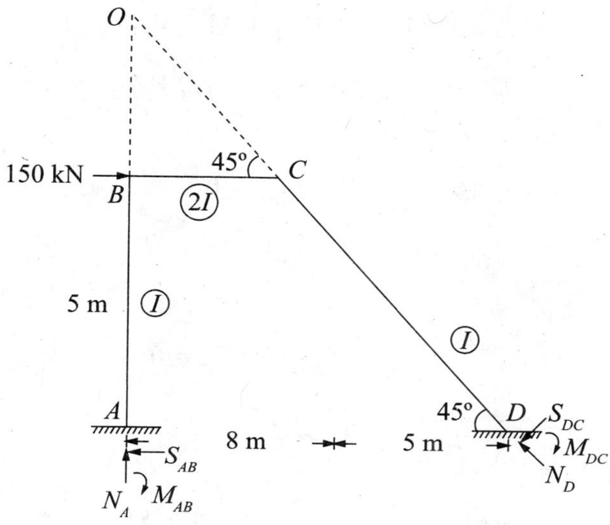

FIG. 1.129

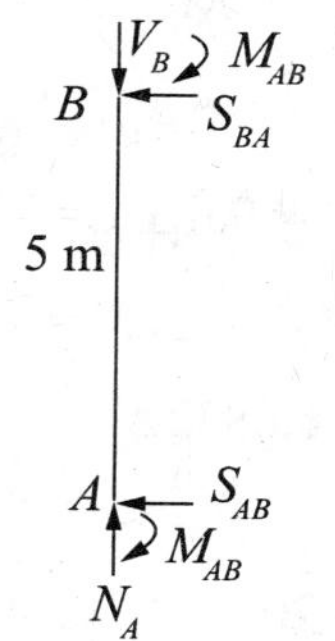

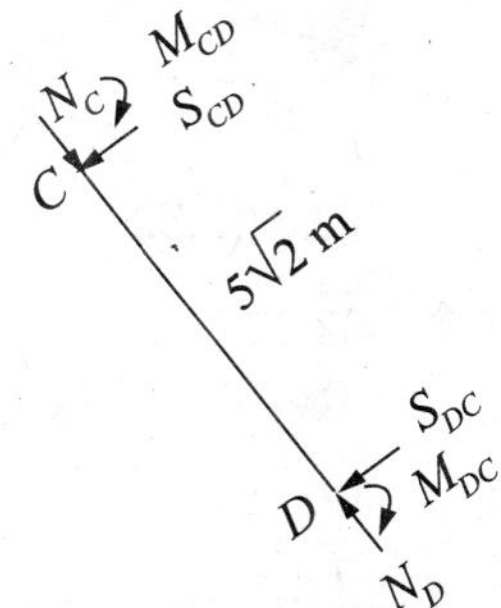

FIG. 1.130

In Δ *OBC*,

$$\tan 45 = \frac{OB}{BC}$$

$$\therefore \quad BC = OB$$

$$\boxed{OB = 8 \text{ m}}$$

$\Sigma M_0 = 0;$

$$-150 \times 8 + 13S_{AB} + M_{AB} + 13\sqrt{2}\, S_{DC} + M_{DC} = 0$$

$$13S_{AB} + 13\sqrt{2}\, S_{DC} + M_{AB} + M_{DC} = 1200 \quad (3)$$

$\Sigma M_B = 0;$

$$5S_{AB} + M_{AB} + M_{BA} = 0$$

$$S_{AB} = -0.2\ (M_{AB} + M_{BA}) \quad (4)$$

$\Sigma M_C = 0;$

$$5\sqrt{2}\, S_{DC} + M_{CD} + M_{DC} = 0$$

$$\therefore \quad S_{DC} = -0.1414\ (M_{CD} + M_{DC}) \quad (5)$$

Substituting (4) and (5) in Eq. (3)

$$-2.6\ (M_{AB} + M_{BA}) - 2.6\ (M_{CD} + M_{DC}) \div M_{AB} + M_{DC} = 1200$$

$$-1.6M_{AB} - 2.6M_{BA} - 2.6M_{CD} - 1.6M_{DC} = 1200$$

$$1.6\ (M_{AB} + M_{DC}) + 2.6(M_{BA} + M_{CD}) = -1200$$

$$1.6\ (1.414\,\theta_B - 0.8484\,\Delta_1 + \theta_C - 0.6\,\Delta_1)$$

$$+2.6\ (2.828\,\theta_B - 0.8484\,\Delta_1 + 2\,\theta_C - 0.6\,\Delta_1) = -1200$$

$$2.2624\,\theta_B + 1.6\,\theta_C - 2.3174\,\Delta_1 + 7.3528\,\theta_B + 5.2\,\theta_C - 3.768\,\Delta_1 = -1200$$

$$9.1652\,\theta_B + 6.8\,\theta_C - 6.0832\,\Delta_1 = -1200 \quad (6)$$

Solving the above

$$\boxed{\theta_B = 7.2352 \quad \theta_C = -4.7741 \quad \Delta_1 = 202.8289}$$

6. End Moments

$$M_{AB} = 1.414\ (7.2352) - 0.8484\ (202.8289) = -161.85 \text{ kNm}$$

$$M_{BA} = 2.828\ (7.2352) - 0.8484\ (202.8289) = -151.62 \text{ kNm}$$

$$M_{BC} = 3.536\ (7.2352) + 1.768\ (-4.7741) + 0.663\ (202.8289) = +151.62 \text{ kNm}$$

$$M_{CB} = 1.768\ (7.2352) + 3.536\ (-4.7741) + 0.663\ (202.8289) = 130.39 \text{ kNm}$$

$$M_{CD} = 2\ (-4.7741) - 0.6\ (202.8289) = -131.25 \text{ kNm}$$

$$M_{DC} = -4.7741 - 0.6\ (202.8289) = -126.47 \text{ kNm}$$

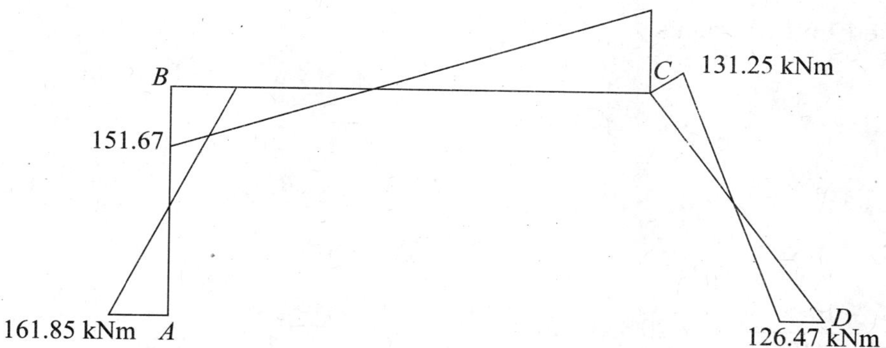

FIG. 1.131 Bending moment diagram

Example 1.35 Analyse the given frame by the slope deflection method. Draw the bending moment diagram.

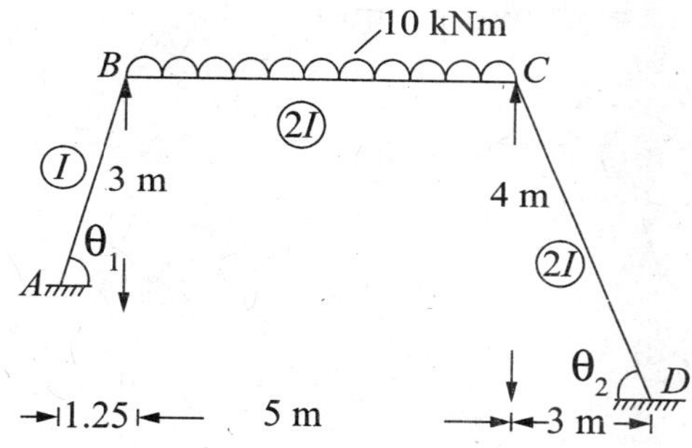

FIG. 1.132

1. Kinematic Redundants

   The kinematic redundants are θ_B, θ_C and the sway Δ. The values of the inclination of the members, viz. θ_1 and θ_2 are found from geometry.

$$AB = \sqrt{1.25^2 + 3^2} = 3.25 \text{ m}$$

$$CD = \sqrt{3^2 + 4^2} = 5.00 \text{ m}$$

$$\sin\theta_1 = 3/3.25 = 0.9231 \qquad \cos\theta_1 = \frac{1.25}{3.25} = 0.3846$$

$$\sin\theta_2 = 4/5 = 0.8 \qquad \cos\theta_2 = \frac{3}{5} = 0.6$$

2. Relative Stiffness Values ($k = I/l$)

Member	I/l	k
AB	$I/3.25$	5
BC	$2I/5$	6.5
CD	$2I/5$	6.5

3. Fixed End Moments

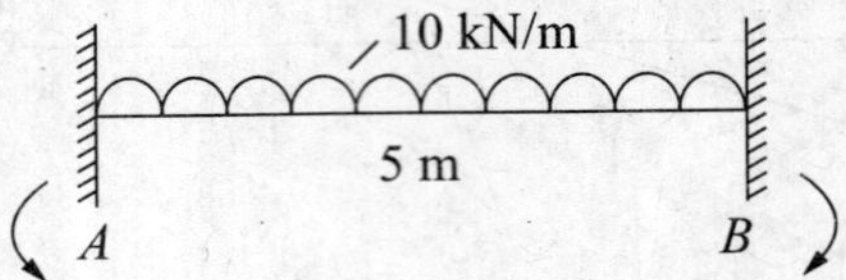

FIG. 1.133

$$\frac{-10 \times 5^2}{12} = -20.83 \text{ kNm} \qquad \frac{+10 \times 5^2}{12} = -20.83 \text{ kNm}$$

4. Displacement Relationship

Let Δ_1 be the perpendicular displacement of B, Δ_2 the displacement of C and Δ the vertical displacement between the displaced position of B and C. The displaced position of the frame is shown in figure below.

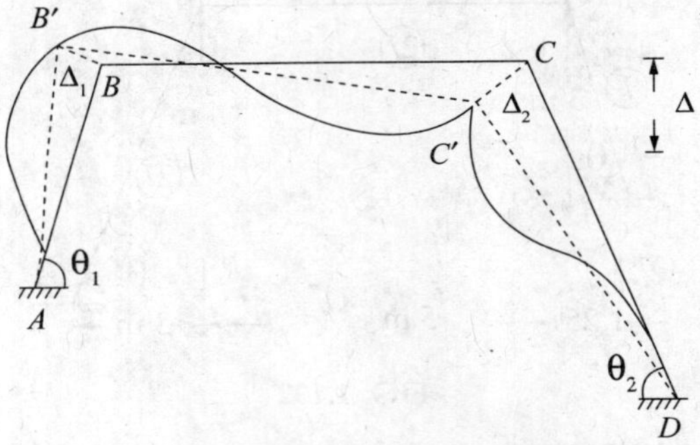

FIG. 1.134 Deflected shape of the frame

It is established from the geometry that

$$\Delta_2 = \Delta_1 \frac{\sin\theta_1}{\sin\theta_2}$$

$$\Delta_2 = \frac{0.9231}{0.8}\Delta_1$$

$$\boxed{\Delta_2 = 1.153\Delta_1}$$

Consider the free body diagram of the inclined member AB

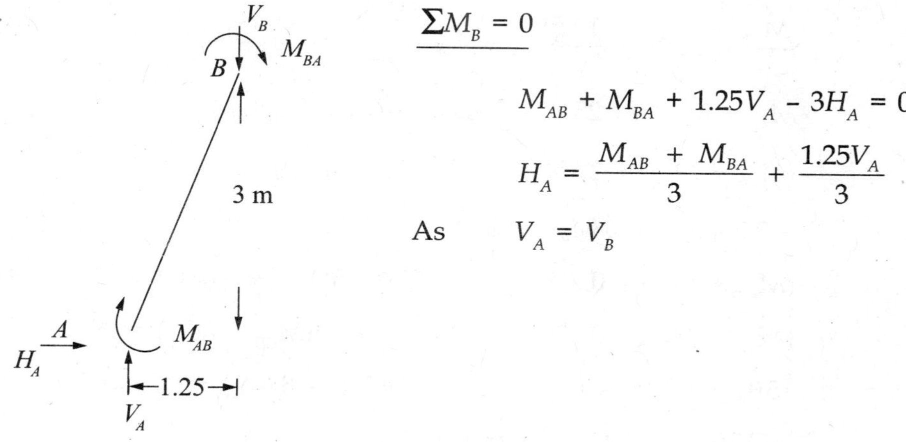

$\underline{\Sigma M_B = 0}$

$$M_{AB} + M_{BA} + 1.25V_A - 3H_A = 0$$

$$H_A = \frac{M_{AB} + M_{BA}}{3} + \frac{1.25V_A}{3}$$

As $\quad V_A = V_B$

FIG. 1.135

$$\boxed{H_A = \frac{(M_{AB} + M_{BA})}{3} + \frac{1.25}{3}\left[25 - \frac{(M_{BC} + M_{CB})}{5}\right]}$$

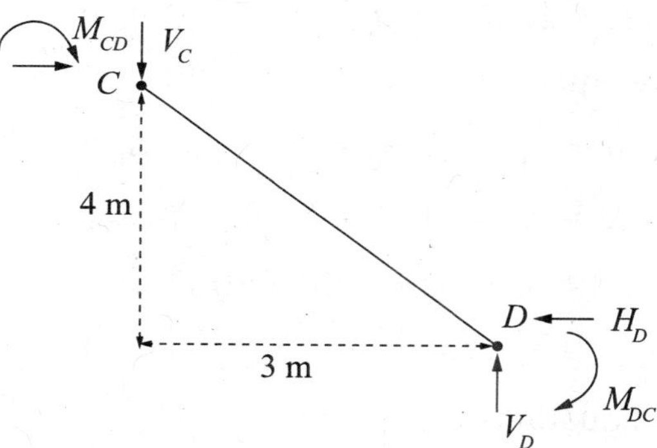

FIG. 1.136

$\Sigma M_C = 0$

$$M_{CD} + M_{DC} + 4H_D - 3V_D = 0$$

$$H_D = \frac{3V_D - (M_{CD} + M_{DC})}{4}$$

$$H_D = 0.75V_D - 0.25(M_{CD} + M_{DC})$$

$$\boxed{H_D = 0.75\left\{25 + \frac{1}{5}(M_{BC} + M_{CB})\right\} - 0.25\,(M_{CD} + M_{DC})}$$

$$H_A = H_D$$

and hence

$$\frac{M_{AB} + M_{BA} + 1.25V_A}{3} = \frac{3V_D - (M_{CD} - M_{DC})}{4}$$

$$M_{AB} + M_{BA} + 1.25V_A = 2.25V_D - 0.75(M_{CD} + M_{DC})$$

$$(M_{AB} + M_{BA}) + 1.25(25 - 0.2\,(M_{BC} + M_{CB}))$$

$$= 2.25(25 + 0.2(M_{BC} + M_{CB})) - 0.75(M_{CD} + M_{DC})$$

$$(M_{AB} + M_{BA}) - 0.2(M_{BC} + M_{CB}) - 0.45(M_{BC} + M_{CB}) + 0.75(M_{CD} + M_{DC}) = 25$$

$$(M_{AB} + M_{BA}) - 0.65(M_{BC} + M_{CB}) + 0.75(M_{CD} + M_{DC}) = 25$$

$$15\theta_B + 9.24\Delta_1 - 0.65(19.5\theta_B + 19.5\theta_C - 8.4\Delta_1) + 0.75(19.5\theta_C + 9\Delta_1) = 25$$

$$2.325\theta_B + 1.95\theta_C + 21.45\Delta_1 = 25$$

The kinematic redundants are determined from

$$23\theta_B + 6.5\theta_C + 0.42\Delta_1 = 20.83 \quad (1)$$

$$6.5\theta_B + 26\theta_C + 0.3\Delta_1 = -20.83 \quad (2)$$

$$2.325\theta_B + 1.95\theta_C + 21.45\Delta_1 = 25 \quad (3)$$

Solving the above equation

$$\theta_B = 1.1998; \quad \theta_C = -1.1142; \quad \Delta_1 = 1.1367$$

$$\Delta = \Delta_1 \cos\theta_1 + \Delta_2 \cos\theta_2$$

$$\Delta = 0.385\Delta_1 + 1.153\Delta_1(0.6)$$

$$\Delta = 1.077\Delta_1$$

5. Slope Deflection Equations

$$M_{AB} = 0 + 5\left(2\theta_A + \theta_B + \frac{3\Delta_1}{3.25}\right) = 5\theta_B + 4.62\Delta_1$$

$$M_{BA} = 0 + 5\left(2\theta_B + \theta_A + \frac{3\Delta_1}{3.25}\right) = 10\theta_B + 4.62\Delta_1$$

$$M_{BC} = -20.83 + 6.5\left(2\theta_B + \theta_C - \frac{3\Delta}{5}\right) = -20.83 + 13\theta_B + 6.5\theta_C - 3.9\Delta$$

$$= -20.83 + 13\theta_B + 6.5\theta_C - 3.9(1.077\Delta_1)$$

$$= -20.83 + 13\theta_B + 6.5\theta_C - 4.2\Delta_1$$

$$M_{BC} = 13\theta_B + 6.5\theta_C - 4.2\Delta_1 - 20.83$$

$$M_{CB} = +20.83 + 6.5\left(2\theta_C + \theta_B - \frac{3\Delta}{5}\right) = 20.83 + 13\theta_C + 6.5\theta_B - 3.9\Delta$$

$$= +20.83 + 6.5\theta_B + 13\theta_C - 3.9\Delta$$

$$= 6.5\theta_B + 13\theta_C - 3.9(1.077\Delta_1) + 20.83$$

$$M_{CB} = 6.5\theta_B + 13\theta_C - 4.200\Delta_1 + 20.83$$

$$M_{CD} = 0 + 6.5(2\theta_C + \cancel{\theta_D} + 3\Delta_2/5)$$

$$M_{CD} = 13\theta_C + 3.9(1.153\Delta_1)$$

$$M_{CD} = 13\theta_C + 4.4967\Delta_1$$

$$M_{CD} = 13\theta_C + 4.5\Delta_1$$

$$M_{DC} = 0 + 6.5(2\cancel{\theta_D} + \theta_C + 3\Delta_2/5)$$

$$M_{DC} = 6.5\theta_C + 3.9(1.153\Delta_1)$$

$$M_{DC} = 6.5\theta_C + 4.4967\Delta_1$$

6. Equilibrium Equations

(i) At joint B ⇒ $M_{BA} + M_{BC} = 0$

$$23\theta_B + 6.5\theta_C + 0.42\Delta_1 = 0 \quad (1)$$

(ii) At joint C ⇒ $M_{CB} + M_{CD} = 0$

$$6.5\theta_B + 26\theta_C + 0.3\Delta_1 = -20.83 \quad (2)$$

(iii) Horizontal shear condition:

Consider the free body diagram of span BC

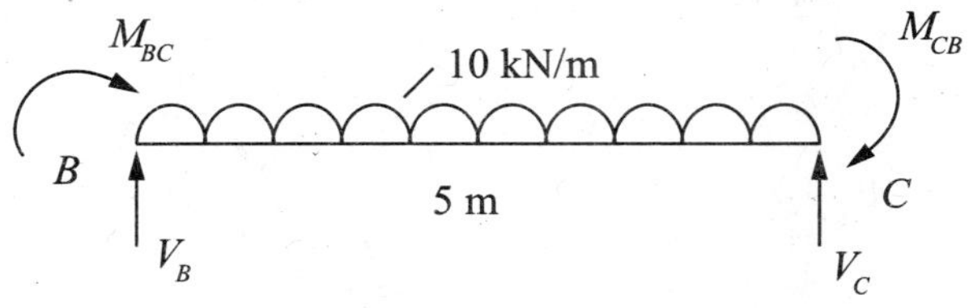

FIG. 1.137

$\Sigma M_B = 0;$

$$M_{BC} + M_{CB} + 5V_B - 10 \times \frac{5^2}{2} = 0$$

$$5V_B = 125 - (M_{BC} + M_{CB})$$

$$\boxed{V_B = 25 - \frac{(M_{BC} + M_{CB})}{5}}$$

$\Sigma M_B = 0;$

$$M_{BC} + M_{CB} - 5V_C + \frac{10 \times 5^2}{2} = 0$$

$$V_C = \frac{M_{BC} + M_{CB} + 125}{5}$$

$$\boxed{V_C = 25 + \frac{1}{5}(M_{BC} + M_{CB})}$$

7. End Moments

$M_{AB} = 5(1.1998) + 4.62(1.1367) = 11.25$ kNm

$M_{BA} = 10(1.1998) + 4.62(1.1367) = 17.249$ kNm

$M_{BC} = 13(1.1998) + 6.5(-1.1142) - 4.2(1.1367) - 20.83 = -17.249$ kNm

$M_{CB} = 6.5(1.1998) + 13(-1.1142) - 4.2(1.1367) + 20.83 = +9.370$ kNm

$M_{CD} = 13(-1.1142) + 4.5(1.1367) = -9.370$ kNm

$M_{DC} = 6.5(-1.1142) + 3.9(1.153 \times 1.1367) = -2.13$ kNm

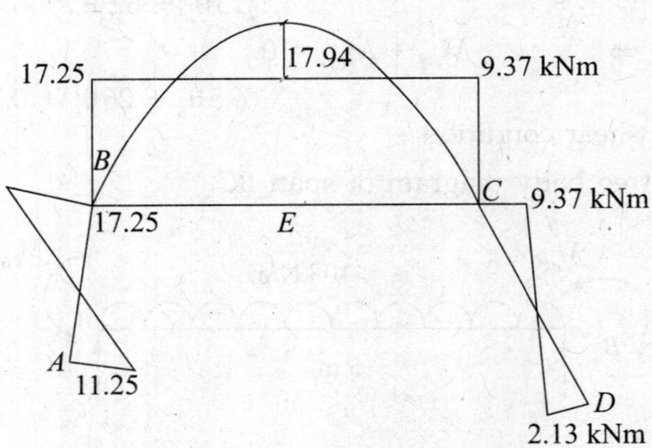

FIG. 1.138 Bending moment diagram

$$M_E = \frac{10 \times 5^2}{8} - \left(\frac{17.25 + 9.37}{2}\right)$$

$= 17.94$ kNm

1.11 ANALYSIS OF GABLE FRAME

Example 1.36 Analyse the Gable frame by the slope deflection method. Draw the shear force and bending moment diagram.

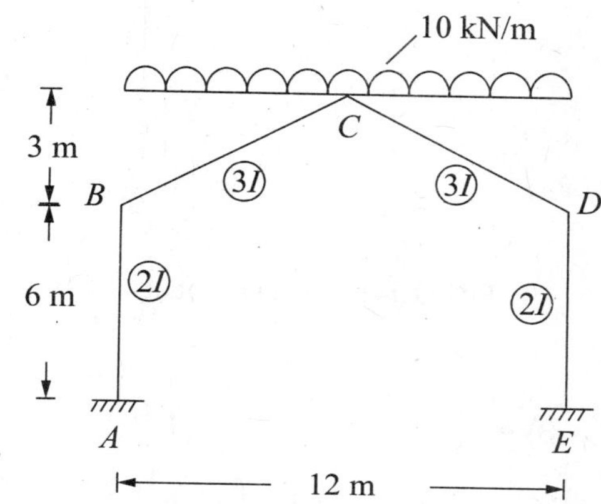

FIG. 1.139 Gable frame

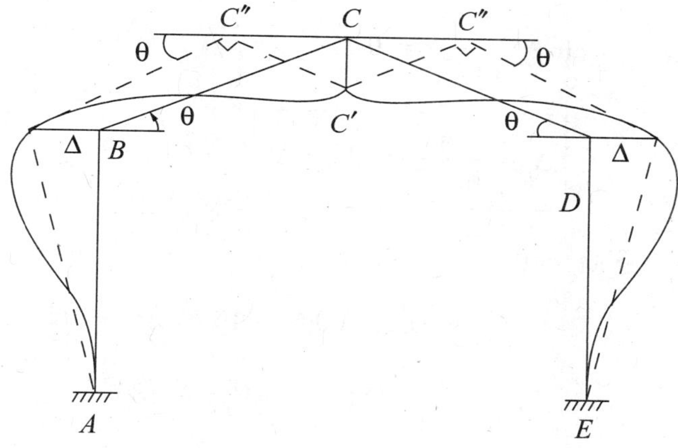

FIG. 1.140 Displacement diagram

1. Kinematic Redundants

   $\theta_A = \theta_E = \theta_C = 0$, due to symmetry $\theta_B = -\theta_D$ and Δ.

2. Relative Stiffness Values

Member	(I/l)	k
AB	$2I/6$	1.34
BC	$3I/6.71$	1.80

3. Displacement Relationship

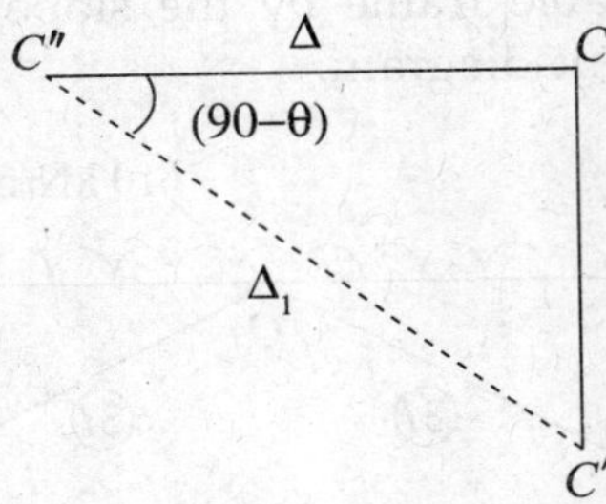

FIG. 1.141 Displacements

Displacements

$$\cos(90 - \theta) = \frac{\Delta}{\Delta_1}$$

$$\sin\theta = \frac{\Delta}{\Delta_1}$$

From geometry, $\sin\theta = \dfrac{3}{6.71} = 0.447$

$$\therefore \quad \boxed{\Delta_1 = \frac{\Delta}{0.447} = 2.236\Delta}$$

$$\Sigma M_C = 0$$

$$M_{BC} + M_{CB} + 60(6) - 10 \times \frac{6^2}{2} - 3H_{BC} = 0$$

$$H_{BC} = \frac{1}{3}\left[(M_{BC} + M_{CB}) + 180\right]$$

$$\boxed{H_{BC} = \frac{M_{BC} + M_{CB}}{3} + 60}$$

$$H_A = +H_{BC}$$

$$\frac{(M_{AB} + M_{BA})}{6} = +\frac{(M_{BC} + M_{CB})}{3} + 60$$

$$\frac{(M_{AB} + M_{BA})}{6} - \frac{(M_{BC} + M_{CB})}{3} = +60$$

$$(M_{AB} + M_{BA}) - 2(M_{BC} + M_{CB}) = +360$$

$$4.02\theta_B + 1.34\Delta - 2(5.4\theta_B - 3.6\Delta) = +360$$

$$\boxed{-6.78\theta_B + 8.54\Delta = +360}$$

Hence,

$$\begin{bmatrix} 6.28 & -1.13 \\ -6.78 & +8.54 \end{bmatrix} \begin{Bmatrix} \theta_B \\ \Delta \end{Bmatrix} = \begin{bmatrix} 30 \\ +360 \end{bmatrix}$$

On solving,

$$\theta_B = 14.422 \text{ radians}$$

$$\Delta = 53.6048$$

4. Fixed End Moments

$$M_{AB}^F = 0, \quad M_{BC}^F = -10 \times \frac{6^2}{12} = -30 \text{ kNm}$$

5. Slope Deflection Equations

$$M_{AB} = 0 + 1.34\left(2\theta_A + \theta_B - \frac{(-3\Delta)}{6}\right) = 1.34\theta_B + 0.67\Delta$$

$$M_{BA} = 0 + 1.34\left(2\theta_B + \theta_A - \frac{(-3\Delta)}{6}\right) = 2.68\theta_B + 0.67\Delta$$

$$M_{BC} = -30 + 1.80\left(2\theta_B + \theta_C - \frac{3\Delta_1}{6.71}\right)$$

$$= -30 + 1.80(2\theta_B + \theta_C - 0.4471\Delta_1)$$

$$= -30 + 1.80(2\theta_B + \theta_C - 0.4471 \times 2.236\Delta)$$

$$= -30 + 1.80(2\theta_B + \theta_C - \Delta) = 3.6\theta_B + 1.80\theta_C - 1.80\Delta - 30$$

$$M_{CB} = +30 + 1.80(2\theta_C + \theta_B - \Delta) = 1.8\theta_B - 1.8\Delta + 30$$

6. Equilibrium Equations

Joint B $\Rightarrow$ $M_{BA} + M_{BC} = 0$

$$6.28\theta_B - 1.13\Delta = 30$$

$\sum M_B = 0$

$$M_{AB} + M_{BA} - 6H_A = 0$$

$$\boxed{H_A = \frac{(M_{AB} + M_{BA})}{6}}$$

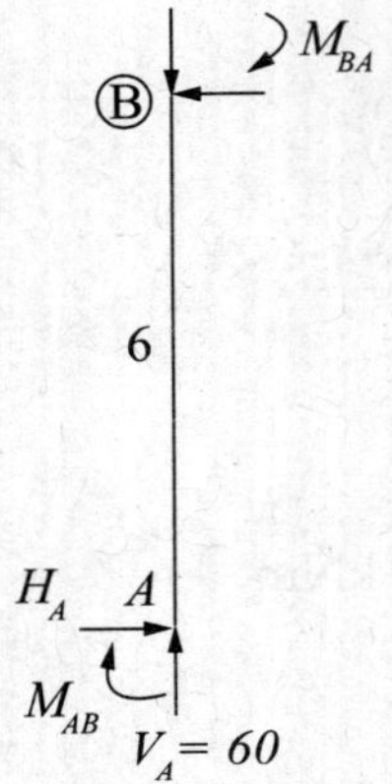

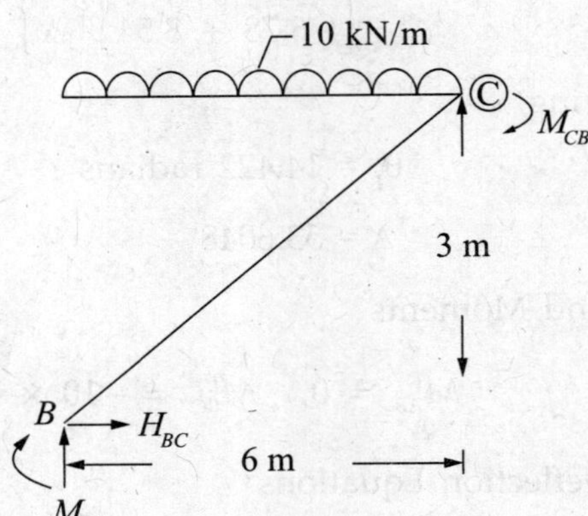

FIG. 1.142

From similar Δ's

$$\frac{x}{(6.71 - x)} = \frac{4.4}{9.68}$$

$$9.68x = 295.24 - 44x$$

$$\boxed{x = 5.5 \text{ m}}$$

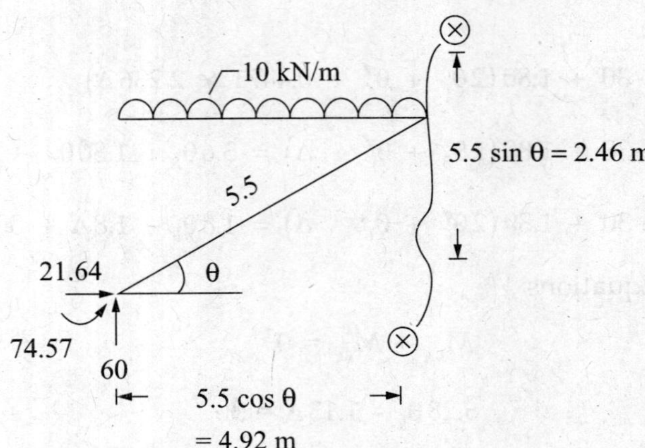

FIG. 1.143

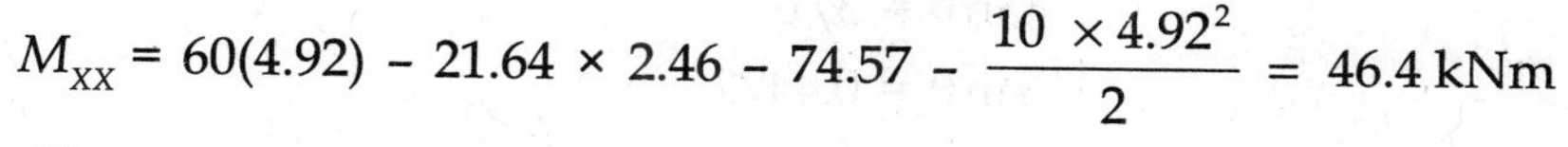

$$M_{XX} = 60(4.92) - 21.64 \times 2.46 - 74.57 - \frac{10 \times 4.92^2}{2} = 46.4 \text{ kNm}$$

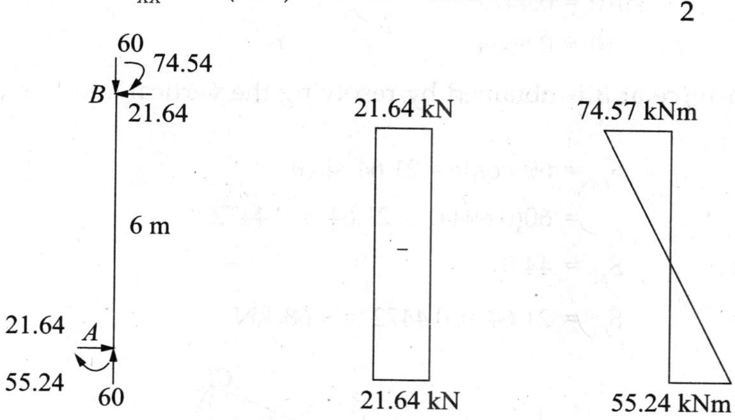

FIG. 1.144

7. End Moments

$$M_{AB} = 1.34(14.422) + 0.67(53.6048) = 55.24 \text{ kNm}$$

$$M_{BA} = 2.68(14.422) + 0.67(53.6048) = 74.57 \text{ kNm}$$

$$M_{BC} = 3.6(14.422) - 1.8(53.6048) - 30 = -74.57 \text{ kNm}$$

$$M_{CB} = 1.8(14.422) - 1.8(53.6048) + 30 = -40.53 \text{ kNm}$$

$$H_A = (55.24 + 74.57)/6 = 21.64 \text{ kNm}$$

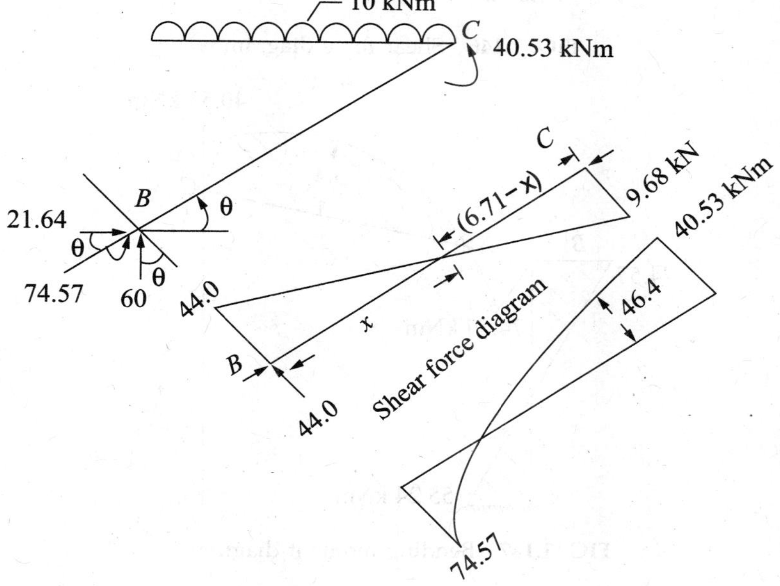

FIG. 1.145

$$\tan\theta = 3/6$$
$$\sin\theta = 0.4472$$
$$\cos\theta = 0.8944$$

The shear force at B is obtained by resolving the vertical and horizontal reaction as

$$S_{BC} = 60\cos\theta - 21.64\sin\theta$$
$$= 60(0.8944) - 21.64 \times 0.4472$$
$$S_{BC} = 44.0$$
$$S_{CB} = 21.64 \times 0.4472 = 9.68 \text{ kN}$$

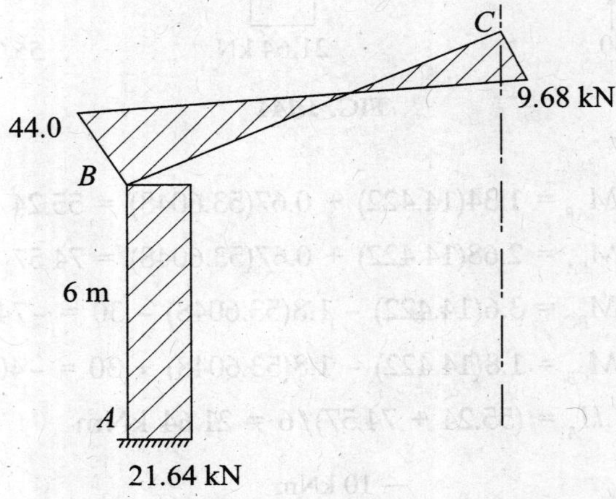

FIG. 1.146 Shear force diagram

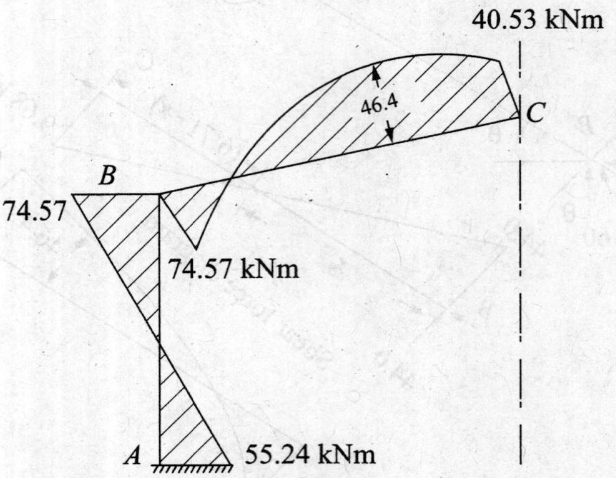

FIG. 1.147 Bending moment diagram

1.12 ADDITIONAL PROBLEMS

Example 1.37 Analyse the non-prismatic beam shown in Fig. 1.148 by the slope deflection method. Draw the bending moment diagram.

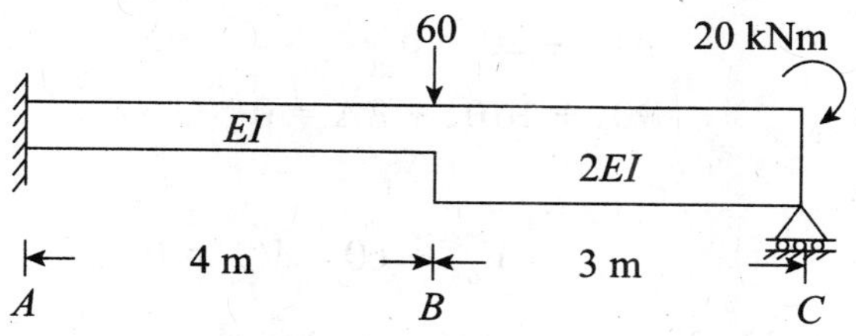

FIG. 1.148

Solution

1. Kinematic Redundants

$$\theta_A = 0; \quad \theta_B \neq 0, \quad \theta_C \neq 0, \quad \Delta_B \neq 0$$

2. Relative Stiffness Values

Member	(I/l)	k
AB	$\left(\frac{}{}\right) \times \frac{12}{}$	3
BC	$\left(\frac{2I}{3}\right) \times \frac{12}{I}$	8

3. Fixed End Moments

   As there is no loading on AB and BC respectively

$$M_{FAB} = M_{FBA} = M_{FBC} = M_{FCB} = 0$$

4. Slope Deflection Equations

$$M_{AB} = 0 + 3\left(2\cancel{\theta_A} + \theta_B - \frac{3\Delta}{4}\right) = 3\theta_B - 2.25\Delta$$

$$M_{BA} = 0 + 3\left(2\theta_B + \cancel{\theta_A} - \frac{3\Delta}{4}\right) = 6\theta_B - 2.25\Delta$$

$$M_{BC} = 0 + 8\left(2\theta_B + \theta_C - \frac{3\Delta}{3}\right) = 16\theta_B + 8\theta_C + 8\Delta$$

$$M_{CB} = 0 + 8\left(2\theta_C + \theta_B - \frac{3\Delta}{3}\right) = 8\theta_B + 16\theta_C + 8\Delta$$

5. Equilibrium Equations

Joint $B \Rightarrow$ $\quad M_{BA} + M_{BC} = 0;$

$$22\theta_B + 8\theta_C + 5.75\Delta = 0 \qquad (1)$$

Joint $C \Rightarrow$ $\quad M_{CB} - 20 = 0$

$$8\theta_B + 16\theta_C + 8\Delta = 0 \qquad (2)$$

Shear condition

$$V_{BA} + 60 - V_{BC} = 0$$

$$\frac{(M_{AB} + M_{BA})}{4} + 60 - \frac{(M_{BC} + M_{CB})}{3} = 0$$

Substituting for moments in terms of slopes and deflections;

$$5.75\theta_B + 8\theta_C + 6.455\Delta = 60 \qquad (3)$$

Solving the above equations;

$$\theta_B = -2.7244, \quad \theta_C = -8.5421, \quad \Delta = 22.309 \qquad (4)$$

6. End Moments

$$M_{AB} = 3(-2.7244) - 2.25(22.308) = -58.37 \text{ kNm}$$

$$M_{BA} = 6(-2.7244) - 2.25(22.308) = -66.54 \text{ kNm}$$

$$M_{BC} = 16(-2.7244) + 8(-8.5421) + 8(22.308) = +66.54 \text{ kNm}$$

$$M_{CB} = 8(-2.7244) + 16(-8.5421) + 8(22.308) = 20.00 \text{ kNm}$$

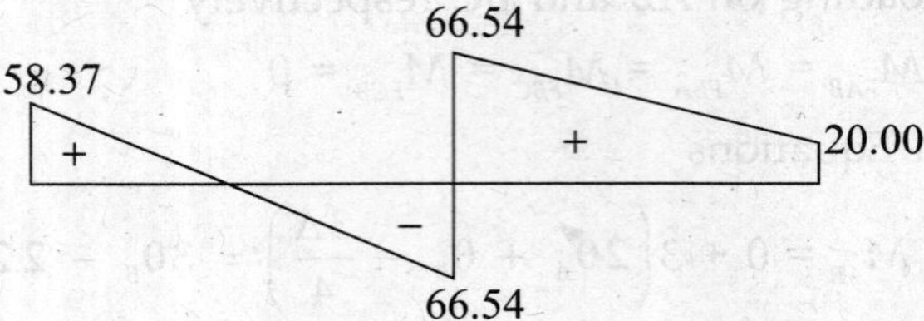

FIG. 1.149 Bending moment diagram

Example 1.38 Analyse the continuous beam shown in Fig. 1.150 by slope deflection method, *EI* is constant. Draw the bending moment diagram.

10 kN/m 30 kN

A 8 m *B* 3 m *C* 3 m *D*

FIG. 1.150

Solution: An internal hinge is provided at *C*. Hence, this problem is solved with the concern of the effect of internal hinge on fixed end moment of span *BD*.

1. Kinematic Redundants

$$\theta_A = 0, \quad \theta_B \neq 0, \quad \theta_C \neq 0, \quad \Delta_{C'} \quad \theta_D = 0$$

2. Fixed End Moments

Due to applied loading

span *AB*

$$M_{FAB} = \frac{-wl^2}{12} = -10 \times \frac{8^2}{12} = -53.33 \text{ kNm}$$

$$M_{FBA} = \frac{+wl^2}{12} = +10 \times \frac{8^2}{12} = +53.33 \text{ kNm}$$

span *BD*

The fixed end moment at *B* due to a load *W* at a distance *x* and with an internal hinge is given by

FIG. 1.151

$$M_{FBD} = -\left(\frac{Wx(2a^3 - 3a^2x + ax^2 + 2b^3)}{2(a^3 + b^3)}\right)$$

$$M_{FDB} = \frac{Wbx^2(3a - x)}{2(a^3 + b^3)}$$

In the above general expression, we substitute $x = a$, $a = b$ it reduces to

$$M_{FBD} = -\frac{Wab^3}{a^3 + b^3} = -\frac{30(3)(3)^3}{3^3 + 3^3} = -45 \text{ kNm}$$

$$M_{FDB} = +\frac{Wa^3b}{a^3 + b^3} = +\frac{30(3)^3(3)}{3^3 + 3^3} = +45 \text{ kNm}$$

The moment due to deformations at *B* is computed separately as

$$M'_{AB} = \frac{2EI}{8}(2\theta_A + \theta_B) = 0.25EI\theta_B$$

$$M'_{BA} = \frac{2EI}{8}(2\theta_B + \theta_A) = 0.5EI\theta_B$$

$$M'_{BD} = 3EIa^2\theta_B/(a^3 + b^3) = 0.5EI\theta_B$$

The final moments are

$$M'_{DB} = 3EIab\theta_B/(a^3 + b^3) = 0.5EI\theta_B$$

Adding the fixed end moments due to loading and due to deformations, the final moments are obtained.

3. Slope Deflection Equations

$$M_{AB} = -53.33 + 0.25EI\theta_B$$

$$M_{BA} = 53.33 + 0.50EI\theta_B$$

$$M_{BD} = -45.00 + 0.5EI\theta_B$$

$$M_{DB} = +45.00 + 0.5EI\theta_B$$

4. Equilibrium Conditions

At joint B;

$$M_{BA} + M_{BD} = 0$$

$$8.33 + EI\theta_B = 0$$

$$\theta_B = -8.33/EI$$

5. End Moments

$$M_{AB} = -53.33 + 0.25(-8.33) = -55.41 \text{ kNm}$$

$$M_{BA} = +53.33 + 0.50(-8.33) = +49.17 \text{ kNm}$$

$$M_{BD} = -45.00 + 0.50(-8.33) = -49.17 \text{ kNm}$$

$$M_{DB} = +45.00 + 0.50(-8.33) = +40.84 \text{ kNm}$$

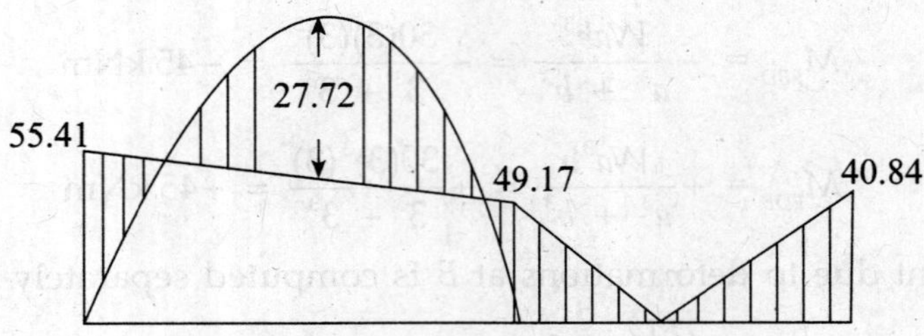

FIG. 1.152 Bending moment diagram

Example 1.39 Analyse the beam shown in Fig. 1.153 by slope deflection method. Draw the bending moment diagram.

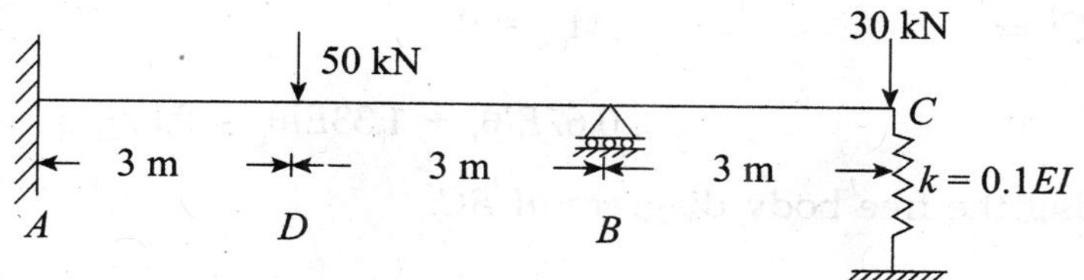

FIG. 1.153

Solution

1. Kinematic Redundants

   A is a fixed support and that $\theta_A = 0$; B is an intermediate support and $\theta_B \neq 0$. C is resting on spring.

   Hence $\theta_C \neq 0$ and $\Delta C \neq 0$.

2. Fixed End Moments

$$M_{FAB} = \frac{-Wl}{8} = \frac{-50 \times 6}{8} = -37.5 \text{ kNm}$$

$$M_{FBA} = \frac{+Wl}{8} = \frac{+50 \times 6}{8} = +37.5 \text{ kNm}$$

3. Slope Deflection Equations

$$M_{AB} = -37.5 + \frac{2EI}{6}(2\theta_A + \theta_B)$$

$$M_{AB} = -37.5 + 0.33EI\theta_B \quad (1)$$

$$M_{BA} = +37.5 + \frac{2EI}{6}(2\theta_B + \theta_A)$$

$$M_{BA} = 37.5 + 0.67EI\theta_B \quad (2)$$

$$M_{BC} = \frac{2EI}{3}\left(2\theta_B + \theta_C - \frac{3\Delta_C}{3}\right)$$

$$M_{BC} = 1.33\ EI\theta_B + 0.67\ EI\theta_C - 0.67\ EI\Delta_C \quad (3)$$

$$M_{CB} = \frac{2EI}{3}\left(2\theta_C + \theta_B - \frac{3\Delta_C}{3}\right)$$

$$M_{CB} = 1.33\ EI\,\theta_C + 0.67\ EI\,\theta_B - 0.67\ EI\Delta_C \quad (4)$$

4. Joint Equilibrium Condition and Shear Condition

   At Joint B $\Rightarrow$ $M_{BA} + M_{BC} = 0$

$$2EI\theta_B + 0.67EI\theta_C + 0.67EI\Delta_C = -37.50 \quad (5)$$

At Joint $C \Rightarrow$ $\qquad M_{CB} = 0$

$$0.67EI\theta_B + 1.33EI\theta_C - 0.67EI\Delta_C = 0 \qquad (6)$$

Considering the free body diagram of BC;

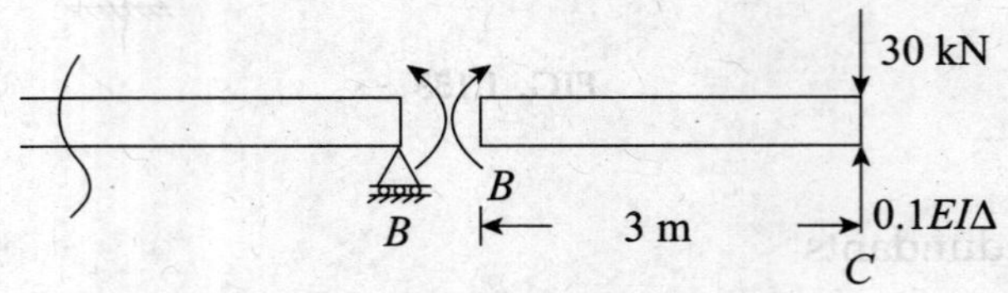

FIG. 1.154

$$\Sigma M_{BC} = 0$$

$$-90 + (0.1EI\Delta_C)3 - M_{BC} = 0$$

$$-90 + 0.3EI\Delta_C - (0.67EI\theta_B + 1.33EI\theta_C - 0.67EI\Delta) = 0$$

$$-0.67EI\theta_B - 1.33EI\theta_C + 0.97EI\Delta_C = 90 \qquad (7)$$

Equations (5) – (7) can be expressed in matrix form as

$$\begin{bmatrix} 2 & 0.67 & -0.67 \\ 0.67 & 1.33 & -0.67 \\ -0.67 & -1.33 & 0.97 \end{bmatrix} \begin{Bmatrix} \theta_B \\ \theta_C \\ \Delta \end{Bmatrix} = \begin{bmatrix} -37.50 \\ 0.00 \\ 90.00 \end{bmatrix}$$

Solving for the kinematic redundants

$$\theta_B = 37.44/EI$$

$$\theta_C = 132.27/EI$$

$$\Delta = 300/EI$$

5. End Moments

$$M_{AB} = -37.5 + 0.33(37.44) = -25.15 \text{ kNm}$$

$$M_{BA} = +37.5 + 0.67(37.44) = +62.58 \text{ kNm}$$

$$M_{BC} = 1.33(37.44) + 0.67(132.27) - 0.67 \times 300 = -62.58 \text{ kNm}$$

$$M_{CB} = 1.33(132.27) + 0.67(37.44) - 0.67 \times 300 = 0$$

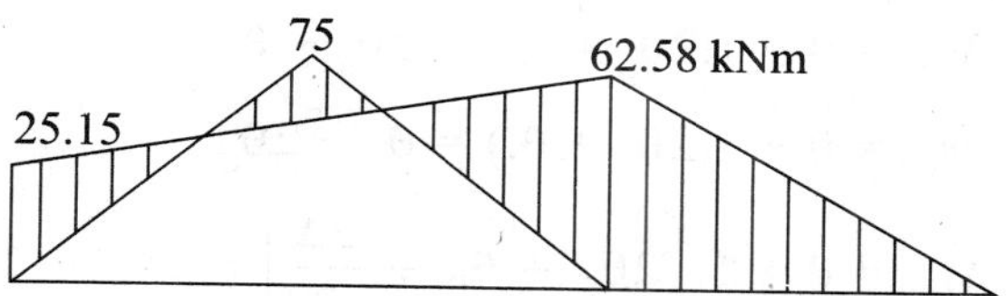

FIG. 1.155 Bending moment diagram

Example 1.40 Analyse the frame shown in Fig. 1.156 by slope deflection method. *EI* is constant.

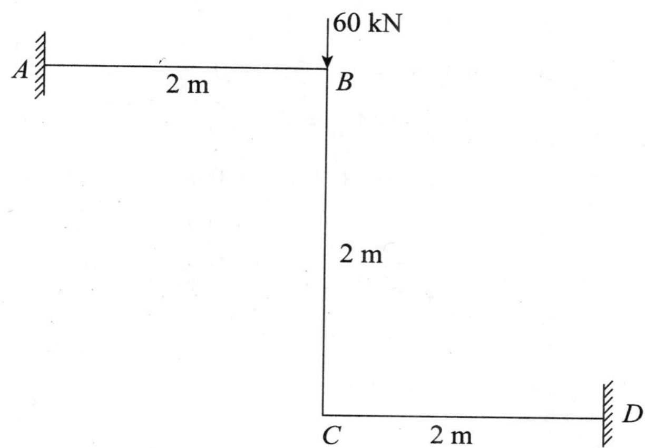

FIG. 1.156

Solution

1. Kinematic Redundants

$$\theta_A = \theta_D = 0; \quad \theta_B \neq 0, \quad \theta_C \neq 0, \quad \Delta \neq 0$$

2. Relative Stiffness Values

Member	(I/l)	k
AB	$I/2$	1
BC	$I/2$	1
CD	$I/2$	1

The value of k is obtained by multiplying the value of (I/l) by $(2/I)$.

3. Fixed End Moments

$$M_{FAB} = M_{FBA} = M_{FBC} = M_{FCB} = M_{FCD} = M_{FDC} = 0$$

4. Slope Deflection Equations

$$M_{AB} = 0 + 1\left(2\theta_A + \theta_B - \frac{3\Delta}{2}\right) = \theta_B - 1.5\Delta$$

$$M_{BA} = 0 + 1\left(2\theta_B + \theta_A - \frac{3\Delta}{2}\right) = 2\theta_B - 1.5\Delta$$

$$M_{BC} = 0 + 1(2\theta_B + \theta_C) = 2\theta_B + \theta_C$$

$$M_{CB} = 0 + 1(2\theta_C + \theta_B) = \theta_B + 2\theta_C$$

$$M_{CD} = 0 + 1\left(2\theta_C + \theta_D + \frac{3\Delta}{2}\right) = 2\theta_C + 1.5\Delta$$

$$M_{DC} = 0 + 1\left(2\theta_D + \theta_C + \frac{3\Delta}{2}\right) = \theta_C + 1.5\Delta$$

5. Joint Equilibrium Equations

At Joint B $\Rightarrow$ $M_{BA} + M_{BC} = 0$

$$4\theta_B + \theta_C - 1.5\Delta = 0 \quad (1)$$

At Joint C $\Rightarrow$ $M_{CB} + M_{CD} = 0$

$$\theta_B + 4\theta_C + 1.5\Delta = 0 \quad (2)$$

Shear condition

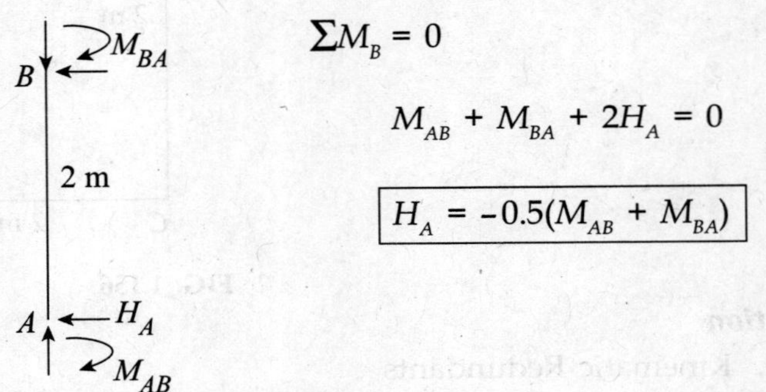

$\Sigma M_B = 0$

$$M_{AB} + M_{BA} + 2H_A = 0$$

$$\boxed{H_A = -0.5(M_{AB} + M_{BA})}$$

FIG. 1.157

$\Sigma M_C = 0$

$$M_{CD} + M_{DC} + 2H_D = 0$$

$$\boxed{H_D = -0.5(M_{CD} + M_{DC})}$$

FIG. 1.158

$\Sigma H = 0,$

$$-0.5(M_{AB} + M_{BA}) + 0.5(M_{CD} + M_{DC}) = 60$$

Using the slope deflection equation and substituting the values of moments;

$$-\theta_B + \theta_C + 2\Delta = 40 \qquad (3)$$

Solving for kinematic redundants;

$$\theta_B = 20, \quad \theta_C = -20, \quad \Delta = 40$$

6. End Moments

$$M_{AB} = 20 - 1.5(40) = -40 \text{ kNm}$$

$$M_{BA} = 2(20) - 1.5(40) = -20 \text{ kNm}$$

$$M_{BC} = 2(20) - 20 = +20 \text{ kNm}$$

$$M_{CB} = 20 + 2(-20) = -20 \text{ kNm}$$

$$M_{CD} = 2(-20) + 1.5(40) = +20 \text{ kNm}$$

$$M_{DC} = -20 + 1.5(40) = +40 \text{ kNm}$$

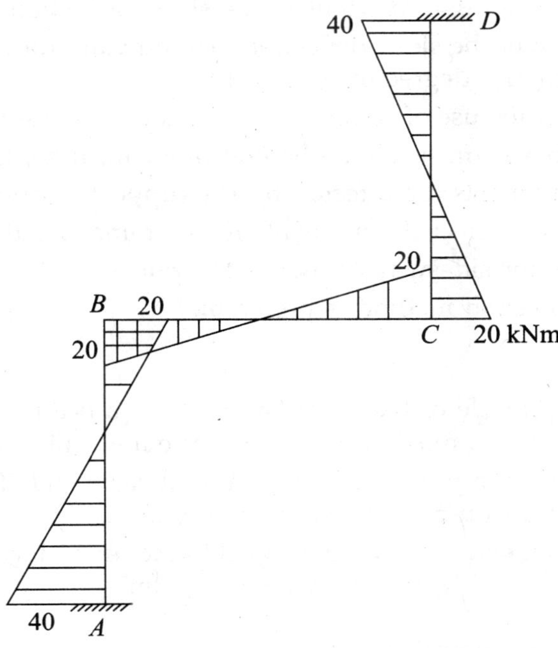

FIG. 1.159 Bending moment diagram

REVIEW QUESTIONS

Remembrance

1.1. What are kinematic redundants?

1.2. Is the slope deflection method applicable for statically determinate beams. If not, why?

1.3. Write down the slope deflection for a beam element?

1.4. What are the assumptions made in the slope deflection method?

1.5. What is sinking of supports? How does it alter the support rotations and settlement?

1.6. How is sinking/settlement of supports can be avoided?

1.7. Are compatibility conditions satisfied in the slope deflection method. If so, how?

1.8. Is equilibrium condition satisfied in the slope deflection method. If so, how?

1.9. What is column shear condition?

1.10. What is the value of the slope along the line of symmetry in a symmetrical frame subjected to symmetrical loading.

Understanding

1.1. What are the primary unknowns in the slope deflection method?

1.2. Why sign convention is important in the slope deflection method?

1.3. Is the application of the slope deflection method valid for any degree of restraint at the ends with any degree of settlement?

1.4. Give examples of the use of equilibrium equations in the slope deflection method?

1.5. Does the member stiffness affect the end moment. If so, how?

1.6. How the end moments are affected by the support rotations and settlement?

1.7. Do the settlements affect the moments if the frame is statically determinate?

1.8. List the reasons for sidesway of the portal frames?

1.9. When slope deflection method is preferable?

Analyse and Apply

1.1. What is the magnitude of fixed end moment for a beam subjected to uniformly varying load which is maximum at the centre and minimum at the supports?

1.2. A fixed beam of span 5 m sinks by 10 mm at right end. Take $I = 3 \times 10^7$ mm^4. $E = 200$ kN/mm^2. Compute the support moments.

1.3. Calculate the fixed end moment developed in the beam due to sinking of supports by 2 mm. $E = 6 \times 10^3$ N/mm^2. $I = 1.6 \times 10^8$ mm^4.

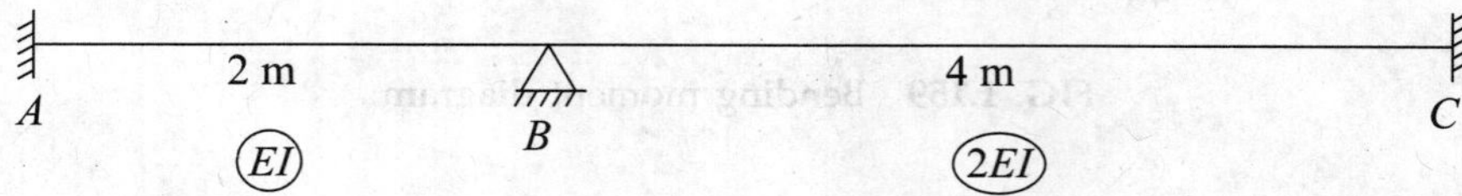

1.4. Calculate the moments developed in span BC for the beam shown in figure EI = 10000 kNm^2. Supports B and C sink by 2 mm and 3 mm respectively?

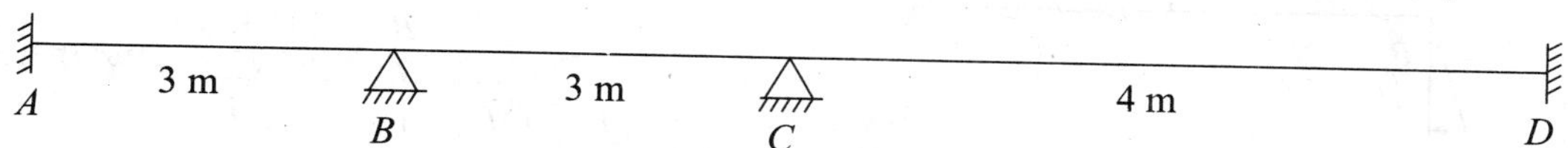

1.5. Why is the slope deflection method called the displacement method?

1.6. What is the angle between the tangents to the various points of the elastic curve meeting at a joint. Will it remain the same with respect to the original undeformed structure.

1.7. In the development of the slope deflection method, are the shear and axial deformations considered?

1.8. Identify the kinematic redundants.

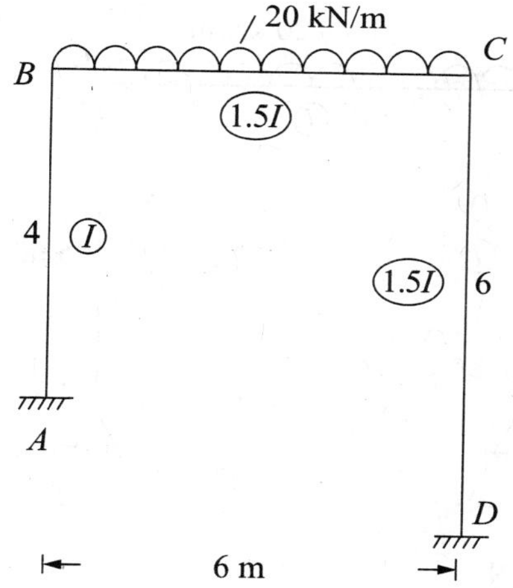

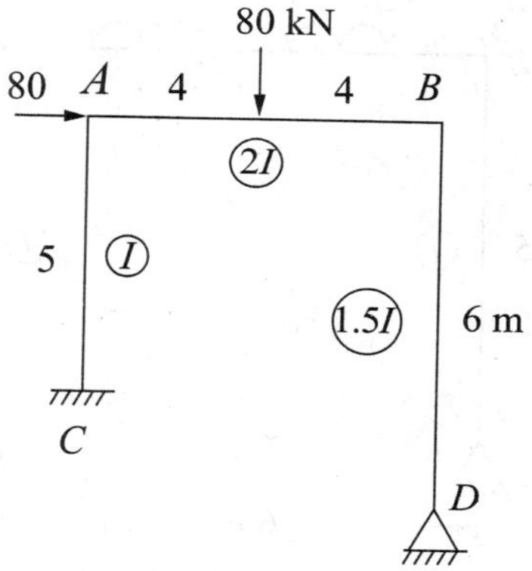

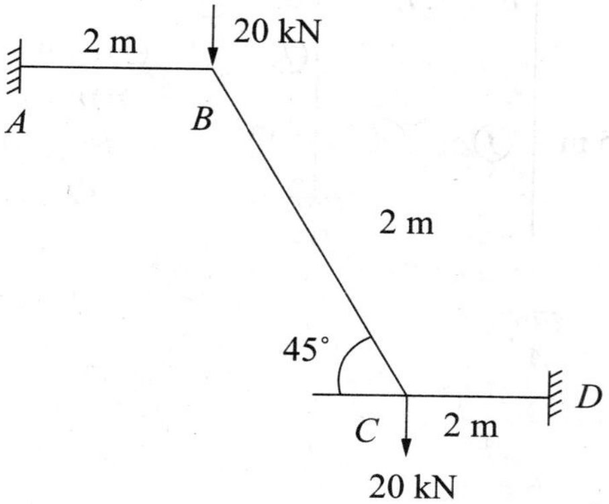

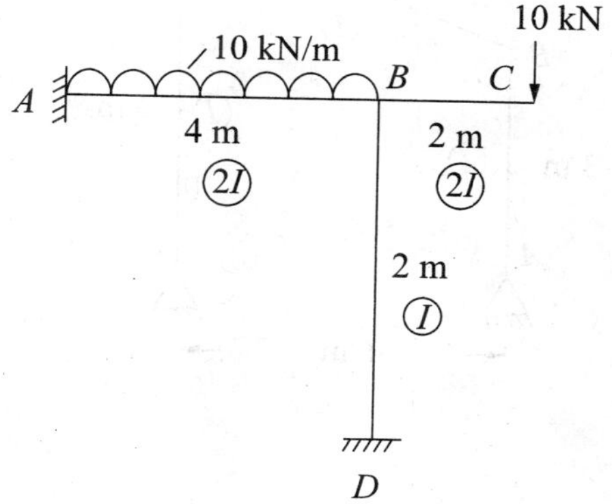

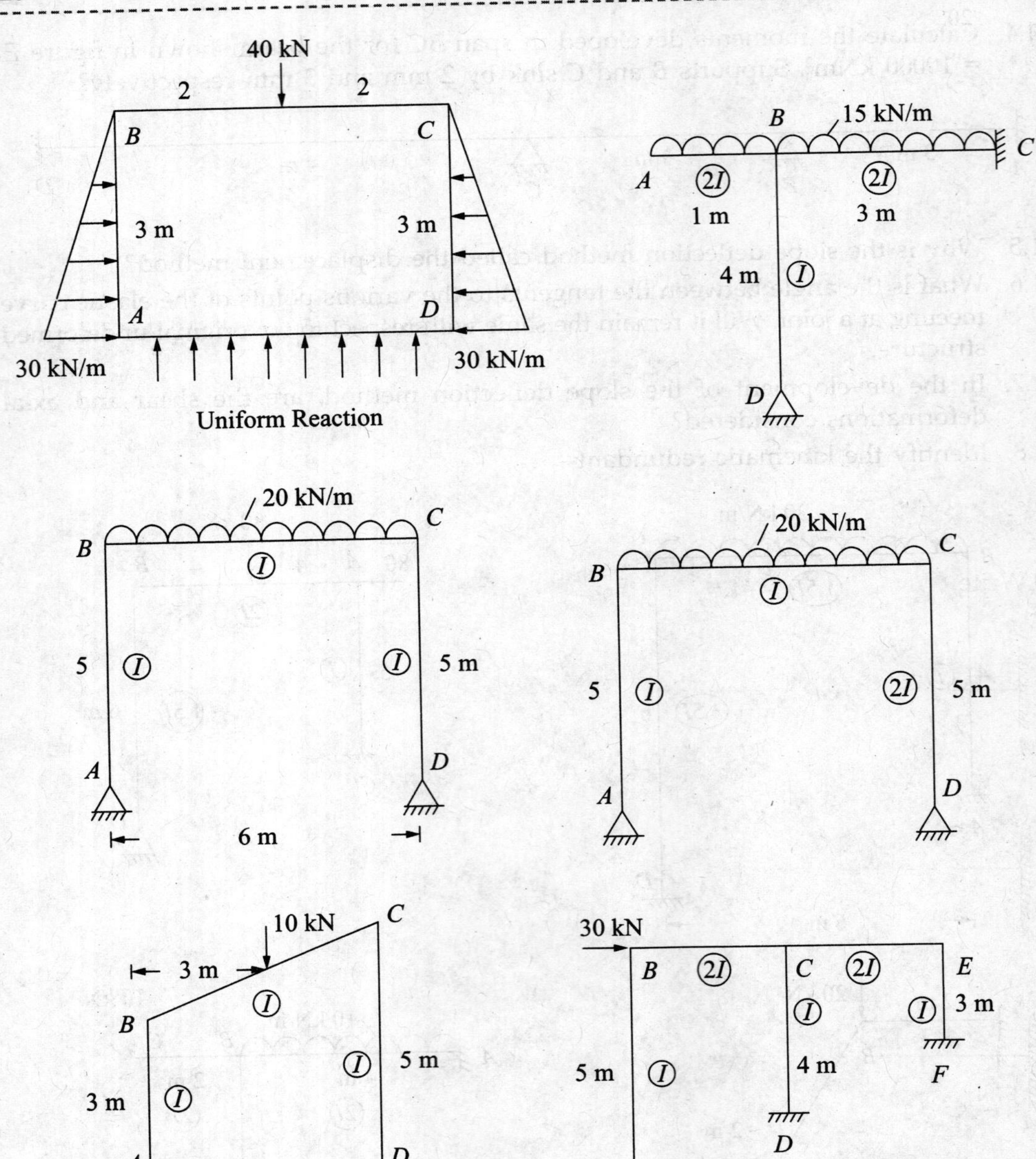
40 kN
2
2
B
C
3 m
3 m
A
D
30 kN/m
30 kN/m
Uniform Reaction
B
15 kN/m
A
2I
2I
C
1 m
3 m
4 m
I
D
20 kN/m
B
C
I
5
I
I
5 m
A
D
6 m
20 kN/m
B
C
I
5
I
2I
5 m
A
D
10 kN
C
3 m
I
B
I
5 m
3 m
I
A
D
6 m
30 kN
B
2I
C
2I
E
I
I
3 m
5 m
I
4 m
F
D
A

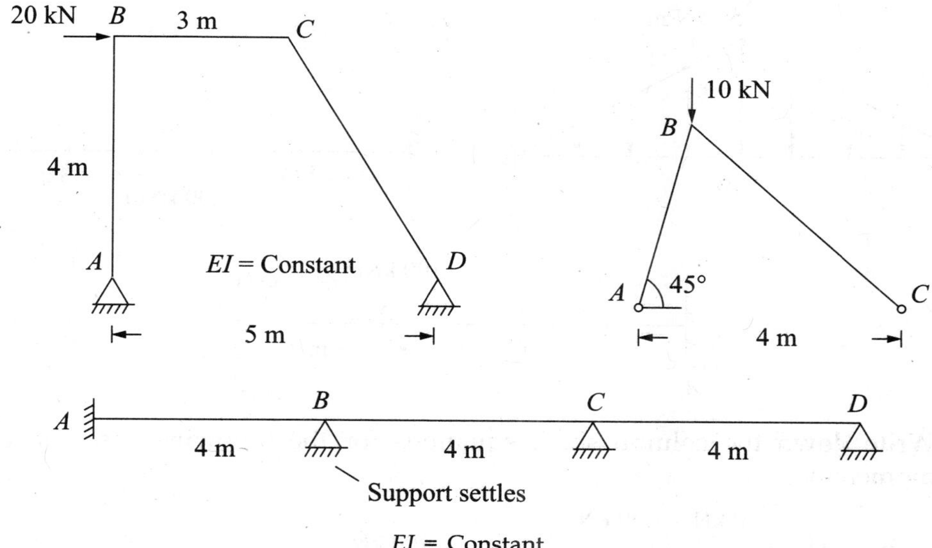

1.9. Write down the fixed end moments for the following cases.

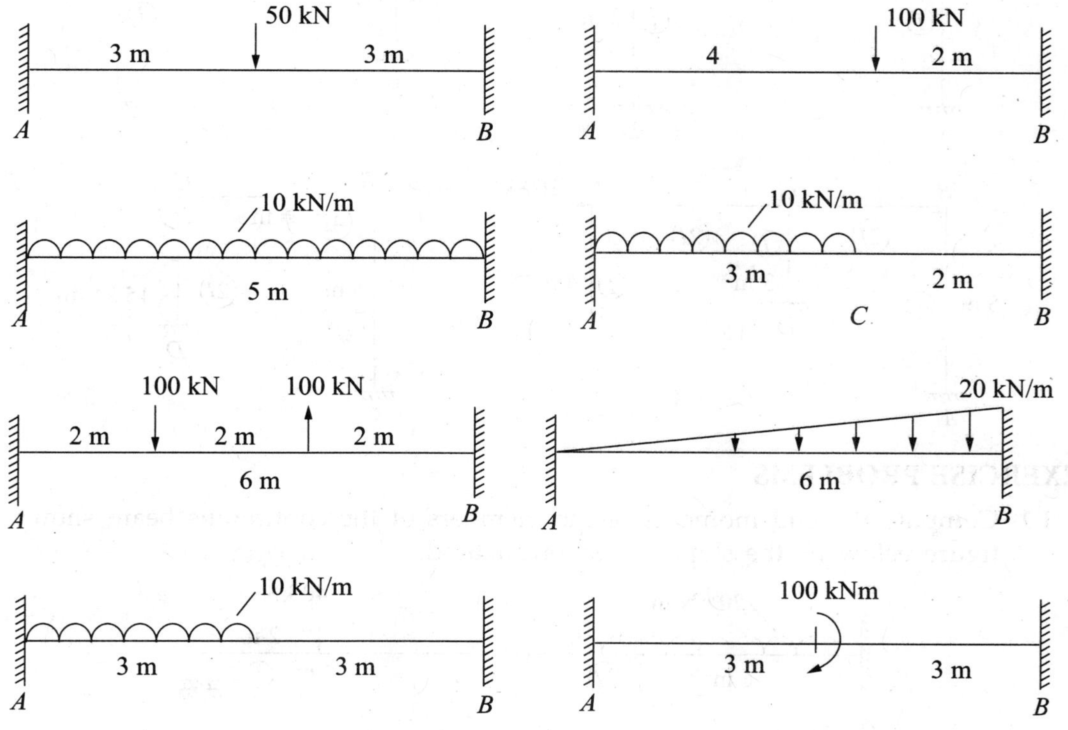

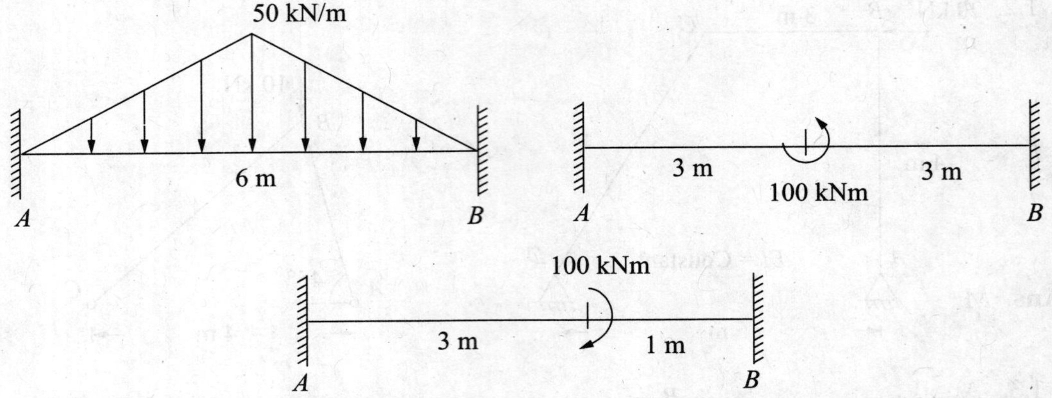

1.10. Write down the column shear equations for the following case (in terms of moments)

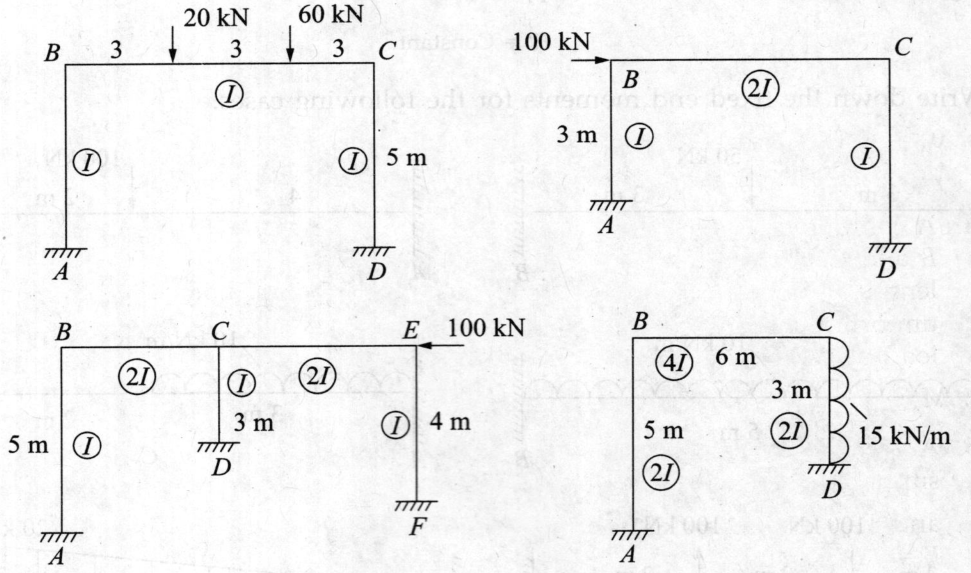

EXERCISE PROBLEMS

1.1 Compute the end moments of the members of the continuous beam shown in figure below by the slope deflection method.

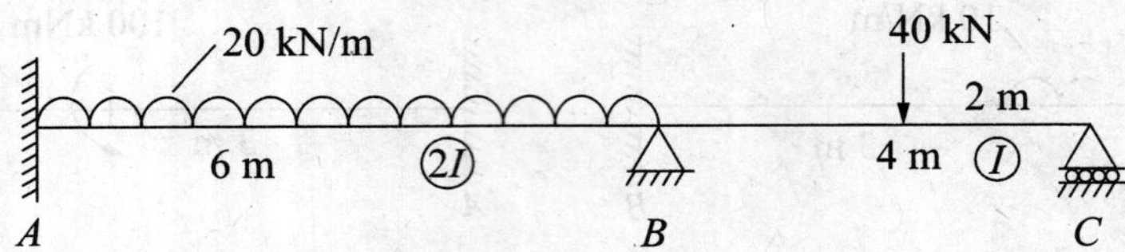

Ans. θ_B = –2.400, θ_C = –2.133, M_A = –69.6 kNm,
M_B = –40.8 kNm, MC = 0

1.2. Analyse the continuous beam shown in figure below by the slope deflection method. Draw bending moment diagram and shear force diagram. EI = Constant.

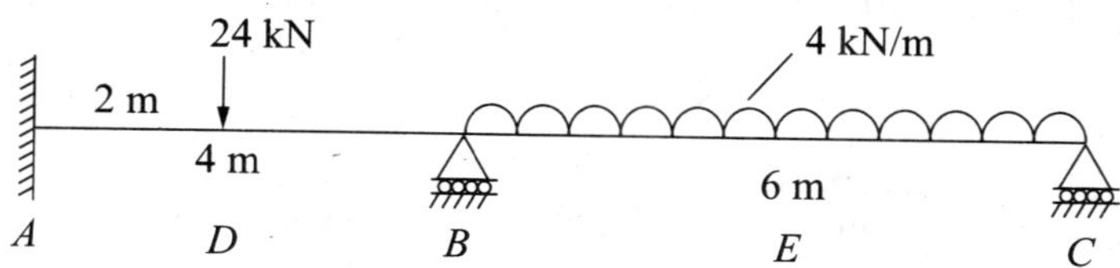

Ans. $M_A = -10.0$ kNm, $M_B = -16.0$ kNm, $M_C = 0$
$V_A = 10.5$ kN, $V_B = 28.2$ kN, $V_C = 9.3$ kN

1.3. Analyse the continuous beam by the slope deflection method. Draw BMD and SFD. EI = Constant

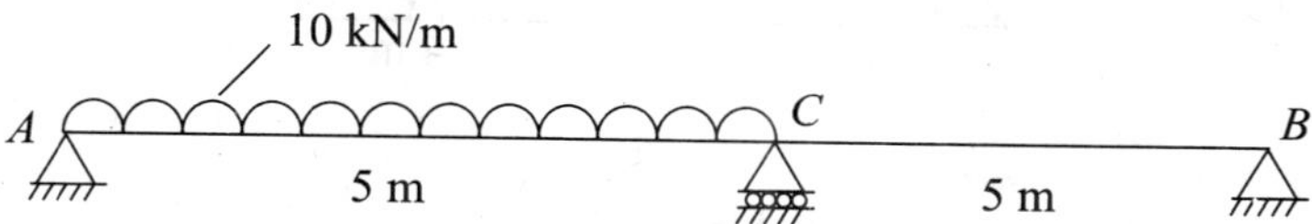

Ans. $\theta_A = 15.623$, $\theta_B = 5.208$, $\theta_C = -10.415$, $M_C = -15.62$ kNm
$V_{AC} = 21.9$ kN, $V_{CA} = 28.1$, $V_{CB} = 3.1$, $V_{BC} = -3.1$ kN

1.4. A continuous beam ABCD is simply supported at A and continuous over spans B and C. AB = 6 m, BC = 6 m respectively. An overhang of CD is of 1 metre length. A concentrated load of 20 kN is acting at 4 metre from support A. A uniformly distributed load of 10 kNm is acting on the span BC. A concentrated load of 10 kN is acting at D. Draw the BMD and SFD.

Ans. $M_A = 0$, $M_B = -28.15$ kNm, $M_C = -10$ kNm

1.5. A continuous beam is loaded as shown in figure below. During loading the support B sinks by 10 mm. Determine the bending moment at the supports and hence sketch the bending moment diagram. $I = 1600 \times 10^4$ mm^4 and $E = 200$ kN/mm^2. Use the slope deflection method.

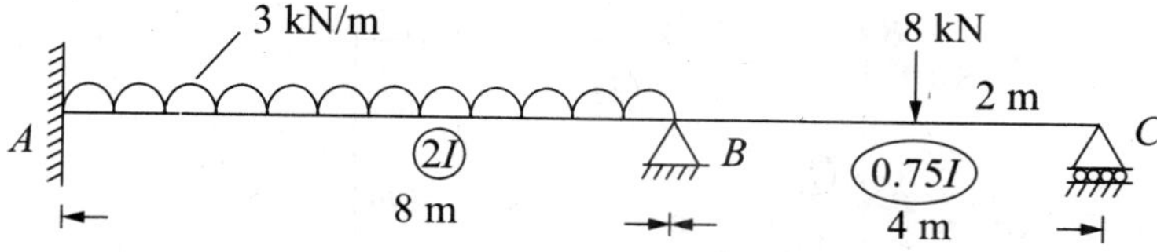

Ans. $M_A = -24.72$ kNm; $M_B = 4.56$ kNm, $M_C = 0$

1.6. For the continuous beam shown in figure below, draw the bending moment diagram and the elastic curve, using slope deflection method. Support B sinks by 20 mm. $E = 2 \times 10^5$ MPa and $I = 400 \times 10^4$ mm^4.

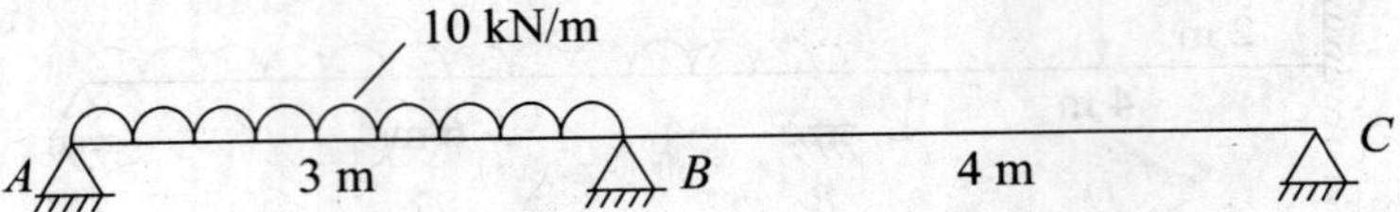

Ans. $M_B = 0.81$ kNm

1.7. Analyse the given frame by the slope deflection method. Draw the BMD and the elastic curve. EI = Constant.

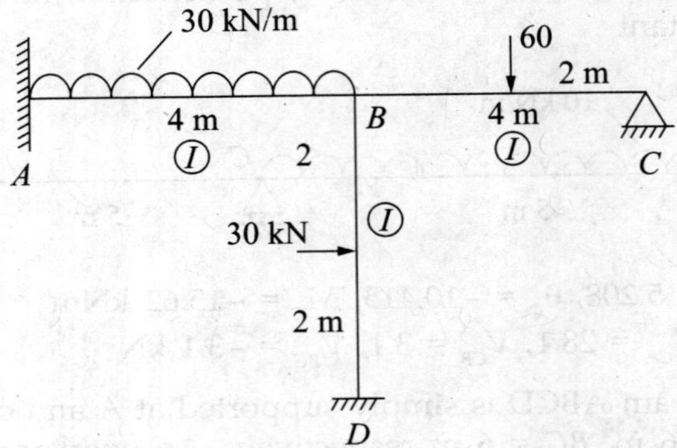

Ans. $\theta_B = -1.951$, $\theta_C = -19.024$, $M_{AB} = -41.95$ kNm,

$M_{BA} = +36.1$, $M_{BC} = -47.2$, $M_{BD} = 11.1$,

$M_{DB} = -17.0$ kNm

1.8. Analyse the frame shown in figure below by the slope deflection method. Draw the bending moment diagram.

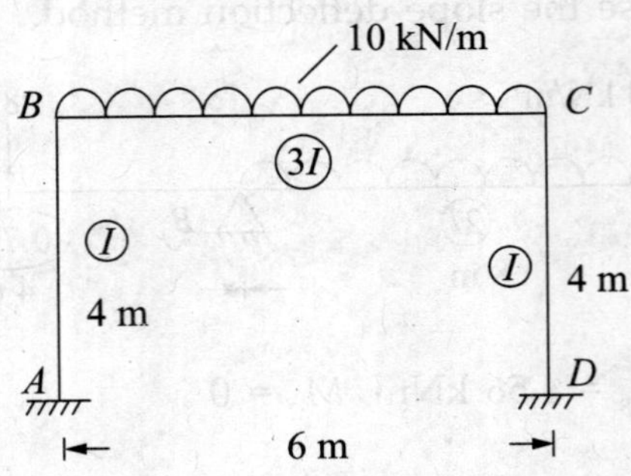

Ans. $M_A = 7.5$ kNm, $M_B = -15$ kNm

1.9. Analyse the frame by the slope deflection method. Draw the bending moment diagram.

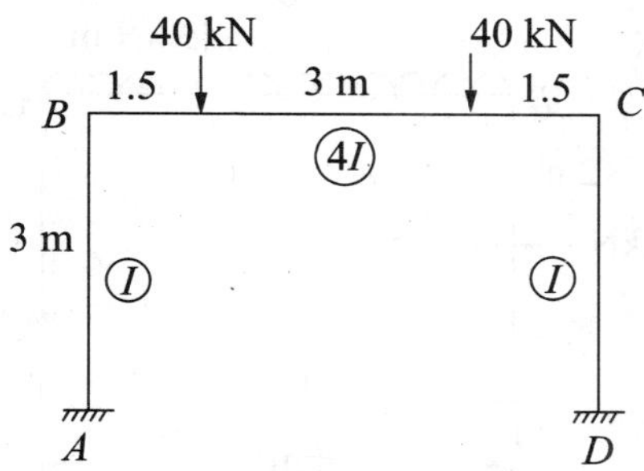

Ans. M_A = 11.25 kNm, M_B = 22.5 kNm.

1.10. Analyse the frame shown in figure by the slope deflection method. Draw the bending moment diagram.

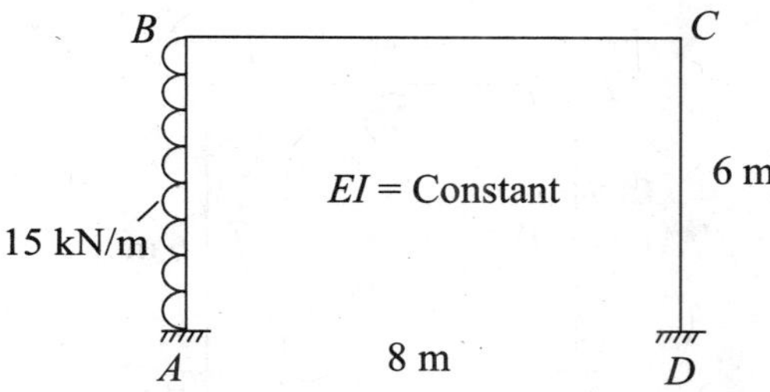

Ans. M_A = 128.9 kNm, M_B = 30.7 kNm, M_C = −42.95 kNm, M_D = −67.5 kNm.

1.11. Analyse the frame shown in figure below and draw the BMD. The frame is of constant cross section.

Ans. M_A = 23.75, M_B = −39.1, M_C = −32.8, M_D = 4.9 kNm

1.12 Using the slope deflection method, determine the end moments of the members of the frame shown in figure below. *EI* is constant throughout.

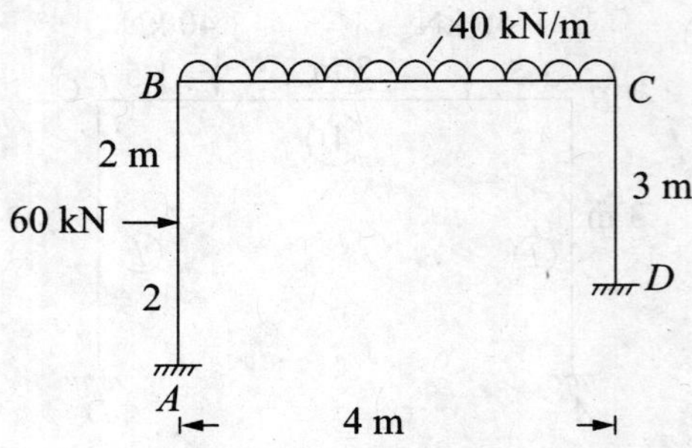

Ans. M_A = –35.3 kNm, M_B = –36.75 kNm, M_C = –50.47 kNm, M_D = –40.6 kNm.

1.13. The rigid frame shown in figure below has $I = 1.6 \times 10^8$ mm^4 $E = 2 \times 10^5$ N/mm^2. Due to partial fixity, the end *A* settles down by 10 mm. Determine the end moments.

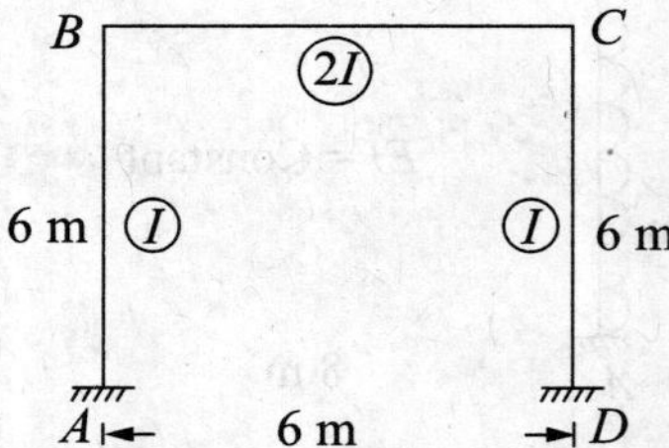

Ans. M_A = 8.21 kNm, M_B = –8.2 kNm, M_C = –8.2 kNm, M_D = 8.2 kNm

1.14. Analyse the box culvert shown below by the slope deflection method.

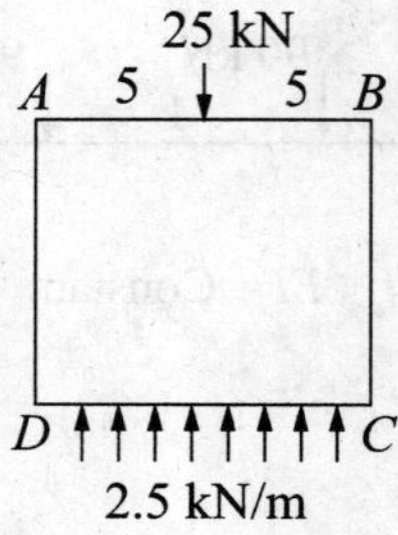

Ans. M_A = 25.1 kNm, M_B = –18.4 kNm, M_C = 5 kNm, M_D = –25.1 kNm

1.15. Analyse the frame shown in figure below by the slope deflection method. Draw the bending moment diagram and the elastic curve.

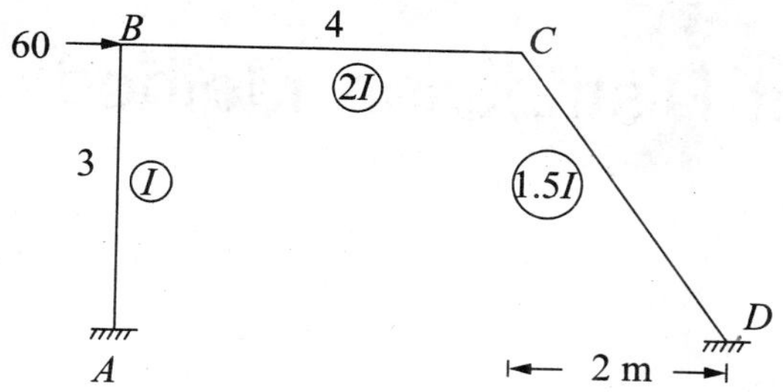

Ans. M_{AB} = –34.4 kNm, M_{BA} = –33.52 kNm, M_{BC} = +33.52 kNm,
M_{CB} = +36.65 kNm, M_{CD} = –36.65 kNm, M_{DC} = 40.34 kNm

1.16. Analyse the given frame shown in figure below by the slope deflection method and sketch the bending moment diagram.

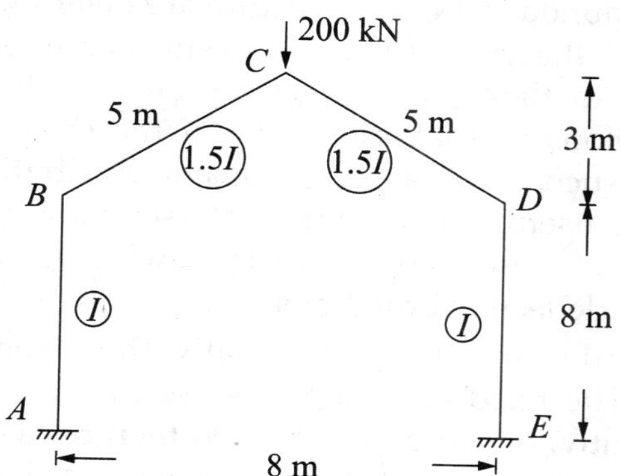

Ans. M_{AB} = 82.4 kNm, M_{BA} = 119.2 kNm, M_{BC} = –119.2 kNm, M_{CB} = –205.2 kNm,
M_{CD} = 205.2 kNm, M_{DC} = +119.2 kNm, M_{DE} = –119.2 kNm, M_{ED} = –82.35 kNm.

2 Moment Distribution Method

Objectives:

Definition of stiffness, carry over factor, distribution factor. Analysis of continuous beams without support yielding – Analysis of continuous beams with support yielding – Analysis of portal frames – Naylor's method of cantilever moment distribution – Analysis of inclined frames – Analysis of Gable frames.

2.1 INTRODUCTION

The end moments of a redundant framed structure are determined by using the classical methods, viz. Clapeyron's theorem of three moments, strain energy method and slope deflection method. These methods of analysis require a solution of set of simultaneous equations. Solving equations is a laborious task if the unknown quantities are more than three in number. In such situations, the moment distribution method developed by Professor Hardy Cross is useful. This method is essentially balancing the moments at a joint or junction. It can be described as a method which gives solution by successive approximations of slope deflection equations.

In the moment distribution method, initially the structure is rigidly fixed at every joint or support. The fixed end moments are calculated for any loading under consideration. Subsequently, one joint at a time is then released. When the moment is released at the joint, the joint moment becomes unbalanced. The equilibrium at this joint is maintained by distributing the unbalanced moment. This joint is temporarily fixed again until all other joints have been released and restrained in the new position. This procedure of fixing the moment and releasing them is repeated several times until the desired accuracy is obtained. The experience of designers points that about five cycles of moment distribution lead to satisfactory converging results.

Basically, in the slope deflection method, the end moments are computed using the slopes and deflection at the ends. Contrarily in the moment distribution method, as a first step – the slopes at the ends are made zero. This is done by fixing the joints. Then with successive release and balancing the joint moments, the state of equilibrium is obtained. The release-balance cycles can be carried out using the following theorems.

2.2 BASIC THEOREMS

2.2.1 Theorem 1

Consider a beam AB as shown in Fig. 2.1(a). The beam AB is of length l, fixed at A and hinged at B. A positive moment is applied at B. Let M_B be the magnitude and by applying M_B a moment of the same nature is induced at the fixed end. Let the moment at A be M_A.

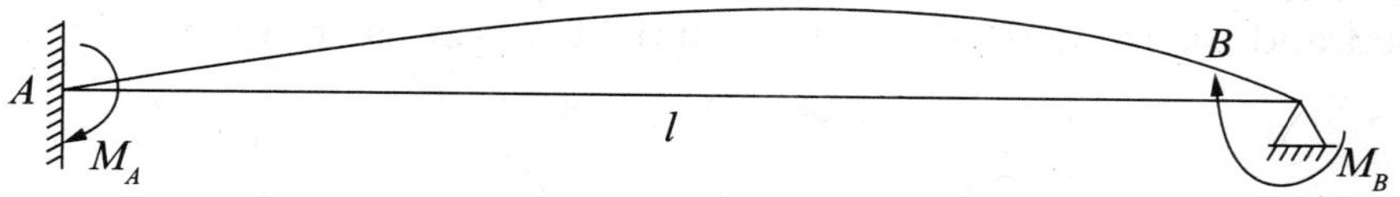

FIG. 2.1(a) Beam with far end fixed

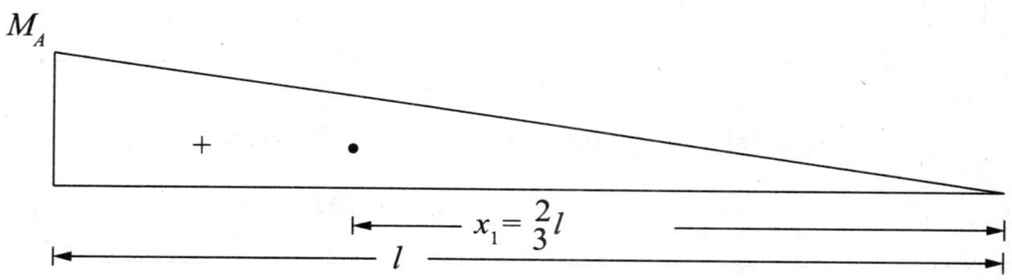

FIG. 2.1(b) Indeterminate moment diagram

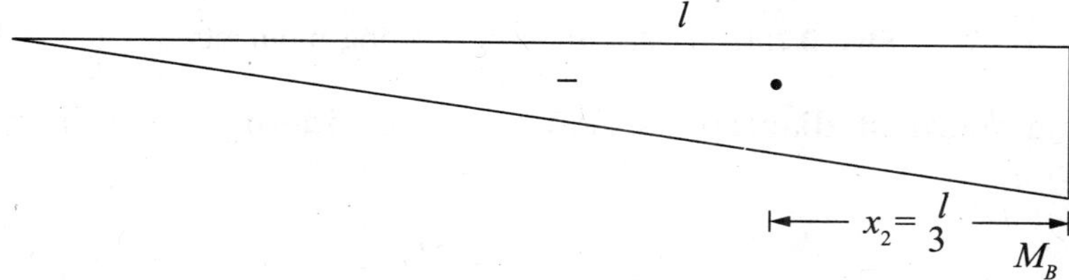

FIG. 2.1(c) Free bending moment diagram

As the support A remains horizontal; and B does not deflect, the tangent at A must pass through B. (Mohr's theorem - refer Basic Structural Analysis - Chapter 10). That is, moments of the areas of the total bending diagram about B must be equal to zero.

$$\left(\frac{1}{2}\right)(l)\,M_A\left(\frac{2l}{3}\right) - \left(\frac{1}{2}\right)(l)\,M_B\left(\frac{l}{3}\right) = 0$$

i.e.

$$2M_A = M_B$$

or

$$\boxed{M_A = \frac{1}{2}M_B}$$

In conclusion, when a positive moment M is applied to the hinged end of a beam a positive moment of $\left(\frac{1}{2}\right) M$ will be transferred to the fixed end.

2.2.2 Theorem 2

Consider a two span continuous beam ACB as shown in Fig. 2.2(a). A and B are fixed supports with a prop at C. A moment is applied at C and it is required to know how much moment is distributed between spans AC and CB. Let this moment M be decomposed and distributed as M_1 to CA and M_2 to CB as shown in Fig. 2.2(b).

i.e. $$M_1 + M_2 = M \tag{1}$$

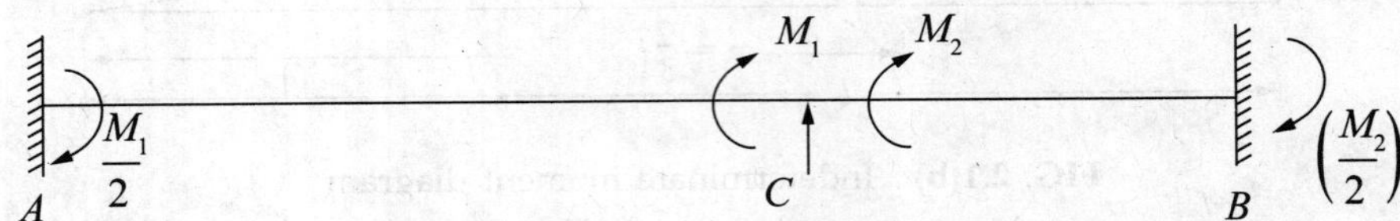

FIG. 2.2(a) Continuous beam ACB with moment M

FIG. 2.2(b) Distribution of bending moments

The bending moment diagram is drawn by considering each span AC and CB respectively.

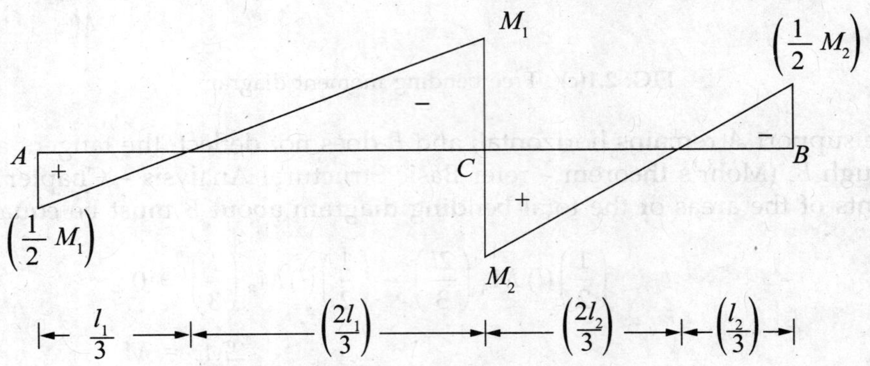

FIG. 2.2(c) Bending moment diagram

As the ends A and B are fixed; the slope between A and B is zero. That is, the area of the bending moment diagram between A and B is zero.

Hence,

$$\text{Area of } \left(\frac{M}{EI}\right)\Bigg|_B^A = \frac{\text{Area of BMD}}{EI_1}\Bigg|_C^A + \frac{\text{Area of BMD}}{EI_2}\Bigg|_B^C$$

$$\frac{\text{Area of BMD}}{EI_1}\Bigg|_C^A = \left\{\left(\frac{1}{2}\right)\left(\frac{l_1}{3}\right)\left(\frac{M_1}{2}\right) - \left(\frac{1}{2}\right)\left(\frac{2l_1}{3}\right)M_1\right\}\Big/EI_1$$

$$\frac{\text{Area of BMD}}{EI_1}\Bigg|_C^A = \frac{-M_1l_1}{4EI_1}$$

$$\frac{\text{Area of BMD}}{EI_2}\Bigg|_B^C = \left\{\left(\frac{1}{2}\right)\left(\frac{2l_2}{3}\right)M_2 - \left(\frac{1}{2}\right)\left(\frac{l_2}{3}\right)\left(\frac{M_2}{2}\right)\right\}\Big/EI_2$$

$$\frac{\text{Area of BMD}}{EI_2}\Bigg|_B^C = \frac{M_2l_2}{4EI_2}$$

$$\therefore \qquad \frac{-M_1l_1}{4EI_1} + \frac{M_2l_2}{4EI_2} = 0$$

$$\frac{M_1l_1}{4EI_1} = \frac{M_2l_2}{4EI_2}$$

Denoting $\frac{I_1}{l_1} = k_1$ and $\frac{I_2}{l_2} = k_2$

where k_1 and k_2 are the relative stiffness of beams AC and CB respectively.

$$\therefore \qquad \boxed{\frac{M_1}{M_2} = \frac{k_1}{k_2}} \qquad (2)$$

Solving the above Eqs. (1) and (2); we obtain

$$M_1 = \frac{Mk_1}{\Sigma k} \qquad (3)$$

and

$$M_2 = \frac{Mk_2}{\Sigma k} \qquad (4)$$

The above equations illustrate that if a moment M is applied to any joint in a structure, that moment will be distributed to other members meeting at that joint in the above ratio.

Consider a joint where four members are meeting.

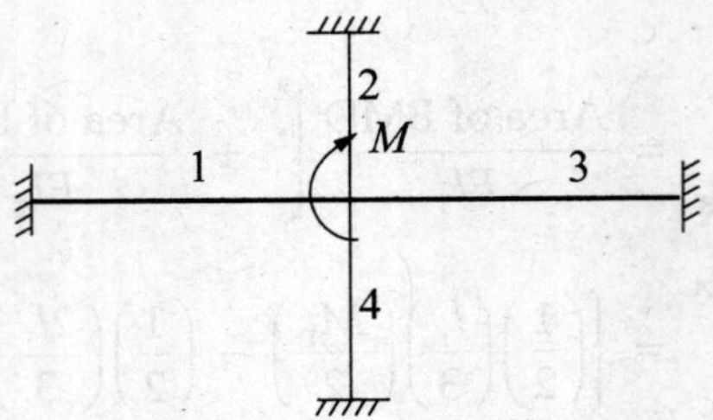

FIG. 2.3

$$M_1 = M \frac{k_1}{(k_1 + k_2 + k_3 + k_4)}$$

$$M_2 = M \frac{k_2}{\Sigma k}$$

$$M_3 = M \frac{k_3}{\Sigma k}$$

$$M_4 = M \frac{k_4}{\Sigma k}$$

2.2.3 Theorem 3

Consider a beam ACB is fixed at A and simply supported at C as shown in Fig. 2.4(a). It is desirable to know how much the applied moment is distributed in spans AC and CB respectively.

FIG. 2.4(a)

Let the moment M_1 be distributed to span CA and the moment M_2 for the span CB respectively.

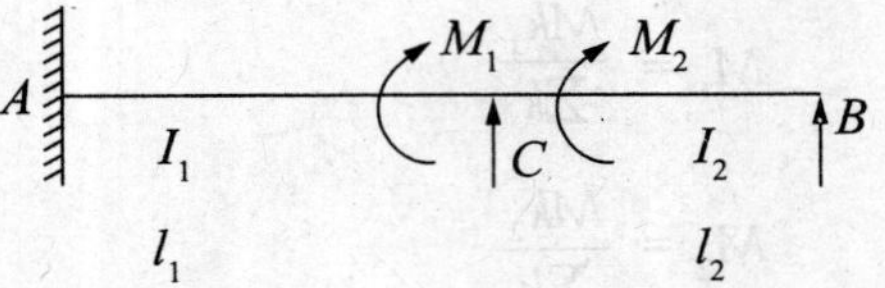

FIG. 2.4(b)

It is known that the moment M_1 applied at C transfers half of the moment to the fixed end.

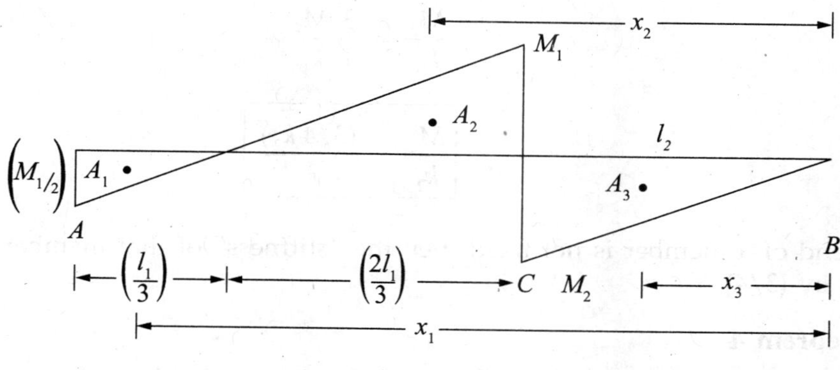

FIG. 2.4(c)

The tangent drawn at A passes through C and B. Hence, from the above figure;

$$\Sigma A_i x_i / EI_i = 0$$

i.e.

$$A_1 = (1/2)\,(l_1/3)\,(M_1/2) = \frac{M_1 l_1}{12}$$

$$x_1 = l_2 + \left(\frac{2}{3}l_1 + \frac{2}{3}\frac{l_1}{3}\right); \qquad \boxed{x_1 = l_2 + \frac{8}{9}l_1}$$

$$A_2 = (-1/2)\left(\frac{2}{3}l_1\right)M_1 = -\frac{M_1 l_1}{3}$$

$$x_2 = l_2 + \left(\frac{1}{3}\right)\left(\frac{2}{3}\right)\left(\frac{l_1}{3}\right); \qquad \boxed{x_1 = l_2 + \frac{2}{9}l_1}$$

$$A_3 = (1/2)\,(l_2)M_2 = \frac{M_2 l_2}{2}$$

$$\boxed{x_3 = \frac{2}{3}l_2}$$

Thus, $\Sigma A_i x_i / EI$ gives

$$\frac{-3M_1 l_1 l_2}{12EI_1} + \frac{M_2 l_2^2}{3EI_2} = 0$$

$$\frac{M_2 l_2^2}{3EI_2} = \frac{-3M_1 l_1 l_2}{12EI_1}$$

$$\frac{l_2 M_2}{I_2} = \frac{3}{4}\frac{M_1 l_1}{I_1}$$

$$\frac{M_2}{k_2} = \frac{3}{4}\frac{M_1}{k_1}$$

i.e.,

$$\boxed{\frac{M_2}{k_2} = \frac{(3/4\,k_2)}{k_1}}$$

If one end of a member is not fixed then the "stiffness" of that member should be multiplied by (3/4).

2.2.4 Theorem 4

Consider a fixed beam AB as shown below. End B has settled by a distance δ. As the ends are fixed, there must develop a fixing moment M at the each end of the beam.

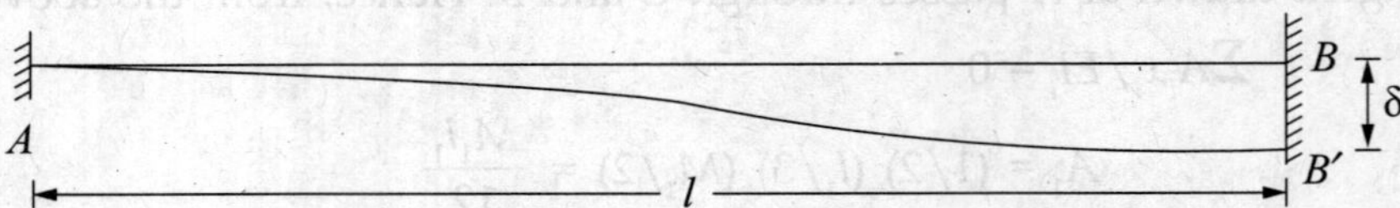

FIG. 2.5(a) Sinking of supports in a fixed beam

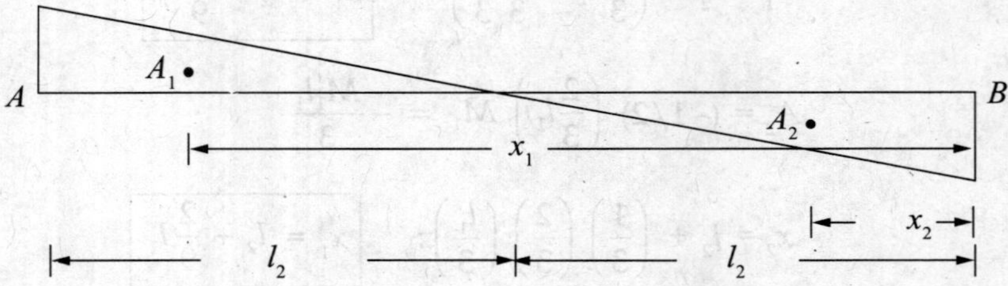

FIG. 2.5(b) Bending moment diagram

Taking moments about B and using the moment area theorem;

$$\Sigma A_i x_i / EI_i = -\delta$$

i.e.

$$A_1 = -\left(\frac{1}{2}\right)\left(\frac{l}{2}\right)M = \frac{-Ml}{4}$$

$$A_2 = \left(\frac{1}{2}\right)\left(\frac{l}{2}\right)M = \frac{Ml}{4}$$

$$x_1 = \frac{l}{2} + \frac{2}{3}\left(\frac{l}{2}\right) = \frac{5l}{6}$$

$$x_2 = \frac{1}{3}\left(\frac{l}{2}\right) = \frac{l}{6}$$

$$\frac{1}{EI}\left[-\left(\frac{Ml}{4}\right)\left(\frac{5l}{6}\right)+\left(\frac{Ml}{4}\right)\left(\frac{l}{6}\right)\right]=-\delta$$

$$\left(\frac{-5Ml^2}{24}+\frac{Ml^2}{24}\right)\frac{1}{EI}=-\delta$$

$$\frac{(-5+1)}{24}\frac{Ml^2}{24}=-\delta$$

$$\boxed{M=\frac{6EI\delta}{l^2}}$$

i.e., when a fixed ended beam settles by an amount δ at one end, the moment required to make the ends horizontal = $6EI\delta/l^2$.

The above four theorems can be summarised as

(1) When the member is fixed at one end and a moment is applied at the other end which is simply supported or hinged, the moment induced at the fixed end is one half of the applied moment. The induced moment at the fixed end is in the same direction as the applied moment.

(2) If a moment is applied in a stiff joint of a structure, the moment is resisted by various members in proportion to their respective stiffnesses (i.e., moment of inertia divided by the length). If the stiffness of the member is more; then it resists more bending moment and it absorbs a greater proportion of the applied moment.

(3) While distributing the moments in a rigid joint, if one end of the member is not restrained then its stiffness should be multiplied by (3/4).

(4) In a fixed beam, if the support settles/subsides/sinks by an amount δ, the moment required to make the ends horizontal is $6EI\delta/l^2$.

2.3 BASIC DEFINITIONS OF TERMS IN THE MOMENT DISTRIBUTION METHOD

(a) Stiffness

Rotational stiffness can be defined as the moment required to rotate through a unit angle (radian) without translation of either end.

(b) Stiffness Factor

(i) It is the moment that must be applied at one end of a constant section member (which is unyielding supports at both ends) to produce a unit rotation of that end when the other end is fixed, i.e. $k = 4EI/l$.

(ii) It is the moment required to rotate the near end of a prismatic member through a unit angle without translation, the far end being hinged is $k = 3EI/l$.

(c) Carry Over Factor

It is the ratio of induced moment to the applied moment (Theorem 1). The carry over factor is always (1/2) for members of constant moment of inertia (prismatic section). If the end is hinged/pin connected, the carry over factor is zero. It should be mentioned here that carry over factors values differ for non-prismatic members. For non-prismatic beams (beams with variable moment of inertia); the carry over factor is not half and is different for both ends.

(d) Distribution Factors

Consider a frame with members *OA, OB, OC* and *OD* rigidly connected at *O* as shown in Fig. 2.6. Let *M* be the applied moment at joint *O* in the clockwise direction. Let the joint rotate through an angle θ. The members *OA,OB,OC* and *OD* also rotate by the same angle θ.

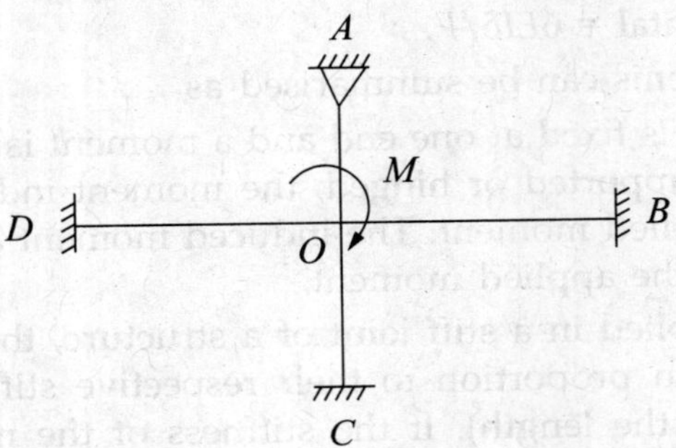

FIG. 2.6

Let k_{OA}, k_{OB}, k_{OC} and k_{OD} be the stiffness values of the members *OA, OB, OC* and *OD* respectively; then

$$M_{OA} = k_{OA}\theta \quad \text{(i)}$$

$$M_{OB} = k_{OB}\theta \quad \text{(ii)}$$

$$M_{OC} = k_{OC}\theta \quad \text{(iii)}$$

$$M_{OD} = k_{OD}\theta \quad \text{(iv)}$$

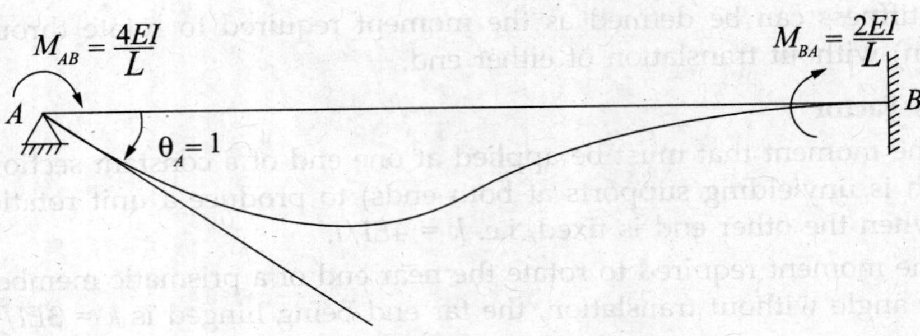

(a) Beam with far end fixed

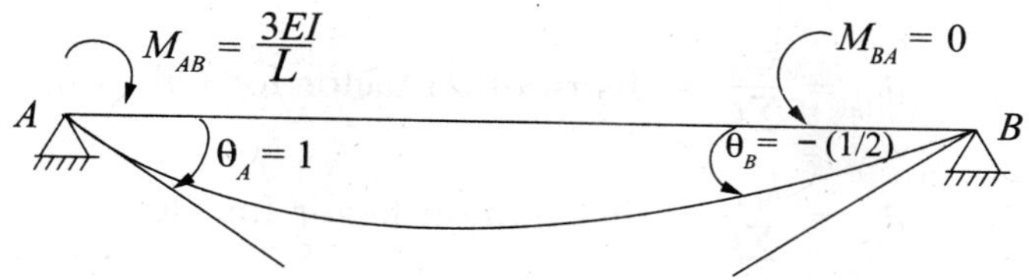

(b) Beam with far end hinged

FIG. 2.7 Stiffness factors for beams with different support conditions

then
$$M = M_{OA} + M_{OB} + M_{OC} + M_{OD}$$
$$M = (k_{OA} + k_{OB} + k_{OC} + k_{OD})\,\theta$$

i.e.
$$M = (\Sigma k)\,\theta \tag{v}$$

where Σk = joint stiffness = $(k_{OA} + k_{OB} + k_{OC} + k_{OD})$

i.e.
$$\boxed{\theta = \frac{M}{\Sigma k}} \tag{vi}$$

Substituting the above equation in Eqs. (i)-(iv); we get

$$M_{OA} = \frac{k_{OA}}{\Sigma k} M \tag{vii}$$

$$M_{OB} = \frac{k_{OB}}{\Sigma k} M \tag{viii}$$

$$M_{OC} = \frac{k_{OC}}{\Sigma k} M \tag{ix}$$

$$M_{OD} = \frac{k_{OD}}{\Sigma k} M \tag{x}$$

Denoting the distribution factor as the ratio of the member stiffness to the joint stiffness; we represent

$$M_{OA} = d_{OA} M$$
$$M_{OB} = d_{OB} M$$
$$M_{OC} = d_{OC} M$$
$$M_{OD} = d_{OD} M$$

and

$$d_{OA} = \frac{k_{OA}}{\Sigma k} = \text{distribution factor for } OA$$

$$d_{OB} = \frac{k_{OB}}{\Sigma k} = \text{distribution factor for } OB$$

$$d_{OC} = \frac{k_{OC}}{\Sigma k} = \text{distribution factor for } OC$$

$$d_{OD} = \frac{k_{OD}}{\Sigma k} = \text{distribution factor for } OD$$

2.4 SIGN CONVENTION

Clockwise moments are considered positive and anticlockwise moments negative.

2.5 BASIC STAGES IN THE MOMENT DISTRIBUTION METHOD

The moment distribution method can be illustrated with the following example.

It is desired to draw the bending moment diagram by computing the bending moments at salient points of the given beam as shown below.

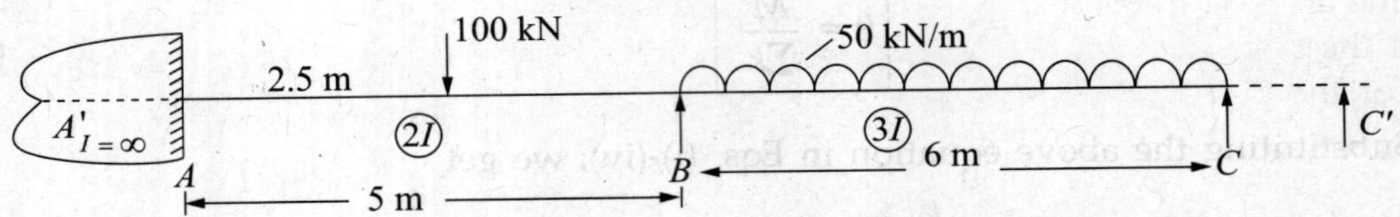

FIG. 2.8 A two span continuous beam

Step 1

Determine the distribution factor at each joint A, B and C respectively. The distribution factor of a member is the ratio of the stiffness of the member divided by the total stiffness of all the members meeting at that joint.

The distribution factor for the fixed support A is determined by assuming an imaginary span $A'A$ (Fig. 2.8). The flexural rigidity of $A'A$ is infinity. Hence, the stiffness is infinite. The stiffness of AB is $2I/5 = 0.4I$. Hence, the total stiffness is infinite. Thus, the distribution factor for $A'A$ is (infinity/infinity) = 1.0 and for AB the distribution factor k_{AB} = ($0.4I$/infinity), i.e. zero. In essence $d_{AA'} = 1.0$ and $d_{AB} = 0.0$

The distribution factor at the support B is determined as follows. The stiffness of the member BA is (0, $4I$). The stiffness of the member BC is taken as three-fourths of its stiffness (refer Theorem 3). Hence, $k_{BC} = \left(\frac{3}{4}\right)\left(\frac{3I}{6}\right) = 0.375I$. The sum of the stiffness at the joint B is $k = (k_{BA} + k_{BC}) = 0.775I$. Therefore, the distribution factors are $d_{BA} = k_{BA}/(k_{BA} + k_{BC})$, i.e. $d_{BA} = 0.4I/(0.4I + 0.375I) = 0.52$; similarly $d_{BC} = k_{BC}/\Sigma k$, i.e. $d_{BC} = (0.375I/0.775I) = 0.48$. The sum of the stiffnesses is $(0.52 + 0.48) = 1.0$.

The distribution factor for the simple support at C is determined by extending the span to CC'. The rigidity of CC' is zero and hence $k_{CC'} = 0$. On the other hand, the stiffness k_{CB} = 0.375*I*. The total stiffness at *C* is $k_{CB} + k_{CC'}$ = 0.375*I*. The distribution factor $d_{CB} = k_{CB}/\sum k = 1.0$ and $d_{CC'} = k_{CC'}/\sum k = 0$.

The above procedure is summarised in the following table for quick understanding.

Table 2.1 Distribution factor at joint *B*

Joint	*Members*	$k = I/l$	$\sum k$	*DF*
	BA	$\frac{2I}{5} = 0.4I$		52
B			0.775*I*	
	BC	$\frac{3}{4}\left(\frac{3I}{6}\right) = 0.3/5I$		0.48

Step 2

Imagine all the three joints *A*, *B* and *C* are rigidly fixed with horizontal tangents. Write down the fixed end moments for the beam *AB* as if it were built in at *A* and *B* and also for the beam *BC* as if it were built in *B* and *C*.

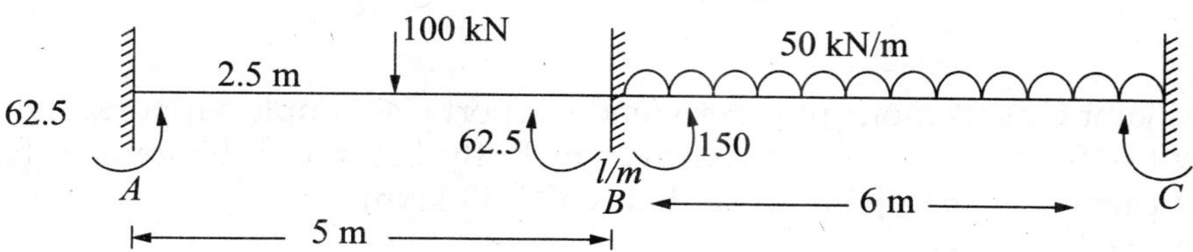

FIG. 2.9 Fixed end moments

$$M_{FAB} = -\frac{100 \times 5}{8} = -62.5 \text{ kNm}$$

$$M_{FBA} = +\frac{100 \times 5}{8} = +62.5 \text{ kNm}$$

$$M_{FBC} = -\frac{50 \times 6^2}{12} = -150.0 \text{ kNm}$$

$$M_{FCB} = +\frac{50 \times 6^2}{12} = +150.0 \text{ kNm}$$

Step 3

Each joint is released in turn and if *B* is released it will be out of balance. T is unbalanced moment (–150.0 + 62.5 = 87.5) is shown in Fig. 2.10.

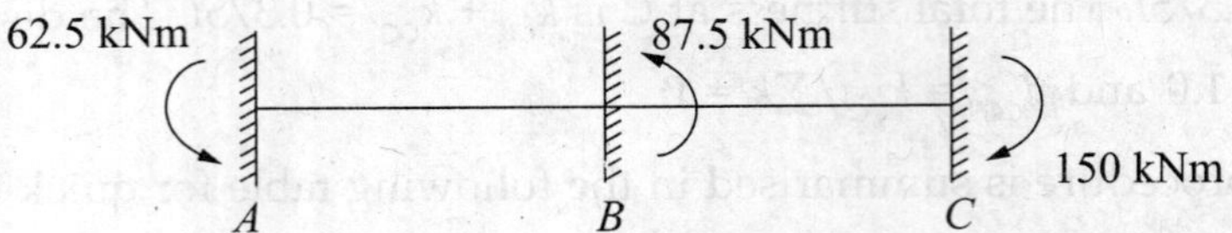

FIG. 2.10 Out of balance moments

Step 4

A moment is applied at B to balance this joint B and it will distribute itself according to the distribution factors. This is shown in the following figure.

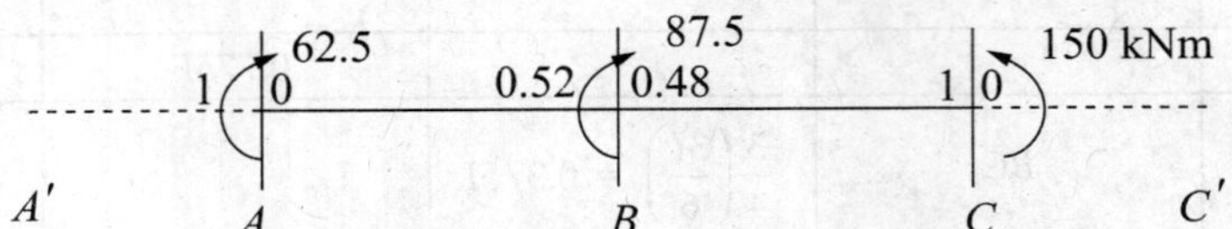

FIG. 2.11 Balancing moments

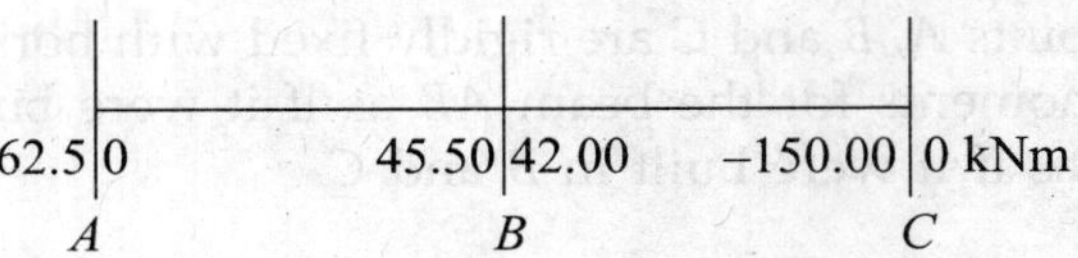

FIG. 2.12 Distributed moments

Step 5

Balance the joint C as its moment is zero (end support C is simple support). By balancing the moment −150, half of it is carried over to B. By balancing the joint B, half of the moment is carried over to joint A, i.e. half of 45.50 kNm.

Step 6

Again the joints become out of balance, and the above procedure is repeated until the moments to be distributed become negligible and can be ignored. This is illustrated in Table 2.2.

Table 2.2 Moment distribution table

Joint	A	B		C
Members	*AB*	*BA*	*BC*	*CB*
DF	0	0.52	0.48	1
FEMS	−62.50	+62.50	−150.00	+150.00
Bal		+45.50	+42.00	−150.00
Co	+22.75		−75.00	
Bal		+39.00	+36.00	
Co	+19.50			
Total	−20.25	+147.00	−147.00	0.00
Nature	↻	↺	↻	*O*

In the above moment distribution table a single vertical line is drawn between the members. Double lines are drawn at the end of each joint. The pure moment diagram can be drawn using the end moments in the moment distribution table. The pure moments are the values just to the left of double line. Thus, $M_A = -20.25$ kNm, $M_B = -147.00$ kNm, $M_C = 0$.

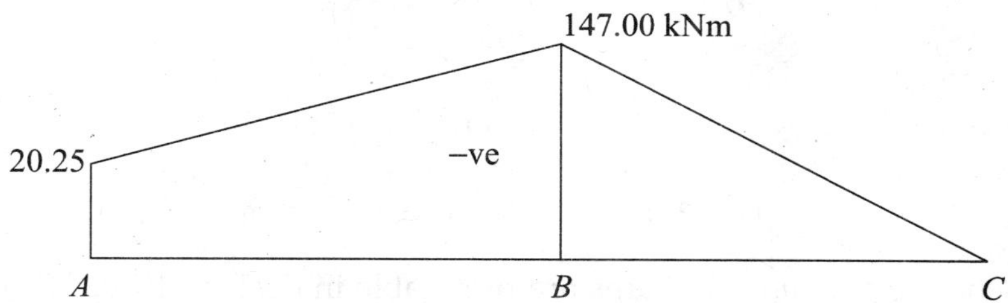

FIG. 2.13 Pure moment diagram

The simple beam moment diagram is drawn by considering each span separately. The simple beam moment diagram is always positive. While the pure moment diagram is negative. The maximum positive bending moment for span *AB* is $(wl/4) = 100 \times 5/4 = 125$ kNm. The maximum bending moment for a simple beam of *BC* is $wl^2/8 = (50 \times 6^2/8) = 225$ kNm.

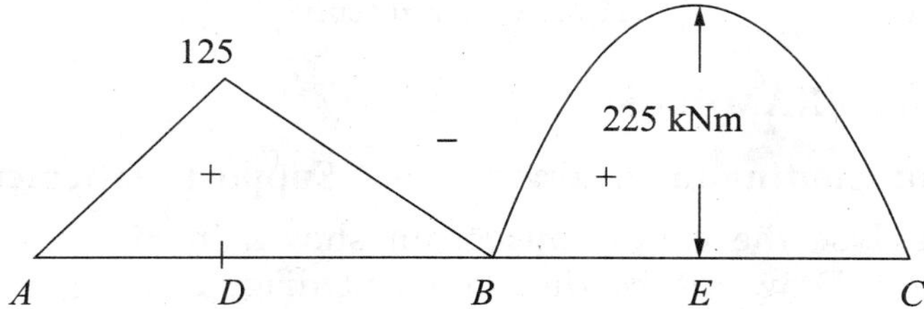

FIG. 2.14 Simple beam moment diagram

The net bending moment diagram is drawn by superimposing the pure moment diagram on the simple beam moment diagram. Thus, the net moment at *D* and *E* are

$$M_D = 125 - \left(\frac{147.00 + 20.25}{2}\right) = +41.38 \text{ kNm}$$

$$M_E = 225 - \left(\frac{147.00 + 0.00}{2}\right) = +151.5 \text{ kNm}$$

The final net bending moment diagram is as follows.

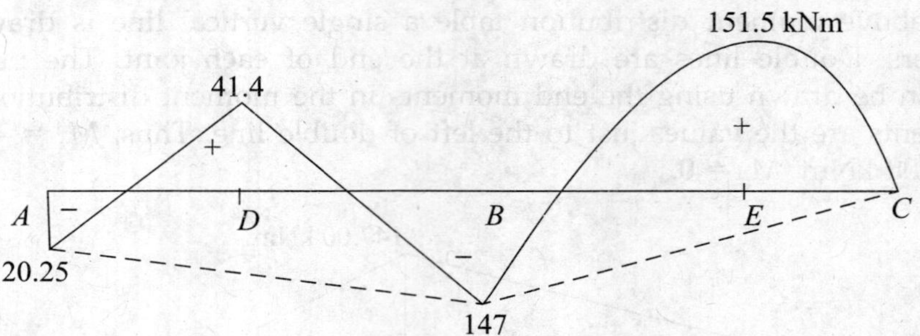

FIG. 2.15 Bending moment diagram

The net bending moment diagrams are preferable in design offices. The elastic curve is drawn using the net bending moment diagram.

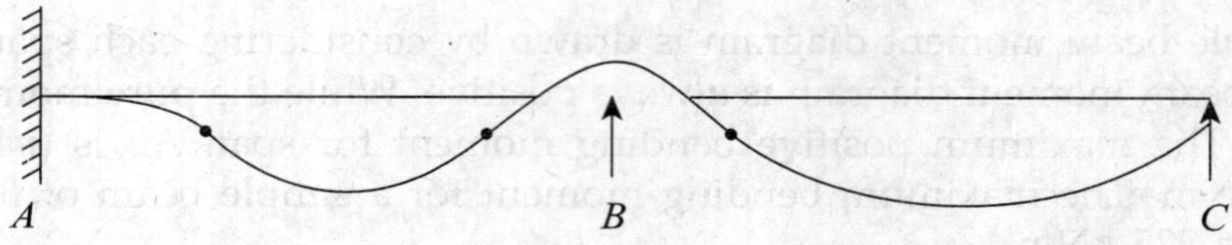

FIG. 2.16 Elastic curve

2.6 NUMERICAL EXAMPLES

2.6.1 Analysis of Continuous Beams without Support Settlement

Example 2.1 Analyse the continuous beam shown in Fig. 2.17 by the moment distribution method. Draw the bending moment diagram and shear force diagram. The beam is of uniform section.

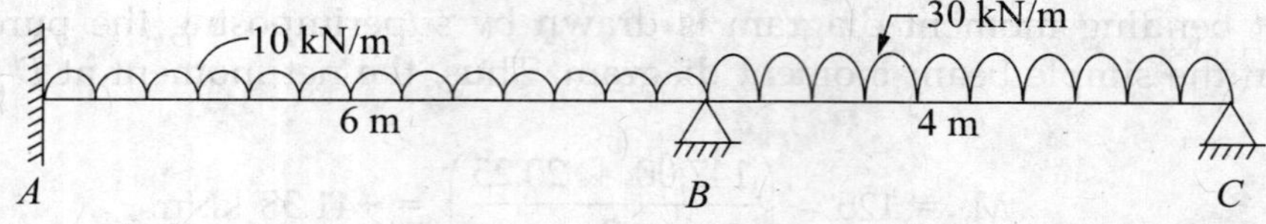

FIG. 2.17

Solution

Step 1

The distribution factors at joint B are evaluated as follows.

Table 2.3 Distribution factors

Joint	*Members*	*Relative Stiffness (k)*	*Sum Σk*	*Distribution Factors (k/Σk)*
	BA	$I/6 = 0.167I$		0.47
B			$0.3555I$	
	BC	$\frac{3}{4} \times \frac{I}{4} = 0.188I$		0.53

Step 2 Fixed End Moments

$$M_{FAB} = \frac{-10 \times 6^2}{12} = -30 \text{ kNm}, M_{FBA} = +30 \text{ kNm}$$

$$M_{FBC} = \frac{-30 \times 4^2}{12} = -40 \text{ kNm}, M_{FCB} = +40 \text{ kNm}$$

Step 3 Moment Distribution Table

As the joint *C* is a hinged end, the moment is zero. Hence, it is balanced first. Then half of this moment is carried over. Then joint *B* is balanced. From the joint *B*, the moment is carried over to *A*.

Table 2.4 Moment distribution table

Joint	*A*	*B*		*C*
Members	*AB* ←	*BA*	*BC* →	*CB*
DF	0	0.47	0.53	1
FEMS	−30.00	+30.00	−40.00	+40.00
Bal	–	–	–	−40.00
Co	–	–	−20.00	
Total	−30.00	+30.00	−60.00	0.00
Bal	–	+14.10	+15.90	–
Co	+7.05	–	–	–
Final	−22.95	+44.10	−44.10	0.00

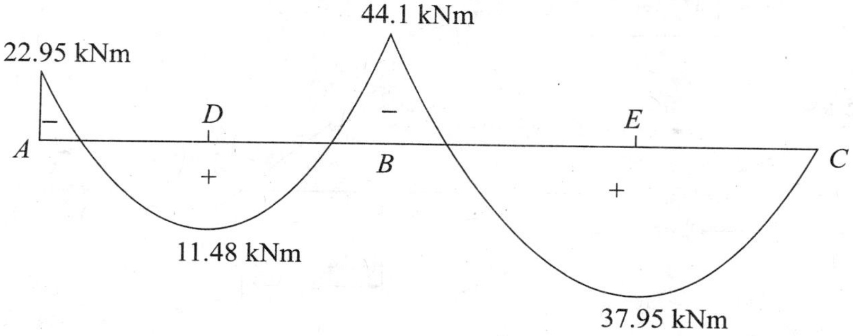

FIG. 2.18 Bending moment diagram

$$M_D = \frac{10 \times 6^2}{8} - \left(\frac{22.95 + 44.1}{2}\right) = 11.48 \text{ kNm}$$

$$M_E = \frac{30 \times 4^2}{8} - \left(\frac{44.1 + 0}{2}\right) = 37.95 \text{ kNm}$$

Step 4 Shear Force Diagrams

Equilibrium of span AB

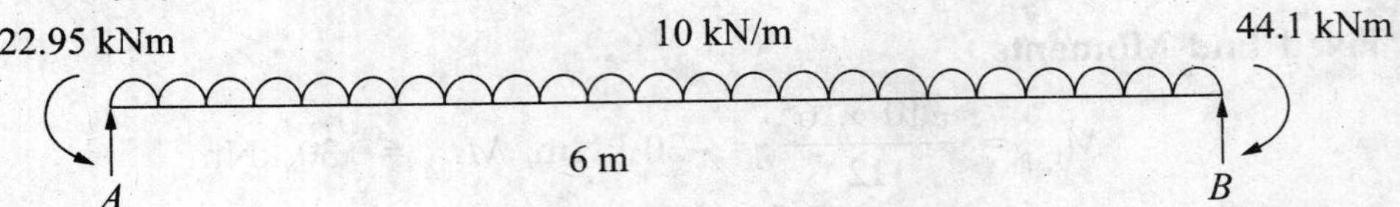

FIG. 2.19

$$\Sigma V = 0; \; V_{AB} + V_{BA} = 60 \quad (1)$$

$$\Sigma M_A = 0; \; -22.95 + 44.1 + 10 \times \frac{6^2}{2} - 6V_{BA} = 0 \quad (2)$$

$$V_{BA} = 33.5 \text{ kN}$$
$$V_{AB} = 26.5 \text{ kN}$$

Equilibrium of span BC

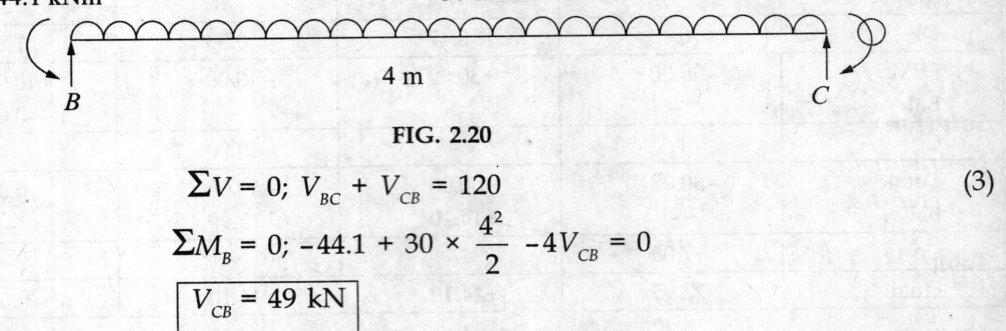

FIG. 2.20

$$\Sigma V = 0; \; V_{BC} + V_{CB} = 120 \quad (3)$$

$$\Sigma M_B = 0; \; -44.1 + 30 \times \frac{4^2}{2} - 4V_{CB} = 0$$

$$V_{CB} = 49 \text{ kN}$$
$$V_{BC} = 71 \text{ kN}$$

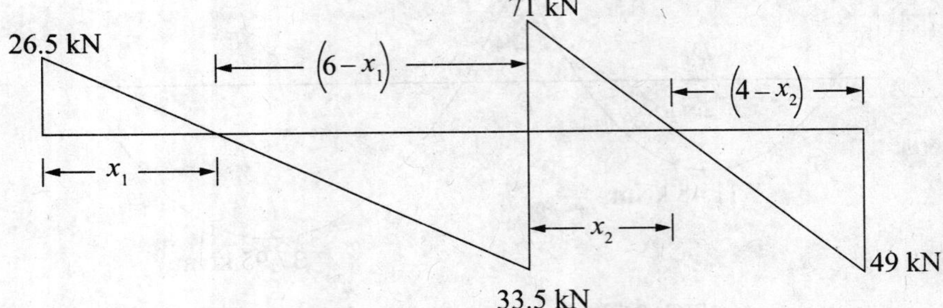

FIG. 2.21 Shear force diagram

From similar triangles

$$\frac{x_1}{(6 - x_1)} = \frac{26.5}{33.5}$$

$$33.5x_1 = 159 - 26.5x_1$$

$$\therefore \quad \boxed{x_1 = 2.65 \text{ m}}$$

$$\frac{x_2}{(4 - x_2)} = \frac{71}{49}$$

$$49x_2 = 284 - 71x_2$$

$$\therefore \quad \boxed{x_2 = 2.37 \text{ m}}$$

Example 2.2 Analyse the continuous beam by moment distribution method. Draw the shear force diagram and bending moment diagram.

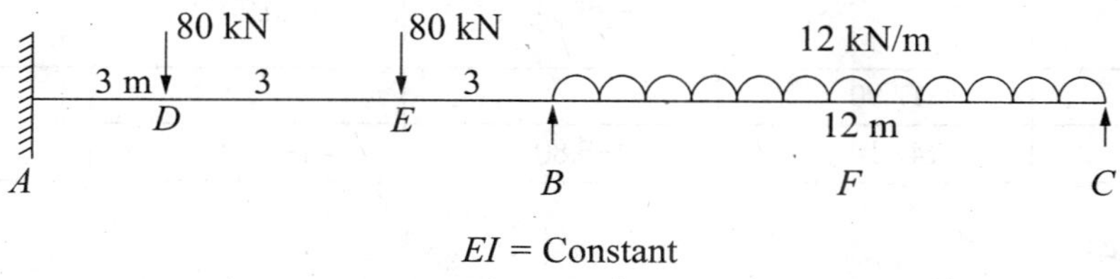

EI = Constant

FIG. 2.22

Solution

Distribution factors

The distribution factors at joint *B* are obtained as follows.

Table 2.5 Distribution factors

Joint	*Members*	*Relative Stiffness (k)*	*Sum Σk*	*Distribution Factors (k/Σk)*
	BA	$I/9 = 0.111I$		0.64
B			$0.174I$	
	BC	$\frac{3}{4} \times \frac{I}{12} = 0.063I$		0.36

Fixed end moments

$$M_{FAB} = -\frac{Wab}{l} = -\frac{80 \times 3 \times 6}{9} = -160 \text{ kNm}$$

$$M_{FBA} = +80 \times 3 \times \frac{6}{9} = +160 \text{ kNm}$$

$$M_{FBC} = -12 \times \frac{12^2}{12} = -144 \text{ kNm}$$

$$M_{FCB} = +144 \text{ kNm}$$

Table 2.6 Moment distribution table

Joint	A	B		C
Members	*AB* ←	*BA*	*BC* ←	*CB*
DF	0	0.64	0.36	1
FEMS	–160.00	+160.00	–144.00	+144.00
Bal				–144.00
Co			–72.00	
Total	–160.00	+160.00	–216.00	0.00
Bal		+35.80	+20.20	
Co	+17.90			
Total	–142.10	+195.80	–195.80	0.00

Shear forces

Equilibrium of span AB

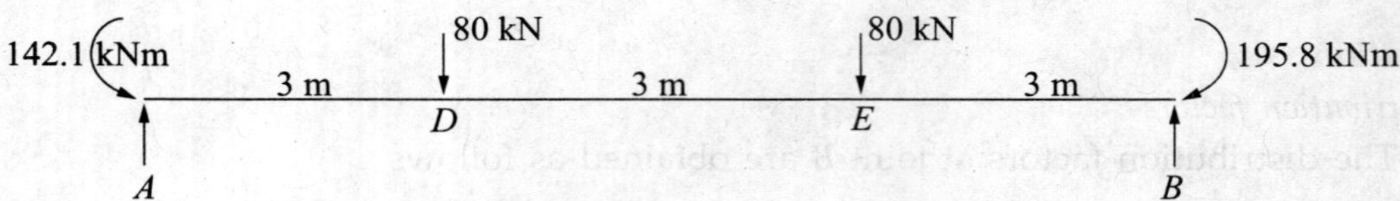

FIG. 2.23

$$\sum V = 0;\ V_{AB} + V_{BA} = 160 \quad (1)$$

$$\sum M_A = 0;\ -142.1 + 195.8 + 80 \times 3 + 80 \times 6 - 9V_{BA} = 0 \quad (2)$$

$$V_{BA} = 86 \text{ kN}$$

$$V_{AB} = 74 \text{ kN}$$

$$M_D = 74(3) - 142.1 = 80 \text{ kNm}$$

$$M_E = 86(3) - 195.8 = 62.2 \text{ kNm}$$

Equilibrium of span BC

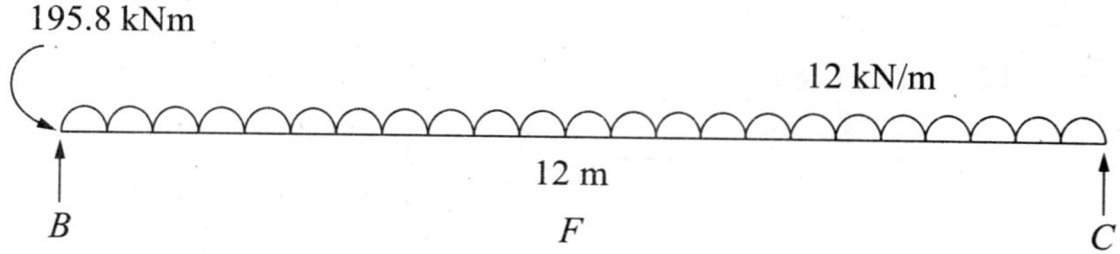

FIG. 2.24

$$\Sigma V = 0;\ V_{BC} + V_{CB} = 144 \tag{3}$$

$$\Sigma M_B = 0;\ -195.8 + 12 \times \frac{12^2}{12} - 12V_{CB} = 0 \tag{4}$$

$$V_{CB} = 55.7 \text{ kN}$$

$$V_{BC} = 88.3 \text{ kN}$$

$$M_F = 55.7 \times 6 - 12 \times \frac{6^2}{2} = 118.2 \text{ kNm}$$

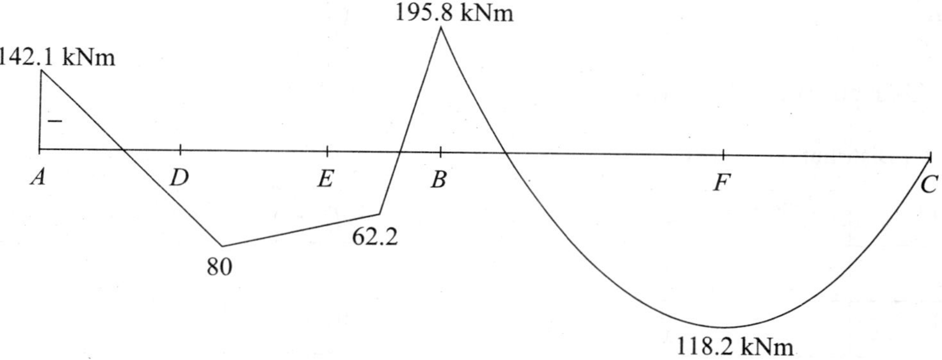

FIG. 2.25 Bending moment diagram

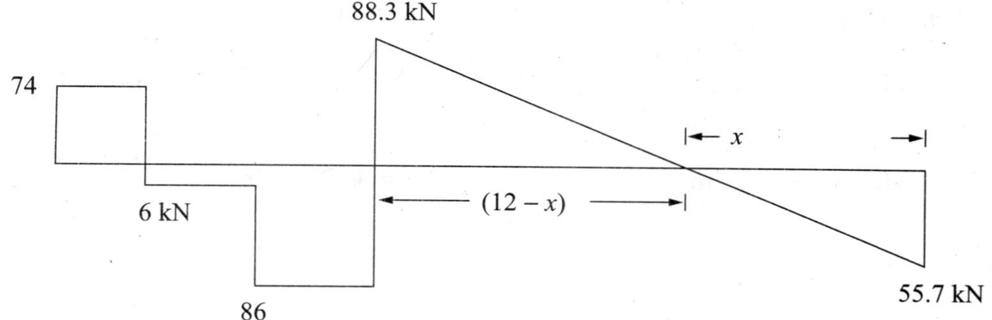

FIG. 2.26 Shear force diagram

From similar Δ's

$$\frac{x}{(12 - x)} = \frac{55.7}{88.3}$$

$$88.3x = 668.4 - 55.7x$$

$$\boxed{x = 4.64 \text{ m}}$$

Example 2.3 Analyse the continuous beam by the moment distribution method. Draw the shear force diagram and bending moment diagram.

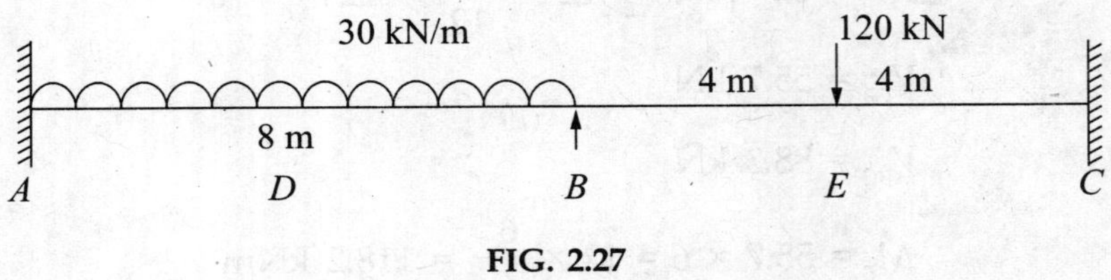

FIG. 2.27

Solution

Distribution factors

The distribution factors at joint B are evaluated as follows.

Table 2.7 Distribution factors

Joint	*Members*	*Relative Stiffness (k)*	*Sum Σk*	*Distribution Factors (k/Σk)*
	BA	$I/8 = 0.125I$		0.5
B			$0.25I$	
	BC	$I/8 = 0.125I$		0.5

Fixed end moments

$$M_{AB} = -30 \times \frac{8^2}{12} = -160 \text{ kNm}; \qquad M_{BC} = -\frac{120 \times 8^2}{8} = -120 \text{ kNm}$$

$$M_{BA} = +160 \text{ kNm}; \qquad M_{CB} = +120 \text{ kNm}$$

Moment Distribution

Table 2.8 Moment distribution table

Joint	*A*	*B*		*C*
Members	*AB* ←	*BA*	*BC* →	*CB*
DF	0	0.5	0.5	0
FEMS	−160	+160	−120	+120
Bal	–	−20	−20	–
Co	−10	–	–	−10
Final	−170	+140	−140	+110

Using the above end moments;

$$M_D = \frac{30 \times 8^2}{8} - \left(\frac{170 + 140}{2}\right) = 85 \text{ kNm}$$

$$M_E = 120 \times \frac{8}{4} - \left(\frac{140 + 110}{2}\right) = 115 \text{ kNm}$$

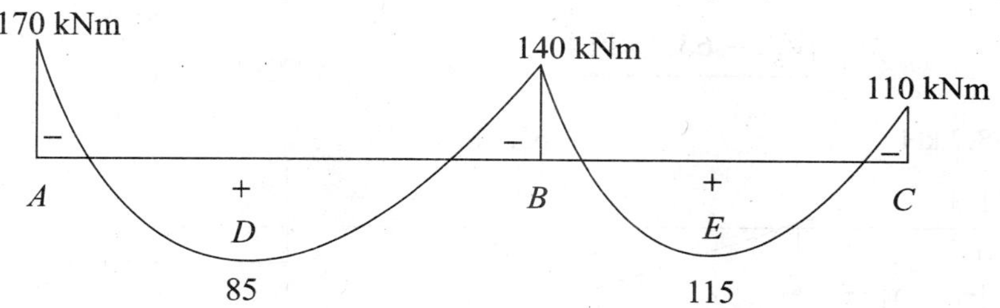

FIG. 2.28 Bending moment diagram

Shear force diagrams
Equilibrium of span AB

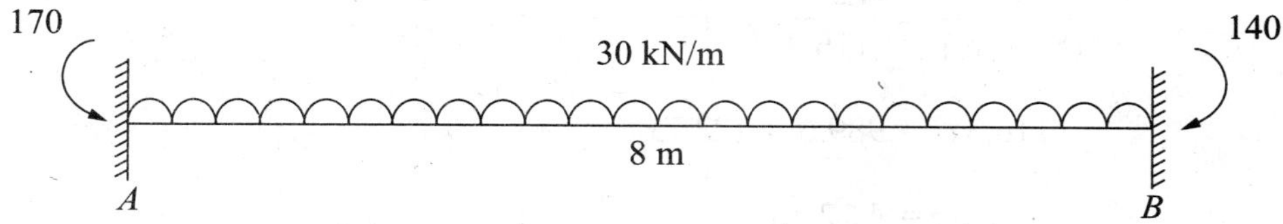

FIG. 2.29

$$\Sigma V = 0;\ V_{AB} + V_{BA} = 240 \qquad (1)$$

$$\Sigma M_A = 0; -170 + 140 + 30 \times \frac{8^2}{2} - 8V_{BA} = 0 \quad (2)$$

$$\boxed{\begin{aligned} V_{BA} &= 116.3 \text{ kN} \\ V_{AB} &= 123.7 \text{ kN} \end{aligned}}$$

Equilibrium of span BC

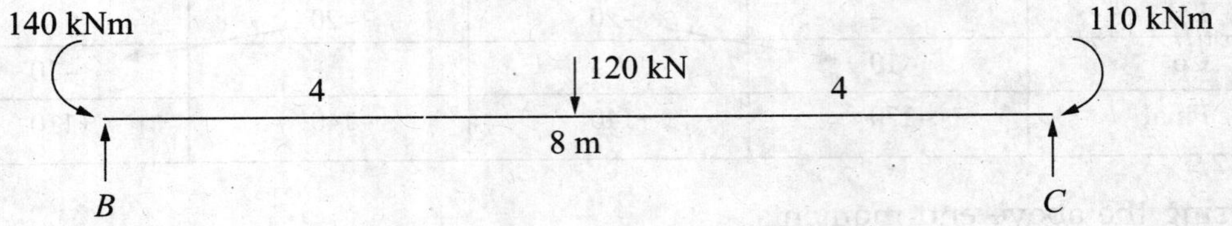

FIG. 2.30

$$\Sigma V = 0; V_{BC} + V_{CB} = 120 \quad (3)$$

$$\Sigma M_B = 0; -140 + 110 + 120(4) - 8V_{CB} = 0$$

$$\boxed{\begin{aligned} V_{CB} &= 56.3 \text{ kN} \\ V_{BC} &= 63.7 \text{ kN} \end{aligned}}$$

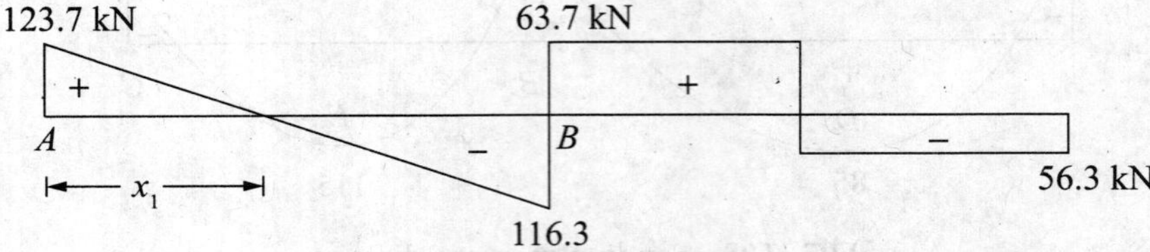

FIG. 2.31 Shear force diagram

From similar Δ's

$$\frac{x_1}{(8 - x_1)} = \frac{123.7}{116.3}$$

$$116.3x_1 = 989.6 - 123.7x_1$$

$$x_1 = 4.12 \text{ m}$$

Example 2.4 Determine the end moments for the continuous beam shown in Fig. 2.32. *EI* is constant. Use the moment distribution method.

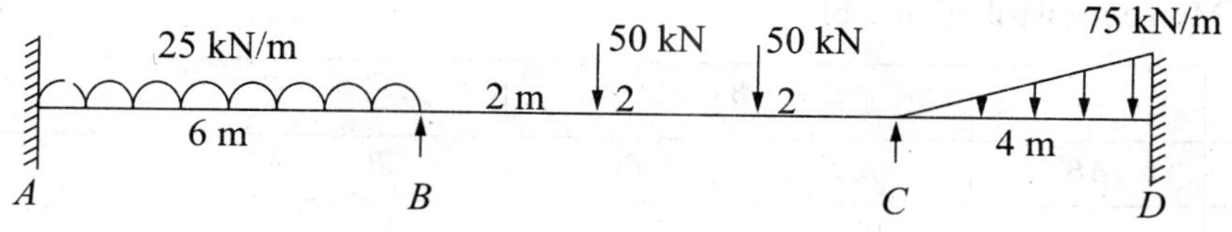

FIG. 2.32

Solution

Distribution factors

The distribution factors at joints *B* and *C* are obtained as follows.

Table 2.9 Distribution factors

Joint	*Members*	*Relative Stiffness Values (k)*	Σk	*Distribution Factors* $k/\Sigma k$
	BA	$I/6$		0.50
B			$2I/6$	
	BC	$I/6$		0.50
	CB	$I/6 = 0.167$		0.40
C			0.417	
	CD	$I/4 = 0.25$		0.60

Fixed end moments

$$M_{FAB} = -25 \times \frac{6^2}{12} = -75 \text{ kNm}$$

$$M_{FBA} = +25 \times \frac{6^2}{12} = +75 \text{ kNm}$$

$$M_{FBC} = -\frac{50 \times 2 \times 4}{6} = -66.67 \text{ kNm}$$

$$M_{FCB} = +50 \times 2 \times \frac{4}{6} = +66.67 \text{ kNm}$$

$$M_{FCD} = -75 \times 4^2/30 = -40.00 \text{ kNm}$$

$$M_{FDC} = +75 \times 4^2/20 = +60.00 \text{ kNm}$$

Table 2.10 Moment distribution table

Joint	*A*	*B*		*C*		*D*
Members	*AB*	*BA*	*BC*	*CB*	*CD*	*DC*
DF	0	0.5	0.5	0.4	0.6	0
FEMS	−75.00	+75.00	−66.67	+66.67	−40.00	+60.00
Bal		−4.17	−4.16	−10.67	+16.00	−
Co	−2.10	−	−5.34	−2.08	−	−8.00
Bal		+2.67	+2.67	+0.83	+1.25	
Co	+1.34		+0.42	+1.34		+0.63
Bal		−0.21	−0.21	−0.54	−0.80	
Co	−0.11		−0.27	−0.11		−0.40
Bal		+0.13	+0.14	+0.04	+0.07	
Co	+.07		+0.02	+0.07		+0.04
Bal		−0.01	−0.01	−0.03	−0.04	
Final	−75.8	+73.41	−73.41	+55.52	−55.52	−52.27

The end moments are

$$M_{AB} = -75.8 \text{ kNm}, \quad M_B = -73.41 \text{ kNm}$$

$$M_c = -55.52 \text{ kNm}, \; M_c = -52.27 \text{ kNm}$$

Example 2.5 Analyse and sketch the bending moment diagram for the beam shown in Fig. 2.33. The values of the second moment area of each span are indicated along the members. Modulus of elasticity is constant.

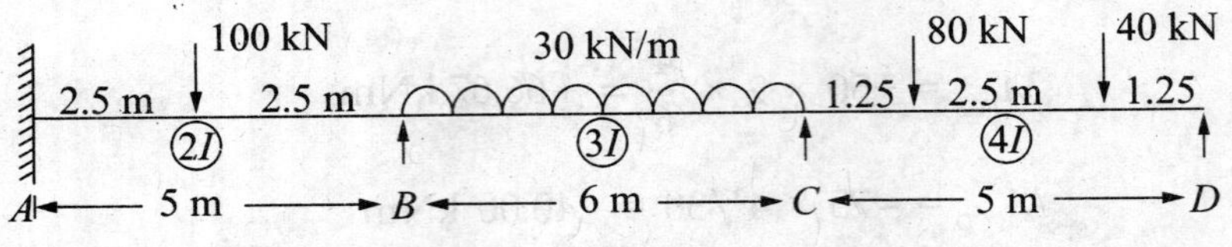

FIG. 2.33

Distribution factors

Table 2.11 Distribution factors

Joint	*Members*	*Relative Stiffness Values (k)*	Σk	*Distribution Factors* $k/\Sigma k$
	BA	$2I/5 = 0.4I$		0.44
B			$0.9I$	
	BC	$3I/6 = 0.5I$		0.56
	CB	$3I/6 = 0.5I$		0.45
C			$1.1I$	
	CD	$\frac{3}{4} \times \frac{4I}{5} = 0.6I$		0.55

Fixed end moments

$$M_{FAB} = -100 \times \frac{5}{8} = -62.5 \text{ kNm}$$

$$M_{FBA} = +100 \times \frac{5}{8} = +62.5 \text{ kNm}$$

$$M_{FBC} = -30 \times \frac{6^2}{12} = -90.0 \text{ kNm}$$

$$M_{FCB} = +30 \times \frac{6^2}{12} = +90.0 \text{ kNm}$$

$$M_{FCD} = -\frac{1.25 \times 80 \times 3.75^2}{5^2} - \frac{3.75 \times 40 \times 1.25^2}{5^2} = -65.625 \text{ kNm}$$

$$M_{FDC} = +\frac{1.25 \times 40 \times 3.75^2}{5^2} + \frac{3.75 \times 80 \times 1.25^2}{5^2} = +46.875 \text{ kNm}$$

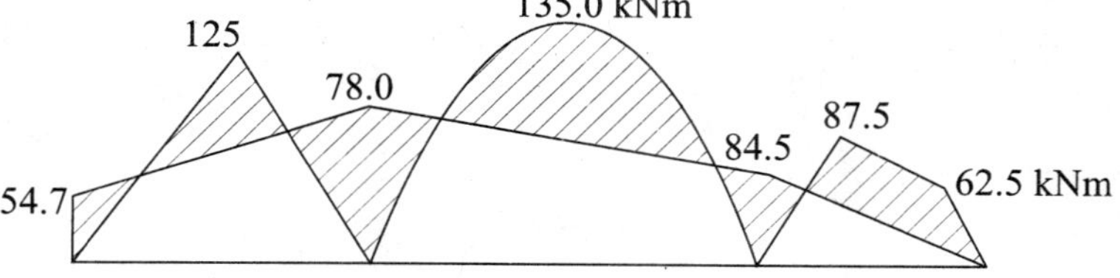

FIG. 2.34 Bending moment diagram

Table 2.12 Moment distribution table

Joint	A	B		C		D
Members	AB	BA	BC	CB	CD	DC
DF	0	0.44	0.56	0.45	0.55	1
FEMS	-62.5	+62.50	-90.00	+90.00	-65.63	+46.88
Bal		+12.10	+15.40	-10.97	-16.4	-46.88
Co	+6.05		-5.49	+7.70		
Bal		+2.42	+3.07	-3.47	-4.23	
Co	+1.21		-1.74	+1.54		
Bal		+0.77	+0.97	-0.69	-0.85	
Co	+0.39		-0.35	+0.49		
Bal		+0.15	+0.20	-0.22	-0.27	
Co	+0.08		-0.11	+0.10		
Bal		0.05	+0.06	-0.05	-0.05	
Co	+0.03		-0.03	+0.03		
Bal		+0.01	+0.02	-0.01	-0.02	
Final	-54.74	+78.0	-78.00	+84.45	-84.45	0.00

2.6.2 Analysis of Continuous Beam with Support Settlement

The settlement of supports causes the bending moment in the members. The settlement of support is mainly due to soil subsidence. In a fixed beam of *AB*, if the support *B* is lower by δ, the induced end moment is $-6EI\delta/l^2$ where δ is the settlement of supports. If the support *B* is higher by δ, the end moment is $+6EI\delta/l^2$. (Refer Basic Structural Analysis – Chapter 11) The moments are computed using the given loading and due to settlement of supports separately and summed up.

Example 2.6 Analyse the beam shown in Fig. 2.35 by the moment distribution method. Support *B* sinks by 10 mm. $E = 200$ kN/mm². $I = 4000 \times 10^4$ mm⁴. Draw *BMD* and *SFD*.

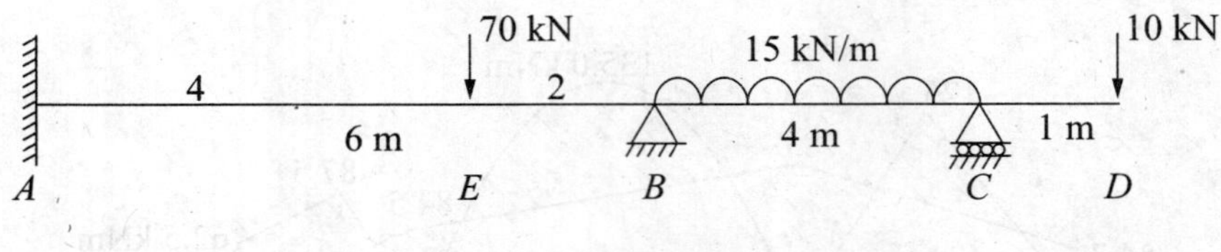

FIG. 2.35

Solution

Distribution factors

Table 2.13 Distribution factors

Joint	*Members*	*Relative Stiffness Values (k)*	Σk	*Distribution Factors* $k/\Sigma k$
	BA	$I/6 = 0.167I$		0.47
B			$0.355I$	
	BC	$\frac{3}{4}\left(\frac{I}{4}\right) = 0.188I$		0.53

Fixed moments

Due to applied loading

$$M'_{AB} = -4(70)\,\frac{2^2}{6^2} = -31.11 \text{ kNm}$$

$$M'_{BA} = +2(70)\,\frac{4^2}{6^2} = +62.22 \text{ kNm}$$

$$M'_{BC} = -15 \times 4^2/12 = -20.00 \text{ kNm}$$

$$M'_{CB} = +15 \times 4^2/12 = +20.00 \text{ kNm}$$

$$M'_{CD} = -10(1) = -10.00 \text{ kNm.}$$

Due to sinking of supports

$$M''_{AB} = M''_{BA} = \frac{-6EI\Delta_B}{l_{AB^2}} = \frac{-6 \times 8000 \times 10}{1000 \times 6^2} = -13.33 \text{ kNm}$$

$$M''_{BC} = M''_{CB} = +\frac{6EI\Delta_B}{l_{CB^2}} = \frac{6 \times 8000 \times 10}{1000 \times 4^2} = +30 \text{ kNm}$$

The actual fixed end moment, is the addition of moment due to applied loading and due to the settlement of supports.

$$M_{FAB} = -31.11 - 13.33 = -44.44 \text{ kNm}$$

$$M_{FBA} = +62.22 - 13.33 = +48.89 \text{ kNm}$$

$$M_{FBC} = -20.00 + 30.00 = +10.00 \text{ kNm}$$

$$M_{FCB} = +20.00 + 30.00 = +50.00 \text{ kNm}$$

$$M_{FCD} = -10.00 \text{ kNm}$$

Table 2.14 Moment distribution table

Joint	A	B		C	
Members	*AB*	*BA*	*BC*	*CB*	*CD*
DF	0	0.47	0.53	1	–
FEMS	–44.44	+48.89	+10.00	+50.00	–10.00
Bal		–27.61	–31.22	–40.00	
Co	–13.84		–20.00		
Bal		+9.40	+10.60		
Co	+4.70				
Final	–53.58	+30.62	–30.62	+10.00	–10.00

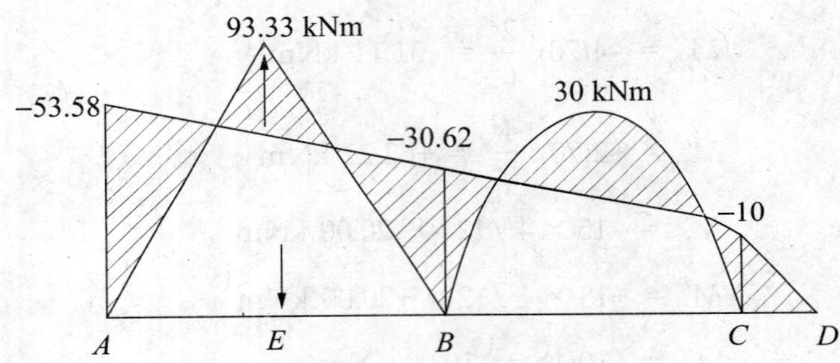

FIG. 2.36 Bending moment diagram

Shear forces
Equilibrium of span AB

53.58 kNm   4   70 kN   2   30.62 kNm

A   B

FIG. 2.37

$$\Sigma V = 0;\ V_{AB} + V_{BA} = 70 \tag{1}$$

$$\Sigma M = 0;\ -53.58 + 30.62 + 70(4) - 6V_{BA} = 0 \tag{2}$$

$$V_{BA} = 42.84 \text{ kN}$$
$$V_{AB} = 27.16 \text{ kN}$$

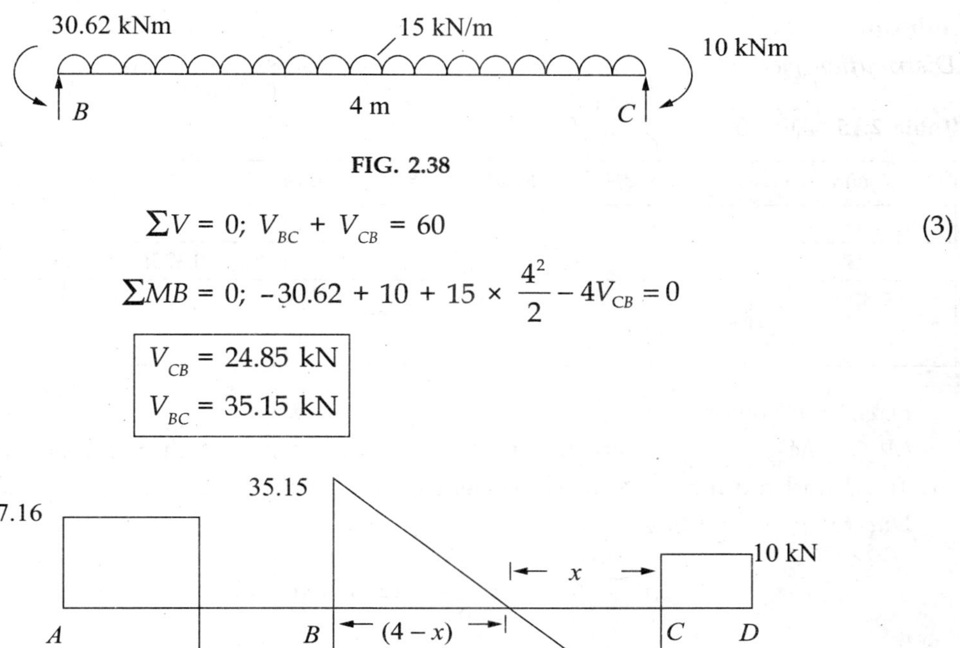

FIG. 2.38

$$\Sigma V = 0;\ V_{BC} + V_{CB} = 60 \tag{3}$$

$$\Sigma MB = 0;\ -30.62 + 10 + 15 \times \frac{4^2}{2} - 4V_{CB} = 0$$

$$V_{CB} = 24.85 \text{ kN}$$
$$V_{BC} = 35.15 \text{ kN}$$

FIG. 2.39 Shear force diagram

The location of zero shear force is located from similar triangles as

$$\frac{x}{4-x} = \frac{24.85}{35.15}$$

$$35.15x = 99.4 - 24.85x$$

$$x = 1.65 \text{ m}$$

Example 2.7 Analyse the continuous beam by the moment distribution. The supports B and C settle by 8 mm and 4 mm respectively. EI = 30000 kNm². Sketch the SFD and BMD.

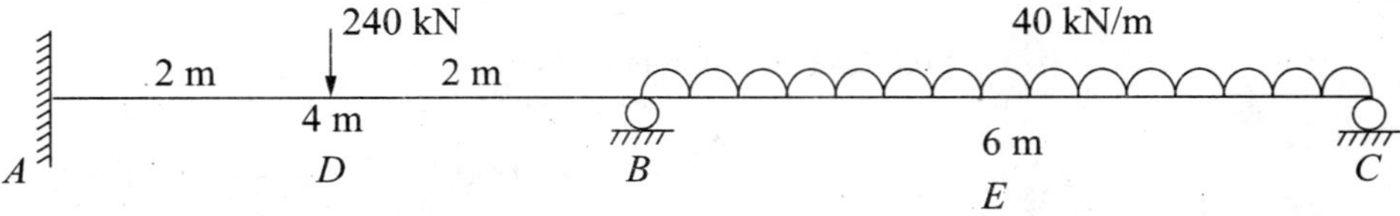

FIG. 2.40

Solution

Distribution factors

Table 2.15 Distribution factors

Joint	*Members*	*Relative Stiff Values (k)*	Σk	$k/\Sigma k$
	BA	$I/4 = 0.25I$		0.67
B			$0.375I$	
	BC	$\frac{3}{4}\times\frac{I}{6}=0.125I$		0.33

Fixed end moments

$M_{FAB} = M'_{AB} + M''_{AB}$ where M'_{AB} is the fixed end moment due to loading and M''_{AB} is the fixed end moment due to settlement of supports.

Due to applied loading

$$M'_{AB} = -240 \times \frac{4}{8} = -120 \text{ kNm}$$

$$M'_{BA} = +240 \times \frac{4}{8} = +120 \text{ kNm}$$

$$M'_{BC} = -40 \times \frac{6^2}{12} = -120 \text{ kNm}$$

$$M'_{CB} = +40 \times \frac{6^2}{12} = +120 \text{ kNm}$$

Due to settlement of supports

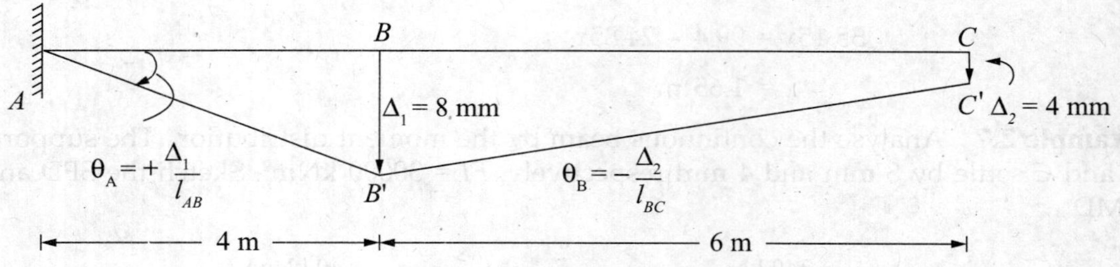

FIG. 2.41 Settlement diagram

$$M''_{AB} = M''_{BA} = \frac{-6EI\Delta_1}{l_1^2} = -6\times 3000 \times \frac{8}{1000}\times\frac{1}{4^2} = -90 \text{ kNm}$$

$$M''_{BC} = M''_{CB} = \frac{-6EI(-\Delta_2)}{l_2^2} = +6\times 3000\times\frac{4}{1000}\times\frac{1}{6^2} = +20 \text{ kNm}$$

Hence, $M_{FAB} = -120 - 90 = 210$ kNm

$M_{FBA} = +120 - 90 = +30$ kNm

$M_{FBC} = -120 + 20 = -100$ kNm

$M_{FCB} = +120 + 20 = +140$ kNm

Moment distribution table

Table 2.16 Moment distribution table

Joint	*A*	*B*		*C*
Members	*AB* ←	*BA*	*BC* ←	*CB*
DF	0	0.67	0.33	1
FEMS	−210.00	+30.00	−100.00	+140.00
Bal		+46.90	+23.10	−140.00
Co	+23.45	–	−70.00	–
Bal		+46.90	+23.10	
Co	+23.45			
Total	−163.10	−123.8	−123.80	0.00

Equilibrium of span AB

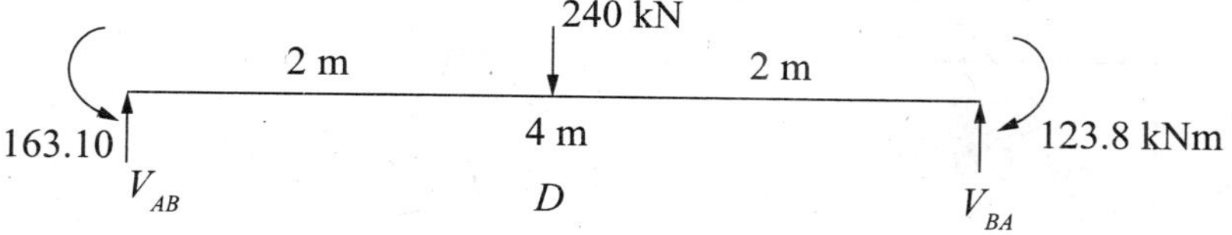

FIG. 2.42

$$\Sigma V = 0;\ V_{AB} + V_{BA} = 240 \tag{1}$$

$$\Sigma M_A = 0;\ -163.10 + 123.8 + 240(2) - 4V_{BA} = 0 \tag{2}$$

$$\boxed{\begin{aligned} V_{BA} &= 110.2 \text{ kN} \\ V_{AB} &= 129.8 \text{ kN} \end{aligned}}$$

$$M_D = -163.10 + 129.8(2) = 96.5 \text{ kNm}$$

Equilibrium of span BC

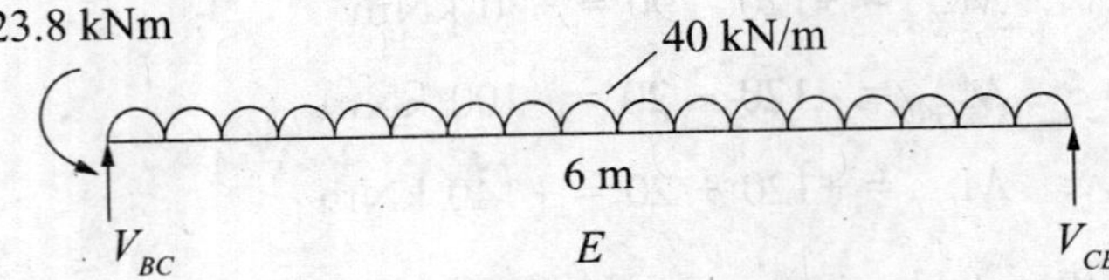

FIG. 2.43

$$\Sigma V = 0; \qquad V_{BC} + V_{CB} = 240 \tag{3}$$

$$\Sigma M_B = 0; \qquad -123.8 - 6V_{CB} + 40 \times \frac{6^2}{2} = 0$$

$$\boxed{\begin{aligned} V_{CB} &= 99.4 \text{ kN} \\ V_{BC} &= 140.6 \text{ kN} \end{aligned}}$$

$$M_E = -123.8 + 140.6 \times 3 - 40 \times \frac{3^2}{2} = 118 \text{ kNm}$$

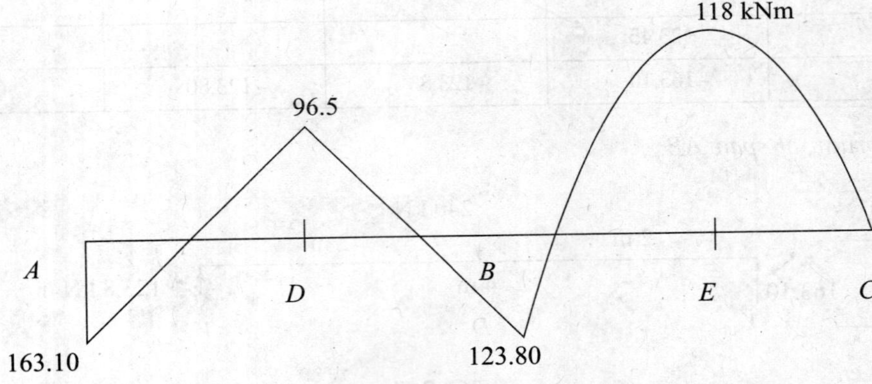

FIG. 2.44 Bending moment diagram

The reader can locate the point of contraflexure and the maximum positive bending moment using basics.

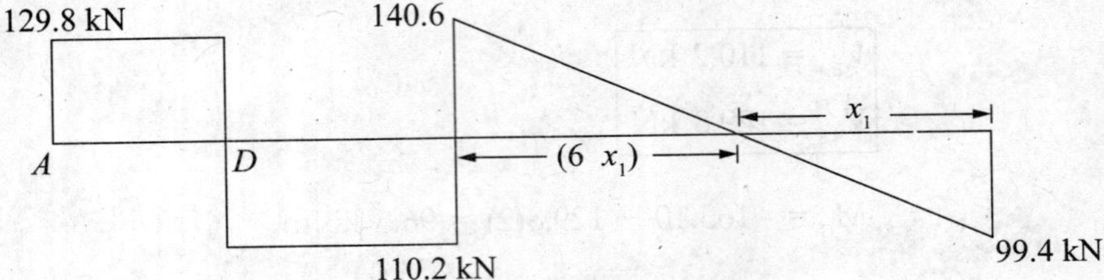

FIG. 2.45

The location of zero shear force can be obtained from similar triangles,

$$\frac{x_1}{(6-x_1)} = \frac{99.4}{140.6}$$

and hence $$x_1 = 2.5 \text{ m}$$

Example 2.8 Analyse the continuous beam by the moment distribution method. Sketch the shear force and bending moment diagrams. Relative to the support *A*, the support *B* sinks by 1 mm and the support *C* raises by 1/2 mm, *EI* = 30000 kNm².

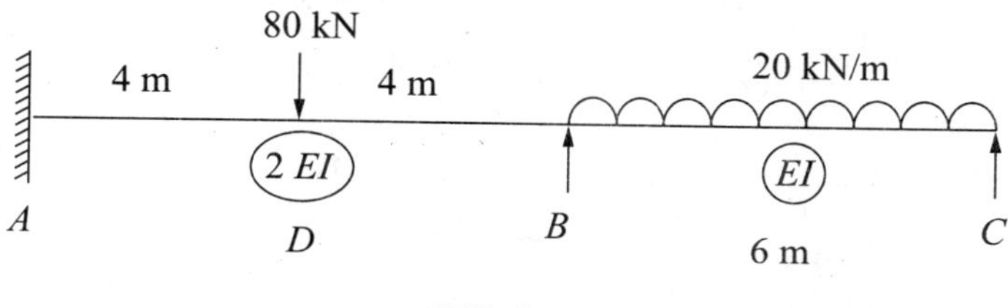

FIG. 2.46

Distribution factors

Table 2.17 Distribution factors

Joint	*Members*	*Relative Stiffness Values (k)*	Σk	$k/\Sigma k$
	BA	$\frac{2I}{8}$		0.67
B			$3I/8$	
	BC	$\frac{3}{4}\times\frac{I}{6}=\frac{I}{8}$		0.33

Fixed end moments

Due to loading

$$M'_{AB} = -80 \times \frac{8}{8} = -80 \text{ kNm}$$

$$M'_{BA} = +80 \text{ kNm}$$

$$M'_{BC} = -20 \times \frac{6^2}{12} = -60 \text{ kNm}$$

$$M'_{CB} = +60 \text{ kNm}$$

Due to settlement of supports

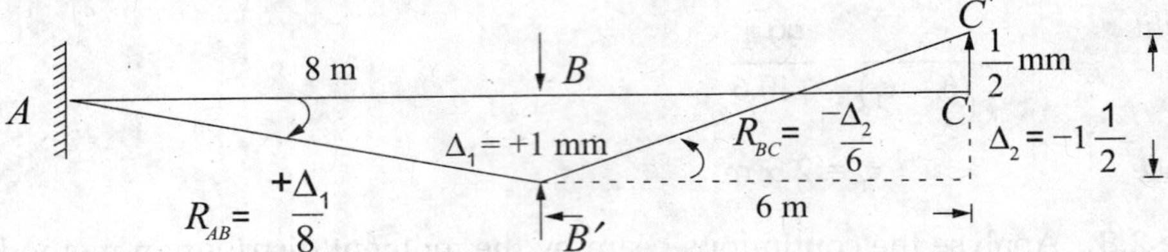

FIG. 2.47 Settlement angles diagram ($R = \Delta/l$)

$$M''_{AB} = \frac{-6E(2I)\Delta_1}{l_1^2} = -\frac{6 \times 30000 \times 1}{1000 \times 8^2} \times 2 = -5.63 \text{ kNm}$$

$$M''_{BA} = \frac{-6E(2I)\Delta_1}{l_1^2} = -5.63 \text{ kNm}$$

$$M''_{BC} = \frac{-6EI(-\Delta_2)}{l_2^2} = +\frac{6 \times 30000 \times 1.5}{1000 \times 6^2} = 7.5 \text{ kNm}$$

It should be noted that the support B settles by 1 mm. In span AB, B settles and hence AB rotates in the clockwise direction. Therefore, (Δ_1/l_1) is positive. In span BC, the support C raises by 1/2 mm from the original level AC. The total displacement is 1(1/2) mm with respect to the displaced position B'. BC displaces to $B'C'$. Hence, the rotation angle (Δ_2/l_2) is negative since the span BC rotates in the anticlockwise direction with respect to B'. The fixed end moment is the sum of the moments due to applied loading and due to the settlement of supports.

$$M_{FAB} = -80 - 5.63 = -85.63 \text{ kNm}$$

$$M_{FBA} = +80 - 5.63 = +74.37 \text{ kNm}$$

$$M_{FBC} = -60 + 7.50 = -52.50 \text{ kNm}$$

$$M_{FCB} = +60 + 7.50 = +67.50 \text{ kNm}$$

Table 2.18 Moment distribution table

Joint	A	B		C
Members	*AB* ←	*BA*	*BC* ←	*CB*
DF	0	0.67	0.33	1
FEMS	−85.63	+74.37	−52.50	+67.50
Bal		−14.65	−7.22	−67.50
Co	−7.33		−33.75	
Bal		+22.61	+11.14	
Co	+11.31			
Final	−81.65	+82.33	−82.33	0.00

Shear force diagrams

Equilibrium of span AB

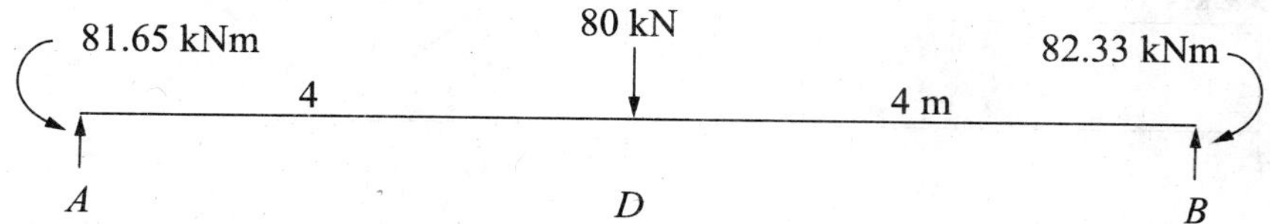

FIG. 2.48

$$\Sigma V = 0, \quad V_{AB} + V_{BA} = 80 \tag{1}$$

$$\Sigma M_A = 0; \quad -81.65 + 82.33 + 80(4) - 8V_{BA} = 0 \tag{2}$$

$$\boxed{\begin{aligned} V_{BA} &= 40.1 \text{ kN} \\ V_{AB} &= 39.9 \text{ kN} \end{aligned}}$$

$$M_D = -81.65 + 39.9 \times 4 = 77.95 \text{ kNm}$$

Equilibrium of span BC

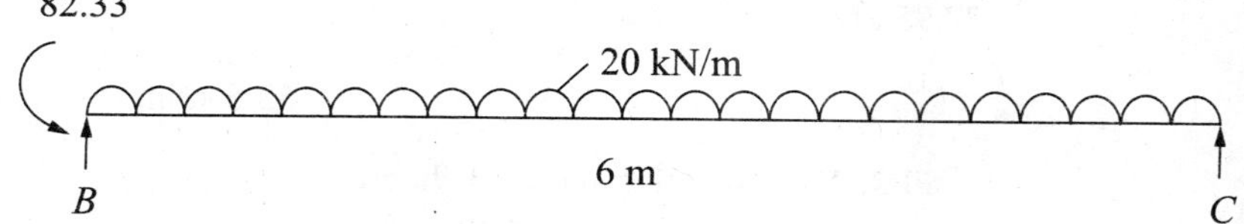

FIG. 2.49

$$\Sigma V = 0; \quad V_{BC} + V_{CB} = 120 \tag{3}$$

$$\Sigma M_B = 0; \quad -82.33 + 20 \times \frac{6^2}{2} - 6V_{CB} = 0$$

$$\boxed{\begin{aligned} V_{CB} &= 46.3 \text{ kN} \\ V_{BC} &= 73.7 \text{ kN} \end{aligned}}$$

The zero shear force location in BC is obtained from

$$\frac{x}{6-x} = \frac{46.3}{73.7}$$

$$73.7x = 277.8 - 46.3x$$

$$\boxed{x = 2.32 \text{ m}}$$

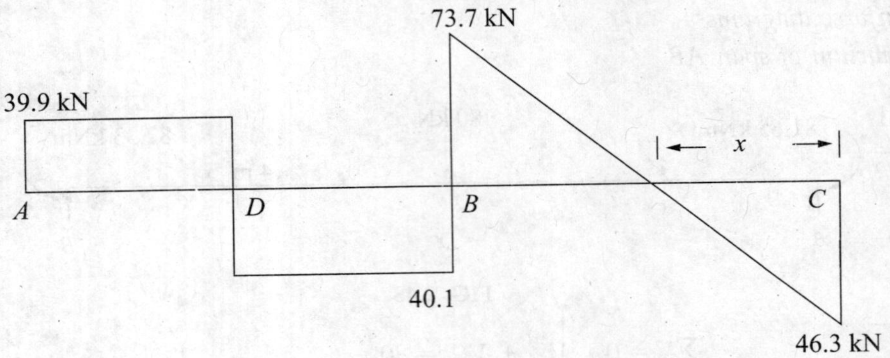

FIG. 2.50 Shear force diagram

Absolute maximum BM is $46.3 \times 2.32 - 20 \times \dfrac{2.32^2}{2} = 53.6$ kNm

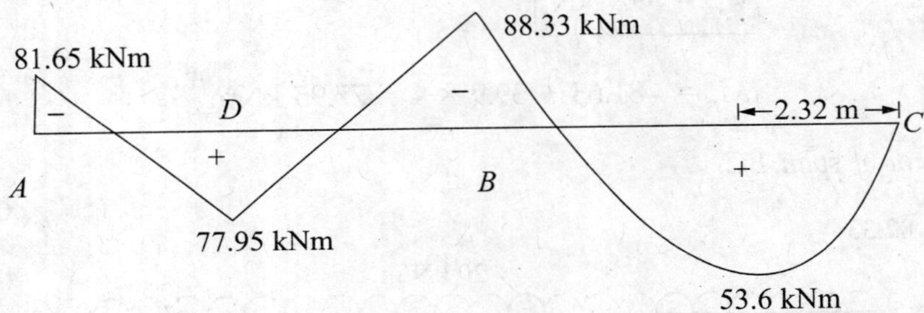

FIG. 2.51 Bending moment diagram

Example 2.9 A continuous beam is loaded as shown in Fig. 2.52. During loading the support B sinks by 10 mm. Determine the bending moments at the supports. Sketch the BMD. Given that $I = 1600(10)^4$ mm^4; $E = 200$ kN/mm^2. Use moment distribution method. Draw SFD also.

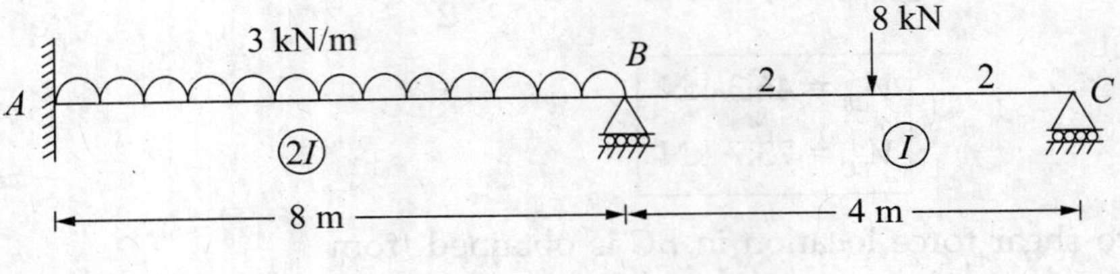

FIG. 2.52

Solution

Distribution factors

Table 2.19 Distribution factors

Joint	*Members*	*Relative Stiffness Values (k)*	Σk	$\frac{k}{\Sigma k}$
	BA	$\frac{2I}{8} = 0.25I$		0.57
B			0.438I	
	BC	$\frac{3}{4} \times \frac{I}{4} = 0.188I$		0.43

Fixed end moments

Due to loading

$$M'_{FAB} = -3 \times \frac{8^2}{12} = -16 \text{ kNm}; \qquad M'_{FBA} = +16 \text{ kNm}$$

$$M'_{FBC} = -8 \times \frac{4}{8} = -4 \text{ kNm}; \qquad M'_{FCB} = +4 \text{ kNm}$$

Due to Settlement

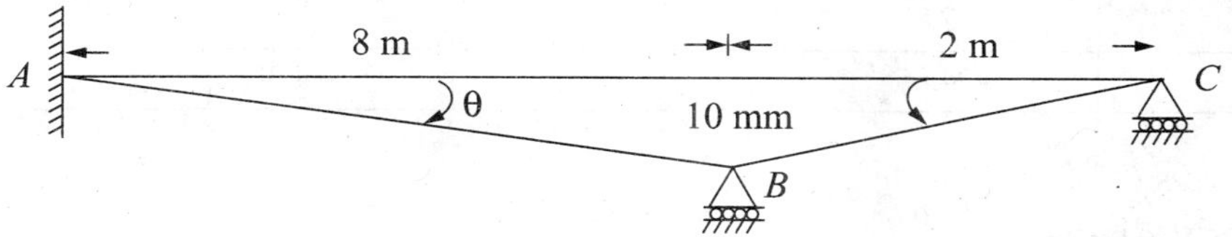

FIG. 2.53 Settlement diagram

$$M''_{FAB} = -\frac{6EI\Delta_1}{l_1^2} = -6 \times 2 \times 3200 \times \frac{10}{1000} \times \frac{1}{8^2} \; (\because \Delta_1/l_1 \text{ is positive})$$

$$M''_{FAB} = -6 \text{ kNm}$$

$$\therefore \quad M''_{FBA} = +6 \text{ kNm}$$

$$M''_{FBA} = -\frac{6EI\Delta_2}{l_2^2} = -\frac{6 \times 3200 \times (-10)}{1000 \times 4^2} = +12 \text{ kNm} \; (\because \Delta_2/l_2 \text{ is negative})$$

$$M''_{FCB} = +12 \text{ kNm}$$

It is to be mentioned here that B settles by 10 mm with respect to A. The settlement angle is positive as AB rotates in clockwise direction with respect to A. The settlement angle for CB is negative as CB rotates in the anticlockwise direction. With respect to C. Thus, the fixed end moments at the joint is the sum of the moments due to loading and the settlement. They are as follows.

Fixed end moments

$$M_{FAB} = -16 - 6 = -22 \text{ kNm}$$

$$M_{FBA} = 16 - 6 = 10 \text{ kNm}$$

$$M_{FBC} = -4 + 12 = 8 \text{ kNm}$$

$$M_{FCB} = +4 + 12 = 16 \text{ kNm}$$

Table 2.20 Moment distribution table

Joint	A	B		C
Members	*AB* ←	*BA*	*BC* ←	*CB*
DF	0	0.57	0.43	1
FEMS	-22.00	+10.00	+8.00	+16.00
Bal	–	–	–	-16.00
Co	–	–	-8.00	–
Total	-22.00	+10.00	0.00	0.00
Bal	–	-5.70	-4.30	–
Co	-2.85	–	–	–
Final	-24.85	+4.30	-4.30	0.00

Shear forces

Free body diagram

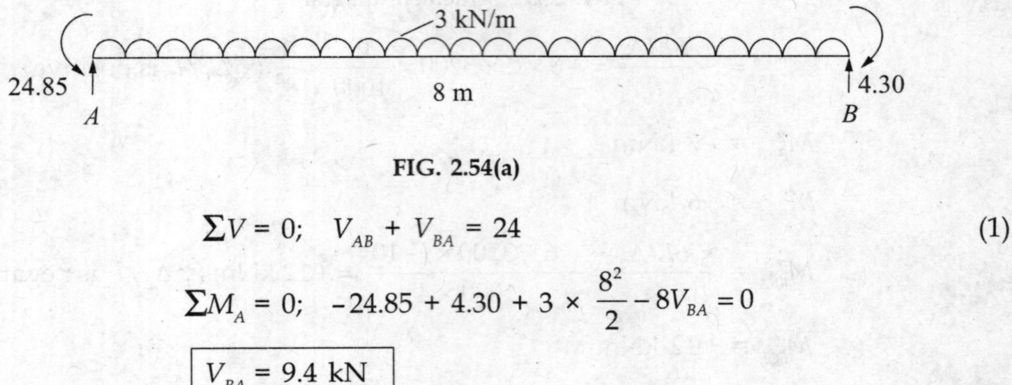

FIG. 2.54(a)

$$\Sigma V = 0; \quad V_{AB} + V_{BA} = 24 \tag{1}$$

$$\Sigma M_A = 0; \quad -24.85 + 4.30 + 3 \times \frac{8^2}{2} - 8V_{BA} = 0$$

$$\boxed{\begin{aligned} V_{BA} &= 9.4 \text{ kN} \\ V_{AB} &= 14.6 \text{ kN} \end{aligned}}$$

$$M_D = -24.85 + 14.6 \times 4 - \frac{3 \times 4^2}{2} = 9.55 \text{ kNm}$$

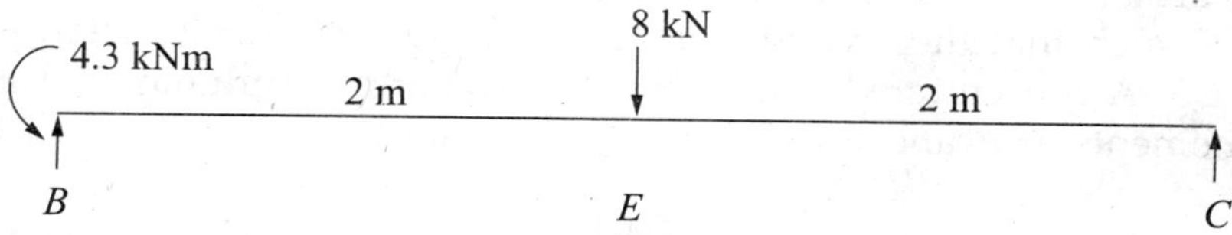

FIG. 2.54(b)

$$\Sigma V = 0; \quad V_{BC} + V_{CB} = 8$$

$$\Sigma M_B = 0; \quad -4.3 + 8(2) - 4V_{CB} = 0$$

$$V_{CB} = 2.9 \text{ kN}$$

$$V_{BC} = 5.1 \text{ kN}$$

$$M_E = 2.9(2) = 5.8 \text{ kNm}$$

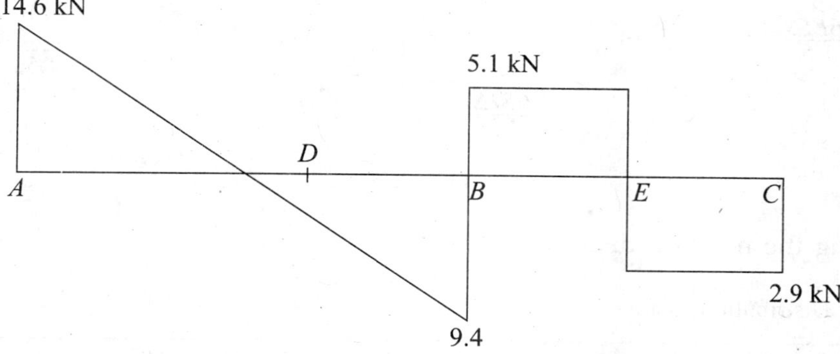

FIG. 2.55 Shear force diagram

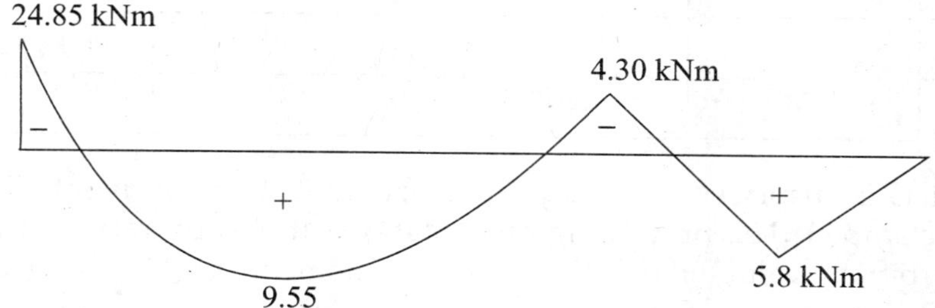

FIG. 2.56 Bending moment diagram

2.6.3 Analysis of Beams with Variable Moment of Inertia

Example 2.10 A horizontal beam *ACB*, three metres long is fixed at both ends *A* and *B*. They are at the same level. The member has a change of section at the centre of the span *C* such that the second moments of area are *I* for a distance *AC* and 2*I* for distance *CB*. A concentrated load of 500 kN acts at the midpoint *C*. Determine the fixing end moments at *A* and *B*.

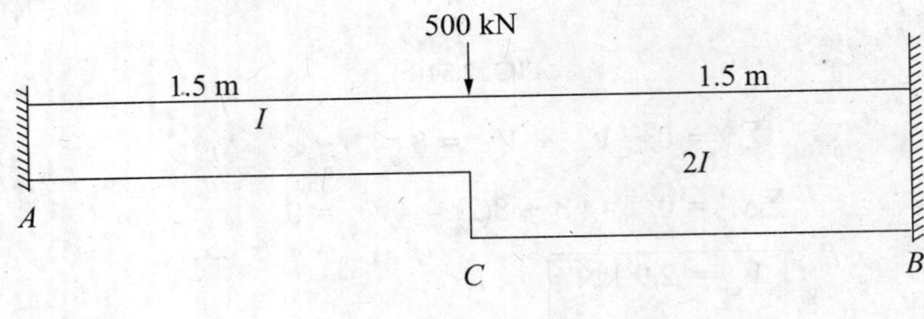

FIG. 2.57

Solution

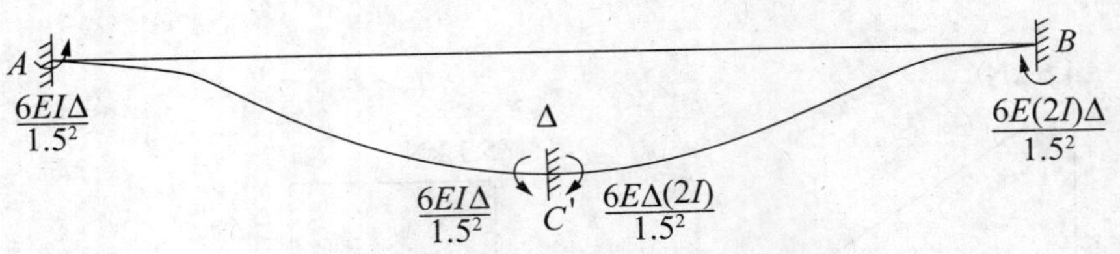

FIG. 2.58

Applying the moment distribution method,

Table 2.21 Distribution factors

Joint	Members	Relative Stiffness (k)	Sum Σk	Distribution Factor (k/Σk)
	CA	I/1.5		0.33
C			2I	
	CB	2I/1.5		0.67

The point *C* is displaced vertically by an amount Δ to the point *C*′ keeping the section at *C* clamped (i.e., preventing any rotation). If an arbitrary end moment M_{CA} is given an arbitrary value of 100, the fixed end moment of M_{CB} will be 200.

Table 2.22 Moment distribution table

Joint	A	C		B
Members	*AC*	*CA*	*CB*	*BC*
DF	1	0.33	0.67	1
FEMS	−100	−100	+200	+200
Bal		−33	−67	
Co	−16.5			−33.5
Final	−116.5	−133	+133	+166.5

Equilibrium of span AC

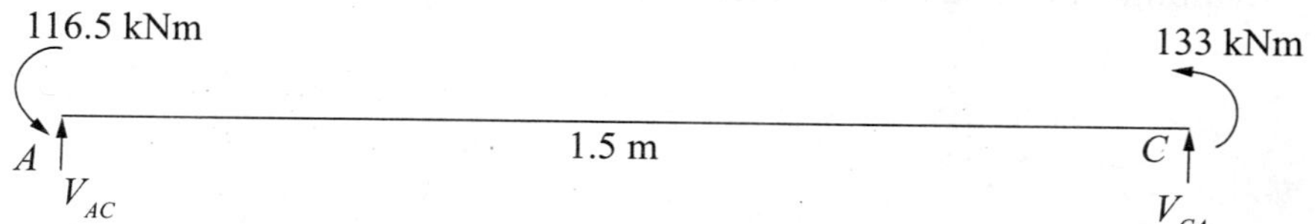

FIG. 2.59

$$\Sigma M_C = 0;$$

$$-116.5 - 133 + 1.5V_{AC} = 0$$

$$V_{AC} = 166.33 \text{ kN}$$

Equilibrium of span CB

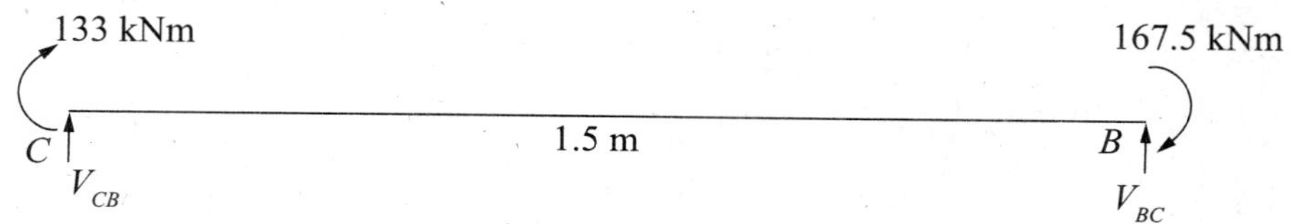

FIG. 2.60

Taking moment about *C*;

$$133 + 167.5 - 1.5V_{BC} = 0$$

$$V_{BC} = 200.33 \text{ kN}$$

Resolving the forces vertically

$$= 166.33 + 200.33 = 366.66 \text{ kN}$$

i.e., if the concentrated load applied at the centre is 366.66 kN then this will yield the moments given in the table. But the applied load is 500 kN. Hence, the moments at A and B are obtained as

$$M_A = -116.5 \times \frac{500}{366.66} = -158.9 \text{ kNm}$$

$$M_B = +167.5 \times \frac{500}{366.66} = +228.4 \text{ kNm}$$

Example 2.11 A vertical column of 8 m height is to carry a crane girder load of 50 kN applied at an eccentricity of 0.2 m. Calculate the moments at A and B due to this load assuming both ends are fixed.

Solution: Evaluate the distribution factor at C.

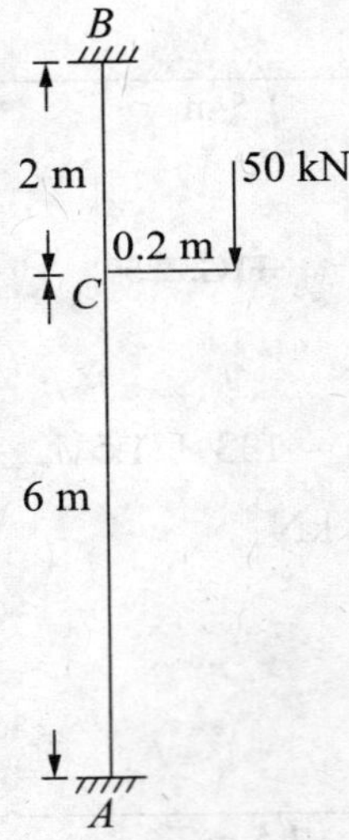

FIG. 2.61

Table 2.23 Distribution factors

Joint	*Members*	*Relative Stiffness (k)*	Σk	$(k/\Sigma k)$
	CA	$I/6$		0.25
C			0.667I	
	CB	$I/2$		0.75

To prevent the sway at C introduce a force X as shown in Fig. 2.62.

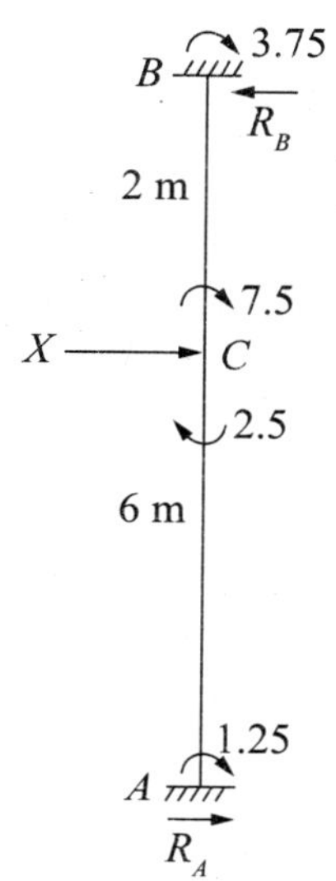

FIG. 2.62

Moment 10 kNm at C is distributed as follows:

$$M'_{CB} = 0.75(10) = 7.5 \text{ kNm}$$

$$M'_{CA} = 0.25(10) = 2.5 \text{ kNm}$$

Carry over moments

$$M'_{BC} = 7.5/2 = 3.75 \text{ kNm}$$

$$M'_{AC} = 2.5/2 = 1.25 \text{ kNm}$$

Equilibrium for the portion CB

Taking moment about C for the portion CB;

$$3.75 + 7.5 - 2R_B = 0$$

$$R_B = 5.625 \text{ kN}$$

Equilibrium for the portion CA

Taking moment about C for the portion CA;

$$2.5 + 1.25 - 6R_A = 0$$

$$R_A = 0.625 \text{ kN}$$

Resolving all the forces horizontally

$$X + 0.625 - 5.625 = 0$$

$$\boxed{X = 5 \text{ kN}}$$

Sway Analysis

Let the column sway by an amount Δ from C to C' as shown in Fig. 2.63. Due to the lateral sway, the moment develops at the support A and B. The moment at A is $6EI\Delta/6^2$ while at B is $-6EI\Delta/2^2$. Hence, the ratio of the moments is taken as

$$M_{CA} : M_{CB}$$

$$\frac{6EI\Delta}{6^2} : -\frac{6EI\Delta}{2^2}$$

$$+4 : -36$$

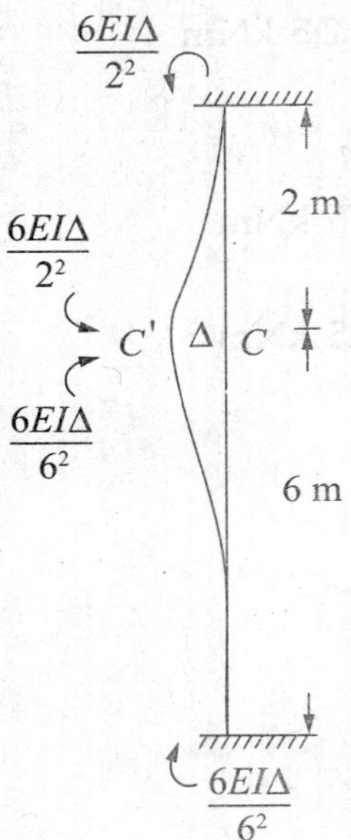

FIG. 2.63

The moment distribution is done with the above assumed moments and the lateral force Y is determined from the free body diagram.

Table 2.24 Moment distribution table

Joint	*A*	*C*		*B*
Members	*AC*	*CA*	*CB*	*BC*
DF	0	0.25	0.75	0
FEMS	+4.00	+4.00	−36.00	−36.00
Bal		+8.00	+24.00	
Co	+4.00	–	–	+12.00
Final	8.00	+12.00	−12.00	−24.00

The above end moments of the members are marked in the free body diagram shown below.

Taking moment about C for the portion CB,

$$2R_B - 24 - 12 = 0$$

$$R_B = 18 \text{ kN}$$

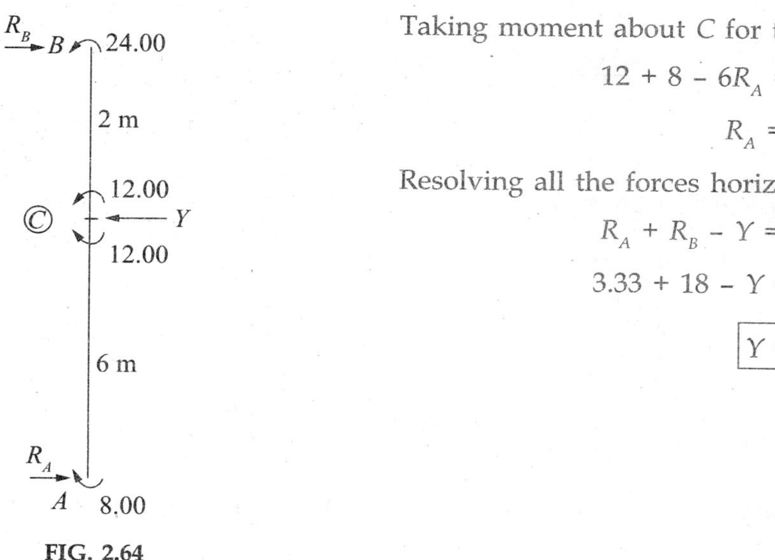

FIG. 2.64

Taking moment about C for the portion AC,

$$12 + 8 - 6R_A = 0$$

$$R_A = 3.33 \text{ kN}$$

Resolving all the forces horizontally

$$R_A + R_B - Y = 0$$

$$3.33 + 18 - Y = 0$$

$$\boxed{Y = 21.33 \text{ kN}}$$

The sway moments corresponding to the sway force 5 kN is obtained by multiplying the moments in the moment distribution table as

$$M''_{AC} = \left(\frac{5}{21.33}\right)8 = 1.88 \text{ kNm}$$

$$M''_{CA} = \left(\frac{5}{21.33}\right)12 = 2.81 \text{ kNm}$$

$$M''_{CB} = \left(\frac{5}{21.33}\right) \times -12 = -2.81 \text{ kNm}$$

$$M''_{BC} = \left(\frac{5}{21.33}\right) \times -24 = -5.63 \text{ kNm}$$

Then the final moments is the sum of sway and nonsway moments as

$$M_{AC} = M'_{AC} + M''_{AC} = 1.25 + 1.88 = 3.13 \text{ kNm}$$

$$M_{BC} = M'_{BC} + M''_{BC} = 3.75 - 5.63 = -1.88 \text{ kNm}$$

2.7 ANALYSIS OF RECTILINEAR FRAMES

Example 2.12 Analyse the frame shown in Fig. 2.65 by moment distribution method. Draw the bending moment diagram.

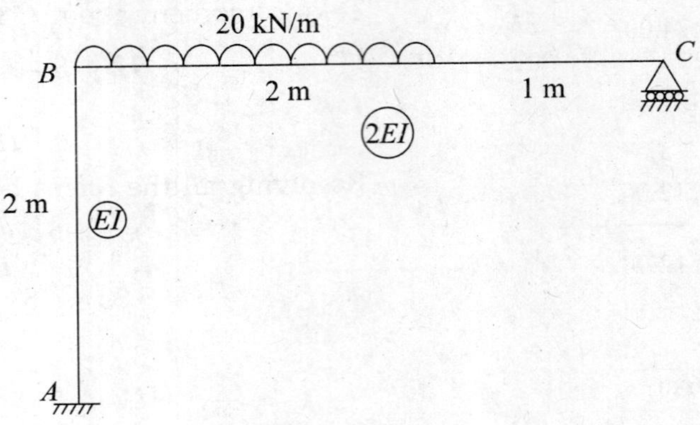

FIG. 2.65

Solution

Distribution factors

Table 2.25 Distribution factors

Joint	*Members*	*Relative Stiffness Values (I/l)*	Σk	$k/\Sigma k$
	BA	$I/2$		0.5
B			I	
	BC	$\frac{3}{4}\left(\frac{2I}{3}\right) = I/2$		0.5

Fixed end moments

$$M_{BC} = \int_0^2 \frac{xwdx(l-x)^2}{l^2} = \int_0^2 \frac{x(20dx)(3-x)^2}{3^2} = 13.33 \text{ kNm}$$

$$M_{CB} = \int_0^2 (I-x)\,wdx\,\frac{x^2}{l^2} = \int_0^2 (3-x)\,20dx\,\frac{x^2}{9} = 8.89 \text{ kNm}$$

Nonsway Analysis

Table 2.26 Moment distribution table

Joint	A	B		C
Members	*AB*	*BA*	*BC*	*CB*
DF	0	0.5	0.5	1.0
FEMS			−13.33	+8.89
Bal		+6.66	+6.67	−8.89
Co	+3.33		−4.44	
Bal		+2.22	+2.22	
Co	+1.11			
	+4.44	+8.88	−8.88	0.00

Sway Analysis

The sway moments are assumed in column *AB*.

Table 2.27 Moment distribution table

Joint	*A*	*B*		*C*
Members	*AB*	*BA*	*BC*	*CB*
DF	0	0.5	0.5	1
FEMS	-100.00	-100.00	0.00	0.00
Bal		+50.00	+50.00	
Co	+25.00			
Total	-75.00	-50.00	+50.00	0.00

Final Moments

$$M_{AB} = 4.44 - 75\,k$$

$$M_{BA} = 8.88 - 50\,k$$

$$M_{BC} = -8.88 + 50\,k$$

$$M_{CB} = 0$$

Column shear condition

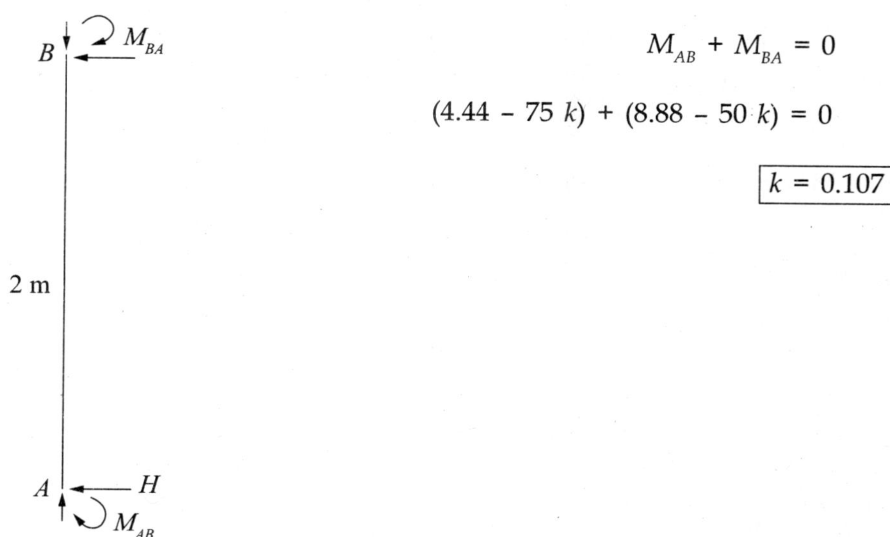

$$M_{AB} + M_{BA} = 0$$

$$(4.44 - 75\,k) + (8.88 - 50\,k) = 0$$

$$\boxed{k = 0.107}$$

FIG. 2.66

Back substitution

$$M_{AB} = 4.44 - 75(0.107) = -3.585 \text{ kNm}$$

$$M_{BA} = 8.88 - 50(0.107) = +3.530 \text{ kNm}$$

$$M_{BC} = -8.88 + 50(0.107) = -3.530 \text{ kNm}$$

$$M_{CB} = 0$$

3.53 B 20 kN/m D C 2 m 1 m

FIG. 2.67

$$\Sigma M_B = 0$$

$$3V_C + 3.53 = \frac{20 \times 2^2}{2}$$

$$\boxed{V_C = 12.157 \text{ kN}}$$

$$\boxed{M_D = 12.157(1) = 12.157 \text{ kNm}}$$

$$\Sigma V = 0$$

$$V_B + V_C = 2(20)$$

$$\boxed{V_B = 40 - 12.157 = 27.843 \text{ kN}}$$

$$(SF)_X = 27.843 - 20x = 0$$

$$\boxed{x = 1.392 \text{ m}}$$

$$\text{Maximum +ve BM} = 27.843(1.392) - \frac{20(1.392)^2}{2} - 3.53$$

$$M_E = 15.851 \text{ kNm}$$

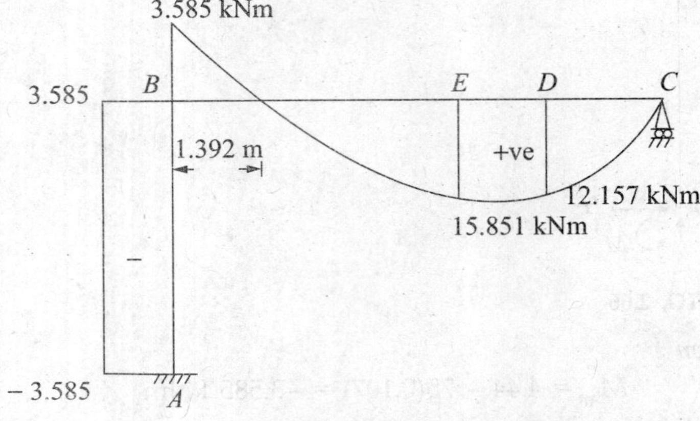

FIG. 2.68 Bending moment diagram

Example 2.13 Analyse the frame shown in Fig. 2.69 by the moment distribution method. Draw the bending moment diagram.

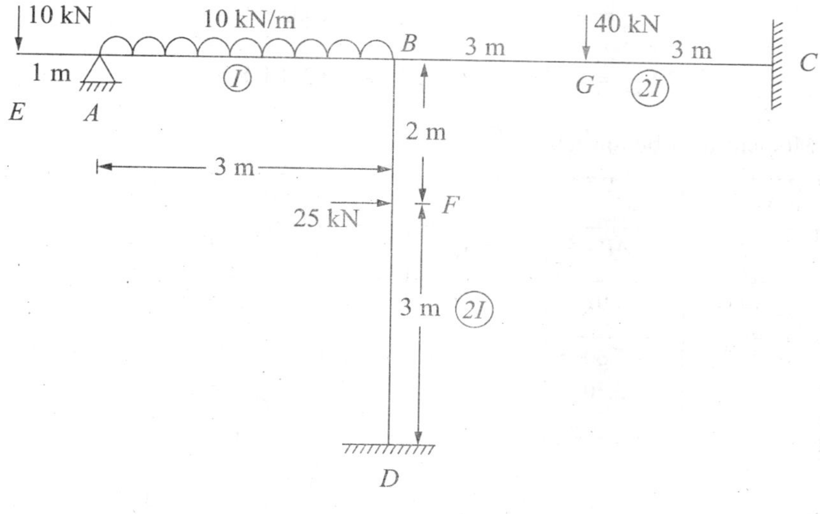

FIG. 2.69

Solution

Distribution factors

Table 2.28 Distribution factors

Joint	*Members*	*Relative Stiffness Values (I/l)*	Σk	$(k/\Sigma k)$
	BA	$\frac{3}{4} \times \frac{I}{3} = \frac{I}{4}$		0.26
B	BD	$(2I/5)$	0.98	0.41
	BC	$2I/6$		0.33

Fixed end moments

$$M_{FAE} = +10 \times 1 = 10 \text{ kNm}$$

$$M_{FAB} = -10 \times \frac{3^2}{12} = -7.5 \text{ kNm}$$

$$M_{FBA} = +10 \times \frac{3^2}{12} = +7.5 \text{ kNm}$$

$$M_{FBC} = -40 \times \frac{6}{8} = -30.0 \text{ kNm}$$

$$M_{FCB} = +40 \times \frac{6}{8} = +30.0 \text{ kNm}$$

$$M_{FBD} = +2 \times 25 \times 3^2/5^2 = +18.0 \text{ kNm}$$

$$M_{FBE} = -3 \times 25 \times 2^2/5^2 = -12.0 \text{ kNm}$$

Table 2.29 Moment distribution table

Joint	*A*		*B*			*C*	*D*
Members	*AE*	*AB* →	*BA*	*BD*	*BC* ↔	*CB*	*DB*
DF	–	1.00	0.26	0.41	0.33	0.00	0.00
FEMS Bal	+10.00	–7.50 –2.50	+7.50 +1.17	+18.00 +1.85	–30.00 +1.48	+30.00	–12.00
Co Bal			–1.25 +0.33	 +0.51	 +0.41	+0.74	+0.93
Co						0.21	
Total	+10.00	–10.00	+7.75	+20.36	–28.11	+30.95	–11.07

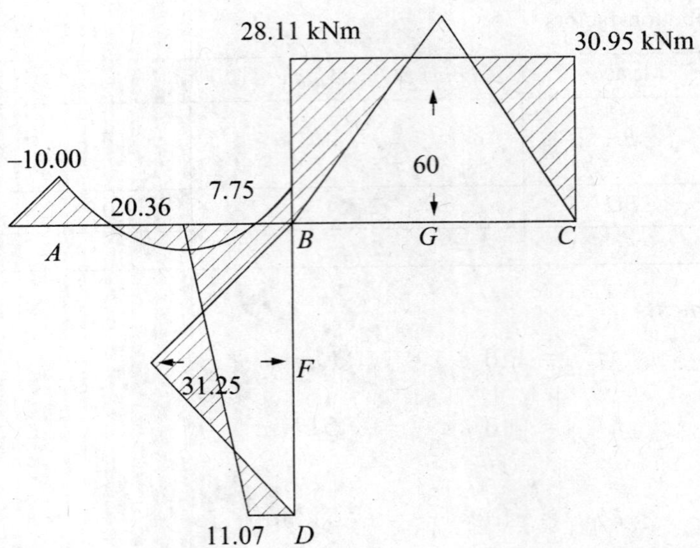

FIG. 2.70 Bending moment diagram

Example 2.14 Analyse the frame by the moment distribution method. Draw the bending moment diagram.

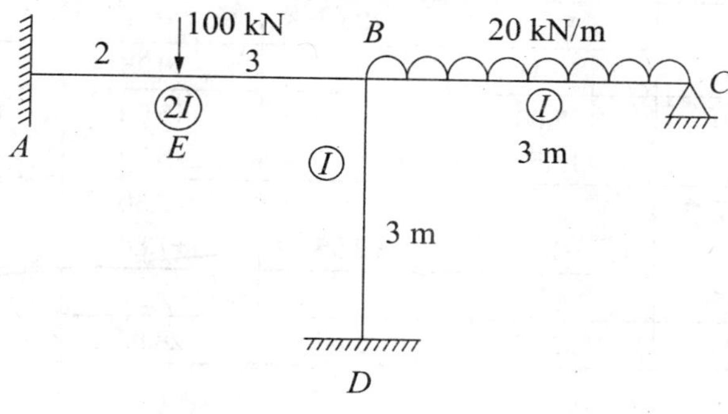

FIG. 2.71

Solution

Distribution factors

Table 2.30 Distribution factors

Joint	*Members*	*Relative Stiffness (k)*	*Σk*	*DF = k/Σk*
	BA	$2I/5 = 0.4I$		0.41
B	BD	$I/3 = 0.33I$	0.98*I*	0.34
	BC	$\frac{3}{4}(I/3) = 0.25I$		0.25

Fixed end moments

$$M_{FAB} = -2 \times 100 \times \frac{3^2}{5^2} = -72 \text{ kNm}$$

$$M_{FBA} = +3 \times 100 \times \frac{2^2}{5^2} = +48 \text{ kNm}$$

$$M_{FBC} = -20 \times \frac{3^2}{12} = -15 \text{ kNm}$$

$$M_{FCB} = +20 \times \frac{3^2}{12} = +15 \text{ kNm}$$

Table 2.31 Moment distribution table

Joint	A		B		C	D
Members	*AB*	*BA*	*BD*	*BC*	*CB*	*DB*
DF	0	0.41	0.34	0.25	1	0
FEMS	−72.00	+48.00	–	−15.00	+15.00	
Bal	–	−13.53	−11.22	−8.25	−15.00	
Co	−6.77			−7.50		−5.61
Bal		+3.08	+2.55	+1.88		
Co	+1.54					+1.28
Total	−77.23	+37.55	−8.67	−28.87	0.00	−4.33

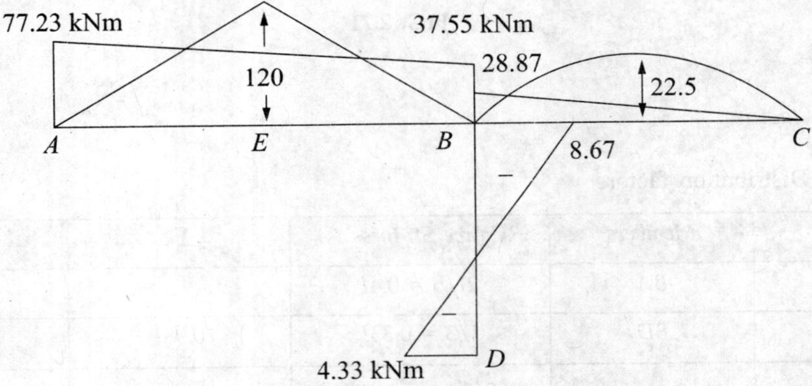

FIG. 2.72 Bending moment diagram

Example 2.15 Analyse the frame by the moment distribution method. Draw the bending moment diagram.

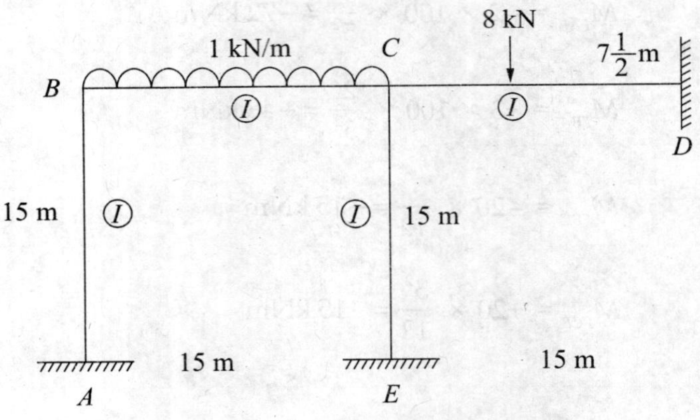

FIG. 2.73

Solution

Distribution factors

Table 2.32 Distribution factors

Joint	*Members*	*Relative Stiffness (k)*	Σk	$k/\Sigma k$
B	BA	I/15	2I/15	0.5
	BC	I/15		0.5
C	CB	I/15	3I/15	0.33
	CE	I/15		0.33
	CD	I/15		0.33

Fixed end moments

$$M_{FBC} = -1 \times 15^2/12 = -18.75 \text{ kNm}$$

$$M_{FCB} = +1 \times 15^2/12 = +18.75 \text{ kNm}$$

$$M_{FCD} = -8 \times 15/8 = -15.00 \text{ kNm}$$

$$M_{FDC} = +8 \times 15/8 = +15.00 \text{ kNm}$$

Table 2.33 Moment distribution table

Joint	A		B			C		D
Members	AB	BA	BC	CB	CE	CD	DC	EC
Distribution Factors	0	0.5	0.5	0.33	0.33	0.33	0	0
FEMS			−18.75	+18.75		−15.00	+15.00	
Balance		+9.38	+9.37	−1.25	−1.25	−1.25		
Co	+4.69		−0.63	+4.69			−0.63	−0.63
Bal		+0.32	+0.31	−1.56	−1.57	−1.56		
Co	+0.16		−0.78	+0.16			−0.78	−0.79
Bal		+0.39	+0.39	−0.05	−0.05	−0.06		
Co	+0.20		−0.03	+0.20			−0.03	−0.03
Bal		+0.02	+0.01	−0.07	−0.06	−0.06		
Final Moments	+5.05	+10.11	−10.11	+20.87	−2.93	−17.93	+13.56	−1.45

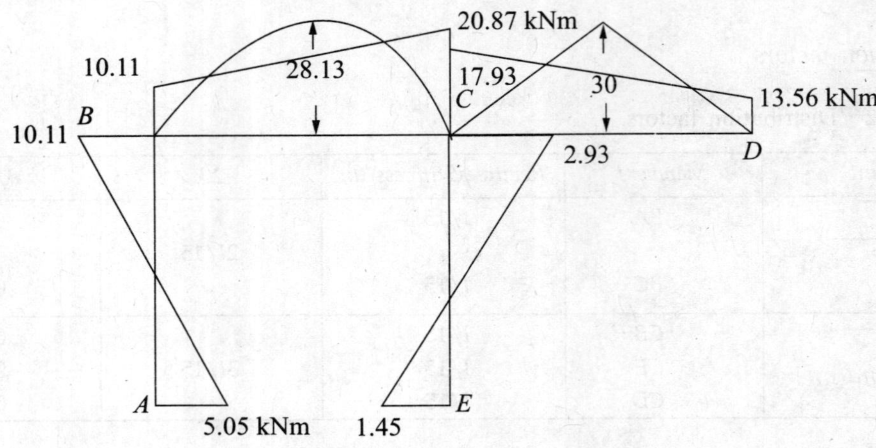

FIG. 2.74 Bending moment diagram

2.8 ANALYSIS OF SYMMETRICAL FRAMES

Example 2.16 Determine the maximum value of W of the frame shown in Fig. 2.75, if the moment at the joint B is 300 kNm. EI is constant throughout. Draw the bending moment diagram subjected to the above loading.

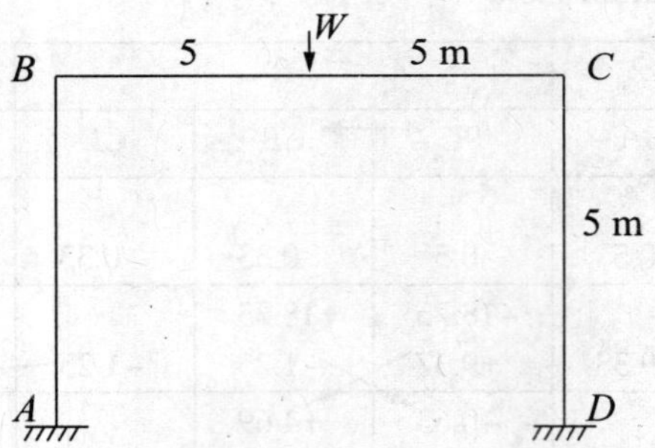

FIG. 2.75

Solution: The above frame is symmetrical and symmetrically loaded. As an effect, there is no lateral horizontal movement with respect to A and B as well as between C and D. It is also to be understood that there is no vertical movement at B and C respectively. Knowing the behaviour of symmetrical frame under symmetrical loading, half frame is analysed. In this shortcut method, the stiffness of the beam BC is taken as half of the original stiffness.

Table 2.34 Distribution factors for symmetrical case

Joint	*Members*	*Relative Stiffness*	Σk	$DF = k/\Sigma k$
	BA	$\frac{I}{5}$		0.8
B			$0.25I$	
	BC	$\frac{1}{2}\left(\frac{I}{10}\right) = 0.05I$		0.2

Fixed end moment

$$M_{FBC} = -\frac{Wl}{8} = -\frac{W(10)}{8} = -1.25W$$

Table 2.35 Moment distribution table

Joint	A	B	
Members	*AB*	*BA*	*BC*
DF	0	0.8	0.2
FEM			$-1.25W$
Bal		$+1.00W$	$+0.25W$
Co	$+0.5W$ ←		
Final	$+0.5W$	$+1.00W$	$-1.00W$

Thus, from the above moment distribution table, joint moment

$$1.00W = 300$$

$$\therefore \quad W = 300 \text{ kN}$$

Hence, $\quad M_{AB} = 0.5 \times 300 = 150$ kNm

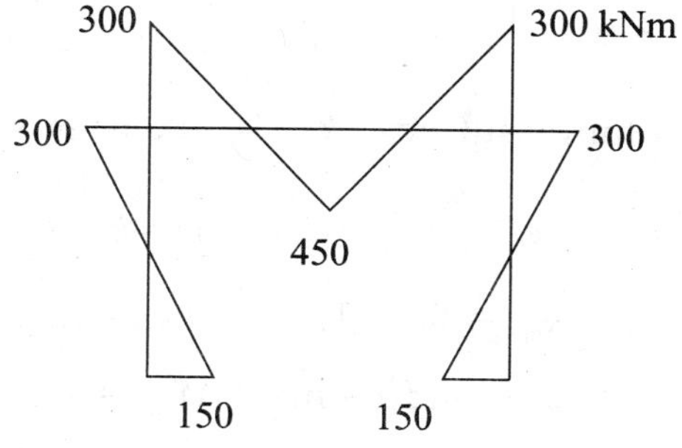

FIG 2.76 Bending moment diagram

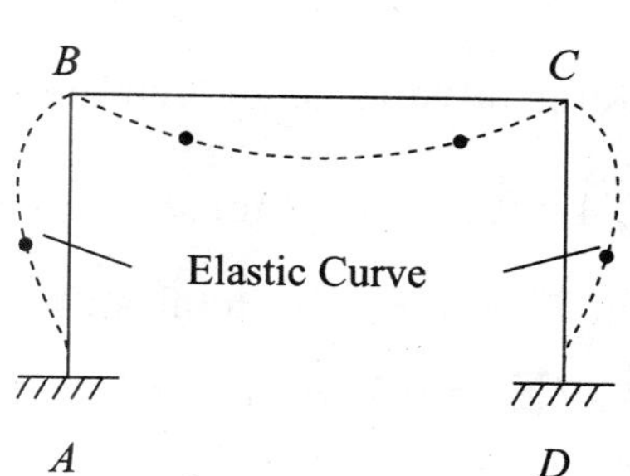

FIG 2.77

Example 2.17 Analyse the given frame by the moment distribution method. Sketch the bending moment diagram.

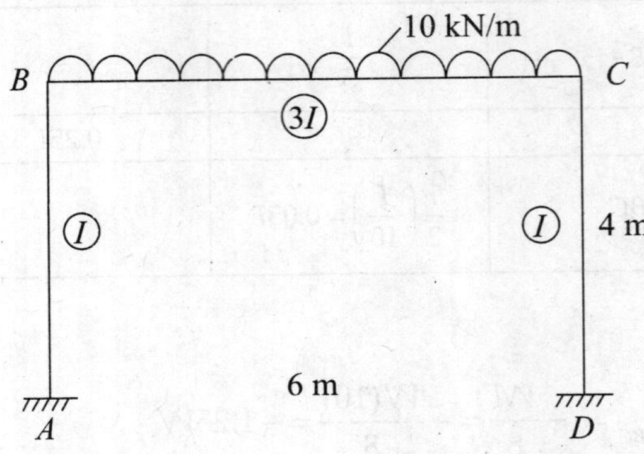

FIG. 2.78

Solution: The above frame is symmetric with respect to geometry, section, boundary conditions and loading. In other words, left half of the frame is mirror image of the right half. These frames do not sway and the joints retain their original positions. However, the joints B and C rotate under the loading by retaining in their original position.

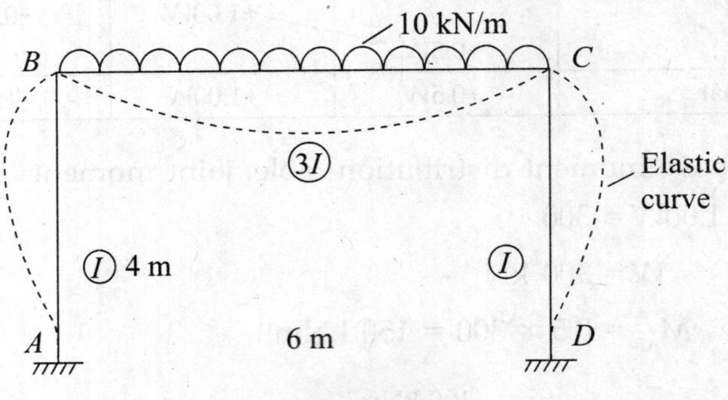

FIG. 2.79

Taking advantage of symmetry; the column stiffness k_{AB} is taken as such, i.e. $k_{AB} = I/4 = 0.25I$, the beam stiffness is taken as half of its original stiffness $k_{BC} = \left(\frac{1}{2}\right)\left(\frac{3I}{6}\right)$ $= 0.25I$. Therefore, the total stiffness $k = k_{AB} + k_{BC} = 0.5I$. The distribution factor for $d_{AB} = k_{AB}/\sum k = 0.5$ and $d_{BC} = k_{BC}/\sum k = 0.5$. The fixed end moment $M_{FBC} = -10 \times \frac{6^2}{12} =$ -30 kNm and $M_{FCB} = +10 \times \frac{6^2}{12} = +30$ kNm.

Table 2.36 Moment distribution table

Joint	*A* ←	*B*	
Members	*AB*	*BA*	*BC*
DF	0.0	0.5	0.5
FEMS Balance		 +15.00	−30.00 +15.00
Co	+7.50		
Final Moments	+7.50	+15.00	−15.00

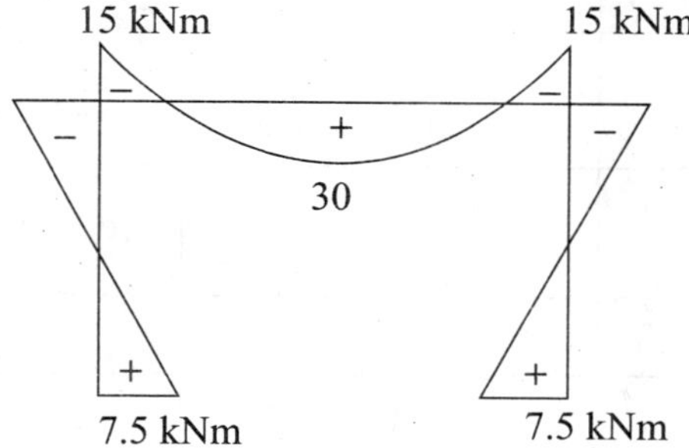

FIG. 2.80 Bending moment diagram

Example 2.18 The culvert shown in Fig. 2.81 is of constant section throughout and the top beam is subjected to a central concentrated load of 25 kN. Assume the base pressure is uniform throughout and analyse the box culvert. Draw the bending moment diagram.

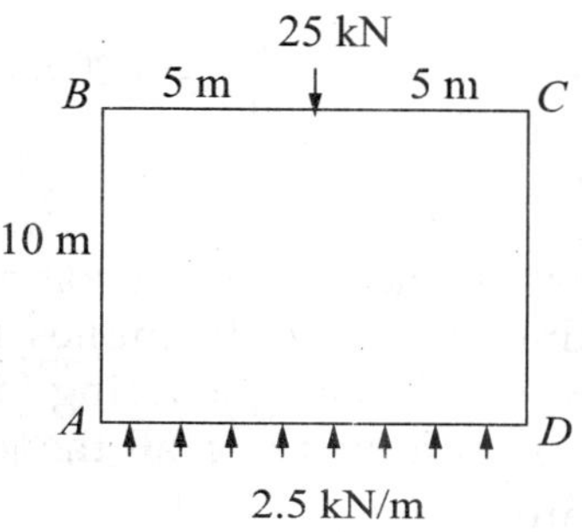

FIG. 2.81

Solution: The above frame is symmetrical and symmetrically loaded. This can be analysed using moment distribution in a straightforward and in a simpler manner by taking advantage of symmetry. The distribution factors are worked out by taking the central member BC as half of its value of stiffness. The process of moment distribution is carried out by considering only half of the frame.

Distribution factors

Table 2.37

Joint	*Members*	*Relative Stiffness Values*	Σk	$k/\Sigma k$
	BA	$\frac{I}{10}=0.1I$		0.67
B			$0.15I$	
	BC	$\frac{1}{2}\left(\frac{I}{10}\right)=0.05I$		0.33
	AB	$\frac{I}{10}=0.1I$		0.67
A			$0.15I$	
	AD	$\frac{1}{2}\left(\frac{I}{10}\right)=0.05I$		0.33

Fixed end moments

$$M_{FBC} = -\frac{Wl}{8}=-25\times\frac{10}{8}=-31.25 \text{ kNm}$$

$$M_{FAD} = +\frac{wl^2}{12}=+2.5\times\frac{10^2}{12}=+20.83 \text{ kNm}$$

Moment Distribution

Due to symmetry analyse half of the frame. The joints A, B, C and D are rigid. This frame is recognised as a continuous frame. It implies that when joint moments are balanced, it is being carried over to the neighbouring joints. This carry over moment is again balanced, the process is continued and all the joints are balanced.

Hence, the joint moments are

$$M_A = M_D = 9.06 \text{ kNm}$$

$$M_B = M_C = 16.95 \text{ kNm}$$

Table 2.38 Moment distribution table

Joint	A		B	
Members	AD	AB ⟷	BA	BC
DF	0.33	0.67	0.67	0.33
FEMS	+20.83			−31.25
Bal	−6.94	−13.89	+20.83	+10.42
Co		+10.42	−6.95	
Bal	−3.47	−6.95	+4.63	+2.32
Co		+2.32	−3.48	
Bal	−0.77	−1.55	+2.32	+1.16
Co		+1.16	−0.78	
Bal	−0.38	−0.78	+0.52	+0.26
Co		+0.26	−0.39	
Bal	−0.17	−0.08	+0.26	+0.13
Co		+0.13	−0.04	
Bal	−0.04	−0.09	+0.03	+.01
Final	9.06	−9.06	16.95	−16.95

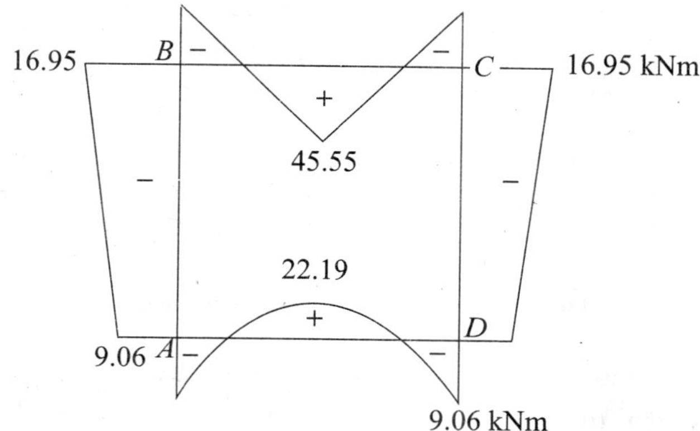

FIG. 2.82 Bending moment diagram

2.9 ANALYSIS OF UNSYMMETRICAL FRAMES

In the previous article, the frame was symmetrical and was symmetrically loaded. Half frame technique was used by taking beam stiffness as half of its original stiffness. It indicates that sways do not occur in the following cases (i) a structure which is held against lateral movement (Ref. Example 2.15) (ii) where both geometry of the frame and the loading are symmetrical (Ref. Examples 2.16 - 2.18).

In reality, the frames are unsymmetrical and the loading could also be unsymmetrical. In such frames, there will be rotation of joints as well as the lateral displacement joints. The lateral displacement causes the frame to sway either to the right or to the left with respect to the centre line of the frame. The sway (side sway) is caused by

(i) Portal frames with unequal columns (Fig. 2.83a)
(ii) Portal frames whose columns are of different moment of inertia (Fig. 2.83b)
(iii) Unsymmetrical loading on the beams (Fig. 2.83c)
(iv) Lateral loading (Fig. 2.83d)

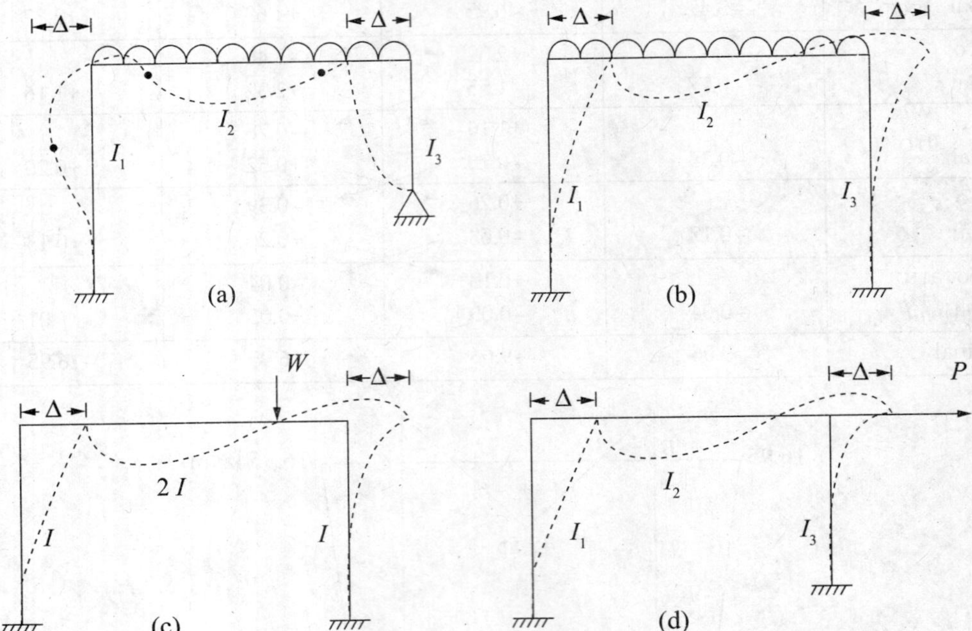

FIG. 2.83 Examples of sway frames

Analysis of sway frames is done in the following way.

(1) The moment distribution was carried out by assuming that the joints do not get displaced. The moments obtained from the moment distribution table are called nonsway moments.
(2) The horizontal reactions at the base of the columns are found out which helps in finding the net out of balance force. A prop is assumed to act opposite to the direction of the above force.
(3) Allow the frame to sway in the direction of sway force which is equal and opposite to the prop force (which acts along the axis of the beam level). Let the actual prop force be X.
(4) Perform the sway moment distribution, by assuming arbitrary moments as per the joint moment ratios.

(5) Calculate the displacement force (Y) from the sway moment distribution.

(6) The correction factor/sway factor $k = X/Y$. The correction factor gives the direction of sway of the frame. If the value of k is positive, then the frame sways in the direction of the sway solution. If the value of k is negative, then the frame sways in the direction opposite to that of assumed sway.

(7) The final end moments are obtained as

Final end moments = Nonsway moments +'Correction factor' × Sway moments

Expressing mathematically;

$$M_{AB} = M'_{AB} + k\, M''_{AB}$$

where M_{AB} is the end moment of the member AB, M'_{AB} is the moment obtained from nonsway moment distribution, M''_{AB} is the computed moment obtained from sway moment distribution k is the correction factor.

2.9.1 Joint Moment Ratios

The various moment ratios for rectangular frames are given below.

Portal frames with both ends fixed

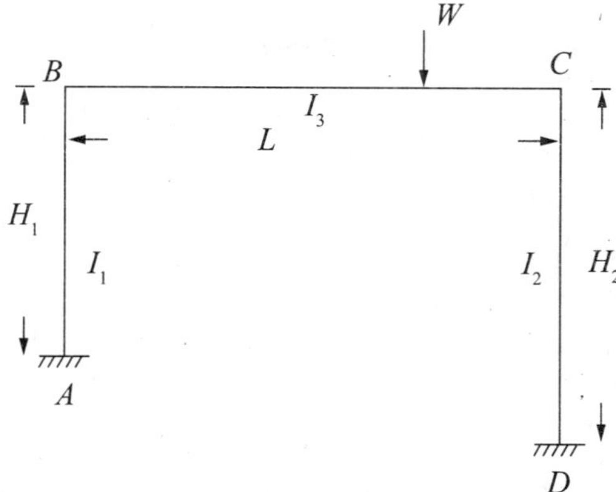

FIG. 2.84(a)

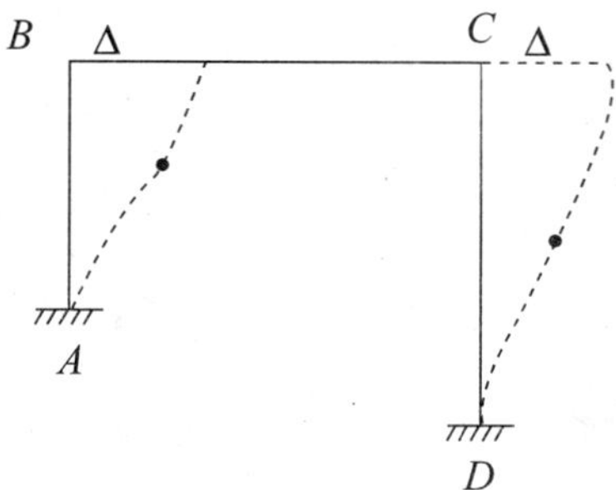

FIG. 2.84(b)

$$M_{AB} = M_{BA} = -\frac{6EI_1\Delta}{H_1^2}$$

and $$M_{CD} = M_{DC} = -\frac{6EI_2\Delta}{H_2^2}$$

$\therefore$ $$\boxed{\frac{M_{AB}}{M_{DC}} = \frac{I_1H_2^2}{I_2H_1^2}}$$

Portal frame with end fixed and other hinged

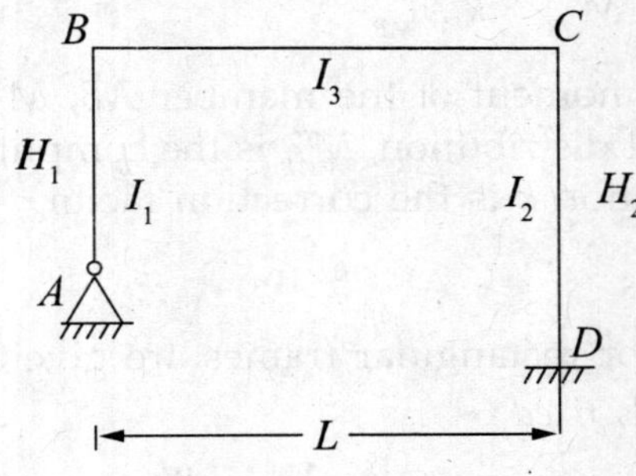

FIG. 2.85(a)

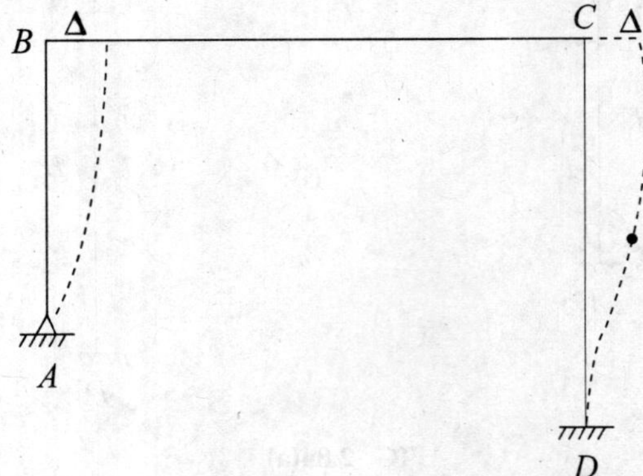

FIG. 2.85(b)

$$M_{AB} = 0 \quad (\because \text{ Hinged})$$

$$M_{BA} = -\frac{3EI_1\Delta}{H_1^2}$$

$$M_{DC} = M_{CD} = -\frac{6EI_2\Delta}{H_2^2}$$

$$\therefore \quad \boxed{\frac{M_{BA}}{M_{DC}} = \frac{I_1 H_2^2}{2 I_2 H_1^2}}$$

Portal frames with bases hinged

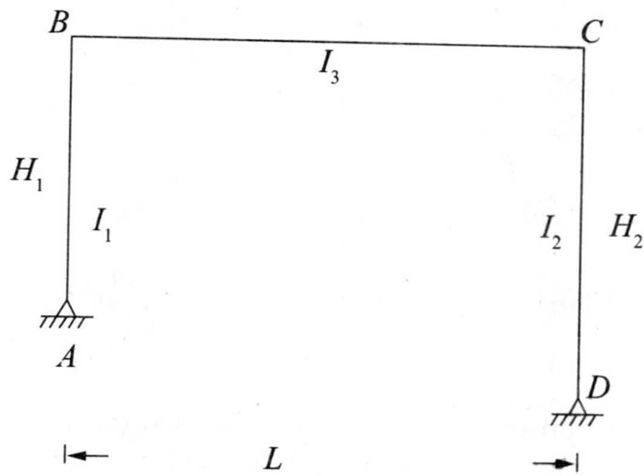

FIG. 2.86(a)

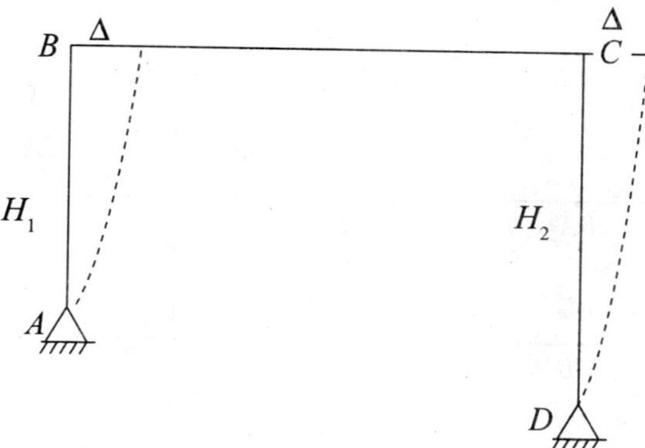

FIG. 2.86(b)

$$M_{AB} = M_{DC} = 0$$

$$\frac{M_{BA}}{M_{CD}} = \left(-\frac{3EI_1\Delta}{H_1^2}\right) \bigg/ \left(-\frac{3EI_2\Delta}{H_2^2}\right)$$

$$\therefore \quad \boxed{\frac{M_{BA}}{M_{CD}} = \frac{I_1 H_2^2}{I_2 H_1^2}}$$

Example 2.19 Analyse the frame shown in Fig. 2.87 by the moment distribution method. Draw BMD and sketch the elastic curve. Joints *B* and *C* are rigid. *A* and *D* are fixed.

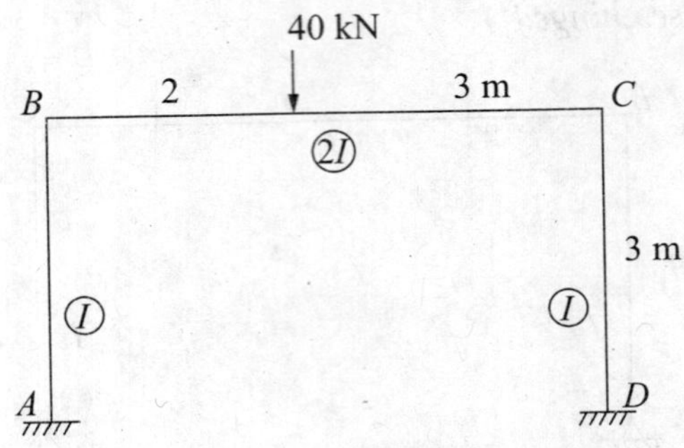

FIG. 2.87

Solution: The frame is symmetrical with respect to geometry and support this case conditions. However, the load is acting on the beam eccentrically. Hence, the frame is subjected to sway. In this case nonsway as well as sway analysis are to be done to determine the end moments of the members and at the supports *A* and *D*.

Distribution factors

Table 2.39

Joint	*Members*	*Relative Stiffness*	Σk	$k/\Sigma k$
B	BA	$I/3 = 0.33I$	$0.66I$	0.5
	BC	$2I/6 = 0.33I$		0.5
C	CB	$2I/6 = 0.33I$	$0.66I$	0.5
	CD	$I/3 = 0.33I$		0.5

Fixed end moments

$$M_{FBC} = -\frac{2 \times 40 \times 3^2}{5^2} = -28.8 \text{ kNm}$$

$$M_{FCB} = +\frac{3 \times 40 \times 2^2}{5^2} = +19.2 \text{ kNm}$$

Nonsway Analysis

Table 2.40 Moment distribution table

Joint	*A*	*B*		*C*		*D*
Members	*AB*	*BA*	*BC*	*CB*	*CD*	*DC*
DF	0 ←	0.45	0.55	0.55	0.45 →	0
FEMS			−28.80	+19.20		
Bal		+12.96	15.84	−10.56	−8.64	
Co	6.48		−5.28	+7.92		−4.32
Bal		+2.38	+2.90	−4.36	−3.56	
Co	1.19		−2.18	+1.45		−1.78
Bal		+0.98	+1.20	−0.80	−0.65	
Co	0.49		−0.40	+0.60		−0.33
Bal		+0.18	+0.22	−0.33	−0.27	
Co	0.09		−0.17	+0.11		−0.14
Bal		+0.08	+0.09	−0.06	−0.05	
Co	0.04		−0.03	+0.05		−0.03
Bal		+0.01	+0.02	−0.03	−0.02	
Total	8.29	16.59	−16.59	13.19	−13.19	−6.60
	M'_{AB}	M'_{BA}	M'_{BC}	M'_{CB}	M'_{CD}	M'_{DC}

Sway Analysis

The assumed sway moments are in the joint ratio

$$\frac{M_{BA}}{M_{CD}} = \frac{-6EI\Delta/3^2}{-6EI\Delta/3^2}$$

i.e. $$M_{BA} : M_{CD} = -1 : -1$$

The moments are arbitrarily assumed in the columns as −100 kNm.

Final moments

These are obtained by adding the nonsway and sway moments as

$$M_{AB} = 8.29 - 82.33k$$

$$M_{BA} = 16.59 - 64.73k$$

$$M_{BC} = -16.59 + 64.73k$$

$$M_{CB} = 13.19 + 64.73k$$

$$M_{CD} = -13.19 - 64.73k$$

$$M_{DC} = -6.60 - 82.33k$$

Table 2.41 Moment distribution table

Joint	*A*	*B*		*C*		*D*
Members	*AB*	*BA*	*BC*	*CB*	*CD*	*DC*
DF	0.00	0.45	0.55	0.55	0.45	0.00
FEMS	−100.00	−100.00			−100.00	−100.00
Bal		+45.00	+55.00	+55.00	+45.00	
Co	+22.50		+27.50	+27.50		+22.50
Bal		−12.38	−15.12	−15.12	−12.38	
Co	−6.19		−7.56	−7.56		−6.19
Bal		+3.40	+4.16	+4.16	+3.40	
Co	+1.70		+2.08	+2.08		+1.70
Bal		−0.94	−1.14	−1.14	−0.94	
Co	−0.47		−0.57	−0.57		−0.47
Bal		+0.26	+0.31	+0.31	+0.26	
Co	+0.13		+0.16	+0.16		+0.13
Bal		−0.07	−0.09	−0.09	−0.07	
Final	−82.33	−64.73	+64.73	64.73	−64.73	−82.33

Column shear condition

The columns are of same height and the ends are fixed. Hence,

$$M_{AB} + M_{BA} + M_{CD} + M_{DC} = 0$$

$$\therefore \qquad 5.09 - 294.12k = 0$$

$$\boxed{k = 0.017}$$

End moments

On substitution of the value of k in the above equations

$$M_{AB} = 8.29 - 82.33 \times 0.017 = 6.89 \text{ kNm}$$

$$M_{BA} = 16.59 - 64.73 \times 0.017 = 15.50 \text{ kNm}$$

$$M_{BC} = -16.59 + 64.73 \times 0.017 = -15.50 \text{ kNm}$$

$$M_{CB} = 13.19 + 64.73 \times .017 = +14.29 \text{ kNm}$$

$$M_{CD} = -13.19 - 64.73 \times .017 = -14.29 \text{ kNm}$$

$$M_{DC} = -6.60 - 82.33 \times 0.017 = -8.00 \text{ kNm}$$

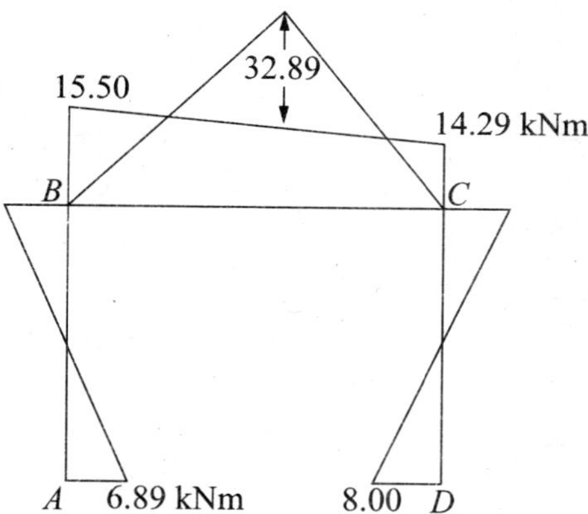

FIG. 2.88 Bending moment diagram

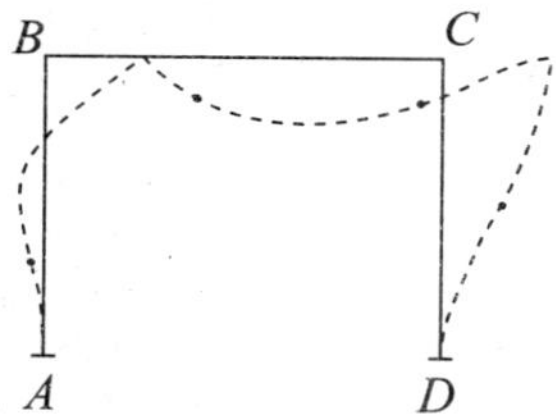

FIG. 2.89 Elastic curve

Example 2.20 Analyse the given frame and draw the bending moment diagram

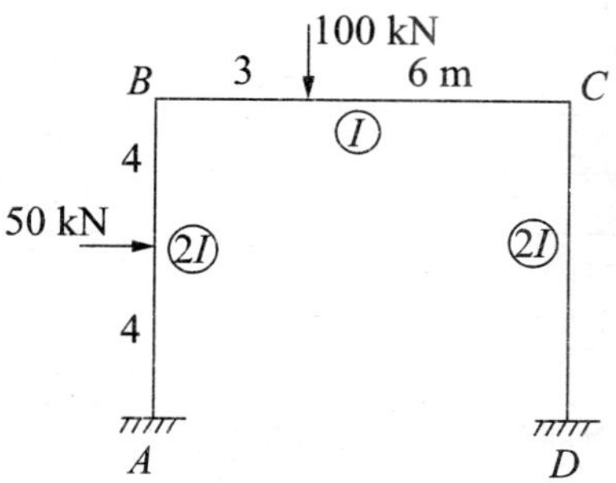

FIG. 2.90

Solution

Distribution factors

Table 2.42

Joint	*Members*	*Relative Stiff Values (I/l)*	Σk	$k/\Sigma k$
	BA	$2I/8 = 0.25I$		0.7
B			$0.36I$	
	BC	$I/9 = 0.11I$		0.3
	CB	$I/9 = 0.11I$		0.3
C			$0.36I$	
	CD	$2I/8 = 0.25I$		0.7

Fixed end moments

$$M_{FAB} = -\frac{Wl}{8} = -50 \times \frac{8}{8} = -50\ \text{kNm}$$

$$M_{FBA} = +\frac{Wl}{8} = +50 \times \frac{8}{8} = +50\ \text{kNm}$$

$$M_{FBC} = -\frac{Wab^2}{l^2} = -100 \times 3 \times \frac{6^2}{9^2} = -133.33\ \text{kNm}$$

$$M_{FCB} = +\frac{Wa^2b}{l^2} = +100 \times 3^2 \times \frac{6}{9^2} = +66.67\ \text{kNm}$$

Nonsway Analysis

Table 2.43 Moment distribution table

Joint	*A*	*B*		*C*		*D*
Members	*AB* ←	*BA*	*BC* ↔	*CB*	*CD* →	*DC*
DF	0	0.70	0.30	0.30	0.70	0
FEMS	−50.00	+50.00	−133.33	+66.67		
Bal		+58.33	+25.00	−20.00	−46.67	
Co	+29.20		−10.00	+12.50		−23.34
Bal		+7.00	+3.00	−3.75	−8.75	
Co	+3.50		−1.88	+1.50		−4.38
Bal		+1.32	+0.56	−0.45	−1.05	
Co	+0.66		−0.23	+0.28		−0.53
Bal		+0.16	+0.07	−0.08	−0.20	
Co	+0.08		−0.04	+0.04		−0.10
Bal		+0.03	+0.01	−0.01	−0.03	
	−16.56	+116.84	−116.84	56.70	−56.70	−28.35
	M'_{AB}	M'_{BA}	M'_{BC}	M'_{CB}	M'_{CD}	M'_{DC}

Sway Analysis

Table 2.44 Moment distribution table

Joint	A	B		C		D
Members	AB ←	BA	BC ↔	CB	CD →	DC
DF	0	0.70	0.30	0.30	0.70	0
FEMS	+100.00	+100.00			+100.00	+100.00
Bal		−70.00	−30.00	−30.00	−70.00	
CO	−35.00		−15.00	−15.00		−35.00
Bal		+10.50	+4.50	+4.50	+10.50	
Co	+5.25		+2.25	+2.25		+5.25
Bal		−1.57	−0.68	−0.68	−1.57	
CO	−0.79		−0.34	−0.34		−0.79
Bal		+0.24	+0.10	+0.10	+0.24	
CO	+0.12		+0.05	+0.05		+0.12
Bal		−0.04	−0.01	−0.02	−0.03	
Final	69.58	39.13	−39.13	−39.13	+39.13	+69.58
	M''_{AB}	M''_{BA}	M''_{BC}	M''_{CB}	M''_{CD}	M''_{DC}

The final moments are the addition of nonsway moments and a constant times sway moments; for example, $M_{AB} = M'_{AB} + k\, M''_{AB}$. Therefore,

End moments

$$M_{AB} = -16.56 + 69.58\, k$$

$$M_{BA} = +116.84 + 39.13\, k$$

$$M_{BC} = -116.84 - 39.13\, k$$

$$M_{CB} = +56.70 - 39.13\, k$$

$$M_{CD} = -56.70 + 39.13\, k$$

$$M_{DC} = -28.35 + 69.58\, k$$

The value of k is determined from the column shear condition as follows.

Column shear condition

Considering the free body diagram of column AB and the value of H_A is determined as

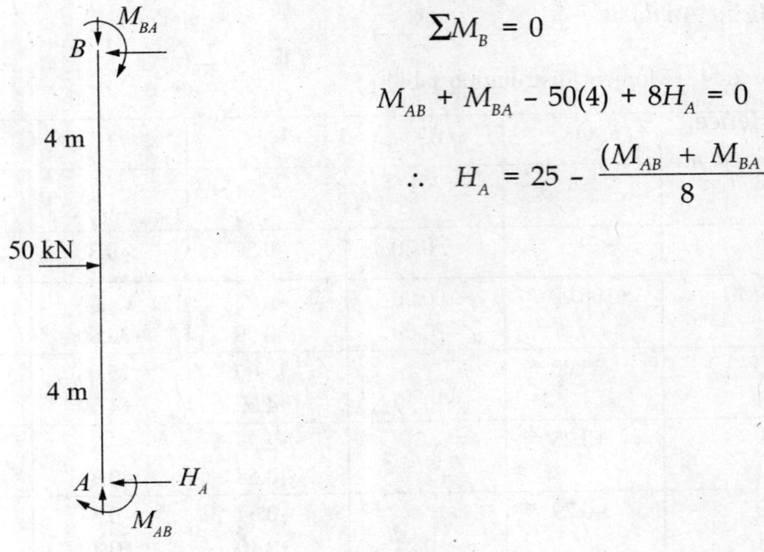

$$\Sigma M_B = 0$$

$$M_{AB} + M_{BA} - 50(4) + 8H_A = 0$$

$$\therefore \quad H_A = 25 - \frac{(M_{AB} + M_{BA})}{8}$$

FIG. 2.91

Substituting the values of M_{AB} and M_{BA} from the above equations

$$H_A = 25 - \frac{1}{8}(100.28 + 108.71\,k)$$

$$\boxed{H_A = 12.465 - 13.59\,k}$$

The horizontal reaction H_D is determined by taking moment C as

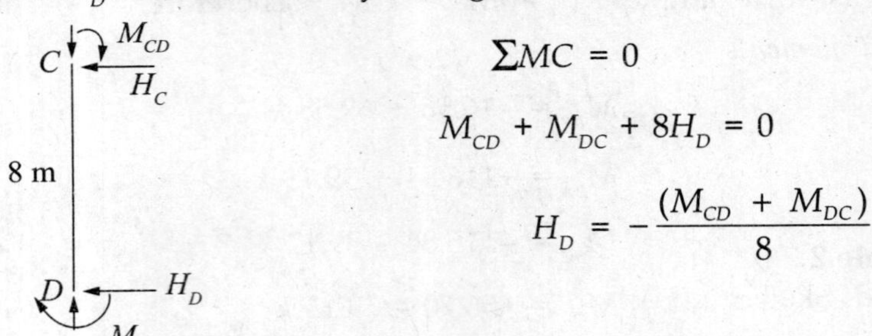

$$\Sigma MC = 0$$

$$M_{CD} + M_{DC} + 8H_D = 0$$

$$H_D = -\frac{(M_{CD} + M_{DC})}{8}$$

FIG. 2.92

The values of M_{CD} and M_{DC} are substituted in the above equation as

$$H_D = \frac{(85.05 - 108.71\,k)}{8}$$

$$\therefore \quad \boxed{H_D = 10.631 - 13.59\,k}$$

Considering the horizontal equilibrium

$$H_A + H_D = 50$$

$$23.096 - 27.18\,k = 50$$

$$\boxed{k = -0.99}$$

Hence,

Final moments

$$M_{AB} = -16.56 + 69.58(-0.99) = -85.4 \text{ kNm}$$

$$M_{BA} = +116.84 + 39.13(-0.99) = +78.1 \text{ kNm}$$

$$M_{BC} = -116.84 - 39.13(-0.99) = -78.1 \text{ kNm}$$

$$M_{CB} = +56.70 - 39.13(-0.99) = +95.4 \text{ kNm}$$

$$M_{CD} = -56.70 + 39.13(-0.99) = -95.4 \text{ kNm}$$

$$M_{DC} = -28.35 + 69.58(-0.99) = -97.2 \text{ kNm}$$

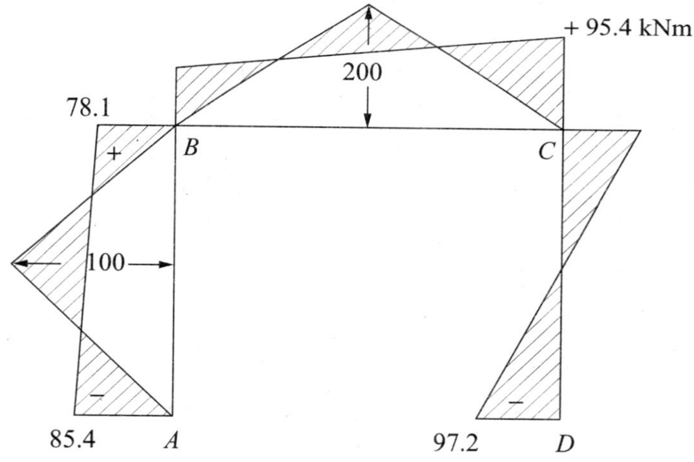

FIG. 2.93 Bending moment diagram

Example 2.21 Analyse the portal frame shown in Fig. 2.94 by the moment distribution method. Sketch the bending moment diagram.

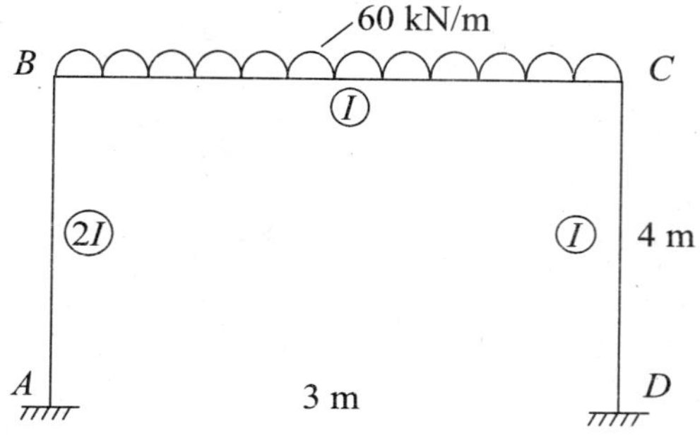

FIG. 2.94

Solution

Distribution factors

Table 2.45

Joint	*Members*	*Relative Stiffness Values*	Σk	$k/\Sigma k$
B	BA	$2I/4$	$0.83I$	0.60
	BC	$I/3$		0.40
C	CB	$I/3$	$0.58I$	0.57
	CD	$I/4$		0.33

Fixed end moments

$$M_{FBC} = -60 \times \frac{3^2}{12} = -45 \text{ kNm}$$

$$M_{FCB} = +60 \times \frac{3^2}{12} = +45 \text{ kNm}$$

Though the boundary conditions and the loading are symmetrical, the frame sways as the columns *AB* and *CD* are having different moments of inertia of the cross-section.

Hence, we have to do nonsway and sway analysis to determine the joint moments.

Nonsway Analysis

Table 2.46 Moment distribution table

Joint	A	B		C		D
Members	AB ←	BA	BC ↔	CB	CD →	DC
DF	0	0.60	0.40	0.57	0.33	0
FEMS			−45.00	+45.00		
Bal		+27.00	+18.00	−25.65	−19.35	
Co	+13.50		−12.83	+9.00		−9.68
Bal		+7.70	+5.13	−5.13	−3.87	
Co	+3.85		−2.56	+2.56		−1.94
Bal		+1.54	+1.02	−1.43	−1.13	
Co	+0.77		−0.72	+0.51		−0.57
Bal		+0.43	+0.29	−0.29	−0.22	
Co	+0.22		−0.15	+0.15		−0.11
Bal		+0.09	+0.06	−0.08	−0.06	
Co	+0.05		−0.04	+0.03		−0.03
Bal		+0.02	+0.02	−0.02	−0.01	
Final	+18.39	+36.78	−36.78	+24.65	−24.65	−12.33
	M'_{AB}	M'_{BA}	M'_{BC}	M'_{CB}	M'_{CD}	M'_{DC}

Sway Analysis

The sway moments are computed using the ratios

$$\frac{M_{BA}}{M_{CD}} = \frac{-6E(2I)\Delta/4^2}{-6EI\Delta/4^2}$$

$$\therefore \qquad M_{BA} : M_{CD} = -2 : -1.0$$

Hence, proportionate an arbitrary moments were assumed –20.00 for M_{BA} and –10.00 for M_{CD}.

Table 2.47 Moment distribution table

Joint	A	B		C		D
Members	*AB* ←	*BA*	*BC* ↔	*CB*	*CD* →	*DC*
DF	0	0.60	0.40	0.57	0.43	0
FEMS	–20.00	–20.00	–	–	–10.00	–10.00
Bal		+12.00	+8.00	+5.700	+4.30	
Co	+6.00		+2.85	+4.00		+2.15
Bal		–1.71	–1.14	–2.28	–1.72	
Co	–0.86		–1.14	–0.57		–0.86
Bal		+0.68	+0.46	+0.32	+0.25	
Co	+0.34		+0.16	+0.23		+0.13
Bal		–0.10	–0.06	–0.13	–0.10	
Co	–0.05		–0.07	–0.03		–0.05
Bal		+0.04	+0.03	+0.02	+0.01	
Final	–14.57	–9.09	+9.09	7.26	–7.26	–8.63
	M''_{AB}	M''_{BA}	M''_{BC}	M''_{CB}	M''_{CD}	M''_{DC}

Corrected final moments

$$M_{AB} = 18.39 - 14.57\,k$$

$$M_{BA} = 36.78 - 9.09\,k$$

$$M_{BC} = -36.78 + 9.09\,k$$

$$M_{CB} = 24.65 + 7.26\,k$$

$$M_{CD} = -24.65 - 7.26\,k$$

$$M_{DC} = -12.33 - 8.63\,k$$

Column shear condition

In the given portal frame, the columns are fixed and there is no horizontal load acting externally. Hence, the horizontal equilibrium gives

$$M_{AB} + M_{BA} + M_{CD} + M_{DC} = 0$$

$$(18.39 + 36.78 - 24.65 - 12.33) - (14.57 + 9.09 + 7.26 + 8.63)\,k = 0$$

$$18.19 - 39.55\,k = 0$$

$$\boxed{k = 0.46}$$

Substituting this value of k in the above equations;

End moments

$$M_{AB} = 18.39 - 14.57(0.46) = 11.7 \text{ kNm}$$

$$M_{BA} = 36.78 - 9.09(0.46) = 32.6 \text{ kNm}$$

$$M_{BC} = -36.78 + 9.09(0.46) = -32.6 \text{ kNm}$$

$$M_{CB} = 24.65 + 7.26(0.46) = +28.0 \text{ kNm}$$

$$M_{CD} = -24.65 - 7.26(0.46) = -28.0 \text{ kNm}$$

$$M_{DC} = -12.33 - 8.63(0.46) = +16.3 \text{ kNm}$$

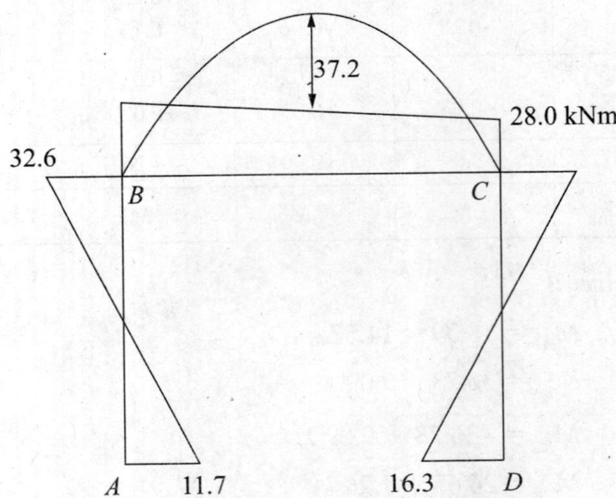

FIG. 2.95 Bending moment diagram

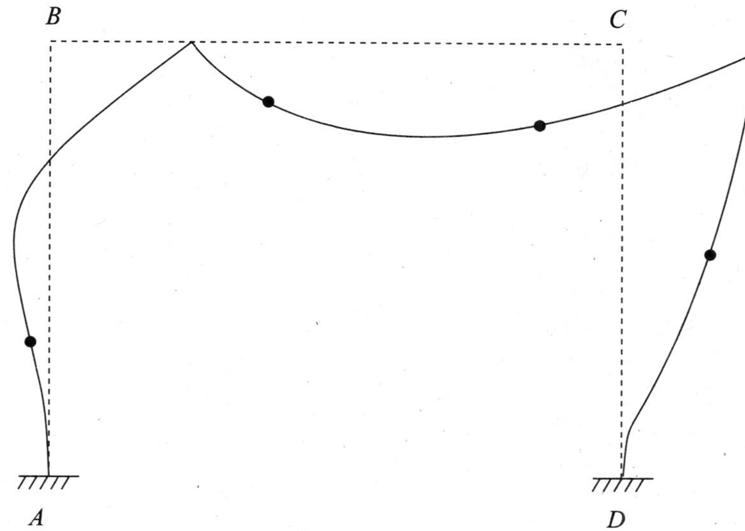

FIG. 2.96 Elastic curve

Example 2.22 Analyse the portal frame shown in Fig. 2.97 by the moment distribution method. Draw the bending moment diagram.

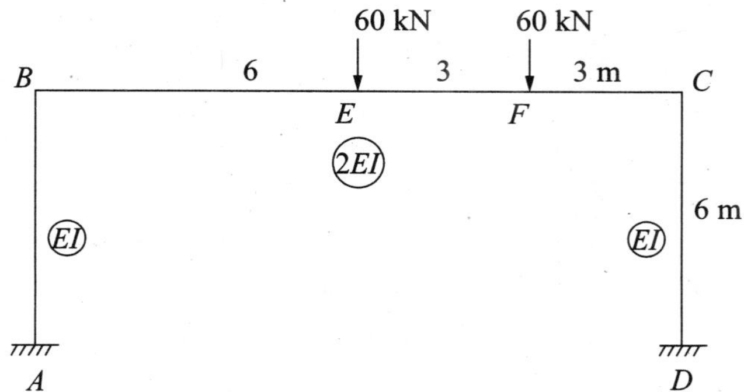

FIG. 2.97

Solution: The frame is subjected to asymmetrical loading. Hence, nonsway and sway analyses were carried out separately and the results are summed up.

Distribution factors

Table 2.48

Joint	*Members*	*Relative Stiff Values (k)*	Σk	*DF*
B	BA	$I/6$	$I/3$	0.5
	BC	$2I/12 = I/6$		0.5
C	CB	$2I/12 = I/6$	$I/3$	0.5
	CD	$I/6$		0.5

Fixed end moments

$$M_{FBC} = -\frac{Wl}{8} - \frac{Wab^2}{l^2} = -60 \times \frac{12}{8} - \frac{60 \times 9 \times 3^2}{12^2} = -123.75 \text{ kNm}$$

$$M_{FCB} = +\frac{Wl}{8} + \frac{Wa^2b}{l^2} = 60 \times \frac{12}{8} + \frac{60 \times 3 \times 9^2}{12^2} = -191.25 \text{ kNm}$$

Nonsway Analysis

Table 2.49 Moment distribution table

Joint	*A*	*B*		*C*		*D*
Members	*AB* ←	*BA*	*BC* ↔	*CB*	*CD* →	*DC*
DF	0.0	0.5	0.5	0.5	0.5	0.0
FEMS			−123.75	+191.25		
Bal		+61.88	+61.87	−95.63	−95.62	
Co	+30.94		−47.82	+30.94		−47.81
Bal		+23.91	+23.91	−15.47	−15.47	
Co	+11.96		−7.74	+11.96		−7.74
Bal		+3.87	+3.87	−5.98	−5.98	
Co	+1.94		2.99	+1.94		−2.99
Bal		+1.50	+1.50	−0.97	−0.97	
Co	+0.75		−0.49	+0.75		−0.49
Bal		+0.24	+0.25	−0.37	−0.38	
Co	+0.12		−0.19	+0.13		−0.19
Bal		+0.10	+0.09	−0.06	−0.07	
Final	45.71	91.50	−91.50	118.49	−118.49	−59.22
	M'_{AB}	M'_{BA}	M'_{BC}	M'_{CB}	M'_{CD}	M'_{DC}

Sway moments

The sway moments are assumed in the joint ratio of

$$\frac{M_{BA}}{M_{DC}} = \frac{-6EI\Delta/6^2}{-6EI\Delta/6^2} = \frac{-1}{-1}$$

$$\therefore \quad M_{BA} : M_{DC} = -100.00 : -100.00$$

Sway Analysis

Table 2.50 Moment distribution table

Joint	A	B		C		D
Members	AB ←	BA	BC ↔	CB	CD →	DC
DF	0.0	0.5	0.5	0.5	0.5	0.0
FEMS	−100.00	−100.00			−100.00	−100.00
Bal		+50.00	+50.00	+50.00	+50.00	
Co	+25.00		+25.00	+25.00		+25.00
Bal		−12.50	−12.50	−12.50	−12.50	
Co	−6.25		−6.25	−6.25		−6.25
Bal		+3.12	+3.13	+3.13	+3.12	
Co	+1.56		+1.57	+1.57		+1.56
Bal		−0.78	−0.79	−0.79	−0.78	
Co	−0.39		−0.40	−0.40		−0.39
Bal		+0.20	+0.20	+0.20	+0.20	
Co	+0.10		+0.10	+0.10		+0.10
Bal		−0.05	−0.05	−0.05	−0.05	
Final	−79.98	−60.03	+60.01	+60.01	−60.01	−79.98
	M''_{AB}	M''_{BA}	M''_{BC}	M''_{CB}	M''_{CD}	M''_{DC}

Final moments

$$M_{AB} = 45.71 - 79.98\ k$$

$$M_{BA} = 91.50 - 60.01\ k$$

$$M_{BC} = -91.50 + 60.01\ k$$

$$M_{CB} = 118.49 + 60.01\ k$$

$$M_{CD} = -118.49 - 60.01\ k$$

$$M_{DC} = -59.22 - 79.98\ k$$

Column shear condition

As the columns are of same height, the column shear condition gives

$$M_{AB} + M_{BA} + M_{CD} + M_{DC} = 0$$

$$(45.71 + 91.50 - 118.49 - 59.22) - (79.98 + 60.01 + 60.01 + 79.98)\ k = 0$$

$$-40.5 - 279.98\ k = 0$$

$$\boxed{k = -0.145}$$

Back substitution

$$M_{AB} = 45.71 - 79.98(-0.145) = 57.31 \text{ kNm}$$

$$M_{BA} = 91.50 - 60.01(-0.145) = 100.2 \text{ kNm}$$

$$M_{BC} = -91.50 + 60.01(-0.145) = -100.2 \text{ kNm}$$

$$M_{CB} = 118.49 + 60.01(-0.145) = 109.8 \text{ kNm}$$

$$M_{CD} = -118.49 - 60.01(-0.145) = -109.8 \text{ kNm}$$

$$M_{DC} = -59.22 - 79.98(-0.145) = -47.6 \text{ kNm}$$

Free body diagram of span BC

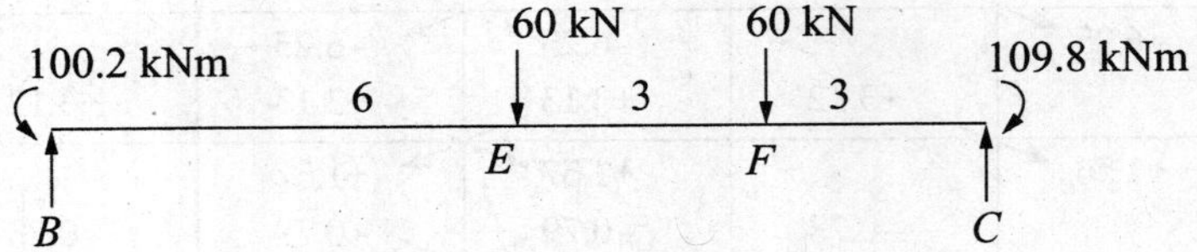

FIG. 2.98

$$\Sigma V = 0; \quad V_B + V_C = 120$$

$$\Sigma M_B = 0; \quad -100.2 + 109.8 + 60(6) + 60(9) - 12V_C = 0$$

$$V_C = 75.8 \text{ kN}$$

$$V_B = 44.2 \text{ kN}$$

$$M_E = 44.2(6) - 100.2 = 165 \text{ kNm}$$

$$M_F = 75.8(3) - 109.8 = 117.6 \text{ kNm}$$

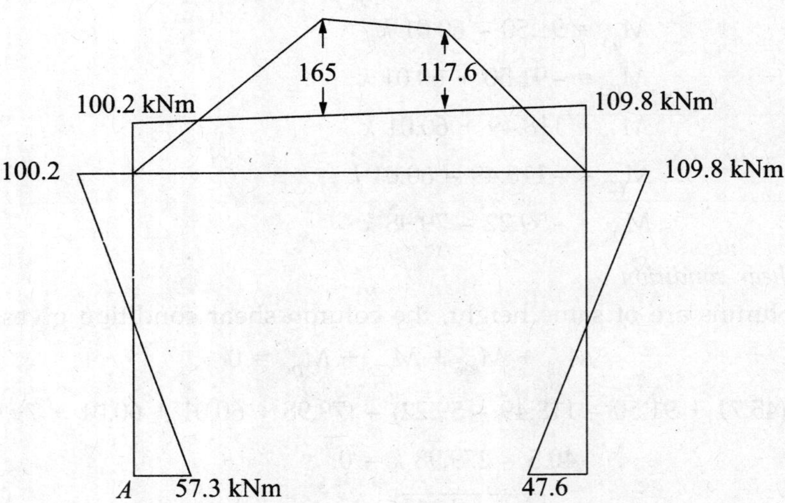

FIG. 2.99 Bending moment diagram

Example 2.23 Analyse the given frame by the moment distribution method. Draw the bending moment diagram and shear force diagram.

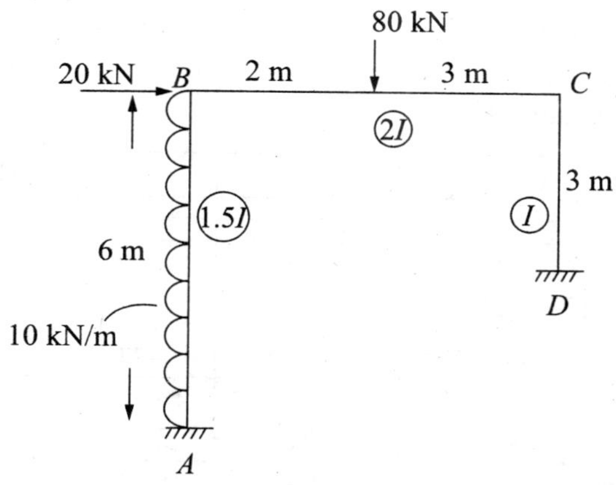

FIG. 2.100

Distribution factors

Table 2.51

Joint	*Members*	*Relative Stiffness*	Σk	$k/\Sigma k$
	BA	$\frac{1.5I}{6} = 0.25I$		0.38
B			$0.65I$	
	BC	$\frac{2I}{5} = 0.4I$		0.62
	CB	$\frac{2I}{5} = 0.4I$		0.55
C			$0.73I$	
	CD	$\frac{I}{3} = 0.33I$		0.45

Fixed end moments

$$M_{FAB} = -10 \times \frac{6^2}{12} = -30 \text{ kNm}$$

$$M_{FBA} = +10 \times \frac{6^2}{12} = +30 \text{ kNm}$$

$$M_{FBC} = -\frac{2(90)3^2}{5^2} = -64.8 \text{ kNm}$$

$$M_{FCB} = +3(90)\ \frac{3^2}{5^2} = +43.2 \text{ kNm}$$

Nonsway Analysis

Table 2.52 Moment distribution table

Joint	*A*	*B*		*C*		*D*
Members	*AB* ←	*BA*	*BC* ↔	*CB*	*CD* →	*DC*
DF	0	0.38	0.62	0.55	0.45	0
FEMS	−30.00	+30.00	−64.80	+43.20	–	–
Bal		+13.22	+21.58	−23.76	−19.44	
Co	+6.61		−11.88	+10.79		−9.72
Bal		+4.51	+7.37	−5.93	−4.86	
Co	+2.26		−2.97	3.69		−2.43
Bal		+1.13	+1.84	−2.03	−1.66	
Co	+0.57		−1.02	+0.92		−0.83
Bal		+0.39	+0.63	−0.51	−0.41	
Co	+0.20		−0.26	0.34		−0.21
Bal		+0.10	+0.16	−0.19	−0.15	
Co	+0.05		−0.10	+0.08		−0.08
Bal		+0.04	+0.06	−0.04	−0.04	
Final	−20.31	+49.39	−49.39	+26.56	−26.56	−12.44
	M'_{AB}	M'_{BA}	M'_{BC}	M'_{CB}	M'_{CD}	M'_{DC}

Sway Analysis

$$M_{BA} : M_{CD} = -6E(1.5I)\ \frac{\Delta}{6^2} : -6E(I)\frac{\Delta}{3^2}$$

$$M_{BA} : M_{CD} = -15.00 : -40.00 \text{ kNm}$$

Correction factor

$$M_{AB} = -20.31 - 14.15\ k$$

$$M_{BA} = +49.39 - 13.34\ k$$

$$M_{BC} = -49.39 + 13.34\ k$$

$$M_{CB} = +26.56 + 22.63\ k$$

$$M_{CD} = -26.56 - 22.63\ k$$

$$M_{DC} = -12.44 - 31.30\ k$$

Table 2.53 Moment distribution table

Joint	A	B		C		D
Members	*AB* ←	*BA*	*BC* ↔	*CB*	*CD* →	*DC*
DF	0	0.38	0.62	0.55	0.45	0
FEMS	−15.00	−15.00	−	−	−40.00	−40.00
Bal		+5.70	+9.30	+22.00	+18.00	
Co	+2.85		+11.00	+4.65		+9.00
Bal		−4.18	−6.82	−2.56	−2.09	
Co	−2.09		−1.28	−3.41		−1.05
Bal		+0.49	+0.79	+1.88	+1.53	
Co	+0.25		+0.94	+0.40		+0.77
Bal		−0.36	−0.58	−0.22	−0.18	
Co	−0.18		−0.11	−0.29		−0.09
Bal		+0.04	+0.07	+0.16	+0.13	
Co	+0.02		+0.08	+0.04		+0.07
Bal		−0.03	−0.05	−0.02	−0.02	
Final	−14.15	−13.34	+13.34	+22.63	−22.63	−31.30
	M''_{AB}	M''_{BA}	M''_{BC}	M''_{CB}	M''_{CD}	M''_{DC}

The value of k is determined from the column shear condition as

Column shear condition

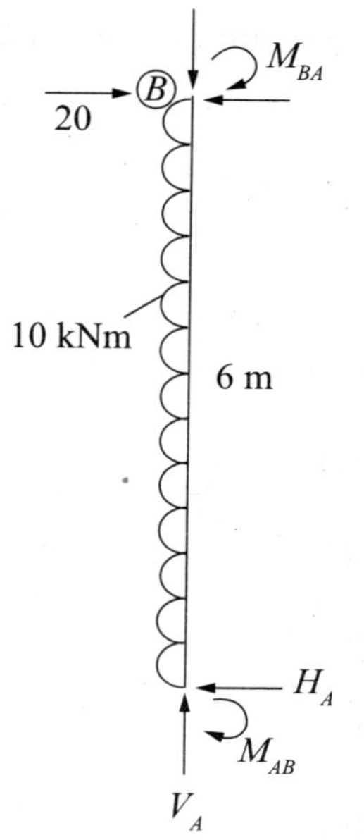

FIG. 2.101

$$\Sigma M_B = 0$$

$$M_{AB} + M_{BA} + 6H_A = 10 \times \frac{6^2}{2}$$

and hence

$$\boxed{H_A = 30 - \frac{(M_{AB} + M_{BA})}{6}}$$

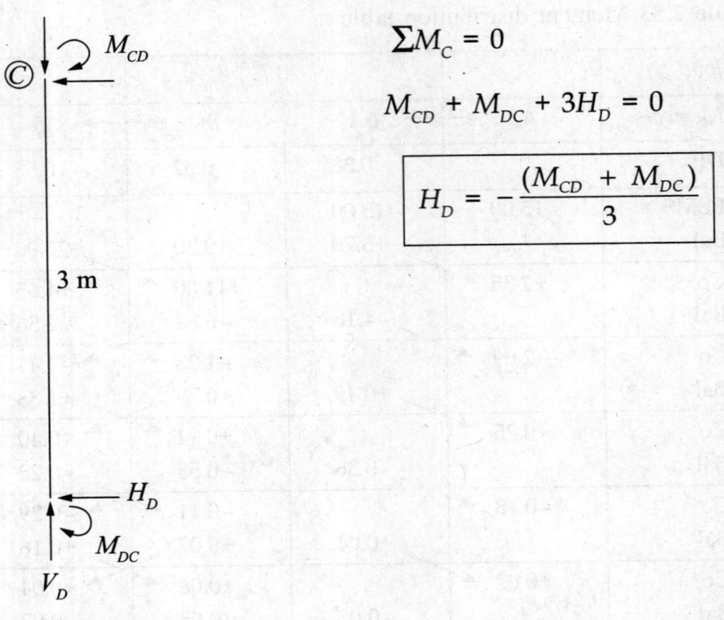

FIG. 2.102

$$\Sigma M_C = 0$$

$$M_{CD} + M_{DC} + 3H_D = 0$$

$$H_D = -\frac{(M_{CD} + M_{DC})}{3}$$

$$H_A + H_D = 6(10) + 20$$

$$30 - \frac{(M_{AB} + M_{BA})}{6} - \frac{(M_{CD} + M_{DC})}{3} = 80$$

Substituting the values of M_{AB}, M_{BA}, M_{CD} and M_{DC} from the above equations; and solving

$$k = 1.843$$

Final moments

$$M_{AB} = -20.31 - 14.15(1.843) = -46.4$$

$$M_{BA} = +49.39 - 13.34(1.843) = +24.8$$

$$M_{BC} = -49.39 + 13.34(1.843) = -24.8$$

$$M_{CB} = +26.56 + 22.63(1.843) = +68.3$$

$$M_{CD} = -26.56 - 22.63(1.843) = -68.3$$

$$M_{DC} = -12.44 - 31.30(1.843) = -70.1$$

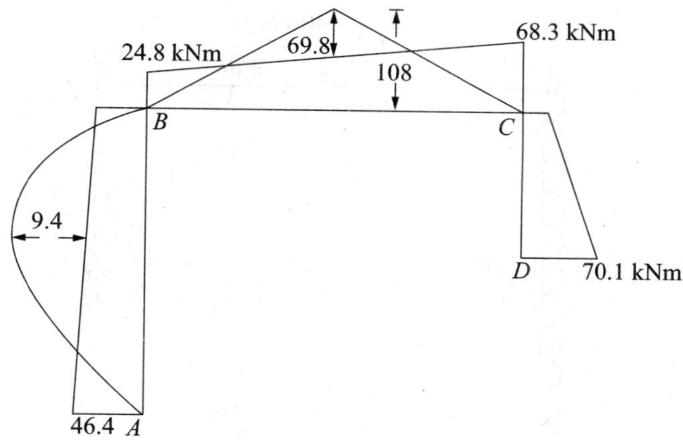

FIG. 2.103 Bending moment diagram

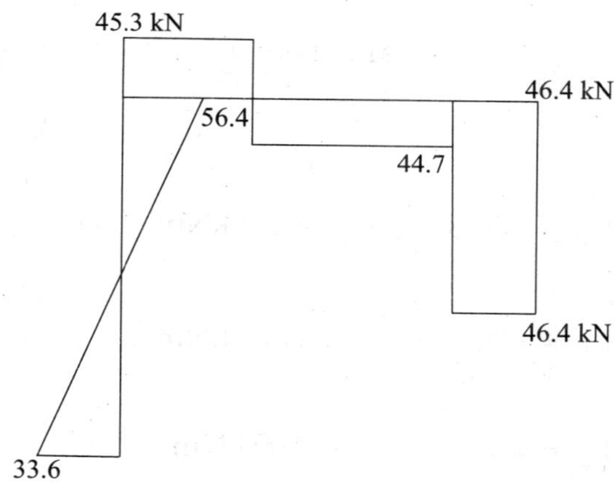

FIG. 2.104 Shear force diagram

Example 2.24 Analyse the rigid portal frame by moment distribution method and hence draw the bending moment diagram (Fig. 2.105).

Solution

Distribution factors

Table 2.54

Joint	*Members*	*Relative Stiffness (I/l)*	Σk	$k/\Sigma k$
B	BA	$2I/5$	$0.9I$	0.44
	BC	$2I/4$		0.56
C	CB	$2I/4$	$0.875I$	0.57
	CD	$\frac{3}{4} \times \frac{1.5I}{3}$		0.43

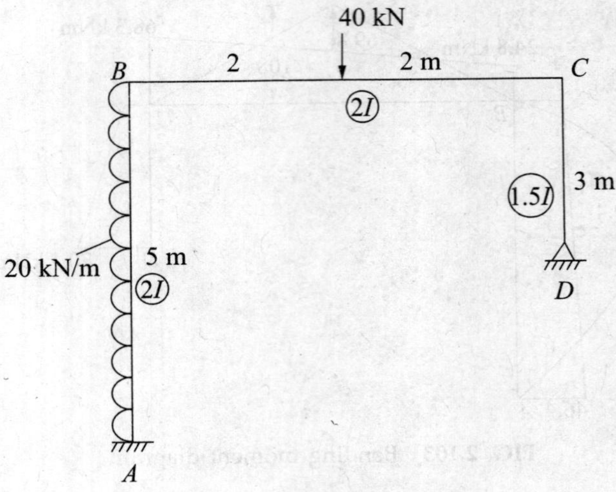

FIG. 2.105

Fixed end moments

$$M_{FAB} = -20 \times \frac{5^2}{12} = -41.67 \text{ kNm}$$

$$M_{FBA} = +20 \times \frac{5^2}{12} = +41.67 \text{ kNm}$$

$$M_{FBC} = -40 \times \frac{4}{8} = -20.00 \text{ kNm}$$

$$M_{FCB} = +40 \times \frac{4}{8} = +20.00 \text{ kNm}$$

Sway Moments

The sway moments are in the following joint ratio

$$\frac{M_{BA}}{M_{CD}} = \frac{-6E(2I)\Delta/5^2}{-3E(1.5I)\Delta/3^2} = \frac{-108}{-112.5}$$

$$M_{BA} : M_{CD} = -108 : -112.5 \text{ kNm}$$

Table 2.55 Moment distribution table (nonsway analysis)

Joint	A	B		C		D
Members	AB	BA	BC	CB	CD	DC
DF	0	0.44	0.56	0.57	0.43	1
FEMS	−41.67	+41.67	−20.00	+20.00	0	0
Bal		−9.53	−12.14	−11.40	−8.60	
Co	−4.77		−5.70	−6.07		
Bal		+2.51	+3.19	+3.46	+2.61	
Co	+1.26		+1.73	+1.60		
Bal		−0.76	−0.97	−0.91	−0.69	
Co	−0.38		−0.46	−0.49		
Bal		+0.20	+0.26	+0.27	+0.22	
Co	+0.10		+0.14	+0.13		
Bal		−0.06	−0.08	−0.07	−0.06	
Co	−0.03		−0.04	−0.04		
Bal		+0.02	+0.02	+0.02	+0.02	
Final	−45.49	+34.05	−34.05	6.50	−6.50	
	M'_{AB}	M'_{BA}	M'_{BC}	M'_{CB}	M'_{CD}	M'_{DC}

Table 2.56 Moment distribution table (sway Analysis)

Joint	A	B		C		D
Members	AB	BA	BC	CB	CD	DC
DF	0	0.44	0.56	0.57	0.43	1
FEMS	−108.00	−108.00			−112.50	−112.50
Bal		+47.52	+60.48	+64.13	+48.37	+112.50
Co	+23.76		+32.07	+30.24	+56.25	
Bal		−14.11	−17.96	−49.30	−37.19	0
Co	−7.06		−24.65	−8.98		
Bal		+10.85	+13.80	+5.12	+3.86	0
Co	+5.43		+2.56	+6.90		
Bal		−1.13	−1.43	−3.93	−2.97	0
Co	−0.57		−1.97	−0.72		
Bal		+0.87	+1.10	+0.41	+0.31	0
Co	+0.44		+0.21	+0.55		
Bal		−0.09	−0.12	−0.31	−0.24	0
	−86.00	−64.09	+64.09	44.11	−44.11	0
	M''_{AB}	M''_{BA}	M''_{BC}	M''_{CB}	M''_{CD}	M''_{DC}

Column shear equation

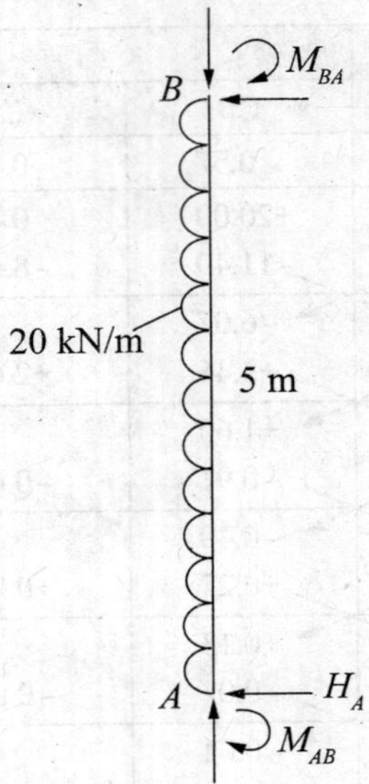

FIG. 2.106

$$\Sigma M_B = 0$$

$$M_{AB} + M_{BA} + 5H_A - 20 \times \frac{5^2}{2} = 0$$

$$\therefore \quad H_A = 50 - 0.2(M_{AB} + M_{BA})$$

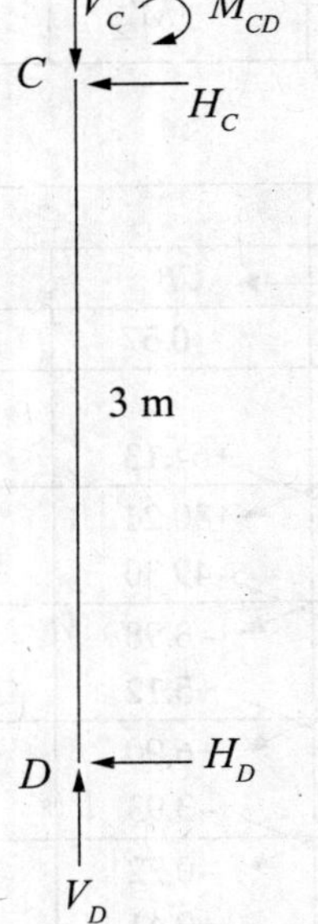

FIG. 2.107

$$\Sigma M_C = 0$$

$$M_{CD} + 3H_D = 0$$

$$\therefore \quad H_D = -0.33M_{CD}$$

$$\Sigma H = 0; \quad H_A + H_D = 20(5)$$

$$\therefore \quad 50 - 0.2(M_{AB} + M_{BA}) - 0.33\, M_{CD} = 100 \qquad (1)$$

The end moments are obtained using the sway and nonsway moment distribution table as

$$M_{AB} = -45.49 - 86.00\,k$$
$$M_{BA} = +34.05 - 64.09\,k$$
$$M_{BC} = -34.05 + 64.09\,k$$
$$M_{CB} = 6.50 + 44.11\,k$$
$$M_{CD} = -6.50 - 44.11\,k$$
$$M_{DC} = 0.00$$

Substituting the above end moments in Eq. (1);

$$\boxed{k = 1.022}$$

Hence, the end moments are obtained by back substitution as

$$M_{AB} = -133.4 \text{ kNm};\ M_{BA} = -31.5 \text{ kNm};\ M_{BC} = +31.5 \text{ kNm};$$
$$M_{CB} = +51.6 \text{ kNm};\ M_{CD} = -51.6 \text{ kNm};\ M_{DC} = 0$$

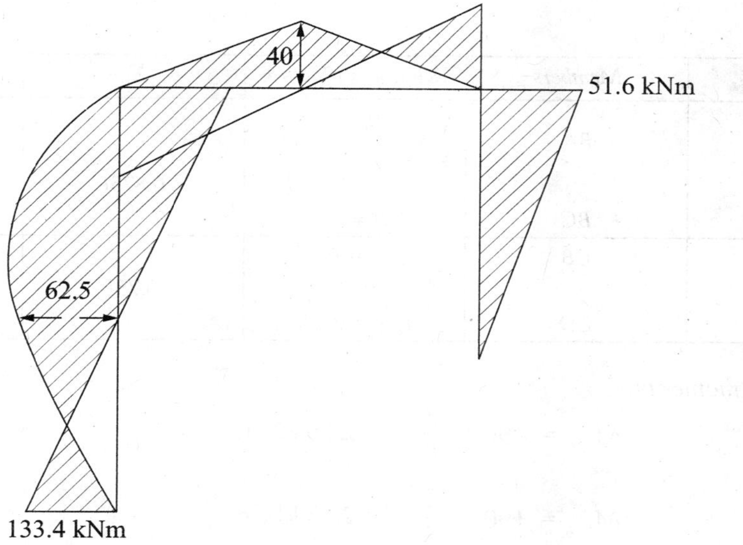

FIG. 2.108 Bending moment diagram

Example 2.25 The frame shown in Fig. 2.109 is hinged at *A*. The end *D* is fixed and the joints *B* and *C* are rigid. Column *CD* is subjected to horizontal loading of 30 kN/m. A concentrated load of 90 kN acts on *BC* at 1 m from *B*. Analyse the frame and sketch the BMD?

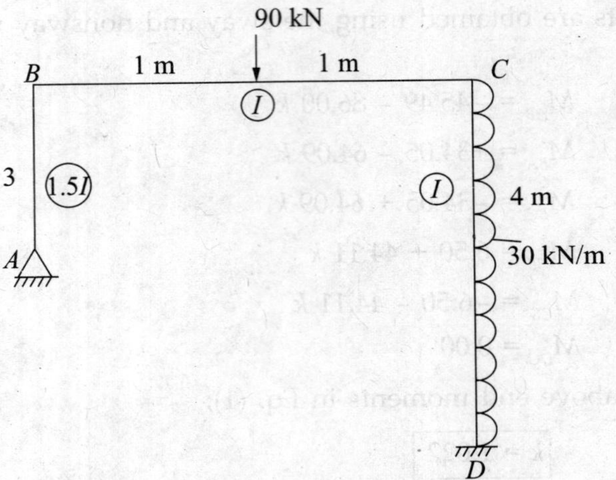

FIG. 2.109

Solution

Distribution factors

Table 2.57

Joint	*Members*	*Relative stiffness k*	Σk	$k/\Sigma k$
B	BA	$\frac{3}{4} \times \frac{1.5I}{3} = 0.375I$	0.875I	0.43
	BC	$I/2 = 0.5I$		0.57
C	CB	$I/2 = 0.5I$	0.75I	0.67
	CD	$I/4 = 0.25I$		0.33

Fixed end moments

$$M_{FBC} = -90 \times \frac{2}{8} = -22.5 \text{ kNm}$$

$$M_{FCB} = +90 \times \frac{2}{8} = +22.5 \text{ kNm}$$

$$M_{FCD} = -30 \times \frac{4^2}{12} = -40 \text{ kNm}$$

$$M_{FDC} = +30 \times \frac{4^2}{12} = +40 \text{ kNm}$$

Nonsway Analysis

Table 2.58 Moment distribution table

Joint	*A*	*B*		*C*		*D*
Members	*AB* ←	*BA*	*BC*	*CB*	*CD* →	*DC*
DF	1	0.43	0.57	0.67	0.33	0
FEMS			−22.50	+22.50	−40.00	+40.00
Bal		+9.68	+12.82	+11.73	+5.77	
Co			+5.87	+6.41		+2.89
Bal		−2.52	−3.35	−4.30	−2.11	
Co			−2.15	−1.68		−1.06
Bal		+0.92	+1.23	+1.12	+0.56	
Co			+0.56	+0.62		+0.28
Bal		−0.24	−0.32	−0.42	−0.20	
Co			−0.21	−0.16		−0.10
Bal		+0.09	+0.12	+0.11	+0.05	
Co			+0.06	+0.06		+0.03
Bal		−0.03	−0.03	−0.04	−0.02	
Final	0	+7.90	−7.90	+35.95	−35.95	+42.04
Members	M'_{AB}	M'_{BA}	M'_{BC}	M'_{CB}	M'_{CD}	M'_{DC}

Sway Analysis

The sway moments are assumed in the following ratios

$$\frac{M_{BA}}{M_{CD}} = \frac{3E(1.5I)\Delta/3^2}{6E(I)\Delta/4^2}$$

$$\frac{M_{BA}}{M_{CD}} = \frac{4}{3}$$

$$\therefore \quad M_{BA} : M_{CD} = +40 : +30 \text{ kNm}$$

Final moments

$$M_{AB} = 0$$

$$M_{BA} = 7.90 + 25.80\,k$$

$$M_{BC} = -7.90 - 25.80\,k$$

$$M_{CB} = +35.95 - 23.22\,k$$

$$M_{CD} = -35.95 + 23.22\,k$$

$$M_{DC} = +42.04 + 26.58\,k$$

Table 2.59 Moment distribution table

Joint	A	B		C		D
Members	*AB*	*BA*	*BC*	*CB*	*CD*	*DC*
DF	1	0.43	0.57	0.67	0.33	0
FEMS	0.00	+40.00			+30.00	+30.00
Bal		−17.20	−22.8	−20.10	−9.90	
Co			−10.1	−11.40		−4.95
Bal		+4.34	+5.76	+7.64	+3.76	
Co			+3.82	+2.88		+1.88
Bal		−1.64	−2.18	−1.93	−0.95	
Co			−0.97	−1.09		−0.48
Bal		+0.42	+0.55	+0.73	+0.36	
Co			+0.37	+0.28		+0.18
Bal		−0.16	−0.21	−0.19	−0.09	
Co			−0.10	−0.11		−0.05
Bal		+0.04	+0.06	+0.07	+0.04	
	0.00	+25.80	−25.80	−23.22	+23.22	+26.58
	M''_{AB}	M''_{BA}	M''_{BC}	M''_{CB}	M''_{CD}	M''_{DC}

The value of k is determined from the column shear equation as

Equilibrium of column AB

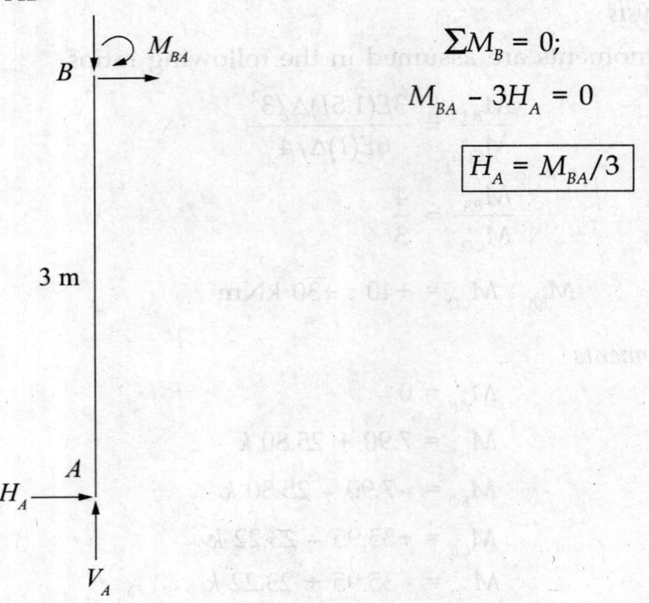

$$\sum M_B = 0;$$

$$M_{BA} - 3H_A = 0$$

$$\boxed{H_A = M_{BA}/3}$$

FIG. 2.110

Equilibrium of column CD

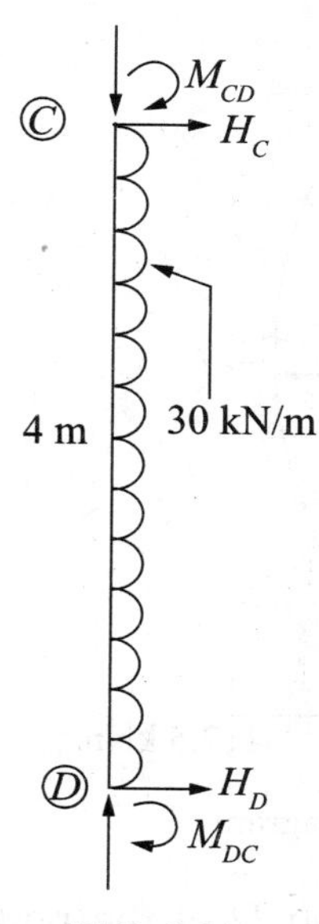

FIG. 2.111

$$\Sigma M_C = 0;$$

$$M_{CD} + M_{DC} + 30 \times \frac{4^2}{2} - 4H_D = 0$$

$$\boxed{H_D = 60 + \frac{1}{4}(M_{CD} + M_{DC})}$$

Considering the horizontal equilibrium of the whole structure;

$$H_A + H_D = 4(30)$$

i.e.
$$\frac{M_{BA}}{3} + 60 + \frac{(M_{CD} + M_{DC})}{4} = 120$$

Substituting the values of the moments from the final moment equations;

$$\boxed{k = 2.65}$$

The end moments were calculated as

$$M_{AB} = 0$$

$$M_{BA} = 7.90 + 25.80 \times 2.65 = 76.3 \text{ kNm}$$

$$M_{BC} = -7.90 - 25.80 \times 2.65 = -76.3 \text{ kNm}$$

$$M_{CB} = +35.95 - 23.22 \times 2.65 = -25.6 \text{ kNm}$$

$$M_{CD} = -35.95 + 23.22 \times 2.65 = +25.6 \text{ kNm}$$

$$M_{DC} = +42.04 + 26.58 \times 2.65 = +112.5 \text{ kNm}$$

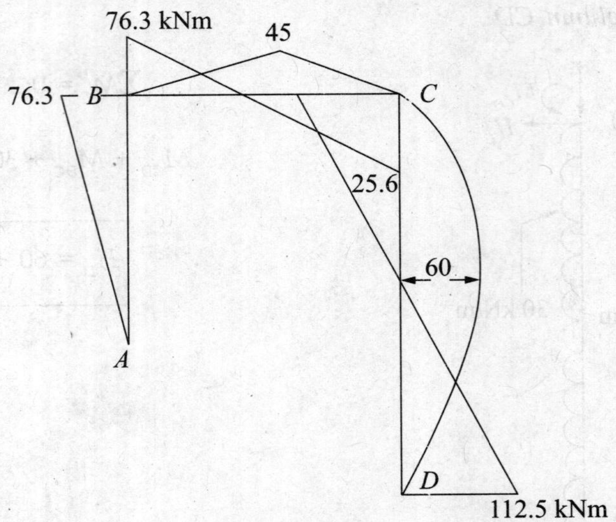

FIG. 2.112 Bending moment diagram

Example 2.26 Determine the moments at *A*, *B*, *C* and *D* of the portal frame shown in Fig. 2.113 by moment distribution method and draw the bending moment diagram.

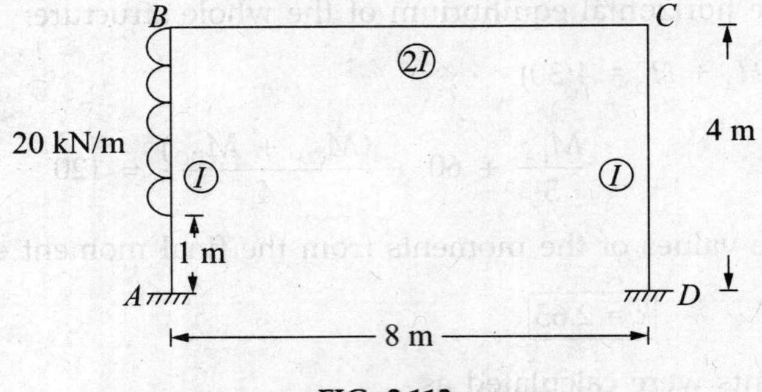

FIG. 2.113

Solution

Distribution factors

Table 2.60

Joint	*Members*	*Relative Stiffness (I/l)*	Σk	$k/\Sigma k$
B	BA	$I/4 = 0.25I$	$0.5I$	0.5
	BC	$2I/8 = 0.25I$		0.5
C	CB	$2I/8 = 0.25I$	$0.5I$	0.5
	CD	$I/4 = 0.25I$		0.5

Fixed end moments

$$M_{FAB} = \int_1^4 \frac{x(20dx)(4 - x)^2}{4^2}$$

$$M_{FAB} = 1.25 \int_1^4 (16x + x^3 - 8x^2)\, dx$$

$$M_{FAB} = 19.66 \text{ kNm}$$

$$M_{FBA} = \int_1^4 (4 - x)\, 20dx \frac{x^2}{4^2} = 25.29 \text{ kNm}$$

Nonsway Analysis

Table 2.61 Moment distribution table

Joint	*A*	*B*		*C*		*D*
Members	*AB*	*BA*	*BC*	*CB*	*CD*	*DC*
	0	0.5	0.5	0.5	0.5	0
FEMS	−19.66	+25.29				
Bal		−12.64	−12.65			
Co	−6.32			−6.32		
Bal				+3.16	+3.16	
Co			+1.58			+1.58
Bal		−0.79	−0.79			
Co	−0.40			−0.40		
Bal				+0.20	+0.20	
Co			+0.10			+0.10
Bal		−0.05	−0.05			
Co	−0.03			−0.03		
				+0.01	+0.02	
Total	−26.41	11.81	−11.81	−3.38	+3.38	+1.68
	M'_{AB}	M'_{BA}	M'_{BC}	M'_{CB}	M'_{CD}	M'_{DC}

Final moments

$$M_{AB} = -26.41 - 79.98\, k$$

$$M_{BA} = 11.81 - 60.01\, k$$

$$M_{BC} = -11.81 + 60.01\, k$$

$$M_{CB} = -3.38 + 60.02\, k$$

$$M_{CD} = +3.38 - 60.02\, k$$

$$M_{DC} = +1.68 - 79.98\, k$$

Sway Analysis

Table 2.62 Moment distribution table

Joint	*A*	*B*		*C*		*D*
Members	*AB* ←	*BA*	*BC* ↔	*CB*	*CD* →	*DC*
DF	0	0.5	0.5	0.5	0.5	0
FEMS	−100.00	−100.00			−100.00	−100.00
Bal		+50.00	+50.00	+50.00	+50.00	
Co	+25.00		+25.00	+25.00		+25.00
Bal		−12.50	−12.50	−12.50	−12.50	
Co	−6.25		−6.25	−6.25		−6.25
Bal		+3.12	+3.13	+3.13	+3.12	
Co	+1.56		+1.56	+1.57		+1.56
Bal		−0.78	−0.78	−0.79	−0.78	
Co	−0.39		−0.40	−0.39		−0.69
Bal		+0.20	+0.20	+0.20	+0.20	
Co	+0.10		+0.10	+0.10		+0.10
Bal		−0.05	−0.05	−0.05	−0.05	
Final	−79.98	−60.01	+60.01	60.02	−60.02	−79.98
	M''_{AB}	M''_{BA}	M''_{BC}	M''_{CB}	M''_{CD}	M''_{DC}

Column shear condition

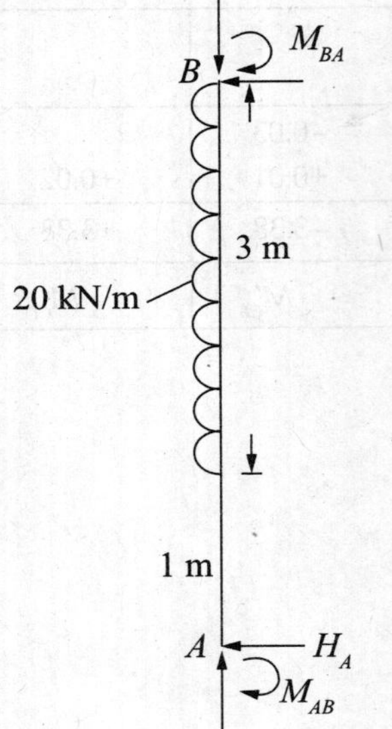

FIG. 2.114

Equilibrium of column AB

$$\Sigma M_B = 0;$$

$$M_{AB} + M_{BA} - 20 \times \frac{3^2}{2} + 4H_A = 0$$

$$\boxed{H_A = \frac{90 - (M_{AB} + M_{BA})}{4}}$$

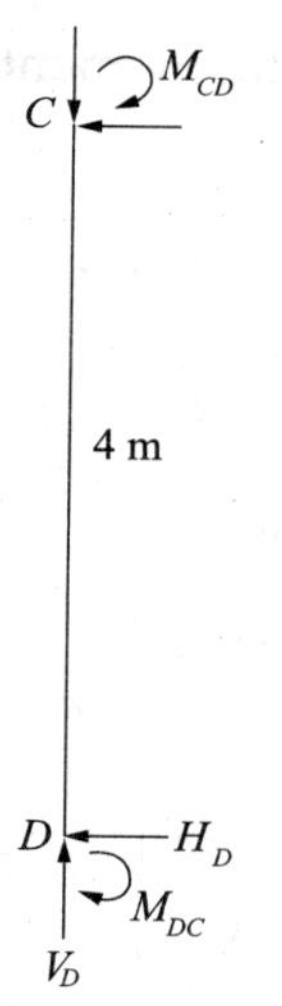

FIG. 2.115

$$\Sigma M_C = 0;$$

$$M_{CD} + M_{DC} + 4H_D = 0$$

$$\boxed{H_D = \frac{-(M_{CD} + M_{DC})}{4}}$$

Considering the horizontal equilibrium
Substituting the values in terms of moments and simplifying

$$H_A + H_D = 20 \times 3$$

$$M_{AB} + M_{BA} + M_{CD} + M_{DC} = -150$$

$$(-26.41 + 11.81 + 3.38 + 1.68) - (79.98 + 60.01 + 60.02 + 79.98)k = -150$$

$$\boxed{k = 0.502}$$

End moments

$$M_{AB} = -26.41 - 79.98(0.502) = 66.6 \text{ kNm}$$
$$M_{BA} = 11.81 - 60.01(0.502) = -18.3 \text{ kNm}$$
$$M_{BC} = -11.81 + 60.01(0.502) = +18.3 \text{ kNm}$$
$$M_{CB} = -3.38 + 60.02(0.502) = +26.8 \text{ kNm}$$
$$M_{CD} = +3.38 - 60.02(0.502) = -26.8 \text{ kNm}$$
$$M_{DC} = 1.68 - 79.98(0.502) = -38.5 \text{ kNm}$$

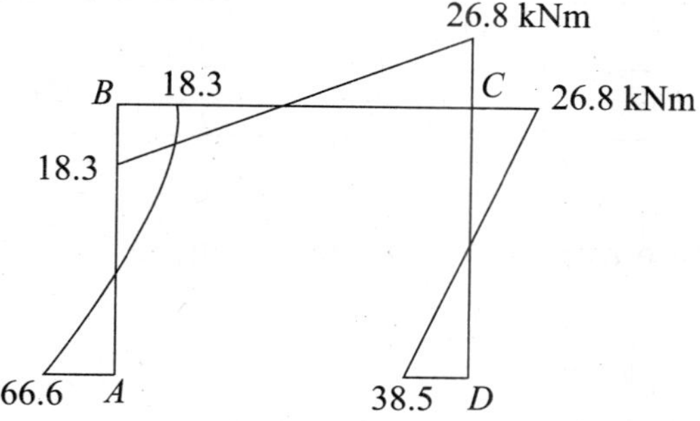

FIG. 2.116 Bending moment diagram

Example 2.27 Analyse the rigid frame shown in Fig. 2.117 by the moment distribution method. Draw the bending moment diagram.

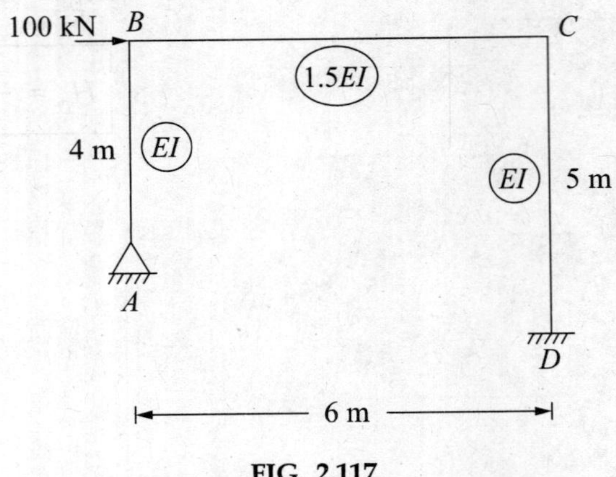

FIG. 2.117

Solution: In the above problem, a lack of symmetry makes the frame to sway to the right. In the sway analysis, it is assumed that the joints B and C have no movement of rotation but there is only lateral translation.

In the first instant, an external force necessary to prevent the lateral translation is assumed. The moments at the supports and joints are computed for the above external force. This is followed by a redistribution of moments allowing for the side sway by removing the assumed external force.

The frame is analysed by considering the sway moments only.

Distribution factors

Table 2.63

Joint	B		C	
Members	BA	BC	CB	CD
k	$\frac{3}{4}\left(\frac{I}{4}\right) = 0.188I$	$\frac{1.5I}{6} = 0.25I$	$\frac{1.5I}{6} = 0.25I$	$\frac{I}{5} = 0.20I$
Σk	0.438I		0.45I	
DF = k/Σk	0.43	0.57	0.56	0.44

Sway moments

The sway moments are assumed in the following ratio

$$\frac{M_{BA}}{M_{CD}} = \frac{-3EI\Delta/4^2}{-6EI\Delta/5^2}$$

$$\therefore \quad M_{BA} : M_{CD} = -25.00 : -32.00 \text{ kNm}$$

Moment distribution table

Table 2.64

Joint	A	B		C		D
Members	AB	BA	BC ⟷	CB	CD ⟶	DC
DF	0	0.43	0.57	0.56	0.44 1	
FEMS	0	−25.00			−32.00	−32.00
Bal		+10.75	+14.25	+17.92	+14.08	
Co			+8.96	+7.13		+7.04
Bal		−3.85	−5.11	−4.00	−3.13	
Co			−2.00	−2.56		−1.57
Bal		+0.86	+1.14	+1.43	+1.13	
Co			+0.72	+0.72		+0.57
Bal		−0.31	−0.41	−0.40	−0.32	
Co			−0.20	−0.21		−0.16
Bal		+0.08	+0.12	+0.12	+0.09	
Co			+0.06	+0.06		+0.05
Bal		−0.03	−0.03	−0.03	−0.03	
Final		−17.50	+17.50	+20.18	−20.18	−26.07
Moment	0					

Column shear condition

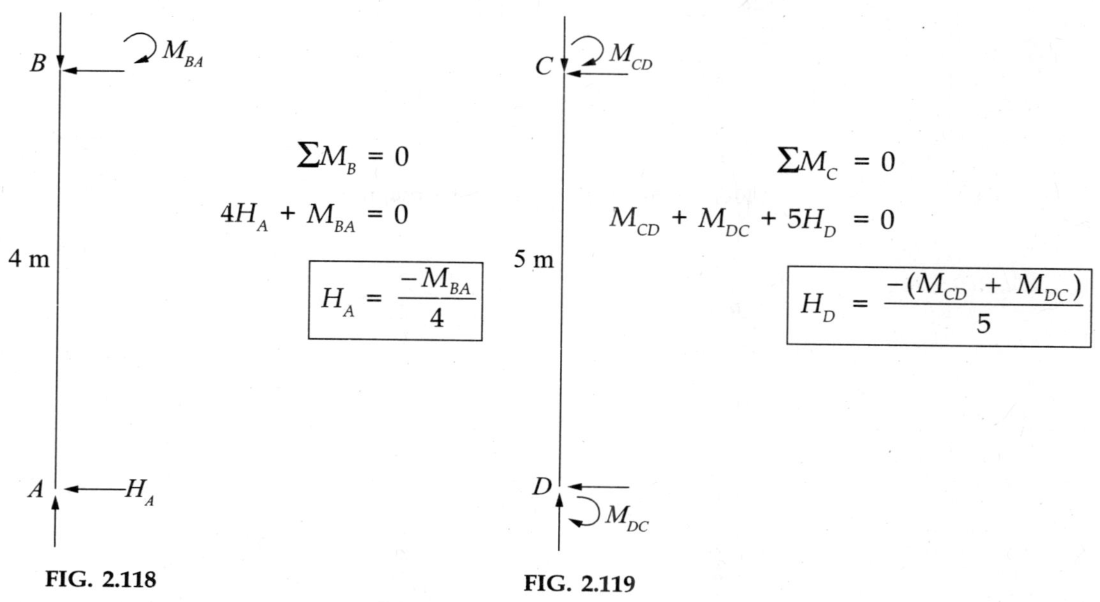

$$\Sigma M_B = 0$$

$$4H_A + M_{BA} = 0$$

$$\boxed{H_A = \frac{-M_{BA}}{4}}$$

$$\Sigma M_C = 0$$

$$M_{CD} + M_{DC} + 5H_D = 0$$

$$\boxed{H_D = \frac{-(M_{CD} + M_{DC})}{5}}$$

FIG. 2.118 **FIG. 2.119**

Considering the horizontal equilibrium, substituting the values of moments, H_A and H_D are calculated as

$$H_A = \frac{-(-17.50)}{4} = 4.38\ \text{kN}$$

$$H_D = \frac{-1}{5}(-20.18 - 26.07) = 9.25\ \text{kN}$$

The addition of H_A and H_D gives the lateral load which produces the assumed moments. Thus, $H_A + H_D$ gives 13.63 kN but the applied load is 100 kN. Hence, the end moments in the table are to be multiplied by the ratio (100/13.63 = 7.34).

The final moments are

$$M_{AB} = 0 \times 7.34 = 0$$

$$M_{BA} = -17.50 \times 7.34 = -128.5\ \text{kNm}$$

$$M_{BC} = +17.50 \times 7.34 = +128.5\ \text{kNm}$$

$$M_{CB} = +20.18 \times 7.34 = +148.1\ \text{kNm}$$

$$M_{CD} = -20.18 \times 7.34 = -148.1\ \text{kNm}$$

$$M_{DC} = -26.07 \times 7.34 = -191.4\ \text{kNm}$$

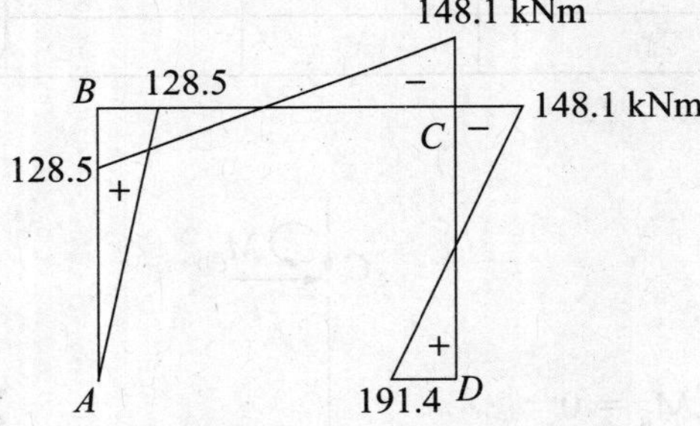

FIG. 2.120 Bending moment diagram

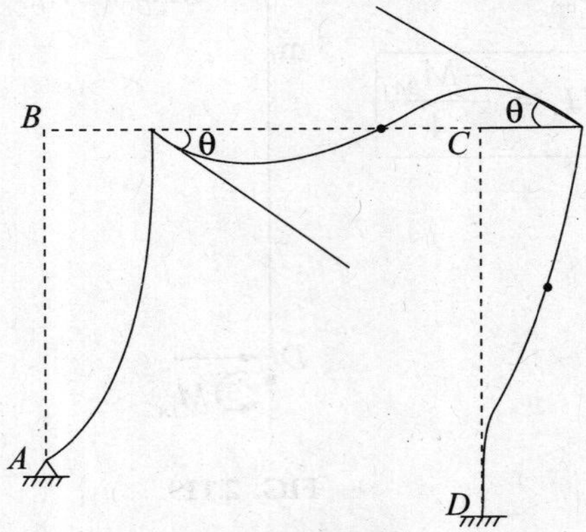

FIG. 2.121 Elastic curve

Example 2.28 Analyse the frame by moment distribution method and draw the shear force diagram and bending moment diagram.

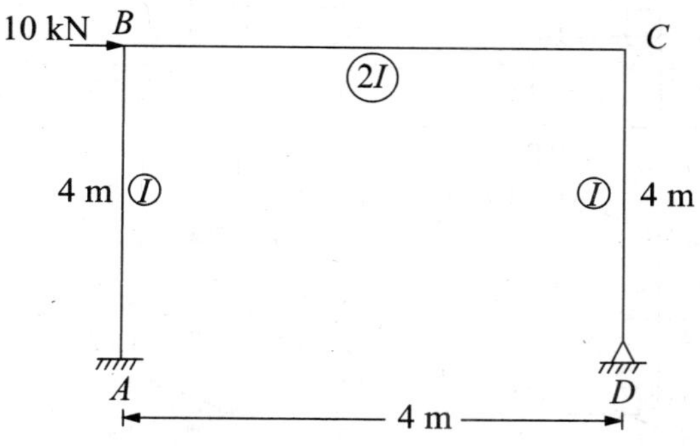

FIG. 2.122

Solution

Distribution factors

Table 2.65

Joint	Members	Relative Stiff (I/l)	Σk	k/Σk
B	BA	$I/4$	$3I/4$	0.33
	BC	$2I/4$		0.67
C	CB	$2I/4$	$0.688I$	0.73
	CD	$\frac{3}{4}\left(\frac{I}{4}\right) = 0.188I$		0.27

Sway moments

The sway moments are assumed in the following ratio:

$$\frac{M_{BA}}{M_{CD}} = \frac{-6EI\Delta/4^2}{-3EI\Delta/4^2} = \frac{-2}{-1}$$

Hence, $M_{BA} : M_{CD} = -20 : -10$ kNm

Table 2.66 Sway moment distribution table

Joint	*A*	*B*		*C*		*D*
Members	*AB* ←	*BA*	*BC* ↔	*CB*	*CD* →	*DC*
DF	1	0.33	0.67	0.73	0.27	0
FEMS	−20.00	−20.00			−10.00	
Bal		+6.60	+13.40	+7.30	+2.70	
Co	+3.30		+3.65	+6.70		
Bal		−1.21	−2.44	−4.89	−1.81	
Co	−0.61		−2.45	−1.22		
Bal		+0.81	+1.64	+0.89	+0.33	
Co	+0.41		+0.45	+0.82		
Bal		−0.15	−0.30	−0.60	−0.22	
Co	−0.08		−0.30	−0.15		
Bal		+0.10	+0.20	+0.11	+0.04	
Co	+0.05		+0.06	+0.10		
Bal		−0.02	−0.04	−0.07	−0.03	
	−16.93	−13.87	+13.87	8.99	−8.99	0.00
	M'_{AB}	M'_{BA}	M'_{BC}	M'_{CB}	M'_{CD}	M'_{DC}

Column shear equation

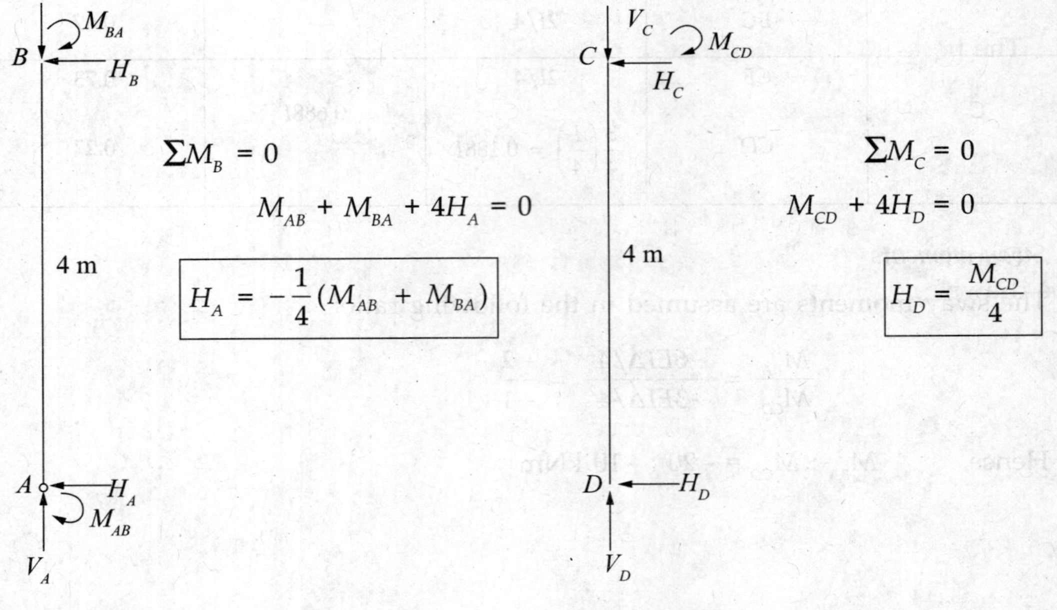

$$\Sigma M_B = 0$$

$$M_{AB} + M_{BA} + 4H_A = 0$$

$$H_A = -\frac{1}{4}(M_{AB} + M_{BA})$$

$$\Sigma M_C = 0$$

$$M_{CD} + 4H_D = 0$$

$$H_D = \frac{M_{CD}}{4}$$

FIG. 2.123 FIG. 2.124

$\Sigma H = 0;$

$$-\frac{(M_{AB} + M_{BA})}{4} - \frac{M_{CD}}{4} - 10 = 0$$

$$\{-16.93 - 13.87 - 8.99\}k = 40$$

$$\boxed{k = 1.005}$$

Final moments

They are obtained by multiplying the sway moments by the value k;

$M_{AB} = -17.0$ kNm, $M_{BA} = -13.9$ kNm, $M_{BC} = -13.9$, $M_{CB} = 9.03$, $M_{CD} = -9.03$ kNm, $M_{DC} = 0$

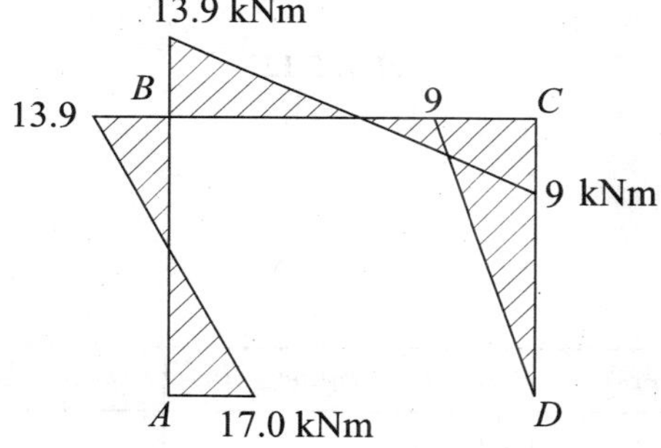

FIG. 2.125 Bending moment diagram

The horizontal reactions are obtained as

$$H_A = +\frac{1}{4}(17 + 13.9) = +7.73\text{ kN}$$

$$H_D = +\frac{1}{4} \times 9 = +2.27\text{ kN}$$

Drawing the free body diagram and evaluating the stress resultants, the SFD can be drawn.

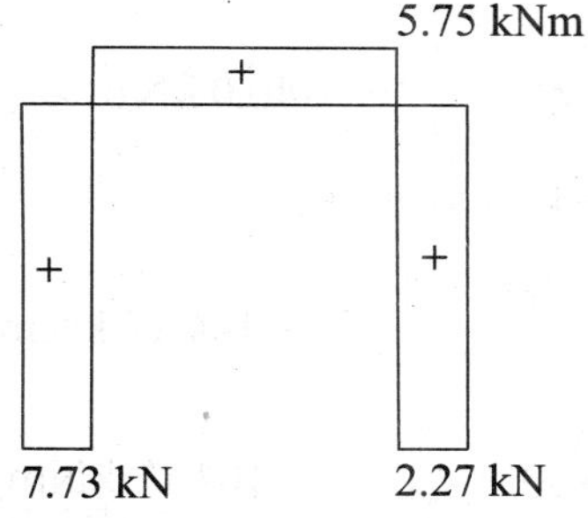

FIG. 2.126 Shear force diagram

Example 2.29 Analyse the frame shown in Fig. 2.127 by the moment distribution method. Draw the bending moment diagram.

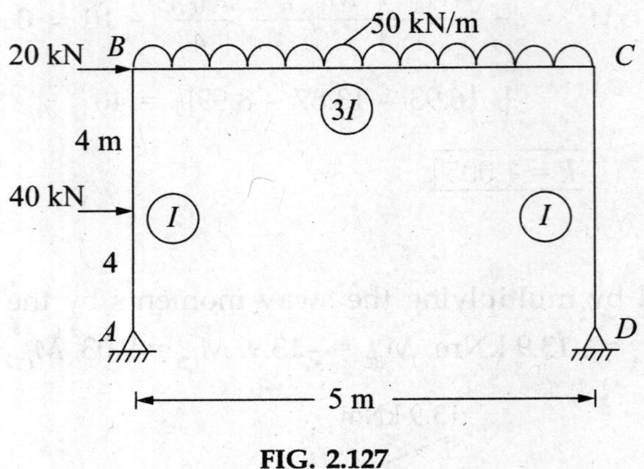

FIG. 2.127

Solution

Distribution factors

Table 2.67

Joint	*Members*	*Relative Stiffness (I/l)*	Σk	$k/\Sigma k$
B	BA	$\frac{3}{4}\left(\frac{1}{8}\right) = .094$	0.694I	0.14
	BC	$\frac{3I}{5} = 0.6I$		0.86
C	CB	$\frac{3I}{5} = 0.6I$	0.694I	0.86
	CD	$\frac{3}{4}\left(\frac{I}{8}\right) = .094I$		0.14

Fixed end moments

$$M_{FAB} = -40 \times \frac{8}{8} = -40.0 \text{ kNm}$$

$$M_{FBA} = +40.0 \text{ kNm}$$

$$M_{FBC} = \frac{-50 \times 5^2}{12} = -104.17 \text{ kNm}$$

$$M_{FCB} = +50 \times \frac{5^2}{12} = +104.17 \text{ kNm}$$

End moments

$$M_{AB} = 0$$

$$M_{BA} = 75.18 - 90.27\,k$$

$$M_{BC} = -75.18 + 90.27\,k$$

$$M_{CB} = +21.07 + 90.27\,k$$

$$M_{CD} = -21.07 - 90.27\,k$$

$$M_{DC} = 0$$

Nonsway Analysis

Table 2.68

Joint	*A*	*B*		*C*		*D*
Members	*AB*	*BA*	*BC* ⟷	*CB*	*CD*	*DC*
DF	1	0.14	0.86	0.86	0.14	1
FEMS	-40.00	+40.00	-104.17	+104.17	0	0
Bal	+40.00	+8.98	+55.19	-89.58	-14.59	
Co		+20.00	-44.79	+27.60		
Bal		+3.47	+21.32	-23.74	-3.86	
Co			-11.87	+10.66		
Bal		+1.66	+10.21	-9.16	-1.50	
Co	0		-4.58	+5.11		0
Bal		+0.64	+3.94	-4.40	-0.71	
Co			-2.20	+1.97		
Bal		+0.31	+1.89	-1.69	-0.28	
Co			-0.85	+0.95		
Bal		+0.12	+0.73	-0.82	-0.13	
		75.18	-75.18	+21.07	-21.07	
	M'_{AB}	M'_{BA}	M'_{BC}	M'_{CB}	M'_{CD}	M'_{DC}

Column shear equations

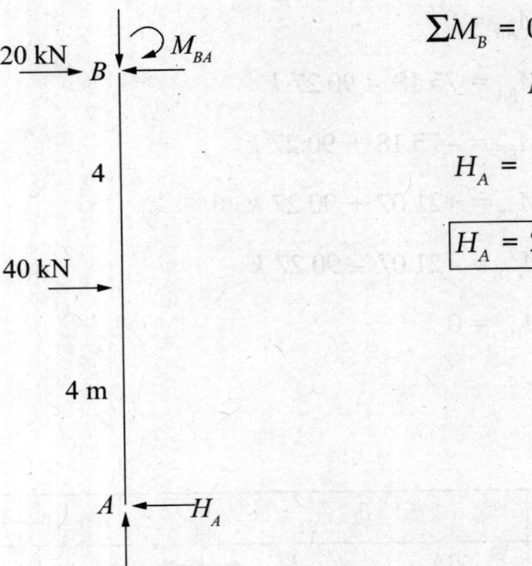

$$\Sigma M_B = 0$$

$$M_{BA} + 8H_A - 40(4) = 0$$

$$H_A = \frac{160 - M_{BA}}{8}$$

$$\boxed{H_A = 20 - 0.125M_{BA}}$$

FIG. 2.128

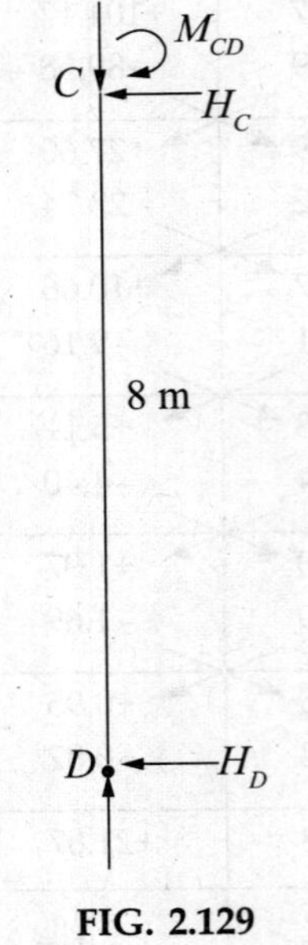

$$\Sigma M_C = 0$$

$$M_{CD} + 8H_D = 0$$

$$\boxed{H_D = M_{CD}/8}$$

$$\Sigma H = 0; \quad H_A + H_D = 60$$

FIG. 2.129

Sway Analysis

The sway moments are assumed in the following ratio:

$$\frac{M_{BA}}{M_{CD}} = \frac{-3EI\Delta/8^2}{-3EI\Delta/8^2} = \frac{-1}{-1}$$

Table 2.69 Moment distribution table

Joint	*A*	*B*		*C*		*D*
Members	*AB* →	*BA*	*BC* ↔	*CB*	*CD*	*DC*
DF	1	0.14	0.86	0.86	0.14	1
FEMS Bal	0	−100.00 +14.00	0 +86.00	0 +86.00	−100.00 +14.00	0
Co Bal	↕	 −6.02	+43.00 −36.98	+43.00 −36.98	 −6.02	↕
Co Bal		 +2.59	−18.49 +15.90	−18.49 +15.90	 +2.59	
Co Bal		 −1.11	+7.95 −6.84	+7.95 −6.84	 −1.11	
Co Bal		 +0.48	−3.42 +2.94	−3.42 +2.94	 +0.48	
Co Bal		 −0.21	+1.47 −1.26	+1.47 −1.26	 −0.21	
	0	−90.27	+90.27	+90.27	−90.27	0
	M''_{AB}	M''_{BA}	M''_{BC}	M''_{CB}	M''_{CD}	M''_{DC}

Hence, $M_{BA} : M_{CD} = -100.00 : -100.00$

$$20 - 0.125M_{BA} - 0.125M_{CD} = 60$$

$$20 - 0.125(75.18 - 90.27k) - 0.125(-21.07 - 90.27k) = 60$$

Solving for $\boxed{k = 2.07}$

Substituting the values of k by back substitution:

$$M_{AB} = 0,\ M_{BA} = -111.86,\ M_{BC} = +111.86,$$

$$M_{CB} = 207.9,\ M_{CD} = -207.9 \text{ kNm},\ M_{DC} = 0$$

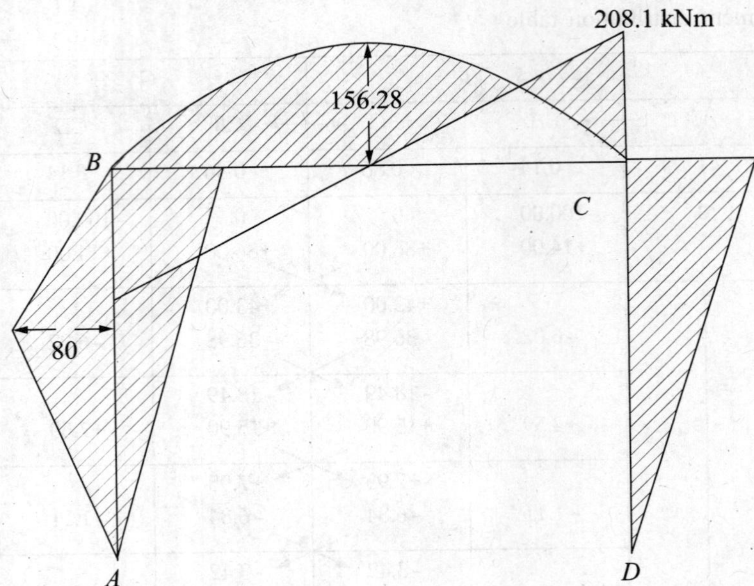

FIG. 2.130 Bending moment diagram

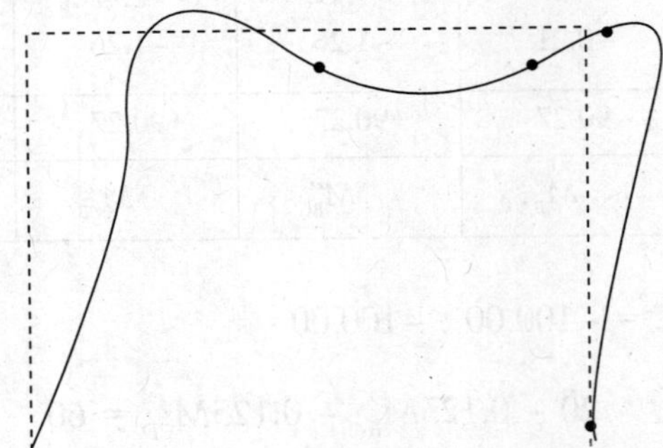

FIG. 2.131 Elastic curve

2.10 NAYLOR'S METHOD FOR SYMMETRICAL FRAMES

This method is an offshoot of Hardy Cross moment distribution method. In 1950, Naylor presented this method with a few interesting and important examples. This method enables joint and sway distribution to be made simultaneously. This method was further developed by Wood, Light Foot and Bolton independently. This method is also referred as the cantilever moment distribution method or no shear method. The important points to be noted are

(i) The rotational stiffness of a cantilever is $k_{AB} = EI/l$.

(ii) The stiffness of a beam of uniform section under antisymmetric loading is $k_{AB} = 6EI/l$.

(iii) The carry over factor is -1.

A few examples are presented in the next section to illustrate the elegance of the Naylor's method. Though, it is a special case of moment distribution, it is advantageous to analyse symmetrical frames with unsymmetrical loading on the beam.

Example 2.30 Analyse the frame shown in Fig. 2.132 by the Naylor's moment distribution method. Sketch the elastic curve and draw the bending moment diagram.

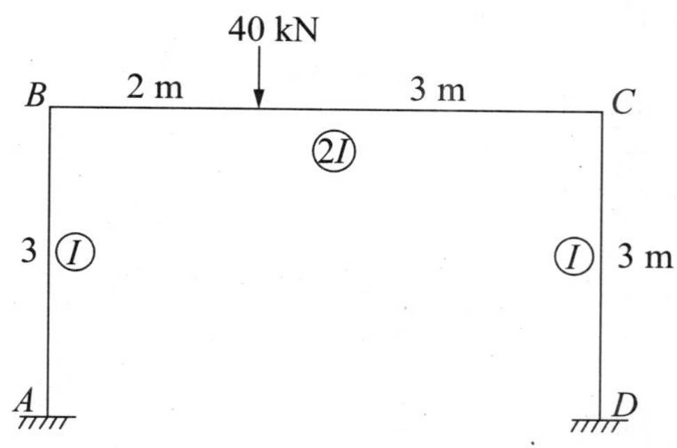

FIG. 2.132

Solution: The given asymmetrical (eccentric) loading is decomposed into symmetric and skew symmetric (antisymmetric) loading respectively. The analysis requires modified relative rotational stiffness and also a different carry over factor for skew symmetric loading.

Symmetric Case

The relative rotational stiffness for column *BA* is $(I/3)$. The beam stiffness of *BC* is taken as half of the actual stiffness. k'_{BC} is taken as $\left(\frac{1}{2} \times \frac{2I}{5}\right) = 0.2I$. Thus, the symmetric case is analysed with actual stiffness of the column and the symmetric stiffness of the beam. The moment distribution is done with one cycle only.

Distribution factors for symmetric case

Table 2.70

Joint	*Members*	*Relative Stiffness*	Σk	$DF = k/\Sigma k$
	BA	$I/3 = 0.33I$		0.62
B			$0.53I$	
	BC	$\frac{1}{2}\left(\frac{2I}{5}\right) = 0.2I$		0.38

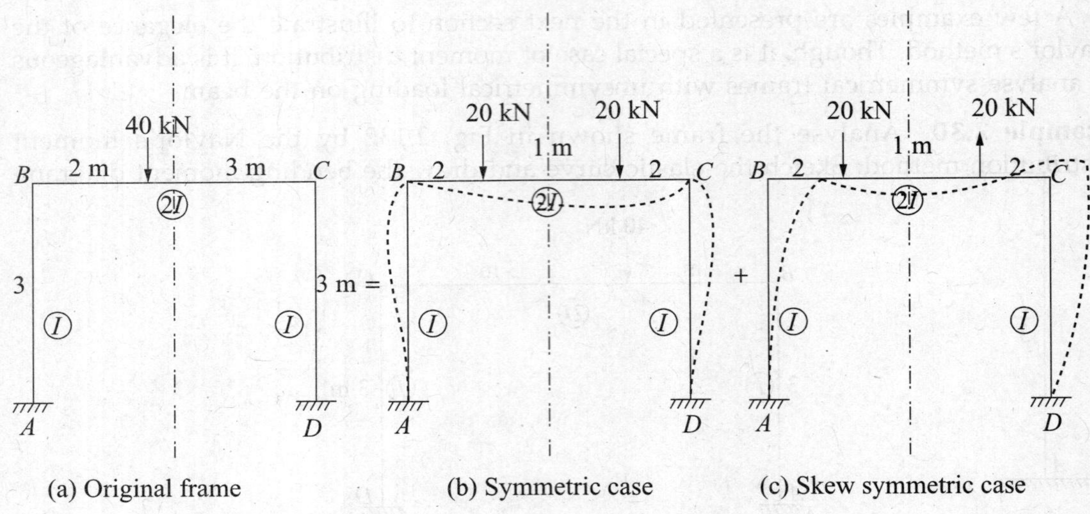

(a) Original frame (b) Symmetric case (c) Skew symmetric case

FIG. 2.133

Decomposition of the structure taking advantage of symmetry.

Fixed end moments in symmetrical case

$$M_{FBC} = -20 \times 2 \times \frac{3}{5} = -24 \text{ kNm}$$

$$M_{FCB} = +20 \times 2 \times \frac{3}{5} = +24 \text{ kNm}$$

Table 2.71 Moment distribution for symmetric case

Joint	*A*	*B*	
Members	*AB*	*BA*	*BC*
DF	0	0.62	0.38
FEM Bal		 +14.88	−24.00 +9.12
Co	7.44		
Final	7.44	14.88	−14.88

Skew–symmetric case

The cantilever stiffness for the column *AB* is taken as ($I/3$). While the skew stiffness of the beam *BC* is taken as $6(2I/5)$. The derivation of the cantilever stiffness and skew stiffness are available in standard references.

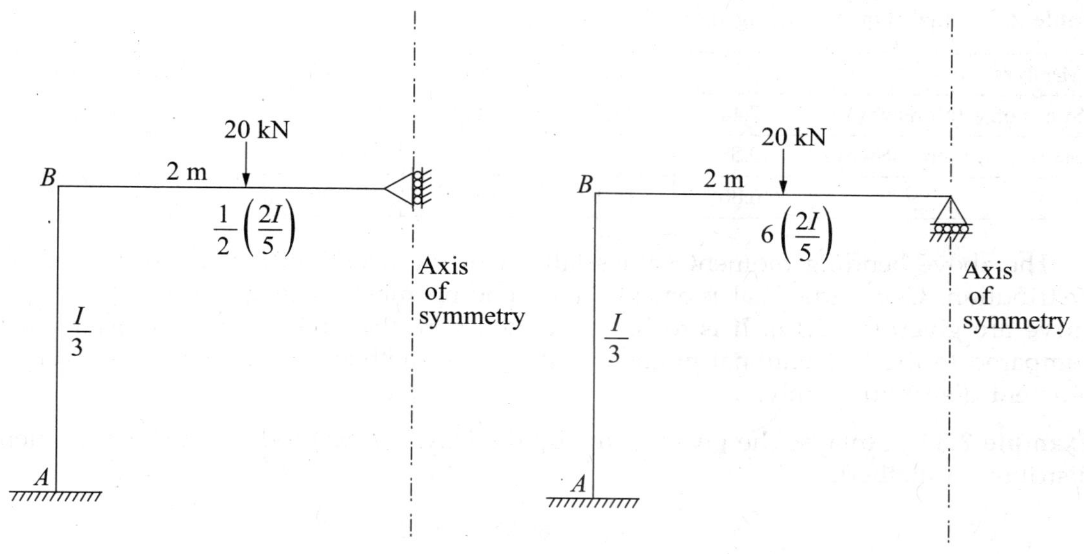

(a) Symmetric case (b) Skew symmetric case

FIG. 2.134

Table 2.72 Distribution factors for skews symmetric case

Joint	*Members*	*Relative Stiffness*	*Σk*	*DF = k/Σk*
	BA	$I/3 = 0.33I$		0.12
B			$2.73I$	
	BC	$6(2I)/5 = 2.4I$		0.88

Fixed end moments in skew symmetric case

$$M_{FBC} = -\frac{2 \times 20 \times 3^2}{5^2} + \frac{3 \times 20 \times 2^2}{5^2} = -4.8 \text{ kNm}$$

$$M_{FCB} = -4.8 \text{ kNm}$$

Table 2.73 Moment distribution for skew symmetric case

Joint	*A*	*B*	
Members	*AB*	*BA*	*BC*
	0	0.12	0.88
FEMS			−4.80
Bal		+0.58	+4.22
Co	−0.58		
Final	−0.58	+0.5	−0.58

Table 2.74 Final moments using moment distribution

Members	*AB*	*BA*	*BC*	*CB*	*CD*	*DC*
Symmetry (Nonsway)	7.44	14.88	−14.88	+14.88	−14.88	−7.44
Skew Symmetry (Sway)	−0.58	+0.58	−0.58	−0.58	+0.58	−0.58
Final	6.86	15.46	−15.46	14.30	−14.30	−8.02

The above bending moment values tally with the results of conventional moment distribution. Using the final moments, the bending moment diagram and the elastic curve are given (Ex 2.19). It is to be mentioned here that this method is faster when compared to the conventional moment distribution method as it requires one cycle of moment distribution only.

Example 2.31 Anlayse the given frame by the Naylor's Method – no shear moment distribution method.

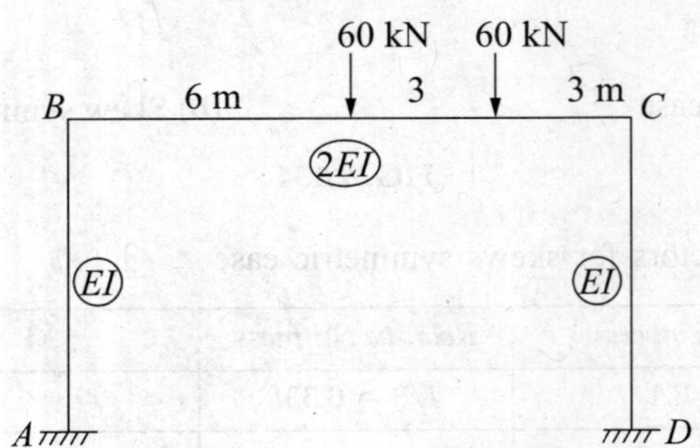

FIG. 2.135

Solution: The given frame is symmetrical with the base fixed. The applied loading is on the beam *BC* only. Hence, the given loading is decomposed into two types of loading.

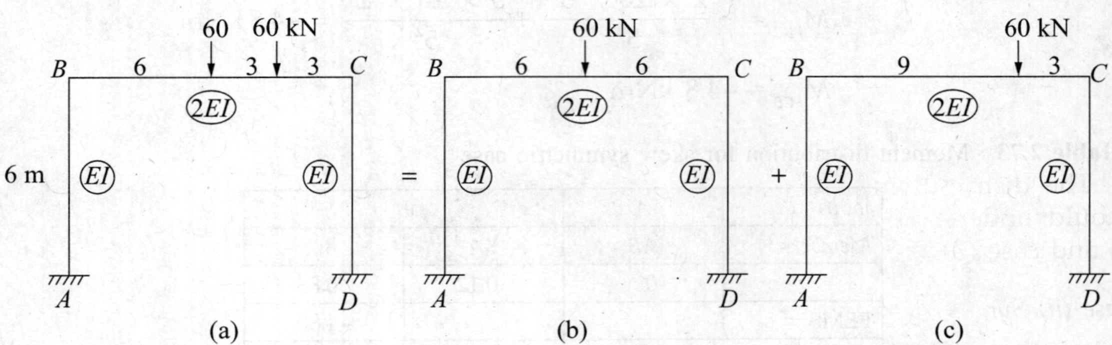

FIG. 2.136 Decomposition of loading

In the first type, a central concentrated load of 60 kN is acting. This is a symmetric case. In the second type, an eccentric loading of 60 kN is acting on span BC. This is further split into a symmetrical loading of 30 kN each acting downwards at 3 metres from B and C respectively. The skew symmetrical loading (antisymmetrical) consists of 30 kN acting up from a distance of 3 mm of B and another 30 kN acting downwards at 3 metres from C. This decomposition of loading on the beam is pictorially shown in Fig. 2.137.

The algebraic sum of the loading systems of symmetric and skew symmetric cases gives the original loading system. The analysis is done for individual cases and the final moments are obtained by adding the results of individual cases.

Case (i) Symmetric case

The distribution factors at joint B are computed for themembers BA and BC. The symmetric stiffness of the beam BC is taken, i.e. half of the beam stiffness k_{BC} = $\frac{1}{2}\left(\frac{2I}{12}\right) = \frac{I}{3}$ = 0.33. The column stiffness is taken as such i.e. $k_{BA} = \frac{I}{6}$ = 0.17. The total stiffness Σk = 0.33 + 0.17 = 0.50. Thus, the distribution factor $d_{BC} = k_{BC}/\Sigma k$ = 0.67 and $d_{BA} = k_{BA}/\Sigma k$ = 0.33. The fixed end moment $M_{FBC} = -Wl/8 = -90$ kNm. Half frame is analysed taking advantage of symmetry.

Table 2.75 Moment distribution table

Joint	A	B	
Members	*AB*	*BA*	*BC*
DF	0.0	0.67	0.33
FEMS Bal		 +60.00	−90.00 +30.00
Co	+30.00 ←	+60.00	−60.00
Final	30.00	60.00	−60.00

The distribution factors obtained in case (i) is used in case (ii) too. The reader should understand obviously the distribution factors do not change between case (i) and case (ii).

Case (ii) Symmetric case

Using the advantage of symmetry, the fixed end moments are $M_{FBC} = -3(30)9/12 =$ −67.5 kNm and M_{FCB} = +67.5 kNm.

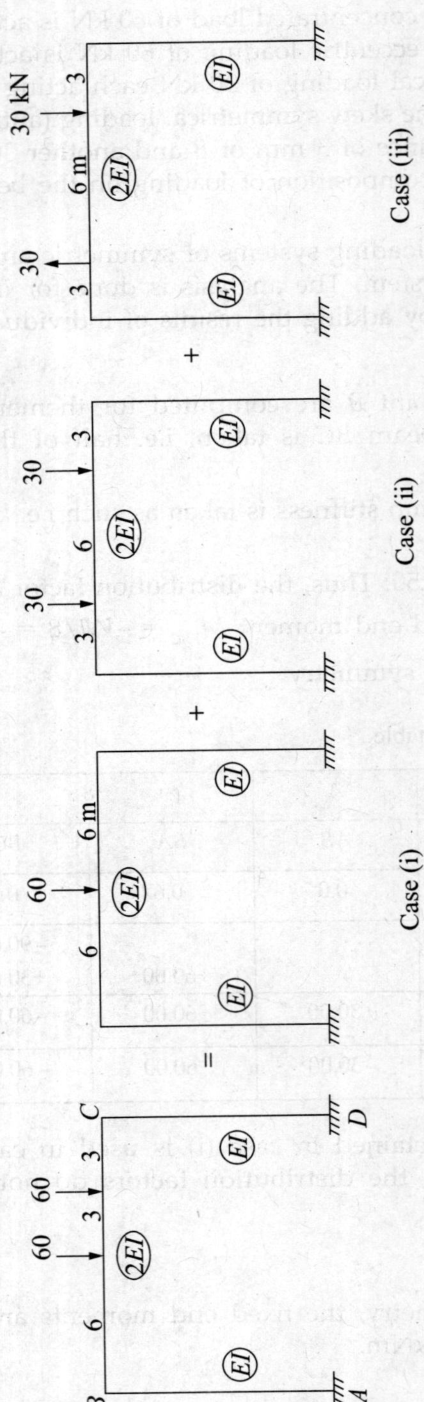

FIG. 2.137 Decomposition of given load system to symmetric and skew symmetric loading systems

Table 2.76

Joint	*A*	*B*	
Members	*AB*	*BA*	*BC*
DF	0.0	0.67	0.33
FEMS Bal		 +45.0	−67.5 +22.5
CO	+22.5		
Total	22.5	45.0	−45.0

Case (iii) Skew symmetric case

The distribution factors at joint B are calculated by considering the cantilever stiffness and taking the skew stiffness of the beam. $M_{FBC} = (3 \times 30 \times 9^2/12^2 - 9 \times 30 \times 3^2/12^2)$ = +33.75 kNm. The cantilever stiffness is taken as $(I/6)$ and for the beam $(6 \times 2I/12)$. Hence,

$$d_{BA} : d_{BC} = (I/6) : (6 \times 2I/12), \quad \text{i.e. } d_{BA} : d_{BC} = 1 : 6$$

i.e. $$d_{BA} = (1/7) \text{ and } d_{BC} = (6/7).$$

Table 2.77

Joint	*A*	*B*	
Members	*AB*	*BA*	*BC*
DF	0	0.14	0.86
FEMS Bal		 −4.72	+33.75 −29.03
CO	+4.72		
Total	+4.72	−4.72	+4.72

The final moments are obtained by adding:

Members	*AB*	*BA*	*BC*	*CB*	*CD*	*DC*
Case (i)	30.00	60.00	−60.00	+60.00	−60.00	−30.00
Case (ii)	22.50	45.00	−45.00	+45.00	−45.00	−22.50
Case (iii)	4.72	−4.72	+4.72	+4.72	−4.72	+4.72
Total	57.22	100.28	−100.28	109.72	−109.72	−47.78

The above moments tally with the moments obtained by classical moment distribution (Ref. Ex. 2.22). The Naylor's method is very much useful if the reader identifies the application of the same to specific problems.

Example 2.32 Analyse the given frame by the Naylor's method and draw the bending moment diagram.

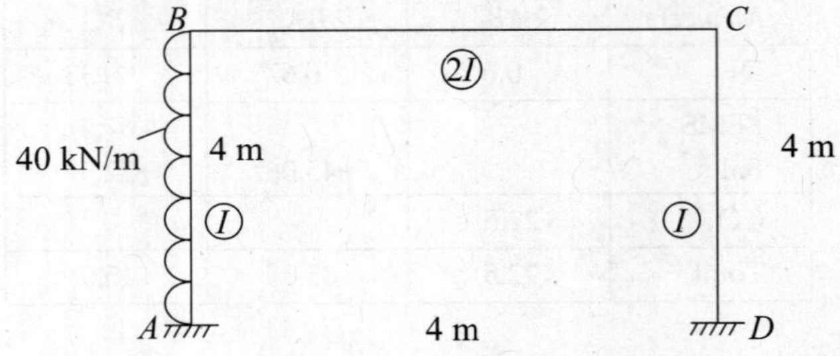

FIG. 2.138

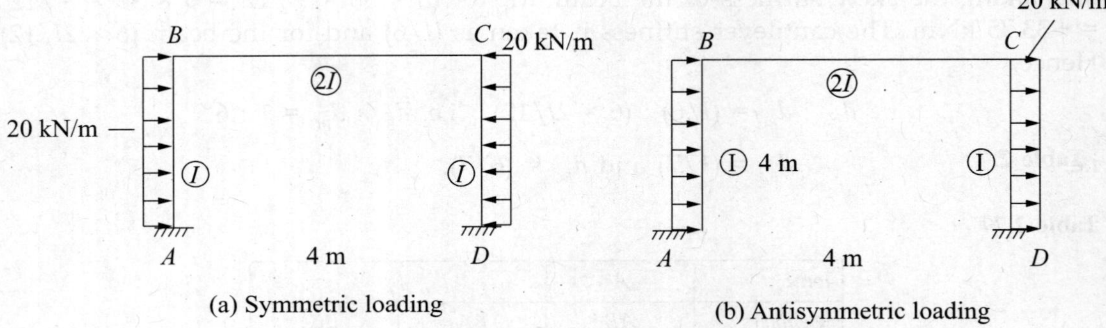

FIG. 2.139

Solution: The given loading is split into symmetric and antisymmetric (skew symmetrical) loading.

In the analysis of symmetrical loading it should be noted that there is no side sway. The distribution is done for one half of the frame. The rotational stiffness for the column *BA* is taken as $(EI/4)$, i.e., the flexural rigidity to the length of the member. The beam stiffness is taken as $\frac{1}{2}\left(\frac{2I}{4}\right) = \frac{I}{4}$. It is half of the original stiffness of the beam. The fixed end moments were obtained in the usual way, i.e.

$$M_F^{AB} = -\frac{20 \times 4^2}{12} = -26.67 \text{ kNm}$$

$$M_F^{BC} = +\frac{20 \times 4^2}{12} = +26.67 \text{ kNm}$$

The distribution factors are obtained as follows.

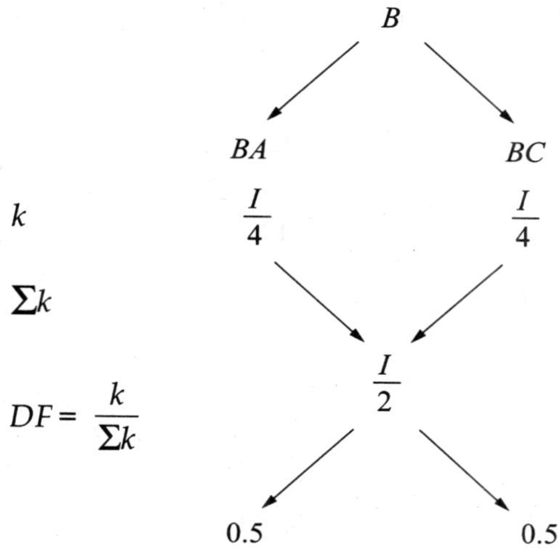

Table 2.78 Member end moments for symmetrical loading

Members	*AB*	*BA*	*BC*
DF	0	0.5	0.5
FEMS Bal	–26.67	+26.67 –13.33	 –13.34
CO	–6.67		
Bal	0.00	0.00	0.00
Total	–33.34	13.34	–13.34

The analysis for antisymmetrical distribution is carried out using cantilever moment distribution. The stiffness for the column is taken as $(EI/4)$ and for the beam it is taken as six times the original stiffness. Hence, $6E(2I)/4 = EI$. The fixed end moments for the columns are taken as

$$M_{FAB} = -\frac{wl^2}{12} - \frac{wl}{2}\left(\frac{l}{2}\right) = -\frac{wl^2}{3} = -20 \times \frac{4^2}{3} = -106.67 \text{ kNm}$$

$$M_{FBA} = +\frac{wl^2}{12} - \frac{wl}{2}\left(\frac{l}{2}\right) = -\frac{wl^2}{6} = -\frac{20 \times 4^2}{6} = -53.33 \text{ kNm}$$

The carryover factor to the column is (–1) in antisymmetrical loading. The distribution factors for the antisymmetrical loading is

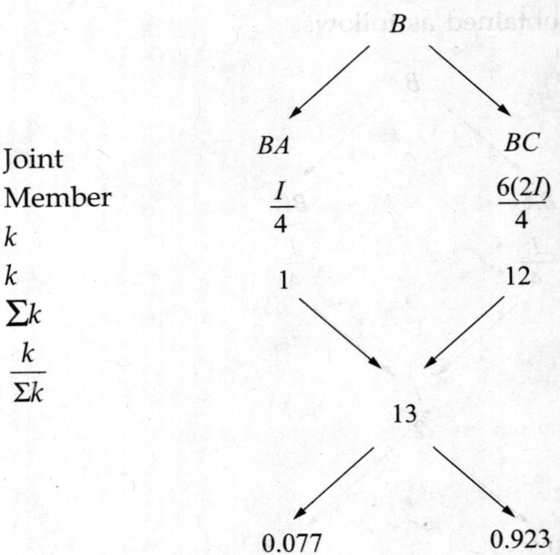

Table 2.79 Member end moments for symmetrical loading

Members	*AB*	*BA*	*BC*
DF	0	0.77	0.923
FEMS Bal	−106.67	−53.33 +4.17	 +49.16
CO	4.17		
Bal	0.00	0.00	0.00
Final	−110.84	−49.16	+49.16

The end moments of the members were obtained by adding the moments due to symmetric loading and antisymmetric loading.

Members	*AB*	*BA*	*BC*	*CB*	*CD*	*DC*
Symmetric Loading	−33.34	+13.34	−13.34	+13.34	−13.34	+33.34
Antisymmetric Loading	−110.84	−49.16	+49.16	+49.16	−49.16	−110.84
Total	−144.18	−35.82	+35.82	+62.5	−62.50	−77.50

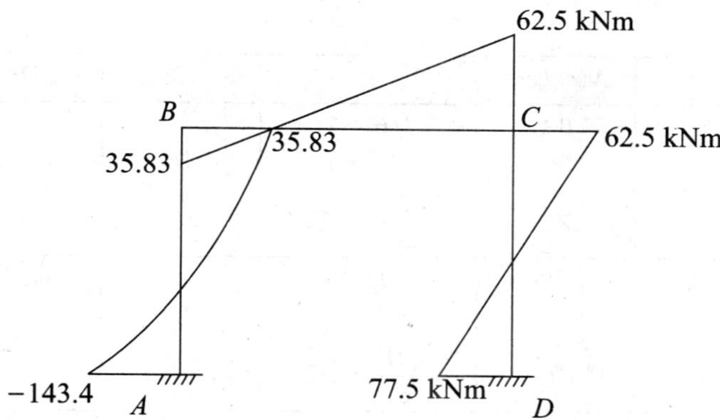

FIG. 2.140 Bending moment diagram

Example 2.33 Analyse the two-storey frame by the Naylor's method. Draw the bending moment diagram.

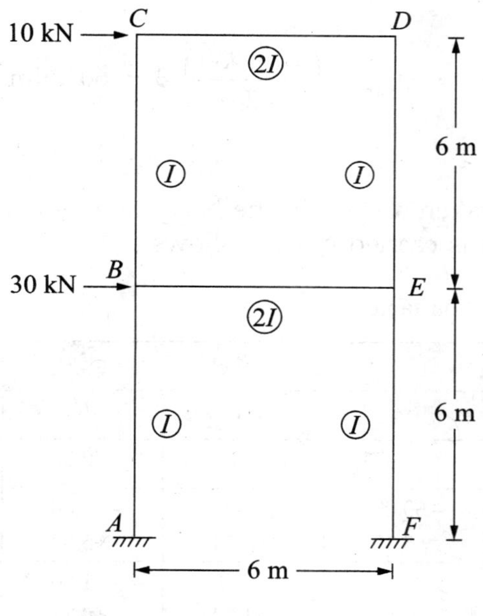

FIG. 2.141

Solution: In the Naylor's method, the relative rotational stiffness for the beam is taken as six times of the actual beam stiffness. The column stiffness is taken as its actual stiffness. Then the distribution factors at joints B and C are determined.

Table 2.80

Joint	Members	k	Σk	DF
B	BA	$I/6 = 1$	14	0.071
	BC	$I/6 = 1$		0.071
	BE	$\frac{6(2I)}{6} = 12$		0.857
C	CB	$I/6$	13	0.077
	CD	$\frac{6(2I)}{6} = 12$		0.923

Fixed end moments

$$M_{FBC} = M_{FCB} = \left(\frac{10}{2}\right) \times 3 = 15\,\text{kNm}$$

$$M_{FBA} = M_{FAB} = \left(\frac{10 + 30}{2}\right) 3 = 60\,\text{kNm}$$

Moment distribution

The carry over factor is taken as (−1) in the Naylor's method of moment distribution. The moment distribution is carried out as follows.

Table 2.81 Member distribution table

Joint	A	B			C	
Members	AB →	BA	BE	BC ↔	CB	CD
DF	0	0.071	0.857	0.071	0.077	0.923
FEMS	−60.00	−60.00	–	−15.00	−15.00	–
Bal		+5.33	+64.34	+5.33	+1.56	+13.44
CO	−5.33			−1.56	−5.33	
Bal		+0.11	+1.34	+0.11	0.41	+4.92
CO	−0.11			0.41	0.11	
Bal		+0.00	+0.39	+0.00	0.00	+0.11
Final	−65.44	−54.56	+66.07	−11.53	−18.47	+18.47

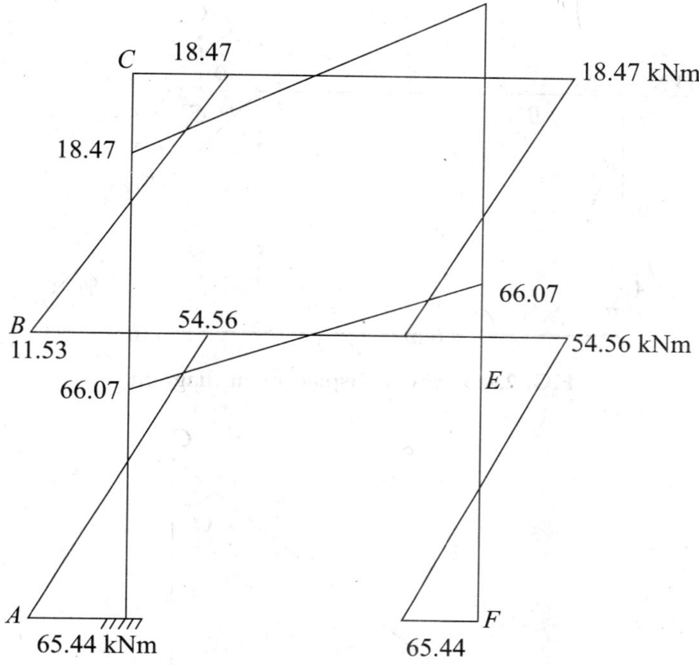

FIG. 2.142 Bending moment diagram

2.11 ANALYSIS OF FRAMES WITH INCLINED LEGS

Example 2.34 Analyse the rigid frame shown in figure by the moment distribution method and draw the BM diagram Support *A* is hinged and support *D* is fixed support.

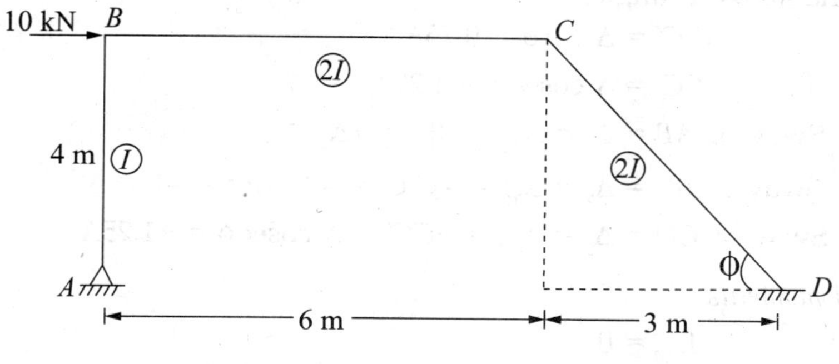

FIG. 2.143

Solution

Determination of sway (Δ) of members

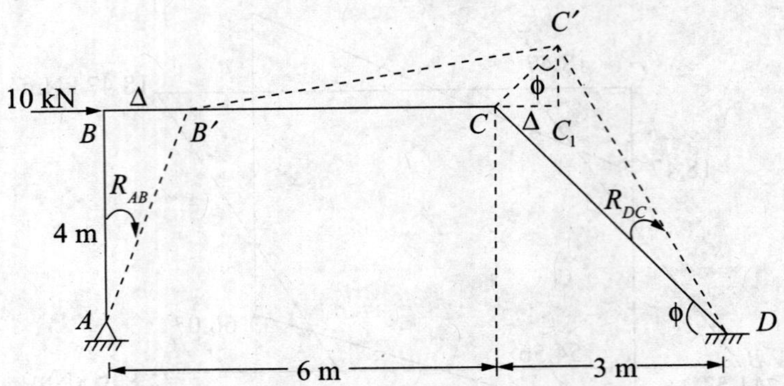

FIG. 2.144 Sway displacement diagram

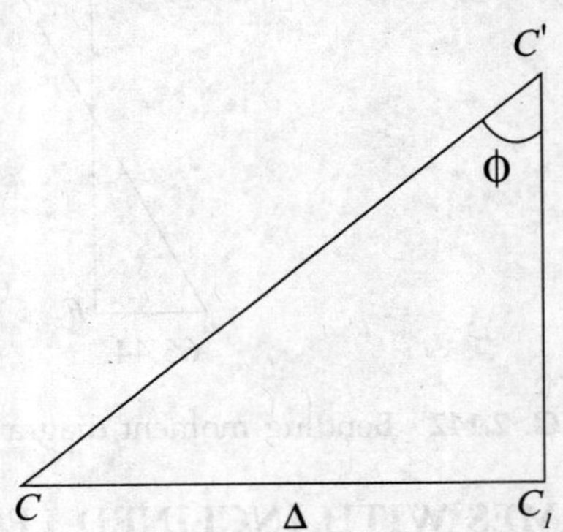

FIG. 2.145

Consider the above triangle,

$$C_1C' = \Delta \cot\phi = 0.75\Delta$$

$$CC' = \Delta \operatorname{cosec}\phi = 1.25\Delta$$

$$\text{Sway of } AB = \Delta_1 = \Delta_{AB} = BB' = +\Delta$$

$$\text{Sway of } BC = \Delta_2 = \Delta_{BC} = -C_1C' = -\Delta \cot\phi = -0.75\Delta$$

$$\text{Sway of } CD = \Delta_3 = \Delta_{CD} = +CC' = \Delta \operatorname{cosec}\phi = +1.25\Delta$$

Sway fixed end moments

$$M_{FAB} = 0$$

$$M_{FAB} = -3EI_1\Delta_1/l_1^2 = -3EI\Delta/4^2 = -\frac{3}{16}EI\Delta$$

$$M_{FBC} = M_{FCB} = -6EI_2\Delta_2/l_2^2 = -6E(2I)(-0.75\Delta)/6^2 = \frac{EI\Delta}{4}$$

$$M_{FCD} = M_{FDC} = -6EI_3\Delta_3/l_3^2 = -6E(2I)(-0.25\Delta)/5^2 = \frac{3EI\Delta}{5}$$

$$\therefore \qquad M_{FBA} : M_{FBC} : M_{FCD} = -\frac{3}{16} : \frac{1}{4} : -\frac{3}{5}$$

$$-15 : +20 : -48$$

Relative Stiffness k Values and Distribution Factors

Table 2.82 Distribution factors

Joint	*Members*	*Relative Stiffness Values I/l*	Σk	$DF = k/\Sigma k$
	BA	$\frac{3}{4}\left(\frac{I}{4}\right) = \frac{3}{16}I$		0.36
B		$= 0.1875I$	$= 0.5175I$	
	BC	$\frac{2I}{6} = 0.33I$		0.64
	CB	$\frac{2I}{6} = 0.33I$		0.45
C			$0.73I$	
	CD	$\frac{2I}{5} = 0.4I$		0.55

Table 2.83 Sway moment distribution table

Joint	*A*	*B*		*C*		*D*
Members	*AB*	*BA*	*BC* ←	→ *CB*	*CD* →	→ *DC*
DF	1	0.36	0.64	0.45	0.55	0
FEMS		−15.00	+20.00	+20.00	−48.00	−48.00
Bal		−1.80	−3.20	+12.60	+15.40	
CO			+6.30	−1.60		+7.70
Bal		−2.27	−4.03	+0.72	+0.88	
CO			+0.36	−2.02		+0.44
Bal		+0.13	−0.23	+0.91	+1.11	
CO			+0.46	−0.12		+0.56
Bal		−0.17	−0.29	+0.05	+0.07	
CO			+0.03	−0.15		+0.04
Bal		−0.01	−0.02	+0.07	+0.08	
CO			+0.04	−0.01		+0.04
Bal		−0.01	−0.03	+0.005	+0.005	
		−19.39	19.39	30.46	−30.46	−39.22

Column shear equation

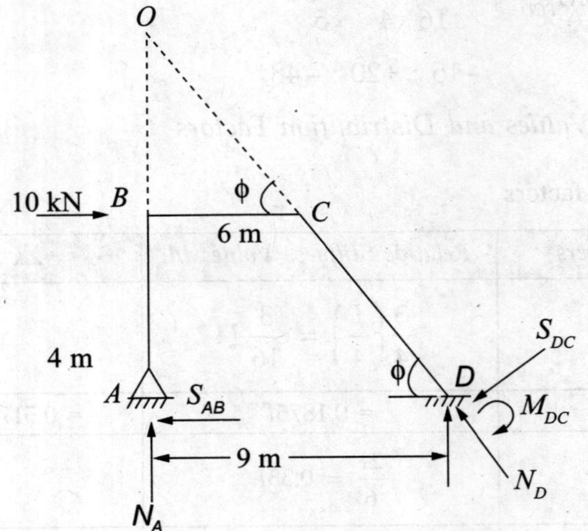

FIG. 2.146(a)

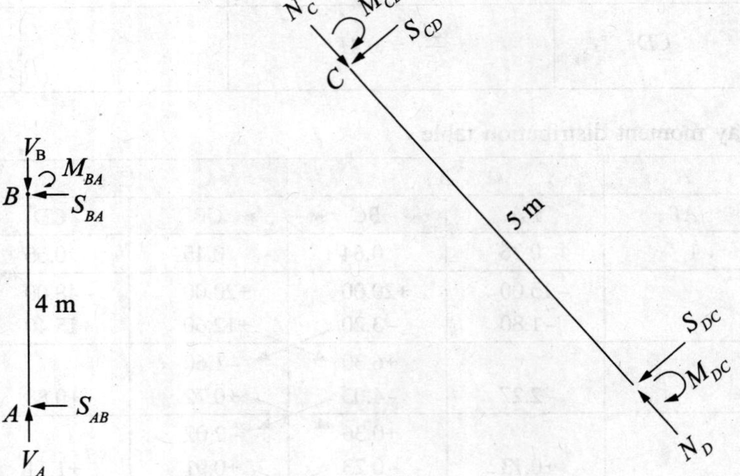

FIG. 2.146(b)

FIG. 2.146(c)

The length of OC in Fig. 2.146(a) can be obtained from geometry as

$$\tan\phi = 4/3$$

i.e.

$$\frac{4}{3} = \frac{OB}{BC}$$

$\therefore$

$$OB = \left(\frac{4}{3}\right) 6 = 8 \text{ m}$$

$$OC = \sqrt{8^2 + 6^2} = 10 \text{ m}$$

Referring to Fig. 2.146 (a),

$$\Sigma M_0 = 0$$

$$12\ S_{AB} + 15\ S_{DC} + M_{DC} - 10(8) = 0$$

$$12\ S_{AB} + 15\ S_{DC} + M_{DC} = 80 \quad (1)$$

Referring to Fig. 2.146 (b),

$$\Sigma M_B = 0$$

$$4\ S_{AB} + M_{BA} = 0 \quad (2)$$

$$\boxed{S_{AB} = -0.25\ M_{BA}}$$

Referring to Fig. 2.146 (c),

$$\Sigma M_C = 0$$

$$5\ S_{DC} + M_{CD} + M_{DC} = 0$$

$$\boxed{S_{DC} = -0.2(M_{CD} + M_{DC})} \quad (3)$$

Substituting S_{DC}, S_{AB} in Eq. (1), and simplifying

$$3\ M_{BA} + 3\ M_{CD} + 2M_{DC} + 80 = 0$$

Substituting the values of moments

$$3(-19.39k) + (-30.46k) + (-30.46k) + 80 = 0$$

$$\boxed{k = 0.38}$$

Final moments

$$M_{AB} = 0.0$$

$M_{BA} = -19.39(0.38) = -7.37$ kNm. Similarly, the sway moments obtained in the table are multiplied by the k value. Hence, $M_{BC} = +7.37$ kNm, $M_{CB} = 11.57$, $M_{CD} = -11.57$, $M_{DC} = -14.9$ kNm

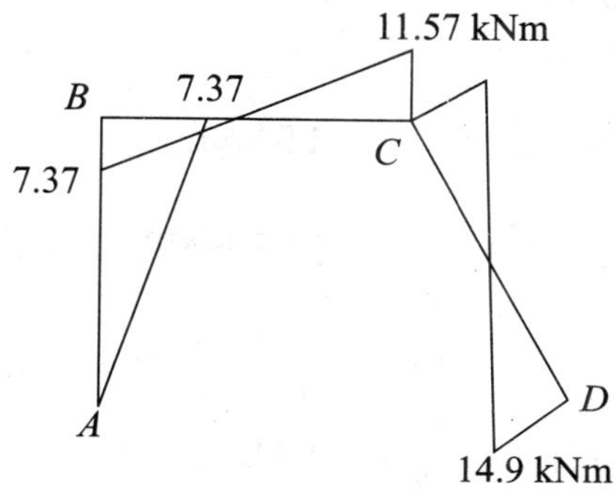

FIG. 2.147 Bending moment diagram

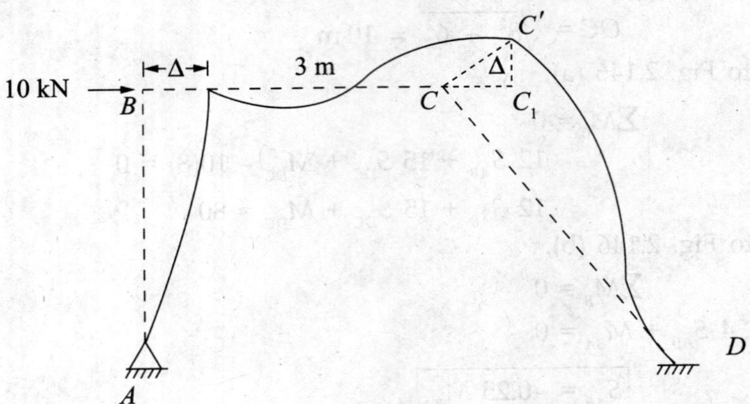

FIG. 2.148 Elastic curve

Example 2.35 Determine the end moments and draw the bending moment diagram.

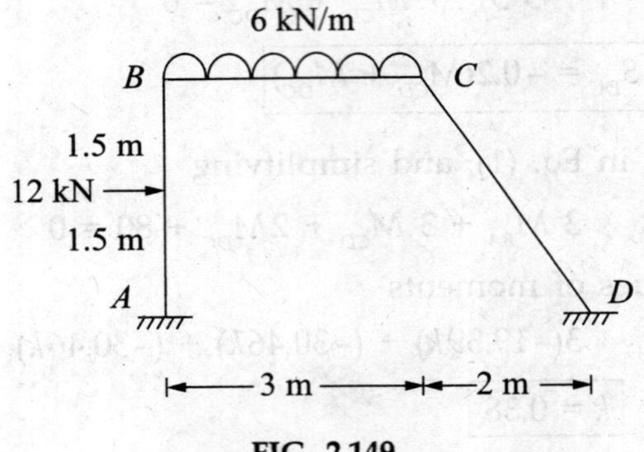

FIG. 2.149

Solution: In inclined frames, the sway of the members is to be determined from geometry. Let the column AB sway by an amount Δ. The beam member BC moves horizontally to $B'C'$ as shown in Fig. 2.150. The displacement CC' is related to sway Δ and the angle of the inclined member CD. Thus, using the geometry of the deflected shape the sway of the members are related.

Fixed end moments due to loading

$$M_{FAB} = -12 \times \frac{3}{8} = -4.5 \text{ kNm};$$

$$M_{FBA} = +12 \times 310 = +4.5 \text{ kNm}$$

$$M_{FBC} = -6 \times \frac{3^2}{12} = -4.5 \text{ kNm}$$

$$M_{FCB} = +6 \times \frac{3^2}{12} = +4.5 \text{ kNm}$$

Determination of sway (Δ) of the member

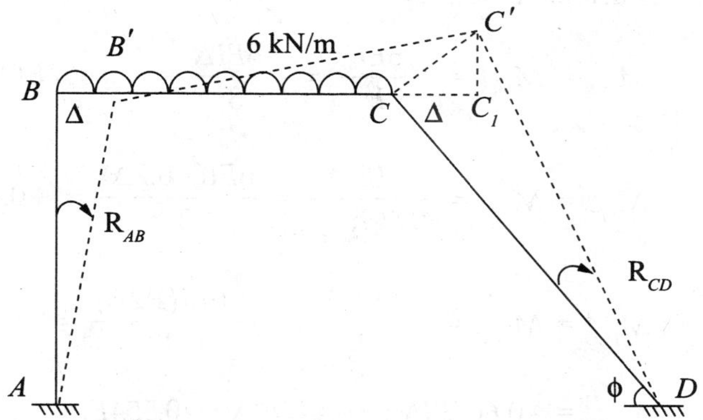

FIG. 2.150 Sway displacement diagram

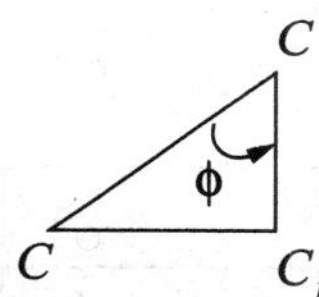

FIG. 2.151

From $\Delta CC_1C'$; $\cot\phi = \dfrac{C_1C'}{\Delta}$

$$\boxed{C_1C' = \Delta\cot\phi}$$

$$\operatorname{cosec}\phi = \frac{CC'}{\Delta}$$

$$\boxed{CC' = \Delta\operatorname{cosec}\phi}$$

Referring to the above given frame

$$\cot\phi = 2/3 = 0.67$$

$$\operatorname{cosec}\phi = \sqrt{13}/3 = 1.20$$

Sway of $AB = \Delta$

Sway of $BC = -0.67\Delta$

Sway of $CD = +1.20\Delta$

Sway moment distribution

The sway moments are assumed as

$$M_{FAB} = M_{FBA} = -\frac{6EI_1}{l_1^2} = -\frac{6EI\Delta}{3^2} = -0.667EI\Delta$$

$$M_{FBC} = M_{FCB} = -\frac{6EI_2\Delta_2}{l_2^2} = \frac{6EI(-0.7\Delta)}{3^2} = +0.447EI\Delta$$

$$M_{FCD} = M_{FDC} = -\frac{6EI_3\Delta_3}{l_3^2} = -\frac{6EI(1.2\Delta)}{13} = -0.554EI\Delta$$

$$\therefore \quad M_{FBA} : M_{FBC} : M_{FCD} = -0.667EI\Delta : +0.447EI\Delta : -0.554EI\Delta$$

$$= -66.70 : +44.70 - 55.40$$

Table 2.84 Moment distribution table

Joint	*A*	*B*		*C*		*D*
Members	*AB* ←	*BA*	*BC* ↔	*CB*	*CD* →	*DC*
	0	0.5	0.5	0.55	0.45	0
FEMS	−66.70	−66.70	+44.70	+44.70	−55.40	−55.40
Bal		+11.00	+11.00	+5.89	+4.82	
CO	+5.50		+2.95	+5.50		+2.41
Bal		−1.48	−1.48	−3.03	−2.47	
CO	−0.74		−1.52	−0.74		−1.24
Bal		+0.76	+0.76	+0.41	+0.33	
CO	+0.38		+0.21	+0.38		+0.17
Bal		−0.10	−0.11	−0.21	−0.17	
CO	−0.05		−0.11	−0.06		−0.09
Bal		+0.06	+0.05	+0.03	+0.03	
CO	+0.03		+0.02	+0.03		+0.02
Bal		−0.01	−0.01	−0.02	−0.01	
Total	−61.58	−56.47	+56.47	+52.88	−52.88	−54.13
	M''_{AB}	M''_{BA}	M''_{BC}	M''_{CB}	M''_{CD}	M''_{DC}

Distribution factors

Table 2.85

Joint	*Members*	*Relative Stiff Values (k)*	Σk	$k/\Sigma k$
B	*BA*	$I/3$	$2I/3$	0.5
	BC	$I/3$		0.5
C	*CB*	$I/3$	$2I/3$	0.55
	CD	$I/\sqrt{3}$		0.45

Table 2.86 Nonsway member distribution table

Joint	A	B		C		D
Members	*AB* ←	*BA*	*BC* ↔	*CB*	*CD* →	*DC*
DF	0	0.5	0.5	0.55	0.45	0
FEMS	-4.50	+4.50	-4.50	+4.50	0	0
Bal		0	0	-2.48	-2.02	
CO	0		-1.24	0		-1.01
Bal		+0.62	+0.62	0	0	
CO	+0.31		0	+0.31		0
Bal		0	0	-0.17	-0.14	
CO	0		-0.09	0		-0.07
Bal		+0.05	+0.04	0	0	
CO	+0.03		0	+0.02		0
Bal		0	0	-0.01	-0.01	
Total	-4.16 M'_{AB}	+5.17 M'_{BA}	-5.17 M'_{BC}	2.17 M'_{CB}	-2.17 M'_{CD}	-108 M'_{DC}
	-61.58 M''_{AB}	-56.47 M''_{BA}	+56.47 M''_{BC}	52.88 M''_{CB}	-52.88 M''_{CD}	-54.13 M''_{DC}

End moments

The end moments are the sum of the moments of the nonsway and sway moments.

$$M_{AB} = -4.16 - 61.58\ k$$

$$M_{BA} = +5.17 - 56.47\ k$$

$$M_{BC} = -5.17 + 56.47\ k$$

$$M_{CB} = +2.17 + 52.88\ k$$

$$M_{CD} = -2.17 - 52.88\ k$$

$$M_{DC} = -1.08 - 54.13\ k$$

Column shear equation

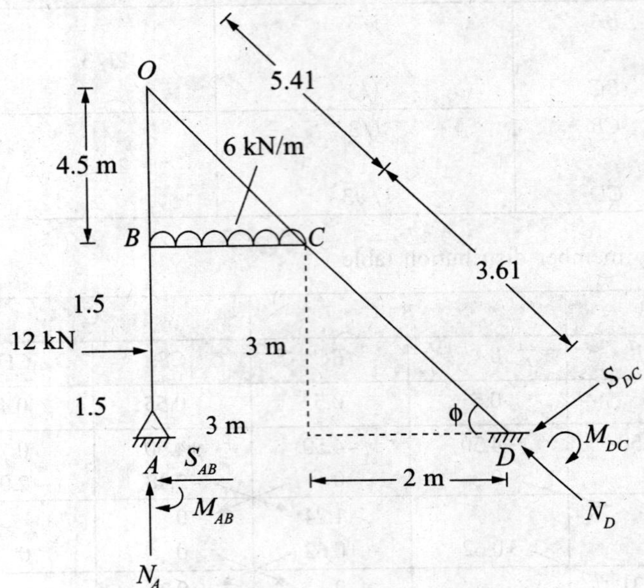

FIG. 2.152(a)

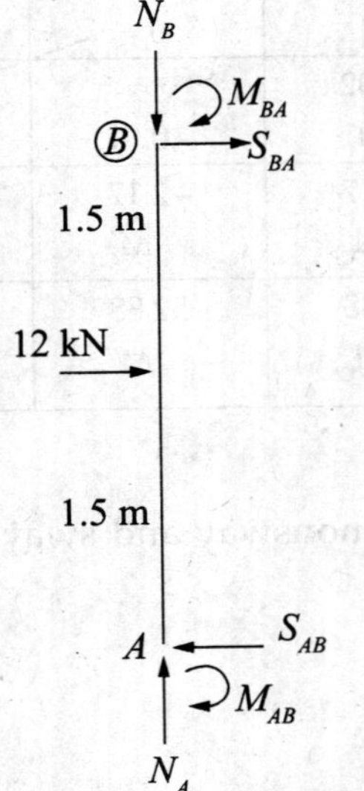

FIG. 2.152(b)

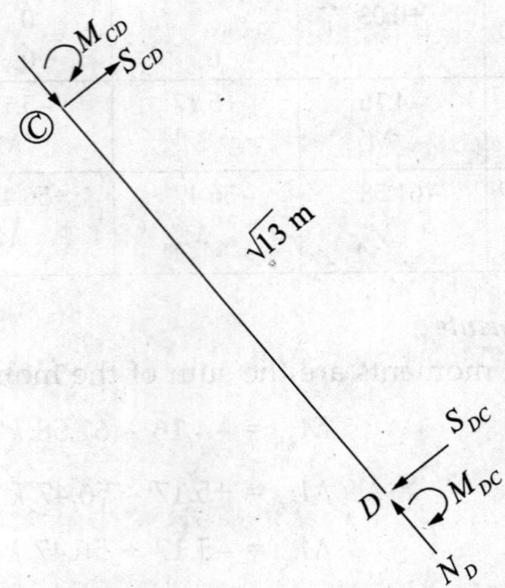

FIG. 2.152(c)

Referring to Fig. 2.152 (a)

$$\Sigma M_0 = 0;$$

$$-12(6) + 6(3)(3/2) + 7.5\ S_{AB} + 9.02\ S_{DC} + (M_{AB} + M_{DC}) = 0$$

$$\boxed{(M_{AB} + M_{DC}) + 7.5\ S_{AB} + 9.02\ S_{DC} = 45} \quad (1)$$

Referring to Fig. 2.152 (b)

$$\Sigma M_B = 0;$$

$$M_{AB} + M_{BA} + 3\ S_{AB} - 12(1.5) = 0$$

$$\boxed{S_{AB} = 6 - \frac{(M_{AB} + M_{BA})}{3}} \quad (2)$$

Referring to Fig. 2.152 (c)

$$\Sigma M_C = 0;$$

$$M_{CD} + M_{DC} + 3.61\ S_{DC} = 0$$

$$\boxed{SDC = -\frac{(M_{CD} + M_{DC})}{3.61}} \quad (3)$$

Substituting (2) and (3) in (1) and with the use of end moments equation;

$$\boxed{k = -0.002}$$

Final moments

$$M_{AB} = -4.16 - 61.58(-0.002) = -4.04\text{ kNm}$$

$$M_{BA} = +5.17 - 56.47(-0.002) = +5.28\text{ kNm}$$

$$M_{BC} = -5.17 + 56.47(-0.002) = -5.28\text{ kNm}$$

$$M_{CB} = 2.17 + 52.90(-0.002) = +2.06\text{ kNm}$$

$$M_{CD} = -2.17 - 52.90(-0.002) = -2.06\text{ kNm}$$

$$M_{DC} = -1.08 - 54.13(-0.002) = -0.97\text{ kNm}$$

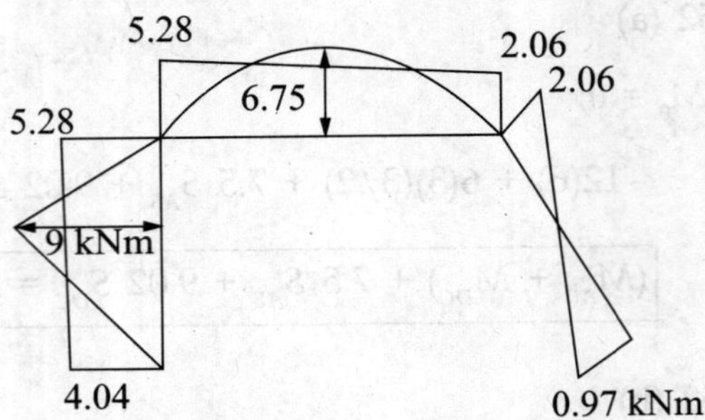

FIG. 2.153 Bending moment diagram

Example 2.36 Analyse the given frame by the moment distribution method. *EI* is same throughout. Draw the bending moment diagram and the elastic curve.

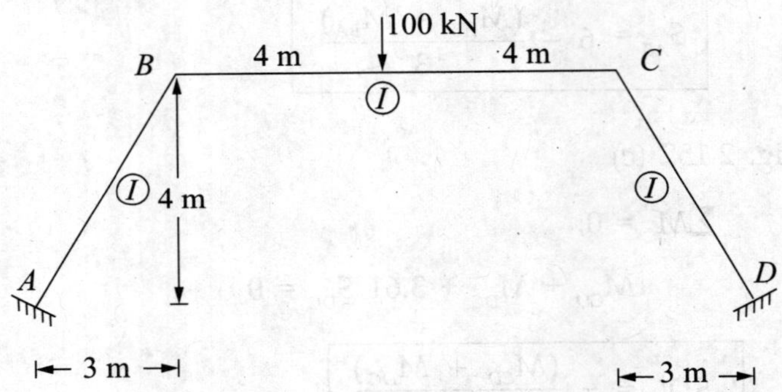

FIG. 2.154

Solution

Determination of sway (Δ) of the members

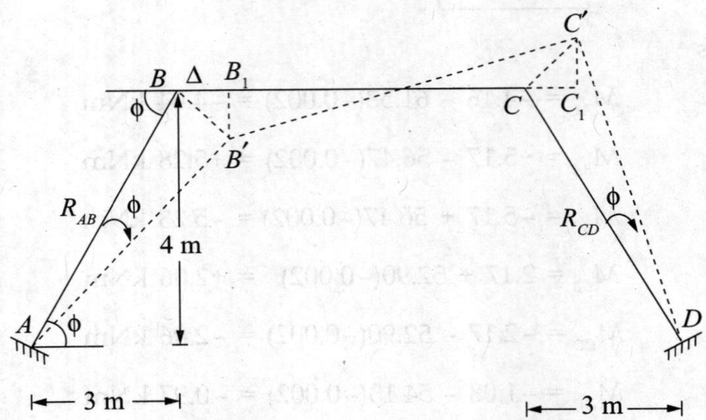

FIG. 2.155 Sway diagram

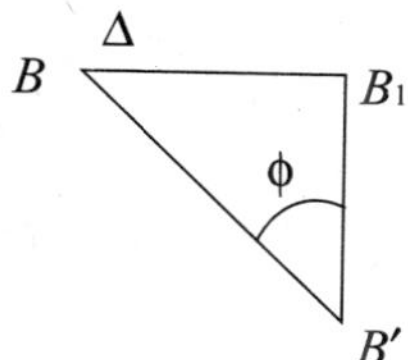

FIG. 2.156

$$\text{cosec}\,\phi = 5/4 = 1.25$$

$$\cot\phi = 3/4 = 0.75$$

From the displacement diagram

$$\cot\phi = \frac{B_1B'}{\Delta}$$

$$\boxed{B_1B' = \Delta\cot\phi}$$

$$\text{cosec}\,\phi = \frac{BB'}{BB_1}$$

$$BB' = BB_1\,\text{cosec}\,\phi$$

$$\boxed{BB' = 1.25\Delta}$$

$$\therefore \quad \boxed{CC' = 1.25\Delta}$$

Vertical displacement between B' and C'

$$= B'B_1 + C_1C' = 2\Delta\cot\phi = 1.5\Delta$$

$$\text{Sway of } AB = \Delta_{AB} = BB' = \Delta\,\text{cosec}\,\phi = +1.25\Delta$$

$$\text{Sway of } BC = \Delta_{BC} = -1.5\Delta$$

$$\text{Sway of } CD = \Delta_{CD} = \Delta\,\text{cosec}\,\phi = +1.25\Delta$$

Nonsway Analysis

$$M_{FBC} = -100 \times \frac{8}{8} = -100\,\text{kNm}$$

$$M_{FCB} = +100 \times \frac{8}{8} = +100\,\text{kNm}$$

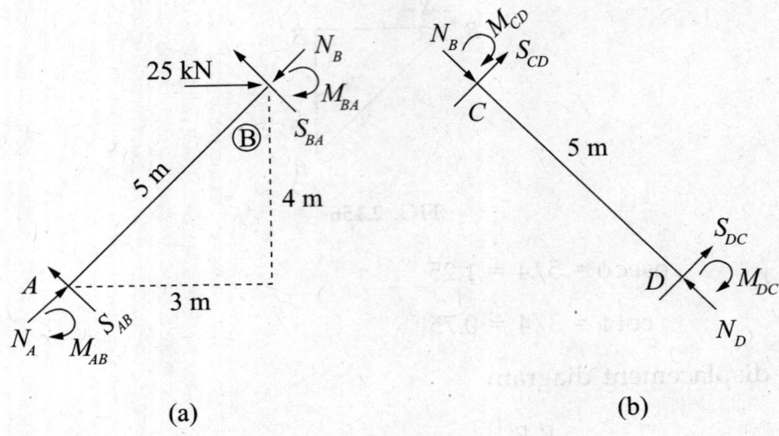

FIG. 2.157

Considering the above free body diagram,

$$\Sigma M_B = 0;$$

$$M_{AB} + M_{BA} + 5\, S_{AB} = 0$$

i.e.
$$\boxed{S_{AB} = -\frac{(M_{AB} + M_{BA})}{5}} \qquad (1)$$

$$\Sigma M_C = 0;$$

$$M_{CD} + M_{DC} - 5\, S_{DC} = 0$$

$$\boxed{S_{DC} = -\frac{(M_{CD} + M_{DC})}{5}} \qquad (2)$$

On simplifying

$$(M_{AB} + M_{DC}) + 1.75(M_{BA} + M_{CD}) + 100 = 0$$

The values of the moments are obtained from the above nonsway moment distribution and sway moment distribution case as

$$\{(38.27 - 26.39k) - (38.27 + 24.78k)\} + 1.75\{(76.55 - 22.74k) + (-76.55 - 24.78k)\} + 100 = 0$$

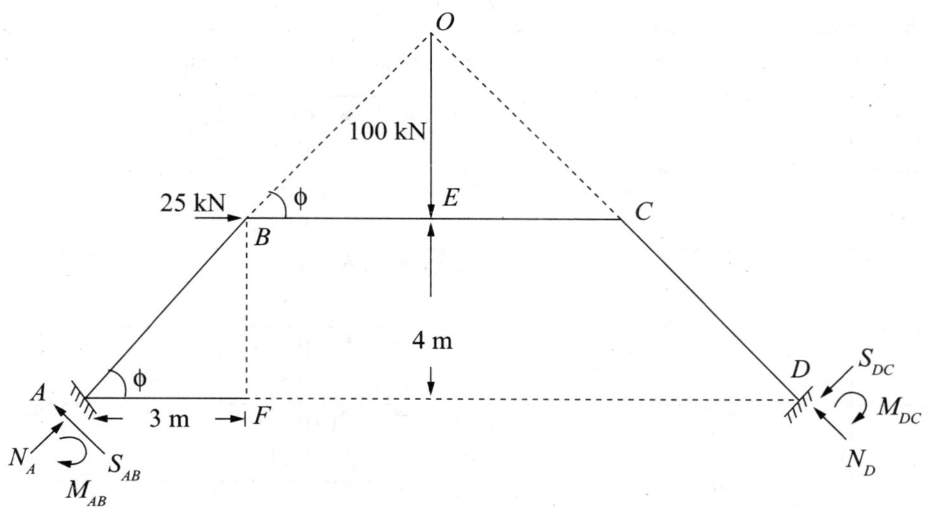

FIG. 2.158

From ΔABF, $\tan\phi = 4/3$

Using the above property in ΔOBE

$$\frac{OE}{BE} = \frac{4}{3} \tag{1}$$

$\therefore$ $$OE = 5.33 \text{ m}$$

Hence, $$OB = \sqrt{4^2 + 5.33^2} = 6.66 \text{ m}$$

$$OA = OB + BA = 6.66 + 5 = 11.66 \text{ m}$$

$$\Sigma M_0 = 0$$

$$11.66\, S_{AB} + 11.66\, S_{DC} + M_{AB} + M_{DC} - 25 \times 5.33 = 0$$

i.e. $$(M_{AB} + M_{DC}) + 11.66(S_{AB} + S_{DC}) = 133.25 \tag{2}$$

Sway Analysis

The sway moment are as follows.

$$M_{FAB} = M_{FBA} = -\frac{6EI_1\Delta_1}{l_1^2} = \frac{6EI(1.25\Delta)}{5^2} = -0.3EI\Delta$$

$$M_{FBC} = M_{FCB} = -\frac{6EI_2\Delta_2}{l_2^2} = \frac{6EI(-1.5\Delta)}{8^2} = +0.14EI\Delta$$

$$M_{FCD} = M_{FDC} = -\frac{6EI_3\Delta_3}{l_3^2} = \frac{6EI(1.25\Delta)}{5^2} = -0.3EI\Delta$$

Taking $EI\Delta = 100$ and the sway moments are assumed as $-30.00 : +14.00 : -30.00$.

Table 2.87 Sway moment distribution table

Joint	*A*	*B*		*C*		*D*
Members	*AB*	*BA*	*BC* ⟷	*CB*	*CD*	*DC*
DF	0	0.62	0.38	0.62	0.38	0
FEM	−30.00	−30.00	+14.00	+14.00	−30.00	−30.00
Bal		+9.92	+6.08	+9.92	+6.08	
Co	+4.96		+4.96	+3.04		+3.04
Bal		−3.10	−1.86	−1.88	−1.16	
Co	−1.55		−0.94	−0.93		−0.58
Bal		+0.58	+0.36	+0.58	+0.35	
Co	+0.29		+0.29	+0.18		+0.18
Bal		−0.18	−0.11	−0.11	−0.07	
Co	−0.09		−0.06	−0.06		−0.04
Bal		+0.04	+0.02	+0.04	+0.02	
Final	−26.39	−22.74	+22.74	+24.78	−24.78	−27.40

Table 2.88 Nonsway moment distribution table

Joint	*A*	*B*		*C*		*D*
Members	*AB* ⟵	*BA*	*BC* ⟷	*CB*	*CD*	*DC*
DF	0	0.62	0.38	0.38	0.62	0
FEMS	0	0	−100.00	+100.00	0	0
Bal		+62.00	+38.00	−38.00	−62.00	
Co	+31.00		−19.00	+19.00		−31.00
Bal		+11.78	+7.22	−7.22	−11.78	
Co	+5.89		−3.61	+3.61		−5.89
Bal		+2.24	+1.37	−1.37	−2.24	
Co	+1.12		−0.69	+0.69		−1.12
Bal		+0.43	+0.26	−0.26	−0.43	
Co	+0.22		−0.13	+0.13		−0.22
Bal		+0.08	+0.05	−0.05	−0.08	
Co	+0.04		−0.03	+0.03		−0.04
Bal		+0.02	+0.01	−0.01	−0.02	
Total	38.27	76.55	−76.55	+76.55	−76.55	−38.27

$$-51.17\,k - 83.16\,k + 100 = 0$$

$$\boxed{k = 0.74}$$

Final moments

$$M_{AB} = 38.27 - 26.39\,k(0.74) = 18.74 \text{ kNm}$$

$$M_{BA} = 76.55 - 22.74\,k(0.74) = 59.72 \text{ kNm}$$

$$M_{BC} = -76.55 + 22.74(0.74) = -59.72 \text{ kNm}$$

$$M_{CB} = +76.55 + 24.78(0.74) = 94.88 \text{ kNm}$$

$$M_{CD} = -76.55 - 24.78(0.74) = -94.88 \text{ kNm}$$

$$M_{DC} = -38.27 - 27.40(0.74) = -58.55 \text{ kNm}$$

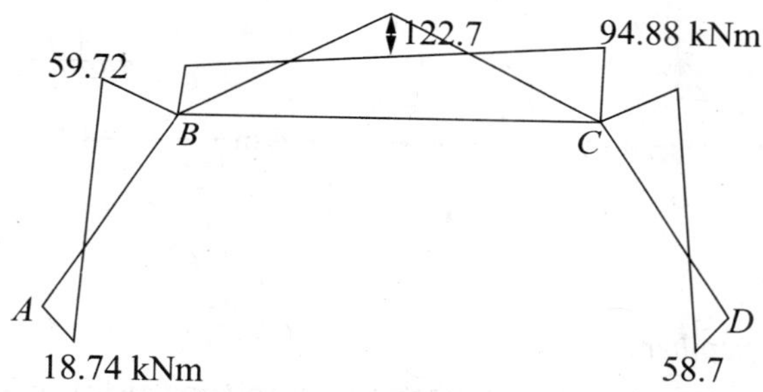

FIG. 2.159 Bending moment diagram

$$M_E = 100 \times \frac{8}{4} - \frac{1}{2}(59.72 + 94.88) = 122.7 \text{ kNm}$$

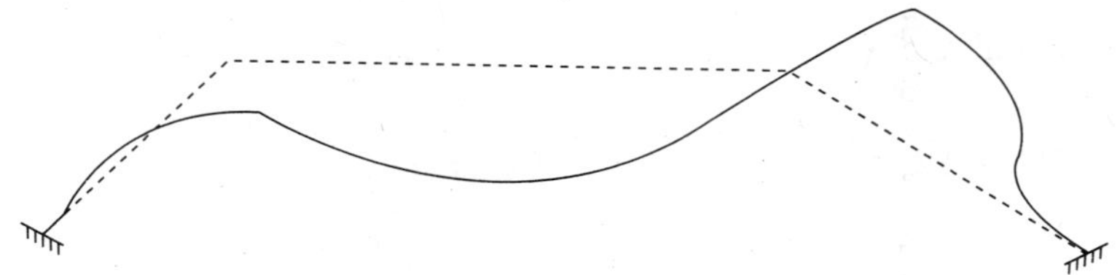

FIG. 2.160 Elastic curve

2.12 ANALYSIS OF GABLE FRAMES

Example 2.37 Analyse the rigid frame shown in Fig. 2.161 by the moment distribution method and sketch the bending moment diagram.

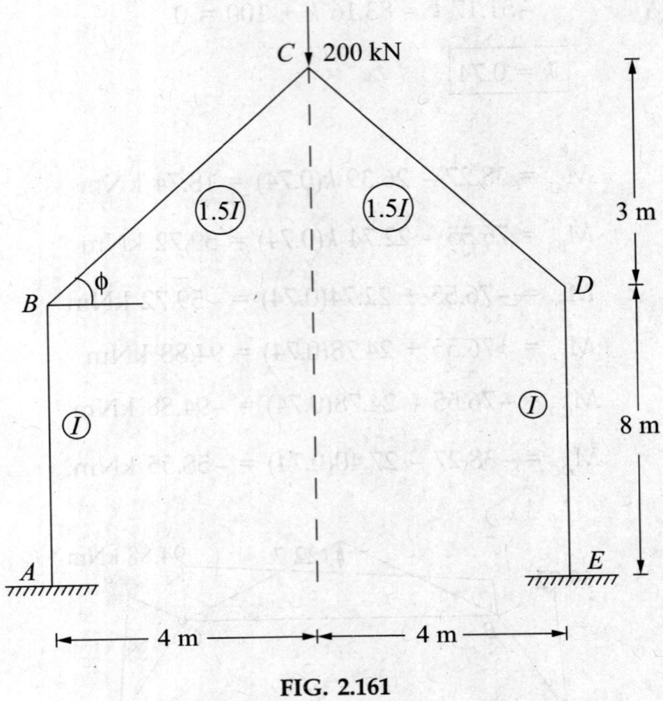

FIG. 2.161

Solution

Sway (Δ) values of the members

The Gable frame sways equally on both sides (either side) at B and D by virtue of symmetry on the frame and the loading on YY axis.

Sway of $AB = \Delta_{AB} = -BB' = -\Delta$ (negative sign because AB sways in the anticlockwise direction with respect to AB)

Sway of $BC = \Delta_{BC} = +C_1C' = +\Delta \operatorname{cosec}\phi$

$$\therefore \quad \Delta_{BC} = +\frac{5}{3}\Delta$$

Sway of $CD = \Delta_{CD} = -C_2C' = -\Delta \operatorname{cosec}\phi$

$$\therefore \quad \Delta_{CD} = -\frac{5}{3}\Delta$$

Sway of $DE = \Delta_{DE} = DD'$

$$\therefore \quad \Delta_{DE} = +\Delta$$

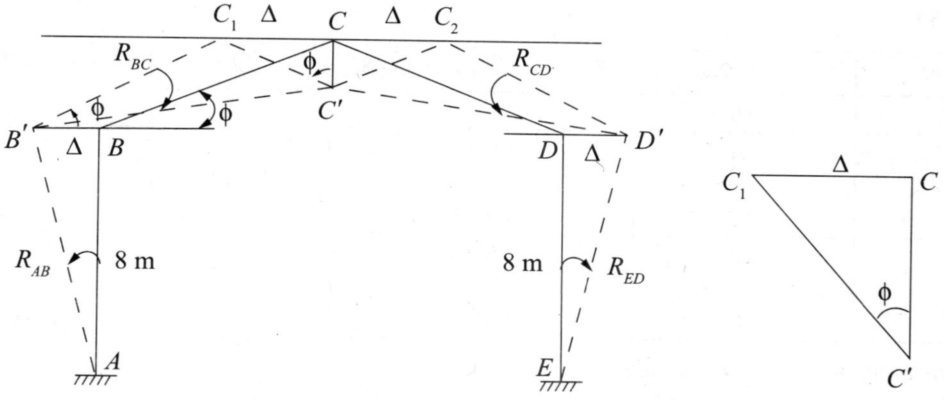

FIG. 2.162 Sway diagram

FIG. 2.163

Sway moment values

(Fixed end moments due to sway)

$$M_{FAB} = M_{FBA} = -\frac{EI_1 \;\; 1}{2} = -\frac{-6EI\Delta}{8^2} = -\frac{6EI(-\Delta)}{64} = +\frac{EI}{64}$$

$$M_{FBC} = M_{FCB} = -\frac{6EI_2\Delta_2}{l_2^2} = -\frac{-6E(1.5\Delta)\,(5/3\Delta)}{5^2} = -\frac{15EI\Delta}{25}$$

$$M_{FCD} = M_{FDC} = -\frac{6EI_3\Delta_3}{l_3^2} = -\frac{-6E(1.5Z)\,(-5/3\Delta)}{5^2} = +\frac{15EI\Delta}{25}$$

$$M_{FDE} = M_{FED} = \frac{6EI_4\Delta_4}{l_4^2} = -\frac{-6EI\Delta}{8^2} = -\frac{6EI\Delta}{64}$$

$$\therefore \qquad \frac{M_{FAB}}{M_{FBC}} = \frac{+6EI\Delta/64}{-15EI\Delta/25} = \frac{+10}{-64}$$

Assume the sway moments in the above proportion as follows:

$$M_{FAB} = M_{FBA} = +10.00 \text{ kNm}$$

$$M_{FBC} = M_{FCB} = -64.00 \text{ kNm}$$

$$M_{FCD} = M_{FDC} = +64.00 \text{ kNm}$$

$$M_{FDE} = M_{FED} = -10.00 \text{ kNm}$$

Distribution factors

Due to symmetry, consider joints B and C only.

Table 2.89

Joint	*Members*	*Distribution Factors (I/l)*	Σk	$DF = k/\Sigma k$
A	*AB*	0	–	0
B	*BA*	$I/8$	$0.425I$	0.30
	BC	$1.5I/5$		0.70
C	*CB*	$1.5I/5$	$0.6I$	0.50
	CD	$1.5I/5$		0.50

Sway moment distribution table

Table 2.90

Joint	*A*	*B*		*C*	
Members	*AB*	*BA*	*BC*	*CB*	*CD*
DF	0	0.30	0.70	0.5	0.5
FEMS	+10.00	+10.00	−64.00	−64.00	+64.00
Bal		+16.20	+37.80	0	0
Co	+8.10	–	0	+18.90	−18.90
Bal		0	0	0	0
Final	+18.10	+26.20	−26.20	−45.10	+45.10
	M''_{AB}	M''_{BA}	M''_{BC}	M''_{CB}	M''_{CD}
	$+18.10k$	$26.20k$	$-26.20k$	$-45.1k$	$+45.1k$

Column shear condition

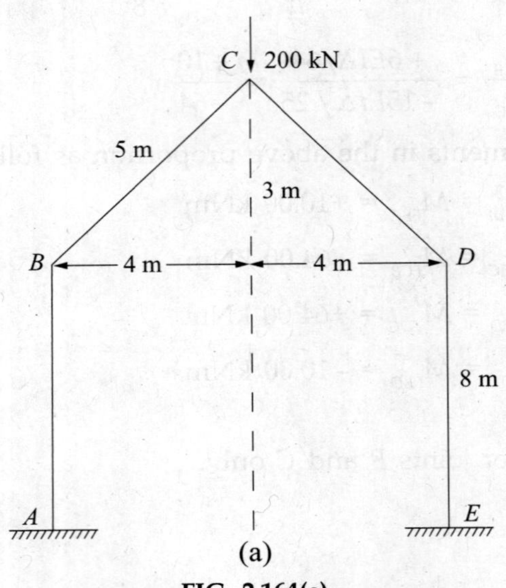

FIG. 2.164(a)

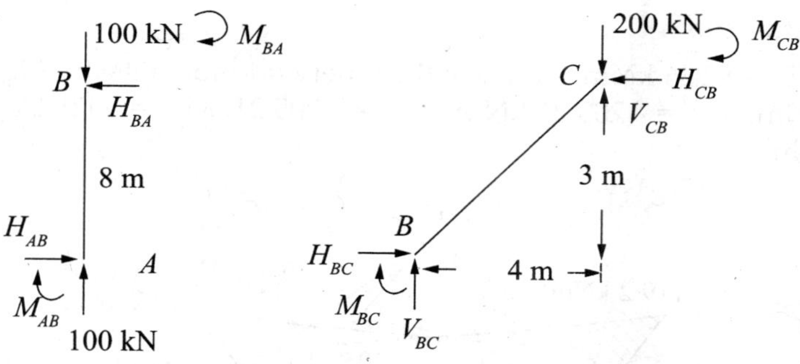

Column AB

FIG. 2.164(b)

Member BC

FIG. 2.164(c)

In Fig. 2.164(b);

$$\Sigma M_B = 0;$$

$$-8H_{AB} + M_{BA} + M_{AB} = 0$$

$$\therefore \quad \boxed{H_{AB} = \frac{(M_{AB} + M_{BA})}{8}}$$

$$H_{BC} = H_{BA} = H_{AB} = (M_{AB} + M_{BA})/8$$

$$\therefore \quad V_{AB} = V_{BC} = V_{BA} = 100 \text{ kN}$$

Referring to Fig. 2.164(c) and taking moment about (c),

$$4\ V_{BC} - 3\ H_{BC} + M_{BC} + M_{CB} = 0$$

Substituting the value of H_{BC} in terms of moments,

$$4(100) - 3\frac{(M_{AB} + M_{BA})}{8} + (M_{BC} + M_{CB}) = 0$$

$$3200 - 3(M_{AB} + M_{BA}) + 8(M_{BC} + M_{CB}) = 0$$

Substituting the end moments from moment distribution table with a multiplier 'k';

$$3200 - 3(18.1k + 26.2k) + 8(-26.2k - 45.1k) = 0$$

$$\therefore \quad \boxed{k = 4.55}$$

On back substitution

$M_{AB} = 18.1(4.55) = 82.36$ kNm, similarly the other end moments are $M_{BA} = 119.21$ kNm, $M_{BC} = -119.21$ kNm, $M_{CB} = -205.21$ kNm, $M_{CD} = +205.21$, $M_{DC} = +119.21$, $M_{DE} = -119.21$, $M_{ED} = -82.36$ kNm.

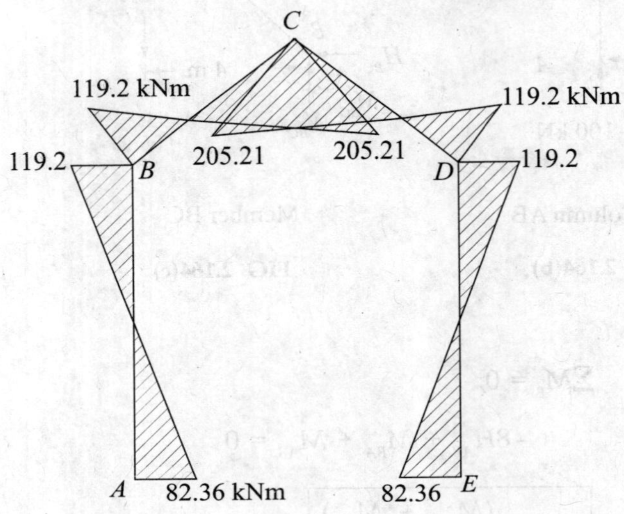

FIG. 2.165 Bending moment diagram

Example 2.38 Analyse the Gable frame shown in Fig. 2.166 by the moment distribution method. Draw the bending moment diagram, shear force diagram and the elastic curve.

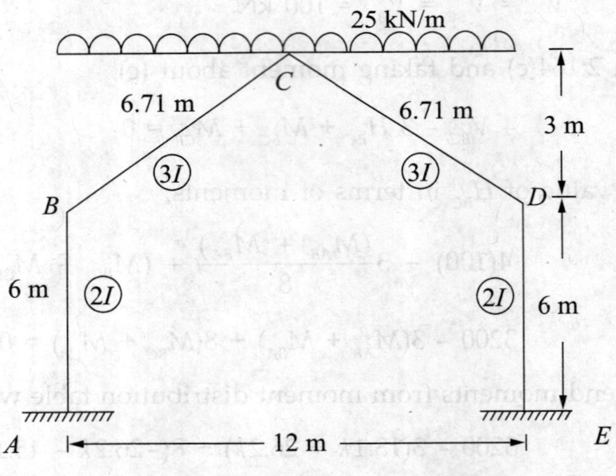

FIG. 2.166

Solution: The Gable frame sways equally on either side *B* and *D* by virtue of symmetry of the frame and loading.

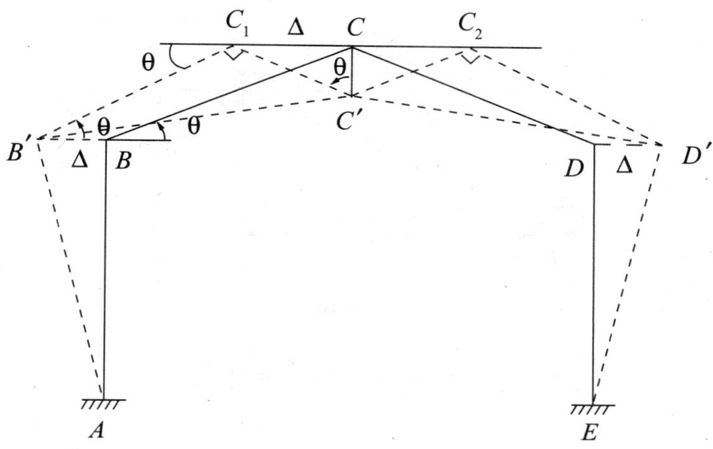

FIG. 2.167 Sway diagram

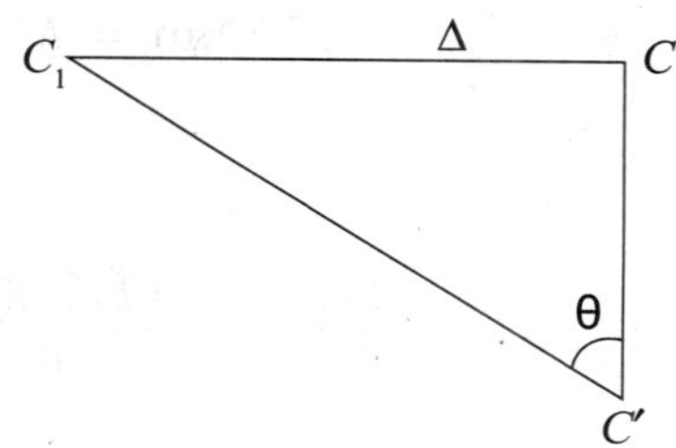

FIG. 2.168

In Fig. 2.166 length $BC = \sqrt{6^2 + 3^2} = 6.71\text{ m}$

$$\operatorname{cosec}\theta = 6.71/3 = 2.237$$

In Fig 2.167 $\operatorname{cosec}\theta = \dfrac{C_1C'}{\Delta}$

$$C_1C' = \Delta \operatorname{cosec}\theta = 2.237\Delta$$

In Fig. 2.167 Sway of $AB = BB' = -\Delta$

Sway of $BC = C_1C' = +2.237\Delta$

Sway of $CD = C_2C' = -2.237\Delta$

Sway of $ED = DD' = +\Delta$

Fixed end moments

Due to loading

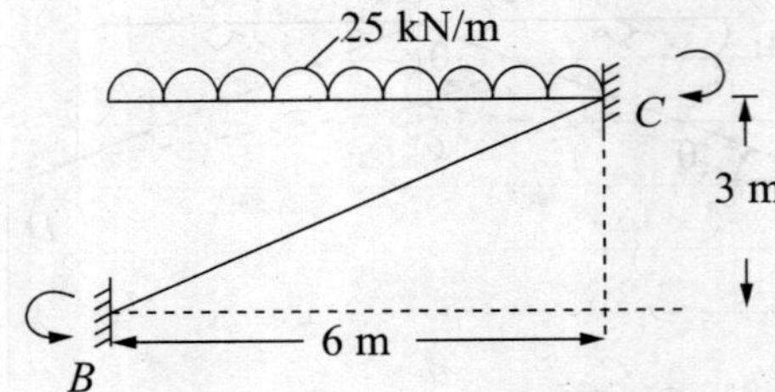

FIG. 2.169

$$M_{FAB} = M_{FBA} = M_{FDE} = M_{FED} = 0$$

$$M_{FBC} = -25 \times \frac{6^2}{12} = -75 \text{ kNm} = M_{FCD}$$

$$M_{FCB} = +25 \times \frac{6^2}{12} = +75 \text{ kNm} = M_{FDC}$$

Sway moment values

$$M_{FAB} = M_{FBA} = -\frac{EI_1 \ _1}{_2} = -\frac{6E(2I)(-\Delta)}{6^2} = \frac{12EI}{36}$$

$$M_{FBC} = M_{FCB} = -\frac{6EI_2\Delta_2}{l_2^2} = -\frac{6E(3I)(2.237\Delta)}{6.71^2} = \frac{-40.27EI\Delta}{45}$$

$$\therefore \quad \frac{M_{FAB}}{M_{FBC}} = \frac{+15}{-40.27}$$

Assume the sway moments in the above proportion as

$$M_{FAB} = M_{FBA} = +15.00 \text{ kNm}$$

$$M_{FBC} = M_{FCB} = -40.27 \text{ kNm}$$

Distribution factors

Due to symmetry, consider joints B and C only.

Table 2.91 Nonsway moment distribution table

Joint	A	B		C
Members	*AB*	*BA*	*BC*	*CB*
DF	0	0.43	0.57	0.50
FEMS			−75.00	+75.00
Bal		+32.25	+42.75	
Co	+16.13			+21.38
Final	+16.13	+32.25	−32.25	+96.38

Table 2.92 Sway moment distribution table

Joint	A	B		C
Members	*AB*	*BA*	*BC*	*CB*
DF	0	0.43	0.57	0.5
FEMS	+15.00	+15.00	−40.27	−40.27
Bal		+10.87	+14.40	
Co	+5.47			+7.20
Final	20.47	25.87	−25.87	−33.07

Final moments

$$M_{AB} = 16.13 + 20.43k$$

$$M_{BA} = 32.25 + 25.87k$$

$$M_{BC} = -32.25 - 25.87k$$

$$M_{CB} = +96.38 - 33.07k$$

Column shear condition

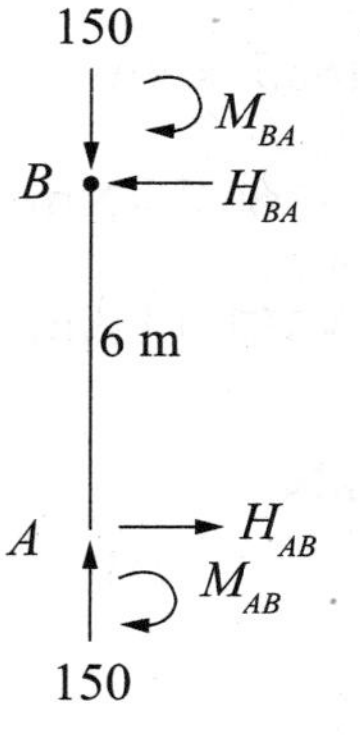

150

M_{BA}

B H_{BA}

6 m

A H_{AB}

M_{AB}

150

FIG. 2.170

$$\Sigma M_B = 0$$

$$M_{AB} + M_{BA} - 6H_{AB} = 0$$

$$M_{AB} = \frac{(M_{AB} + M_{BA})}{6}$$

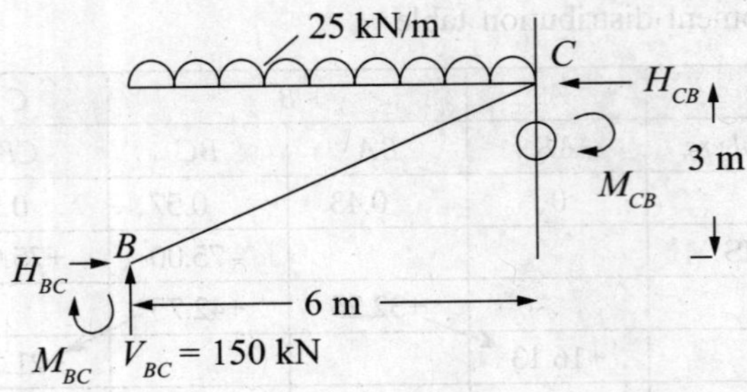

FIG. 2.171

$$\Sigma M_C = 0$$

$$150(6) + M_{BC} + M_{CB} - 3H_{BC} - 25 \times \frac{6^2}{2} = 0$$

As $H_{BC} = H_{BA} = H_{AB}$ and replacing in terms of moments as above.

$$150 \times 6 + M_{BC} + M_{CB} - \frac{3}{6}(M_{AB} + M_{BA}) - 450 = 0$$

$$900 + (64.13 - 58.94k) - 0.5(48.38 + 46.3k) - 450 = 0$$

$$489.94 - 82.09k = 0$$

$$\boxed{k = 5.97}$$

End moments

The end moments are obtained by substituting the value of k in the above equations as

$$M_{AB} = 16.13 + 20.43(5.97) = 138.1 \text{ kNm}$$

$$M_{BA} = 32.25 + 25.87(5.97) = 186.7 \text{ kNm}$$

$$M_{BC} = -32.25 - 25.87(5.97) = -186.7 \text{ kNm}$$

$$M_{CB} = +96.38 - 33.07(5.97) = -101.0 \text{ kNm}$$

$$H_{AB} = \frac{M_{AB} + M_{BA}}{6} = 54.13 \text{ kN}$$

From geometry; $\cos\theta = 6/6.71 = 0.894$ and

$$\sin\theta = 3/6.71 = 0.447$$

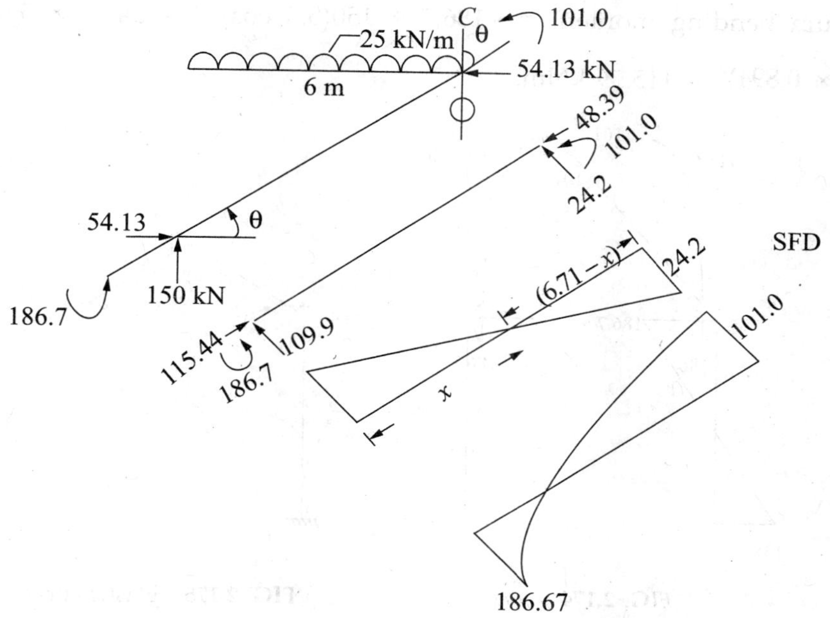

FIG. 2.172 Bending moment diagram (BMD)

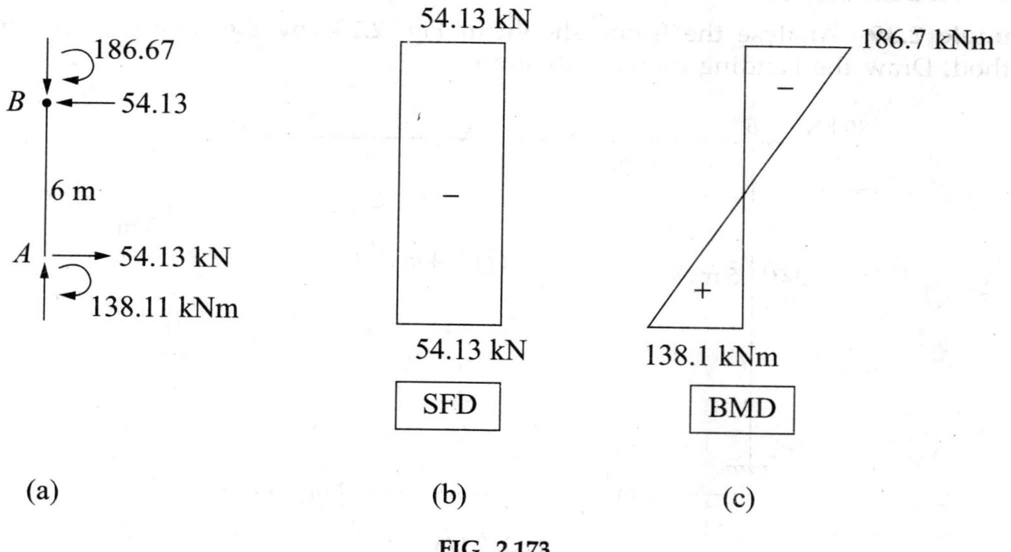

FIG. 2.173

Referring to shear force diagram,

$$\frac{x}{6.71 - x} = \frac{109.9}{24.2}$$

$$\boxed{x = 5.5 \text{ m}}$$

Maximum bending moment = $-186.7 + 150(5.5 \cos\theta) - 54.13 \times 5.5 \times 0.447 - \frac{25}{2} \times (5.5 \times 0.894)^2 = 115.59$ kNm.

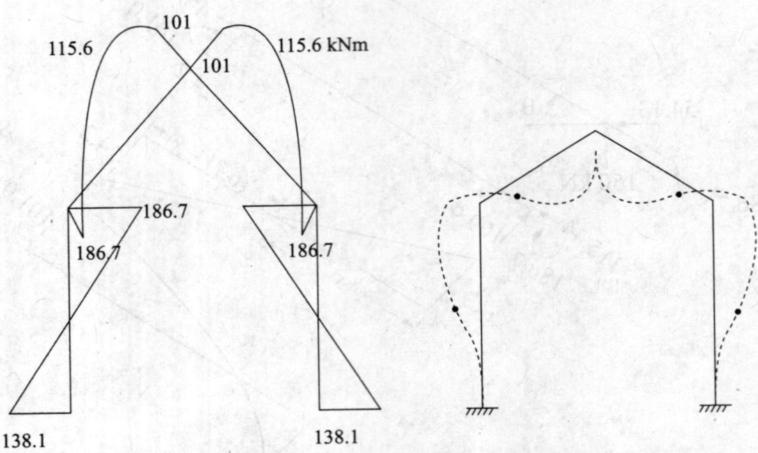

FIG. 2.174

FIG. 2.175 Elastic curve

2.13 ANALYSIS OF DOUBLE BAY PORTAL FRAME

Example 2.39 Analyse the frame shown in Fig. 2.176 by the moment distribution method. Draw the bending moment diagram.

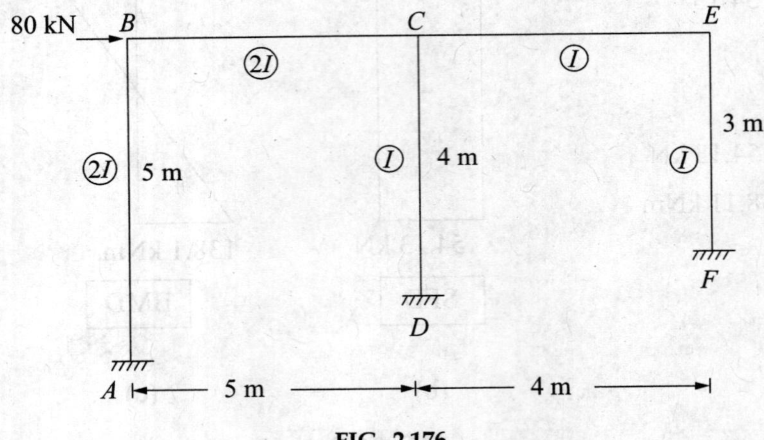

FIG. 2.176

Solution: As the nonsway fixed end moments are zero's, only sway moment distribution is carried out.

Sway fixed end moments

$$M_{FAB} = M_{FBA} = -\frac{6EI_1\Delta_1}{l_1^2} = -\frac{6E(2I)\,\Delta}{5^2} = -\frac{12}{25}EI\Delta$$

$$M_{FCD} = M_{FDC} = -\frac{6EI_2\Delta_2}{l_2^2} = -\frac{6EI\Delta}{4^2} = \frac{3}{8}EI\Delta$$

$$M_{EF} = M_{FE} = -\frac{6EI_3\Delta_3}{l_3^2} = -\frac{6EI\Delta}{3^2} = \frac{6EI\Delta}{9}$$

$$M_{FBC} = M_{FCB} = 0$$

$$M_{FCE} = M_{FEC} = 0$$

$$\therefore \quad M_{FBA} : M_{FCD} : M_{FEF} = -\frac{12EI\Delta}{25} : -\frac{3EI\Delta}{8} : -\frac{6EI\Delta}{9}$$

$$M_{FAB} : M_{FCD} : M_{FEF} = -0.48\Delta : -0.375\Delta : -0.67\Delta$$

Hence,

$$M_{FAB} : M_{FCD} : M_{FEF} = -48.00 : -37.50 : -66.7 \text{ kNm}$$

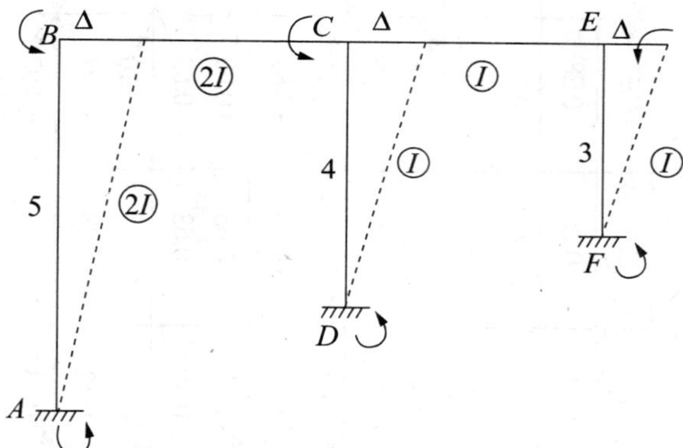

FIG. 2.177

Distribution factors

Table 2.93

Joint	*Members*	*Relative Stiffness Values (k)*	*ΣI/l*	*k/Σk*
B	BA	2I/5	4I/5	0.50
	BC	2I/5		0.50
C	CB	2I/5	0.9I	0.44
	CD	I/4		0.28
	CE	I/4		0.28
E	EC	I/4	0.58I	0.43
	EF	I/3		0.57

Table 2.94 Moment distribution table

Joint	A	B		C			E		F	D
Members	*AB*	*BA*	*BC*	*CB*	*CD*	*CE*	*EC*	*EF*	*FE*	*DC*
DF	0	0.50	0.50	0.44	0.28	0.28	0.43	0.57	0	0
FEMS	–48.00	–48.00	0	0	–37.50	0	0	–66.70	–66.70	–37.50
Bal		+24.00	+24.00	16.50	10.50	10.50	28.68	38.02		
Co	+12.00		+8.25	12.00		14.34	5.25		19.01	+5.25
Bal		–4.12	–4.13	–11.59	–7.38	7.38	–2.25	–3.00		
Co	–2.06		–5.80	–2.07		–1.13	–3.69		–1.50	–3.69
Bal		+2.90	+2.90	+1.41	+0.90	+0.89	+1.59	+2.10		
Co	+1.45		+0.71	+1.45	0.63	+0.80	+0.45		+1.05	+0.45
Bal		–0.36	–0.35	–0.99	–0.63	–0.63	–0.19	–0.26		
Co	–0.18		–0.50	–0.18		–0.10	–0.32		–0.13	–0.32
Bal		+0.25	+0.25	+0.12	+0.08	+0.08	+0.14	+0.18		
Co	+0.13		+0.06	+0.13		+0.07	+0.04		+0.09	+0.04
Bal		–0.03	–0.03	–0.08	–0.056	–0.06	–0.02	–0.02		
	–36.66	–25.36	+25.36	16.70	–34.09	+17.38	29.68	–29.68	–48.18	–35.77

Column shear equation

$$\Sigma M_B = 0$$

$$M_{AB} + M_{BA} + 5H_A = 0$$

$$\boxed{H_A = -0.2(M_{AB} + M_{BA})}$$

FIG. 2.178

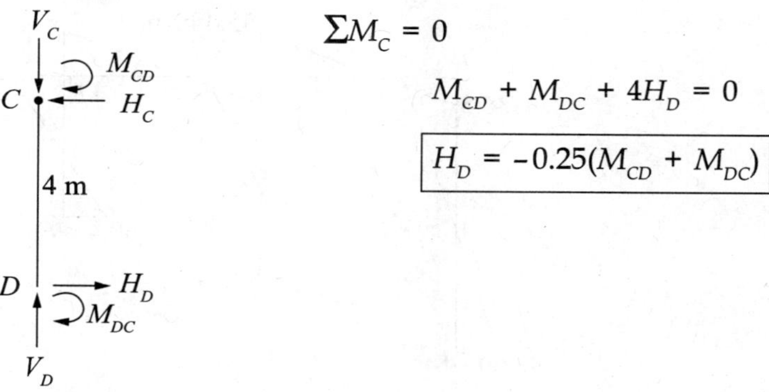

$$\Sigma M_C = 0$$

$$M_{CD} + M_{DC} + 4H_D = 0$$

$$\boxed{H_D = -0.25(M_{CD} + M_{DC})}$$

FIG. 2.179

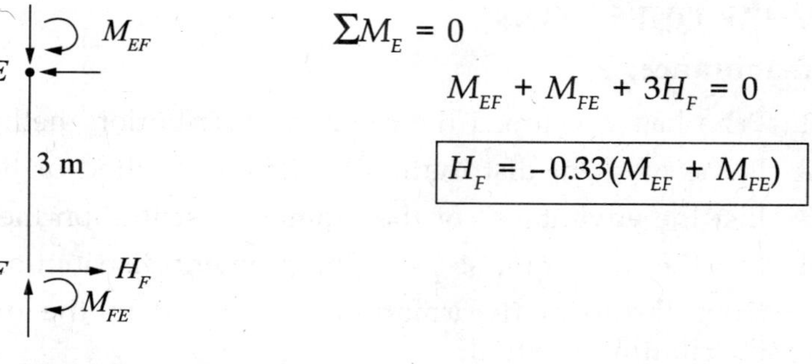

$$\Sigma M_E = 0$$

$$M_{EF} + M_{FE} + 3H_F = 0$$

$$\boxed{H_F = -0.33(M_{EF} + M_{FE})}$$

FIG. 2.180

Considering the horizontal equilibrium

$$80 - H_B - H_D - H_F = 0$$

$$80 + 0.2(M_{AB} + M_{BA}) + 0.25(M_{CD} + M_{DC}) + 0.33(M_{EF} + M_{FE}) = 0$$

$$80 + \{0.2(-36.66 - 25.36) + 0.25(-34.09 - 35.77) + 0.33(-29.68 - 48.18)\}k = 0$$

$$\boxed{k = 1.44}$$

Final moments

The moments obtained at the end of moment distribution table is multiplied by k and hence

$$M_{AB} = -52.8 \text{ kNm}, M_{BA} = -36.5 \text{ kNm}, M_{BC} = -36.5 \text{ kNm}$$

$$M_{CB} = +24.1 \text{ kNm}, M_{CD} = -49.1 \text{ kNm}, M_{DC} = -51.5 \text{ kNm}$$

$$M_{CE} = +35.4 \text{ kNm}, M_{EC} = +42.7 \text{ kNm}, M_{EF} = -42.7 \text{ kNm}$$

$$M_{FE} = -69.4 \text{ kNm}$$

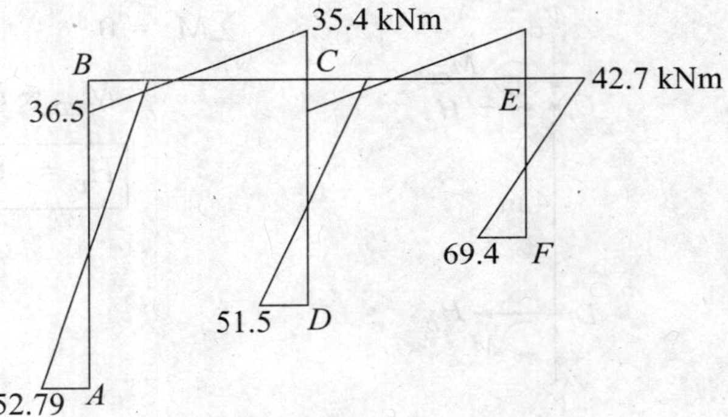

FIG. 2.181 Bending moment diagram

REVIEW QUESTIONS

Remembrance

2.1. Who has developed the moment distribution method?

2.2. Is the moment distribution a stiffness method or flexibility method?

2.3. List the advantages of the moment distribution method?

2.4. List the important steps in the moment distribution method?

2.5. Does the axial deformation considered in the development of the moment distribution method?

2.6. Define rotational stiffness?

2.7. Explain distribution factor?

2.8. Define carry over factor?

2.9. Which moment distribution is preferable for symmetric frames subjected to lateral loads as storey heights?

2.10. What are the other names for cantilever moment distribution method?

2.11. What are the advantages of Naylor's method?

2.12. What is the rotational stiffness of a cantilever?

2.13. What is the magnitude of a stiffness of a member under antisymmetric bending in Naylor's method?

2.14. List the values of symmetric stiffness factor, skew symmetric stiffness factor and cantilever stiffness factor?

Understanding

2.1. In a member *AB*, if a moment of 10 kNm is applied at *A*, what is the moment carried over to the fixed end *B*?

2.2. What is sinking of supports? What is its effect on the end moments of the member?

2.3. Calculate the M_{FBA} and M_{FBC} for the beam shown in figure below due to sinking of support by 2 mm. E = 6000 N/mm² and I = 1.6(10)⁸ mm⁴.

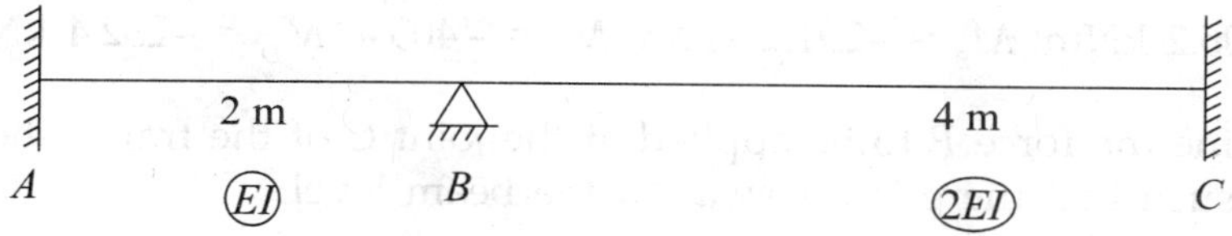

2.4. Calculate the M_{FBC} and M_{FCB} for the beam shown in figure below $EI = 10 \times 10^{11}$ Nmm². Support *B* and *C* sinks by 2 mm and 3 mm respectively.

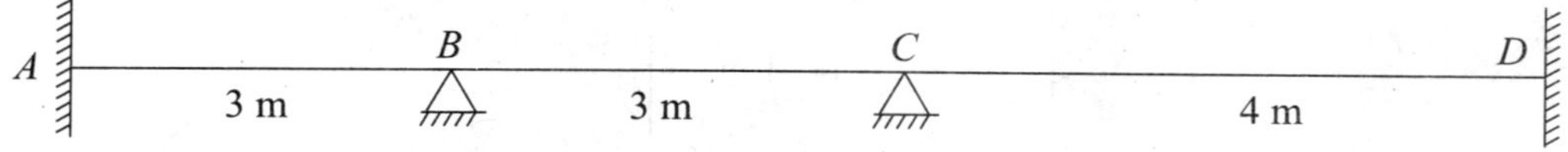

2.5. Is it possible to determine the beam deflections in a continuous beam by moment distribution method?

2.6. List the reasons for sidesway of the portal frame.

2.7. What is balancing at a joint?

2.8. While applying the moment distribution method, a designer remembers that "nothing comes back from the fixed end". Justify.

2.9. What we can understand from the computed sway correction factor, i.e., if it is positive what does it indicate as well as if it is negative what does it point out?

2.10. Why sway correction is required when we analyse unsymmetrical frames?

2.11. Does the carry over factor is half for non-prismatic members?

2.12. When a symmetrical structure is subjected to symmetrical loading, it does not sway. Does this statement applicable to Gable frames? If not, why?

EXERCISE PROBLEMS

2.1. Analyse the frame by the moment distribution method and draw the BMD.

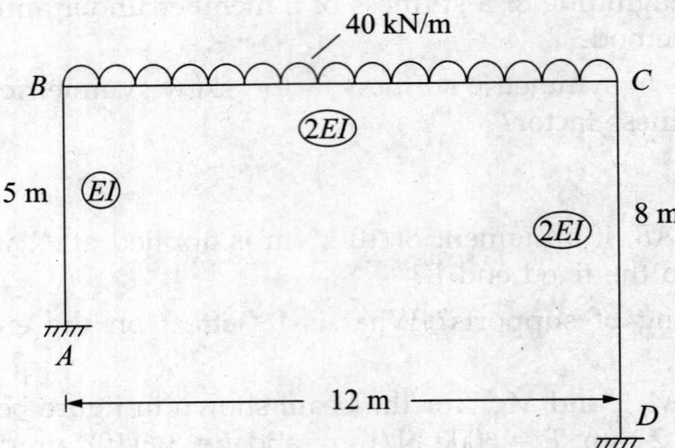

Ans. M_{AB} = 106.2 kNm, M_B = –291.2 kNm, M_C = –403.4, M_D = –232.4 kNm

2.2. Determine the force *P* to be applied at the joint *C* of the frame shown in figure below, such that there is nonsway at the beam level.

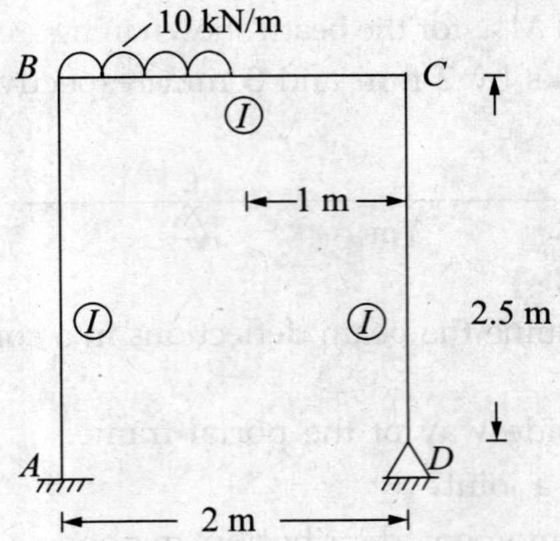

Ans. P = 0.5 kN

2.3. Analyse the portal frame *ABCD* subjected to loading as shown in figure below. Use the moment distribution method and draw BMD.

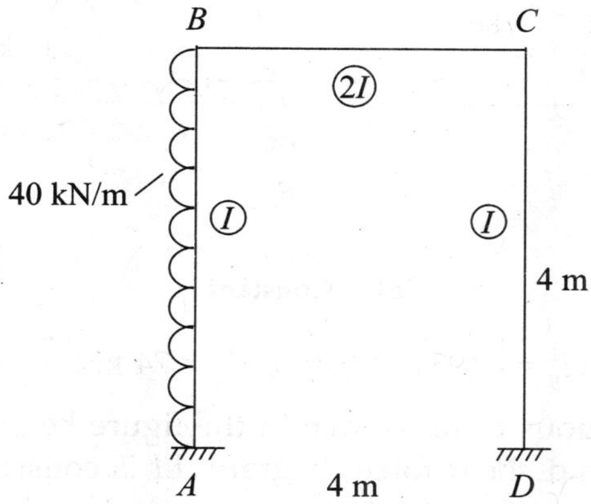

Ans. $M_A = -143.9$ kNm, $M_B = -35.8$ kNm, $M_C = -62.7$, $M_D = -77.4$ kNm

2.4. Analyse the given frame by the moment distribution method and draw the bending moment diagram.

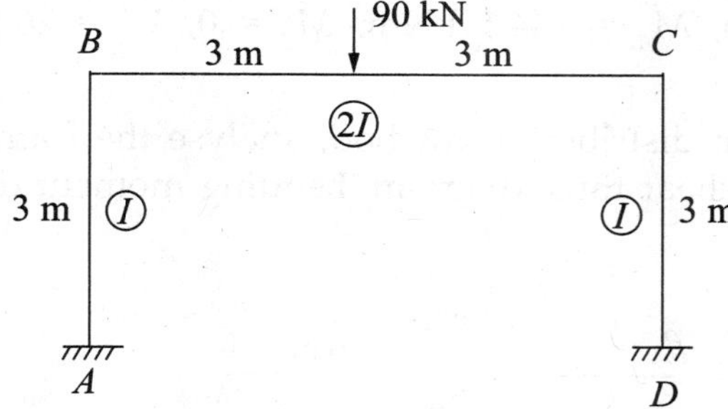

Ans. $M_A = -22.6$, $M_B = -45.3$, $M_C = -45.3$, $M_D = -22.6$ kNm

2.5. Analyse the given frame by the moment distribution method.

75 kN
B 6 3 C
2I E
6 m
I I
A D

Ans. $M_A = -25.3$ kNm, $M_B = -42.3$, $M_C = -47.9$, $M_D = -19.7$

2.6. Analyse the continuous beam shown in the figure below. Draw the shear force diagram and bending moment diagram.

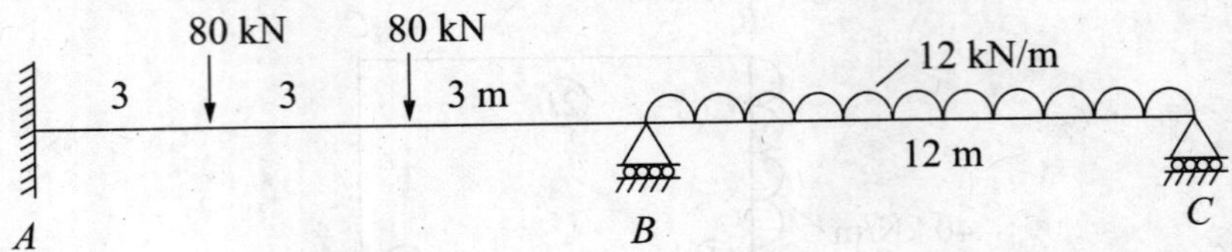

***EI* = Constant**

Ans. $M_A = -142.10$ kNm, $M_B = -195.8$, $M_C = 0$, $V_A = 74$ kN, $V_B = 174.3$ kN, $V_C = 55.7$ kN

2.7. Analyse the continuous beam shown in the figure below and draw the bending moment diagram and shear force diagram. *EI* is constant.

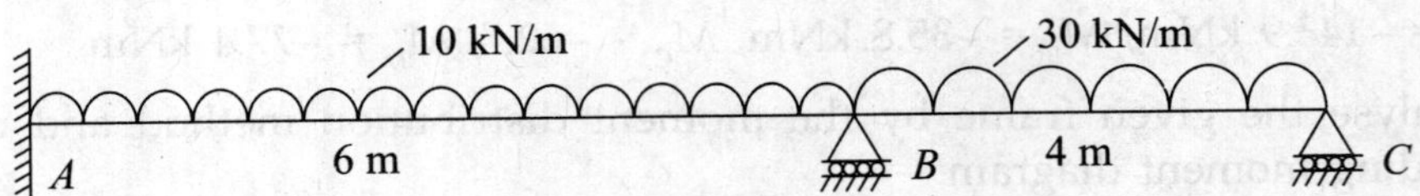

Ans. $M_A = -22.95$ kNm, $M_B = -44.1$ kNm, $M_C = 0$, $V_A = 26.5$ kN, $V_B = 104.5$ kN, $V_C = 49$ kN

2.8. Using the moment distribution method, analyse the frame shown in the figure below. Draw the shear force diagram, bending moment diagram and the elastic curve.

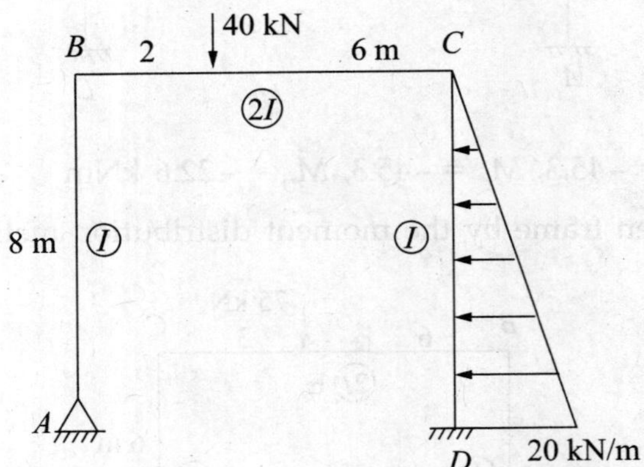

Ans. $M_A = 0$, $M_B = -50.5$ kNm, $M_C = -22.5$, $M_D = -140.6$ kNm

2.9. Analyse the frame by the moment distribution method. Draw the bending moment diagram.

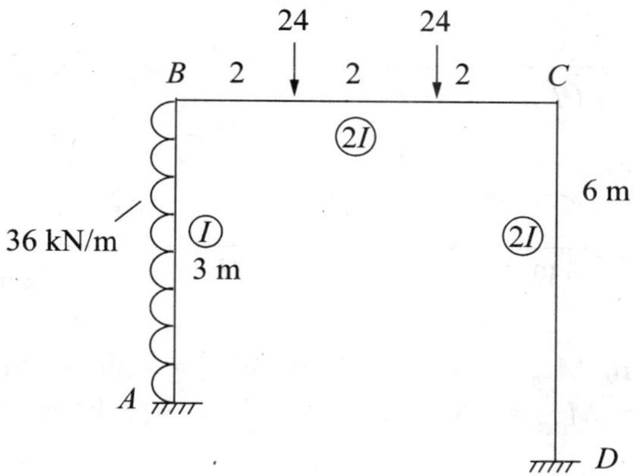

Ans. $M_A = -50.8$ kNm, $M_B = -85.9$, $M_C = -17.65$, $M_D = -33$ kNm

2.10. Analyse the given frame by the moment distribution method. The flexural rigidity *EI* is constant throughout. Draw the bending moment diagram and the elastic curve.

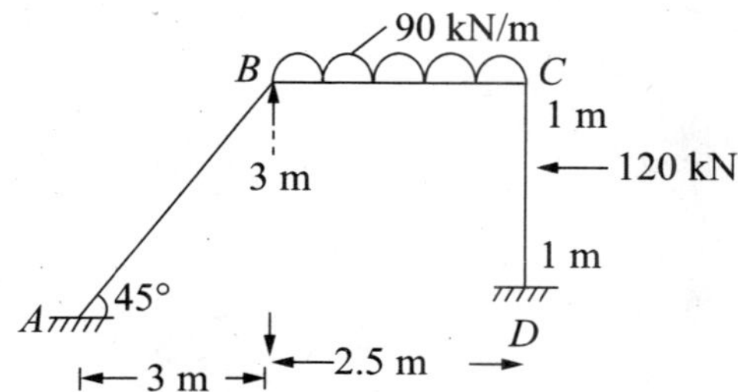

Ans. $M_A = 0.76$ kNm, $M_B = 8.922$, $M_C = 66.03$, $M_D = -0.226$ kNm

2.11. Analyse the portal frame shown in the figure below by moment distribution method.

B D F 60 kN

2I I

3 m 2I 3I I

A 6 m C 3 m E

Ans. $M_{AB} = -35.9$ kNm, $M_{BA} = -21.0$, $M_{BD} = +21.0$, $M_{DB} = 20.1$, $M_{DC} =$ -`37.7, $M_{DF} = 17.6$, $M_{FD} = 15.9$, $M_{FE} = -15.9$, $M_{CD} = -56.9$, $M_{EF} = -20.6$ kNm

2.12. Analyse the Vierendeel girder by moment distribution method.

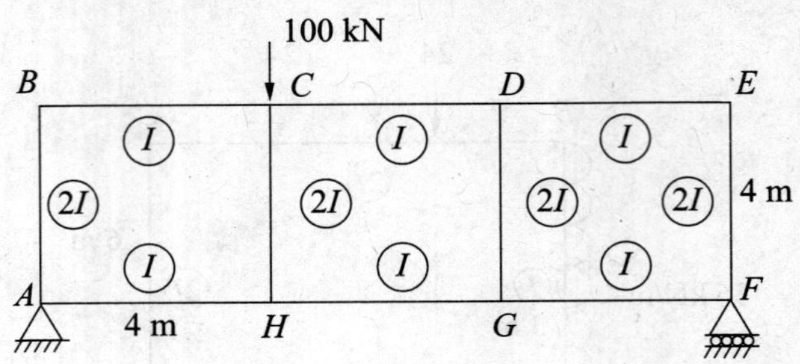

Ans. M_{BA} = –63.5 kNm, M_{CB} = +68.89 kNm, M_{CH} = –35.2 kNm, M_{CD} = –33.6 kNm, M_{DC} = –29.7 kNm, M_{DG} = +60.5 kNm, M_{DE} = –30.8 kNm, M_{ED} = –35.0 kNm

3 Kani's Method

Objectives:

Concept - Relationship between bending moment and deformations – analysis of beams with and without settlement of supports – analysis of frames with and without sway.

3.1 INTRODUCTION

In 1947, Gasper Kani of East Germany proposed an iterative method for the analysis of multistoreyed frames. It involves repeated distribution of moments at successive joints in beams and frames. This method was preferred by the designers in late eighties. With the advent of computer; its charm started diminishing. This method is similar to Gauss-Seidel Iteration method and is self corrective. It does not require the solution of simultaneous equations as in case of the slope deflection method. The essential difference between the Hardy Cross moment distribution method and that of Kani's rotation contribution method is that in Kani's method the total joint moment is distributed while in moment distribution, the unbalanced joint moments are distributed. Kani's method has been developed on the basis of slope deflection method with a different form of moment distribution.

This method is advantageous due to its simplicity, speed and accuracy. It requires only a single table calculation unlike in the moment distribution method, we require one for nonsway analysis and the other for sway analysis. However, Kani's method is advantageous in cross verifying the results obtained from software packages. Kani's method becomes more laborious if the number of joints in the frame exceeds twenty-five.

3.2 FUNDAMENTAL EQUATIONS IN KANI'S METHOD

3.2.1 Frames Without Sway

The frames do not sway in the following cases: (i)When they are restrained at the end joints in rectilinear frames and (ii) symmetric frames subjected to symmetric loading. The reader can recall that a structure is said to be symmetric, when half of the structure is mirror image of the other; identical members match in orientation and also the supports are to be identical. Consider the frame shown in Fig. 3.1.

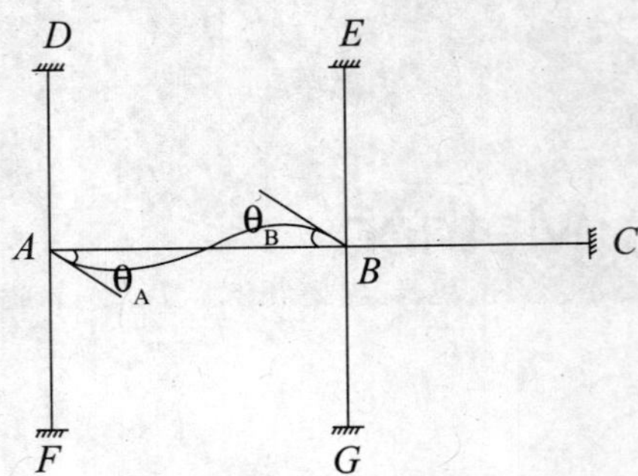

FIG. 3.1 A typical frame

The moment at A for the member AB can be written as the sum of

$$M_{AB} = M_{FAB} + \left(\frac{4EI}{l}\right)\theta_A + \left(\frac{2EI}{l}\right)\theta_B \tag{1}$$

where θ_A and θ_B are the rotations at A and B respectively. The above equation is nothing but the slope deflection of the member AB. It can be recast as

$$M_{AB} = M_{FAB} + 2\left(\frac{2EI}{l}\theta_A\right) + \left(\frac{2EI}{l}\theta_B\right) \tag{1a}$$

Denoting the terms in the brackets as m_{ab} and m_{ba} respectively;

$$M_{AB} = M_{FAB} + 2m_{ab} + m_{ba} \tag{2}$$

where m_{ab} is the rotational moment due to θ_A and m_{ba} is the rotational moment due to θ_B.

The joint A is in equilibrium, when the sum of all moments meeting at a joint should be zero.

i.e.
$$\Sigma(M_{FAB} + 2m_{ab} + m_{ba}) = 0 \tag{3}$$

and hence
$$\Sigma m_{ab} = -\frac{1}{2}\left[\Sigma M_{FAB} + \Sigma m_{ba}\right] \tag{4}$$

Thus, the above equation has been derived based on the slope deflection method. According to Theorem 2 in the moment distribution method, if a moment M is applied to any joint in a structure, that moment will be distributed to other members meeting at that joint in the ratio; stiffness of member to the sum of the stiffnesses of the members. Hence,

$$m_{ab} = \frac{k_{ab}}{\Sigma k}\,\Sigma m_{ab} \tag{5}$$

Thus, combining the basis of the slope deflection method along with the basics of moment distribution method, i.e., substituting Eq. 4 in Eq. 5;

$$m_{ab} = \frac{k_{ab}}{\Sigma k}\left\{-\frac{1}{2}\left[\Sigma M_{Fab} + \Sigma m_{ba}\right]\right\} \tag{6}$$

i.e.

$$\boxed{m_{ab} = - - -\,{}^{ab}\left[\Sigma M_{Fab} + \Sigma m_{ba}\right]} \tag{7}$$

where k_{ab} is the stiffness of AB and Σk is the sum of the stiffnesses of the members meeting at the joint. The term $\left(-\frac{1}{2}\frac{k_{ab}}{\Sigma k}\right)$ is called rotation factor. In other words, rotation factor is −(1/2) of the 'distribution factor' in the moment distribution method.

The above equation can be rewritten as

$$m_{ab} = +R_{AB}\left(\Sigma M_{Fab} + \Sigma m_{ba}\right) \tag{8}$$

where

R_{AB} = Rotation factor for AB element

ΣM_{Fab} = Sum of restraint moment at joint A

Σm_{ba} = Sum of rotation moments

In the above equation the sum of the restraint moment can be found out using standard formulae for fixed beams. Σm_{ba} is an unknown quantity to start with.

The application of Kani's method is illustrated in the following example.

3.2.1.1 *Analysis of Continuous Beam without Support Settlement*

Example 3.1 Determine the support moments for the continuous beam shown in Fig. 3.2 by Kani's method. The relative I values are indicated along the member in each span. EI is constant. Sketch the BMD and SFD. (VTU, June 2009)

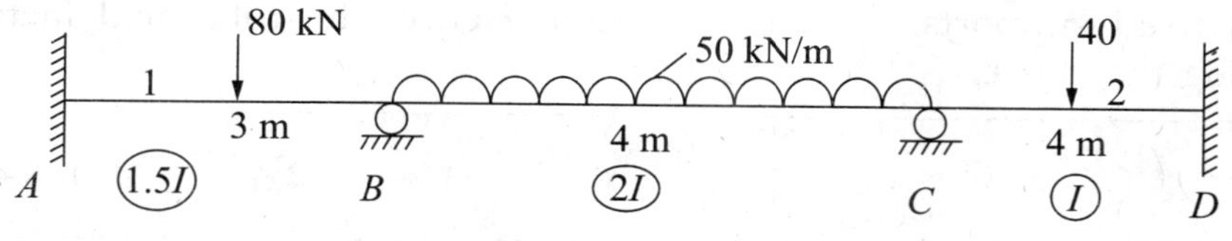

FIG. 3.2

Solution

Step 1: Calculate the rotation factor at each joint and enter in the rotation factor table. Joint A is fixed support. Hence, the rotation is zero. Therefore, rotational moment/factor is zero.

Joint *B*

$$k_{BA} = \frac{1.5I}{3} = 0.5I$$

$$k_{BC} = \frac{2I}{4} = 0.5I$$

$$\Sigma k = k_{BA} + k_{BC} = I$$

$$\therefore \quad R_{BA} = -\frac{1}{2}\frac{k_{BA}}{\Sigma k} = -\frac{1}{2}\left(\frac{0.5I}{I}\right) = -0.25$$

$$R_{BC} = -\frac{1}{2}\frac{k_{BC}}{\Sigma k} = -\frac{1}{2}\left(\frac{0.5I}{I}\right) = -0.25$$

The sum of the rotation factor at joint B is (–1/2). The reader should remember that the sum of the rotation factors at a joint is (–1/2).

Joint C

$$k_{CB} = \frac{2I}{4}$$

$$k_{CD} = \frac{I}{4}$$

$$\Sigma k = \frac{3I}{4}$$

$$\therefore \quad R_{CB} = -\frac{1}{2}\left(\frac{2I/4}{3I/4}\right) = -0.33$$

$$R_{CD} = -\frac{1}{2}\left(\frac{I/4}{3I/4}\right) = -0.17$$

Joint D

As D is a fixed supports, the slope is zero. Hence, the rotational factor is zero. Summarising the above,

Joint	Members	k	Σk	$R = -\frac{1}{2}\frac{k}{\Sigma k}$
B	BA	+0.5I	I	–0.25
	BC	+0.5I		–0.25
C	CB	+0.5I	0.75I	–0.33
	CD	+0.25I		–0.17

Step 2: The fixed end moments are

$$M_{FAB} = -1(80)2^2/3^2 = -35.6 \text{ kNm}$$

$$M_{FBA} = +2(80)1^2/3^2 = +17.8 \text{ kNm}$$
$$M_{FBC} = -50(4)^2/12 = -66.7 \text{ kNm}$$
$$M_{FCB} = +50(4)^2/12 = +66.7 \text{ kNm}$$
$$M_{FCD} = -40(4)/8 = -20.0 \text{ kNm}$$
$$M_{FDC} = +40(4)/8 = +20.0 \text{ kNm}$$

Step 3: The restraint moments are obtained at joints B and C are;

Restraint moment at $B = M_{FBA} + M_{FBC} = -48.9$ kNm

Restraint moment at $C = M_{FCB} + M_{FCD} = +46.7$ kNm

Step 4: The rotation factors and the restraint moments are entered as follows. The rotational factors are in the annular box and the restraint moments are in the joint location.

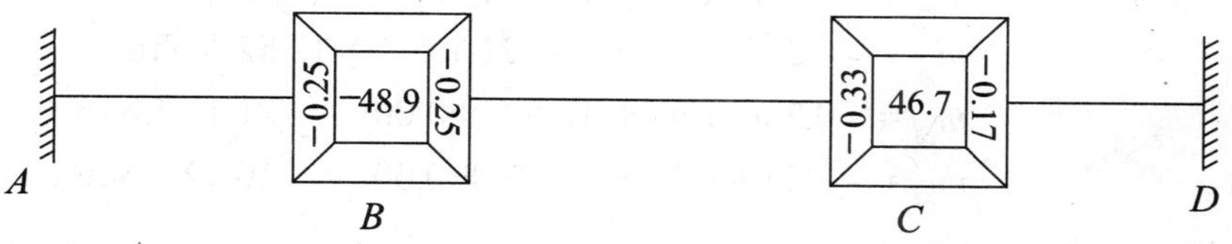

FIG. 3.3

Step 5:

Iteration 1

$$m_{ba} = R_{AB}\left(M_{FB} + \Sigma m_{ba}\right)$$

The rotational moment m_{ba} is obtained from the above equation as

$$m_{ba} = -0.25(-48.9 + 0) = +12.23 \text{ kNm}$$
$$m_{bc} = -0.25(-48.9 + 0) = +12.23 \text{ kNm}$$

Similarly, the rotational moments at C is obtained as

$$m_{cb} = R_{CB}\left(M_{FC} + \Sigma m_{cb}\right)$$

i.e.

$$m_{cb} = -0.33(46.7 + 12.23 + 0) = -19.45 \text{ kNm}$$
$$m_{cd} = -0.17(46.7 + 12.23 + 0) = -10.02 \text{ kNm}$$

Iteration 2

$$m_{ba} = -0.25(-48.9 + 0 - 19.45) = +17.09 \text{ kNm}$$
$$m_{bc} = -0.25(-48.9 + 0 - 19.45) = +17.09 \text{ kNm}$$
$$m_{cb} = -0.33(46.7 + 17.09 + 0.0) = -21.05 \text{ kNm}$$
$$m_{cd} = -0.17(46.7 + 17.09 + 0.0) = -10.84 \text{ kNm}$$

Iteration 3

$$m_{ba} = -0.25(-48.9 + 0 - 21.05) = +17.49 \text{ kNm}$$
$$m_{bc} = -0.25(-48.9 + 0 - 21.05) = +17.49 \text{ kNm}$$
$$m_{cb} = -0.33(46.7 + 17.49 + 0.0) = -21.18 \text{ kNm}$$
$$m_{cd} = -0.17(46.7 + 17.49 + 0.0) = -10.91 \text{ kNm}$$

Iteration 4

$$m_{ba} = -0.25(-48.9 + 0 - 21.18) = +17.52 \text{ kNm}$$
$$m_{bc} = -0.25(-48.9 + 0 - 21.18) = +17.52 \text{ kNm}$$
$$m_{cb} = -0.33(46.7 + 17.52 + 0.00) = -21.19 \text{ kNm}$$
$$m_{cd} = -0.17(46.7 + 17.52 + 0.00) = -10.92 \text{ kNm}$$

Iteration 5

$$m_{ba} = -0.25(-48.9 + 0 - 21.19) = +17.52 \text{ kNm}$$
$$m_{bc} = -0.25(-48.9 + 0 - 21.19) = +17.52 \text{ kNm}$$
$$m_{cb} = -0.33(46.70 + 17.52 + 0.00) = -21.19 \text{ kNm}$$
$$m_{cd} = -0.17(46.70 + 17.52 + 0.00) = -10.92 \text{ kNm}$$

The iterations 4 and 5 gave the same value of rotational moments to a desired accuracy of two decimals. Two-decimal accuracy is enough from the practical point of view.

Step 6:

The end moments are obtained from

$$M_{AB} = M_{FAB} + 2m_{ab} + m_{ba} = -35.6 + 2(0) + 17.52 = -18.08 \text{ kNm}$$
$$M_{BA} = M_{FBA} + 2m_{ba} + m_{ab} = +17.58 + 2(17.52) + 0.00 = +52.62 \text{ kNm}$$
$$M_{BC} = M_{FBC} + 2m_{bc} + m_{cb} = -66.7 + 2(17.52) - 21.19 = -52.85 \text{ kNm}$$
$$M_{CB} = M_{FCB} + 2m_{cb} + m_{bc} = +66.7 + 2(-21.19) + 17.52 = +41.84 \text{ kNm}$$
$$M_{CD} = M_{FCD} + 2m_{cd} + m_{dc} = -20.0 + 2(-10.92) + 0.00 = -41.84 \text{ kNm}$$
$$M_{DC} = M_{FDC} + 2m_{dc} + m_{cd} = +20.0 + 2(0.00) - 10.92 = +9.08 \text{ kNm}$$

These end moments are used to draw the pure moment diagram.

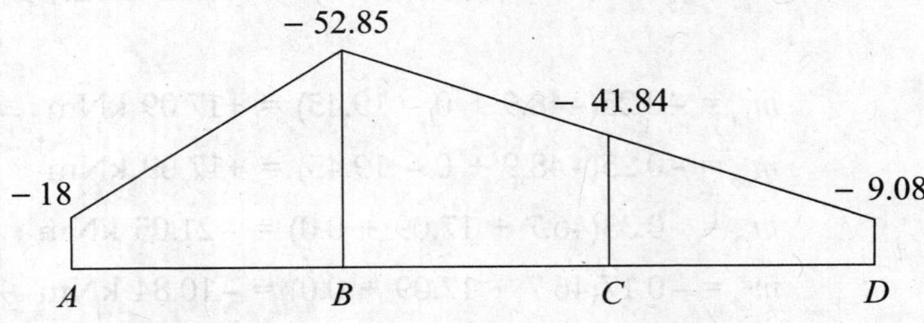

FIG. 3.4 Pure moment diagram (kNm)

The net bending moment diagram can be drawn by computing the bending moments in the span points.

Equilibrium of span AB

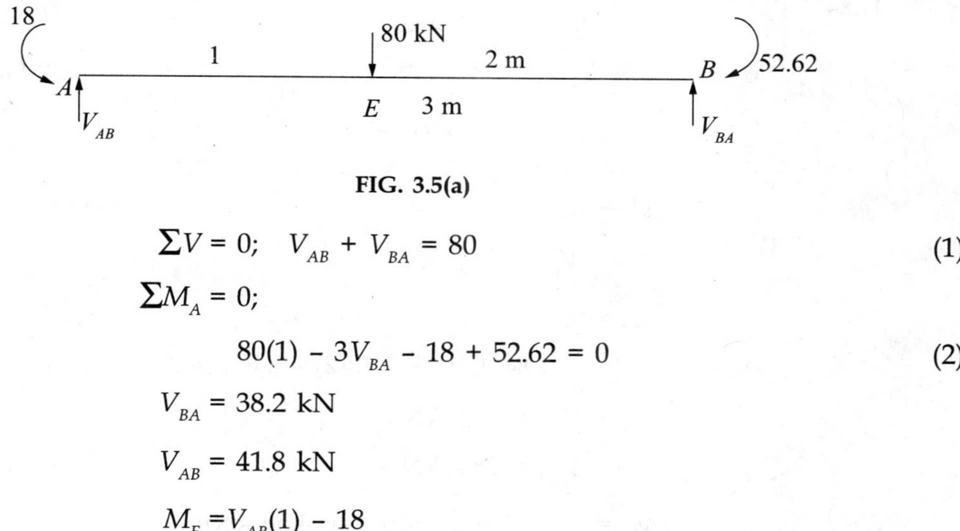

FIG. 3.5(a)

$$\Sigma V = 0; \quad V_{AB} + V_{BA} = 80 \tag{1}$$

$$\Sigma M_A = 0;$$

$$80(1) - 3V_{BA} - 18 + 52.62 = 0 \tag{2}$$

$$V_{BA} = 38.2 \text{ kN}$$

$$V_{AB} = 41.8 \text{ kN}$$

$$\therefore \quad M_E = V_{AB}(1) - 18$$

$$M_E = 23.8 \text{ kNm}$$

Equilibrium of span BC

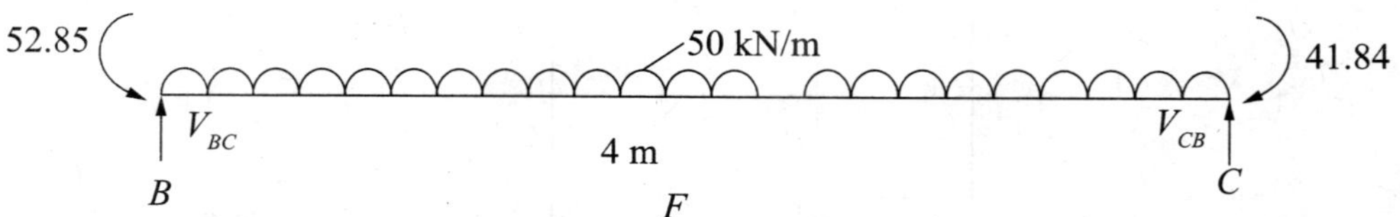

FIG. 3.5(b)

$$\Sigma V = 0; \quad V_{BC} + V_{CB} = 4(50) = 200 \tag{3}$$

$$\Sigma M_B = 0; \quad -52.85 + 41.84 + \frac{50 \times 4^2}{2} - 4V_{CB} = 0 \tag{4}$$

$$V_{CB} = 97.25 \text{ kN}$$

$$V_{BC} = 102.75 \text{ kN}$$

$$M_F = -52.85 + 102.75(2) - 50 \times \frac{2^2}{2} = 52.65 \text{ kNm}$$

Equilibrium of span CD

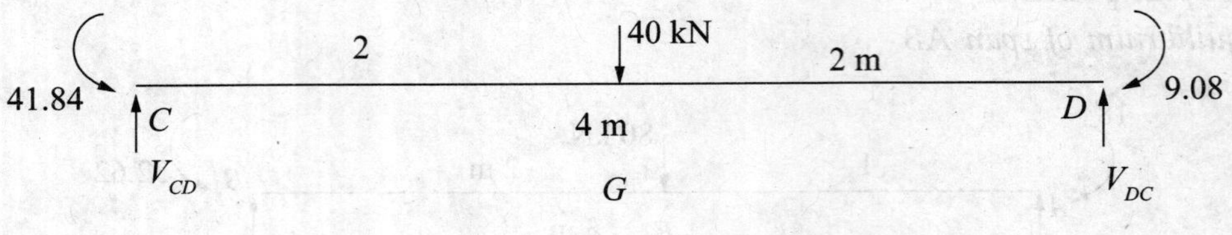

FIG. 3.5

$$\sum V = 0;$$

$$V_{CD} + V_{DC} = 40 \qquad (5)$$

$$\sum M_C = 0;$$

$$-41.84 + 9.08 + 40(2) - 4V_{DC} = 0$$

$$V_{DC} = 11.8 \text{ kN}$$

$$\therefore \quad V_{CD} = 28.2 \text{ kN}$$

$$M_G = -41.84 + 28.2(2) = +14.56 \text{ kNm}$$

The bending moment diagram and shear force diagram are drawn as

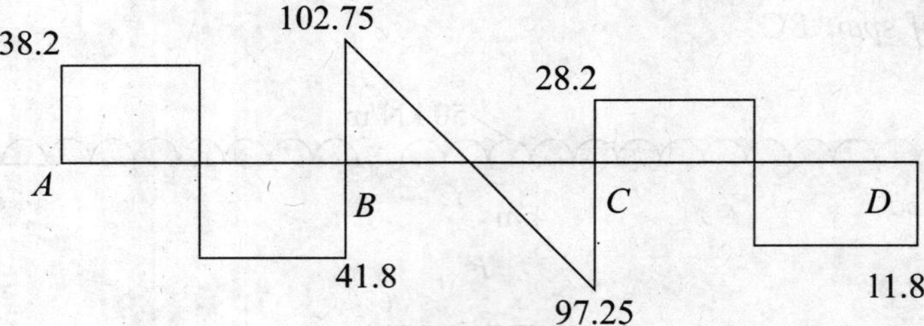

FIG. 3.6 Shear force diagram

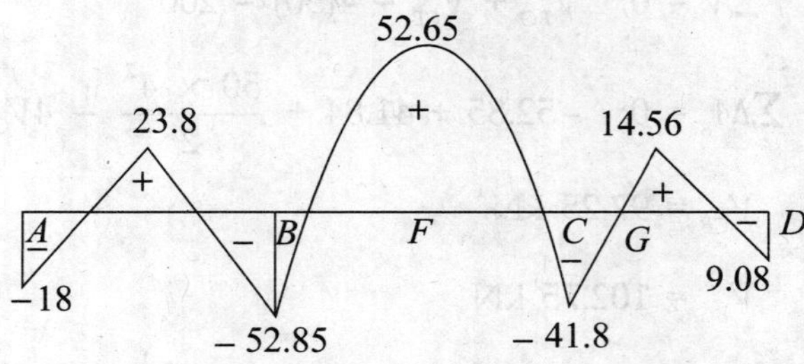

FIG. 3.7 Bending moment diagram

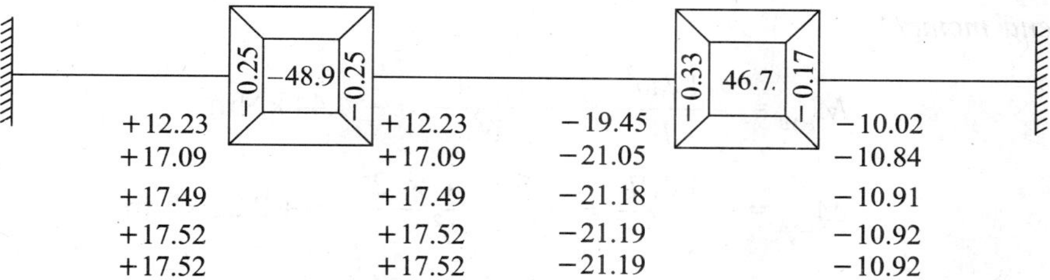

FIG. 3.8 Rotational moments

A −35.60 +17.80 B −66.70 +66.70 C −20.00 +20.00

0	+17.52	+17.52	−21.19	−10.92	0
0	+17.52	+17.52	−21.19	−10.92	0
+17.52	0.00	−21.19	+17.52	0.00	−10.92
−18.08	52.84	−52.85	+41.84	−41.84	9.08

FIG. 3.9 Computation of end moments

Example 3.2 Determine the support moments for the continuous beam shown in Fig. 3.10, EI = constant.

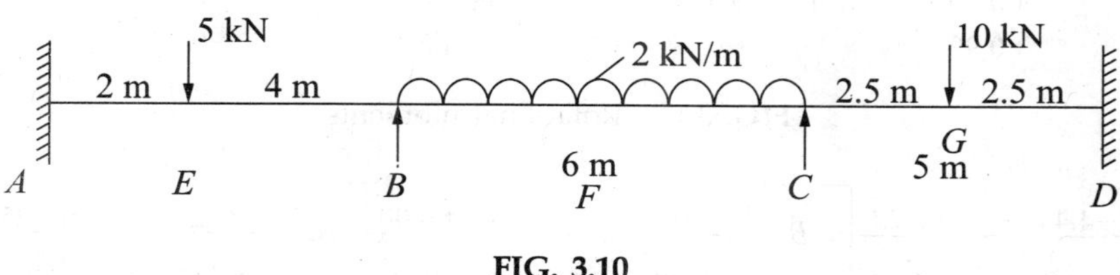

FIG. 3.10

Use Kani's method.

Solution

Rotational stiffness factors

Joint	*Members*	*Relative Stiffness k*	Σk	$R = \frac{-1}{2}\left(\frac{k}{\Sigma k}\right)$
B	BA	$I/6$	$2I/6$	−1/4
	BC	$I/6$		−1/4
C	CB	$I/6$	$11I/30$	−5/22
	CD	$I/5$		−6/22

Fixed end moments

$$M_{FAB} = -\frac{Wab^2}{l^2} = -\frac{5(2)4^2}{6^2} = -4.44 \text{ kNm}$$

$$M_{FBA} = +\frac{Wa^2b}{l^2} = +\frac{5 \times 4 \times 2^2}{6^2} = +2.22 \text{ kNm}$$

$$M_{FBC} = -\frac{wl^2}{12} = -2 \times \frac{6^2}{12} = -6.00 \text{ kNm}$$

$$M_{FCB} = +\frac{wl^2}{12} = +2 \times \frac{6^2}{12} = +6.00 \text{ kNm}$$

$$M_{FCD} = -\frac{Wl}{8} = -\frac{10(5)}{8} = -6.25 \text{ kNm}$$

$$M_{FDC} = +\frac{Wl}{8} = +\frac{10(5)}{8} = +6.25 \text{ kNm}$$

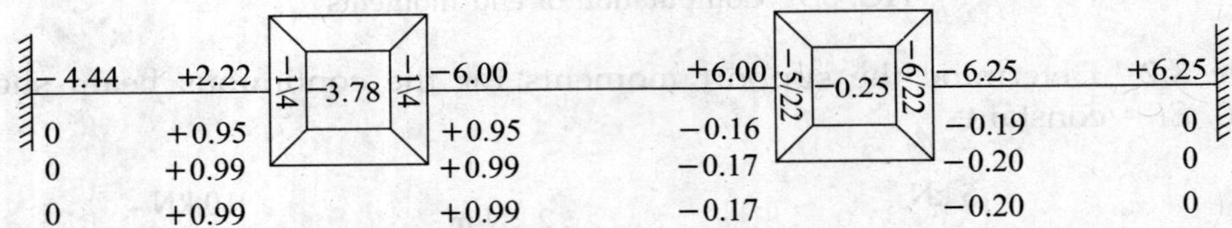

FIG. 3.11 Rotational moments

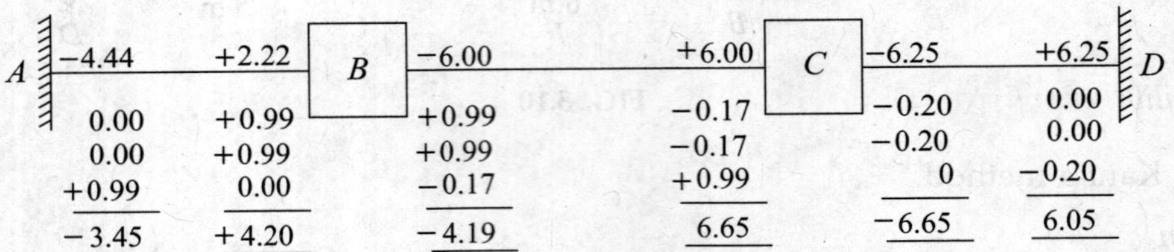

FIG. 3.12 Computation of end moments

Equilibrium of span AB

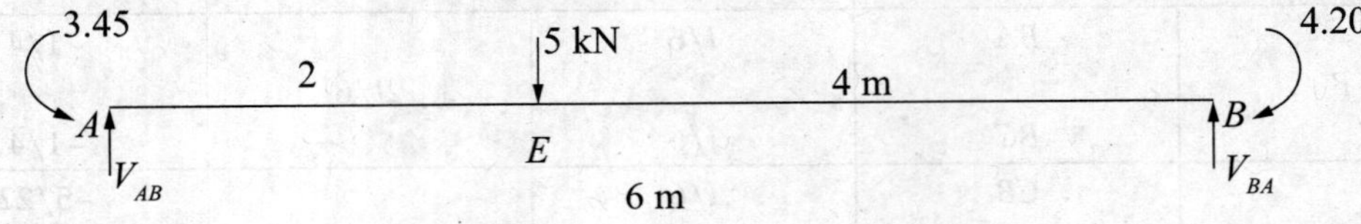

FIG. 3.13

$$\Sigma V = 0; \quad V_{AB} + V_{BA} = 5 \tag{1}$$

$$\Sigma M_A = 0; \quad -3.45 + 5(2) + 4.20 - 6V_{BA} = 0 \tag{2}$$

$\therefore$

$$V_{BA} = 1.79 \text{ kN}$$

$$V_{AB} = 3.21 \text{ kN}$$

$$M_E = 3.21(2) - 3.45 = 2.97 \text{ kNm}$$

Equilibrium of span BC

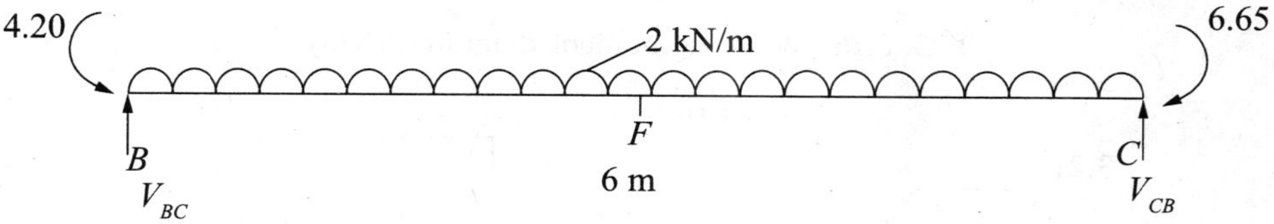

FIG. 3.14

$$\Sigma V = 0; \quad V_{BC} + V_{CB} = 2(6) = 12 \tag{3}$$

$$\Sigma M_B = 0; \quad -4.20 + 2 \times \frac{6^2}{2} + 6.65 - 6V_{CB} = 0 \tag{4}$$

$$V_{CB} = 6.41 \text{ kN}$$

$$V_{BC} = 5.59 \text{ kN}$$

$$M_F = -4.20 + 5.59(3) - 2(3)^2/2 = 3.57 \text{ kNm}$$

Equilibrium of span CD

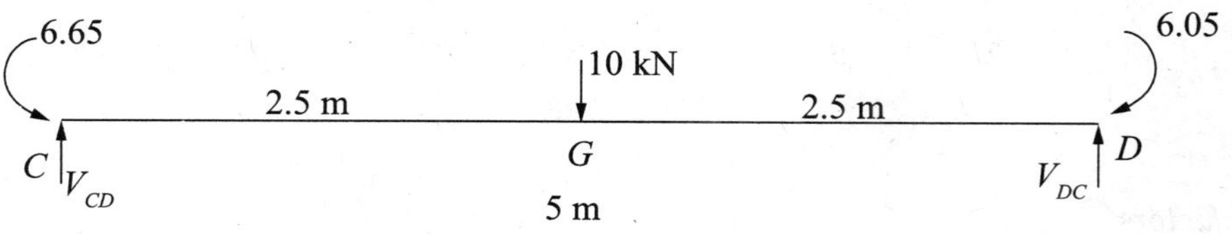

FIG. 3.15

$$\Sigma V = 0; \quad V_{CD} + V_{DC} = 10 \tag{5}$$

$$\Sigma M_C = 0; \quad -6.65 + 10(2.5) + 6.05 - 5V_{DC} = 0$$

$$V_{DC} = 4.88 \text{ kN}$$

$$V_{CD} = 5.12 \text{ kN}$$

$$M_G = -6.65 + 5.12(2.5) = 6.15 \text{ kNm}$$

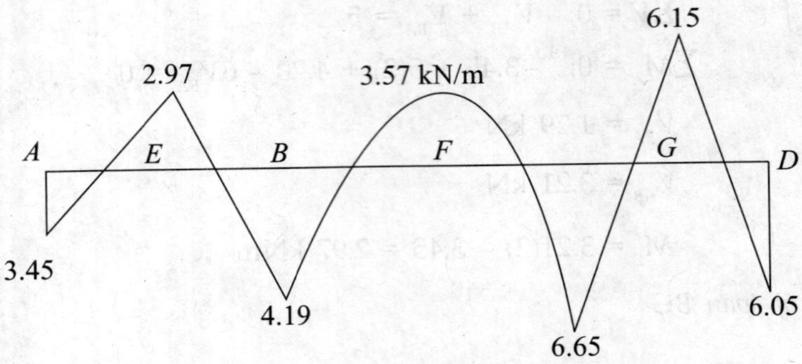

FIG. 3.16 Bending moment diagram (kNm)

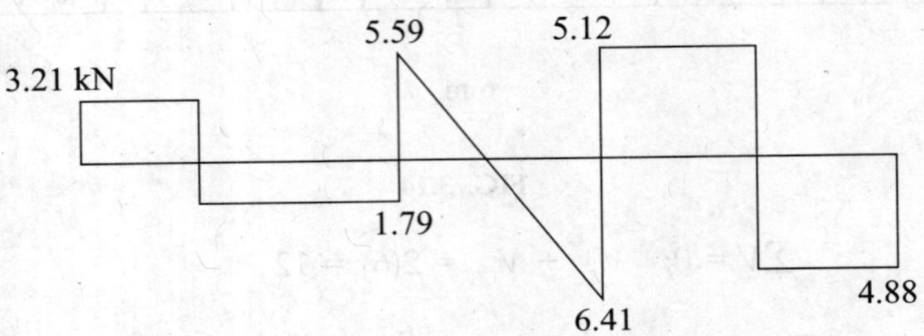

FIG. 3.17 Shear force diagram (kN)

Example 3.3 Analyse the three span continuous beam using Kani's method. The values of second moment area of each span are indicated along the members. *EI* is constant. Calculate the end moments.

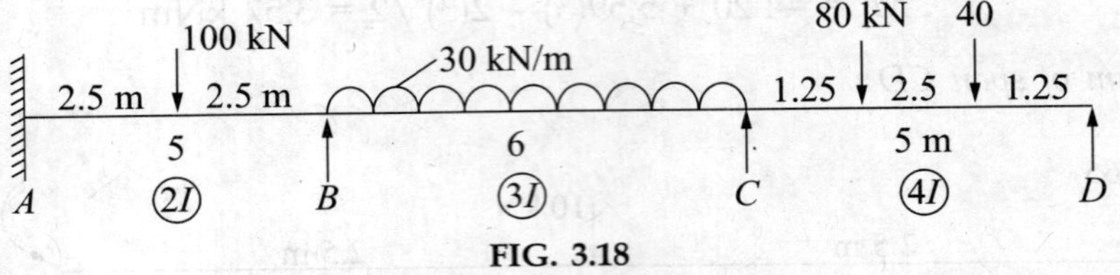

FIG. 3.18

Solution

Rotation factors

Joint	*Members*	*Relative Stiffness*	Σk	$R = -\frac{1}{2}\left(\frac{k}{\Sigma k}\right)$
B	*BA*	$2I/5 = 0.4I$	$0.9I$	-0.22
	BC	$3I/6 = 0.5I$		-0.28
C	*CB*	$3I/6 = 0.5I$	$1.1I$	-0.23
	CD	$\frac{3}{4} \times \frac{4I}{5} = 0.6I$		-0.27

The support D is simple support which is an end support. The relative span stiffness is taken as three-fourths of its original stiffness. Also the fixed end moments is to be modified as the moment at D is zero.

Fixed end moments

$$M_{FAB} = -100 \times \frac{5}{8} = -62.5 \text{ kNm}$$

$$M_{FBA} = +100 \times \frac{5}{8} = +62.5 \text{ kNm}$$

$$M_{FBC} = -30 \times \frac{6^2}{12} = -90.00 \text{ kNm}$$

$$M_{FCB} = +30 \times \frac{6^2}{12} = +90.00 \text{ kNm}$$

$$M_{FCD} = -\frac{1.25 \times 80 \times 3.75^2}{5^2} - \frac{3.75 \times 40 \times 1.25^2}{5^2} = -65.63 \text{ kNm}$$

$$M_{FDC} = +\frac{1.25^2 \times 80 \times 3.75}{5^2} + \frac{1.25 \times 40 \times 3.75^2}{5^2} = +46.88 \text{ kNm}$$

Modified fixed end moments

AB	*BA*	*BC*	*CB*	*CD*	*DC*
−62.50	+62.50	−90.00	+90.00	−65.63	+46.88
–	–	–	–	−23.44 ←	−46.88
−62.50	+62.50	−90.00	+90.00	−89.67	0

The modified fixed end moments and the rotation factors are given in Fig. 3.19.

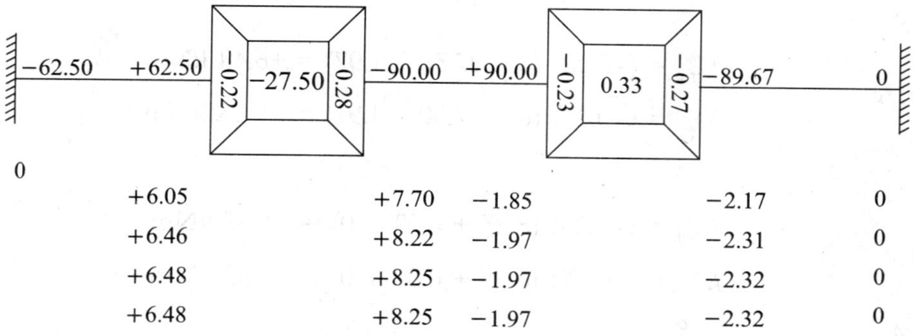

FIG. 3.19

The fixed end moments and the rotation factors are computed. The above values are entered in the appropriate places, i.e., the fixed end moment values are entered in the end of the member. The rotation factors are entered in the annular space in the above square boxes. The iterations are made as follows.

Iteration 1

Joint *B*

The rotation moments are assumed to be zero at the ends

Therefore,

$$M'_{BA} = (-0.22)(0 - 27.50 + 0) = +6.05 \text{ kNm}$$

$$M'_{BC} = (-0.28)(0 - 27.50 + 0) = +7.70 \text{ kNm}$$

Joint *C*

$$M'_{CB} = (-0.23)(7.70 + 0.33 + 0) = -1.85 \text{ kNm}$$

$$M'_{CD} = (-0.27)(7.70 + 0.33 + 0) = -2.17 \text{ kNm}$$

Iteration 2

Joint *B*

The approximate rotation moments obtained in cycle 1 is further improved as follows.

$$M'_{BA} = (-0.22)\ (0 - 27.50 - 1.85) = +6.46 \text{ kNm}$$

$$M'_{BC} = (-0.28)\ (0 - 27.50 - 1.85) = +8.22 \text{ kNm}$$

Joint *C*

$$M'_{CB} = (-0.23)(8.22 + 0.33 + 0) = -1.97 \text{ kNm}$$

$$M'_{CD} = (-0.27)(8.22 + 0.33 + 0) = -2.31 \text{ kNm}$$

Iteration 3

$$M'_{BA} = (-0.22)\ (0 - 27.50 - 1.97) = +6.48 \text{ kNm}$$

$$M'_{BC} = (-0.28)\ (0 - 27.50 - 1.97) = +8.25 \text{ kNm}$$

Joint *C*

$$M'_{CB} = (-0.23)\ (8.25 + 0.33 + 0) = -1.97 \text{ kNm}$$

$$M'_{CD} = (-0.27)\ (8.25 + 0.33 + 0) = -2.32 \text{ kNm}$$

Iteration 4

Joint *B*

$$M'_{BA} = -0.22\ (0 - 27.50 - 1.97) = 6.48 \text{ kNm}$$

$$M'_{BC} = -0.28\ (0 - 27.50 - 1.97) = 8.25 \text{ kNm}$$

Joint *C*

$$M'_{CB} = (-0.23)\ (8.25 + 0.33 + 0) = -1.97 \text{ kNm}$$

$$M'_{CD} = (-0.27)\ (8.25 + 0.33 + 0) = -2.32 \text{ kNm}$$

Comparing the values of the rotational moments in cycles 3 and 4; the values converge.

Knowing the rotational end moments, the final moments are computed as follows.

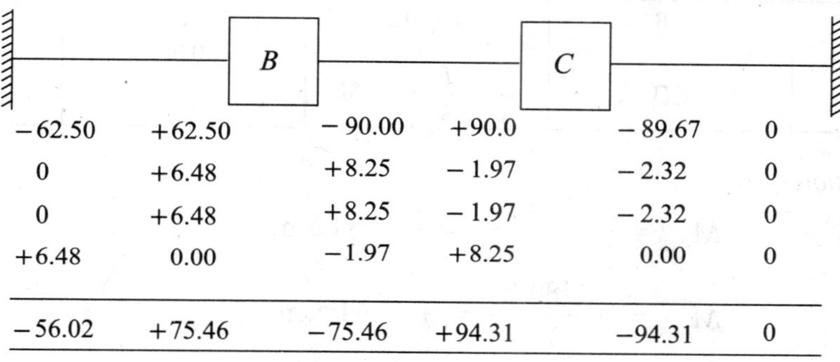

FIG. 3.20 Computation of end moments

Example 3.4 A horizontal beam *ABCD* of constant moment of inertia is 10 metres long. It is hinged at *A* and on rollers at *D*. It is continuous over *B* and *C* such that *AB* = 3 m, *BC* = 4 m and *CD* = 3 m. If the beam is loaded as shown in Fig. 3.21, draw the BMD and SFD. Use Kani's method.

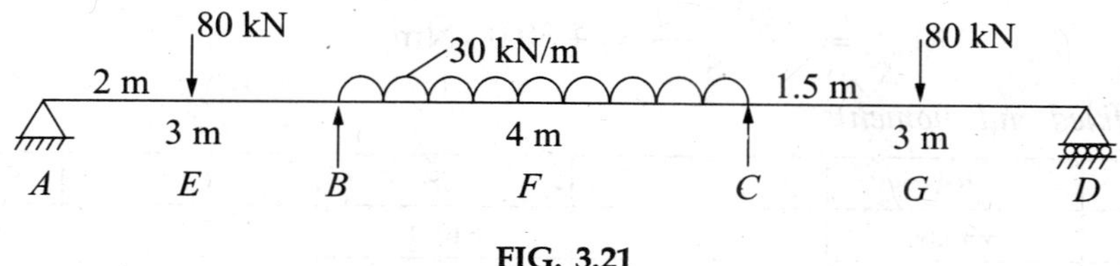

FIG. 3.21

Solution: The beam is supported on hinge at *A* and the roller at *D*. They are the end supports and therefore the end moments are zeros. Hence, it requires modification on end moments which are initially taken as fixed ends. Also the stiffness of the element. *AB* and *CD* are to be taken as three-fourths of the original stiffness. In summary:

(1) Take the stiffness of member *AB* and *CD* as $(3/4)k_{AB}$ and $(3/4)k_{CD}$

(2) Release the fixed end moments at *A* and *D* respectively and make them as zeros.

Rotation factors

Joint	*Members*	*Relative Stiffness*	Σk	$R = -\frac{1}{2}\left(\frac{k}{\Sigma k}\right)$
B	*BA*	$\frac{3}{4} \times \frac{I}{3} = 0.25I$	0.5*I*	–0.25
	BC	$I/4 = 0.25I$		–0.25
C	*CB*	$I/4 = 0.25I$	0.5*I*	–0.25
	CD	$\frac{3}{4} \times \frac{I}{3} = 0.25I$		–0.25

Fixed end moments

$$M_{FAB} = -\frac{2(80)1}{} = -17.8 \text{ kNm}$$

$$M_{FBA} = \frac{+1(80)2^2}{3^2} = +35.6 \text{ kNm}$$

$$M_{FBC} = -\frac{30(4)^2}{12} = -40.0 \text{ kNm}$$

$$M_{FCB} = +\frac{30(4)^2}{12} = +40.0 \text{ kNm}$$

$$M_{FCD} = -\frac{80 \times 3}{8} = -30.0 \text{ kNm}$$

$$M_{FDC} = +\frac{80 \times 3}{8} = +30.0 \text{ kNm}$$

Modified fixed end moments

AB	*BA*	*BC*	*CB*	*CD*	*DC*
–17.8	+35.60	–40.0	+40.0	–30.0	+30.0
+17.8 ⊖	+8.90	–	–	–15.0 ⊖	–30.0
0.0	+44.50	–40.0	+40.0	–45.0	0.0

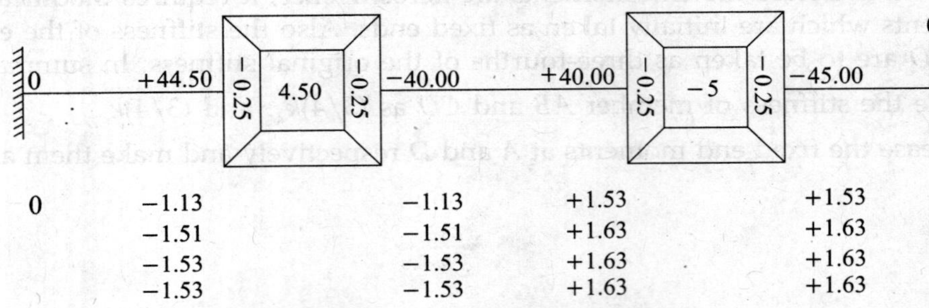

FIG. 3.22 Rotational moments

The above rotational moments are used to determine the end moments.

A		B		C		D
0	+44.50		−40.0 +40.0		−45.0	0.0

AB	BA	BC	CB	CD	DC
0	+44.50	−40.0	+40.0	−45.0	0.0
0	−1.53	−1.53	+1.63	+1.63	0
0	−1.53	−1.53	+1.63	+1.63	0
−	0.00	+1.63	−1.53	0	0
0	+41.44	−41.44	41.73	−41.73	0

FIG. 3.23

Shearing forces at the end of the members AB, BC and CD

Member AB

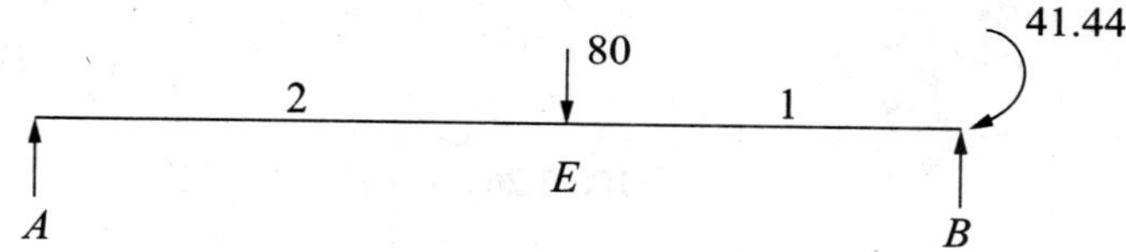

FIG. 3.24

$\Sigma V = 0;$

$$V_{AB} + V_{BA} = 80 \qquad (1)$$

$\Sigma M_A = 0;$

$$80(2) + 41.44 - 3V_{BA}$$

$$V_{BA} = 67.15 \text{ kN} \qquad (2)$$

$$\therefore \quad V_{AB} = 12.85 \text{ kN}$$

$$M_E = 12.85(2) = 25.7 \text{ kNm}$$

Member BC

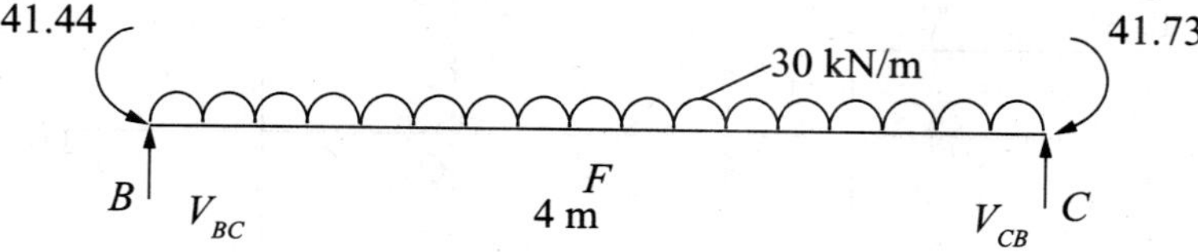

FIG. 3.25

$\sum V = 0;$

$$V_{BC} + V_{CB} = 30(4) = 120 \quad (3)$$

$\sum M_B = 0;$

$$-41.44 + 41.73 + 30 \times \frac{4^2}{2} - 4V_{CB} = 0 \quad (4)$$

$$V_{CB} = 60.1 \text{ kN}$$

$$V_{BC} = 59.9 \text{ kN}$$

$$M_F = -41.44 + 59.9(2) = 78.36 \text{ kNm}$$

Member CD

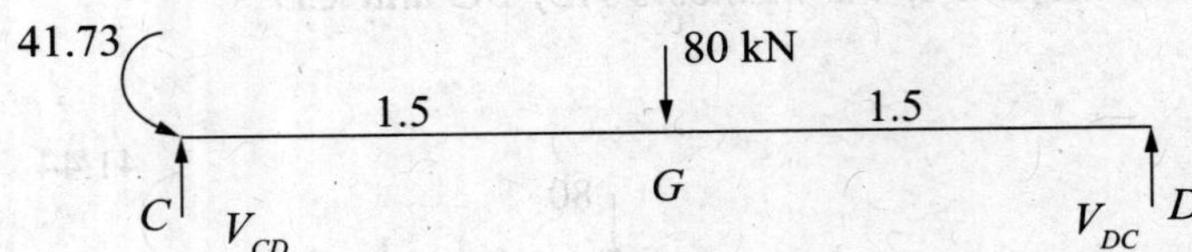

FIG. 3.26

$$\sum V = 0; \quad V_{CD} + V_{DC} = 80 \quad (5)$$

$$\sum M_C = 0; -41.73 + 80(1.5) - 3V_{DC} = 0$$

$$V_{DC} = 26 \text{ kN}$$

$$V_{CD} = 54 \text{ kN}$$

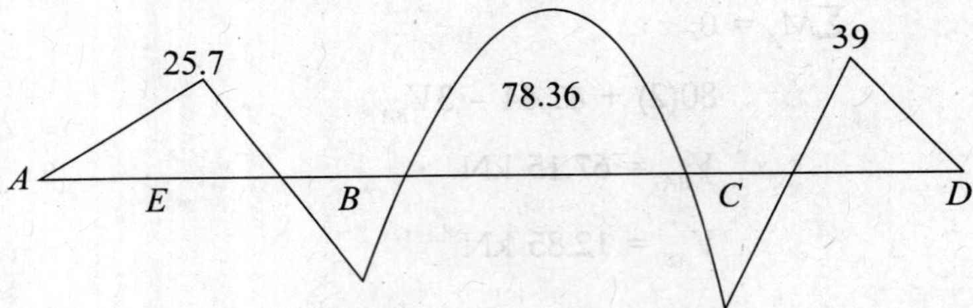

FIG. 3.27 Bending moment diagram (kNm)

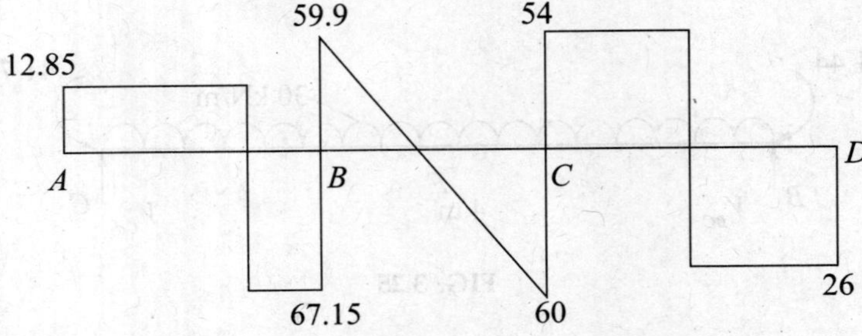

FIG. 3.28 Shear force diagram (kN)

3.2.1.2 *Analysis of Continuous Beam with Support Settlement*

Example 3.5 A continuous beam is loaded as shown in Fig. 3.29. During loading the support *B* sinks by 10 mm. Determine the bending moments at the supports and hence sketch the bending moment diagram. Given that $I = 1600(10)^4$ mm^4; $E = 200$ kN/mm^2. Use Kani's method.

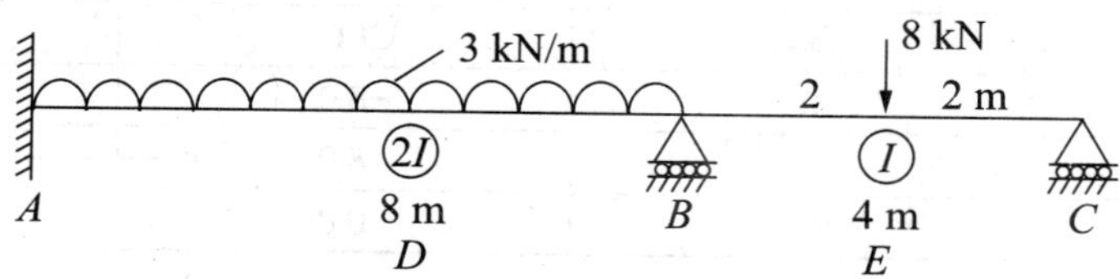

FIG. 3.29

Solution

Rotational factors

Joint	*Members*	*Relatie Stiffness (k)*	Σk	$R = -\frac{1}{2}\left(\frac{k}{\Sigma k}\right)$
	BA	$\frac{2I}{8} = 0.25I$		–0.28
B			$0.44I$	
	BC	$\frac{3}{4}\left(\frac{I}{4}\right) = 0.19I$		–0.22

Fixed end moments

$$M_{FAB} = -3(8)^2/12 = -16 \text{ kNm}$$

$$M_{FBA} = +3(8)^2/12 = +16 \text{ kNm}$$

$$M_{FBC} = -8 \times 4/8 = -4 \text{ kNm}$$

$$M_{FCB} = +8 \times 4/8 = +4 \text{ kNm}$$

Fixed end moments due to settlement of supports

$$M'_{AB} = M'_{BA} = -\frac{6EI\Delta}{l_1^2} = -\frac{6(200)\,1600(10)^4(2)(10)}{8000^2 \times 1000}$$

$$\therefore \quad M'_{AB} = M'_{BA} = -6 \text{ kNm}$$

$$M'_{BC} = M'_{CB} = +\frac{6EI\Delta}{l_2^2} = +\frac{6(200)\,1600 \times 10^4 \times 10}{8000^2}$$

$$\therefore \quad M'_{BC} = M'_{CB} = +12 \text{ kNm}$$

Final moments due to loading and settlement of supports

$$M_{AB} = -16 - 6 = -22.0 \text{ kNm}$$

$$M_{BA} = +16 - 6 = +10.0 \text{ kNm}$$

$$M_{BC} = -4 + 12 = +8.0 \text{ kNm}$$

$$M_{CB} = +4 + 12 = +16.0 \text{ kNm}$$

Modified fixed end moments for hinged ends

AB	BA	BC	CB
−22.0	+10.0	+8.0	+16.0
–	–	−8.0 ←	−16.0
−22.0	+10.0	0.0	0.0

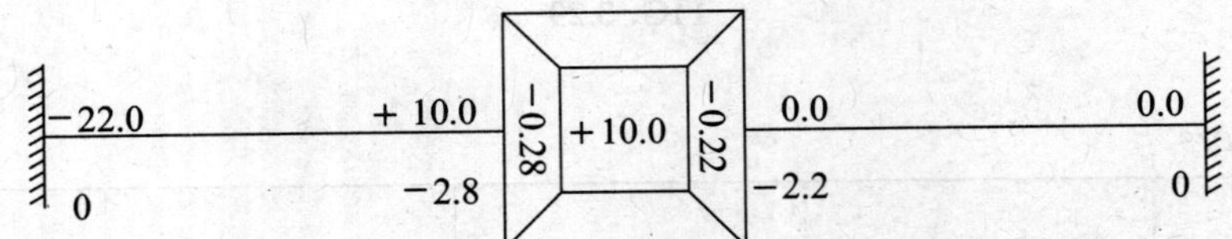

FIG. 3.30

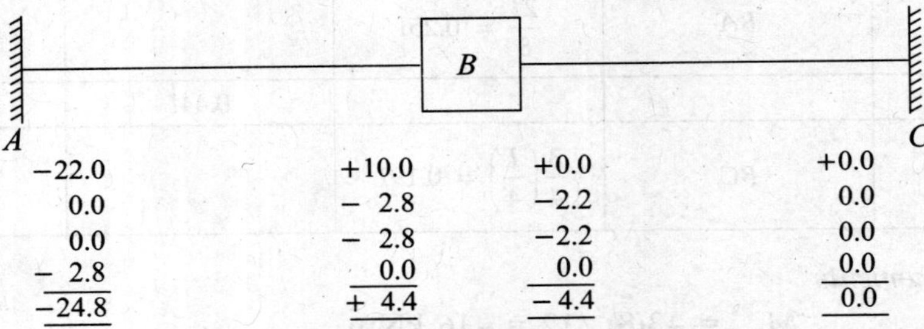

FIG. 3.31 End moments

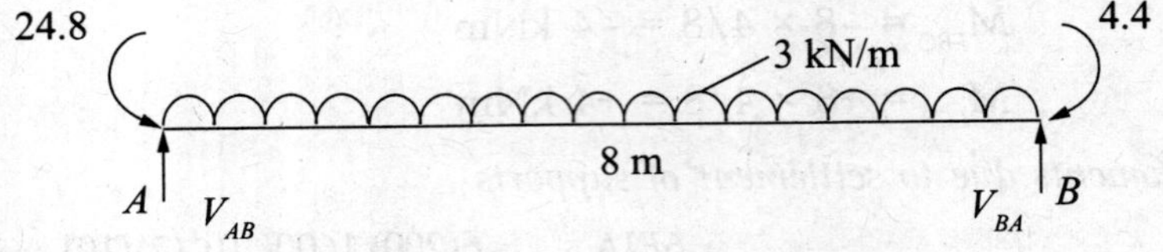

FIG. 3.32

Equilibrium of span AB $\sum V = 0;$

$$V_{AB} + V_{BA} = 24 \tag{1}$$

$$\sum M_A = 0;$$

$$-24.8 + 3 \times \frac{8^2}{2} + 4.4 - 8V_{BA} = 0 \tag{2}$$

$$V_{BA} = 9.45 \text{ kN}$$

$$V_{AB} = 14.55 \text{ kN}$$

$$M_D = 14.55(4) - 3(4)^2/2 = 34.2 \text{ kNm}$$

Equilibrium of span BC

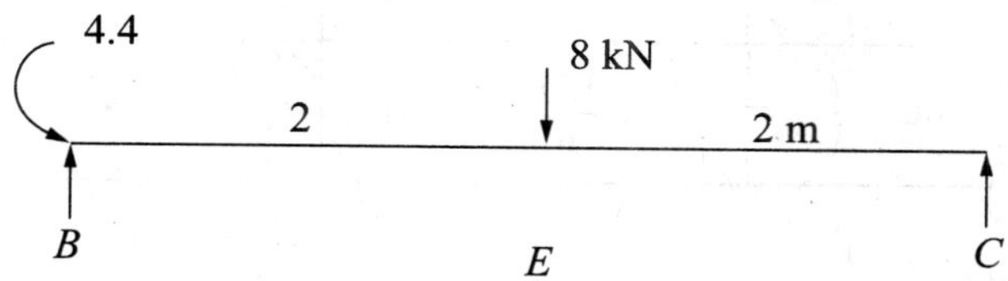

FIG. 3.33

$$\Sigma V = 0; \quad V_{BC} + V_{CB} = 8 \tag{3}$$

$$\Sigma M_B = 0; -4.4 + 8(2) - 4V_{CB} = 0$$

$$V_{CB} = 2.9 \text{ kN}$$

$$V_{BC} = 5.1 \text{ kN}$$

$$M_E = 5.1(2) - 4.4 = 5.8 \text{ kNm}$$

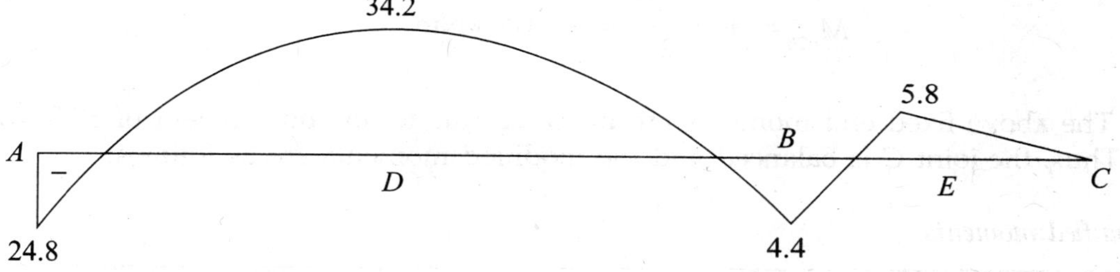

FIG. 3.34 Bending moment diagram (kNm)

Example 3.6 Determine the reactions at the rigid supports. *A*, *B* and *C* of a continuous beam are shown in Fig. 3.35. Use Kani's method of analysis. Sketch the shear force and bending moment diagrams.

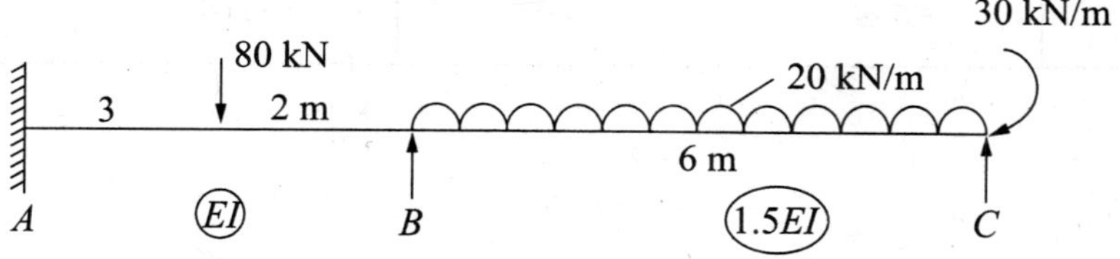

FIG. 3.35

Rotational stiffness factors

Joint	*Members*	*Relative Stiffness (k)*	Σk	$R = -\frac{1}{2}\left(\frac{k}{\Sigma k}\right)$
	BA	$I/5 = 0.2I$		-0.26
B			$0.39I$	
	BC	$\frac{3}{4} \times \frac{1.5I}{6} = 0.19I$		-0.24

Fixed end moments

$$M_{FAB} = -\frac{3(80)2^2}{5^2} = -38.40 \text{ kNm}$$

$$M_{FBA} = +\frac{2(80)3^2}{5^2} = +57.60 \text{ kNm}$$

$$M_{FBC} = -\frac{20(6)^2}{12} = -60.00 \text{ kNm}$$

$$M_{FCB} = +\frac{20(6)^2}{12} = +60.00 \text{ kNm}$$

The above fixed end moments are modified due to an end moment of 30 kNm at *C*. Thus, the joint *C* is balanced and the modified moments are as follows.

Modified moments

AB	*BA*	*BC*	*CB*	*CE*
-38.40	+57.60	-60.00 ←	+60.00	+30.00
-	-	-45.00	-90.00	
-38.40	+57.60	-105.00	-30.00	+30.00

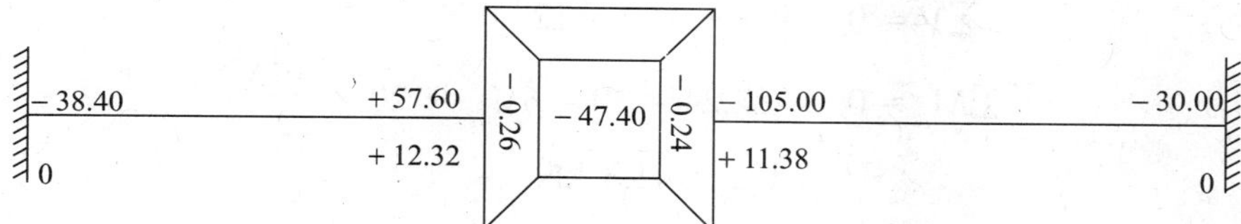

FIG. 3.36

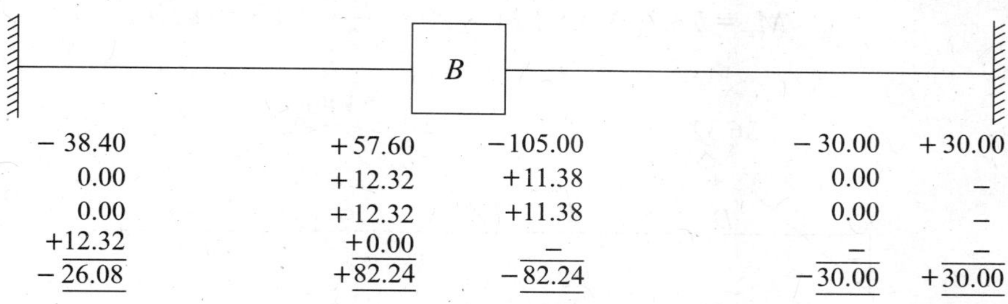

FIG. 3.37

Shearing forces and bending moments in members AB and BC

Considering the free body diagram of each span separately

Member AB

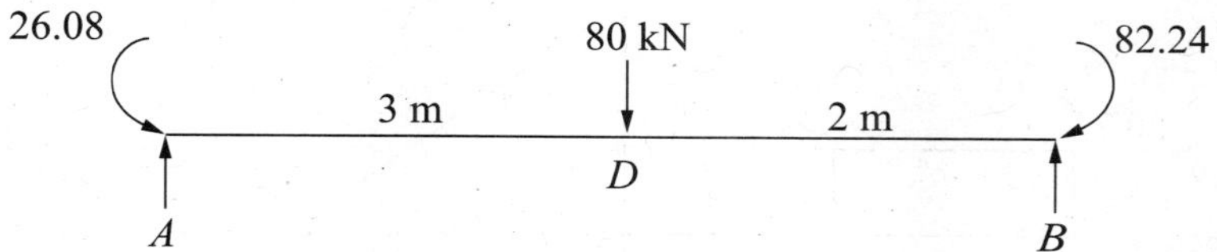

FIG. 3.38

$$\Sigma V = 0; \quad V_{AB} + V_{BA} = 80 \tag{1}$$

$$\Sigma M_A = 0; \quad -26.08 + 80(3) + 82.24 - 5V_{BA} = 0 \tag{2}$$

$$V_{BA} = 59.2 \text{ kN}$$

$$V_{AB} = 20.8 \text{ kN}$$

$$M_D = -26.08 + 20.8(3) = 36.32 \text{ kNm}$$

Member BC

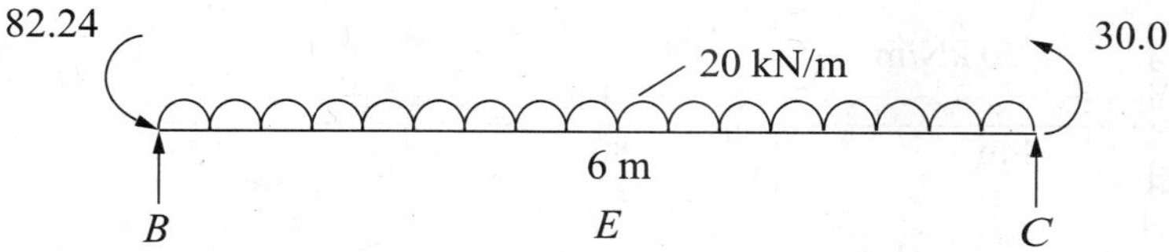

FIG. 3.39

$$\Sigma V = 0; \quad V_{BC} + V_{CB} = 120 \tag{3}$$

$$\Sigma M_B = 0; \quad -82.24 - 30 - 6V_{CB} + 20 \times \frac{6^2}{2} = 0$$

$$V_{CB} = 41.3 \text{ kN}$$

$$V_{BC} = 78.7 \text{ kN}$$

$$M_E = 78.7(3) - 82.24 - 20 \times \frac{3^2}{2} = 63.86 \text{ kNm}$$

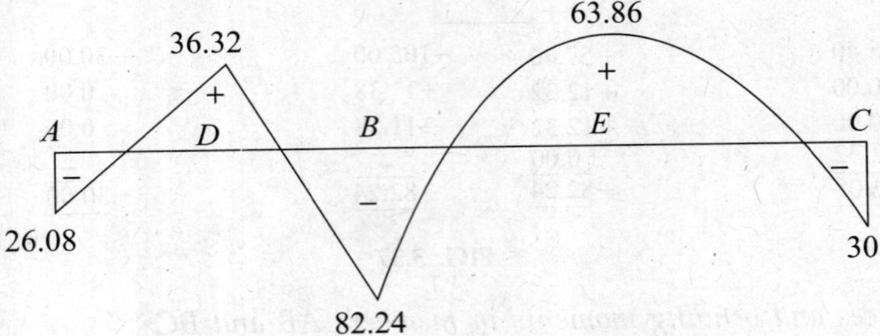

FIG. 3.40 Bending moment diagram (kNm)

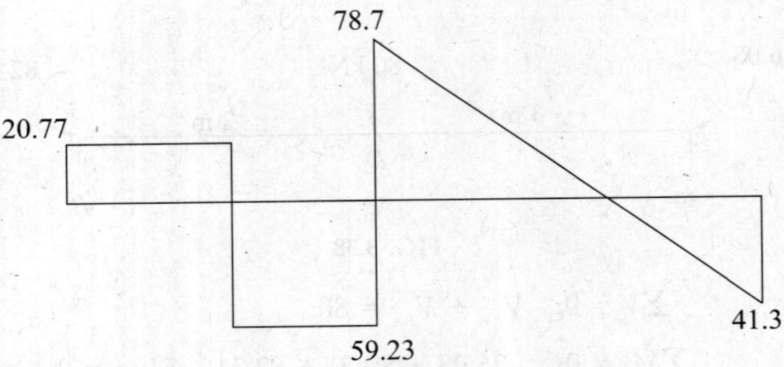

FIG. 3.41 Shear force diagram (kN)

Example 3.7 A continuous beam *ABCD* subjected to the loading shown in Fig. 3.42 is fixed at *A* and freely supported at *B* and *C*. The support *C* sinks by 20 mm. *A* and *B* remain in the same level. Calculate the reactions *A*, *B* and *C*. $I_{AB} = 170(10)^6$ mm^4; $I_{BC} = I_{CD} = 340(10)^6$ mm^4. $E = 200$ kN/mm^2.

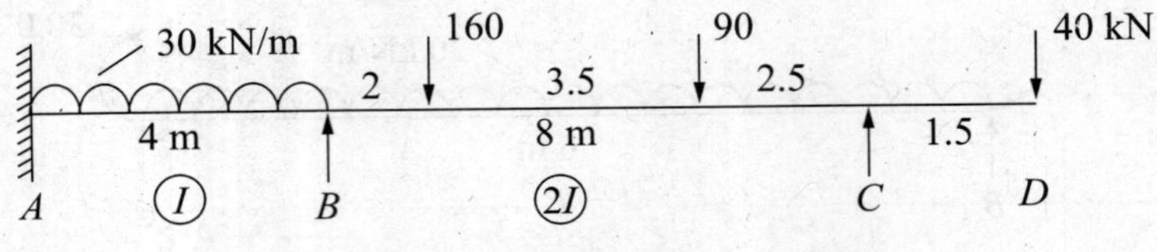

FIG. 3.42

Rotation factors

Joint	*Members*	*Relative Stiffness*	Σk	$R = -\frac{1}{2}\left(\frac{k}{\Sigma k}\right)$
	BA	$I/4 = 0.25I$		−0.29
B			$0.438I$	
	BC	$\frac{3}{4}\left(\frac{2I}{8}\right) = 0.188I$		−0.21

Fixed end moments due to applied loading

$$M'_{AB} = -\frac{30(4)^2}{12} = -40 \text{ kNm}$$

$$M'_{BA} = +\frac{30(4)^2}{12} = +40 \text{ kNm}$$

$$M'_{BC} = -\frac{2(160)6^2}{8^2} - \frac{5.5(90)2.5^2}{8^2} = -228.34 \text{ kNm}$$

$$M'_{CB} = +\frac{6(160)2^2}{8^2} + \frac{2.5(90)5.5^2}{8^2} = +166.35 \text{ kNm}$$

Fixed end moments due to settlement of supports C

$$M''_{BC} = -\frac{6EI\Delta}{l^2} = -\frac{6 \times 200 \times 340 \times 10^6 \times 20}{8000^2 \times 1000} = -127.5 \text{ kNm}$$

$$M''_{CB} = -\frac{6EI\Delta}{l^2} = -127.5 \text{ kNm}$$

Total fixed moments

$$M_{FAB} = -40 \text{ kNm}$$

$$M_{FBA} = +40 \text{ kNm}$$

$$M_{FBC} = -228.34 - 127.5 = -355.84 \text{ kNm}$$

$$M_{FCB} = +166.35 - 127.5 = +38.85 \text{ kNm}$$

$$M_{FCD} = -60.0 \text{ kNm}$$

Modified moments

AB	*BA*	*BC*	*CB*	*CD*
−40.00	+40.00	−355.84	+38.85	−60.00
–	–	+10.58 ⟵	+21.15	
−40.00	+40.00	−345.26	+60.00	+60.00

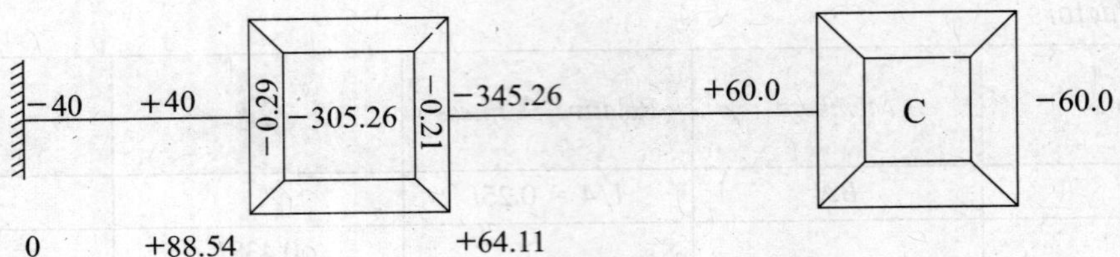

FIG. 3.43

In Fig. 3.43, the moments are to be balanced at joint C. This has been done in the above modified moment table. As the rotational moment is to be distributed to only joint *B* (as already joint *C* is balanced).

This has been done as follows.

Cycle 1

$$m_{ba} = -0.29(0 - 305.26 + 0) = 88.53 \text{ kNm}$$

$$m_{bc} = -0.21(0 - 305.26 + 0) = 64.11 \text{ kNm}$$

The final end moments are computed as follows.

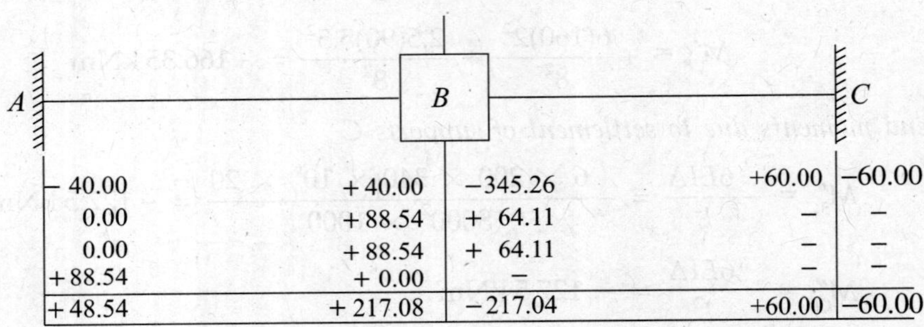

FIG. 3.44

Shearing forces at the ends of members AB, BC

Member AB

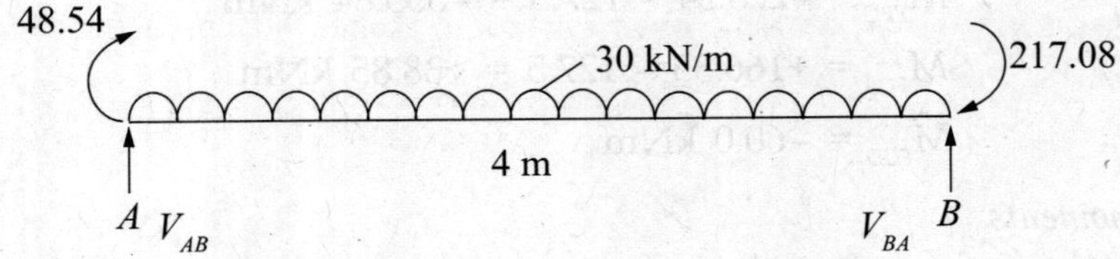

FIG. 3.45

$$\Sigma V = 0; \quad V_{AB} + V_{BA} = 30(4) = 120 \tag{1}$$

$$\Sigma M_A = 0; \quad 48.54 + \frac{30(4)^2}{2} + 217.08 - 4V_{BA} = 0 \tag{2}$$

$$V_{BA} = 126.4 \text{ kN}$$

$$V_{AB} = -6.4 \text{ kN}$$

Member BC

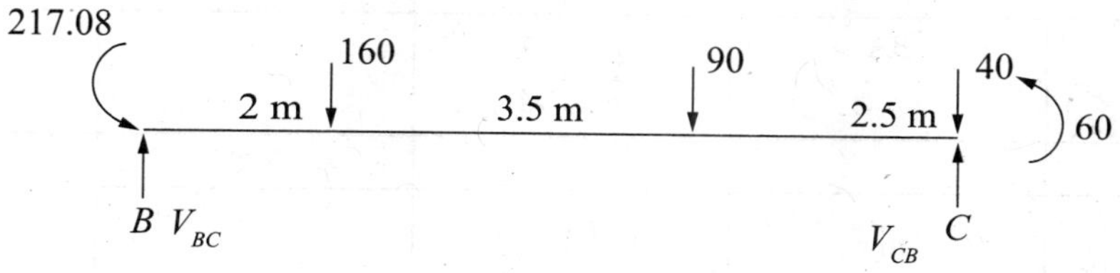

FIG. 3.46

$$\Sigma V = 0; \quad V_{BC} + V_{CB} = 250 \qquad (3)$$

$$\Sigma M_B = 0; \quad -217.08 + 160(2) + 90(5.5) - 60 - 8V_{CB} = 0$$
$$+ 40 \times 8$$

$$V_{CB} = 107.24 \text{ kN}$$

$$V_{BC} = 182.76 \text{ kN}$$

Reactions

$$R_A = 6.4 \text{ kN}(\downarrow)$$

$$R_B = 309.2 \text{ kN} \uparrow$$

$$R_C = 107.2 \uparrow$$

Check

$$R_A + R_B + R_C = 30(4) + 160 + 90 + 40$$

$$-6.4 + 309.2 + 107.2 = 410$$

Example 3.8 Analyse the continuous beam shown in Fig. 3.47 by Kani's method. Sketch the shear force and bending moment diagrams. Relative to the support *A*, the support *B* sinks by 1 mm and the support C rises by 1/2 mm. $EI = 30000$ kNm².

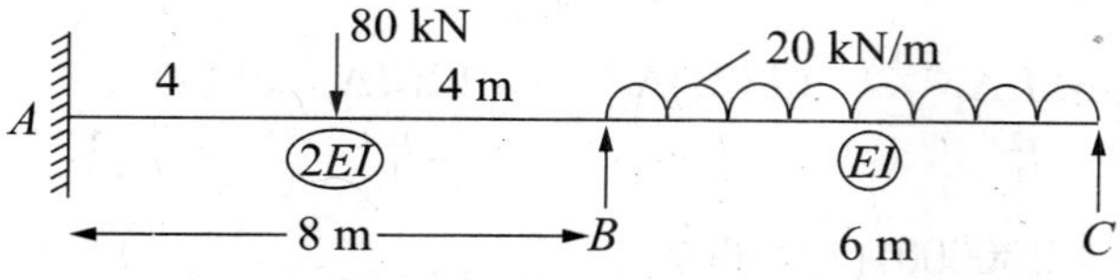

FIG. 3.47

Solution

Rotation factors

Joint	*Members*	*k*	Σk	$R = -\frac{1}{2}\frac{k}{\Sigma k}$
	BA	$\frac{2I}{8} = \frac{I}{4}$		−1/3
B			3/8	
	BC	$\frac{3}{4}\left(\frac{I}{6}\right) = \frac{I}{8}$		−1/6

Fixed end moments

$$M'_{FAB} = -\frac{Wl}{8} = -\frac{80(8)}{8} = -80 \text{ kNm}$$

$$M'_{FBA} = +\frac{Wl}{8} = +\ 80 \times \frac{8}{8} = +80 \text{ kNm}$$

$$M'_{FBC} = -\frac{wl^2}{12} = -\frac{20 \times 6^2}{12} = -60 \text{ kNm}$$

$$M'_{FCB} = +\frac{wl^2}{12} = +\frac{20 \times 6^2}{12} = +60 \text{ kNm}$$

FEM'S due to settlement of supports

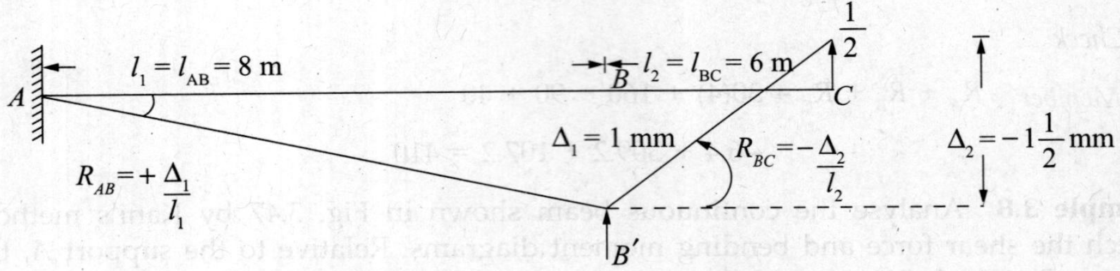

FIG. 3.48 Settlement angles diagram ($R = \Delta/l$)

$$M''_{FAB} = M''_{FBA} = -\frac{6EI_1\Delta_1}{l_1^2} = -\frac{6E(2I)\Delta_1}{l_1^2} = -\frac{12EI\Delta_1}{l_1^2}$$

$$= -\frac{12(30000)\,(1/1000)}{8^2} = -\,5.63 \text{ kNm}$$

$$M''_{FBC} = M''_{FCB} = -\frac{6EI_2\Delta_2}{l_2^2} = -\frac{6EI\Delta_2}{l_2^2} = -\frac{6(30000)\,(-1.5/1000)}{6^2} = +7.5 \text{ kNm}$$

Total fixed end moments

The total fixed end moments is found out by adding the fixed end moments due to the applied loading as well as due to settlement of supports.

$$M_{FAB} = -80 - 5.63 = -85.63 \text{ kNm}$$

$$M_{FBA} = +80 - 5.63 = +74.37 \text{ kNm}$$

$$M_{FBC} = -60 + 7.5 = -52.50 \text{ kNm}$$

$$M_{FCB} = +60 + 7.5 = +67.50 \text{ kNm}$$

Shear force and bending moment in members AB and BC

Member AB

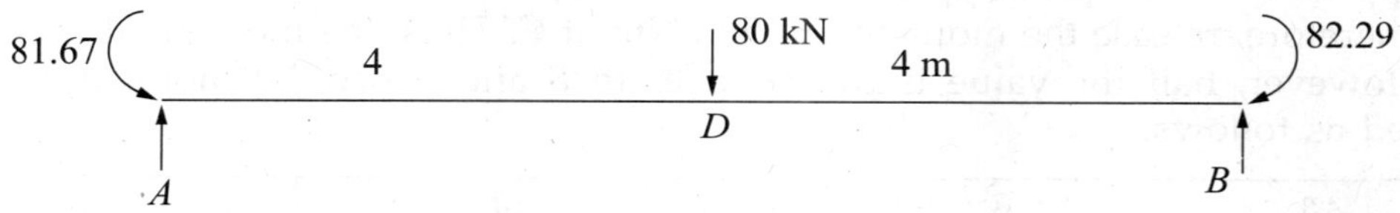

FIG. 3.49

$$\Sigma V = 0; \quad V_{AB} + V_{BA} = 80 \qquad (1)$$

$$\Sigma M_A = 0; \quad -81.67 + 82.29 + 80(4) - 8V_{BA} = 0 \qquad (2)$$

$$V_{BA} = 40.08 \text{ kN}$$

$$V_{AB} = 39.92 \text{ kN}$$

$$MD = -81.67 + 39.92(4) = 78.01 \text{ kNm}$$

Member BC

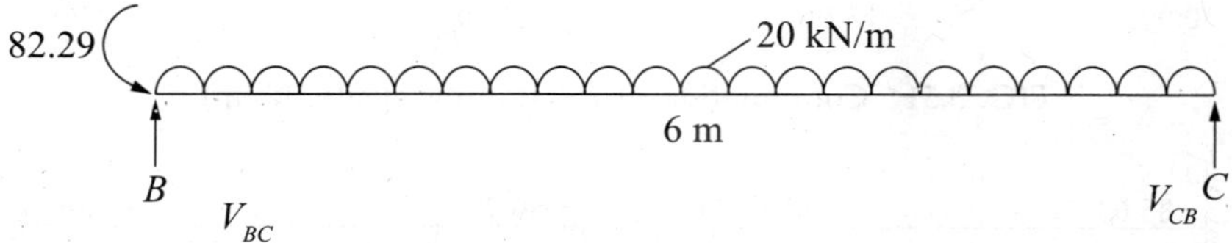

FIG. 3.50

$$\Sigma V = 0; \quad V_{BC} + V_{CB} = 120 \qquad (3)$$

$$\Sigma M_B = 0; \quad -82.29 + 20 \times \frac{6^2}{2} - 6V_{CB} = 0$$

$$V_{CB} = 46.29 \text{ kN}$$

$$V_{BC} = 73.71 \text{ kN}$$

$$M_X = 46.29x - 20\frac{x^2}{2}$$

$$\frac{dM_x}{dx} = 46.29 - 20x = 0$$

$$x = 2.31 \text{ m}$$

$$\text{Max + ve BM} = 46.29(2.31) - 20(2.31)^2/2 = 53.56 \text{ kNm}$$

Modified fixed end moments

The support *C* is a simple support and an end support. Hence, the moment should be zero. Therefore, release the moment −67.50 kNm at *C*. Thus, the moment at *C* becomes zero. However, half the value is carried over to *B* and hence the moment at *B* gets modified as follows.

AB	*BA*	*BC*	*CB*
−85.63	+74.37	−52.50	+67.50 −67.50
–	–	−33.75	
−85.63	+74.37	−86.25	0.00

Using the above 'modified fixed end moments', the rotation moments are computed as follows.

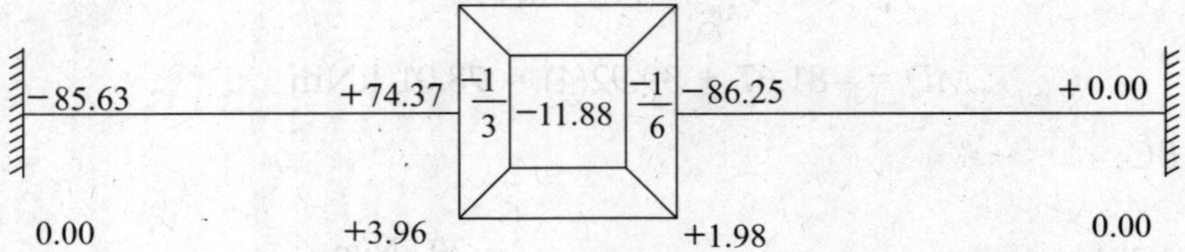

FIG. 3.51 Computation of rotation moments (kNm)

−85.63	+74.37	−86.25	+0.00
0.00	+3.96	+1.98	+0.00
0.00	+3.96	+1.98	+0.00
+3.96	+0.00	+0.00	—
−81.67	+82.29	−82.29	0.00

FIG. 3.52

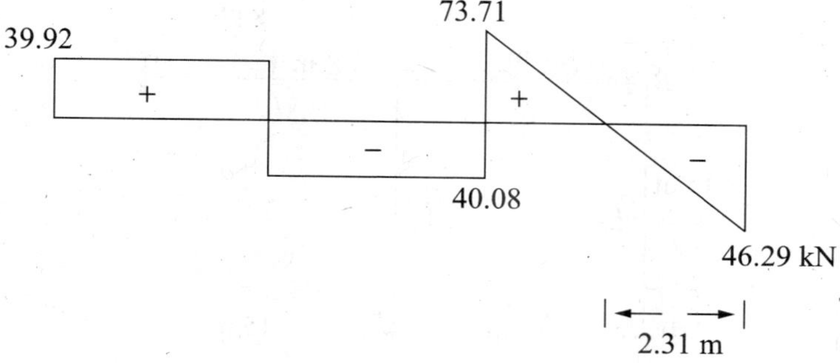

FIG. 3.53 Shear force diagram (kN)

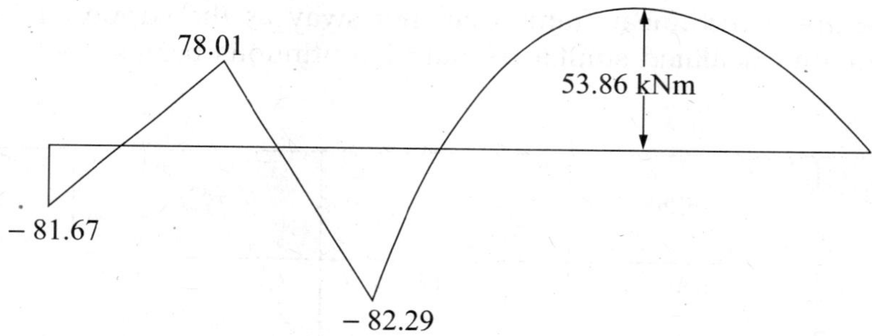

FIG. 3.54 Bending moment diagram (kNm)

Note on sinking of supports

The settlement/sinking/subsidence/uneven supports cause serious changes in the bending moment. This effects in bending stresses. This sinking of supports is a demerit to recommend the use of continuous beams. The practical problem is the difficulty in continuity during construction and also the estimation of degree of fixity.

When the loads are applied on a continuous beam, the bending moment diagram gives the indication of two points of contraflexure occurring between the supports. At these points if the beam is hinged, the bending moment at such location is zero. In such cases, the sinking of supports does not affect the bending moment. This leads to the principle of cantilever bridge construction. However, construction detailing has to be taken care of for providing internal hinges. Hence, the settlement of supports is an important factor for determining the internal forces and moments.

3.2.1.3 *Analysis of Rectilinear Frames*

Example 3.9 The rigid jointed frame shown in Fig. 3.55 is encastre at *A*, *D* and *E*. All the members are assumed to be inextensible and to have the same flexural rigidity. Determine the axial forces in *BC* and *CD*. Use Kani's method.

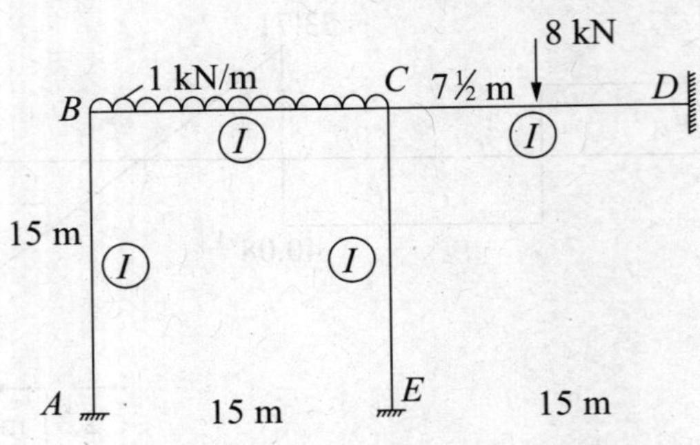

FIG. 3.55

Solution: The given rectilinear frame does not sway as the support D is fixed. The rotation factors are calculated similar to that of continuous beams.

Rotation factors

Joint	*Members*	k	Σk	$R = -\frac{1}{2}\frac{k}{\Sigma k}$
B	BA	$I/15$	$2I/15$	−0.25
	BC	$I/15$		−0.25
C	CB	$I/15$	$3I/15$	−1/6
	CD	$I/15$		−1/6
	CE	$I/15$		−1/6

Fixed end moments

$$M_{FBC} = -\frac{1(15)^2}{12} = -18.75 \text{ kNm}$$

$$M_{FCB} = +\frac{1(15)}{12} = +18.75 \text{ kNm}$$

$$M_{FCD} = -\frac{8 \times 15}{8} = -15.00 \text{ kNm}$$

$$M_{FDC} = +\frac{8 \times 15}{8} = +15.00 \text{ kNm}$$

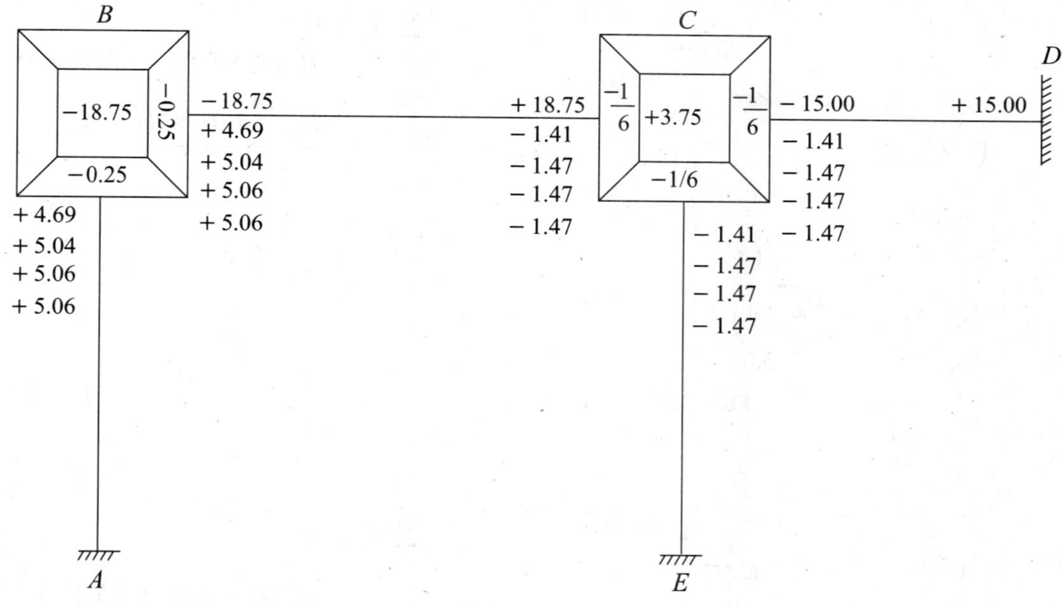

FIG. 3.56 Rotational moments

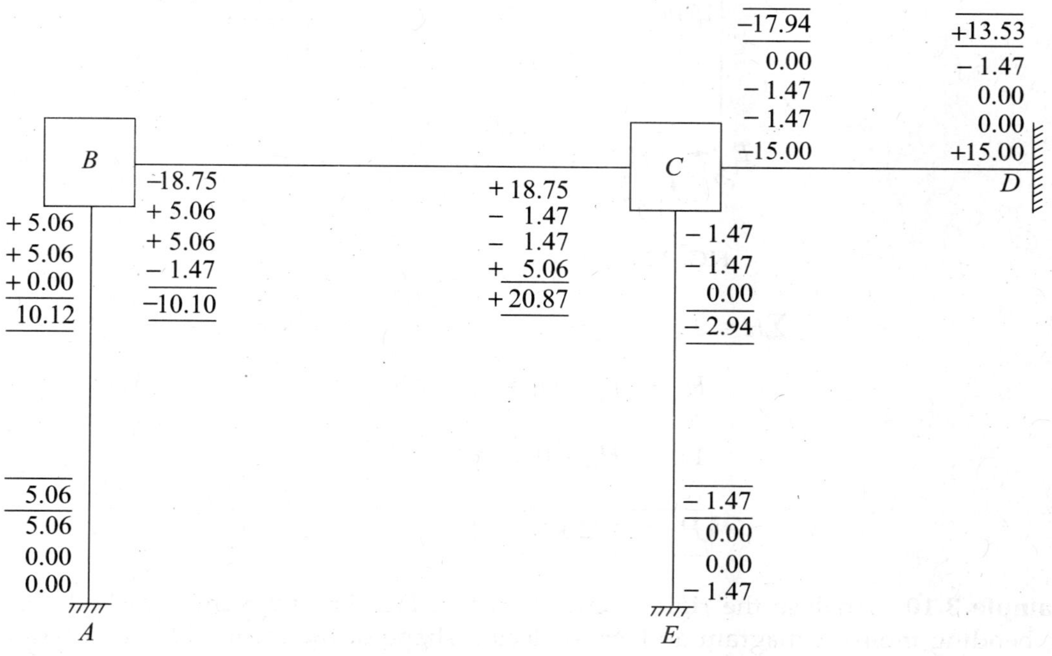

FIG. 3.57 End moments

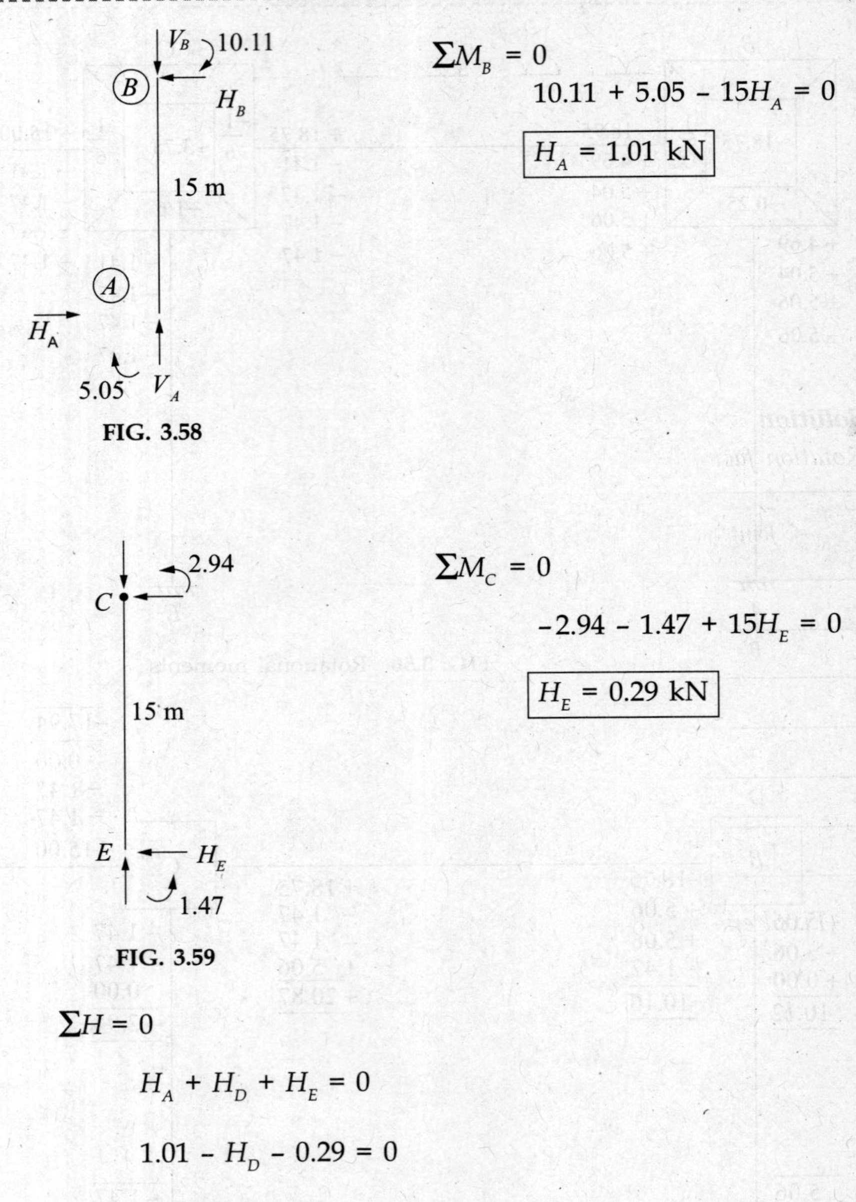

FIG. 3.58

$$\Sigma M_B = 0$$

$$10.11 + 5.05 - 15H_A = 0$$

$$\boxed{H_A = 1.01 \text{ kN}}$$

FIG. 3.59

$$\Sigma M_C = 0$$

$$-2.94 - 1.47 + 15H_E = 0$$

$$\boxed{H_E = 0.29 \text{ kN}}$$

$$\Sigma H = 0$$

$$H_A + H_D + H_E = 0$$

$$1.01 - H_D - 0.29 = 0$$

$$\boxed{H_D = 0.72 \text{ kN}}$$

Example 3.10 Analyse the rigid frame shown in Fig. 3.60 by Kani's method. Draw the bending moment diagram and the deflected shape of the frame. *EI* is constant for all the members.

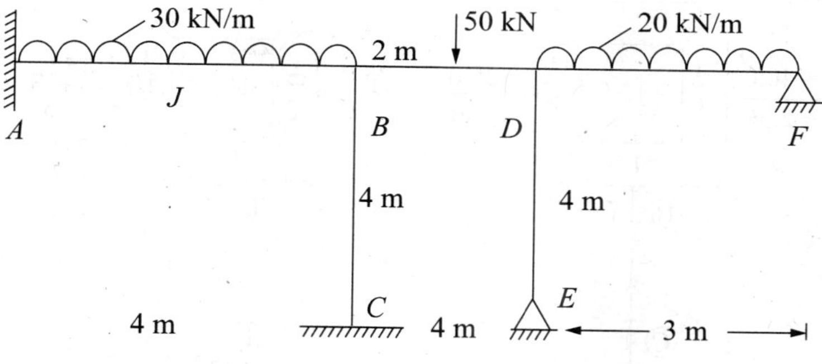

FIG. 3.60

Solution

Rotation factors

Joint	*Members*	*Relative Stiffness*	Σk	$R = -\frac{1}{2}\frac{k}{\Sigma k}$
	BA	$I/4$		$-1/6$
B	*BC*	$I/4$	$3I/4$	$-1/6$
	BD	$I/4$		$-1/6$
	DB	$I/4 = 0.25I$		-0.18
D	*DF*	$3/4(I/4) = 0.19I$	$0.69I$	-0.14
	DE	$3/4(I/3) = 0.25I$		-0.18

Fixed end moments

$$M_{FAB} = -\frac{30(4)^2}{12} = -40.0 \text{ kNm}$$

$$M_{FBA} = +\frac{30(4)^2}{12} = +40.0 \text{ kNm}$$

$$M_{FBD} = -\frac{50(4)}{8} = -25.0 \text{ kNm}$$

$$M_{FDB} = +\frac{50(4)}{8} = +25.0 \text{ kNm}$$

$$M_{FDF} = -\frac{20(3)^2}{12} = -15.0 \text{ kNm}$$

$$M_{FFD} = +\frac{20(3)^2}{12} = +15.0 \text{ kNm}$$

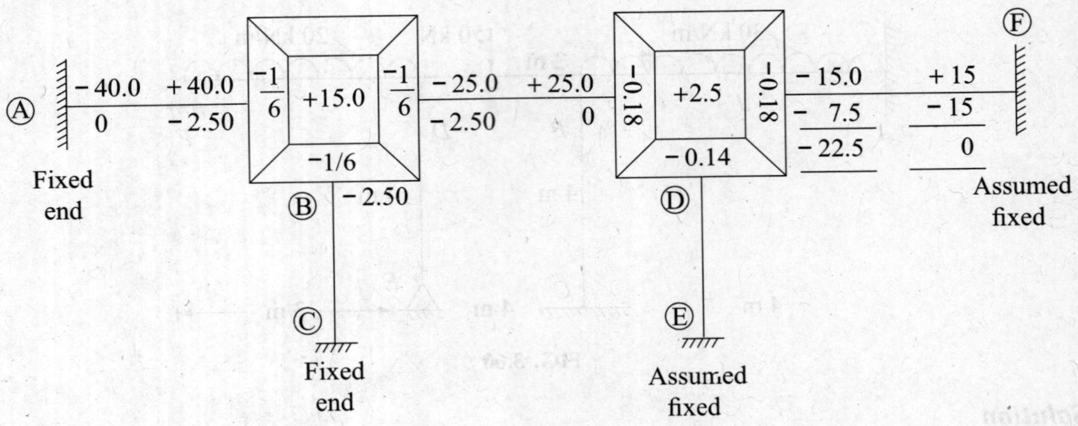

FIG. 3.61

Computation of end moments

This is done by using Kani's formulation as

Moment at near end = FEM at near end + 2(Rotation contribution at near end) + 1(Rotation contribution at far end)

$$M_{AB} = -40 + 2(0) + (-0.25) = -42.5 \text{ kNm}$$

$$M_{BA} = +40 + 2(-2.5) + 0 = +35.0 \text{ kNm}$$

$$M_{BC} = 0 + 2(-2.5) + 0 = -5.0 \text{ kNm}$$

$$M_{CB} = 0 + 2(0) - 2.5 = -2.5 \text{ kNm}$$

$$M_{BD} = -25.0 + 2(-2.5) + 0 = -30.0 \text{ kNm}$$

$$M_{DB} = +25.0 + 2(0) - 2.5 = +22.5 \text{ kNm}$$

$$M_{DE} = 0 + 2(0) + 0 = 0$$

$$M_{DF} = -22.5 + 2(0) + 0 = -22.5 \text{ kNm}$$

$$M_{ED} = 0$$

$$M_{FD} = 0$$

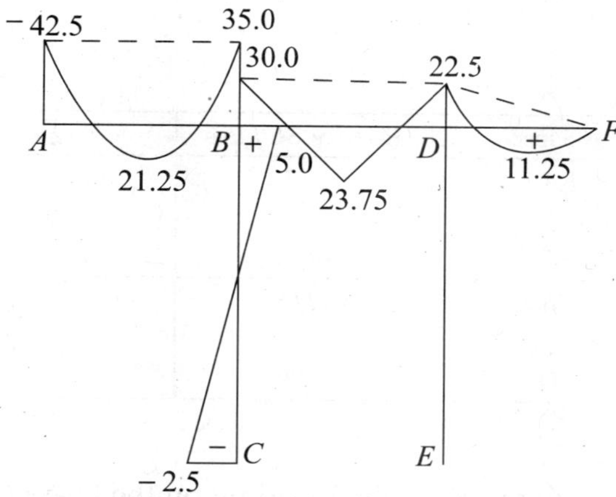

FIG. 3.62 Bending moment diagram

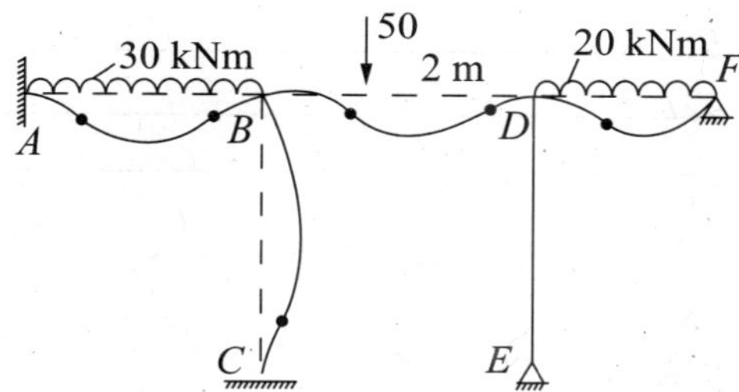

FIG. 3.63 Elastic curve of the frame

3.2.1.4 *Analysis of Frames with no Lateral Translation of Joints*

The Kani's method of moment distribution is done similar to that of continuous beams. The lateral translation is prevented due to symmetry in loading and boundary conditions.

Example 3.11 Analyse the given frame by Kani's method. Draw the bending moment diagram.

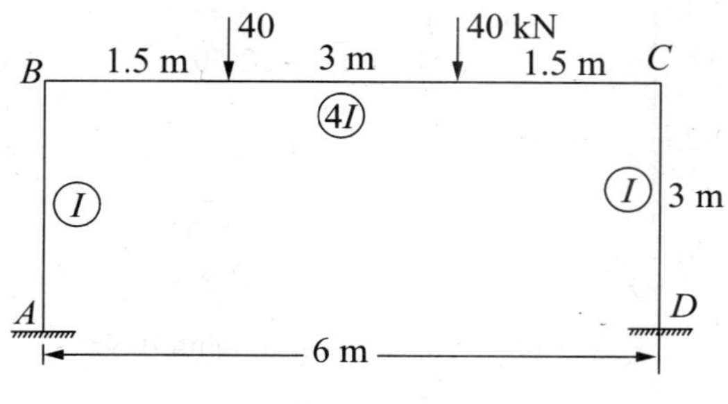

FIG. 3.64

Solution

Rotation factors

Joint	*Member*	*Relative Stiffness*	Σk	*Rotation Factor*
B	BA	I/3	I	-1/6
	BC	4I/6		-1/3
C	CB	4I/6	I	-1/3
	CD	I/3		-1/6

Fixed end moments

The fixed end moments due to the applied loading on the beam (transom) is obtained as.

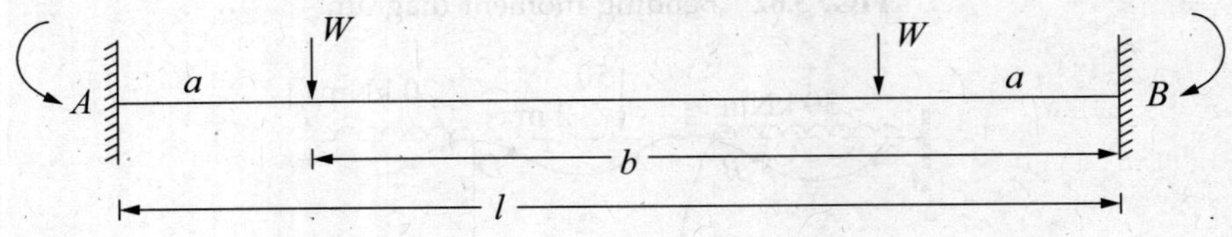

FIG. 3.65

$$M_{FAB} = -\frac{Wab}{l}$$

$$\therefore \quad M_{FAB} = -\frac{40(1.5)(4.5)}{6} = -45.00 \text{ kNm}$$

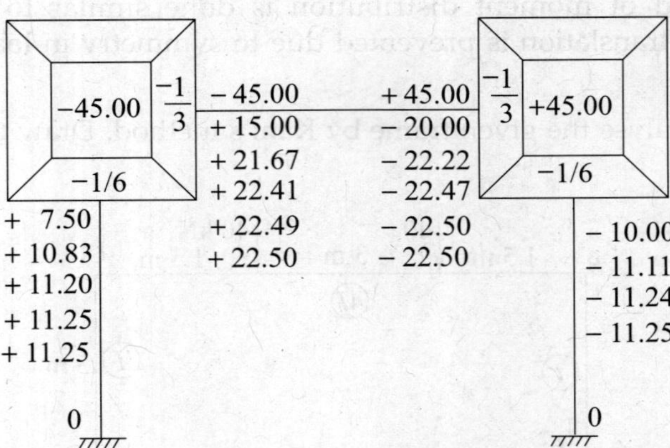

FIG. 3.66 Rotational moments (kNm)

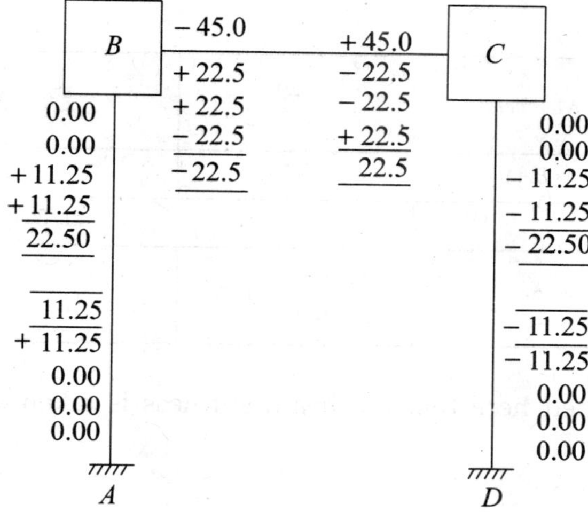

FIG. 3.67 Final moments (kNm)

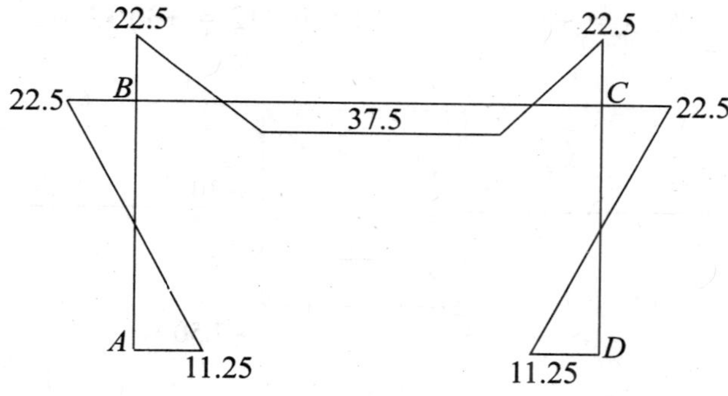

FIG. 3.68 Bending moment diagram (kNm)

Example 3.12 Analyse the given frame by Kani's method. Sketch the bending moment diagram.

(Jan 2010, VTU)

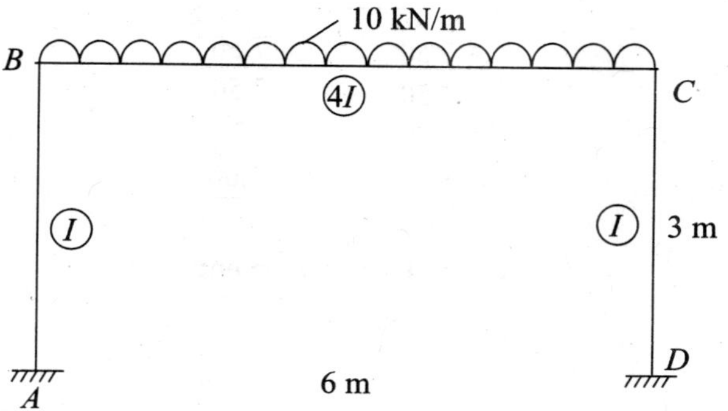

FIG. 3.69

Rotation factors

Joint	*Members*	*k*	Σk	$R = -\frac{1}{2}\left(\frac{k}{\Sigma k}\right)$
	BA	*I*/3		−0.25
B			2*I*/3	
	BC	$\frac{1}{2}\left(\frac{4I}{6}\right) = I/3$		−0.25

It is to be mentioned here that the beam stiffness is taken as half of the original stiffness.

Fixed end moments

$$M_{FBC} = -wl^2/12 = -10 \times 6^2/12 = -30 \text{ kNm}$$

$$M_{FCB} = +wl^2/12 = +10 \times 6^2/12 = +30 \text{ kNm}$$

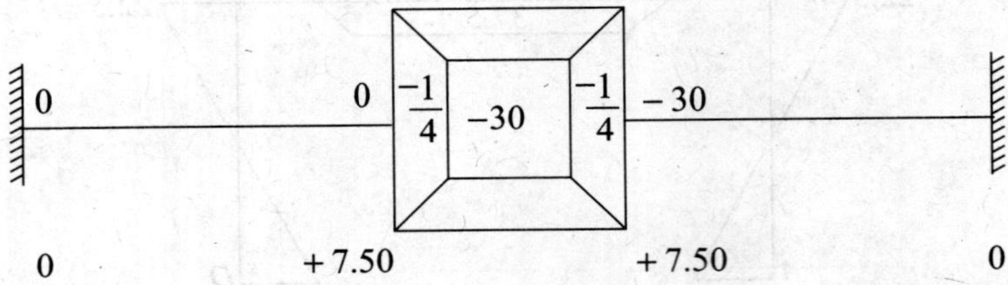

FIG. 3.70

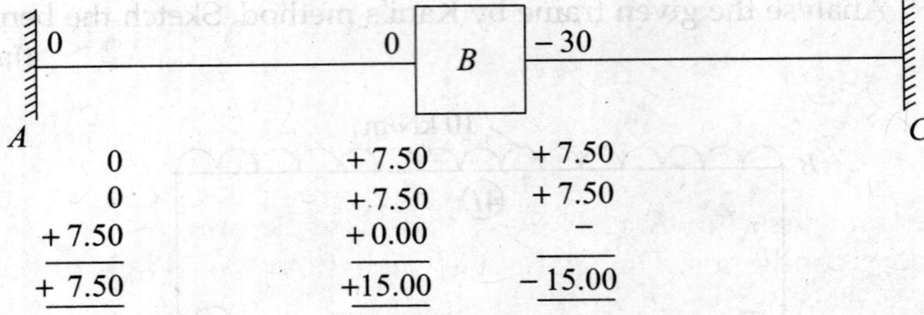

FIG. 3.71 End moments

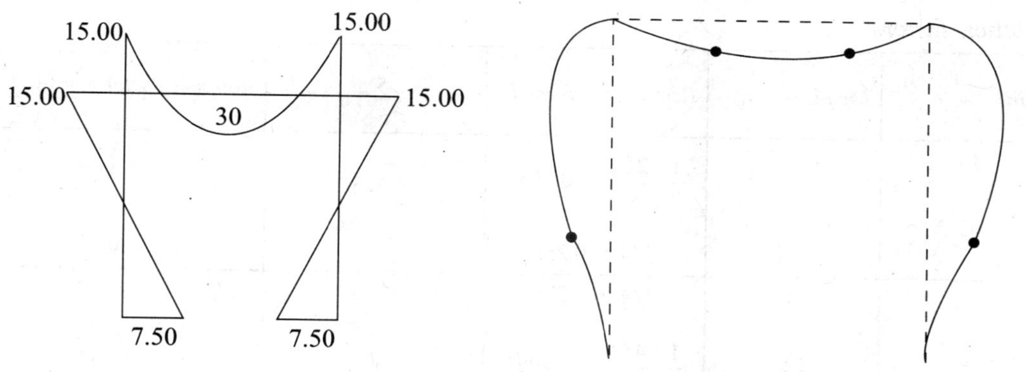

FIG. 3.72 Bending moment diagram (kNm)

FIG. 3.73 Elastic curve

Example 3.13 The frame shown in Fig. 3.74 is fixed at *A* and *F*, and all the other joints are rigid. The second moment of the areas and the loads are shown. Analyse the frame by Kani's method.

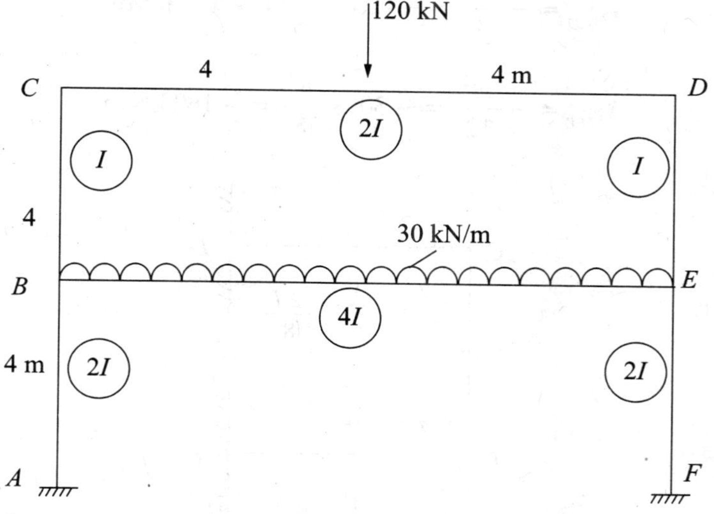

FIG. 3.74

Solution: The frame is symmetrically loaded and is symmetric with respect to geometry and the boundary conditions. The analysis is carried out for an substitute frame whose beam stiffness is half of the actual beam stiffness. The rotation factors are computed and entered in the annular space inside the box. The rotation moments are computed using the rotation factors and the fixed end moments due to applied loading. Due to symmetry, the frame does not sway and the advantage of symmetry, has been used in the following solution.

Rotation factors

Joint	Members	Relative Stiffness k	$\sum k$	Rotation Factor $R = -\frac{1}{2}\frac{k}{\sum k}$
C	CD	$\frac{1}{2}\left(\frac{2I}{8}\right) = 0.125I$	$0.375I$	−0.17
	CB	$I/4 = 0.25I$		−0.33
B	BC	$I/4 = 0.25I$		−0.13
	BE	$\frac{1}{2}\left(\frac{4I}{8}\right) = 0.125I$	I	−0.13
	BA	$2I/4 = 0.5I$		−0.24

Fixed end moments

$$M_{FCD} = -\frac{Wl}{8} = -120 \times \frac{8}{8} = -120 \text{ kNm}$$

$$M_{FBE} = -\frac{wl^2}{12} = -30 \times \frac{8^2}{8} = -160 \text{ kNm}$$

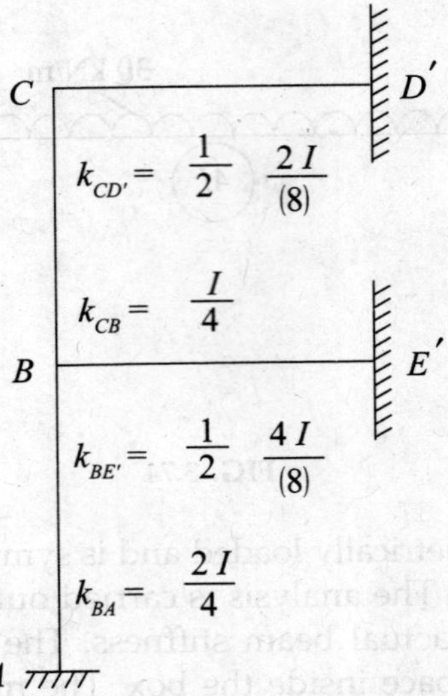

FIG. 3.75 Substitute frame

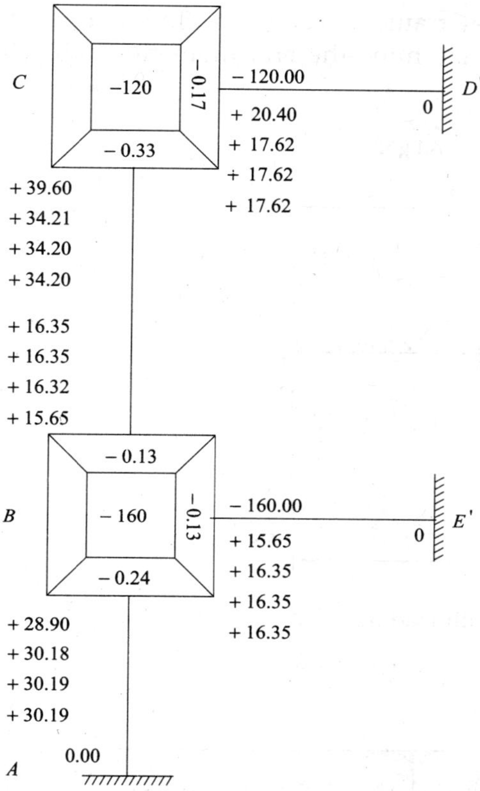

C
−120
−0.17
−0.33
− 120.00
0
D'
+ 20.40
+ 17.62
+ 17.62
+ 17.62
+ 39.60
+ 34.21
+ 34.20
+ 34.20
+ 16.35
+ 16.35
+ 16.32
+ 15.65
− 0.13
B
− 160
−0.13
− 160.00
0
E'
+ 15.65
+ 16.35
+ 16.35
+ 16.35
− 0.24
+ 28.90
+ 30.18
+ 30.19
+ 30.19
0.00
A

FIG. 3.76 Rotational moments (kNm)

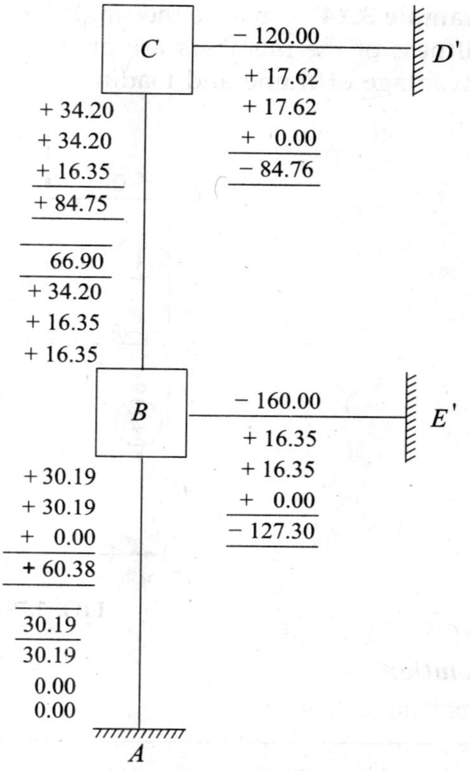

C
− 120.00
D'
+ 17.62
+ 17.62
+ 0.00
− 84.76
+ 34.20
+ 34.20
+ 16.35
+ 84.75
66.90
+ 34.20
+ 16.35
+ 16.35
B
− 160.00
E'
+ 16.35
+ 16.35
+ 0.00
− 127.30
+ 30.19
+ 30.19
+ 0.00
+ 60.38
30.19
30.19
0.00
0.00
A

FIG. 3.77 Computation of end moments (kNm)

−84.76
+84.76
84.75
−84.75
66.90
− 127.30
+ 127.3
−66.90
60.38
−60.38
30.19
−30.19

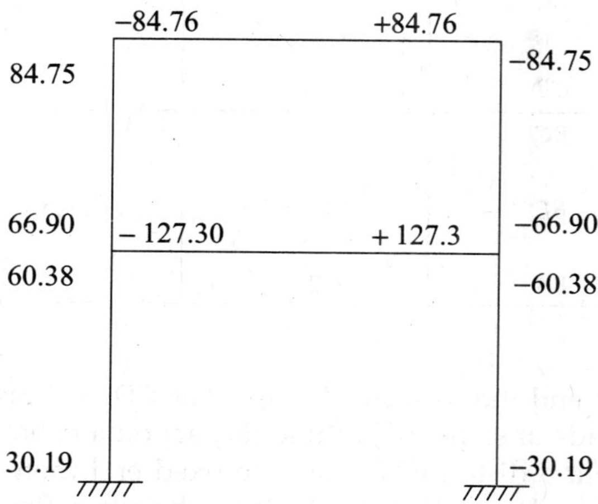

FIG. 3.78 End moments

Example 3.14 Analyse the single bay two-storey frame by Kani's method. The relative stiffness of the members are given in circles. Determine the end moments taking the advantage of frame and loading.

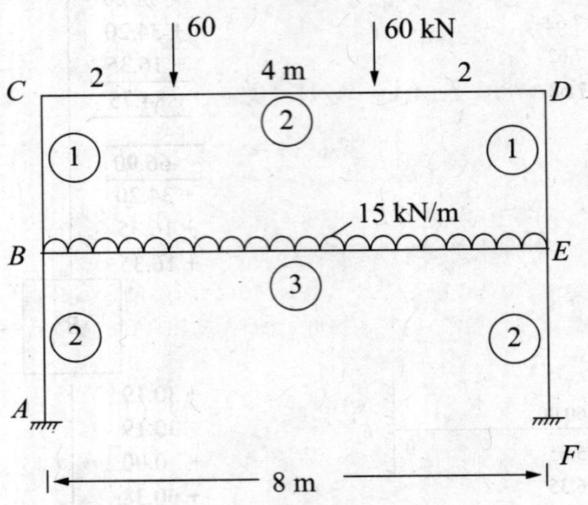

FIG. 3.79 Frame with loading

Solution

Rotation factors

Joint	*Members*	*Relative Stiffness*	Σk	$R = -\frac{1}{2}\left(\frac{k}{\Sigma k}\right)$
C	CD	$\left(\frac{1}{2}\right)2 = 1$	2	-0.25
	CB	1		-0.25
B	BC	1	4.5	-0.11
	BE	$\frac{1}{2}(3) = 1.5$		-0.17
	BA	2.0		-0.22

Fixed end moments

In calculating the fixed end moment for the member *CD*; a simplied formula has been used. That is, if two loads of same magnitude (*W*) act on a fixed beam, the distance for one symmetric load is at a distance '*a*' from the fixed end then the fixed end moment (due to two equal loads of *W* at a distance '*a*' from the either fixed end) is $M_F = -Wab/l$ where $b = (l - a)$. Therefore, for the beam element *CD*, $a = 2$ m and $b = (6 - 2)$ m.

$$M_{FCD} = -2 \times 60 \times \frac{6}{8} = -90.00 \text{ kNm}$$

$$M_{FBE} = -15 \times \frac{8^2}{12} = -80.00 \text{ kNm}$$

Method of Computations in Kani's Method

Iteration 1

The computed rotation factors at *B* and *C* are obtained from the table above. These values are entered in the appropriate annular boxes. For example, $R_{BC} = -0.11$, $R_{BE} = -0.17$ and $R_{BA} = -0.22$ are entered in the appropriate places in the annular boxes first.

Next, the restraint (algebraic sum of fixed end moments at that joint) moments to be balanced are entered in *B* and *C* as –80.00 and –90.00 respectively. Using the rotation factors and the moment to be distributed are obtained as

$$m_{ba} = R_{BA}(-80.00) = +17.60 \text{ kNm}$$

$$m_{bb'} = R_{BE}(-80.00) = +13.60 \text{ kNm}$$

$$m_{bc} = R_{BC}(-80.00) = +8.80 \text{ kNm}$$

They are written adjacent to the box at *B*.

Then

$$m_{cb} = -(0.25)\ (-90.00 + 8.80 + 0.00) = +20.30$$

$$m_{cc'} = -0.25\ (-90.00 + 8.80 + 0.00) = +20.30$$

This completes Iteration 1.

Iteration 2

Knowing the rotation contribution moment C, $m_{cb} = 20.30$ kNm,

$$m_{bc} = -0.11(-80.00 + 20.30 + 0.00) = +6.57 \text{ kNm}$$

$$m_{bb'} = -0.17(-80.00 + 20.30 + 0.00) = +10.15 \text{ kNm}$$

$$m_{ba} = -0.22(-80.00 + 20.30 + 0.00) = +13.13 \text{ kNm}$$

Hence,

$$m_{cb} = -0.25(-90.00 + 6.57 + 0.00) = +20.86 \text{ kNm}$$

$$m_{cc'} = -0.25(-90.00 + 6.57 + 0.00) = +20.86 \text{ kNm}$$

Iteration 3

$$m_{bc} = -0.11(-80.00 + 20.86 + 0.00) = +6.51 \text{ kNm}$$

$$m_{bb'} = -0.17(-80.00 + 20.86 + 0.00) = +10.05 \text{ kNm}$$

$$m_{ba} = -0.22(-80.00 + 20.86 + 0.00) = +13.01 \text{ kNm}$$

$$m_{cb} = -0.25(-90.00 + 6.51 + 0.00) = +20.87 \text{ kNm}$$

$$m_{cc'} = -0.25(-90.00 + 6.51 + 0.00) = +20.87 \text{ kNm}$$

The rotational moments of Iteration 2 and Iteration 3 show that they will converge in the next iterations. Hence, the fourth iteration is carried out.

Iteration 4

$$m_{bc} = -0.11(-80.00 + 20.87 + 0.00) = +6.50 \text{ kNm}$$

$$m_{bb'} = -0.17(-80.00 + 20.87 + 0.00) = +10.05 \text{ kNm}$$

$$m_{ba} = -0.22(-80.00 + 20.87 + 0.00) = +13.01 \text{ kNm}$$

$$m_{cb} = -0.25(-90.00 + 6.50 + 0.00) = +20.88 \text{ kNm}$$

$$m_{cc'} = -0.25(-90.00 + 6.50 + 0.00) = +20.88 \text{ kNm}$$

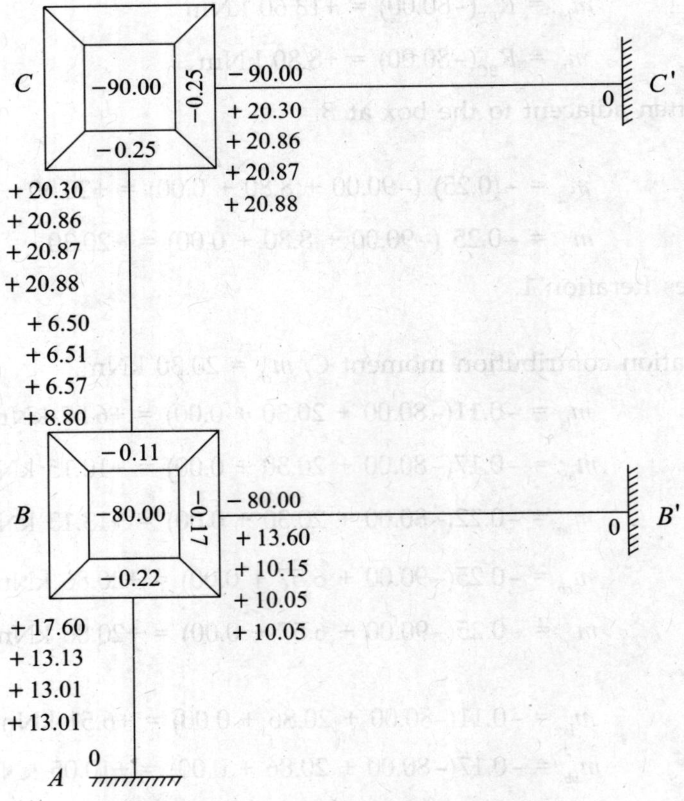

FIG. 3.80 Rotational moments (kNm)

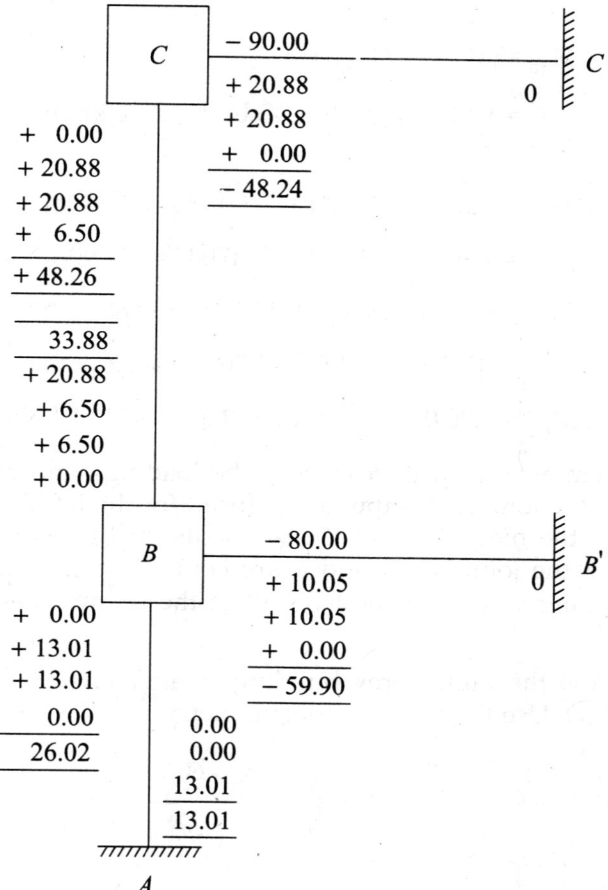

FIG. 3.81 End moments in half frame (kNm)

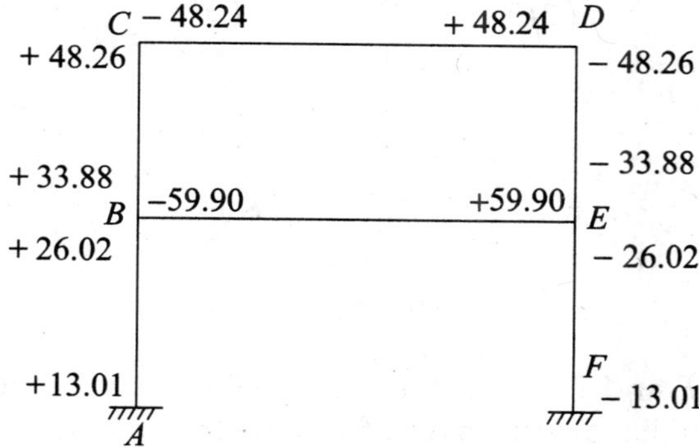

FIG. 3.82 Final end moments (kNm)

End moments

$$M_{AB} = M_{FAB} + 2M'_{AB} + M'_{BA}$$

$$= 0.00 + 2(0.00) + 13.01 = 13.01 \text{ kNm}$$

Similarly,

$$M_{BA} = 0.00 + 2(13.01) + 0.00 = 26.02 \text{ kNm}$$

$$M_{BE} = -80.00 + 2(10.05) + 0.00 = -59.90 \text{ kNm}$$

$$M_{BC} = 0.00 + 2(6.50) + 20.88 = +33.88 \text{ kNm}$$

$$M_{CB} = 0.00 + 2(20.88) + 6.50 = +48.26 \text{ kNm}$$

$$M_{CD} = -90.00 + 2(20.88) + 0.00 = -48.24 \text{ kNm}$$

As the frame is symmetrical with respect to the loading, boundary conditions and geometry of the cross section, the computations (used for the left half of the frame) can be used to determine the magnitude of the moments on the right half of the frame. Thus, the moments at the joints *D*, *E* and *F* are equal in magnitude but opposite in sign to the corresponding values on the left half of the frame. These values are given in Fig. 3.82.

Example 3.15 Analyse the multistorey building frame shown in Fig. 3.83 by Kani's method and draw BMD. Use the principle of symmetry. (Dec. 2010, VTU)

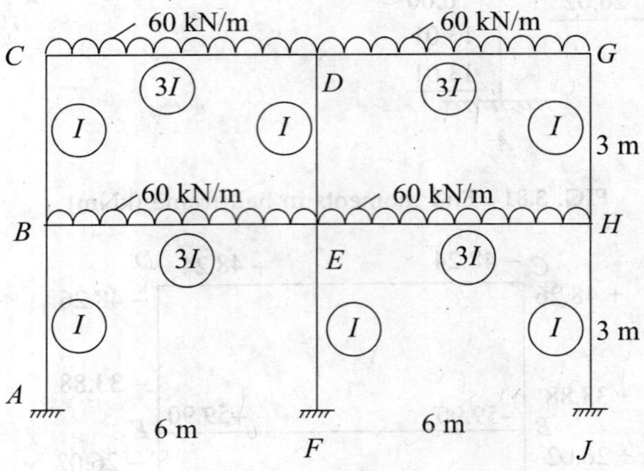

FIG. 3.83

Solution: The joints *D*,*E* and *F* are on the axis of symmetry of the above symmetrical frame. The joints which lie on the axis of symmetry do not rotate and hence are considered fixed. Hence, only half frame is considered. This half frame with joints *D*,*E* and F are considered fixed and such frame is called substitute frame. The reader should not confuse this with substitute frame technique which is being used by designers for the analysis of multistorey frame. The substitute frame in the present chapter is only a half frame.

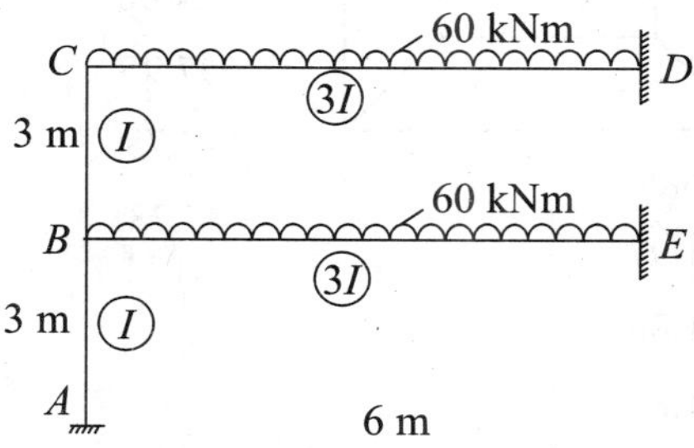

FIG. 3.84 Substitute frame

Fixed end moments

$$M_{FCD} = -60 \times \frac{6^2}{12} = -180 \text{ kNm}$$

$$M_{FDC} = +180 \text{ kNm}$$

$$M_{FBE} = -60 \times \frac{6^2}{12} = -180 \text{ kNm}$$

$$M_{FEB} = +180 \text{ kNm}$$

Rotation factors

Joints	*Members*	$k = I/l$	Σk	$R = -\frac{1}{2}\left(\frac{k}{\Sigma k}\right)$
C	CD	$3I/6 = I/2$	$0.83I$	−0.30
	CB	$I/3 = 0.33I$		−0.20
B	BC	$I/3 = 0.33I$	$1.16I$	−0.14
	BE	$3I/6 = 0.50I$		−0.22
	BA	$I/3 = 0.33I$		−0.14

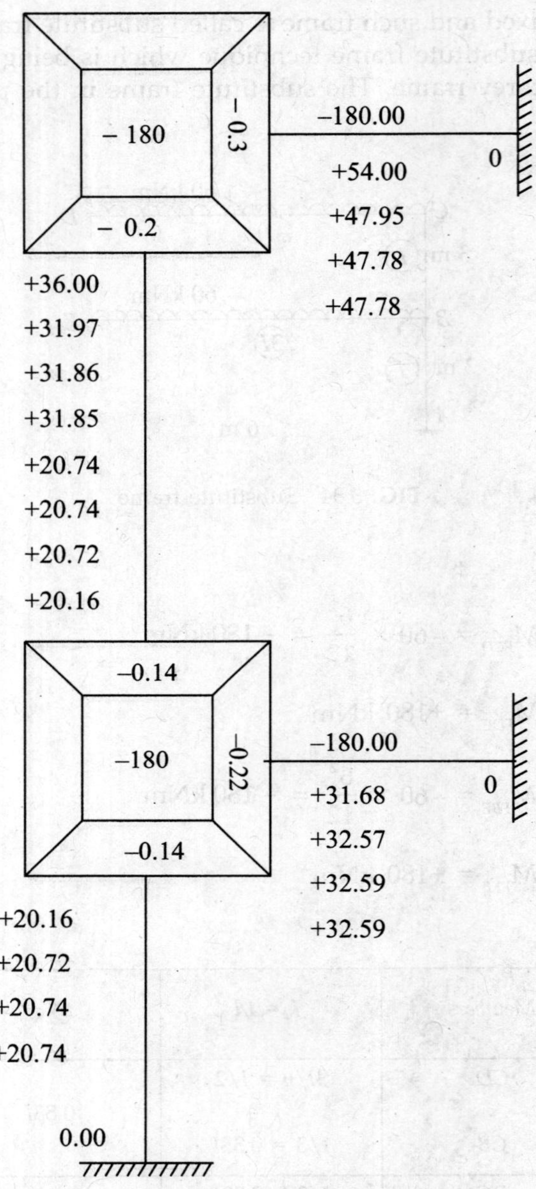

FIG. 3.85 Rotational moments (kNm)

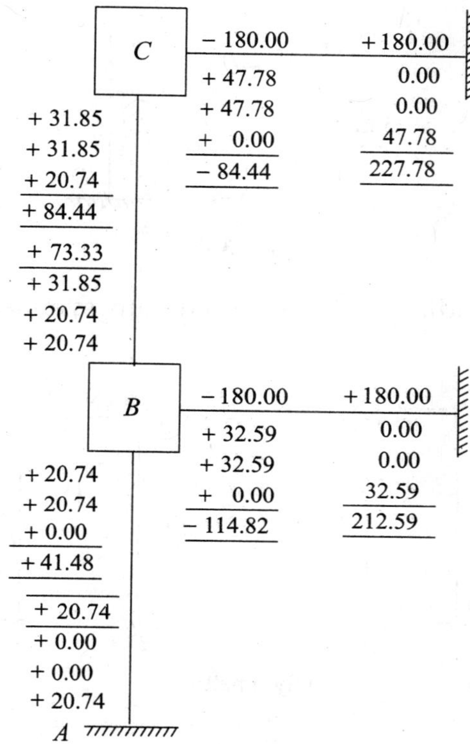

FIG. 3.86 Calculation end moments (kNm)

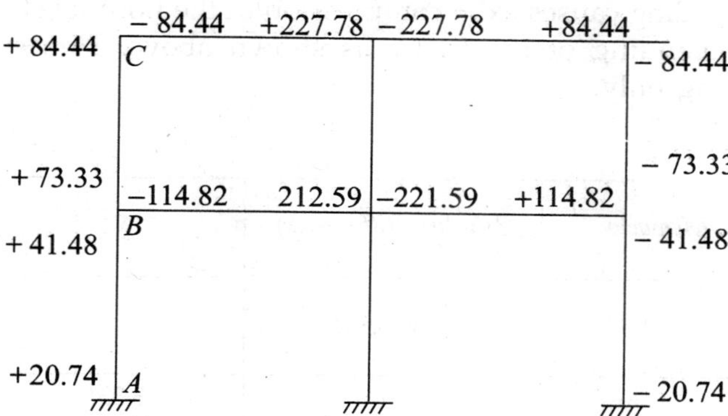

FIG. 3.87 Final end moments (kNm)

3.2.1.5 *Analysis of Frames with Lateral Loads*

Example 3.16 Analyse the portal frame by Kani's method. Compute the lateral drift at the floor level. Take $EI = 1.5(10)^4$ kNm^2.

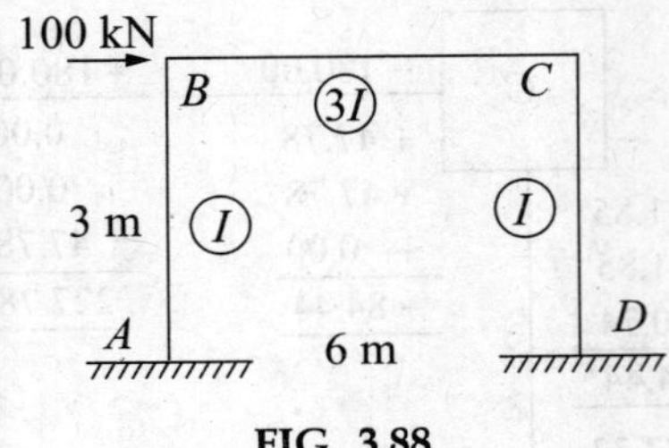

FIG. 3.88

Solution: The given loading is decomposed into symmetric and antisymmetric loading as

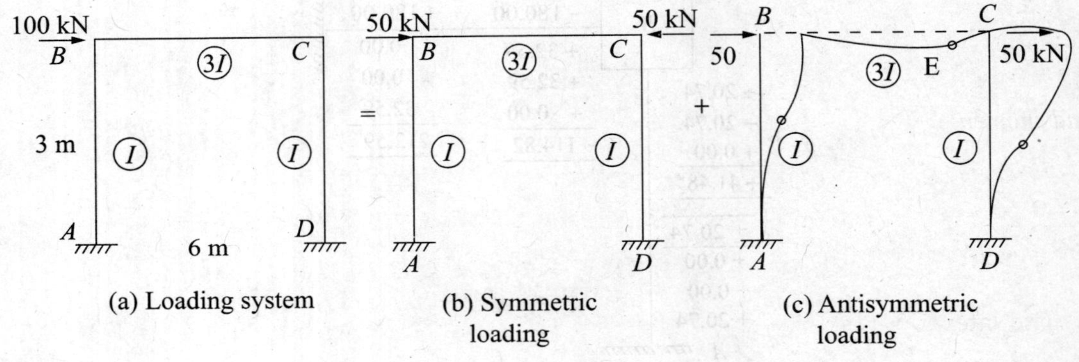

FIG. 3.89

The symmetric loading causes axial compression in the floor level. The antisymmetric loading causes the bending of the frame as shown above. Hence, analyse only the antisymmetric loading only.

Rotation factor

Joint	*Members*	*Relative Stiffness* (*k*)	Σk	$R = -\frac{1}{2}\frac{k}{\Sigma k}$
	BA	$\frac{I}{3} = 0.33I$		−0.05
B			$3.33I$	
	BC	$\frac{6(3I)}{6} = 3I$		−0.45

Fixed end moments

$$M_{FB} = -\frac{Wh}{2} = -\frac{50 \times 3}{2} = -75 \text{ kNm}$$

Rotational moments

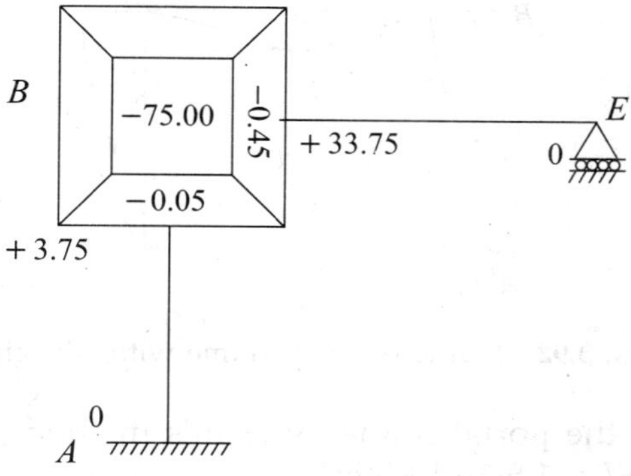

FIG. 3.90 Rotational moments (kNm)

End moments

$$M_{AB} = M_{FAB} - 2M'_{AB} = -75.00 - 2(3.75) = -82.5 \text{ kNm}$$

$$M_{BA} = M_{FBA} + 2M'_{BA} = -75.00 + 2(3.75) = -67.5 \text{ kNm}$$

$$M_{BC} = M_{FBC} + 2M'_{BC} = 0.00 + 2(33.75) = +67.5 \text{ kNm}$$

The lateral drift in the floor level

$$\Delta = \frac{Wh^3}{12EI} + \frac{Ml}{3(3EI)}\left(\frac{h}{2}\right)$$

$$\Delta = \frac{50(3)^3}{12EI} + \frac{67.5 \times 3}{3 \times 3EI \times 2} \times 3$$

$$\Delta = \frac{146.25}{EI}$$

$$\Delta = \frac{146.25}{1.5(10)^4} \times 1000 = 9.75 \text{ mm}$$

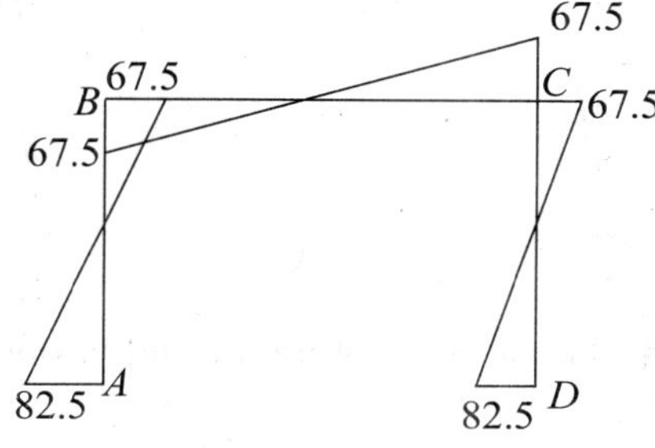

FIG. 3.91 Bending moment diagram (kNm)

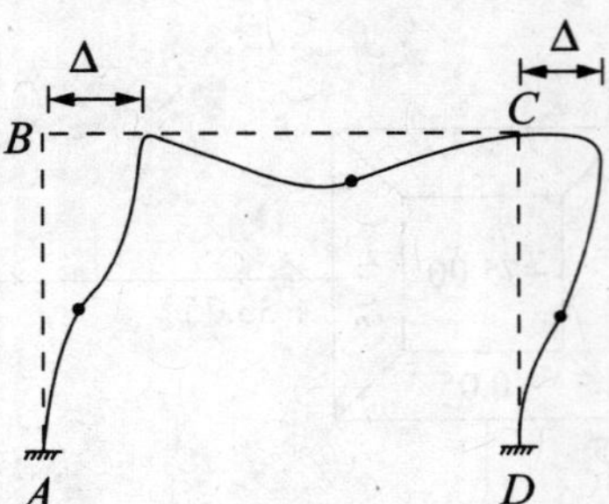

FIG. 3.92 Elastic curve of frame with stiff girder

Example 3.17 Analyse the portal frame by Kani's method. Compute the lateral drift at the floor level. Take $EI = 1.5(10)^4$ kNm².

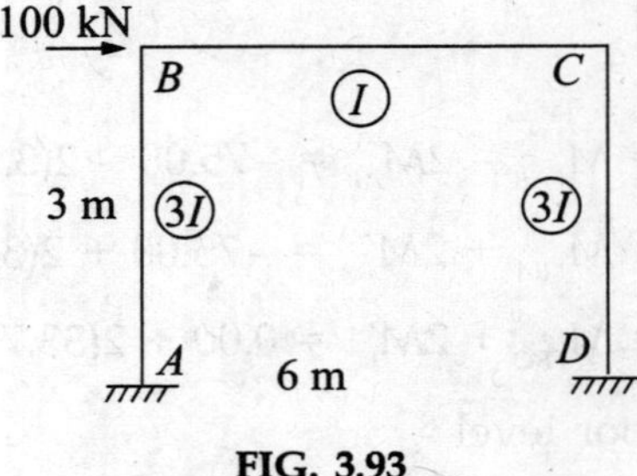

FIG. 3.93

Solution: In the above frame, the column is stiff (3*I*) and the girder is flexible (*I*). This kind of frame is preferable in the areas prone to earthquakes. This frame behaviour is different when compared with the previous problem, wherein the girder is stiff (3*I*) and the column is flexible (*I*). The analysis is done on similar lines with the earlier problem.

Considering the antisymmetric loading only which causes the bending of frame.

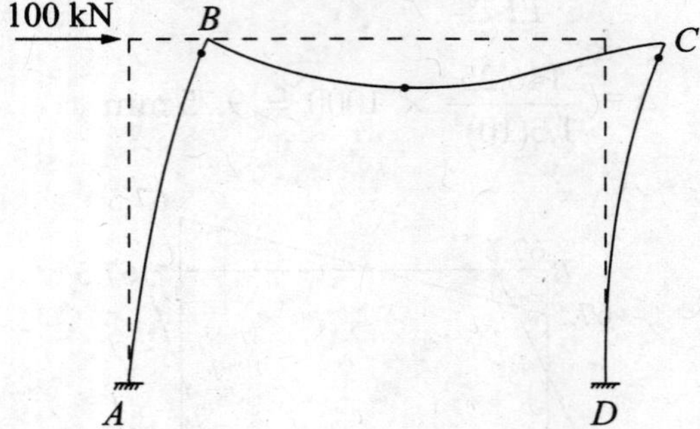

FIG. 3.94 Elastic curve of frame with flexible girder

Rotation factors

Joint	*Members*	*Relative Stiffness (k)*	Σk	$R = -\frac{1}{2}\frac{k}{\Sigma k}$
B	*BA*	$\frac{3I}{3}$	$2I$	−0.25
	BC	$6I/6$		−0.25

Fixed end moments

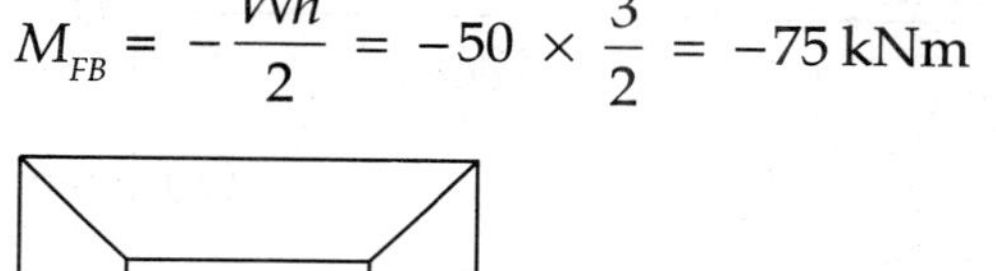

$$M_{FB} = -\frac{Wh}{2} = -50 \times \frac{3}{2} = -75 \text{ kNm}$$

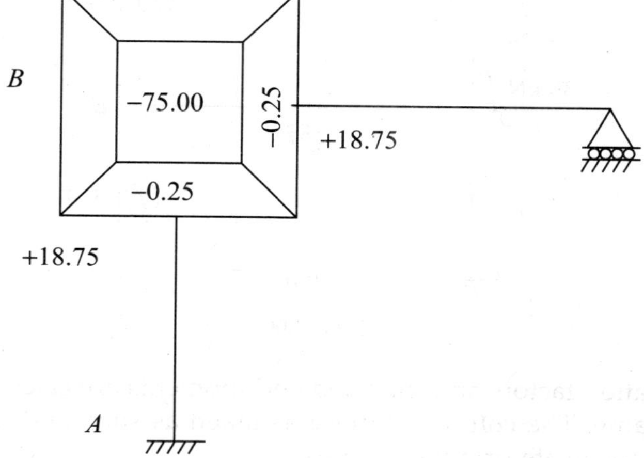

FIG. 3.95

End moments

$$M_{AB} = M_{FAB} - 2M'_{AB} = -75.00 - 2(18.75) = -112.50$$

$$M_{BA} = M_{FBA} + 2M'_{BA} = -75.00 + 2(18.75) = -37.50$$

$$M_{BC} = M_{FBC} + 2M'_{BC} = 0.00 + 2(18.75) = +37.50$$

The lateral drift in the floor level is

$$\Delta = \frac{Wh^3}{12E(3I)} + \frac{Ml}{3(EI)} \times \frac{h}{2}$$

$$\Delta = \frac{50 \times 3^3}{36EI} + \frac{37.5 \times 3 \times 3}{3EI \times 2} = \frac{93.75}{EI}$$

$$\Delta = 6.25 \text{ mm}$$

The lateral drift is less when compared to a companion frame where the girder is stiff and the column is flexible. Also the point of contraflexure in columns is not at the mid height (which usually occurs nearly in frames where the column is flexible and the girder is stiff); but it is closer to the girder-column junction. This is advantageous when structure fails at ultimate load by the formation of plastic hinges nearer to the column girder junction rather than at mid height of the column. In other words, the frame with stiff column and flexible girder under lateral loads leads to the initiation of failure of girders rather than column failure immediately.

Example 3.18 Analyse the given frame subjected to lateral loads by Kani's method. Draw the bending moment diagram.

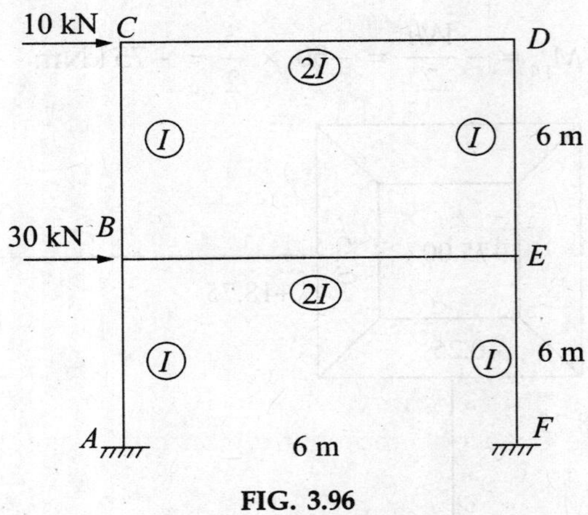

FIG. 3.96

Solution: The rotation factors and the fixed end moments are calculated by considering only half of the frame. The column stiffness is taken as such and the beam stiffness is modified as six times of its original stiffness.

Rotation factors

Joint	*Members*	k	Σk	$R = -\frac{1}{2}\frac{k}{\Sigma k}$
C	CD	$\frac{6(2I)}{6} = 2I$	2.167I	−0.46
	CB	$I/6 = 0.167I$		−0.04
B	BC	$\frac{I}{6} = 0.167I$		−0.04
	BE	$\frac{6(2I)}{6} = 2I$	2.334I	−0.43
	BA	$I/6 = 0.167I$		−0.03

Fixed end moments

Half frame technique is employed taking into antisymmetry account:

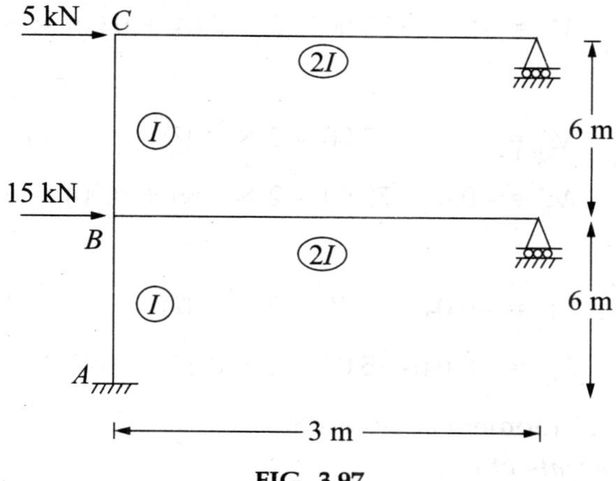

FIG. 3.97

The horizontal shear in the top storey is 5 kN. The height being 6 m,

$$M_{FCB} = M_{FBC} = -\frac{5 \times 6}{2} = -15.00 \text{ kNm}$$

The horizontal shear in the bottom storey is the sum of the lateral loads acting above and therefore the fixed end moment is computed as

$$M_{FBA} = M_{FAB} = -(5 + 15)\ \frac{6}{2} = -60.00 \text{ kNm}$$

Rotational Moments

The restraint moment at C is M_{FCB} and it is written in the box, while the restraining moment at B is the sum of $(M_{FBC} + M_{FBA})$, i.e. -75.00 kNm. The rotational moments ($M'_{AB} = M'_{HB} = M'_{GC} = 0.$) at A, H and G are taken as zeros. Taking cantilever action into account, the modified Kani's equation for a beam element AB can be written as

$$M_{AB} = M_{FAB} + 2M'_{AB} - 2M'_{BA}$$

The iterations are carried out using the above equation. The computation of rotational moments is done for columns only:

Iteration 1

Joint C to start with;

$$M'_{cb} = -15(-0.04) = 0.60 \text{ kNm}$$

$$M'_{bc} = -0.04(-75.00 - 2 \times 0.60 + 0.00) = 3.05 \text{ kNm}$$

Iteration 2

$$M'_{cb} = -0.04(-15.00 - 2 \times 3.05) = +0.84 \text{ kNm}$$

$$M'_{bc} = -0.04(-75.00 - 2 \times 0.84 + 0.00) = +3.07 \text{ kNm}$$

Iteration 3

$$M'_{cb} = -0.04(-15.00 - 2 \times 3.07) = +0.85 \text{ kNm}$$

$$M'_{bc} = -0.04(-75.00 - 2 \times 0.85 + 0.00) = +3.07 \text{ kNm}$$

Iteration 4

$$M'_{cb} = -0.04(-15.00 - 2 \times 3.07) = +0.85 \text{ kNm}$$

$$M'_{bc} = -0.04(-75.00 - 2 \times 0.85 + 0.00) = +3.07 \text{ kNm}$$

The values of rotational moments converge.

Computation of end moments of columns

$$M_{CB} = -15.00 + 2(0.85) - 2(3.07) = -19.44 \text{ kNm}$$
$$M_{BC} = -15.00 + 2(3.07) - 2(0.85) = -10.56 \text{ kNm}$$

Taking m'_{ba} as 0.75 times m'_{bc},

$$m'_{ba} = 0.75 \times 3.07 = 2.30 \text{ kNm}$$

$$m'_{ab} = 0$$

Thus, $$M_{BA} = -60.00 + 2(230) = -55.4 \text{ kNm}$$

$$M_{AB} = -60.00 - 2(2.30) = -64.6 \text{ kNm}$$

Computation of end moments in beams

The beam moments are obtained by summing up the moments at the joints and equate them to zero.

Joint $C \Rightarrow$ $$M_{CB} + M_{CD} = 0$$

$\therefore$ $$M_{CD} = -M_{CB}$$
$$M_{CD} = +19.44 \text{ kNm}$$

Joint $B \Rightarrow$ $$M_{BC} + M_{BE} + M_{BA} = 0$$

$$-10.56 + MBE - 55.4 = 0$$

$$M_{BE} = 65.96 \text{ kNm}$$

The end moments on the right half can be written with same magnitude and same nature as shown below.

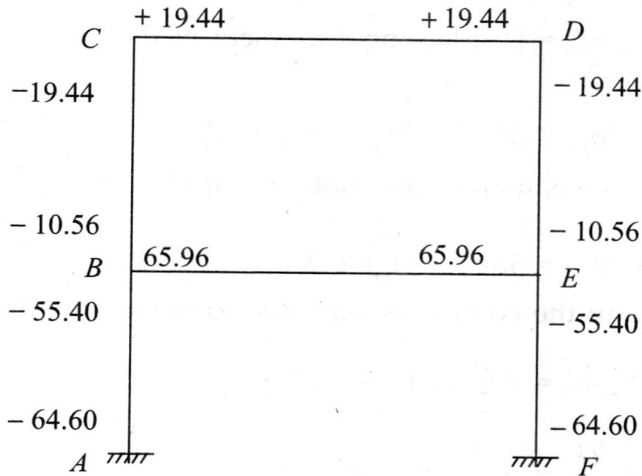

FIG. 3.98 End moments

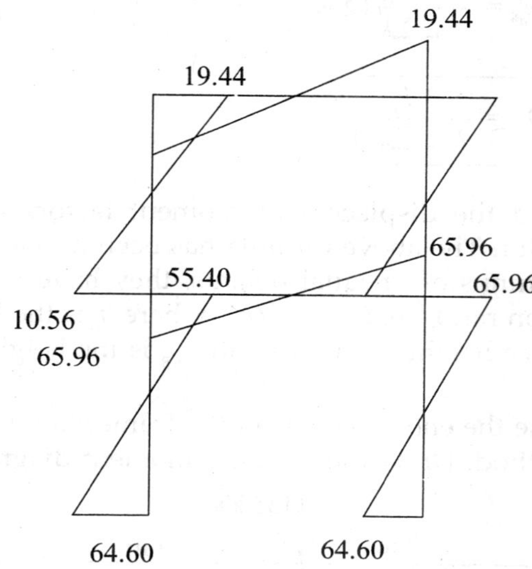

FIG. 3.99 BMD (kNm)

3.2.2 Frames with Sway

3.2.2.1 Columns of Same Height and Subjected to Vertical Loads

The slope deflection equation is written in another form as

$$M_{AB} = M_{FAB} + 2m_{ab} + m_{ba} + m'_{ab} \tag{1}$$

where M_{AB} = Total moment at A of column AB

M_{FAB} = Fixed moment in column due to applied loads

m_{ab}, m_{ba} = Rotational moments

m'_{ab} = Displacement moment at the joint A

Similarly,

$$M_{BA} = M_{FBA} + 2m_{ba} + m_{ab} + m'_{ba} \quad (2)$$

Adding the above two equations and noting that $M_{FAB} = M_{FBA} = 0$ for vertical loads in columns;

$$M_{AB} + M_{BA} = 3(m_{ab} + m_{ba}) + 2m'_{ab} \quad (3)$$

The total moments in the column is zero due to horizontal shear condition,

$$\Sigma m'_{ab} = -\frac{3}{2}(\Sigma m_{ab} + m_{ba}) \quad (4)$$

i.e.

$$m'_{ab}\frac{\Sigma k}{k_{ab}} = -\frac{3}{2}(\Sigma m_{ab} + m_{ba}) \quad (5)$$

$$m'_{ab} = -\frac{3}{2}\frac{k_{ab}}{\Sigma k}(\Sigma m_{ab} + m_{ba}) \quad (6)$$

and

$$\boxed{D_{ab} = -\frac{3}{2}\frac{k_{ab}}{\Sigma k}} \quad (7)$$

The term D_{ab} is called the displacement moment factor. This is also called the translational moment factor. The above formula has been derived for columns of equal height. However, for columns of unequal heights, they have to be modified with the reduction factor/correction ratio, i.e. $C_{AB} = h'_r/h_{AB}$ where h'_r is the height of the rth storey which is taken as reference height in metres and h_{AB} is the height of the column AB in metres.

Example 3.19 Determine the end moments of the frame shown in Fig. 3.100 by Kani's rotation contribution method. Draw the bending moment diagram.

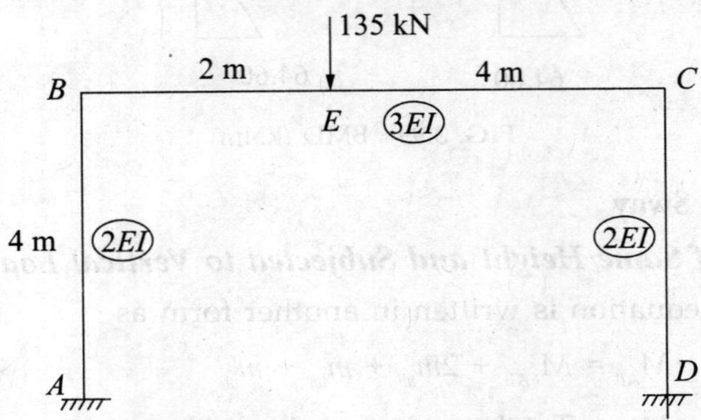

FIG. 3.100

Solution

Rotation factor

Joints	*Members*	*k*	*Σk*	$R = -\frac{1}{2}\left(\frac{k}{\Sigma k}\right)$
B	BA	$2I/4 = 0.5I$	I	−0.25
	BC	$3I/6 = 0.5I$		−0.25
C	CB	$3I/6 = 0.5I$	I	−0.25
	CD	$2I/4 = 0.5I$		−0.25

Fixed end moments

$$M_{FAB} = M_{FBC} = 0$$

$$M_{FBC} = -\frac{2(135)4^2}{6^2} = -120\text{ kNm}$$

$$M_{FCB} = +\frac{4(135)2^2}{6^2} = +60\text{ kNm}$$

Displacement factors

$$\left(D = -\frac{3}{2}\frac{k_c}{\Sigma k_c}\right)$$

$$D_{AB} = -\frac{3}{2}\left(\frac{0.5I}{0.5I + 0.5I}\right) = -\frac{3}{4}$$

$$D_{CD} = -\frac{3}{2}\left(\frac{0.5I}{0.5I + 0.5I}\right) = -\frac{3}{4}$$

Rotation moments

$$M_{AB} = R_{AB}\,[\Sigma M'_{FAB} + \Sigma M'_{BA} + \Sigma M''_{BA(c)}] \tag{1}$$

where

$$M''_{BA(c)} = D_{AB(c)}\,[\Sigma M'_{AB(c)} + \Sigma M'_{BA(c)}] \tag{2}$$

End moment

$$M_{AB} = M_{FAB} + 2M'_{AB} + M'_{BA} + M''_{BA(c)} \tag{3}$$

The rotation moments are computed using the above equations in a sequential manner. The following is the sequence, joint B and joint C followed by column AB and column CD.

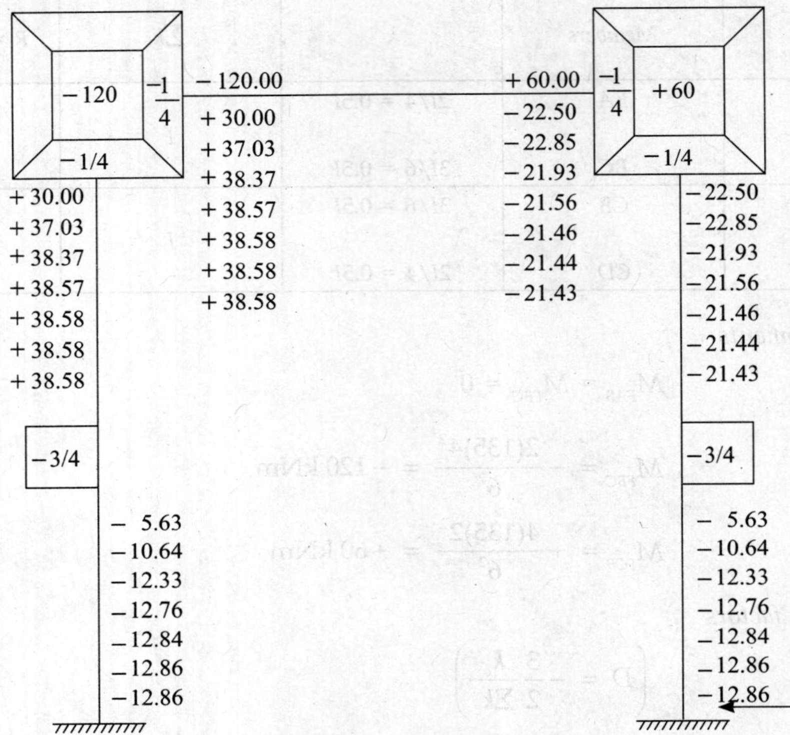

FIG. 3.101 Computation of rotation moments (kNm)

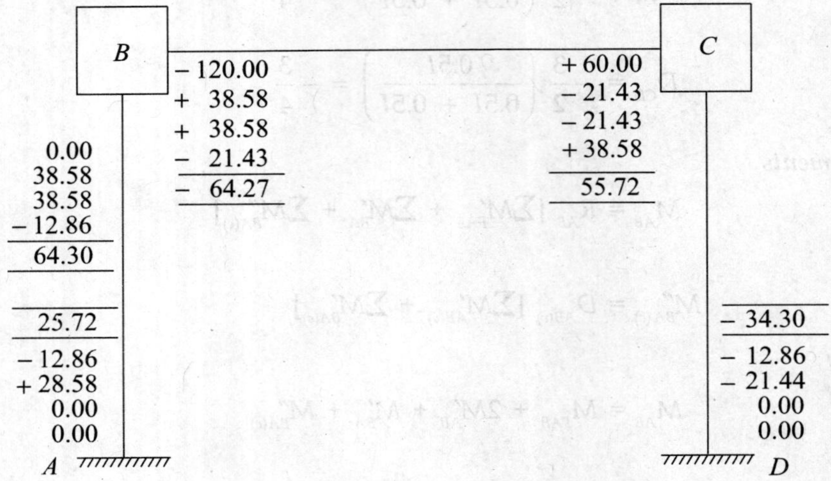

FIG. 3.102 End moments (kNm)

The above pictorial representation of computation of end moments is explained in the following section for more clarity.

End moment calculation of the above table

$$M_{AB} = M_{FAB} + 2M'_{AB} + M'_{BA} + M''_{AB(c)}$$

$$M_{AB} = 0 + 2(0) + 38.58 - 12.86 = 25.72 \text{ kNm}$$

$$M_{BA} = 0 + 2(38.58) + 0 - 12.86 = 64.30 \text{ kNm}$$

$$M_{BC} = -120.00 + 2(38.58) - 21.43 = -64.30 \text{ kNm}$$

$$M_{CB} = +60.00 + 2(-21.43) + 38.58 = +55.72 \text{ kNm}$$

$$M_{CD} = 0.00 + 2(-21.44) + 0 - 12.86 = -55.74 \text{ kNm}$$

$$M_{DC} = 0.00 + 2(0) - 21.44 - 12.86 = -34.30 \text{ kNm}$$

Bending moment at E

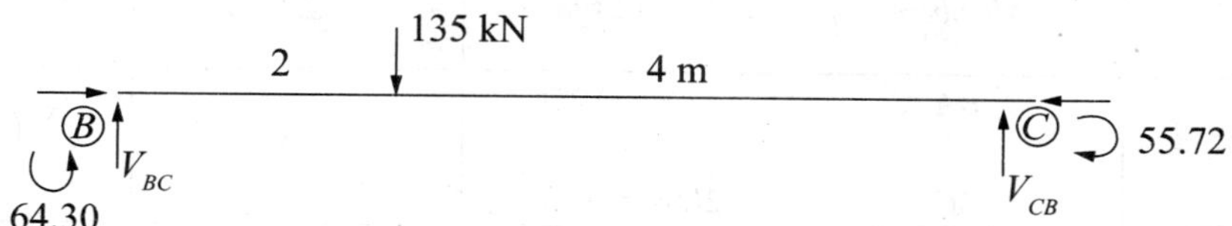

FIG. 3.103

$$\Sigma M_C = 0; \quad 55.72 - 64.3 + 6V_{BC} - 135(4) = 0$$

$$V_{BC} = 91.43 \text{ kN}$$

$$M_E = -64.30 + 91.43(2) = 118.56 \text{ kNm}$$

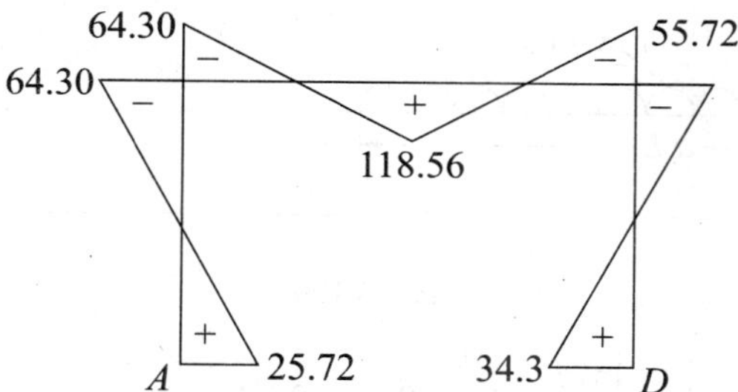

FIG. 3.104 Bending moment diagram (kNm)

Example 3.20 Using Kani's rotation contribution method, analyse the rigid frame and draw the bending moment diagram. Determine the maximum positive bending moment in the beam.

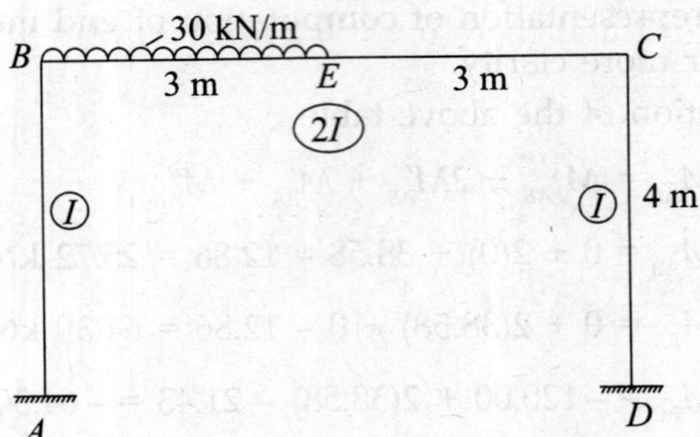

FIG. 3.105

Solution

Rotation factors

Joint	*Members*	*Relative Stiffness*	Σk	$R = -\frac{1}{2}\frac{k}{\Sigma k}$
B	BA	$I/4 = 0.25I$	0.58I	−0.22
	BC	$2I/6 = 0.33I$		−0.28
C	CB	$2I/6 = 0.33I$	0.58I	−0.28
	CD	$I/4 = 0.25I$		−0.22

Fixed end moments

Figure 3.106 shows a fixed beam subjected to partial uniformly distributed load. The fixed end moment is given by a formula.

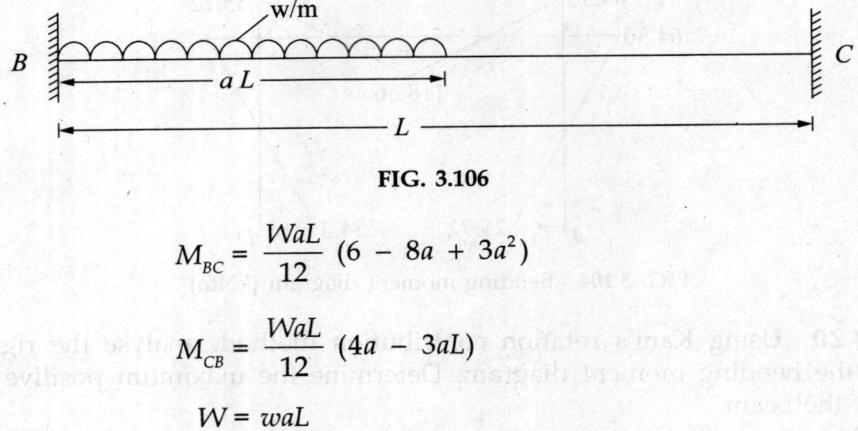

FIG. 3.106

$$M_{BC} = \frac{WaL}{12}(6 - 8a + 3a^2)$$

$$M_{CB} = \frac{WaL}{12}(4a - 3aL)$$

where $W = waL$

In this problem, $a = L/2$ and therefore the above expression reduces to

$$M_{BC} = -\frac{11}{192}wl^2 \qquad M_{FBC} = -\frac{11}{192}(30)6^2 = -61.88\text{ kNm}$$

$$M_{CB} = +\frac{5}{192}wl^2 \qquad M_{FCB} = +\frac{5}{192}(30)6^2 = +28.13\text{ kNm}$$

Displacement factor

The columns have the same relative stiffness.

i.e. $k_{AB} = k_{DC}$ and hence

$$D_{AB} = -\frac{3}{2}\frac{k_{AB}}{k_{AB} + k_{DC}} = -\frac{3}{2}\left(\frac{k_{AB}}{2k_{AB}}\right) = -0.75$$

Hence, $D_{CD} = -0.75$

In simple words for symmetrical frames having the same column stiffness with fixed support, the displacement factor is −0.75.

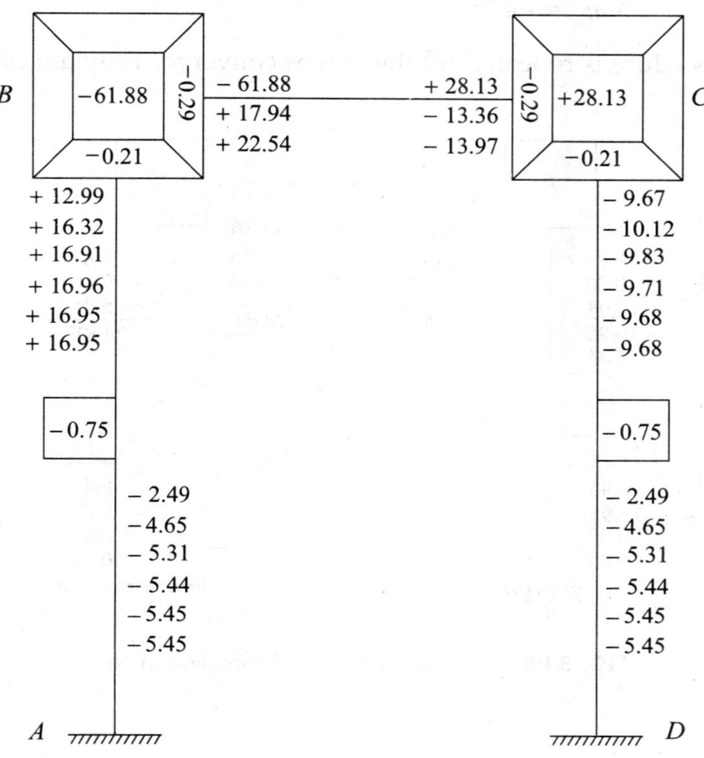

FIG. 3.107 Rotational moments (kNm)

Iteration 1

$$m_{ba} = -61.88(-0.21) = +12.99$$

$$m_{bc} = -61.88(-0.29) = +17.94$$

$$m_{cb} = (17.94 + (28.13)) - 0.29 + 0 = -13.36$$

$$m_{cd} = (17.94 + 28.13)(-0.21) = -9.67$$

$$m_{ab} = -0.75(12.99 - 9.67) = -2.49$$

$$m_{dc} = -2.49$$

Iteration 2

$$m_{ba} = -0.21(-2.49 - 61.88 - 13.36) = +16.32$$

$$m_{bc} = -0.29(-2.49 - 61.88 - 13.36) = +22.54$$

$$m_{cb} = -0.29(22.54 + 28.13 - 2.49) = -13.97$$

$$m_{cd} = -0.21(22.54 - 2.49 + 28.13) = -10.12$$

$$m_{ab} = -0.75(16.32 - 10.12) = -4.65$$

$$m_{dc} = -4.65$$

The same procedure is repeated till the values converge. They are pictorially shown below.

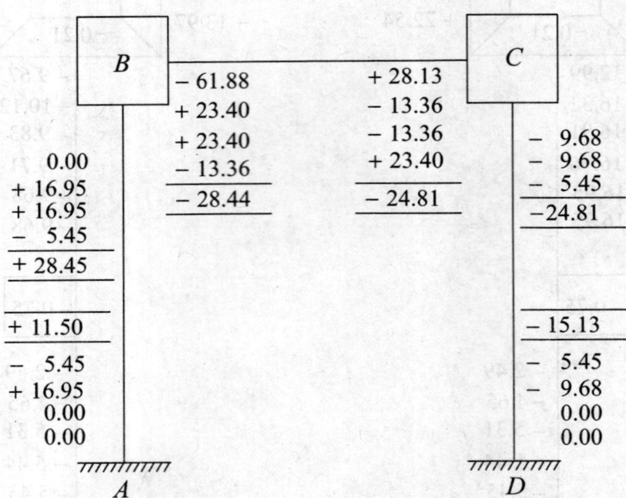

FIG. 3.108 Computation of end moments (kNm)

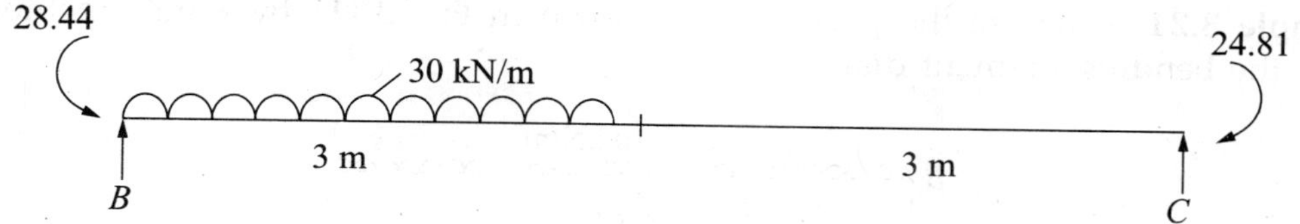

FIG. 3.109

$$\Sigma M_B = 0$$

$$-28.44 + 24.81 - 6V_{CB} + 30 \times \frac{3^2}{2} = 0$$

$$V_{CB} = 21.9 \text{ kN}$$

$$\Sigma V = 0;$$

$$V_{BC} + V_{CB} = 30(3) = 90 \text{ kN}$$

i.e. $$V_{BC} = 68.1 \text{ kN}$$

Bending moment at a distance x from B;

$$68.1x - 28.44 - 30 \times \frac{x^2}{2} = 0$$

$$-15x^2 + 68.1x - 28.44 = 0$$

$$x = 0.47 \text{ m}$$

$\therefore$ Maximum positive bending moment $= -28.44 + 68.1(0.47) - 30 \times \dfrac{0.47^2}{2}$

$= 0.25$ kNm

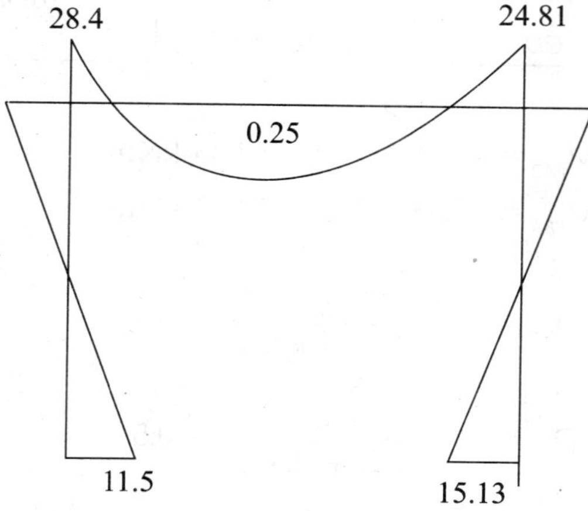

FIG. 3.110 Bending moment diagram (kNm)

Example 3.21 Analyse the portal frame shown in Fig. 3.111 by Kani's method and draw the bending moment diagram.

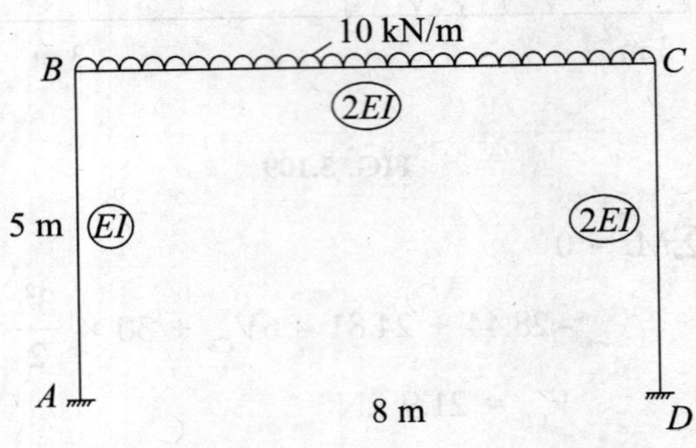

FIG. 3.111

Solution: The frame is symmetrical with respect to loading and support conditions. But the column *AB* and column *CD* have different flexural rigidity, viz. *EI* and 2*EI*. Therefore, the frame is asymmetrical with respect to the geometry of the cross section of columns. Hence, the frame is prone to sway due to geometric asymmetry.

Rotation factors

Joint	*Members*	*Relative Stiffness*	Σk	$R = -\frac{1}{2}\frac{k}{\Sigma k}$
B	BA	$I/5 = 0.2I$	$0.45I$	−0.22
	BC	$2I/8 = 0.25I$		−0.28
C	CB	$2I/8 = 0.25I$	$0.65I$	−0.19
	CD	$2I/5 = 0.40I$		−0.31

Fixed end moments

$$M_{FAB} = -10(8)^2/12 = -53.33 \text{ kNm}$$

$$M_{FBA} = +10(8)^2/12 = +53.33 \text{ kNm}$$

Displacement factors

$$D_{AB} = -\frac{3}{2}\left(\frac{k_{AB}}{k_{AB} + k_{CD}}\right)$$

$$D_{AB} = -\frac{3}{2}\left(\frac{0.2}{0.2 + 0.4}\right) = -0.5$$

Similarly,

$$D_{CD} = -\frac{3}{2}\left(\frac{0.4}{0.2 + 0.4}\right) = -1.0$$

It is to be noted that the fixed end columns *AB* and *CD* have different flexural stiffnesses. Hence, the displacement factors are not same.

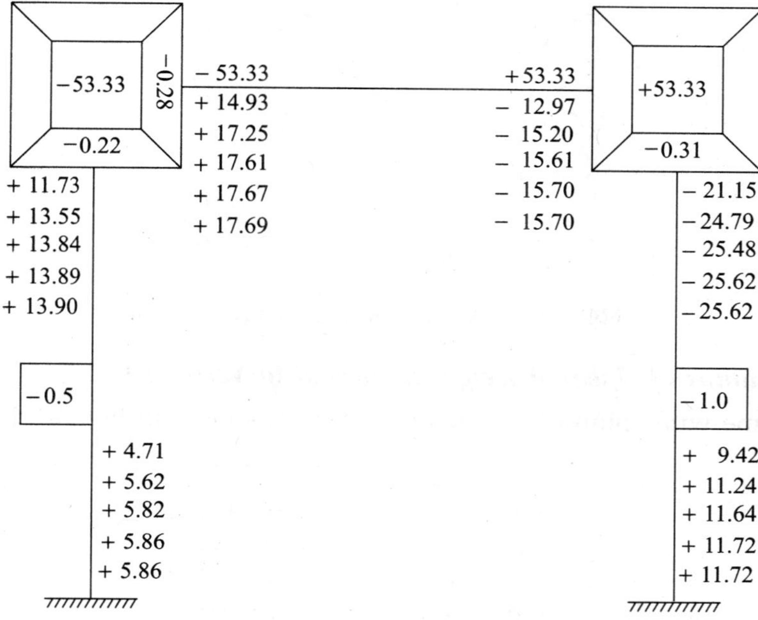

FIG. 3.112 Rotation moments (kNm)

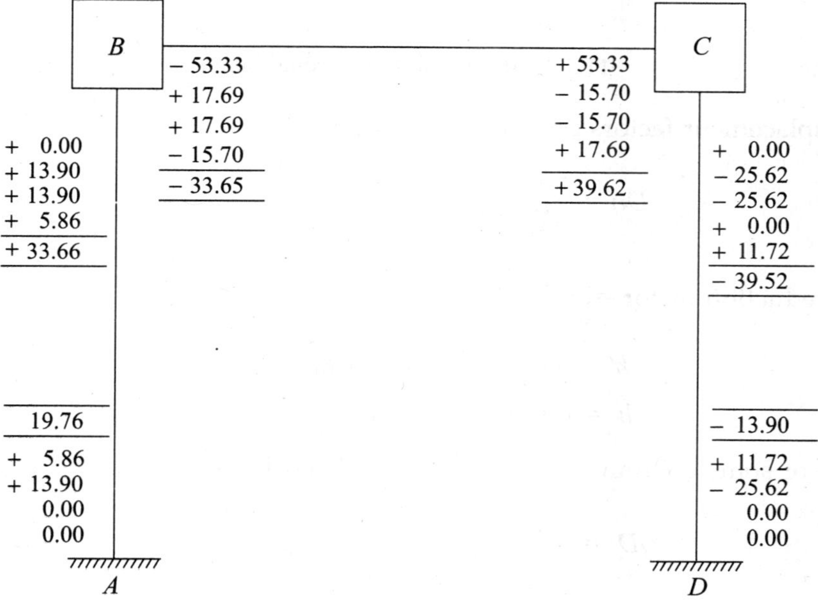

FIG. 3.113 Computation of end moments (kNm)

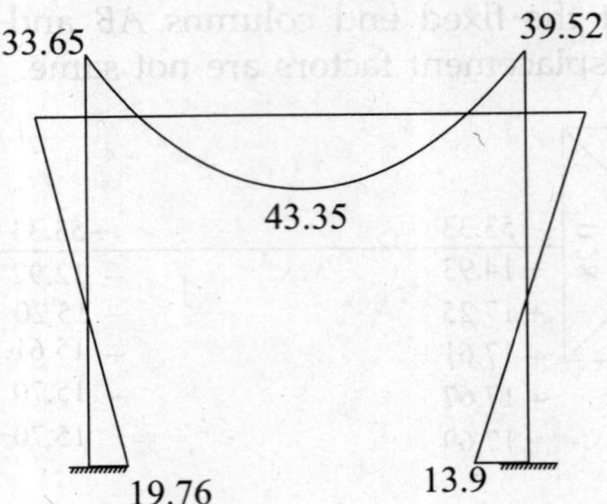

FIG. 3.114 Bending moment diagram (kNm)

3.2.2.2 *Columns of Unequal Height Subjected to Vertical Loads*

A typical frame with columns of unequal height is shown in Fig. 3.115.

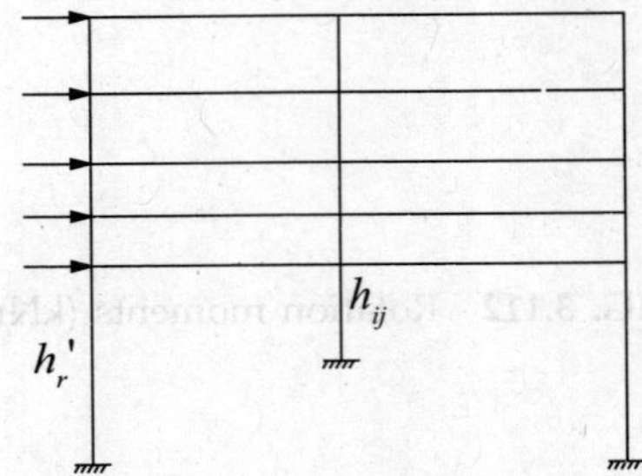

FIG. 3.115 Frames of unequal height

The displacement factors are computed from

$$Dij = \frac{-1.5\,C_{ij}\,K_{ij}}{\Sigma\, C_{ij}^2\,K_{ij}}$$

and C_{ij} = reduction factor = $\dfrac{h_r'}{h_{ij}}$

where h_r' = height of 'r'th storey

h_{ij} = reference height

In case of some columns hinged and some fixed it can be shown that

$$D_{ij} = \frac{-1.5\,C_{ij}\,K_{ij}}{\Sigma m\, C_{ij}^2\,K_{ij}}$$

where $m = 3/4$ for hinged ends; $m = 1$ for columns with fixed ends. If all the columns are hinged then $m = 3/4$ for all columns. Hence,

$$D_{ij} = \frac{-2\,C_{ij}\,K_{ij}}{\Sigma C_{ij}^2\,K_{ij}}$$

Example 3.22 Analyse the rigid frame shown in Fig. 3.116 by Kani's rotation contribution method.

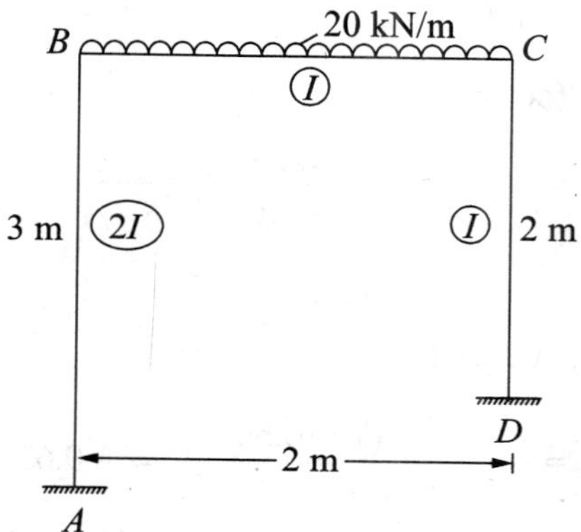

FIG. 3.116

Rotation factors

Joint	*Members*	*Stiffness k*	Σk	$R = -\frac{1}{2}\frac{k}{\Sigma k}$
	BA	2I/3		−0.29
B			1.167I	
	BC	I/2		−0.21
	CB	I/2		−0.25
C			I	
	CD	I/2		−0.25

Displacement factors

$$D_{ij} = \frac{-1.5\,C_{ij}\,K_{ij}}{\Sigma C_{ij}^2\,K_{ij}}$$

The column reduction factor (correction ratio) is computed from the relation

$$C_{ij} = \frac{h_r'}{h_{ij}}$$

and taking BA as the reference height hr

$$C_{BA} = \frac{h_r'}{h_{BA}} = \frac{3}{3} = 1.00$$

$$C_{CD} = \frac{h_r'}{h_{CD}} = \frac{3}{2} = 1.50$$

Thus, the displacement factors are

$$D_{BA} = \frac{-1.5\,C_{BA}\,K_{BA}}{(C_{AB})^2\,K_{AB} + (C_{CD})^2\,K_{CD}}$$

$$D_{BA} = \frac{-1.5(1)0.67}{1^2(0.67) + 1.5^2(0.5)} = -0.56$$

$$D_{CD} = \frac{-1.5(1.5)0.50}{1^2(0.67) + 1.5^2(0.5)} = -0.63$$

The method of computation of rotation contribution and displacement contribution are as follows.

Fixed end moments

$$M_{FBC} = -20 \times \frac{2^2}{12} = -6.67\text{ kNm}$$

$$M_{FCB} = +20 \times \frac{2^2}{12} = +6.67\text{ kNm}$$

Iteration 1

Rotation contribution

$$m_{ij} = R_{ij}\left[M_{Fi} + \sum_j (m_{ji} + M_{ij}')\right]$$

At *B*

$$m_{bc} = -0.21[-6.67 + 0 + 0 + 0] = 1.41\text{ kNm}$$

$$m_{ba} = -0.29[-6.67 + 0 + 0 + 0] = 1.93\text{ kNm}$$

At *C*

$$m_{cb} = -0.25[+6.67 + 1.41 + 0 + 0] = 2.02\text{ kNm}$$

$$m_{cd} = -0.25[6.67 + 1.41 + 0 + 0] = -2.02\text{ kNm}$$

Displacement contribution

$$m'_{ij} = D_{ij}[C_{BA}(m_{BA} + \cancel{m}_{AB}) + C_{CD}(m_{CD} + \cancel{m}_{DC})]$$

$$m'_{ij} = D_{ij}[C_{BA}\ m_{BA} + C_{CD}\ m_{CD}]$$

$$m'_{ba} = -0.56[1(1.93) + 1.5(-2.02)] = 0.62 \text{ kNm}$$

$$m'_{cd} = -0.63[1(1.93) + 1.5(-2.02)] = 0.69 \text{ kNm}$$

Iteration 2

Rotation contribution

At *B*;

$$m_{bc} = -0.21[-6.67 - 2.02 + 0.62] = +1.70 \text{ kNm}$$

$$m_{cd} = -0.29[-6.67 - 2.02 + 0.62] = +2.34 \text{ kNm}$$

At *C*;

$$m_{cb} = -0.25[+6.67 + 1.70 + 0.69] = -2.27 \text{ kNm}$$

$$m_{cd} = -0.25[6.67 + 1.70 + 0.69] = -2.27 \text{ kNm}$$

Displacement contribution

$$m'_{ba} = -0.56[1(2.34) + 1.5(-2.27)] = 0.60 \text{ kNm}$$

$$m'_{cd} = -0.63[1(2.34) + 1.5(-2.27)] = 0.67 \text{ kNm}$$

Iteration 3

Rotation contribution

At *B*;

$$m_{bc} = -0.21[-6.67 + 0.60 - 2.27] = 1.75 \text{ kNm}$$

$$m_{cd} = -0.29[-6.67 + 0.60 - 2.27] = 2.42 \text{ kNm}$$

At *C*;

$$m_{cb} = -0.25[6.67 + 1.75 + 0.67] = -2.27 \text{ kNm}$$

$$m_{cd} = -0.25[6.67 + 1.75 + 0.67] = -2.27 \text{ kNm}$$

Displacement contribution

$$m'_{ba} = -0.56[1(2.42) + 1.5(-2.27)] = 0.55 \text{ kNm}$$

$$m'_{cd} = -0.63[1(2.42) + 1.5(-2.27)] = 0.62 \text{ kNm}$$

The operations are repeated till the values converge.

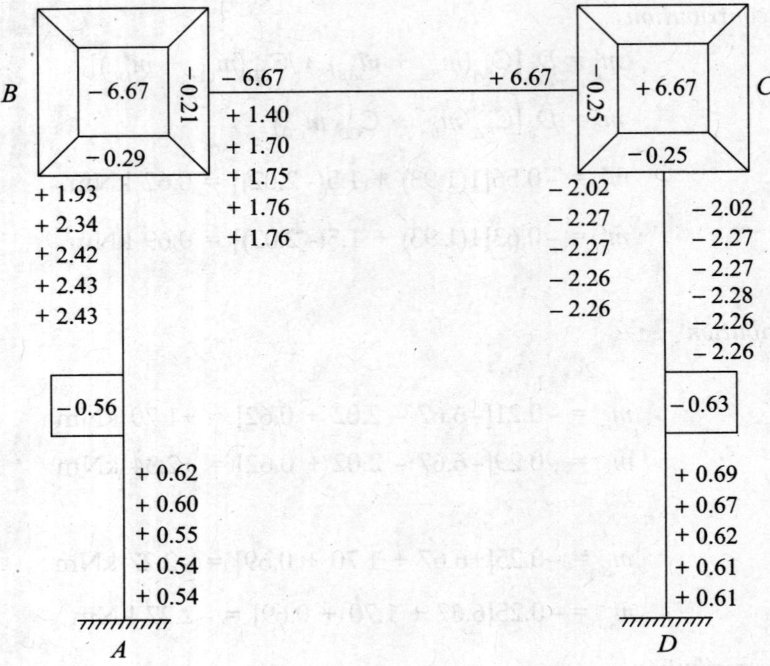

FIG. 3.117 Rotational and translational moments (kNm)

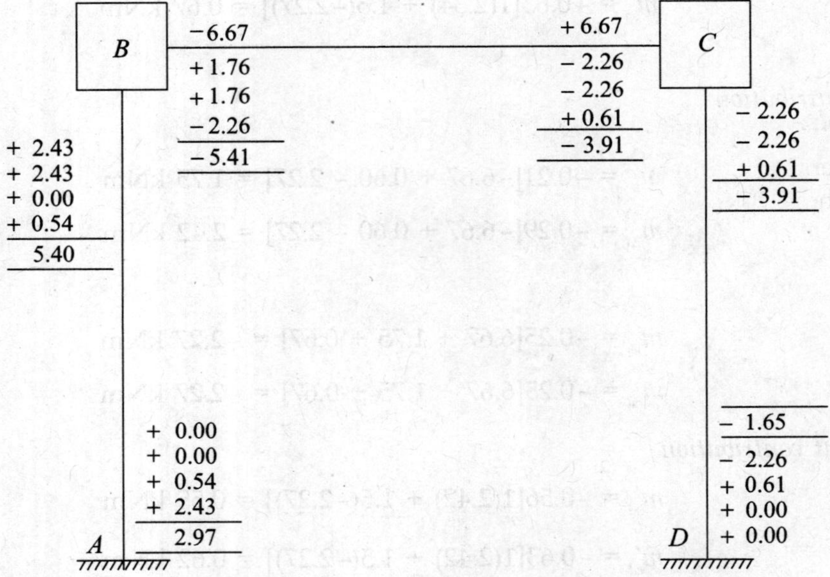

FIG. 3.118 Computation of end moments (kNm)

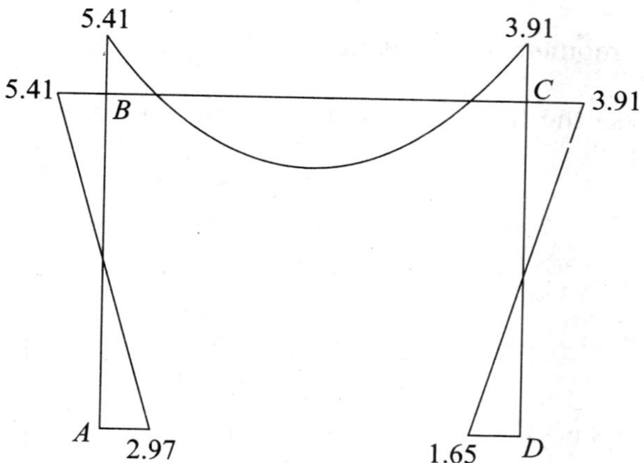

FIG. 3.119 Bending moment diagram (kNm)

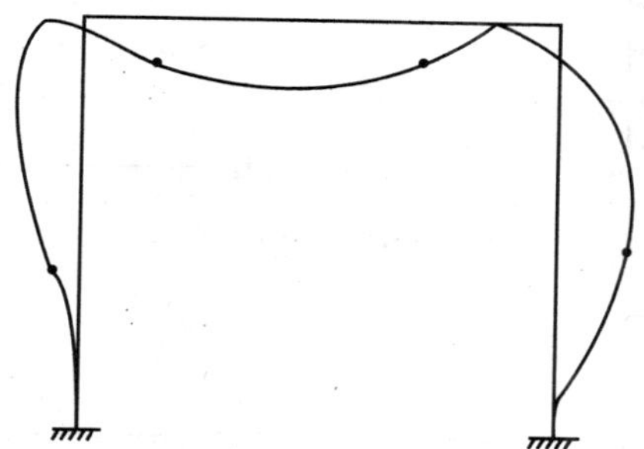

FIG. 3.120 Elastic curve

3.2.2.3 *Columns of Same Height Subjected to Horizontal Loads*

Let a fixed ended column *AB* of height *h* be subjected to horizontal load *P*; then the horizontal shear condition gives

$$\frac{\Sigma(M_{AB} + M_{BA})}{h} + P = 0 \tag{1}$$

$$\Sigma(M_{AB} + M_{BA}) = -Ph \tag{2}$$

$$\Sigma(2m_{ab} + m_{ba} + m'_{ab} + 2m_{ba} + m_{ab} + m'_{ba}) + Ph = 0 \tag{3}$$

$$\Sigma m'_{ab} = -\frac{3}{2}\left[\frac{Ph}{3} + \Sigma(m_{ab} + m_{ba})\right] \tag{4}$$

i.e.

$$\boxed{m'_{ab} = -\frac{3}{2}\frac{k_{ab}}{\Sigma k}\left[\frac{Ph}{3} + \Sigma(m_{ab} + m_{ba})\right]} \tag{5}$$

where $-\frac{Ph}{2}$ = Storey moment which is taken into consideration for sway.

Example 3.23 Analyse the frame by Kani's method and draw the bending moment diagram.

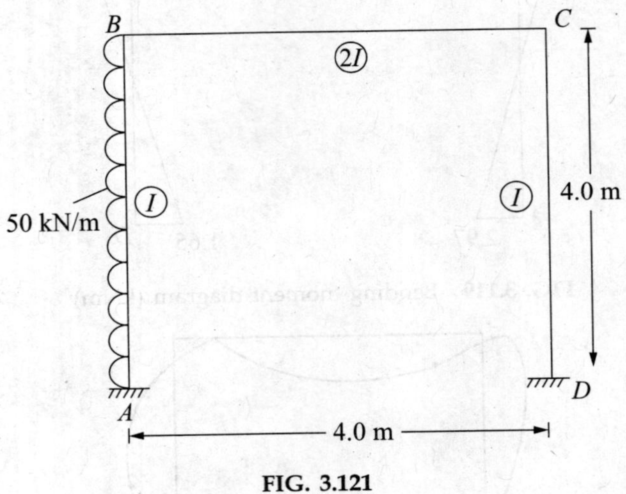

FIG. 3.121

Solution

Rotational factors

Joint	*Members*	*Relative Stiffness k*	Σk	$R = -\frac{1}{2}\frac{k}{\Sigma k}$
B	BA	I/4	3I/4	−0.17
	BC	2I/4		−0.33
C	CB	2I/4	3I/4	−0.33
	CD	I/4		−0.17

Displacement factors

The displacement factor is given by

$$D_{ij} = \frac{-1.5\,C_{ij}\,K_{ij}}{\Sigma m\,C_{ij}^2\,K_{ij}}$$

As the ends of the columns AB and DC are fixed $m = 1$, the reduction factor (correction ratio) for columns is computed from

$$C_{ij} = \frac{h'_r}{h_{ij}}$$

and the relative height h'_r is taken as 4.0 m. Therefore,

$$C_{BA} = \frac{h'_r}{h_{BA}} = \frac{4}{4} = 1.0$$

$$C_{CD} = \frac{h'_r}{h_{CD}} = \frac{4}{4} = 1.0$$

Thus, the displacement factor reduces to

$$D_{ij} = \frac{-1.5(1)K_{ij}}{\Sigma 1(1)^2 K_{ij}}$$

i.e.

$$D_{ij} = \frac{-1.5K_{ij}}{\Sigma K_{ij}}$$

$$D_{BA} = \frac{-1.5K_{BA}}{K_{BA} + K_{CD}}$$

As $K_{BA} = K_{CD}$ the ratio reduces to

$$D_{BA} = -\frac{1.5}{2} = -0.75$$

Similarly, $D_{CD} = -0.75$

The above reasoning of computation of displacement factors leads to the conclusion that:

In a single bay, single-storey frame with fixed ends and the columns have the same values of moment of inertia, the displacement factor for the columns is –0.75

Fixed end moments

$$M_{FAB} = -\frac{50 \times 4^2}{12} = -66.67 \text{ kNm}$$

$$M_{FBA} = +\frac{50 \times 4^2}{12} = -66.67 \text{ kNm}$$

Storey moment

$$\text{Storey shear } P_r = \frac{50 \times 4}{2} = 100 \text{ kNm}$$

$$\text{Storey moment } M_r = \frac{P_r h_r}{3} = \frac{100 \times 4}{3}$$

$$M_r = 133.33 \text{ kNm}$$

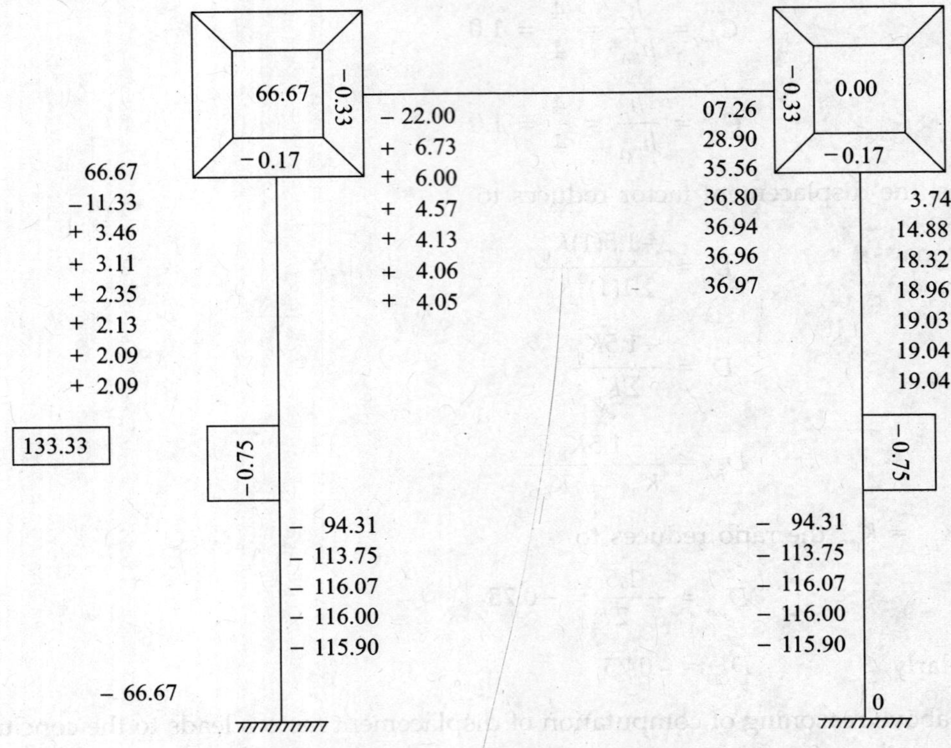

FIG. 3.122 Rotational translational moment (kNm)

The fixed end moments, rotation factors and displacement factors are entered in the appropriate places.

Iteration 1

$$m'_{ba} = -0.17[66.67 + 0 + 0] = -11.33 \text{ kNm}$$

$$m'_{bc} = -0.33[66.67 + 0 + 0] = -22.00 \text{ kNm}$$

$$m'_{cb} = -0.33[-22.0 + 0 + 0] = +7.26 \text{ kNm}$$

$$m'_{cd} = -0.17[-22.0 + 0 + 0] = +3.74 \text{ kNm}$$

$$m''_{ba} = -0.75[133.33 - 11.33 + 3.74] = -94.31 \text{ kNm}$$

$$m''_{cd} = -0.75[133.33 - 11.33 + 3.74] = -94.31 \text{ kNm}$$

Iteration 2

$$m'_{ba} = -0.17[66.67 - 94.31 + 7.26] = 3.46 \text{ kNm}$$

$$m'_{bc} = -0.33[66.67 - 94.31 + 7.26] = 6.73 \text{ kNm}$$

$$m'_{cb} = -0.33[6.73 - 94.31 + 0] = 28.90 \text{ kNm}$$

$$m'_{cd} = -0.17[6.73 - 94.31 + 0] = 14.88 \text{ kNm}$$

$$m''_{ba} = -0.77[133.33 + 3.46 + 14.88] = -113.75 \text{ kNm}$$

$$m''_{cd} = -0.75[133.33+3.46+14.88] = -113.75 \text{ kNm}$$

It is to be remembered here that m'_{ba} represents rotational moments and m''_{ba} represents translational (displacement) moments. The iterations are continued till two successive iterations converge.

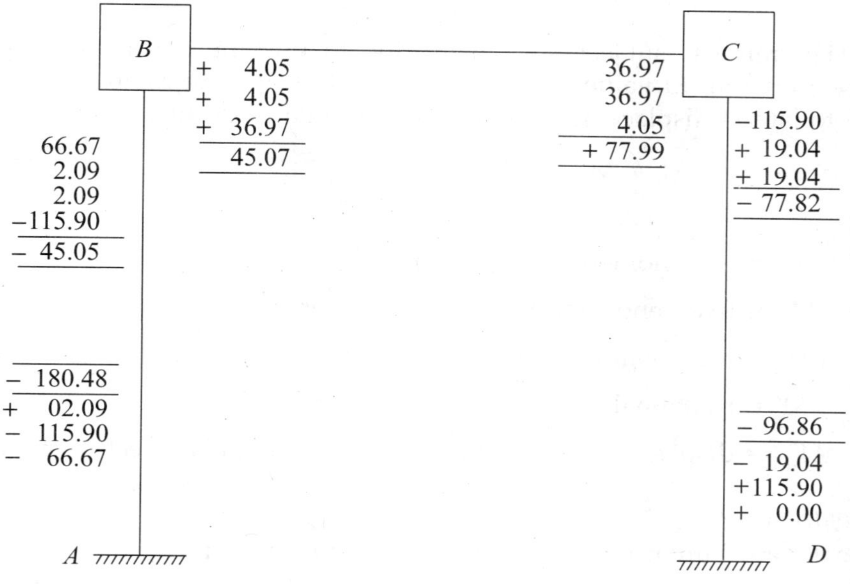

FIG. 3.123 Computation of end moments (kNm)

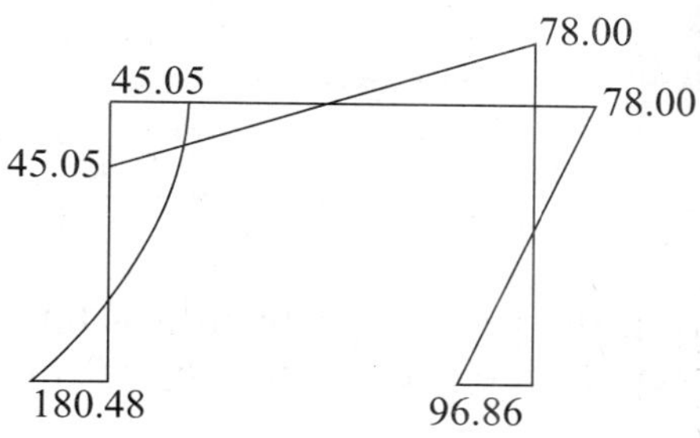

FIG. 3.124 Bending moment diagram (kNm)

Example 3.24: Analyse the rigid portal frame by Kani's method of moment distribution. Draw the bending moment diagram.

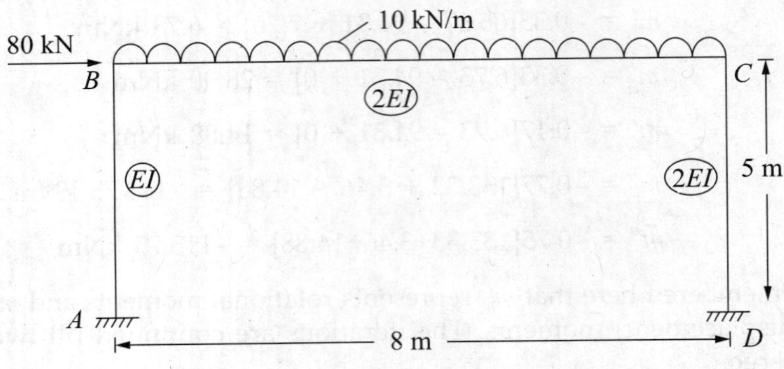

FIG. 3.125

Solution: The frame is subjected to sway due to unsymmetrical loading and the columns have different *EI* values. Hence, there will be effect of rotational moments as well as due to lateral displacement moments. The final moment is evaluated as

$$M_{ij} = M_{Fij} + 2M'_{ij} + M'_{ji} + M''_{ij(c)} \tag{1}$$

where

M_{ij} = end moment of the member $i - j$

M_{Fij} = fixed end moment of the member

M_{ij} = rotational moment at *i*th moment

M'_{ji} = rotational moment at *j*th end

$M''_{ji(c)}$ = displacement moment due to lateral loads/effect of sway

Rotational contribution

The rotational contribution moment in the member $i - j$ is given by

$$M'_{ij} = R_{ij}\left[\Sigma M_{Fij} + \Sigma M'_{ji} + M''_{ij}\right] \tag{2}$$

Displacement contribution

This moment due to lateral displacement is given by

$$M''_{ij} = D_{ij}\left[\Sigma M_{ij(c)} + \Sigma M'_{ji(c)} + \frac{P_r h_r}{3}\right] \tag{3}$$

where P_r = storey shear and in the present problem it is 80 kN and h_r = height of the column and r = level of storey. In the present case h_r = 5 m.

D_{ij} = displacement factor for the column and in general term

$$D = -\frac{3}{2}\frac{k_c}{\Sigma k_c} \tag{4}$$

Thus, in the present case, the end moment can be obtained by using the fixed end moments due to external loading, displacement contribution moments and rotation contribution moments.

Rotation factors (R_{ij})

Joint	*Members*	*Relative Stiffness k*	Σk	$R = -\frac{1}{2}\left(\frac{k}{\Sigma k}\right)$
B	BA	$I/5 = 0.2I$	$0.45I$	–0.22
	BC	$2I/8 = 0.25I$		–0.28
C	CB	$2I/8 = 0.25I$	$0.65I$	–0.19
	CD	$2I/5 = 0.4I$		–0.31

Fixed end moments (MF_{ij})

$$M_{FBC} = -\frac{wl^2}{12} = -\frac{10 \times 8^2}{12} = -53.33 \text{ kNm}$$

$$M_{FCB} = +\frac{wl^2}{12} = +10 \times \frac{8^2}{12} = +53.33 \text{ kNm}$$

Displacement factor (D_{ij})

$$D_{AB} == -\frac{3}{2}\frac{k_{AB}}{k_{AB} + k_{CD}} = -\frac{3}{2}\left(\frac{0.2}{0.2 + 0.4}\right) = -0.5$$

$$D_{CD} == -\frac{3}{2}\frac{k_{CD}}{k_{AB} + k_{CD}} = -\frac{3}{2}\left(\frac{0.4}{0.2 + 0.4}\right) = -1.0$$

These displacement factors for each column are entered in the appropriate box at mid height of the column (shown in Fig. 3.126). The sum of the displacement factor is –1.5 and the sum of the rotation factor is –0.5 at a joint.

Storey moment ($P_r\, h_r/3$)

The storey shear is the lateral load applied, i.e. 80 kN. Height of the storey is 5 m. The storey moment

$$M_r = \frac{P_r\, h_r}{3} = \frac{80 \times 5}{3} = 133.33 \text{ kNm}$$

This storey moment is positive if P_r acts from left to right. This storey moment is recorded in a box on the mid height towards the left of the column.

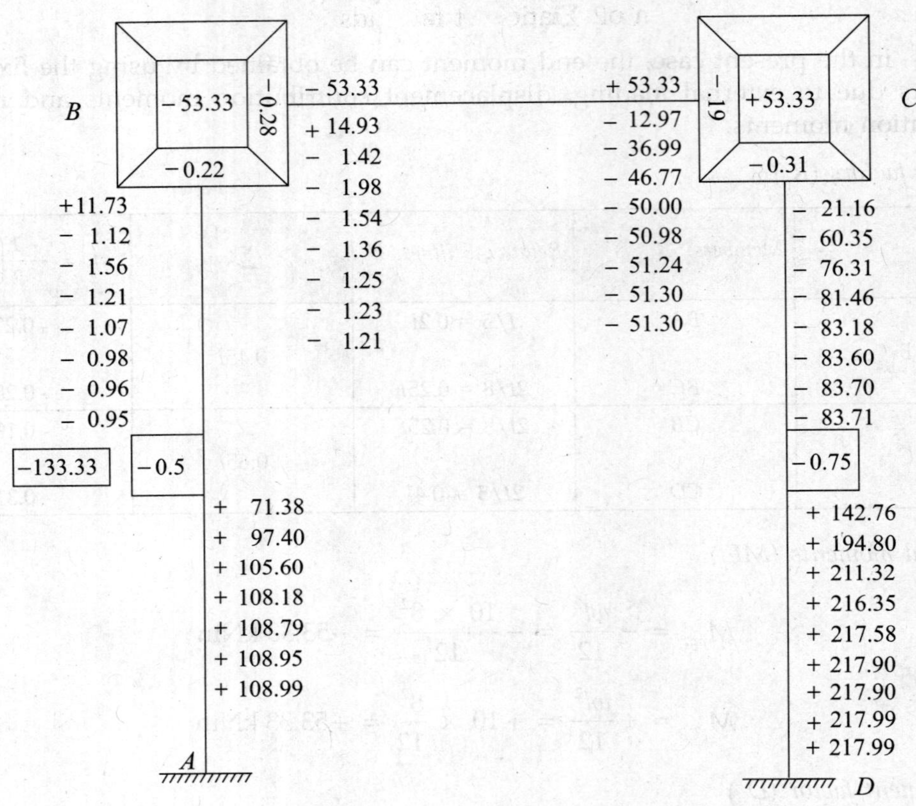

FIG. 3.126 Rotational and translational moments (kNm)

Iteration 1

Joint *B*

The rotation moments and translational moments are initially assumed to be zero.

Sum of the joint moments = −53.33 kNm

Sum of the rotation moments

at the ends A = 0 (fixed end)

C = 0 (assumed)

Column BA translational moments = 0 (assumed)

Total = −53.33 kNm

$$\text{Rotational moment } M'_{BA} = -0.22(-53.33) = +11.73 \text{ kNm}$$

$$M'_{BC} = -0.28(-53.33) = +14.93 \text{ kNm}$$

Joint C

$$\text{Sum of joint moments} = +53.33 \text{ kNm}$$

Sum of rotation at far ends

$$B = +14.93 \text{ kNm}$$

$$D = 0$$

Sum of the translation moments

$$\text{column } CD = 0$$

$$\text{Total moments} = +68.26 \text{ kNm}$$

$$\text{Rotational moments } M'_{CB} = -0.19(68.26) = -12.97 \text{ kNm}$$

$$M'_{CD} = -0.31(68.26) = -21.16 \text{ kNm}$$

After knowing the rotational moments the translational moment is computed as follows.

$$\text{Storey moment} = -133.33 \text{ kNm}$$

Rotation moments at ends of columns

$$BA = +11.73 \text{ kNm}$$

$$CD = -21.16 \text{ kNm}$$

$$\text{Total} = -142.76 \text{ kNm}$$

Using the displacement factors $D_{AB}(= -0.5)$ and $D_{CD}(= -1.0)$. The displacement moments are obtained as

$$M''_{AB} = -0.5(-142.76) = +71.38 \text{ kNm}$$

$$M''_{DC} = -1.0(-142.76) = +142.76 \text{ kNm}$$

Iteration 2

Joint B

$$\text{Sum of the joint moments} = -53.33 \text{ kNm}$$

Sum of the rotation moments

$$A = 0 \text{ (fixed)}$$

$$C = -12.97 \text{ kNm}$$

$$\text{Column } BA\text{, translational moment} = +71.38 \text{ kNm}$$

$$\text{Total} = +5.08 \text{ kNm}$$

$$\text{Rotational moments } M'_{BA} = -0.22(5.08) = -1.12 \text{ kNm}$$

$$M'_{BC} = -0.28(5.08) = -1.42 \text{ kNm}$$

Joint *C*

$$\text{Sum of the joint moments} = +53.33 \text{ kNm}$$

Sum of the rotation moments at far ends

$$B = -1.42 \text{ kNm}$$

$$D = 0.00$$

$$\text{Displacement moment} = 142.76$$

$$\text{Total} = 194.67 \text{ kNm}$$

$$\text{Rotational moment } M'_{CB} = 194.67 \times (-0.19) = -36.99 \text{ kNm}$$

$$M'_{CD} = -0.31(194.67) = -60.35 \text{ kNm}$$

Displacement moments

$$M''_{AB} = (-133.33 - 1.12 - 60.35)(-0.5) = 97.4 \text{ kNm}$$

$$M''_{CD} = (-133.33 - 1.12 - 60.35)(-1) = 194.8 \text{ kNm}$$

This completes the second iteration. The procedure is repeated till the convergence is obtained.

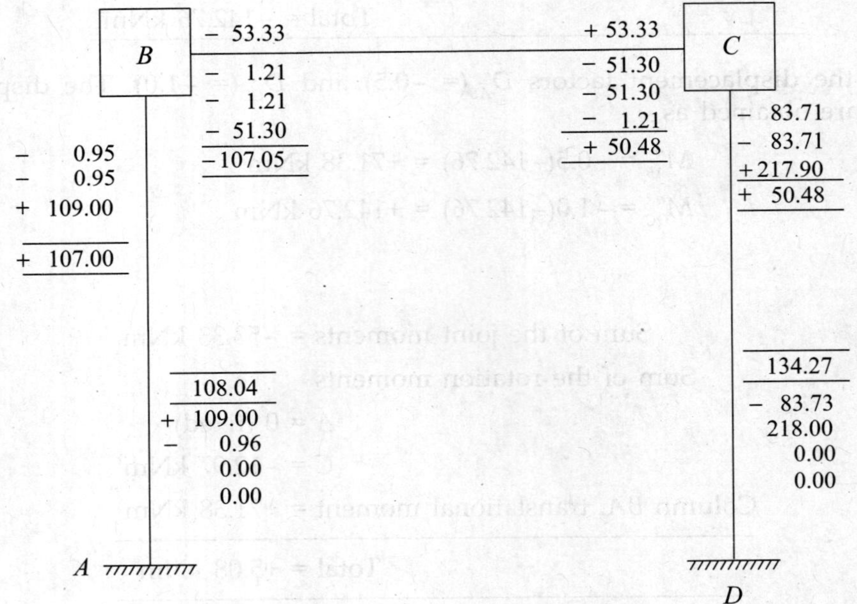

FIG. 3.127 Final end moments (kNm)

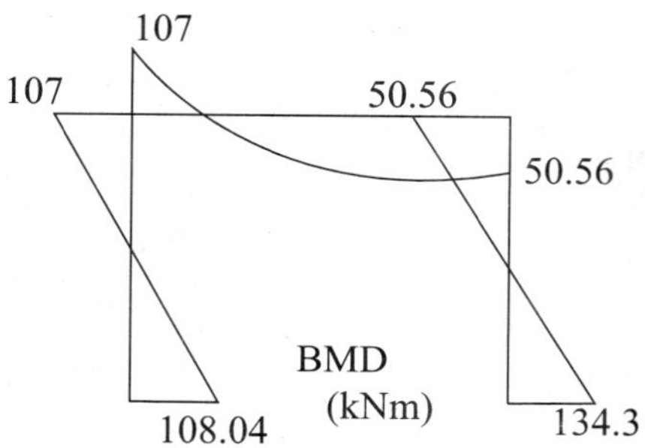

FIG. 3.128

Example 3.25 Analyse the frame shown in Fig. 3.129 by Kani's method. Draw the bending moment diagram, shear force diagram and the elastic curve.

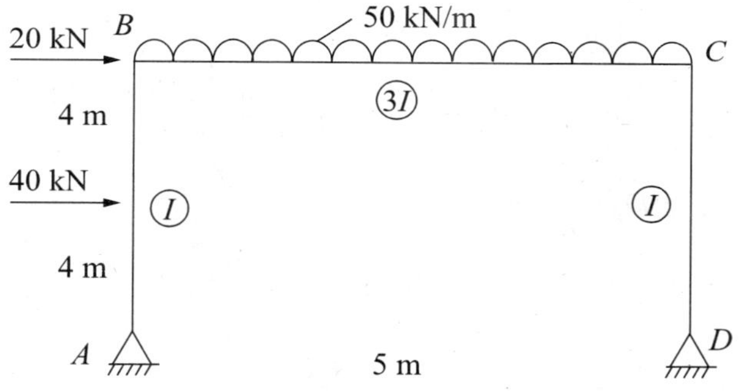

FIG. 3.129

Solution

Rotation factors

Joint	*Members*	*Stiffness (k)*	Σk	$R = -\frac{1}{2}\frac{k}{\Sigma k}$
B	BA	$3/4(I/8) = 0.0938I$	$0.6938I$	−0.068
	BC	$3I/5 = 0.6I$		−0.432
C	CD	$\frac{3}{4}\left(\frac{I}{8}\right) = 0.0938I$	$0.6938I$	−0.068
	CB	$\frac{3I}{5} = 0.60I$		−0.432

Fixed end moments

$$M_{FAB} = -\frac{40 \times 8}{8} = -40 \text{ kN}$$

$$M_{FBA} = +\frac{40 \times 8}{8} = +40 \text{ kNm}$$

$$M_{FBC} = -\frac{50 \times 5^2}{12} = -104.17 \text{ kNm}$$

$$M_{FCB} = +\frac{50 \times 5^2}{12} = +104.17 \text{ kNm}$$

Displacement factor

$$D_{ij} = \frac{-2C_{ij}\, K_{ij}}{\Sigma C_{ij}^2\, K_{ij}}$$

$$D_{ij} = \frac{-1.5C_{ij}\, K_{ij}}{C_{ij}[C_{ij}\, K_{ij} + C_{ij}\, K_{ij}]} = \frac{1.5K_{ij}}{2K_{ij}} = -0.75$$

Taking constant 2; $\boxed{D_{ij} = -1.50}$

Storey moments

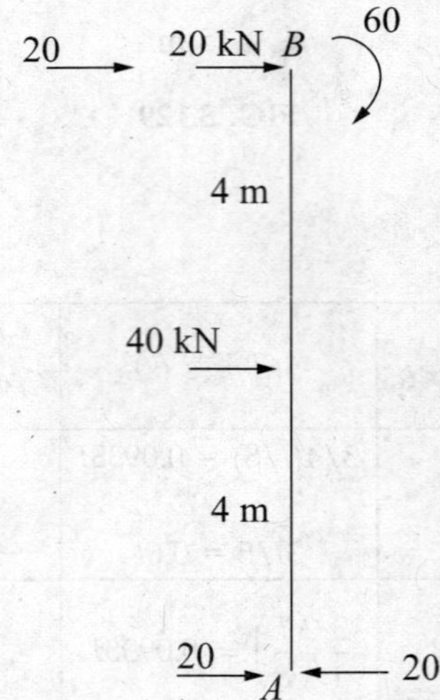

FIG. 3.130

$$\text{Top shear} = 20 + 20 + \frac{60}{8} = 47.5 \text{ kN}$$

Top shear was calculated as follows. The central concentrated load 40 kN is distributed as 20 kN at top and 20 kN at bottom. Also at the top there is a horizontal load of 20 kN. The support A is hinged and hence the moment is zero. The fixed end moment at A and B are $-\frac{Wl}{8} = -40 \times \frac{8}{8} = -40\ \text{kNm}$. To make the moment zero at the hinged end, the moment is released and half of the moment is carried over to B. Therefore, the shear at the top due to the end moment 60 kNm is 60/8 = 7.5 kN. Hence, the total shear force is 20 + 20 + 7.5 = 47.5 kN at top. Hence, the storey moment is

$$M_r = \frac{Ph_r}{3}$$

$$M_r = \frac{47.5 \times 8}{3} = 126.67\ \text{kNm}$$

$$\boxed{M_r = 126.67\ \text{kNm}}$$

Iteration 1

Rotation contribution

$$m_{ij} = R_{ij}\left[M_{Fi} + \sum_j \left(M_{ji + m'_{ji}}\right)\right]$$

At Ⓑ,

$$m_{BC} = R_{BC}[M_{FB} + m_{CB} + m_{BA} + m'_{BA}]$$

$$m_{BC} = -0.432[-44.17 + 0 + 0 + 0] = 19.08$$

$$m_{BA} = -0.068[-44.17 + 0 + 0] = 3.00$$

At Ⓒ,

$$M_{CB} = R_{CB}[M_{FC} + m_{BC} + m_{DC} + m'_{CD}]$$

$$= -0.432[104.17 + 19.081 + 0 + 0] = -53.244$$

$$M_{CD} = R_{CD}[M_{FC} + m_{BC} + m_{DC} + m'_{CD}]$$

$$= -0.068[104.17 + 19.081 + 0 + 0] = -8.38$$

Displacement contribution

$$m'_{ab} = D_{ab}\left[\frac{Ph_r}{3} + \frac{2}{3}\Sigma M_{ba}\right]$$

$$= -1.5\left[126.67 + \frac{2}{3}(3 - 8.38)\right] = -184.59$$

Iteration 2

Rotation contribution

$$m_{bc} = -0.432[-44.17 - 53.244 + 0 - 184.59] = 121.825$$

At *B*,

$$m_{ba} = -0.068[-44.17 - 53.244 + 0 - 184.59] = -19.18$$

At *C*,

$$m_{cb} = -0.432[104.17 + 121.82 + 0 - 184.59] = -17.88$$

$$m_{cd} = -0.068[104.17 + 121.82 + 0 - 184.59] = -2.82$$

Displacement contribution

$$m'_{ab} = D_{ab}\left[\frac{Ph_r}{3} + \frac{2}{3}\Sigma M_{ba}\right]$$

$$= -1.5\left[126.67 + \frac{2}{3}(19.18 - 2.82)\right] = -206.45$$

$$m'_{cd} = -206.45$$

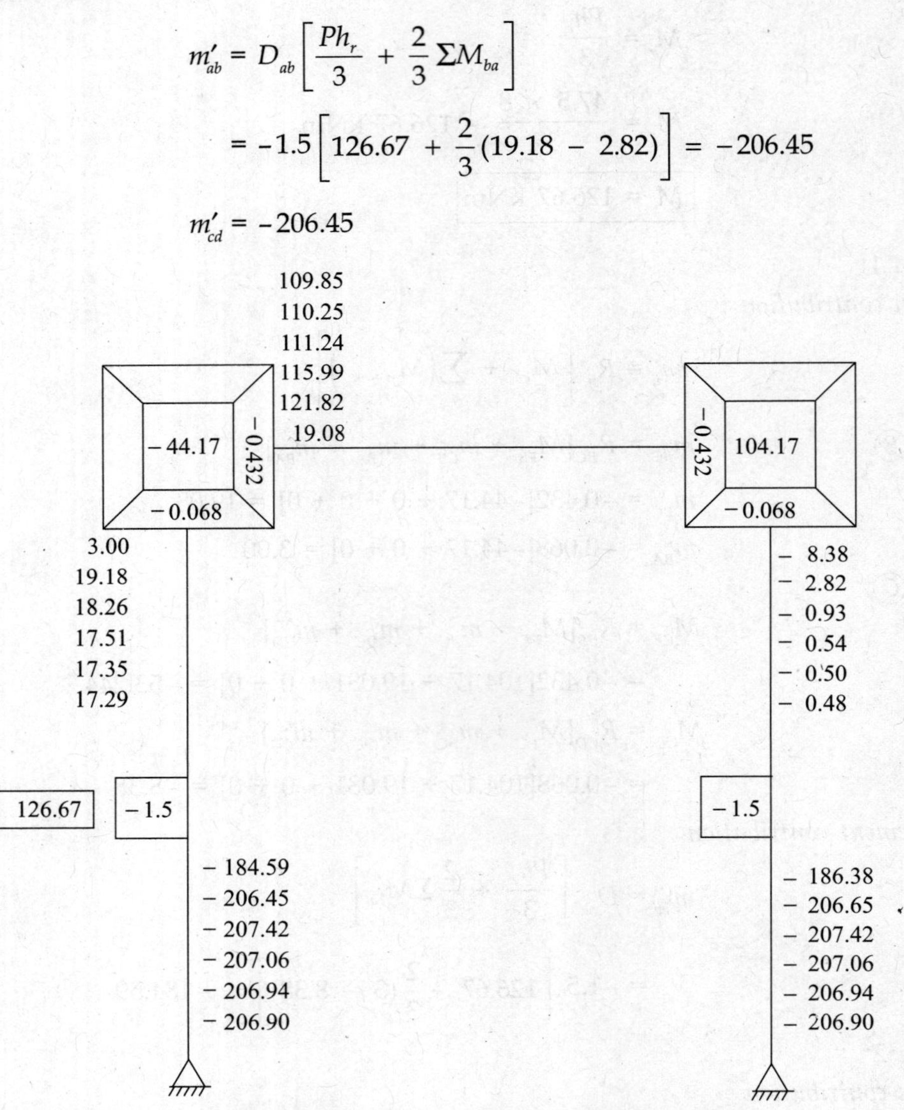

FIG. 3.131 Rotational and displacement moments (kNm)

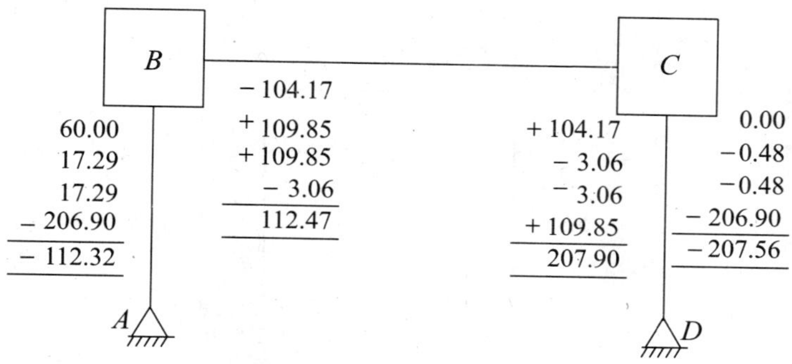

FIG. 3.132 End moments (kNm)

$$H_D = \frac{M_{CD}}{8} = \frac{207.56}{8} = 25.95 \text{ kN}$$

$$H_A + H_D = 60$$

$$\therefore \quad H_A = 60 - 25.95 = 34.05 \text{ kN}$$

$$\Sigma M_A = 0;$$

$$40(4) + 20(8) + 50 \times \frac{5^2}{2} - 5R_D = 0$$

$$R_D = 189 \text{ kN}$$

$$V_A = 250 - 189 = 61 \text{ kN}$$

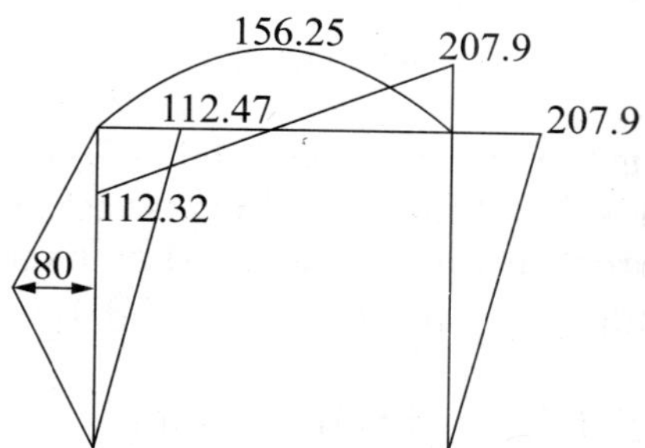

FIG. 3.133 Bending moment diagram (kNm)

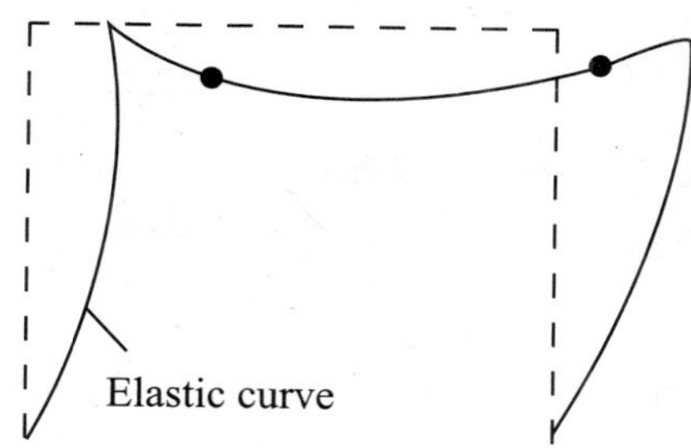

FIG. 3.134

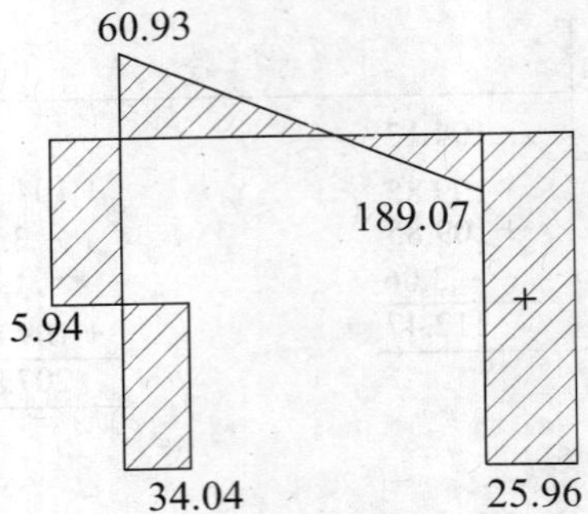

FIG. 3.135 Shear force diagram (kNm)

Example 3.26 Analyse the given frame and draw the bending moment diagram.

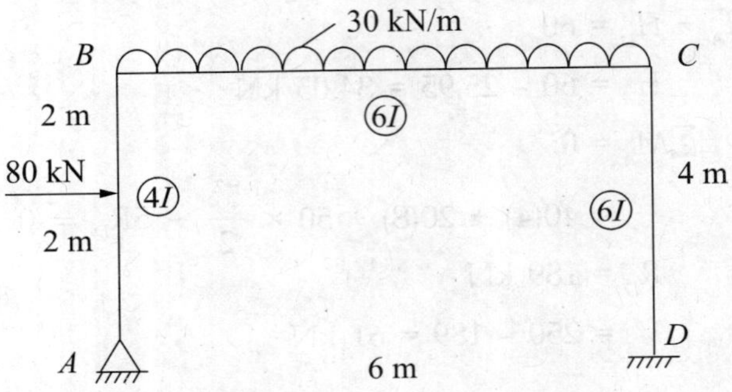

FIG. 3.136

Solution: The given frame is hinged at A and fixed at D. In reality, the hinged condition arises in case of columns on isolated footing which is resting on soil. On the other hand, if the column is resting on rock it can be idealised as fixed condition.

In Kani's method, the hinged column is replaced by a column fixed at the base having

$$h'_{AB} = 1.5h_{AB} = 1.5(4) = 6.0 \text{ m}$$

$$K'_{AB} = \frac{3}{4}k_{AB} = \frac{3}{4}\left(\frac{4I}{4}\right) = 0.75I$$

Rotation factors

Joint	*Members*	*Stiffness k*	Σk	$R = -\frac{1}{2}\frac{k}{\Sigma k}$
B	BA	$\frac{3}{4} \times \frac{4}{4} = 0.75$	1.75	−0.21
	BC	$\frac{6}{6} = 1.0$		−0.29

C	CB	6/6 = 1.0	2.00	-0.25
	CD	6/6 = 1.0		-0.25

Fixed end moments

$$M_{FBC} = -\frac{wl^2}{12} = -30 \times \frac{6^2}{12} = -90.00 \text{ kNm}$$

$$M_{FCB} = +\frac{wl^2}{12} = +30 \times \frac{6^2}{12} = +90.00 \text{ kNm}$$

$$M_{FAB} = -\frac{Wl}{8} = -80 \times \frac{4}{8} = -40.0 \text{ kNm}$$

$$M_{FBA} = +\frac{Wl}{8} = +80 \times \frac{4}{8} = +40.0 \text{ kNm}$$

Storey Moment

The top shear is calculated as follows. The lateral load 80 kN is replaced at top and bottom nodes as 40 kN. The end *A* is hinged. Therefore, the fixed end moment of 40.0 kNm is released. Half of the moment is transferred to the joint *B*. Thus, the moment at *B* is $40 + \frac{40}{2} = 60$ kNm. Therefore, this causes the storey shear of $\frac{60}{4} = 15$ kN in addition to 40 kN. Thus, the value of P_r is 40 + 15 = 55 kN. Therefore,

$$M_r^F = \frac{P_r h_r}{3} = \frac{55(4)}{3} = 73.33 \text{ kNm}$$

Translation factor

The translation factor is given by

$$D_{ij} = \frac{-1.5\, C_{ij}\, K_{ij}}{\Sigma m'\, C_{ij}^2\, K_{ij}}$$

The correction ratio is given as

$$C_{ij} = \frac{h'_r}{h_{ij}}$$

Taking the reference height h'_r as 4.0 m.

$$\therefore \quad C_{BA} = \frac{4}{6} = 0.67$$

$$C_{CD} = \frac{4}{4} = 1.00$$

The displacement factor (translational factor)

Column *AB*

$$K_{AB} = 3/4; \quad C_{AB} = 2/3 \quad \text{and} \quad m'_{AB} = 0.75 \quad (\because \text{ Hinged})$$

$$\therefore \quad C_{AB}\, K_{AB} = 0.5; \quad m'_{AB}\, C^2_{AB}\, K_{AB} = 0.25$$

Column *CD*

$$K_{CD} = 1.0; \quad C_{CD} = 1.0 \quad \text{and} \quad m'_{CD} = 1.0 \quad (\because \text{ Fixed})$$

$$m'_{CD}\, C^2_{CD}\, K_{CD} = 1.0;$$

$$\therefore \quad D_{AB} = -1.5 \times 0.5/(1.0 + 0.25) = -0.6$$

$$D_{CD} = -1.5 \times 1.0/(1.0 + 0.25) = -1.2$$

Iteration 1

Rotational moments

Joint *B*

$$m_{ij} = R_{ij}\, [M_{Fi} + \Sigma(m_{ji} + m'_{ij})]$$

$$m_{BC} = R_{BC}\, [M_{FB} + m_{CB} + m_{AB} + m'_{BA}]$$

$$= -0.29[(-90 + 60.0) + 0 + 0 + 0] = 8.70$$

$$m_{BA} = RBA\, [M_{FB} + m_{CB} + m_{AB} + m'_{BA}]$$

$$= -0.21[(-90 + 60) + 0 + 0 + 0] = 6.30$$

Joint *C*

$$m_{CB} = R_{CB}\, [M_{FC} + m_{BC} + m_{dc} + m'_{cd}]$$

$$= -0.25[90 + 8.70 + 0 + 0] = -24.68$$

$$m_{cd} = -0.25[90 + 8.70 + 0 + 0] = -24.68$$

Translation moments

$$m'_{ab} = -0.6 \left[68.75 + \frac{2}{3}(6.30 - 24.68) \right] = -33.90 \text{ kNm}$$

$$m'_{cd} = -1.2 \left[68.75 + \frac{2}{3}(6.30 - 24.68) \right] = -67.80$$

Iteration 2

Similar computations are shown in the computation figure.

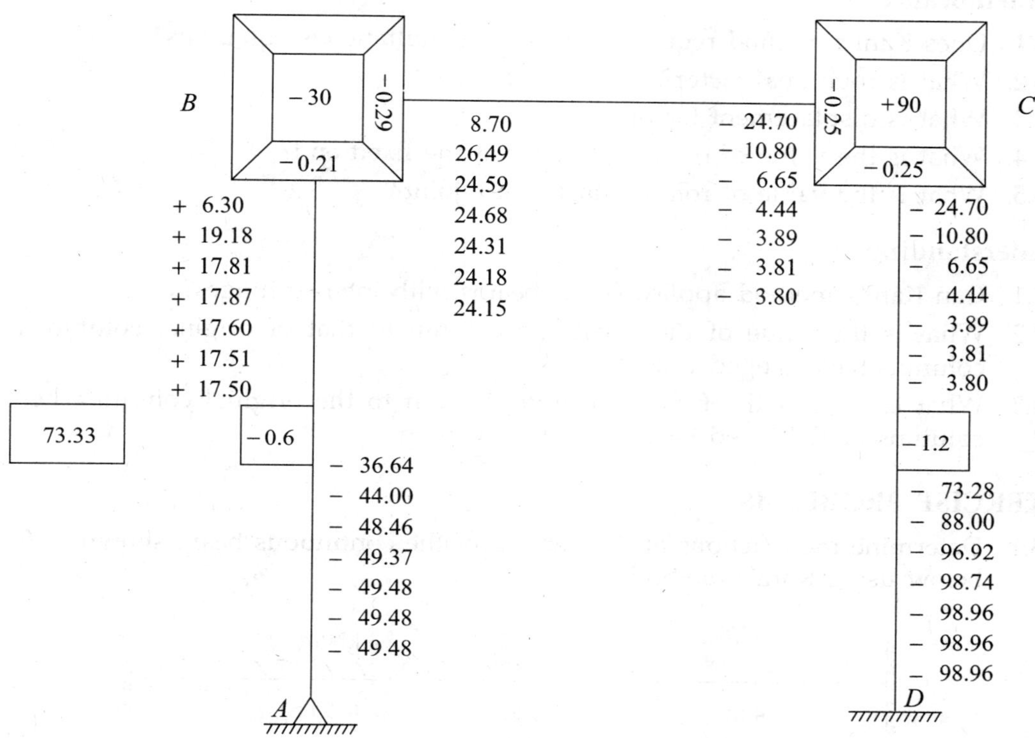

FIG. 3.137 Rotation and displacement moments (kNm)

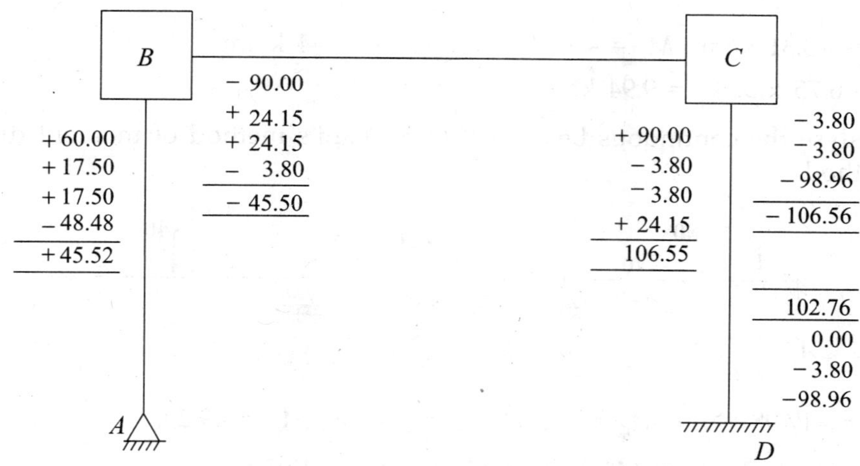

FIG. 3.138 Computation of end moments (kNm)

REVIEW QUESTIONS

Remembrance

3.1. Does Kani's method require solution of simultaneous equations?

3.2. What is rotational factor?

3.3. What is displacement factor?

3.4. What is the value of rotational factor at the fixed end?

3.5. What is the value of rotation factor at a joint?

Understanding

3.1. Can Kani's method applicable for beams with interior hinges?

3.2. What is the value of the substitute column to that of original column if the columns have hinged bases?

3.3. What is the length of the substitute column to the original column when the columns have hinged bases in Kani's method.

EXERCISE PROBLEMS

3.1. Determine the reactions at *A*, *B* and *C* of the continuous beam shown in figure below using Kani's method.

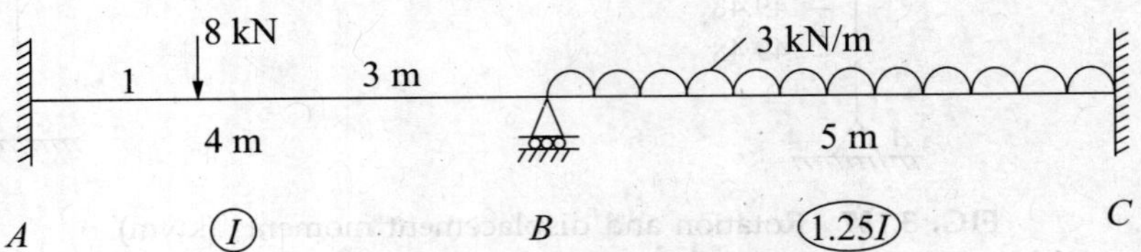

Ans: $M_A = -3.31$ kNm, $M_B = -3.87$ kNm, $M_C = -7.44$ kNm

$V_A = 6.75$ kN, $V_B = 9.94$ kN, $V_C = 6.31$ kN

3.2. Analyse the continuous beam *ABCD* by Kani's method of moment distribution method.

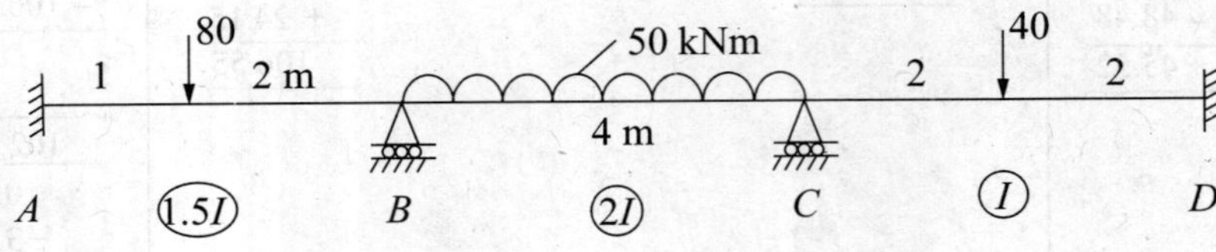

Ans: $M_A = -17.98$ kNm, $M_B = -52.93$, $M_C = -41.46$, $M_D = -9.29$,

$R_A = 41.68$ kN, $R_B = 141.2$, $R_C = 125.15$, $R_D = 11.97$ kN

3.3. Analyse the three span continuous beam by Kani's method. Draw the bending moment diagram. EI is constant.

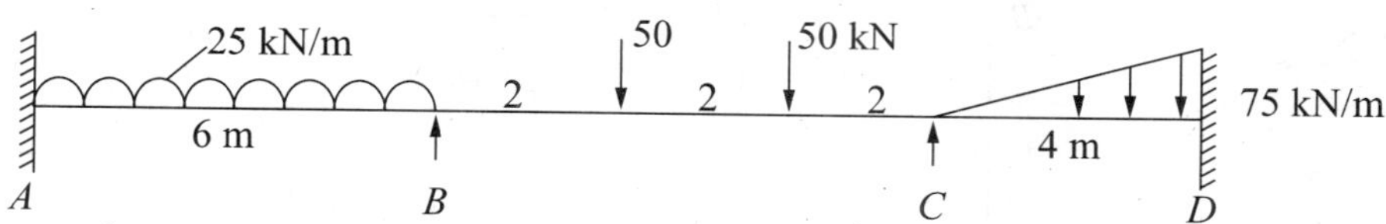

Ans: M_A = −75.8 kNm, M_B = −73.4 kNm, M_C = −55.5 kNm, M_D = −52.2 kNm

3.4. Determine the reactions at A, B and C of the continuous beam shown in figure below. Adopt Kani's method of analysis. The supports A, B and C are rigid and at the same level.

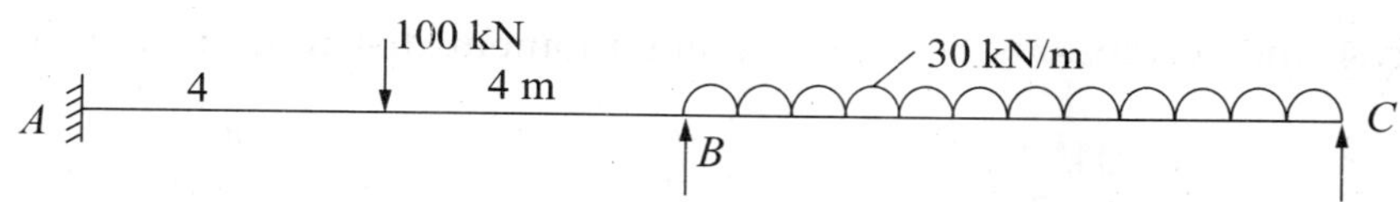

Ans: M_A = −8.83 kNm, M_B = −12.33 kNm, M_C = 0
R_A = 45 kN, R_B = 165.5 kN, R_C = 69.5 kN

3.5. Sketch the bending moment diagram for the rigid frame shown in the figure below using Kani's moment distribution method.

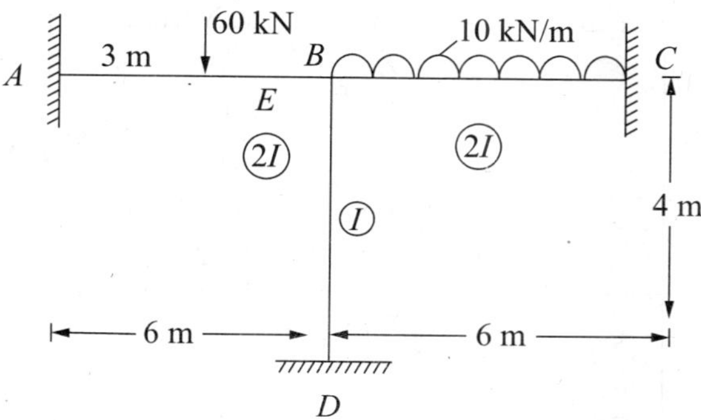

Ans: M_{AB} = −47.7 kNm, M_{BA} = +39.6 kNm, M_{BC} = −35.4 kNm, M_{CB} = +27.3 kNm,
M_{BD} = −4.2 kNm, M_{DB} = −2.1 kNm

3.6. Analyse the rigid frame by Kani's method.

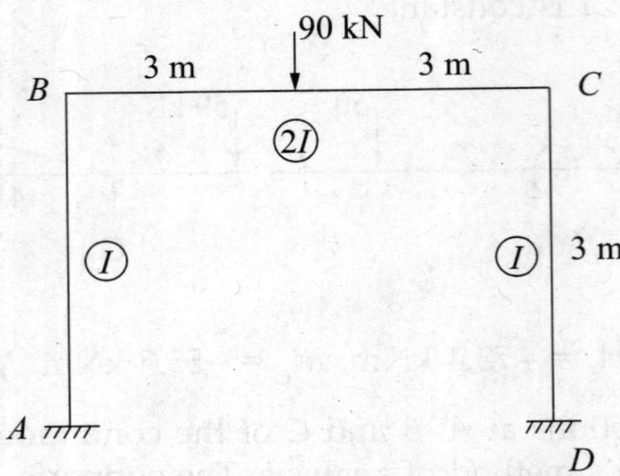

Ans: M_A = −22.5 kNm, M_B = −45.0, M_C = +45.0, M_D = −22.5 kNm

3.7. Analyse the continuous beam by Kani's moment distribution method.

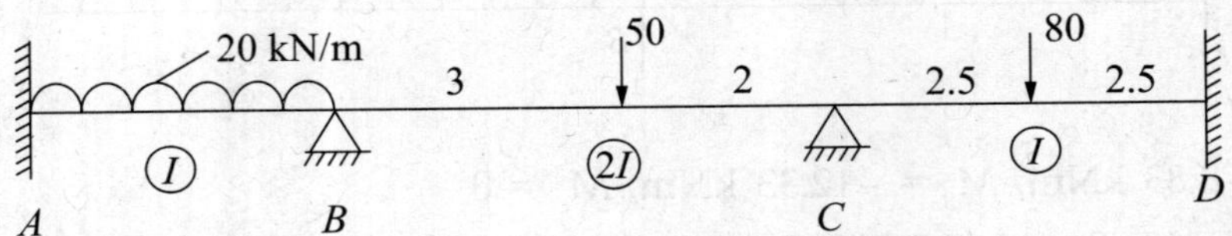

Ans: M_A = −66.88 kNm, M_B = −46.22, M_C = −39.78, M_D = −55.11 kNm

3.8. Analyse the given frame by Kani's method.

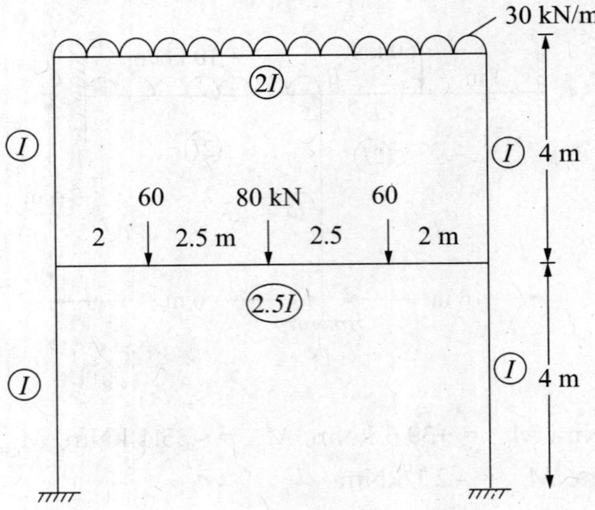

Ans: M_{AB} = 24.18 kNm, M_{BA} = 48.36, M_{BE} = −159.15,
M_{BC} = 110.77, M_{CB} = 149.0, M_{CD} = −149.0

3.9. A continuous beam of uniform section *ABCD* is supported and loaded as shown in figure below. If the support *B* sinks by 10 mm, determine the resultant moments and reactions at the supports. Assume $I = 6(10)^7$ mm^4; $E = 2(10)^5$ N/mm^2.

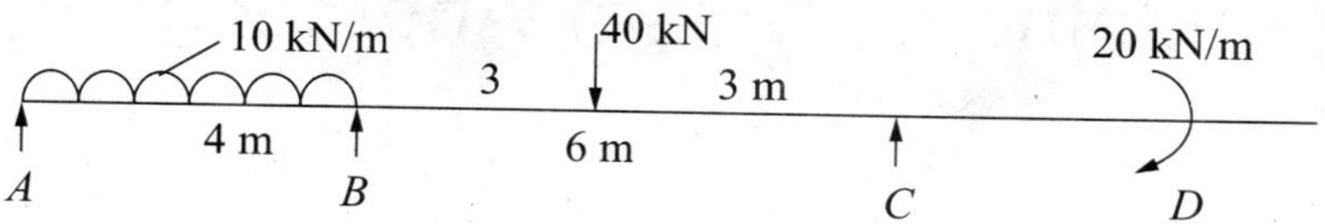

Ans: $M_B = -14.0$ kNm, $R_A = 16.5$ kN, $R_B = 42.5$ kN, $R_C = 21.0$ kN

3.10. Analyse the portal frame shown in figure below by Kani's method.

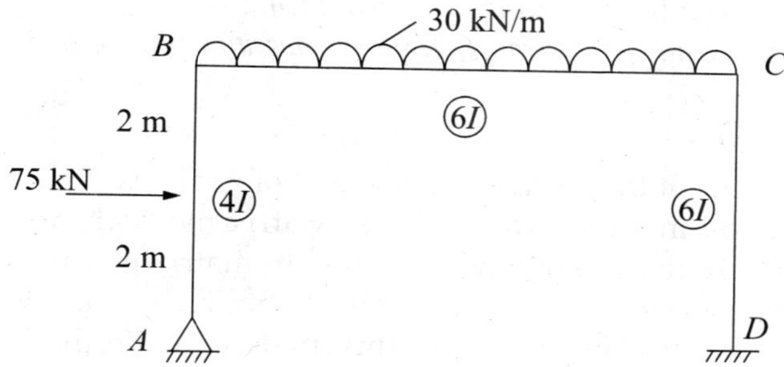

Ans: $M_B = -47.3$ kNm, $M_C = -101.9$ kNm, $M_D = -95.5$ kNm.

4 Flexibility Method: System Approach

Objectives:
To understand the relation between the flexibility and stiffness approach. Define flexibility coefficients. To explain the concept of system approach. Development of governing equations – settlement of supports – choice of redundants, flexibility analysis of propped cantilever beams, fixed beams, continuous beams, bent frame, rigid frames and pin jointed plane trusses.

4.1 INTRODUCTION

The earlier chapters dealt with the classical method of analysis. The advent of computers has changed the outlook in structural analysis. Matrix methods are ideally suitable for complex problems where the unknowns are large. In matrix methods, analysis is carried out by expressing the forces and displacements in matrix form. The displacements are expressed in terms of forces through flexibility matrix, while the forces are expressed in terms of displacement through stiffness matrix. Matrix formats facilitate writing of computer codes and commercially available software packages are based on matrix methods of analysis.

The dual nature of force-displacement relationship gives way to:

- Flexibility or force approach
- Stiffness or displacement approach

Consider an axially loaded bar shown in Fig. 4.1.

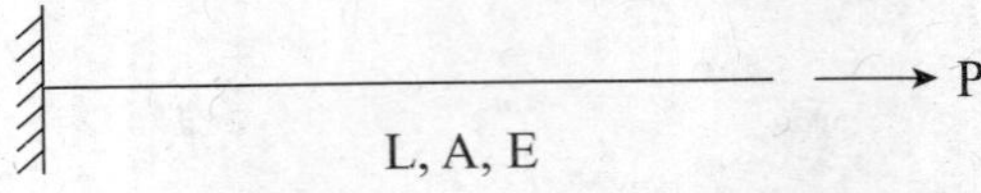

FIG. 4.1 Axially loaded bar

The elongation at the end is given by

$$\delta = \frac{PL}{AE} \qquad (1)$$

where

δ = elongation of the bar
P = load at the end

L = length of the bar

A = area of cross section of the bar

E = modulus of elasticity of the material

In the above equation if P = 1 (unity), then the elongation of the bar is $[L/AE]$, which is known as the flexibility of the element. *Flexibility is the deformation at a point due to a unit load at the same or different point.* Equation (1) can also be written as

$$\delta = fP \tag{2}$$

where f = flexibility of the bar

In general structural system with more joints, loads and displacements can be expressed as

$$\{\delta\} = \{f\}\{P\} \tag{3}$$

where

$\{\delta\}$ = displacement vector

$\{f\}$ = flexibility matrix for the structure

$\{P\}$ = force vector

Equation (1) can also be written as

$$P = [AE/L]\delta \tag{4}$$

In the above equation if δ =1 (unity), then the force in the bar is $[AE/L]$, which is known as the stiffness of the element. Thus, the stiffness can be defined as:

Stiffness is the force developed at a point due to a unit deformation at the same or at a different point.

Equation (4) can also be written as

$$P = k\delta \tag{5}$$

where k is the stiffness of the bar.

In general structural system with more joints, loads and displacements, Eq. (5) can be expressed as

$$\{P\} = [k]\{\delta\} \tag{6}$$

where $[k]$ = stiffness matrix of the structure.

It can be seen from the above discussion that the flexibility and stiffness are inverse of each other. This is proved in the next section.

4.2 RELATION BETWEEN FLEXIBILITY AND STIFFNESS MATRICES

Consider the basic equations of flexibility and stiffness approaches,

$$\{\delta\} = [f]\{P\} \tag{7}$$

$$\{P\} = [k]\{\delta\} \tag{8}$$

From the above equations;

$$\{\delta\} = [k]^{-1}\,\{P\} \tag{9}$$

Comparing Eqs. (3) and (9)

$$[f] = [k]^{-1} \tag{10}$$

and
$$[k] = [f]^{-1} \tag{11}$$

Equations (10) and (11) indicate that flexibility and stiffness matrices are inverse of each other. This statement requires the same set of coordinates to be used in flexibility and stiffness formulations.

4.3 FLEXIBILITY COEFFICIENTS

A flexibility coefficient represents a displacement at a point caused by a unit force or unit moment applied at the same or different point. Units – m/kN or mm/N.

For example, f_{11} is the flexibility coefficient representing a displacement at point 1 due to unit force at the same point 1.

As a second example, f_{23} is the flexibility coefficient representing a displacement at point 2 due to unit force at 3.

In general, f_{ij} is the flexibility coefficient representing a displacement at point i due to unit force at point j.

4.4 FLEXIBILITY MATRICES ARE SYMMETRICAL

Consider the external work done equation expressed in

$$W_E = \left(\frac{1}{2}\right)[P]^T\,\{\delta\} \tag{12}$$

But
$$\delta = [f]\{P\} \tag{13}$$

Substituting the value of δ in W_E

$$W_E = \left(\frac{1}{2}\right)[P]^T\,\{f\}\{P\} \tag{14}$$

Similarly, we know

$$W_E = \left[\frac{1}{2}\right][\delta]^T\,\{P\} \tag{15}$$

But

$$[\delta]^T = [P]^T\,\{f\}^T \tag{16}$$

Substituting (16) in (15)

$$W_E = \left[\frac{1}{2}\right][P]^T \{f\}^T \{P\} \tag{17}$$

Comparing Eqs. (14) and (17)

$$[f] = [f]^T \tag{18}$$

The above equation indicates that flexibility matrices are symmetrical.

4.5 FLEXIBILITY METHODS IN ANALYSIS OF STRUCTURES

In this method, the analysis is carried out by using

$$\{\delta\} = [f]\{P\} \tag{19}$$

with usual notations.

This method considers the redundant (statically indeterminate) forces as unknown and is also called the force method. Using the compatibility of displacements simultaneous equations are formulated and the unknown redundant forces are determined. Flexibility methods are not suitable for general-purpose computer coding since the redundant forces (internal moments or external reactions) can be chosen arbitrarily and not automatically.

Flexibility methods can be carried out using

- System approach
- Element approach

4.6 SYSTEM APPROACH

Statically indeterminate structures can be analysed by writing the compatibility equations in terms of flexibility coefficients and selected redundants. This method involves the calculations of displacements due to external loads and the selected redundants in the released structure. Released structure or statically determinate structure is obtained by releasing the selected redundants and shall be stable. The analyst has the choice to select the redundants.

4.6.1 Development of Equations

Consider the two-span continuous beam shown in Fig. 4.2. The beam has a static indeterminacy of 2 which can be two external vertical reactions at B and C or external moment at A and internal moment at B. Figure 4.2 shows two adjacent spans AB and BC of lengths l_1 and l_2 respectively of a continuous beam with I_1 and I_2 beginning the corresponding moment of inertia.

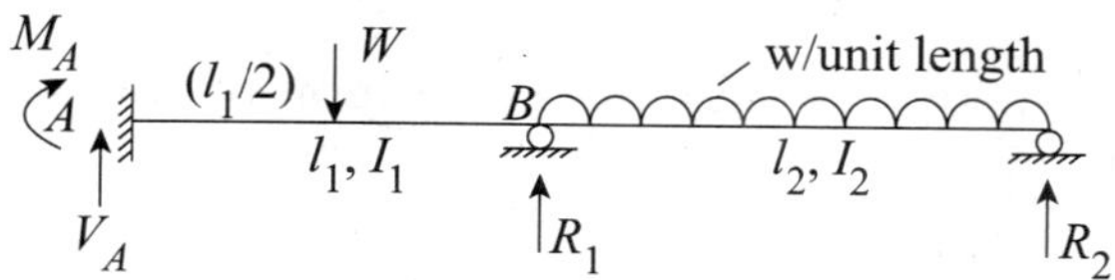

FIG. 4.2 Two-span continuous beam

The selected redundants are the two vertical reactions R_1 and R_2. The released structure is a cantilever beam with all the external loads but without the reactions R_1 and R_2 (Fig. 4.3). Let δ_{L1} and δ_{L2} be the displacements at B and$C$ due to external loads (where the redundants are acting). The displacements are positive if they are in the same direction of the redundants assumed, otherwise negative. The cantilever released structure is stable.

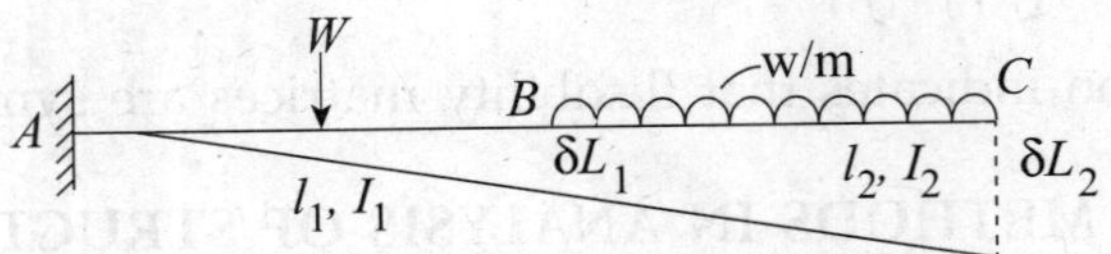

FIG. 4.3 Released structure with external loads

The flexibility coefficients are calculated in the released structure by applying a unit force at B and C now called points 1 and 2. The unit forces are applied one at a time and in the direction of the redundants assumed. The flexibility coefficients are indicated in Figs. 4.4 and 4.5. These coefficients are calculated using the moment area methods in case of beams and the unit load method in case of trusses and rigid frames.

In Fig. 4.4, f_{11} and f_{21} are the flexibility coefficients at points 1 and 2 due to unit load at point 1, while in Fig. 4.5, f_{12} and f_{22} are the flexibility coefficients at points 1 and 2 due to unit load at point 2.

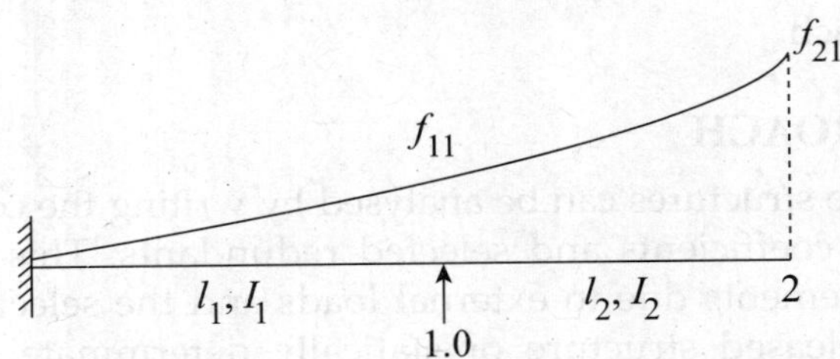

FIG. 4.4 Unit load at 1

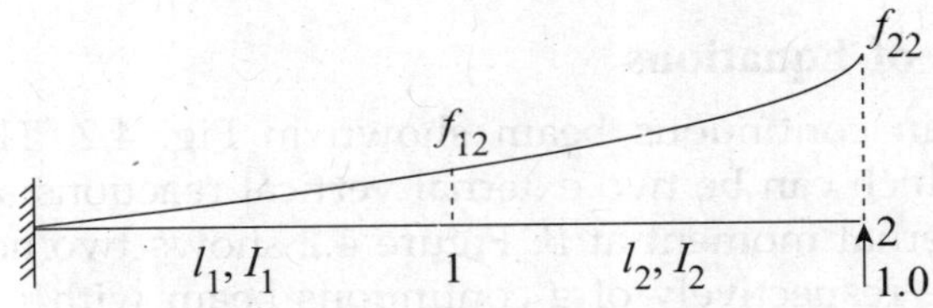

FIG. 4.5 Unit load at 2

Using the principle of superposition, the condition of compatibility at joints B and C can be expressed as

$$\delta_{L1} + f_{11}R_1 + f_{12}R_2 = 0 \tag{20}$$

$$\delta_{L2} + f_{21}R_1 + f_{22}R_2 = 0 \tag{21}$$

The above equations can also be expressed in matrix form and solution for R_1 and R_2 can be obtained.

The above force method or flexibility method or compatibility method was developed by James Clerk Maxwell in 1864. Later Otto Mohr and Muller-Breslau refined the above force method. Essentially, principle of superposition, force displacement, compatibility and equilibirium equations form the basis of the force method.

This method can be conveniently used for indeterminate beams, frames, arches, trusses and trussed beams. It is to be noted that the force or flexibility method can be used to analyse the indeterminate truss up to second degree. The hand calculation is tedious if the degree of redundancy is greater than 2. In such cases, commercial software package can be used.

4.6.2 Settlement of Supports

Sinking of supports and rotation of fixed supports are converted into equivalent loads by calculating the fixed end moments (reactions) due to sinking and rotation and reversing the sign of moments (action).

Consider a beam with settlements and rotations as shown below.

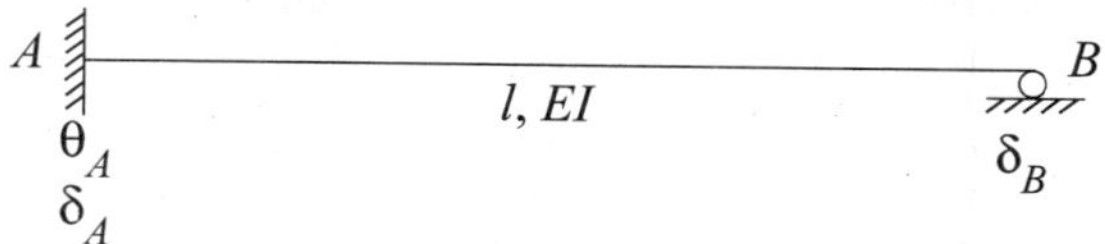

FIG. 4.6 Beam with rotation and settlement at supports

In the above propped cantilever beam, let θ_A be the rotation at A, say clockwise at A. δ_A is the settlement at support A.

Let δ_B be the settlement at support B and $\delta_B > \delta_A$

Relative displacement between A and B =

$$\delta = \delta_B - \delta_A$$

Fixed end moments at A and B = $6EI\delta/l^2$ both anticlockwise are indicated in Fig. 4.7. The moments are clockwise if $\delta_A > \delta_B$. These moments are due to relative displacements between A and B.

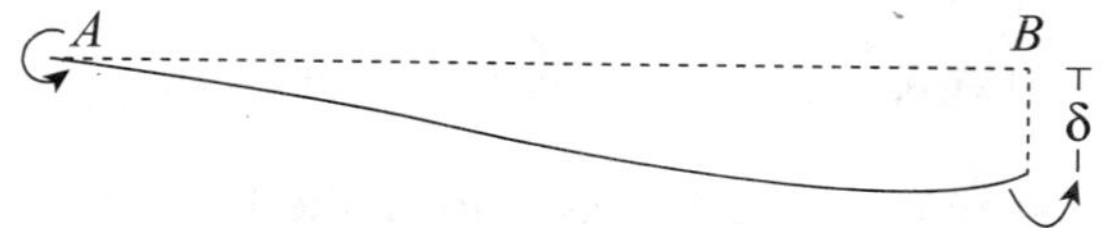

FIG. 4.7 Beam with relative displacement at the supports

Due to the rotation at A; the fixed end moments developed at A and B will be $4EI\theta_A/l$ and $2EI\theta_A/l$. The nature of moments at A and B are both clockwise as shown in Fig. 4.8.

If the rotation at A is anticlockwise,

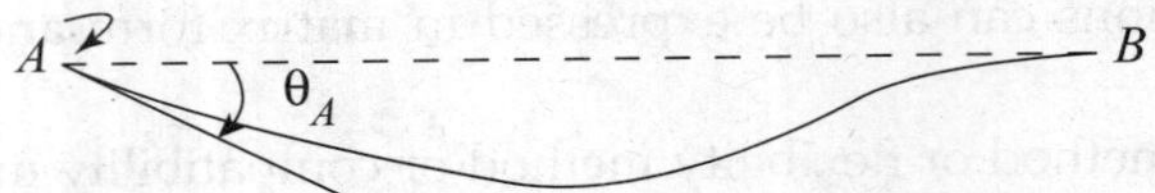

FIG. 4.8 Beam with clockwise rotations at support A

then the fixed end moment will also be anticlockwise moments. The above moments are reversed and applied at the joints to determine the displacements along with external loads. The other steps are similar as in the case of non-settlement problems.

4.6.3 Choice of Redundants

The choice of the redundants shall be such that the calculation of displacements will be simple. For example, in Fig. 4.2, the structure was made determinate by releasing the reactions R_1 and R_2. However, the calculations for estimating the displacements are tedious, if the two spans have different moment of inertia. Again, for the same problem in Fig. 4.2, the structure can be made determinate by releasing the external moment at A and vertical reaction at B, so that the beam is simply supported at A and C as shown in Fig. 4.9. The calculation for estimation of deflections is tedious, if the spans have uniform moment of inertia or different moment of inertia.

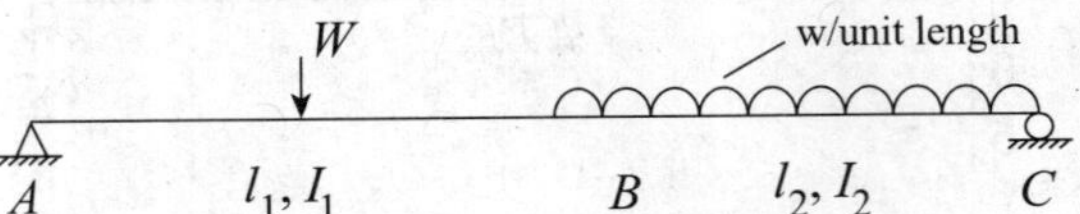

FIG. 4.9 Released structure with external loads (Simply supported at A and C)

The outcome of the above discussion is as follows: The simplest way to get the released or statically determinate structure will be to release the external moment at A and the internal moment at B so that the released structure will be two simply supported beams AB and BC as shown in Fig. 4.10. The calculations involving the displacements will be relatively simple as it involves the rotations at A and B is the direction of the redundants.

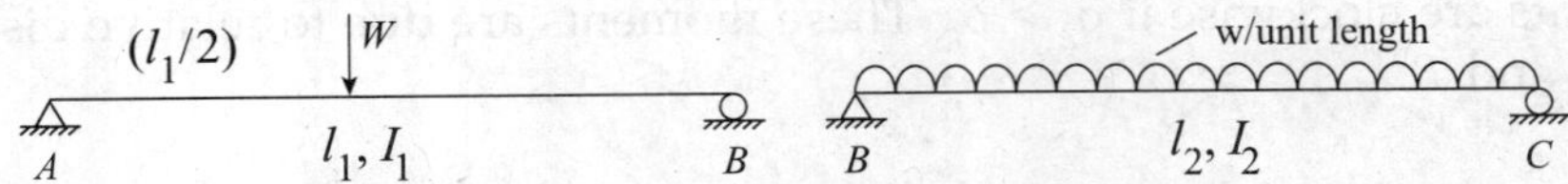

FIG. 4.10 Released structure with external loads (Two simply supported beams)

In rigid frames, the supports are released to make the structure determinate, while in pin jointed frames, either the excess members or the supports are released depending on the internal and external redundancies.

4.6.4 Basic Concepts in Flexibility Method

- A suitable number of releases are made to make statically indeterminate skeletal structure to a determinate structure.
- The number of releases required is equal to statical indeterminacy, i.e., number of redundants equal to the number of reaction components minus number of equilibrium equations.
- The insertion of releases cause displacement discontinuities at the location of releases due to the applied loads.
- To restore the continuity, unknown biactions (forces and moments) are applied at these releases.
- The compatibility equations in terms of displacements, (a set of linear simultaneous equations) are formed. The solution of these equations give the values of biactions redundants.
- After computing the redundants biactions, the other reactions are obtained using the free body diagram of the members and the equilibrium equations.
- The required displacements can be found out by displacement calculations.

4.7 STANDARD FORMULAE FOR CALCULATING ROTATIONS IN SIMPLY SUPPORTED BEAMS

The calculations of rotations in simply supported beams for standard loadings are given below.

(a) Simply supported beams with udl throughout

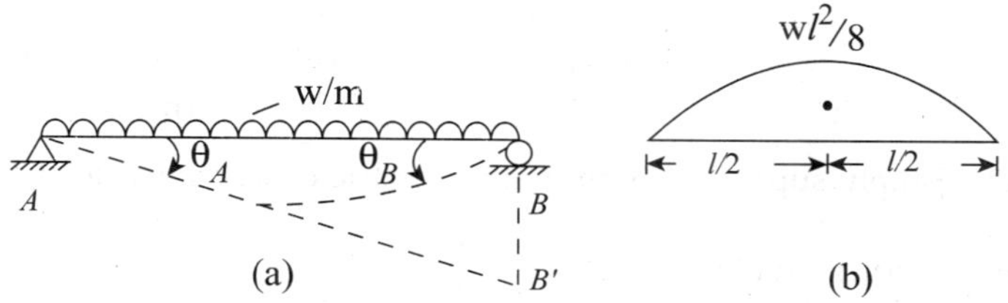

FIG. 4.11 Simply supported beam with udl and BMD

Using the moment area method;

$$\theta_A = \frac{BB'}{l} = \text{Moment of BMD between } A \text{ and } B \text{ about } B/EIl$$

$$\theta_A = \left(\frac{2l}{3}\right)\left(\frac{wl^2}{8}\right)\left(\frac{l}{2}\right)\left(\frac{1}{EI}\right)\left(\frac{1}{l}\right) = \frac{wl^3}{24EI} = \theta_B$$

The value of θ_B is equal to θ_A due to symmetry.

(b) Simply supported between beam with central concentrated load

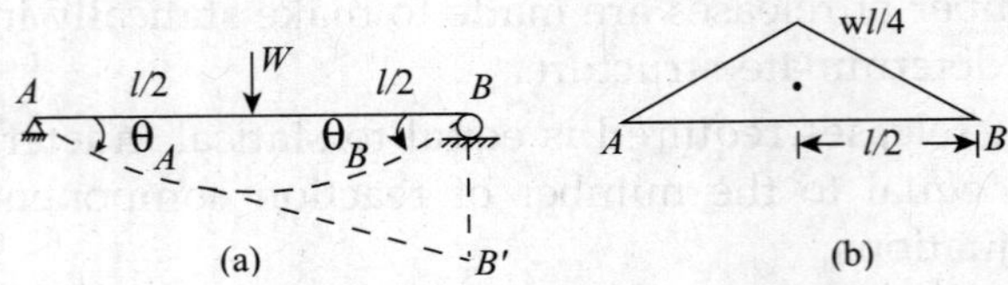

FIG. 4.12 Simply supported beam with central concentrated load and BMD

Using the moment area method,

$$\theta_A = \frac{BB'}{l}$$

$$= \frac{\text{Moment of BMD between A and B about B}}{1EI}$$

$$\theta_A = \left(\frac{1}{2}\right)(l)\,(wl/4)\left(\frac{l}{2}\right)/lEI = Wl^2/16EI$$

Due to symmetry; $\theta_A = \theta_B$

(c) Simply supported beam with eccentric concentrated load

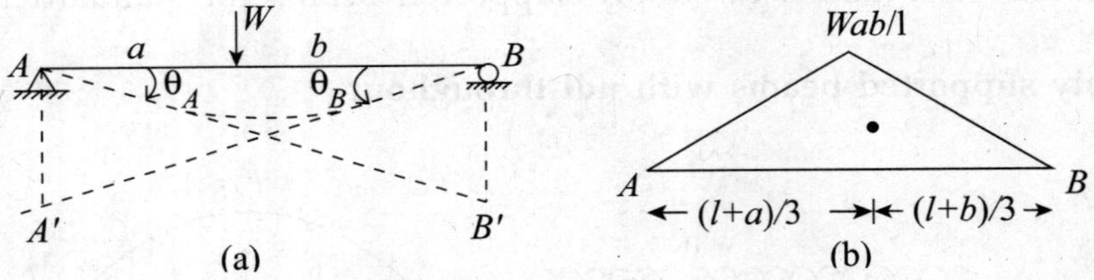

FIG. 4.13 Simply supported beam with eccentric concentrated load and BMD

Using the moment area method;

$$\theta_A = \frac{BB'}{l}$$

$$= \left(\frac{1}{2}\right)(l)\left(\frac{Wab}{l}\right)\left(\frac{l+b}{3}\right)/EIl$$

$$\boxed{\theta_A = \frac{Wab(l+b)}{6EIl}}$$

Similarly,

$$\theta_B = \frac{AA'}{l}$$

$$\theta_B = \frac{\text{Moment of the BMD between A and B about A}}{EIl}$$

$$\theta_B = \left(\frac{1}{2}\right)(l)\left(\frac{Wab}{l}\right)\left(\frac{l+a}{3}\right)/EIl$$

$$\boxed{\theta_B = \frac{Wab(l+a)}{6EIl}}$$

(d) Simply supported beam with end moment

This type of loading occurs in overhang continuous beams and is also required for calculating the flexibility coefficients.

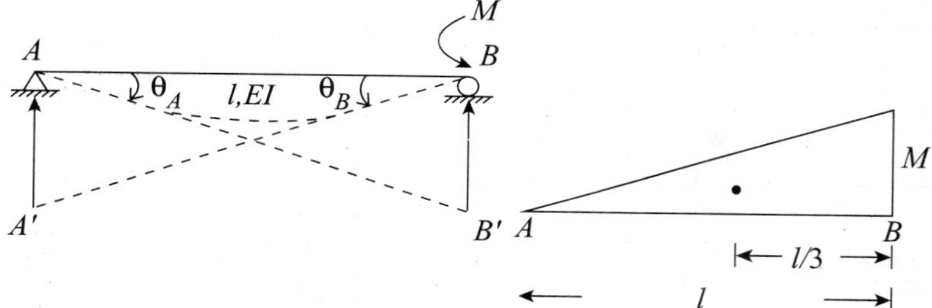

FIG. 4.14 Simply supported beam with end moment at B and BMD

Using the moment area method

$$\theta_A = \frac{BB'}{l}$$

$$\theta_A = \left(\frac{1}{2}\right)(M)(l)\left(\frac{l}{3}\right)/EIl$$

$$\boxed{\theta_A = \frac{Ml}{6EI}}$$

Similarly,

$$\theta_B = \frac{AA'}{l}$$

$$\theta_B = \left(\frac{1}{2}\right)(M)(l)\left(\frac{2l}{3}\right)/EIl$$

$$\boxed{\theta_B = \frac{Ml}{3EI}}$$

If the applied moment is unity, i.e., $M = 1$; the flexibility coefficients at A and B are as follows:

$$\text{Flexibility coefficients at } A = l/6EI$$

$$\text{Flexibility coefficients at } B = l/3EI$$

The rotations for other loadings can be obtained by using moment area method.

Table 4.1 Slopes and deflection of simple beams

	Beam and Loading	Slope at ends	Maximum deflection
(1)	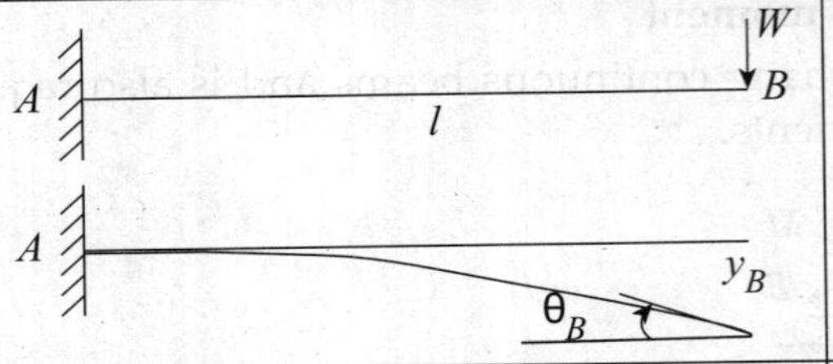 Cantilever AB, length l, point load W at B	$\theta_A = 0$ $\theta_B = Wl^2/2EI$	$y_B = Wl^3/3EI$
(2)	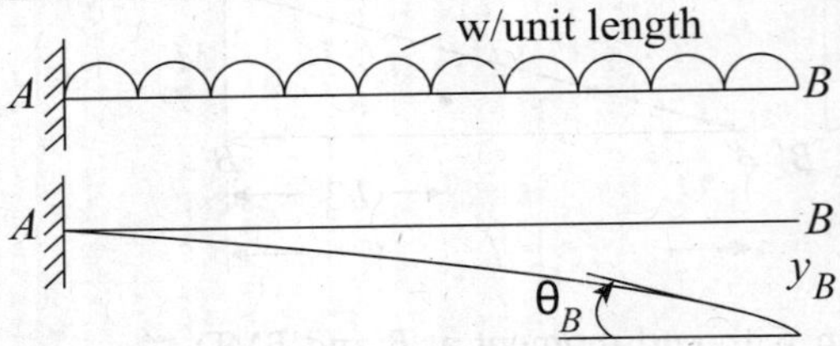 Cantilever AB, w/unit length	$\theta_A = 0$ $\theta_B = Wl^3/6EI$	$y_B = Wl^4/8EI$
(3)	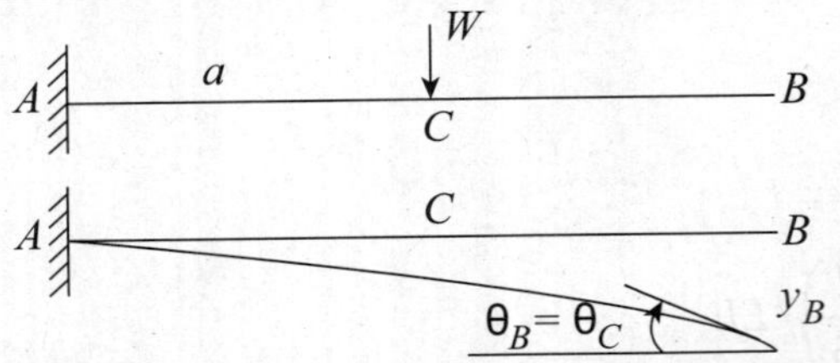 Cantilever AB, point load W at C, distance a	$\theta_A = 0$ $\theta_C = \theta_B = Wa^2/2EI$	$y_B = \dfrac{Wa^2}{6EI}(3l - a)$
(4)	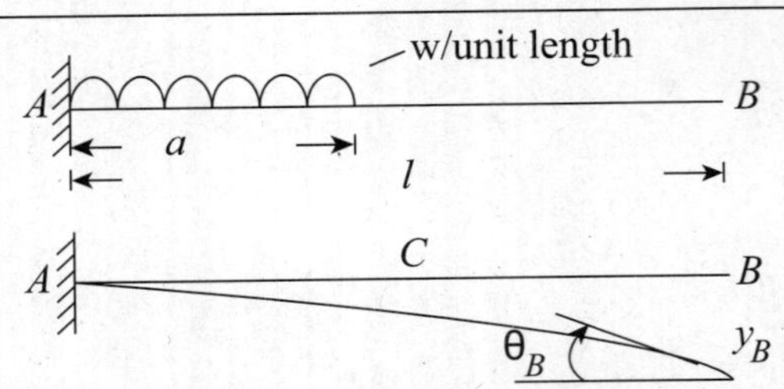 Cantilever AB, length l, w/unit length over a	$\theta_A = 0$ $\theta_C = \theta_B = \dfrac{Wa^3}{6EI}$	$y_B = \dfrac{Wa^3}{24EI}(4l - a)$
(5)	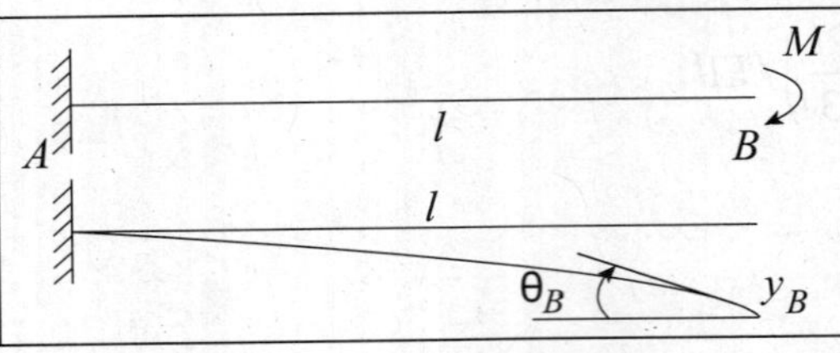 Cantilever AB, length l, moment M at B	$\theta_A = 0$ $\theta_B = \dfrac{Ml}{EI}$	$y_B = \dfrac{Ml^2}{2EI}$

	Beam and Loading	Slope at ends	Maximum deflection
(6)	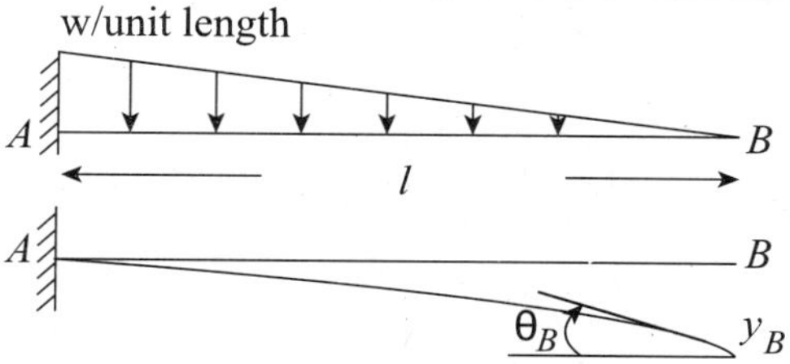 w/unit length; A, B, l; θ_B, y_B	$\theta_A = 0$ $\theta_B = \dfrac{Wl^3}{24EI}$	$y_B = \dfrac{Wl^4}{30EI}$
(7)	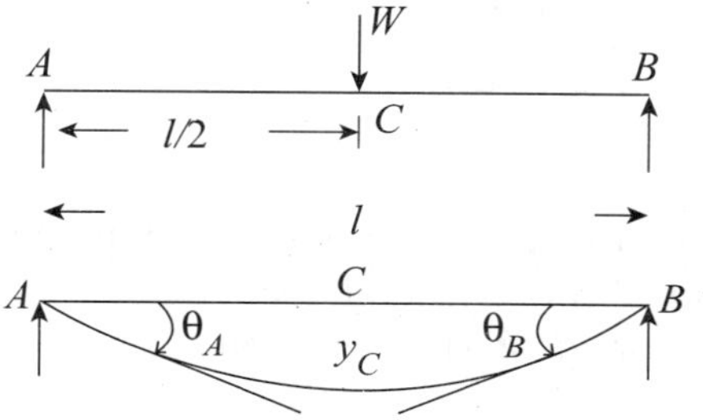 W; A, B, C; $l/2$; l; θ_A, y_C, θ_B	$\theta_A = -\theta_B = \dfrac{Wl^2}{16EI}$	$y_C = \dfrac{Wl^3}{48EI}$
(8)	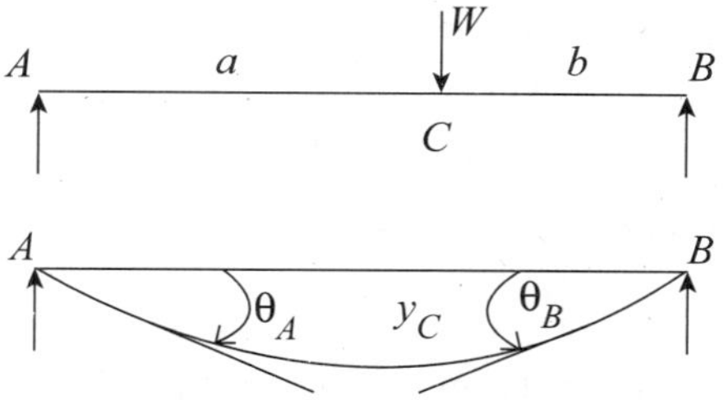 W; A, B, C; a, b; θ_A, y_C, θ_B	$\theta_A = \dfrac{Wab(l+b)}{6EIl}$ $\theta_B = \dfrac{Wab(l+a)}{6EIl}$	$y_{max} = \dfrac{Wa^{3/2}(2l-a)^{3/2}(l-a)}{9\sqrt{3}\,EIl}$ at $x = [a(2l - 3)/3]^{1/2}$ $y_C = \dfrac{Wa^2b^2}{3EIl}$
(9)	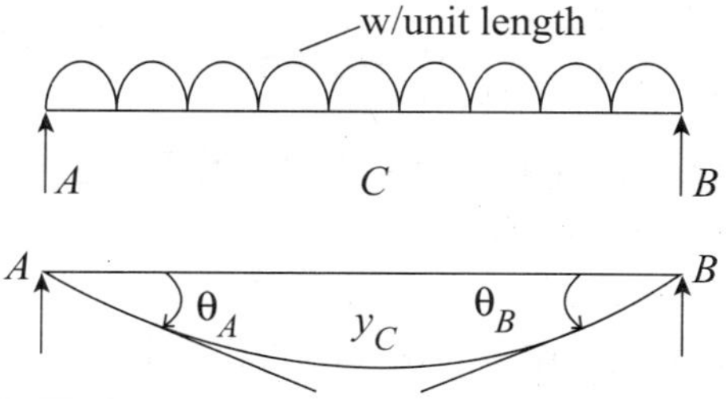 w/unit length; A, B, C; θ_A, y_C, θ_B	$\theta_A = -\theta_B = \dfrac{Wl^3}{24EI}$	$y_C = \dfrac{5}{384}\dfrac{Wl^4}{EI}$
(10)	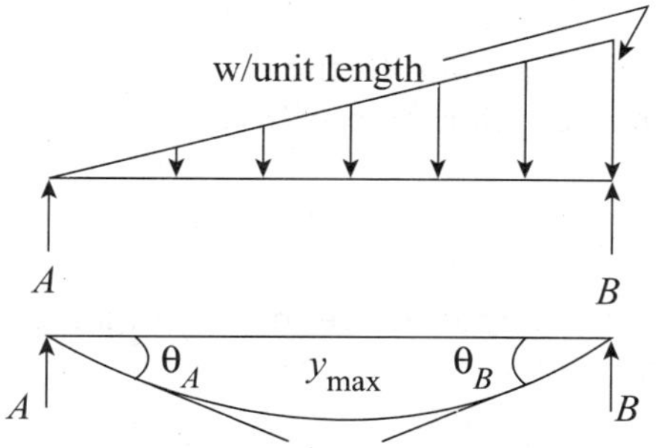 w/unit length; A, B; θ_A, y_{max}, θ_B	$\theta_A = \dfrac{7Wl^3}{360EI}$ $\theta_B = -\dfrac{Wl^3}{45EI}$	$y_{max} = 0.006522\,\dfrac{Wl^4}{EI}$ at $x = 0.519l$

Contd...

	Beam and Loading	Slope at ends	Maximum deflection
(11)	w/unit length; A, B, C, $l/2$; A, B, θ_A, y_C, θ_B	$\theta_A = -\theta_B = \dfrac{5Wl^3}{192EI}$	$y_C = \dfrac{Wl^4}{120EI}$

4.8 NUMERICAL EXAMPLES

Example 4.1 Analyse the propped cantilever of span 8 m subjected to uniformly distributed load of 30 kN/m throughout and a concentrated load of 300 kN at the centre. Sketch the bending moment diagram and shear force diagram. The beam is of uniform flexural rigidity.

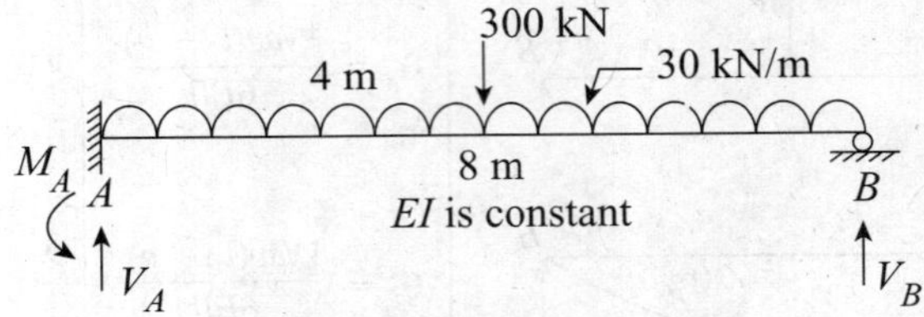

FIG. 4.15 Propped cantilever beam

Solution: The unknown external reactions are V_A, V_B and M_A. Hence, the unknown are three and the equilibrium equations are 2.

Note: In beams subjected to transverse loads (vertical loads) only $\Sigma V = \Sigma M = 0$ is valid and $\Sigma H = 0$ is of no consequence.

Hence, the static redundancy (redundancy)

= Number of unknown reactions – Available equations of equilibrium

= 3 – 2 = 1

Thus, the propped cantilever is statically indeterminate to the first degree. This can be made determinate by removing any one of the external reactions, i.e., M_A, V_A and V_B.

First Method: Releasing the moment M_A

The propped cantilever beam becomes a simply supported beam by remaining fixity at *A*. Figure 4.16 shows the released structure.

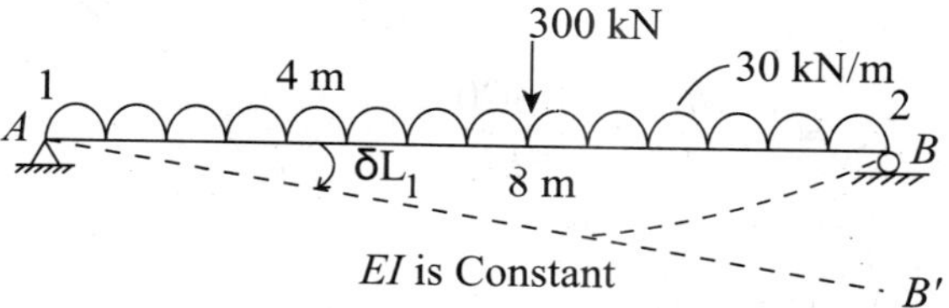

FIG. 4.16 Released structure with external loads

The redundant chosen is M_A and hence rotation at A is only required.

Let BB' is the tangential deviation at B and is computed using the moment area method.

$$BB' = \frac{\text{Moment of BMD between } A \text{ and } B \text{ about } B}{EI}$$

$$\delta L_1 = \frac{BB'}{l}$$

The rotation δL_1 is negative as the direction of M_A is anticlockwise, while the rotation of A with respect to AB is clockwise. The value of BB' is obtained using the bending moment diagrams.

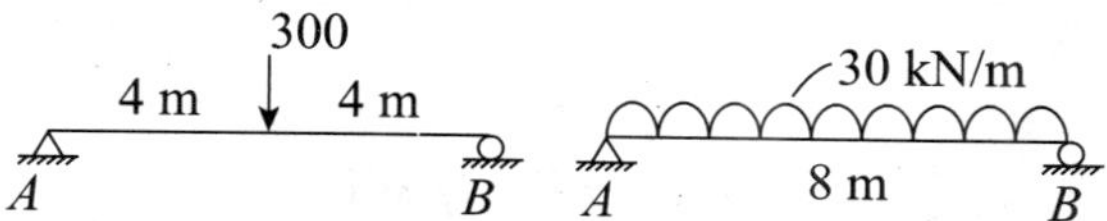

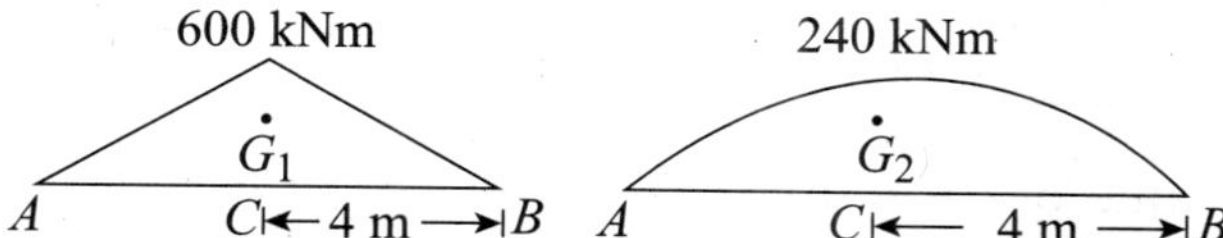

FIG. 4.17 BMD for each load of the released structure

$$\begin{aligned} BB' &= \text{area of } M/EI \text{ of the above diagrams} \times \text{centroidal distance from } B \\ &= \text{area of } \Delta \text{ BMD} \times BG_1 + \text{area of BMD} \times BG_2 \\ &= \frac{1}{2} \times \text{base} \times \text{height} \times BG_1 + \frac{2}{3} \times \text{base} \times \text{height} \times BG_2 \\ &= \left(\frac{1}{2} \times 8 \times 600 \times 4 + \frac{2}{3} \times 8 \times 240 \times 4\right) / EI \\ &= 14920/EI \end{aligned}$$

Hence,

$$\delta L_1 = -\frac{BB'}{l} = \frac{14920}{8EI}$$

$$\boxed{\delta L_1 = -1840/EI}$$

Flexibility Coefficients

Since the redundancy is one, only one flexibility coefficient f_{11} is required. A unit moment at 1 is applied and the bending moment diagram is as shown in Fig. 4.18.

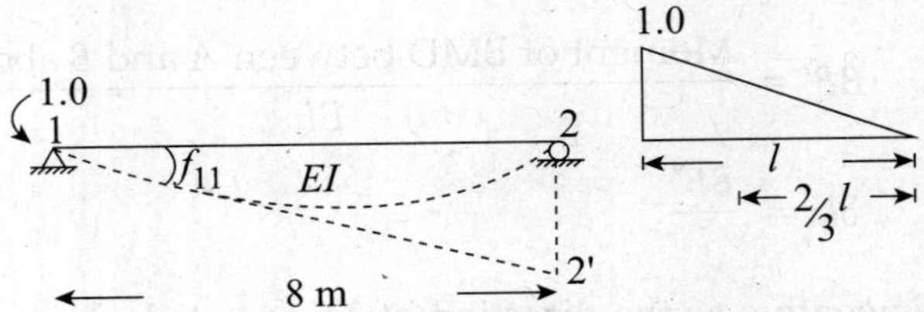

FIG. 4.18 Unit moment at 1 and bending moment diagram

$$f_{11} = \frac{2-2'}{l} = \frac{\text{(Moment of the BMD between 1 and 2 about 2)}}{(EI/l)}$$

$$f_{11} = \frac{1}{2}l(1.0)\left(\frac{2}{3}l\right)/EIl = l/3EI$$

$$\boxed{f_{11} = \frac{l}{3EI}}$$

Substituting the value of $l = 8$ m;

$$\boxed{f_{11} = 8/3EI}$$

As there are no external displacements at A;

$$\delta L_1 + f_{11}M_A = 0$$

substituting for δL_1 and f_{11}; we have

$$-1840/EI + (8/3EI)M_A = 0$$

i.e. $$\boxed{M_A = 690 \text{ kNm}}$$

The assumed direction is correct and the bending moment is hogging knowing the value of M_A, the other reactions V_A and V_B are found out using equilibrium equations. Referring to Fig. 4.11.

$\Sigma M_A = 0$

$$-690 + 300(4) + 30(8)^2/2 - 8V_B = 0$$

$$\boxed{V_B = 183.75 \text{ kN}}$$

$\Sigma V = 0$

$$V_A + V_B = 300 + 30(8)$$

Substituting the value of V_B

$$\boxed{V_A = 356.25 \text{ kN}}$$

2nd Method: Releasing the vertical reaction, V_A

The released structure is a beam supported as shown in Fig. 4.19 with a gliding support (guide support) at A.

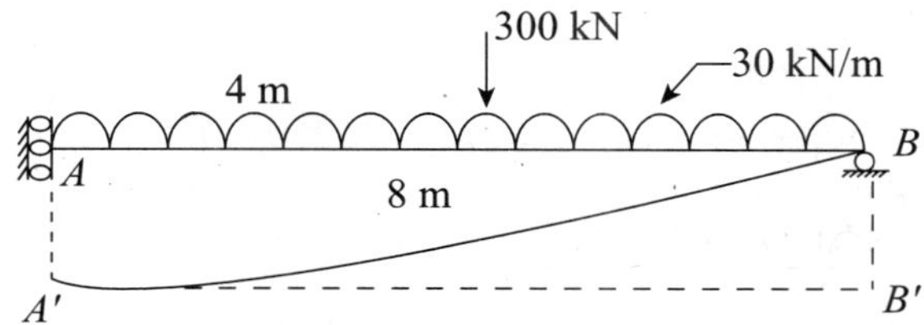

FIG. 4.19 Released structure with external loads

Bending moment diagram for two external loads and the vertical reaction VB are drawn separately for the released structure as shown in Fig. 4.20.

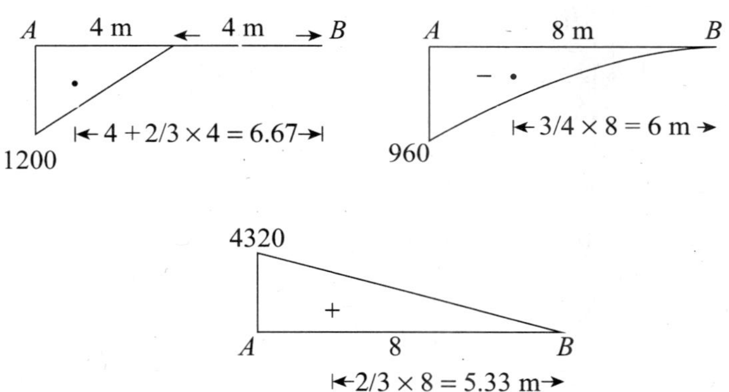

FIG. 4.20 BMD for each load in released structure

The analysis of the released structure is as follows. As 'A' is a gliding support; the entire vertical load is borne by the reaction R_B. Hence,

$$R_B = 30(8) + 300$$

$$\boxed{R_B = 540 \text{ kN}}$$

Taking moment about A and assuming in the clockwise direction.

$$M_A + 300(4) + 30(8)^2/2 - 540(8) = 0$$

$$\boxed{M_A = 2160 \text{ kNm}}$$

and

$$\delta L_1 = BB'/EI$$

$$\delta L_1 = -\left\{\frac{1}{2} \times 8 \times 4320 \times \frac{2}{3} \times 8 - \frac{1}{2}(4)(1200)6.67 - \left(\frac{1}{3}\right)(8)(960)(6)\right\}/EI$$

$$\delta L_1 = -92160 - 16008 - 15360/EI$$

$$\boxed{\delta L1 \simeq -60800/EI}$$

Flexibility Coefficients

As the redundancy is one; only one flexibility coefficient is required. A unit force is applied in the direction of V_A.

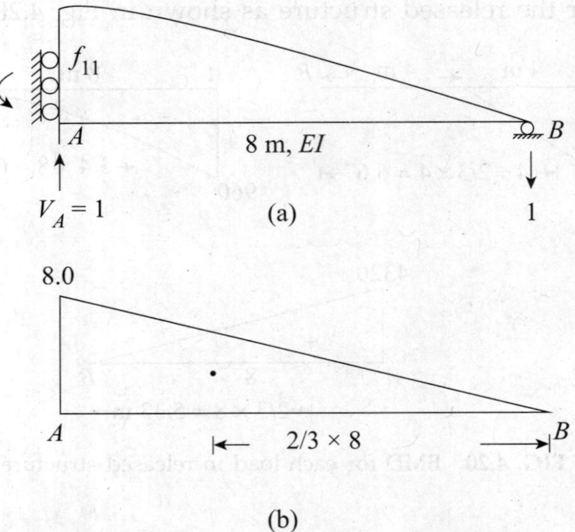

FIG. 4.21 Unit force at A and BMD

Referring to Fig. 4.21;

$$f_{11} = \text{Moment of BMD between } A \text{ and } B \text{ about } B/EI$$

$$f_{11} = \left(\frac{1}{2}\right)8\,(8)\left(\frac{2}{3} \times 8\right)/EI = 512/3EI$$

As there are no external displacements,

$$\delta L_1 + f_{11}V_A = 0$$

$$-60800/EI + (512/3EI)V_A = 0$$

$$\boxed{V_A = 356.25 \text{ kN}}$$

$$\Sigma V = 0,$$

$$V_A + V_B = 300 + 30(8)$$

$$356.25 + V_B = 540$$

$$\boxed{V_B = 183.75 \text{ kN}}$$

In Fig. 4.11

$$\Sigma M_A = 0;$$

$$-M_A + 300(4) + 30 \times \frac{8^2}{2} - 8(193.75) = 0$$

$$\boxed{M_A = 690 \text{ kNm}}$$

3rd Method: Releasing the Vertical Reaction V_B

The released structure is a cantilever beam with the removal of support *B*.

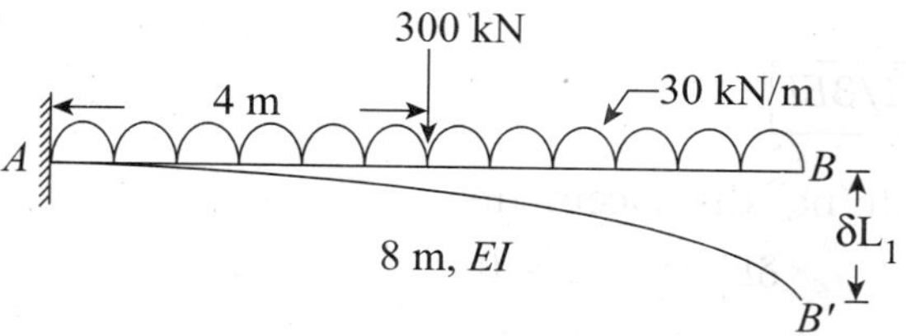

FIG. 4.22 Released structure with external loads

In this method, the redundant chosen is V_B. Hence, the deflection at *B* is required. Since the slope at *A* is zero, the tangent at *A* is horizontal and *BB′* is the tangent deviation at *B* and can be obtained by using the moment area method.

$BB' = $ [Moment of BMD between A and B about B]$/EI$

$\delta L_1 = $ Deflection at B due to external loads

$= BB'$ and this is negative as the direction of the redundant chosen is upwards, while the deflection at B is downwards.

Bending moment diagram for the two loads are drawn separately for the released structure as shown in Fig. 4.23.

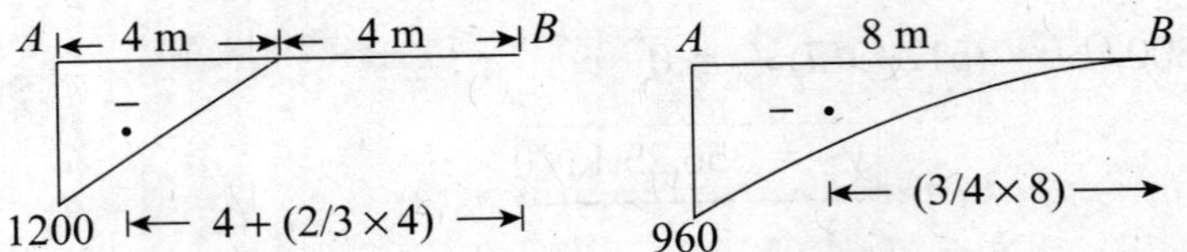

FIG. 4.23 BMD for each load in released structure

Flexibility Coefficients

As the redundancy is 1, only one flexibility coefficient at B is required. A unit force at B is applied and BMD for the same is shown in Fig. 4.24.

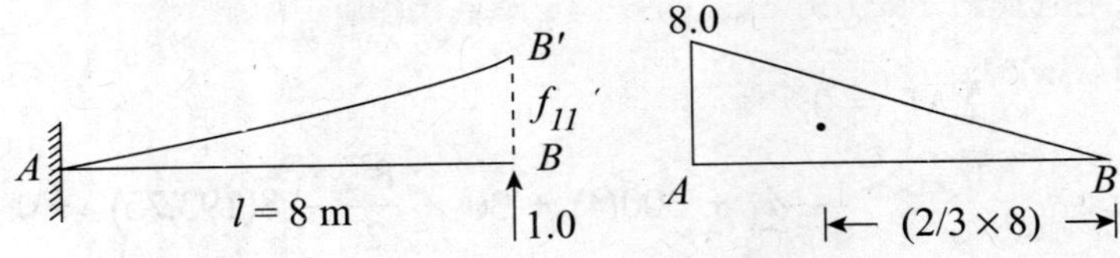

FIG. 4.24 Unit load at B and BMD

$$f_{11} = BB' = \text{Moment of the BMD between } A \text{ and } B \text{ about } B/EI$$

$$f_{11} = \left(\frac{1}{2}\right)(8)(8)\left(\frac{2}{3} \times 8\right)/EI$$

$$\boxed{f_{11} = 512/3EI}$$

As there are no external displacements,

$$\delta L_1 + f_{11}V_B = 0$$

Substituting the obtained values in the above equation,

$$-31360/EI + (512/3EI)V_B = 0$$

$$\boxed{V_B = 183.75 \text{ kN}}$$

The reactions V_A and M_A are found using equilibrium equations as;

$\Sigma V = 0;$

$$V_A + V_B = 300 + 30(8)$$

$$\boxed{V_A = 356.25 \text{ kN}}$$

$\Sigma M = 0$;

Referring to Fig. 4.15;

$$-M_A - 183.75(8) + 300(4) + 30(8)^2/2 = 0$$

$$\boxed{M_A = 690 \text{ kNm}}$$

Final Note

The propped cantilever beam subjected to uniformly distributed load throughout and a central point load was analysed in three different ways. The essential difference in the method of analysis is the choice of redundant. In the first method, the indeterminate moment M_A was chosen as redundant. This method is found to be simple for calculating the displacements. The vertical reaction V_A is taken as redundant in the second method. The calculation of displacements is tedious and quite involved. In the third method, V_B is taken as redundant. This method is simple for calculating the displacements. Hence, all propped cantilever beams can be analysed by the third method while it is preferable to use the first method for the analysis of continuous beams.

Shear Force and Bending Moment Diagram

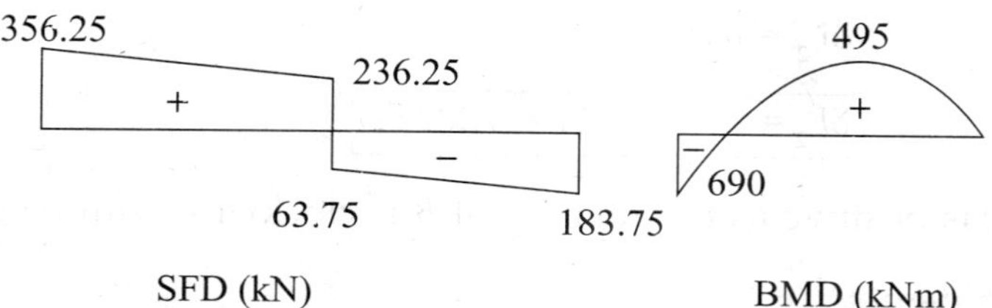

FIG. 4.25 Shear force and bending moment diagram

Example 4.2 Analyse the fixed beam *AB* shown in Fig. 4.26 by the flexibility method, *EI* is constant.

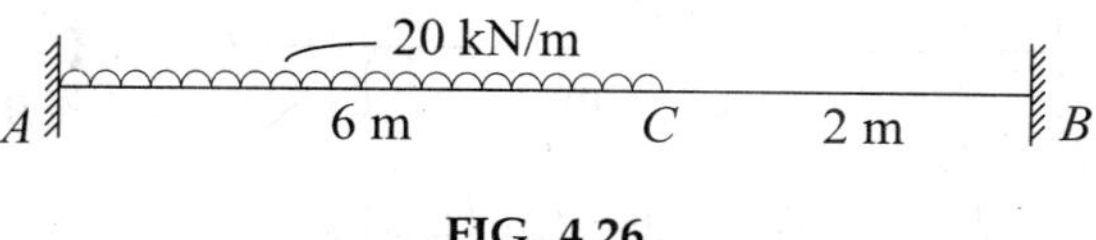

FIG. 4.26

Solution: The fixed beam is statically indeterminate to the second degree. R_B and M_B are taken as redundants and the beam is treated as a cantilever beam with above redundants. R_B is taken as upwards (positive) and M_B is in clockwise direction (positive). The displacements are assumed positive if they are acting in the direction of the redundants.

Released Structure

The removal of VB and MB make the released structure as cantilever beam with fixed end at A. Let δL_1 be the vertical displacement at the free end B and δL_2 is the rotation at B.

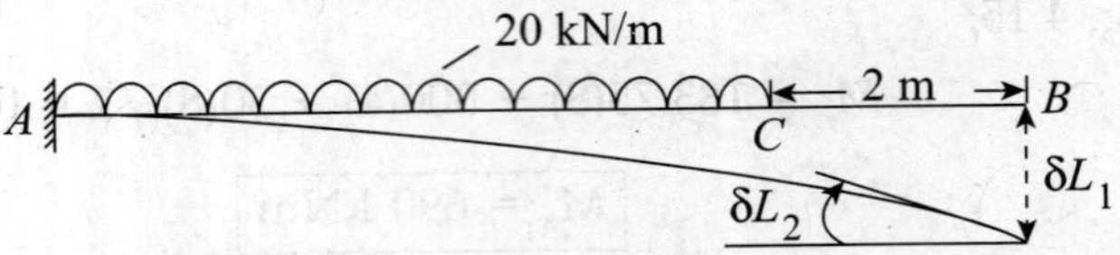

FIG. 4.27 Released structure with external loads

$$\delta L_1 = \left(\frac{wl^4}{8EI}\right) + \left(\frac{wl^3}{6EI}\right) l_1$$

$$\delta L_1 = -\left[\frac{20(6)^4}{8EI} + \frac{20(6)^3 2}{6EI}\right]$$

$$\boxed{\delta L_1 = -4680/EI}$$

The negative sign is due to the vertical downward displacement; as upward displacement in the direction of RB is positive.

$$\delta L_2 = wl^3/6EI$$

$$\boxed{\delta L_2 = 20(6)^3/6EI = 720/EI}$$

The slope δL_2 is positive as the rotation at B is clockwise with respect to horizontal.

Flexibility Coefficients

A unit load is applied vertically upwards at B; then

f_{11} = Vertical displacement at B due to unit vertical load at B

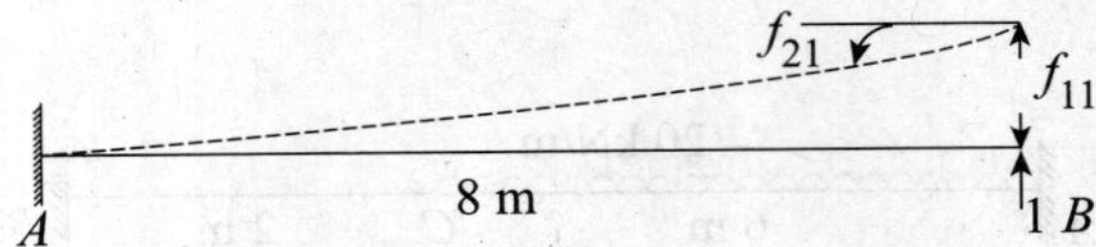

FIG. 4.28 Unit load at B

f_{21} = Rotation at B due to unit load at B

Hence,

$$f_{11} = l^3/3EI = 8^3/3EI = 170.67/EI$$

$$f_{21} = -l^2/2EI = -8^2/2EI = -32/EI$$

The negative is because of the rotation in the anticlockwise direction.

A unit moment is applied in the clockwise direction at B. Referring to the figure given below;

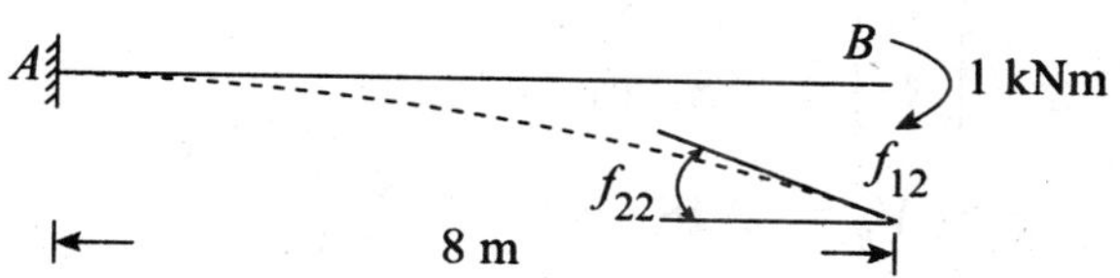

FIG. 4.29 Unit moment is applied at B

f_{12} = Vertical displacement at B due to unit moment at B

f_{22} = Rotation at B due to unit moment at B

$$f_{12} = -Ml^2/2EI = -(1)8^2/2EI = -32/EI$$

The negative sign is due to downward vertical displacement.

$$f_{22} = \frac{Ml}{EI} = +1(8)/EI = 8/EI$$

The positive sign is due to clockwise rotation at B.

The flexibility matrix

$$[F] = \begin{bmatrix} f_{11} & f_{12} \\ f_{21} & f_{22} \end{bmatrix} = \frac{1}{EI}\begin{bmatrix} 170.67 & -32 \\ -32 & +8 \end{bmatrix}$$

and
$$[F]^{-1} = \frac{EI}{f_{11}f_{22} - f_{21}f_{12}}\begin{bmatrix} f_{22} & -f_{12} \\ -f_{21} & f_{11} \end{bmatrix}$$

Thus,
$$[F]^{-1} = \frac{EI}{170.67(8) - (-32)(-32)}\begin{bmatrix} 8 & 32 \\ 32 & 170.67 \end{bmatrix}$$

$$[F]^{-1} = \frac{EI}{341.36} = \begin{bmatrix} 8 & 32 \\ 32 & 170.67 \end{bmatrix}$$

Compatibility Equation

As the deflection at B is zero;

$$\delta L_1 + f_{11}R_1 + f_{12}R_2 = 0$$

$$\delta L_2 + f_{21}R_1 + f_{22}R_2 = 0$$

$$[\delta L] + [F][R] = 0$$

$$[R] - [F] - 1[\delta L]$$

$$[R] = -\frac{EI}{341.36} = \begin{bmatrix} 8 & 32 \\ 32 & 170.67 \end{bmatrix} \times \frac{I}{EI} \begin{bmatrix} -4680 \\ 720 \end{bmatrix}$$

$$\begin{bmatrix} R_1 \\ R_2 \end{bmatrix} = \begin{bmatrix} 42.18 \\ 78.74 \end{bmatrix}$$

i.e.

$$R_1 = V_B = 42.18 \text{ kN}$$

$$R_2 = M_B = 78.74 \text{ kNm}$$

It is to be mentioned here that much time could be saved by solving the simultaneous equations using the calculator directly. As both V_B and M_B have positive signs, the assumed directions are correct. Using the free body diagram, the other reactions V_A and M_A are computed. The resulting shear force and bending moment diagrams are given below.

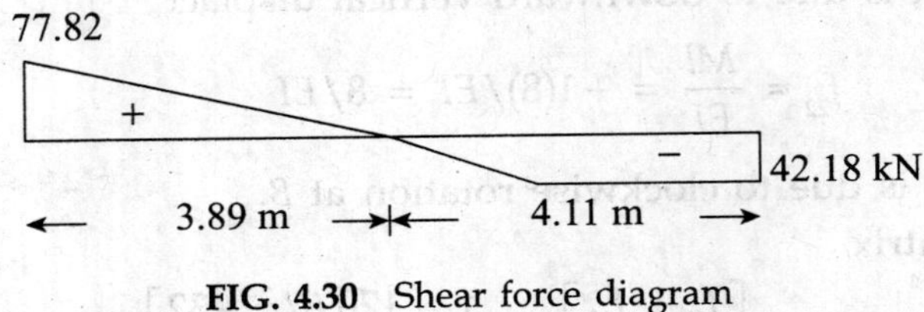

FIG. 4.30 Shear force diagram

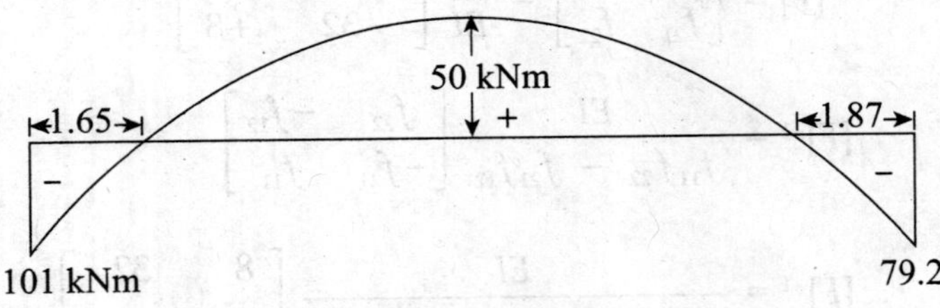

FIG. 4.31 Bending moment diagram

Example 4.3 Analyse the continuous beam shown in Fig. 4.32. by the flexibility method. Draw the shear force and bending moment diagram.

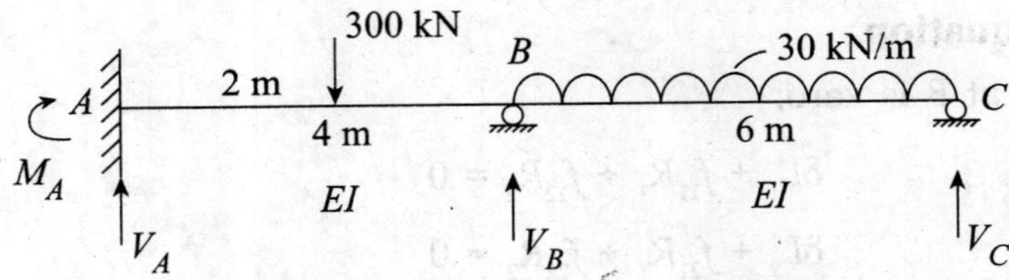

FIG. 4.32 Continuous beam with reaction components

Solution: The redundants are V_A, V_B, V_C and M_A. The static indeterminacy is 4 – 2 = 2. Thus, the above beam is statically indeterminate to the second degree as the number of redundants and the available equations of equilibrium are 2. It is to be noted that for the beams subjected to transverse loads (vertical loads), $\Sigma H = 0$ is of no consequence and only $\Sigma V = \Sigma M = 0$ is valid.

The continuous beam is statically indeterminate to the second degree. The beam is made determinate by removing the external moment M_A and the internal moment at B. [B is a continuous support, while C is a simple support].

The redundants R_1 and R_2 are shown in Fig. 4.27 and are assumed sagging.

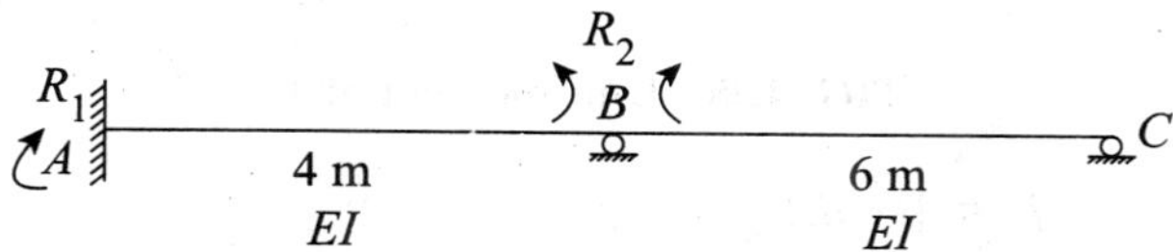

FIG. 4.33 Redundants

The released structure consists of two simply supported beams with the same external loads as shown in Fig. 4.34.

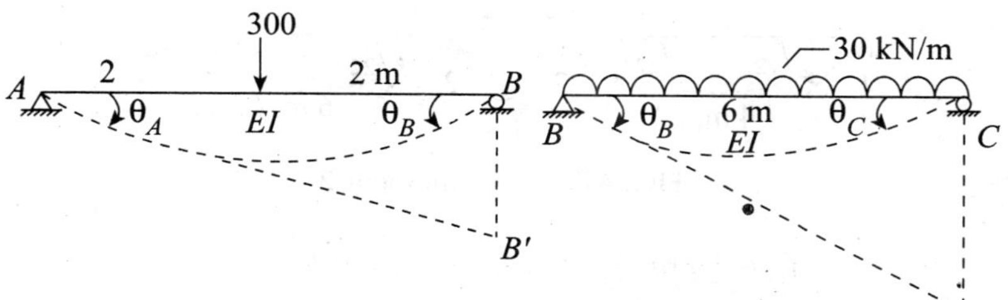

FIG. 4.34 Released structure

The rotations at A and B due to the external loads are +ve, as the directions assumed A and B is same as in Fig. 4.34.

$$\delta L_1 = \theta_A = \frac{Wl^2}{16EI} = \frac{300(4)^2}{16EI} = \frac{300}{EI}$$

$$\delta L_2 = \left\{\theta_B\big|_{BA} + \theta_B\big|_{BC}\right\} = \left\{\frac{Wl^2}{16EI}\bigg|_{BA} + \frac{Wl^3}{24EI}\bigg|_{BC}\right\}$$

$$\delta L_2 = \frac{300(4)^2}{16EI}\bigg|_{BA} + \frac{30 \times 6^3}{24EI}\bigg|_{BC} = \frac{570}{EI}$$

Flexibility Coefficients

Since the redundancy is 2, the flexibility coefficients at 1 and 2 are required. A unit moment at 1 is applied as shown in Fig. 4.35 and the flexibility coefficients, f_{11} and f_{21} are calculated.

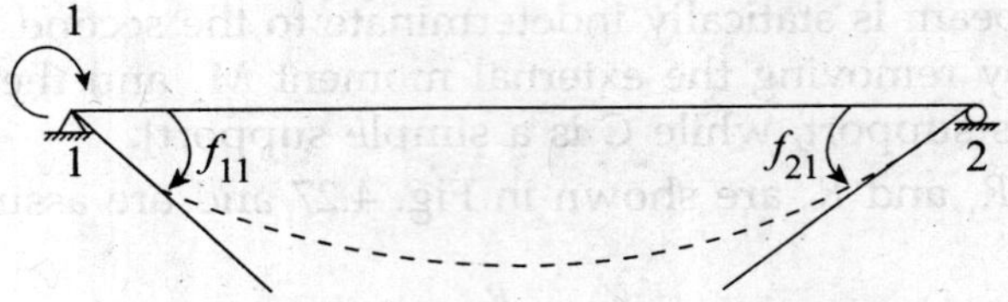

FIG. 4.35 Unit moment at 1

$$f_{11} = [l/3EI]_{AB} = 4/3EI$$

$$f_{21} = [l/6EI]_{BA} = 4/6EI = 2/3EI$$

A unit moment at 2 is applied and the flexibility coefficients, f_{12} and f_{22} are calculated.

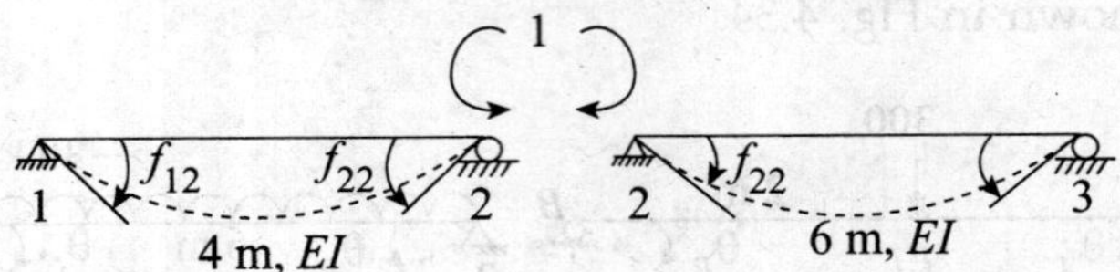

FIG. 4.36 Unit moment 2

$$f_{12} = [l/6EI]_{AB} = 4/6EI = 2/3EI$$

$$f_{22} = [l/3EI]_{BA} + [l/3EI]_{BC} = [4/3EI]_{BA} + [6/3EI]_{BC}$$

i.e. $$f_{22} = 10/3EI$$

It shall be remembered that the flexibility coefficients are given by ($l/3EI$) for the support where the unit moment is applied and ($l/6EI$) for the other support where the unit moment is not applied. Using the Eqs. (20 & 21)

$$\delta L_1 + f_{11}R_1 + f_{12}R_2 = 0$$

$$\delta L_2 + f_{21}R_1 + f_{22}R_2 = 0$$

On substitution;

$$(300/EI) + (4/3EI)R_1 + (2/3EI)R_2 = 0 \quad (1)$$

$$(570/EI) + (2/3EI)R_1 + (10/3EI)R_2 = 0 \quad (2)$$

Multiplying the Eqs. (1) and (2) by $3EI$,

$$900 + 4R_1 + 2R_2 = 0 \quad (3)$$

$$1710 + 2R_1 + 10R_2 = 0 \quad (4)$$

Solving the above equations

$$R_1 = -155 \text{ kNm}$$
$$R_2 = -140 \text{ kNm}$$

The negative sign indicates that the moments are hogging and not sagging as assumed.

Drawing the free body diagram for each span along with $M_A(R_1)$ and $M_B(R_2)$ respectively.

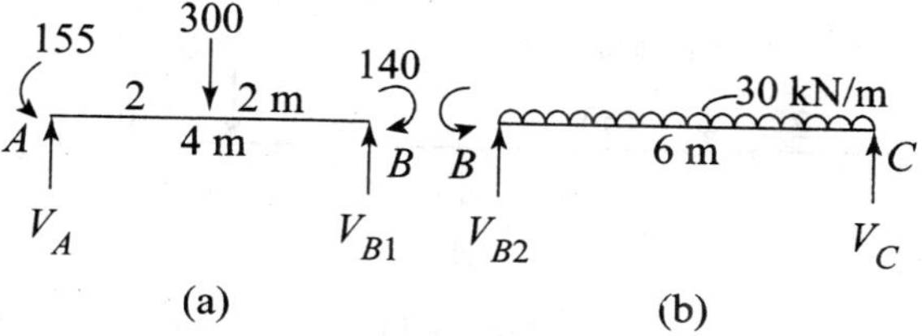

FIG. 4.37 Free body diagrams

$\Sigma V = 0;$

$$V_A + V_{B1} = 300 \tag{5}$$

$\Sigma M_A = 0$

$$-155 + 140 + 300(2) - 4V_{B1} = 0 \tag{6}$$

$$V_{B1} = 146.25 \text{ kN}$$

$$V_A = 153.75 \text{ kN}$$

$\Sigma V = 0$

$$V_{B2} + V_C = 30(6) \tag{7}$$

$\Sigma M_B = 0$

$$-140 + 30(6)^2/2 - 6V_C = 0 \tag{8}$$

$$V_C = 66.67 \text{ kN}$$

$$V_{B2} = 113.33 \text{ kN}$$

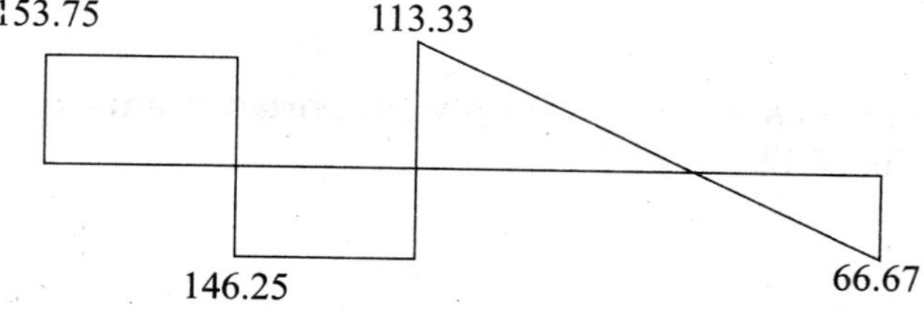

FIG. 4.38 Shear force diagram

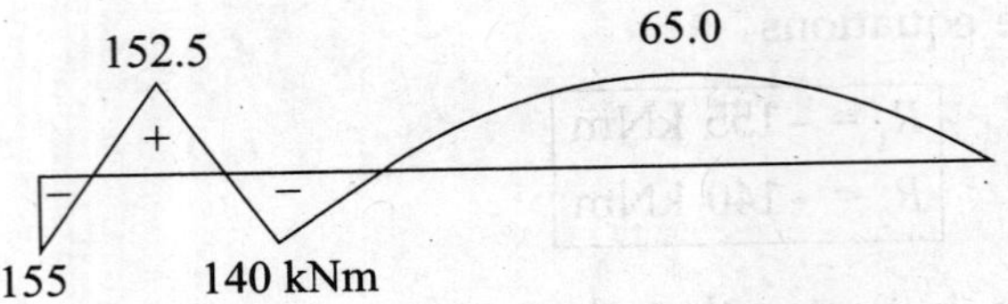

FIG. 4.39 Bending moment diagram

Example 4.4 Analyse the continuous beam shown in Fig. (4.40). Draw the BMD. *EI* is constant.

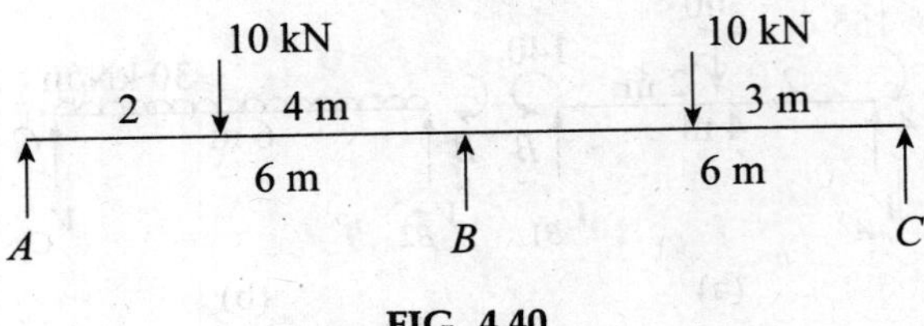

FIG. 4.40

Solution: The number of unknown reactions to be determined are V_A, V_B and V_C. The available equilibrium equations are two, i.e., sum of all the vertical forces zero and sum of all the moments is zero. Hence, the statical indeterminacy (redundancy) is 3 – 2 = 1. Thus, the above continuous beam is statically indeterminate to the first degree. The internal moment at *B* is taken as the redundant. The problem reduces to find a solution consisting of (i) rotation at *B* due to the applied loads and (ii) computing the flexibility coefficient at *B*.

Redundant (R_1)

The redundant R_1 (internal moment at *B*, i.e. M_B) is shown in Fig. 4.41. and it is assumed as sagging.

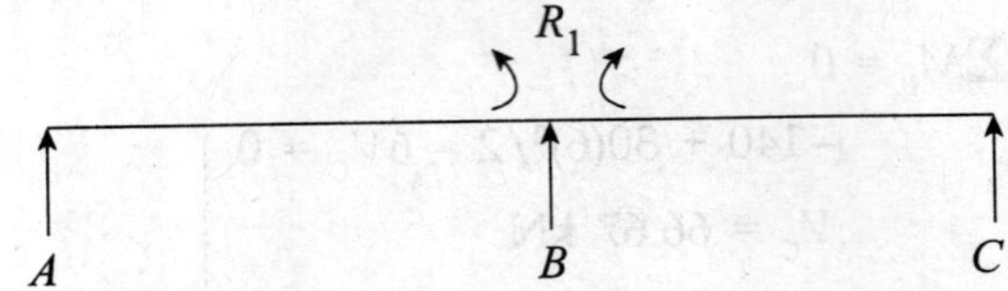

FIG. 4.41 Redundant R_1 (internal moment *MB*)

Released structure

The released structure consists of two simply supported beams with the same external loads as shown in Fig. 4.42.

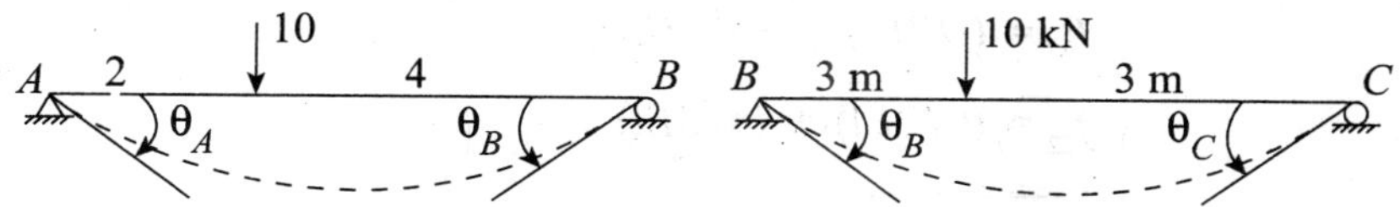

FIG. 4.42 Released structure

Rotation at B

$$\delta L_1 = \theta_B\big|_{BA} + \theta_B\big|_{BC} \tag{1}$$

The rotation at B, i.e., δL_1 is the summation of rotations at B in span BA and BC respectively. (Eq. 1)

$$\theta_B\big|_{BA} = \frac{Wab(l + a)}{6EIl} \qquad \text{(Refer Table 4.1)}$$

$$= \frac{10(2)(4)(6 + 2)}{6EI(6)}$$

$$\boxed{\theta_B\big|_{BA} = 17.78/EI}$$

$$\theta_B\big|_{BC} = \frac{Wl^2}{16EI}$$

$$\theta_B\big|_{BC} = \frac{10(6)^2}{16EI}$$

$$\theta_B\big|_{BC} = 22.5/EI$$

Hence,

$$\delta L_1 = \frac{17.78 + 22.5}{EI} = \frac{40.28}{EI} \tag{2}$$

Flexibility Coefficient at B

As the redundancy is 1, the flexibility coefficient at B is required. A unit moment is applied at the intermediate support and f_{22} is calculated as follows.

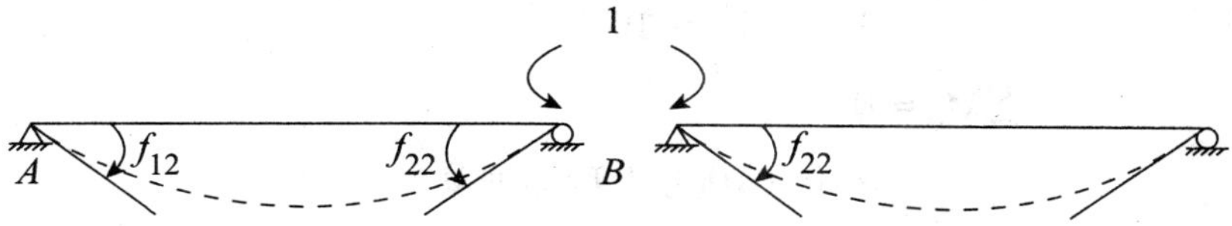

FIG. 4.43 Unit moment at B

$$f_{22} = (l/3EI)_{BA} + (l/3EI)_{BC}$$

$$f_{22} = (6/3EI) + (6/3EI)$$

$$\boxed{f_{22} = 4/EI} \quad (3)$$

Compatibility Equation

$$\delta L_1 + f_{22}R_1 = 0$$

$$(40.28/EI) + (4/EI)R_1 = 0$$

$$\boxed{R_1 = -10 \text{ kNm}}$$

The negative sign indicates that the moment is hogging and not sagging as assumed (Ref. Fig. 4.41).

Bending Moment Diagram

Drawing the free body diagram for each span along with indeterminate moment M_B

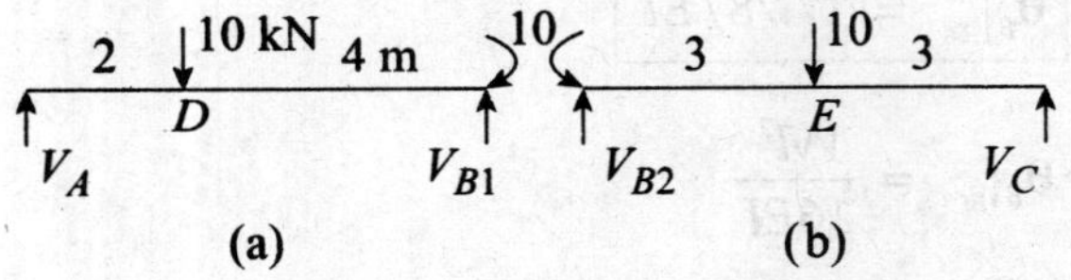

(a) Span AB: V_A at A; 10 kN load at D, 2 m from A; 4 m from D to B; moment 10 and V_{B1} at B.

(b) Span BC: moment 10 and V_{B2} at B; 10 kN load at E, 3 m from B; 3 m from E to C; V_C at C.

FIG. 4.44 Free body diagram

$$\Sigma V = 0$$

$$V_A + V_{B1} = 10 \quad (4)$$

$$\Sigma M_A = 0$$

$$10(2) + 10 - 6V_{B1} = 0 \quad (5)$$

$$V_{B1} = 5 \text{ kN}$$

$$\therefore \quad V_A = 5 \text{ kN}$$

$$\Sigma V = 0$$

$$V_{B2} + V_C = 10 \quad (6)$$

$$\Sigma M_C = 0$$

$$-10 - 10(3) + 6V_{B2} = 0$$

$$V_{B2} = 6.67 \text{ kN}$$

$$V_C = 3.33 \text{ kN}$$

Bending moments at salient points

$$M_A = 0; \quad M_B = -10 \text{ kNm}; \quad M_C = 0$$
$$M_D = 5(2) = 10 \text{ kNm}$$
$$M_E = 3.33(3) = 10 \text{ kNm}$$

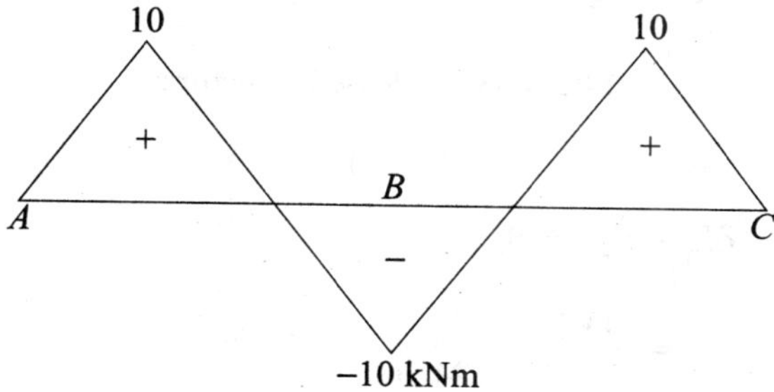

FIG. 4.45 Bending moment diagram

Example 4.5 Analyse the continuous beam shown in Fig. 4.46 by the flexibility method. Use system approach. (VTU, Dec. 2010)

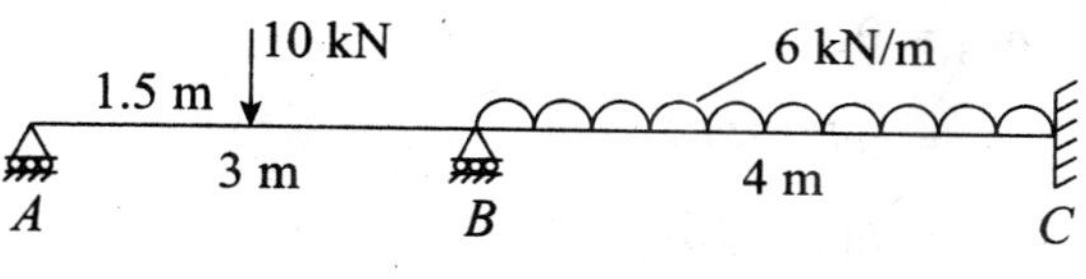

FIG. 4.46

Solution: The number of unknown reactions are V_A, V_B, V_C and M_C. The available equilibrium equations are two, i.e., $\Sigma V = \Sigma M = 0$; Hence, the statical indeterminacy is $4 - 2 = 2$. Hence, the two redundants chosen are internal moment M_B and external moment M_C. The unknown redundants are chosen such that the given continuous beam turns out to be a combination of two simple beams AB and BC respectively.

Redundants

The redundant R_1 is (internal moment at B, i.e., M_B) shown in Fig. 4.47 and is assumed sagging. The moment at C is taken as R_2 and is again assumed sagging.

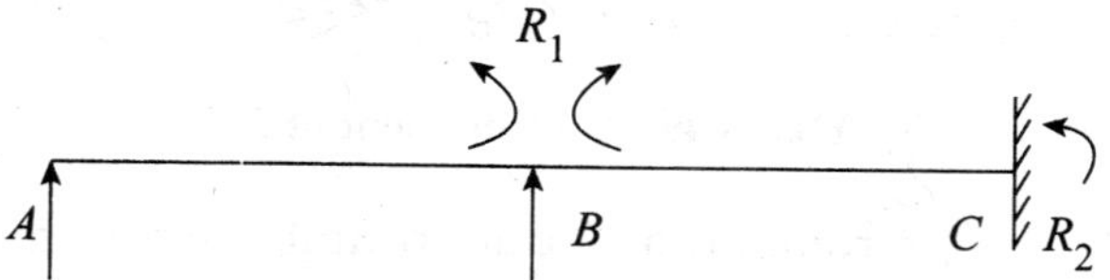

FIG. 4.47 Redundants R_1 and R_2

Released Structure

The released structure consists of two simply supported beams with the external loads.

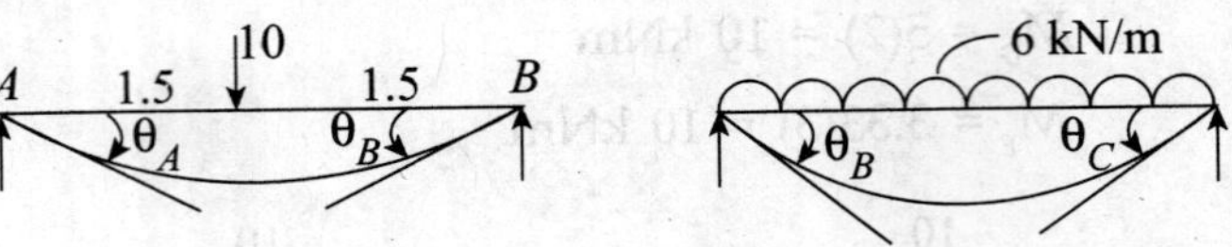

FIG. 4.48 Released structure

Deformations at B and C

$$\delta L_1 = \theta_{BA} + \theta_{BC}$$

$$= \frac{Wl^2}{16EI} + \frac{Wl^3}{24EI}$$

$$\delta L_1 = \frac{10(3)^2}{16EI} + \frac{6 \times 4^3}{24EI}$$

$$\delta L_1 = -21.63/EI$$

$$\delta L_2 = \theta_{CB}$$

$$= wl^3/24EI$$

$$= 6(4)^3/24EI$$

$$\delta L_2 = 16/EI$$

Analysis of Structures for Unit Redundants

Unit moment at B

The flexibility coefficient at *B* is calculated as follows. A unit moment is applied at *B* and the rotations at *B* and *C* are calculated as follows.

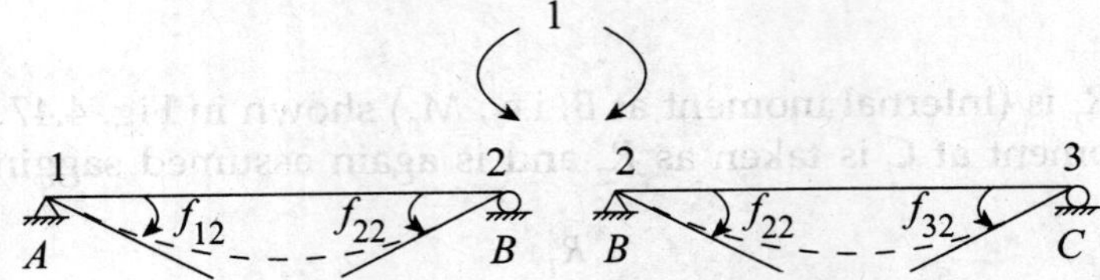

FIG. 4.49 Unit moment at *B*

f_{22} = rotation at 2 due to unit sagging moment at 2

f_{22} = (3/3*EI*) + (4/3*EI*) = 7/3*EI*

The first term corresponds to span BA and the second span corresponds to span BC respectively. It is again reminded that the rotation at the point where unit moment is applied is equal to $(l/3EI)$; and the rotation at the other end is $(l/6EI)$

$$f_{32} = \text{rotation at 3 due to unit sagging moment at 2}$$

$$f_{32} = 4/6EI = 2/3EI$$

Unit moment at C

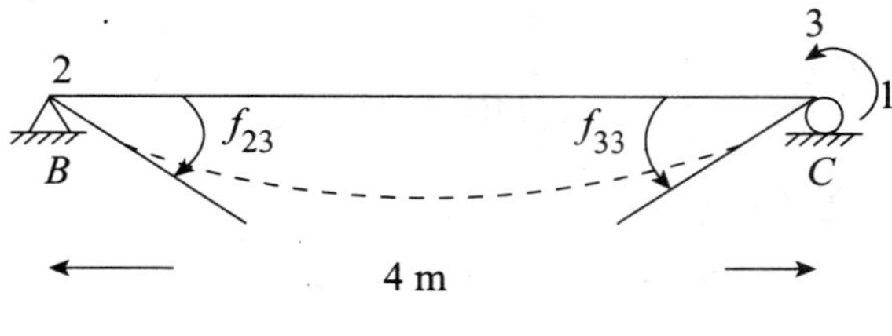

FIG. 4.50

Apply a unit moment as shown above.

$$f_{23} = \text{rotation at 2 due to unit sagging moment at 3}$$

$$f_{23} = 4/6EI = 2/3EI$$

$$f_{33} = \text{rotation at 3 due to unit sagging moment at 3}$$

$$f_{33} = 4/3EI$$

Compatibility Equation

Since the net deflection at B and C are zeros,

$$\delta L_1 + f_{22}R_1 + f_{32}R_2 = 0$$

$$\delta L_2 + f_{23}R_1 + f_{33}R_2 = 0$$

Substituting the values of deformations at B and C $(\delta L_1, \delta L_2)$ and the flexibility coefficients $(f_{22}, f_{32}, f_{23}, f_{33})$ the redundants R_1 and R_2 are computed.

$$21.63/EI + (7/3EI)R_1 + (2/3EI)R_2 = 0 \quad (1)$$

$$16/EI + (2/3EI)R_1 + (4/3EI)R_2 = 0 \quad (2)$$

Multiplying the above equations and rearranging

$$7R_1 + 2R_2 = -64.89 \quad (3)$$

$$2R_1 + 4R_2 = -48.00 \quad (4)$$

on solving;

$$R_1 = M_B = -6.81 \text{ kNm}$$

$$R_2 = M_C = -8.59 \text{ kNm}$$

The negative sign indicates that the above moments are hogging in nature.

Determination of other reactions

Drawing the free body diagram of each span AB and BC with M_B and M_C.

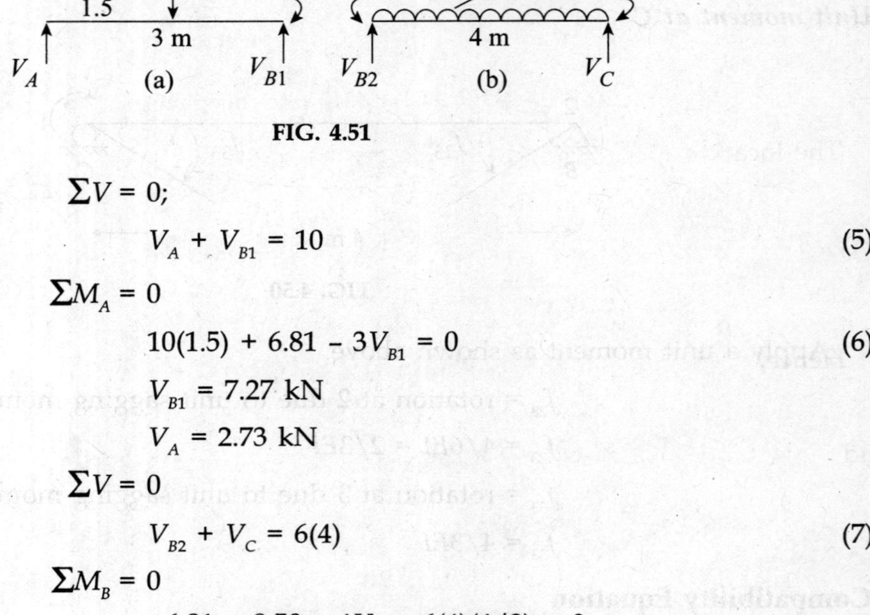

FIG. 4.51

$$\Sigma V = 0;$$

$$V_A + V_{B1} = 10 \qquad (5)$$

$$\Sigma M_A = 0$$

$$10(1.5) + 6.81 - 3V_{B1} = 0 \qquad (6)$$

$$V_{B1} = 7.27 \text{ kN}$$

$$V_A = 2.73 \text{ kN}$$

$$\Sigma V = 0$$

$$V_{B2} + V_C = 6(4) \qquad (7)$$

$$\Sigma M_B = 0$$

$$-6.81 + 8.59 - 4V_C + 6(4)(4/2) = 0$$

$$V_C = 12.45 \text{ kN}$$

$$\therefore \quad V_{B2} = 11.55 \text{ kN}$$

Hence, the reactions are

$$V_A = 2.73 \text{ kN}, V_B = 18.82, V_C = 12.45 \text{ kN}$$

Numerical check

$$\Sigma \text{ upward reactions} = \Sigma \text{ downward loads}$$

$$V_A + V_B + V_C = 10 + (6 \times 4)$$

$$2.73 + 18.82 + 12.45 = 34$$

Hence; ok

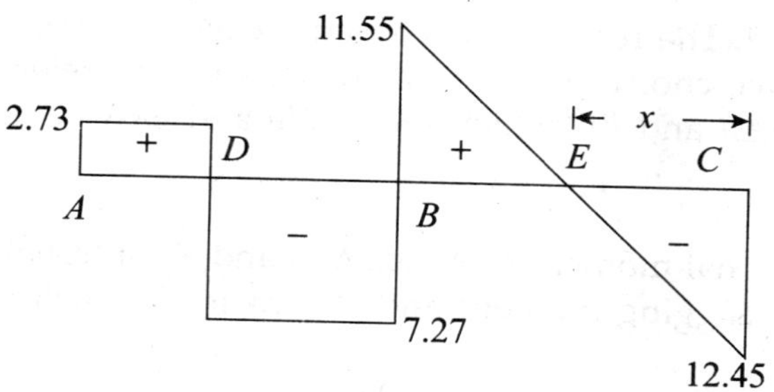

FIG. 4.52 Shear force diagram

The location E is found out by similar triangles

$$\frac{x}{(4-x)} = \frac{12.45}{11.55}$$

$$\boxed{x = 2.075 \text{ m}}$$

Hence,

$$M_E = 12.45(2.075) - 6(2.075)^2/2 - 8.59$$

$$M_E = 12.92 \text{ kNm}$$

and

$$M_D = 2.73(1.5) = 4.1 \text{ kNm}$$

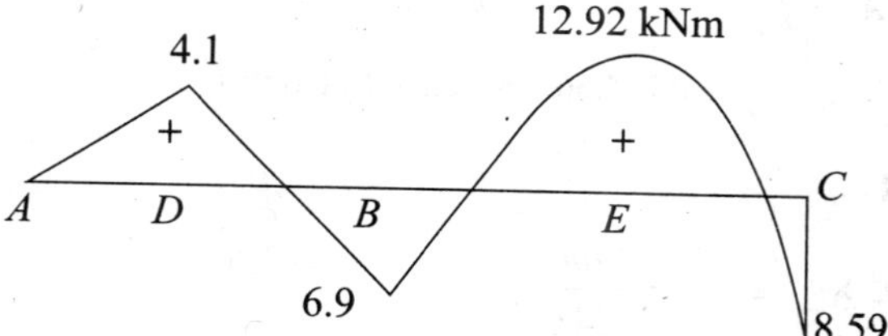

FIG. 4.53 Bending moment diagram

Example 4.6 Analyse the continuous beam shown in Fig. 4.54 is fixed at A. Analyse and draw the bending moment diagram using flexibility matrix method. Assume $EI = 10$ MNm²

(VTU, June 2010)

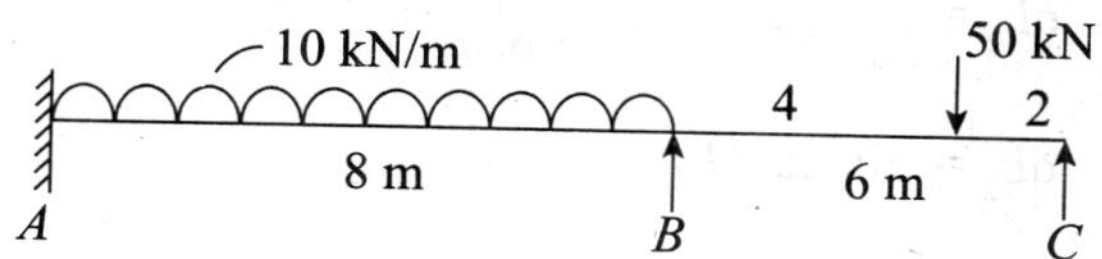

FIG. 4.54

Solution: The reaction components of the above beam are M_A, V_A, V_B and V_C. Thus, there are four unknowns and the equilibrium equations are two. Hence, the degree of redundancy is 4 – 2 = 2. The redundants chosen are M_A and the indeterminate internal moment MB. The above choice has been made so that the released structure consists of two simple beams AB and BC which are stable and easy to analyse.

Redundants

The redundant R_1 (external moment at A, i.e., M_A) and R_2 (internal moment at B, i.e. M_B) are assumed to act as sagging moment and shown in Fig. 4.55.

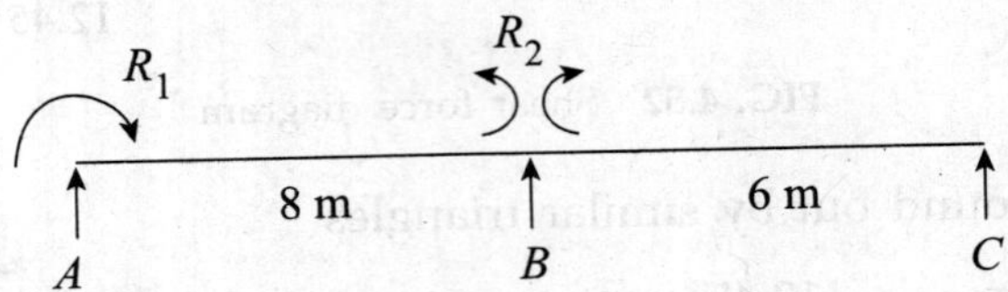

FIG. 4.55 Redundants

Released Structure

The released structure consists of two simply supported beams with external loads.

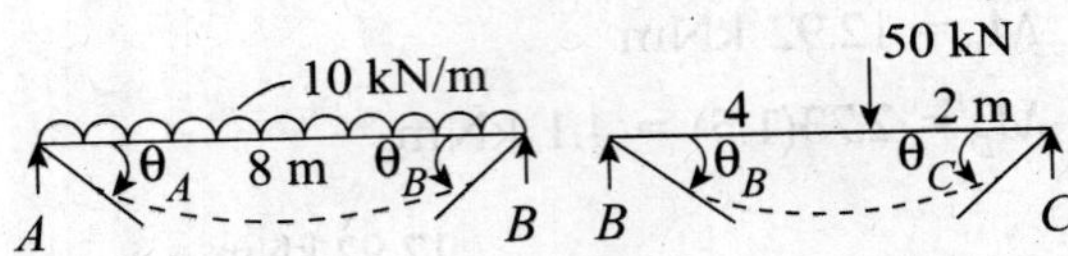

FIG. 4.56 Released structure

Rotations at A and B

$$\delta L_1 = \theta_A = \frac{wl^3}{24EI}$$

$$\delta L_1 = 10(8)^3/24EI = 213.33/EI$$

$$\delta L_2 = \theta_B = \frac{wl^3}{24EI} + \frac{Wab}{6EIl}(l + b) \quad \text{(Refer Table 4.1)}$$

$$\delta L_2 = \frac{10(8)^3}{24EI} + \frac{50(4)(2)(6 + 2)}{6EI(6)}$$

$$\delta L_2 = 302.22/EI$$

Analysis of Structures for Unit Redundants

Unit moment at A

The flexibility coefficients at A, viz., f_{11} and f_{21} are calculated as follows.

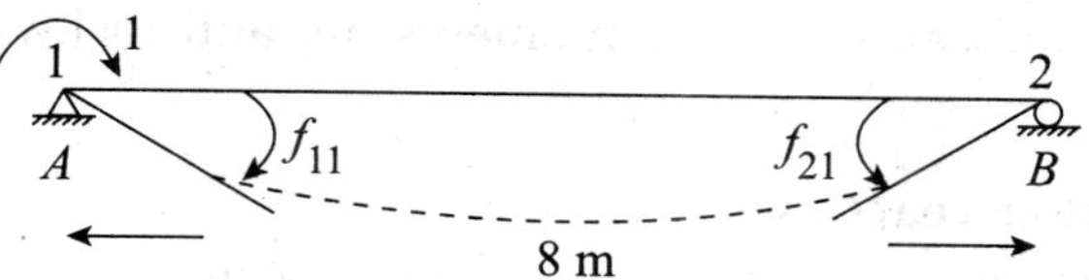

FIG. 4.57(a) Unit moment at *A*

$$f_{11} = l/3EI = 8/3EI$$

$$f_{21} = l/6EI = 8/6EI$$

Unit moment at B

A unit moment is applied at *B* and the rotations are calculated at *B* and *A* respectively.

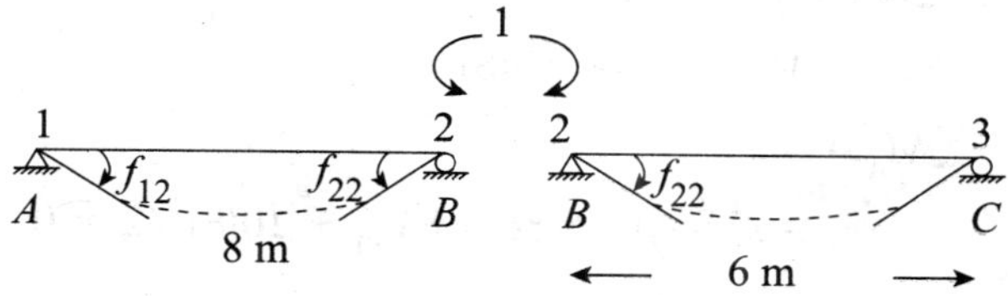

FIG. 4.57(b) Unit moment at *B*

f_{22} = rotation at 2 due to unit sagging moment at 2

$$f_{22} = 8/3EI + 6/3EI = 14/3EI$$

f_{12} = rotation at 1 due to unit sagging moment at 2

$$f_{12} = 8/6EI = 4/3EI$$

Compatibility Equations

Since the net deformations at *A* and *B* are zeros, the compatibility equations can be written as

$$\delta L_1 + f_{11}R_1 + f_{12}R_2 = 0 \qquad (1)$$

$$\delta L_2 + f_{21}R_1 + f_{22}R_2 = 0 \qquad (2)$$

Substituting the values;

$$213.33/EI + (8/3EI)R_1 + (4/3EI)R_2 = 0 \qquad (3)$$

$$302.22/EI + (4/3EI)R_1 + (14/3EI)R_2 = 0 \qquad (4)$$

Solving the above

$$R_1 = M_A = -55.56 \text{ kNm}$$

$$R_2 = M_B = -48.86 \text{ kNm}$$

The negative sign indicates that the moments are anticlockwise and not clockwise moment as assumed.

Determination of other reactions

Drawing the free body diagram of each span AB and BC.

55.56 10 kN/m 48.86 48.86 50
8 m 4 2 m 6 m
V_A (a) V_{B1} V_{B2} (b) V_C

FIG. 4.58

$$\Sigma V = 0;$$

$$V_A + V_{B1} = 10(8) \qquad (5)$$

$$\Sigma M_A = 0;$$

$$-55.56 + 48.86 - 8V_{B1} + 10(8)^2/2 = 0 \qquad (6)$$

$$V_{B1} = 39.2 \text{ kN}$$

$$\therefore \quad V_A = 40.8 \text{ kN}$$

$$\Sigma V = 0$$

$$V_{B2} + V_C = 50 \qquad (7)$$

$$\Sigma M_B = 0$$

$$-48.86 + 50(4) - 6V_C = 0$$

$$V_C = 25.2 \text{ kN}$$

$$V_{B2} = 24.8 \text{ kN}$$

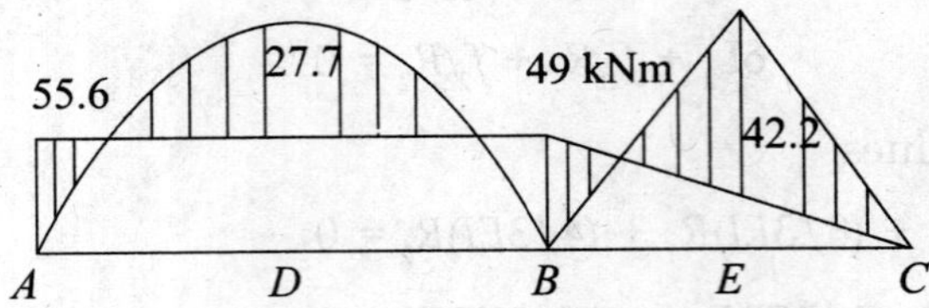

FIG. 4.59 Bending moment diagram

Example 4.7 Analyse the continuous beam by the flexibility matrix method. EI = constant. Draw the bending moment diagram. (July 2008, VTU)

FIG. 4.60

Solution: The given two-span continuous beam is statically indeterminate to the second degree. Here M_A and M_B are treated as redundants. The choice of choosing the above redundants makes the beam as two simply supported beams of *AB* and *BC* as the released structure.

Redundants

The redundant R_1 is M_A and is assumed to be sagging. R_2 is the moment at the interior support at *B*. This is also assumed to be sagging moment as shown in Fig. 4.61.

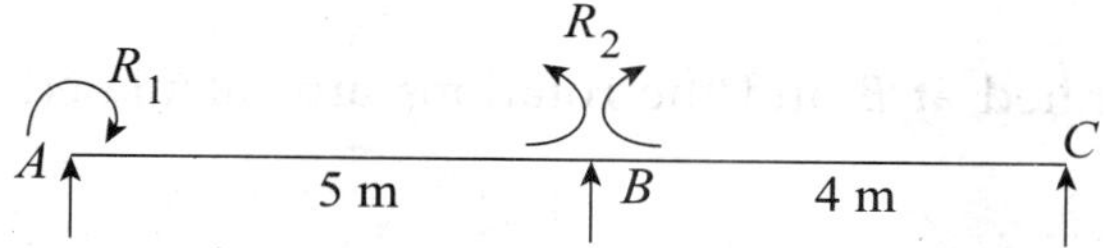

FIG. 4.61 Redundants

Released Structure

The released structure with the external loads is shown below.

FIG. 4.62 Released structure

Rotations at *A* and *B*

$$\delta L_1 = \theta_A = \frac{Wab}{6EIl}(l + b) = \frac{50(2)(3)(5 + 3)}{6EI(5)} = 80/EI$$

$$\delta L_2 = \theta_{BA} + \theta_{BC} = \frac{Wab}{6EIl}(l + a) + \frac{Wl^2}{16EI}$$

$$\delta L_2 = \frac{50(2)(3)(5 + 2)}{6EI(5)} + \frac{40(4)^2}{16EI} = 110/EI$$

Analysis of Structure for Unit Redundants

Unit moment at A

The flexibility coefficients at *A*, viz., f_{11} and f_{21} are calculated as

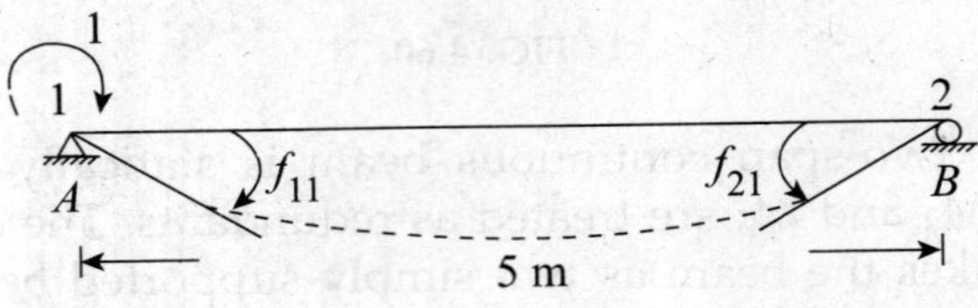

FIG. 4.63

$$f_{11} = l/3EI = 5/3EI$$
$$f_{21} = l/6EI = 5/6EI$$

Unit moment at B

A unit moment is applied at *B* and the rotations are calculated at *B* and *A*.

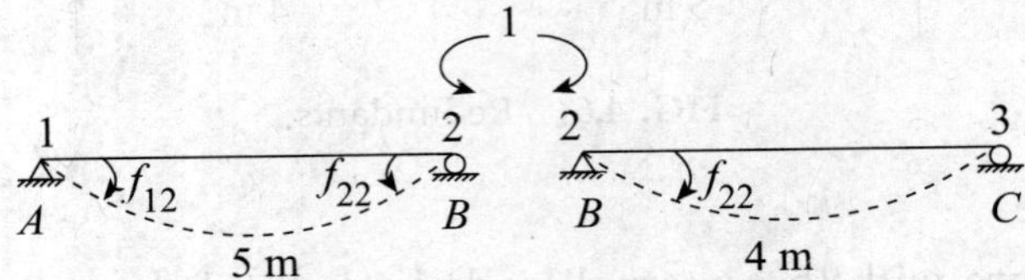

FIG. 4.64 Unit moment at *B*

f_{22} = rotation at 2 due to unit sagging moment at 2

$$f_{22} = 5/3EI + 4/3EI = 9/3EI$$

f_{12} = rotation at 1 due to unit sagging moment at 2

$$f_{12} = 5/6EI$$

Compatibility Equations

As the net deformations at *A* and *B* are zeros; the compatibility equations are of the form

$$\delta L_1 + f_{11}R_1 + f_{12}R_2 = 0 \quad (1)$$
$$\delta L_2 + f_{21}R_1 + f_{22}R_2 = 0 \quad (2)$$

Substituting the computing values,

$$(80/EI) + (5/3EI)R_1 + (5/6EI)R_2 = 0 \quad (3)$$
$$(110/EI) + (5/6EI)R_1 + (9/3EI)R_2 = 0 \quad (4)$$

The above equations are multiplied throughout by $6EI$ and solving for R_1 and R_2;

$$R_1 = -34.5 \text{ kNm} = M_A$$
$$R_2 = -27.1 \text{ kNm} = M_B$$

The negative sign indicates that both M_A and M_B are hogging in nature.

Determination of other reactions

The free body diagrams are drawn along with M_A and M_B.

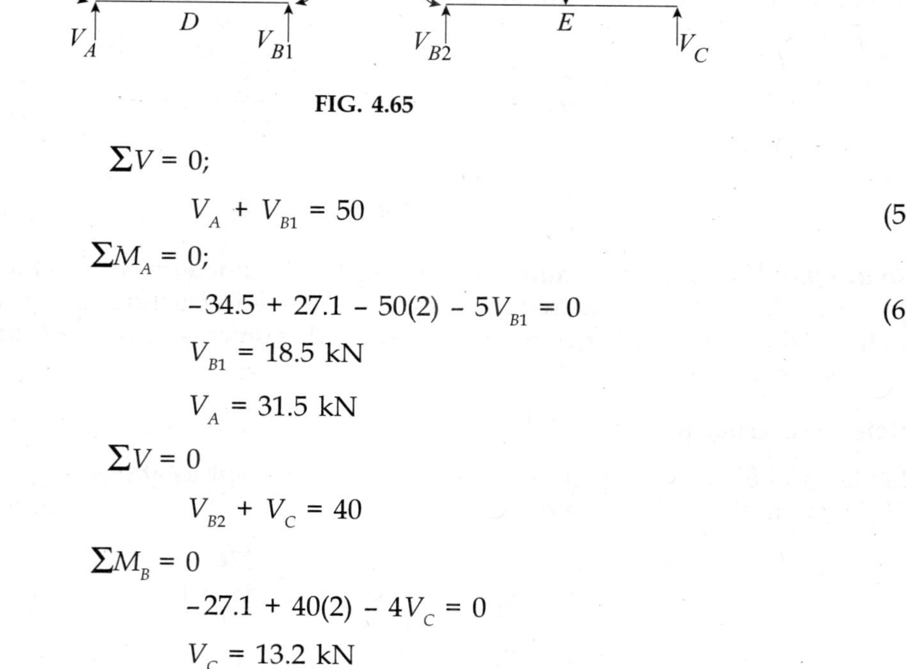

FIG. 4.65

$\Sigma V = 0;$

$$V_A + V_{B1} = 50 \tag{5}$$

$\Sigma M_A = 0;$

$$-34.5 + 27.1 - 50(2) - 5V_{B1} = 0 \tag{6}$$

$$V_{B1} = 18.5 \text{ kN}$$
$$V_A = 31.5 \text{ kN}$$

$\Sigma V = 0$

$$V_{B2} + V_C = 40$$

$\Sigma M_B = 0$

$$-27.1 + 40(2) - 4V_C = 0$$
$$V_C = 13.2 \text{ kN}$$
$$V_{B2} = 26.8 \text{ kN}$$

Bending moment values

$$M_A = -34.5 \text{ kNm}$$
$$M_D = -34.5 + 31.5(2) = 28.5 \text{ kNm}$$
$$M_B = -27.1 \text{ kNm}$$
$$M_E = 13.2(2) = 26.4 \text{ kNm}$$
$$M_C = 0$$

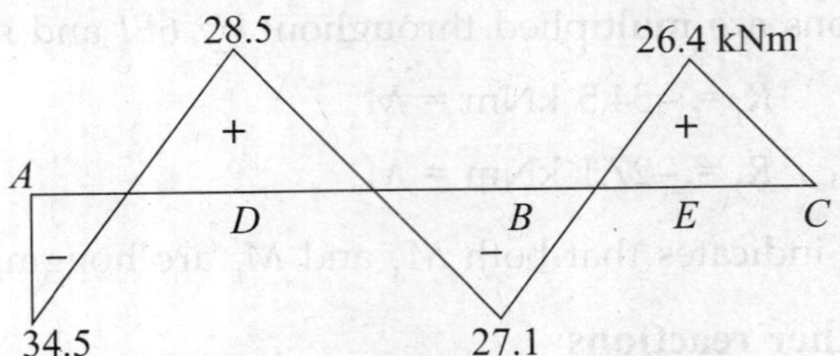

FIG. 4.66 Bending moment diagram

Example 4.8 Analyse the beam shown in Fig. 4.67 by the flexibility method and compute the indeterminate moments. (VTU, Dec. 2011)

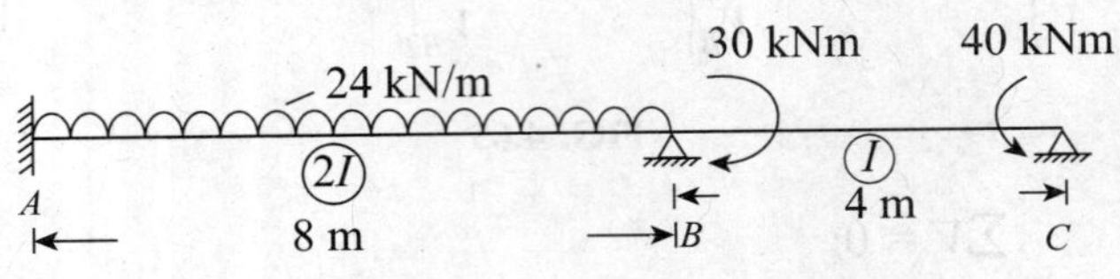

FIG. 4.67

Solution: The given continuous beam is statically indeterminate to the second degree. M_A and M_B are treated as redundants. The released structure consists of two simple beams AB and BC and at B and C; the applied moments at B and C are reversed and applied in direction.

Released structure

Let δL_1 and δL_2 be the rotations at the chosen redundant locations. δL_1 corresponds to θ_A. δL_2 refers to the sum of the rotations at B; i.e., $\delta L_2 = \theta_{BA} + \theta_{BC}$. The computed values are

$$\delta L_1 = \frac{24(8)^3}{24E(2I)} + \frac{30(8)}{6E(2I)} = \frac{276}{EI}$$

$$\delta L_2 = \theta_{BA} + \theta_{BC}$$

$$\theta_{BA} = \frac{24(8)^3}{24E(2I)} + \frac{30 \times 8}{3E(2I)} = \frac{296}{EI}$$

$$\theta_{BC} = \frac{30(4)}{3EI} - \frac{40(4)}{6EI} = \frac{13.33}{EI}$$

$$\therefore \quad \delta L_2 = 296/EI + 13.33/EI = 309.33/EI$$

Flexibility coefficients

$$f_{11} = 8/3E(2I) = 4/3EI$$

$$f_{21} = 8/6E(2I) = 2/3EI$$

$$f_{12} = 8/6E(2I) = 2/3EI$$

$$f_{22} = 8/3E(2I) + 4/3(EI) = 8/3EI$$

Compatibility equations

$$\delta L_1 + f_{11}R_1 + f_{12}R_2 = 0 \quad (1)$$

$$\delta L_2 + f_{21}R_1 + f_{22}R_2 = 0 \quad (2)$$

Substituting the computed values in the above equation

$$276/EI + (4/3EI)R_1 + (2/3EI)R_2 = 0 \quad (3)$$

$$309.33/EI + (2/3EI)R_1 + (8/3EI)R_2 = 0 \quad (4)$$

on simplification

$$4R_1 + 2R_2 = -828 \quad (5)$$

$$2R_1 + 8R_2 = -928 \quad (6)$$

Solving the above equations for the redundants

$$R_1 = M_1 = -170.3 \text{ kNm}$$

$$R_2 = M_2 = -73.4 \text{ kNm}$$

Example 4.9 Analyse the continuous beam shown in Fig. 4.68 by the force method. Draw the bending moment diagram and shear force diagram. During loading, the supports B and C sink by 10 and 5 mm respectively with E = 200 GPa and I = 80E06 mm^2.

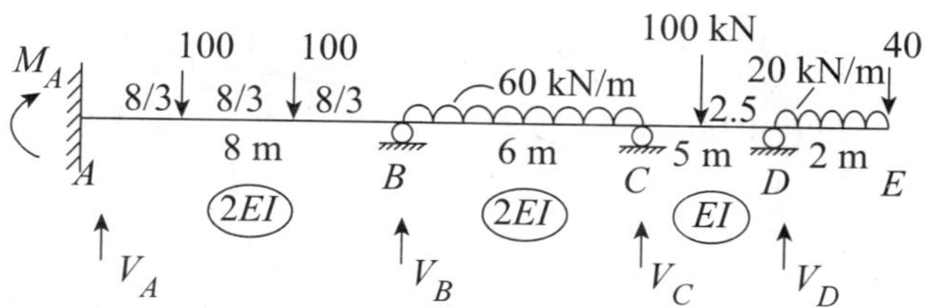

FIG. 4.68 Three-span continuous beam with the reaction components

Solution: The number of unknown reaction components are five, viz., V_A, V_B, V_C, V_D and M_A. The number of equilibrium equations are two. Hence, the static indeterminacy is 5 – 2 = 3. The redundants taken are M_A, M_B and M_C as M_D is known. In other words, the beam is made determinate by removing the external moment M_A and the internal moments at B and C, viz., M_B and M_C.

Redundants

The redundants R_1, R_2 and R_3 are assumed as sagging and shown in Fig. 4.69.

FIG. 4.69

Released Structure

The released structure consists of three simply supported beams with same external loads and the settlement moments applied in reverse direction for each span. The effect of overhang is to apply a clockwise moment of (40 × 2 + 20 × 22/2) – 120 kNm. But in the analysis we have to consider the net external moment taking into account settlement moments.

Effect of sinking/settlement/subsidence of supports

The sinking of supports at *B* and *C* are converted into equivalent joint loads at *A*, *B*, *C* and *D* by calculating the fixed end moments (reaction) due to sinking and reversing the sign of moments (action). Fixed end moments due to sinking of supports at *B* andC affecting the spans are shown in the following figures. Relative displacements between the supports are only considered. $EI = 200(10)^3 \times 80(10)^6$ Nmm², i.e. $EI = 16000$ kNm²

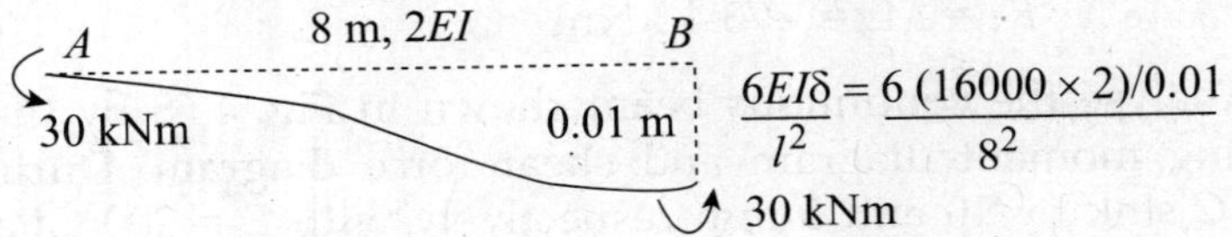

FIG. 4.70 Fixed end moments due to sinking of supports at *B*

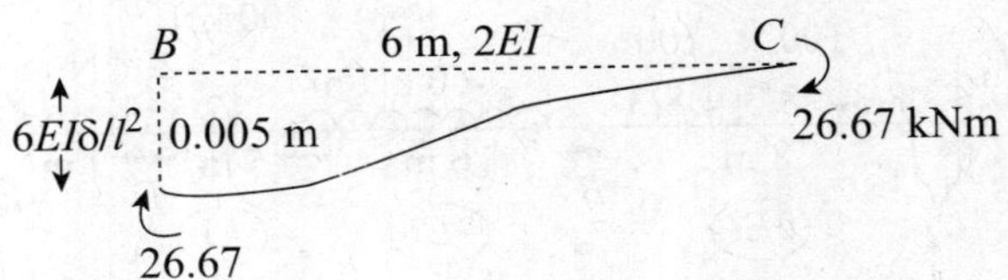

FIG. 4.71 Fixed end moments due to sinking of supports at *B* and *C*

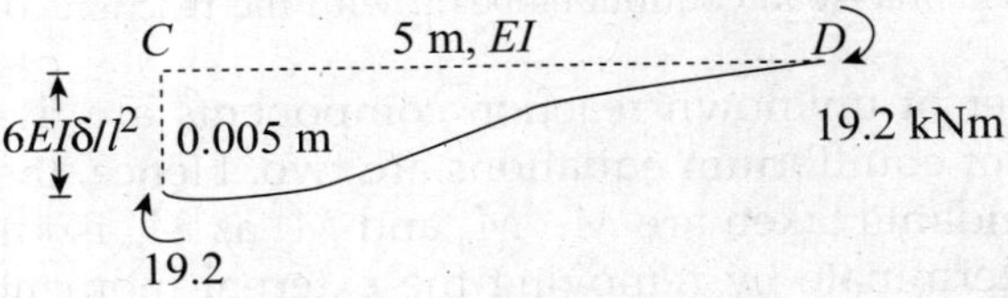

FIG. 4.72 Fixed end moments due to sinking of supports at *C*

The net moment at *D* is obtained by referring to the above diagram, i.e., 120 – 19.2 = 100.8 kNm clockwise at *D*.

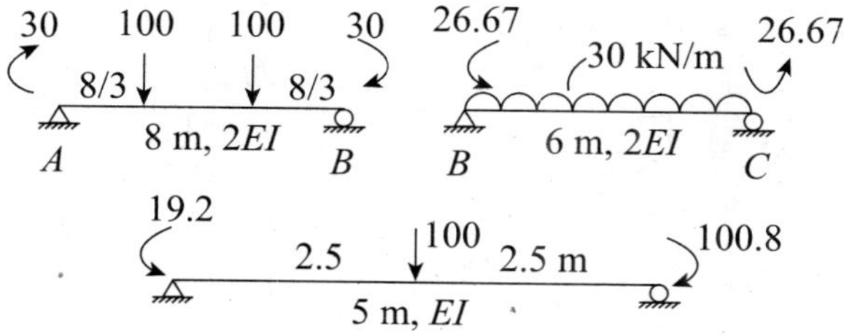

30 100 100 30 26.67 26.67
8/3 8/3 30 kN/m
8 m, 2EI 6 m, 2EI
A B B C
19.2
2.5 100 2.5 m 100.8
5 m, EI

FIG. 4.73 Released structures

The rotations at A, B and C due to external loads and moments are calculated and considered positive, if they are in the same directions of the redundant at A, B and C as indicated in the above figure.

$$\delta L_1 = \theta_A = \left\{\frac{Wab(l+b)}{6EIl} \text{ for each concentrated load } + Ml/3EI - Ml/6EI\right\}_{AB}$$

$$\theta_A = \{100 \times 8/3 \times 16/3 \times (8 + 16/3)\}/(6 \times 2EI \times 8)$$

$$+\{100 \times 16/3 \times 8/3 \times (8 + 8/3)\}/(6 \times 2EI \times 8)$$

$$+\{30 \times 8/(3 \times 2EI) - 30 \times 8/(6 \times 2EI)\}_{AB}$$

$$\boxed{\theta_A = 3380/9EI}$$

$$\delta L_2 = \{\theta_{BA} + \theta_{BC}\}$$

$$= \left\{\frac{Wab(l+a)}{6EI} \text{ for each load } - \frac{Ml}{3EI} + \frac{Ml}{6EI}\right\}_{BA}$$

$$= +\left\{\frac{Wl^3}{24EI} \text{ for udl } - \frac{Ml}{3EI} + \frac{Ml}{6EI}\right\}_{BC}$$

$$\delta L_2 = \left\{100 \times \frac{8}{3} \times \frac{16}{3} \times \left(8 + \frac{8}{3}\right)\Big/(6 \times 2EI \times 8) + 100 \times 16/3 \times 8/3\right.$$

$$\times (8 + 16/3)/[6 \times 2EI \times 8] - (30 \times 8)/3 \times 2EI) + (30 \times 8)/$$

$$\left.(6 \times 2EI)\right\} + \{(60 \times 6^3)/(24 \times 2EI) - (26.67 \times 6)/(3 \times 2EI)$$

$$+ (26.67 \times 6)/(6 \times 2EI)\}_{BC}$$

$$\boxed{\delta L_2 = 5330/9\text{EI}}$$

$$\delta L_3 = \{\theta_{CB} + \theta_{CD}\}$$

$$\delta L_3 = \left\{\frac{wl^3}{24EI} \text{ for udl } + \frac{Ml}{3EI} - \frac{Ml}{6EI}\right\}_{CB}$$

$$+ \left\{\frac{Wl^2}{16EI} - \frac{Ml}{3EI} - \frac{Ml}{6EI}\right\}_{CD}$$

$$\delta L_3 = \{60/(6)^3/24(2EI) + 26.67(6)/3(2EI)\}_{CB}$$

$$+ \left\{\frac{100(5)^2}{16EI} - \frac{19.2(5)}{3EI} - \frac{100.8(5)}{6EI}\right\}_{CD}$$

$$\boxed{\delta L_3 = 11649/36EI}$$

Flexibility coefficient

Unit moment at A

Since the redundancy is 3, the flexibility coefficient at 1, 2, and 3 are required. A unit moment at 1 is applied as shown in Fig. 4.74 and the flexibility coefficients f_{11}, f_{21} and f_{31} are calculated.

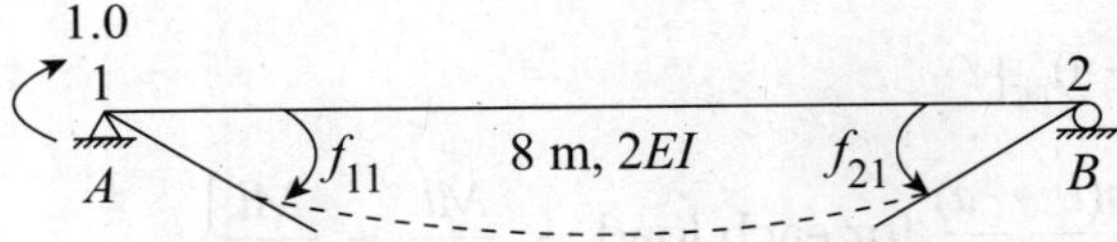

FIG. 4.74 Unit moment at 1

It is to be noted that the flexibility coefficients are given by $(l/3EI)$ for the support where the unit moment is applied and $(l/6EI)$ for the other support.

$$f_{11} = (l/3EI)_{AB} = 8/(3 \times 2EI) = 4/3EI$$

$$f_{21} = (l/6EI)BA = 8/(6{\times}2EI) = 2/3EI$$

$$f_{31} = 0(\because \text{ span } BC \text{ and } CD \text{ not affected})$$

Unit moment at B

A unit moment at B is applied and the flexibility coefficients are calculated.

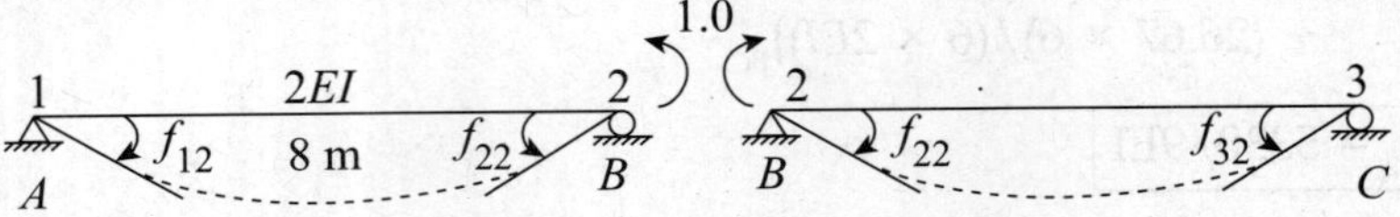

FIG. 4.75 Unit moment at B(2)

$f_{12} = (l/6EI)_{AB} = 8/(6 \times 2EI) = 2/3EI$

$f_{22} = (l/3EI)_{BA} + (l/3EI)_{BC} = \{8/(3 \times 2EI)\}_{BA} + \{6/(3 \times 2EI)\}_{BC} = 7/3EI$

$f_{32} = (l/6EI)_{BC} = 6/(6 \times 2EI) = 1/2EI$

Unit moment at C

A unit moment at *C* is applied and the flexibility coefficients f_{13}, f_{23} and f_{33} are calculated.

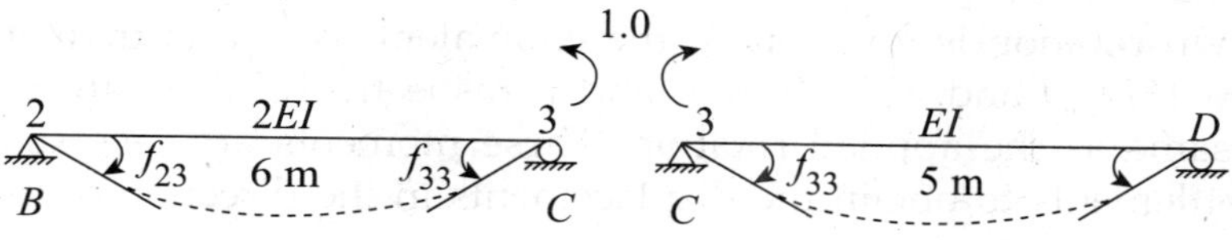

FIG. 4.76 Unit moment at C(3)

$f_{13} = 0$ (span *AB* not affected)

$f_{23} = (l/6EI)_{BC} = 6/(6 \times 2EI) = 1/2EI$

$f_{33} = (l/3EI)_{CB} + (l/3EI)_{CD} = \{6/(3 \times 2EI)\}_{CB} + \{5/(3 \times 2EI)\}_{CD} = 8/3EI$

Compatibility equations

They are

$$\delta L_1 + f_{11}R_1 + f_{12}R_2 + f_{13}R_3 = 0 \quad (1)$$

$$\delta L_2 + f_{21}R_1 + f_{22}R_2 + f_{23}R_3 = 0 \quad (2)$$

$$\delta L_3 + f_{31}R_1 + f_{32}R_2 + f_{33}R_3 = 0 \quad (3)$$

Substituting the computed values

$$3380/9EI + (4/3EI)(R_1) + (2/3EI)R_2 = 0 \quad (4)$$

$$5330/9EI + (2/3EI)R_1 + (7/3EI)R_2 + (1/2EI)R_3 = 0 \quad (5)$$

$$11649/36EI + (1/2EI)R_2 + (8/3EI)R_3 = 0 \quad (6)$$

Solving the above equations;

$$R_1 = -191.5 \text{ kNm},\ R_2 = -180.34 \text{ kNm},\ R_3 = -87.53 \text{ kNm}$$

The negative sign indicates that the moments are hogging in nature and not sagging as assumed. The shear force diagram and bending moment diagram are given below.

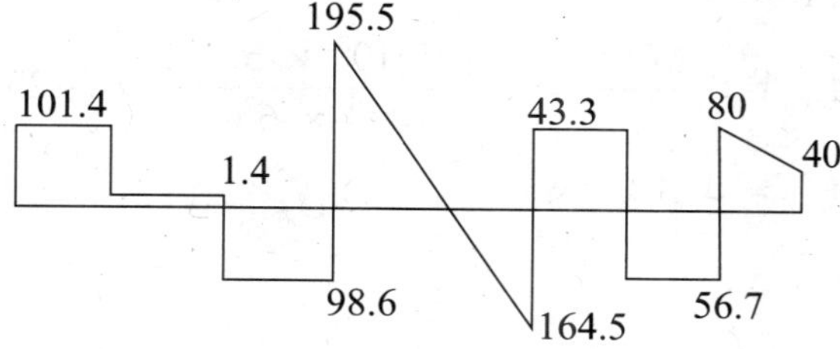

FIG. 4.77 Shear force diagram

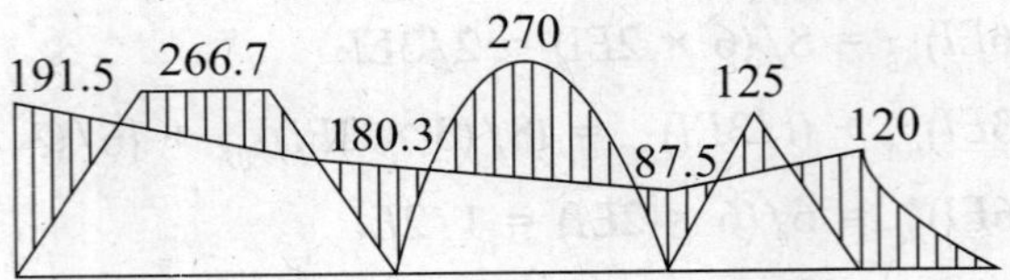

FIG. 4.78 Bending moment diagram

Note:

1. If a known rotation is given at A, the equivalent fixed end moments at A and B will be $4EI\theta_A/l$ and $2EI\theta_A/l$ at A and B respectively. The nature of moments will be same as the applied rotation. These moments are reversed and added to the settlements to obtain the displacements in the direction of the redundant chosen.
2. If support A is made simply supported, then the number of redundants will be 2. [Number of reactions − available equations of equilibrium = 4 − 2 = 2, as the number of reactions at A will be only V_A]. The redundants will be R_1 and R_2 at B and C.

Example 4.10 Analyse the continuous beam shown in Fig. 4.79. by the flexibility method using system approach. Support B sinks by 5 mm, sketch the bending moment diagram and the elastic curve. $EI = 15(10)^3$ kNm² (VTU, Jan. 2008)

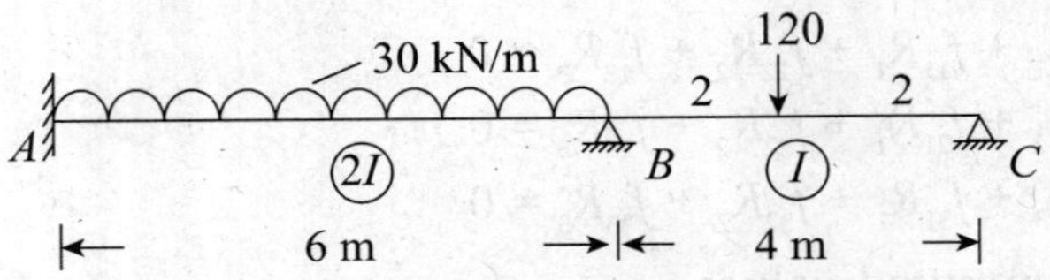

FIG. 4.79

Solution: The given beam is statically indeterminate to the second degree. The redundants chosen are M_A and M_B respectively. They are denoted as R_1 and R_2 and assumed to be clockwise. The released structure consists of two simply supported beams with external loads and the settlement moments are applied in the reverse direction for each span.

Fixed end moments

Due to settlement at support B, the induced moments are

$$M_{FAB} = \frac{-6EI\Delta}{l^2} = \frac{-6(15)10^3 \times 5 \times 2}{1000 \times 6^2} = -25 \text{ kNm} \circlearrowleft$$

$$M_{FBA} = -25 \text{ kNm} \circlearrowleft$$

$$M_{FBC} = \frac{+6EI\Delta}{l^2} = \frac{+6(15)10^3 \times 5}{1000 \times 4^2} = +28.13 \text{ kNm} \circlearrowleft$$

$$M_{FCB} = +28.13 \text{ kNm} \circlearrowleft$$

The above moments are reversed in sign and applied at the ends of the member.

Released structure along with settlement moments

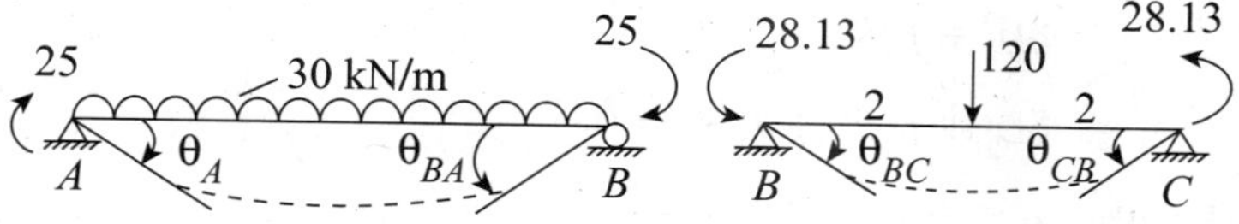

FIG. 4.80

$$\delta L_1 = \theta_A = \frac{wl^3}{24EI} + \frac{Ml}{3EI} - \frac{Ml}{6EI}$$

$$\delta L_1 = \frac{30(6)^3}{24E(2I)} + \frac{25 \times 6}{3E(2I)} - \frac{25(6)}{6E(2I)}$$

$$\boxed{\delta L_1 = 147.5/EI}$$

$$\delta L_2 = \theta_{BA} + \theta_{BC}$$

$$\theta_{BA} = \frac{wl^3}{24E(2I)} + \frac{Ml}{3E(2I)} - \frac{Ml}{6E(2I)}$$

$$\theta_{BA} = \frac{30(6)^3}{24E(2I)} - \frac{25 \times 6}{3E(2I)} + \frac{25 \times 6}{6E(2I)}$$

$$\theta_{BA} = 122.5/EI$$

$$\theta_{BC} = \frac{Wl^2}{16EI} - \frac{Ml}{3EI} + \frac{Ml}{6EI}$$

$$\theta_{BC} = \frac{120(4)^2}{16EI} - \frac{28.13(4)}{3EI} + \frac{28.13(4)}{6EI}$$

$$\theta_{BC} = 101.25/EI$$

Hence,

$$\delta L_2 = 122.5/EI + 101.25/EI$$

$$\boxed{\delta L_2 = 223.75/EI}$$

Flexibility coefficients

$$f_{11} = (l/3EI)_{AB} = 6/3E(2I) = 1/EI$$

$$f_{21} = (l/6EI)_{BA} = 6/6E(2I) = 1/2EI$$

$$f_{22} = (l/3EI)_{BA} + (l/3EI)_{BC} = 7/3EI$$

$$f_{12} = (l/6EI)_{AB} = 6/6EI(2) = 1/EI(2)$$

Compatibility equations

$$\delta L_1 + f_{11}R_1 + f_{12}R_2 = 0 \quad (1)$$

$$\delta L_2 + f_{21}R_1 + f_{22}R_2 = 0 \quad (2)$$

Substituting the above values,

$$147.5/EI + (1/EI)R_1 + (1/2EI)R_2 = 0 \quad (3)$$

$$223.75/EI + (1/2EI)R_1 + (7/3EI)R_2 = 0 \quad (4)$$

Solving the above

$$R_1 = -111.4 \text{ kNm}$$
$$R_2 = -72.1 \text{ kNm}$$

The final bending moment diagram is given below.

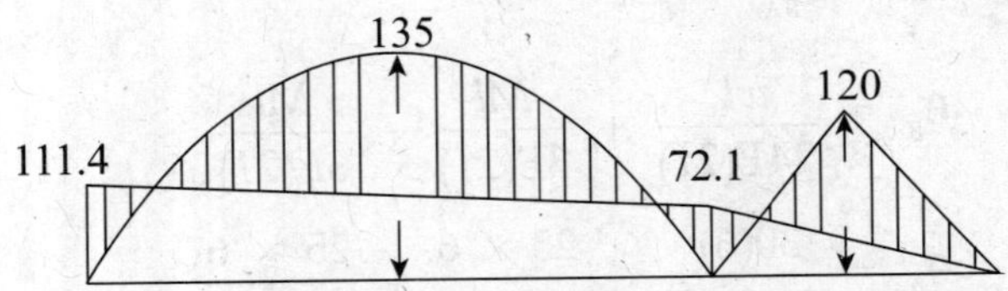

FIG. 4.81 Bending moment diagram

Example 4.11 Analyse the continuous beam shown in Fig. 4.82 by the flexibility method. Draw the shear force and bending moment diagram.

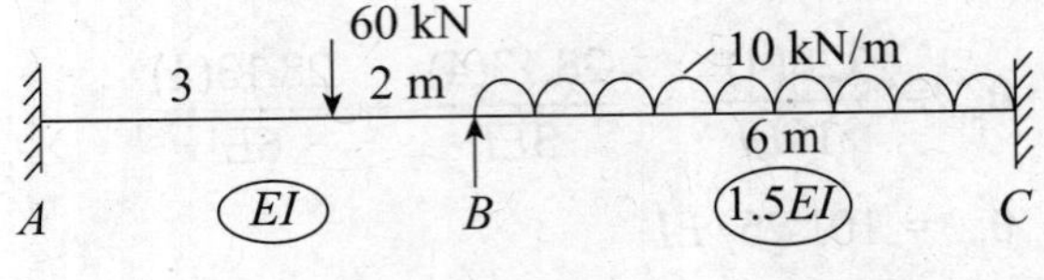

FIG. 4.82

Solution: The beam is statically indeterminate to the third degree. The redundants removed are M_A, M_B and M_C. They are designated as R_1, R_2 and R_3.

FIG. 4.83 Redundants

FIG. 4.84 Released structure

$$\delta L_1 = \theta_A = \frac{Wab(l + b)}{6EIl} = \frac{60(3)(2)(5 + 2)}{6EI(5)} = 84/EI$$

$$\delta L_2 = \theta_{BA} + \theta_{BC} = \frac{Wab(l + a)}{6EIl} + \frac{wl^3}{24E(1.5I)}$$

$$\delta L_2 = \frac{60(3)(2)(5 + 3)}{6EI(5)} + \frac{10(6)^3}{24E(1.5I)} = \frac{156}{EI}$$

$$\delta L_3 = \frac{wl^3}{24E(1.5I)} = \frac{10(6)^3}{24E(1.5I)} = \frac{60}{EI}$$

Flexibility coefficients

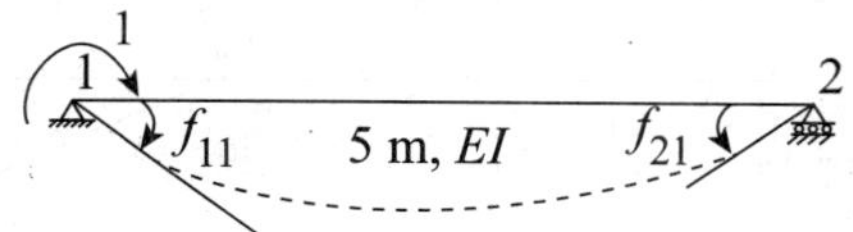

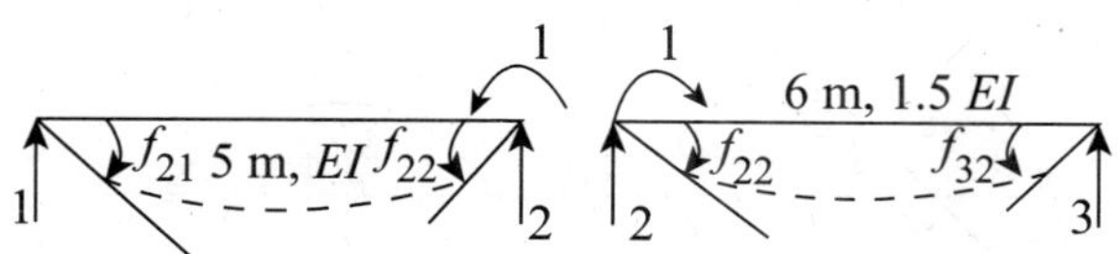

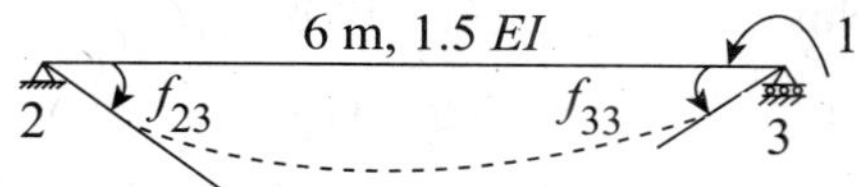

FIG. 4.85 Flexibility coefficients

$f_{11} = 5/3EI$ $\quad f_{12} = 5/6EI$ $\quad f_{13} = 0$

$f_{21} = 5/6EI$ $\quad f_{22} = 5/3EI + 6/(3 \times 1.5EI)$ $\quad f_{23} = 6/(6 \times 1EI)$

$f_{31} = 0$ $\quad f_{22} = 9/3EI$ $\quad f_{23} = 2/3EI$

$f_{32} = 6/6(1.5EI)$ $\quad f_{33} = 6/3(1.5EI)$

$f_{32} = 2/3EI$ $\quad f_{33} = 4/3EI$

Compatibility conditions

$$\delta L_1 + f_{11}R_1 + f_{12}R_2 + f_{13}R_3 = 0 \qquad (1)$$

$$\delta L_2 + f_{21}R_1 + f_{22}R_2 + f_{23}R_3 = 0 \qquad (2)$$

$$\delta L_3 + f_{31}R_1 + f_{32}R_2 + f_{33}R_3 = 0 \qquad (3)$$

Substituting the values

$$(84/EI) + (5/3EI)R_1 + (5/6EI)R_2 + 0R_3 = 0 \qquad (4)$$

$$(156/EI) + (5/6EI)R_1 + (9/3EI)R_2 + (2/3EI)R_3 = 0 \qquad (5)$$

$$(60/EI) + 0R_1 + (2/3EI)R_2 + (4/3EI)R_3 = 0 \qquad (6)$$

Solving the above equations

$$R_1 = M_A = -31.73 \text{ kNm}$$

$$R_2 = M_B = -37.33 \text{ kNm}$$

$$R_3 = M_C = -26.33 \text{ kNm}$$

Using the static equilibrium conditions, the shear force and bending moment diagrams are drawn.

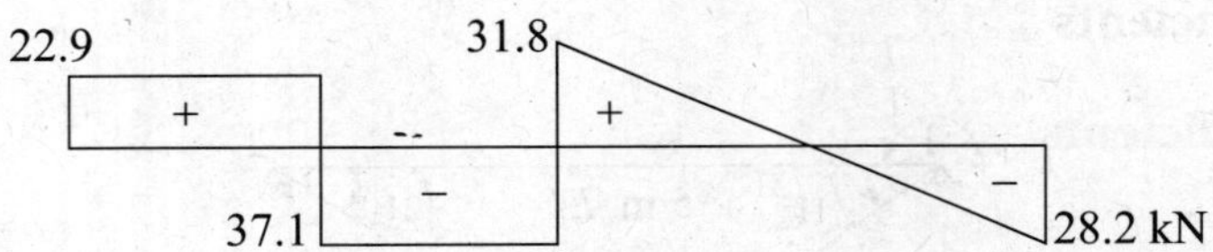

FIG. 4.86

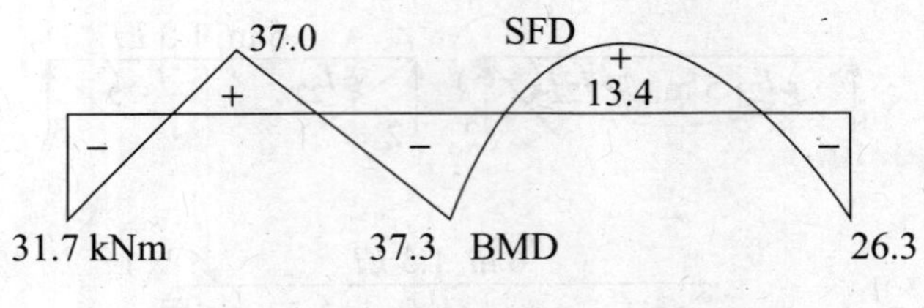

FIG. 4.87

Example 4.12 Analyse the continuous beam shown in Fig. 4.88 by the flexibility method. Draw BMD. (VTU, July 2006)

10 kN/m, 100; A, B, C, D; I, $2I$, I; 4 m, 6 m, 2, 2, 4 m

FIG. 4.88

Solution: The given beam is statically indeterminate to the second degree. M_B and M_C are taken as redundants. They are designated as R_2 and R_3 respectively. Their directions are assumed to be clockwise. The released structure consists of three simple beams and are given below.

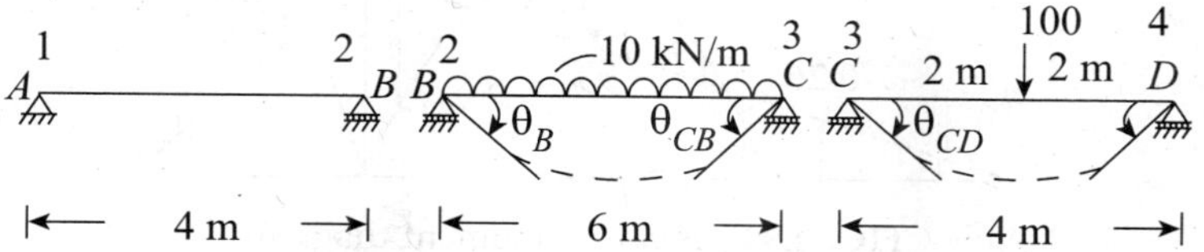

FIG. 4.89 Released structure

$$\delta L_2 = \theta_B = \frac{wl^3}{24EI} = \frac{10(6)^3}{24EI} = 45/EI$$

$$\delta L_3 = \theta_{CB} + \theta_{CD}$$

$$= \frac{wl^3}{24EI} + \frac{Wl^2}{16EI}$$

$$\delta L_3 = \frac{10(6)^3}{24(2EI)} + \frac{100(4)^2}{16EI} = 145/EI$$

Flexibility coefficients

$$f_{22} = 4/3EI + 6/3E(2I) = 7/3EI$$

$$f_{32} = 6/6E(2I) = 1/2EI$$

$$f_{23} = 6/6E(2I) = 1/2EI$$

$$f_{33} = 6/3(2EI) + 4/3EI = 7/3EI$$

Compatibility equations

$$\delta L_2 + f_{22}R_1 + f_{32}R_2 = 0 \quad (1)$$

$$\delta L_3 + f_{23}R_1 + f_{33}R_2 = 0 \quad (2)$$

Substituting the values obtained,

$$45/EI + (7/3EI)R_1 + (1/2EI)R_2 = 0 \quad (3)$$

$$145/EI + (1/2EI)R_1 + (7/3EI)R_2 = 0 \quad (4)$$

On solving, we get

$$R_1 = -6.26 \text{ kNm} = M_B$$

$$R_2 = -60.8 \text{ kNm} = M_C$$

The final BMD is shown below.

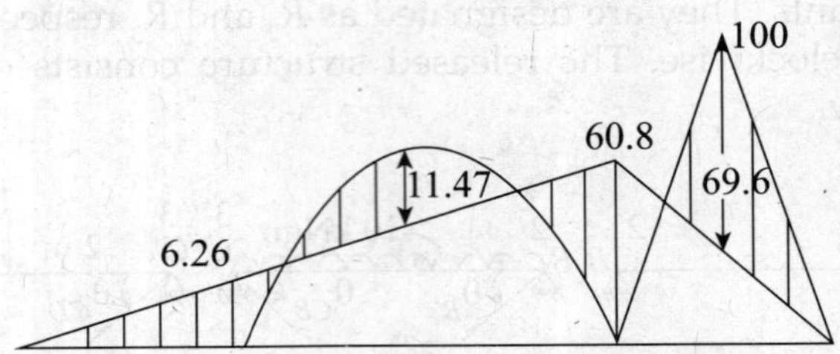

FIG. 4.90 Bending moment diagram

Example 4.13 Analyse the continuous beam shown in Fig. 4.91 and find the moments at the supports *B* and *C*. The supports *B* and *C* sink down by (150/*EI*) kNm units and (75/*EI*) kNm units respectively. Apply the flexibility method.

FIG. 4.91

Solution: The given beam is statically indeterminate to the second degree. M_B and M_C are treated as redundants. They are denoted as R_2 and R_3 and assumed to be clockwise. The released structure consists of three simple beams as along with settlement moments.

Effect of sinking of supports

The sinking of supports at *B* and *C* are converted into joint loads at *A, B, C* and *D*. The fixed end moments due to sinking of supports are given in the following figures.

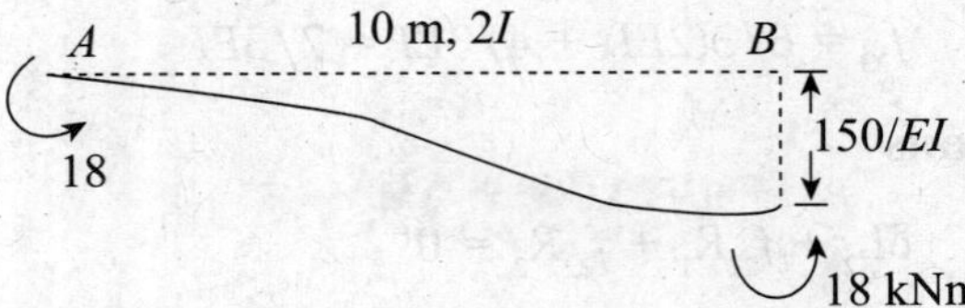

FIG. 4.92 Fixed end moments due to sinking of support at *B*

$$M'_{AB} = M'_{BA} = -6E(2I)\left(\frac{150}{EI}\right)/10^2 = -18\text{ kNm}$$

The fixed end moments due to settlement at *B* and *C* are as follows. Support *B* sinks by (150/*EI*) while *C* sinks by (75/*EI*). Hence, the net settlement between *C* and *B* with respect to *C* is given by (150/*EI*) – (75/*EI*) = 75/*EI*. This induces the fixed end moment as

$$M'_{BC} = M'_{CB} = +6E(3I)\left(\frac{75}{EI}\right)/10^2 = +13.5\ \text{kNm}$$

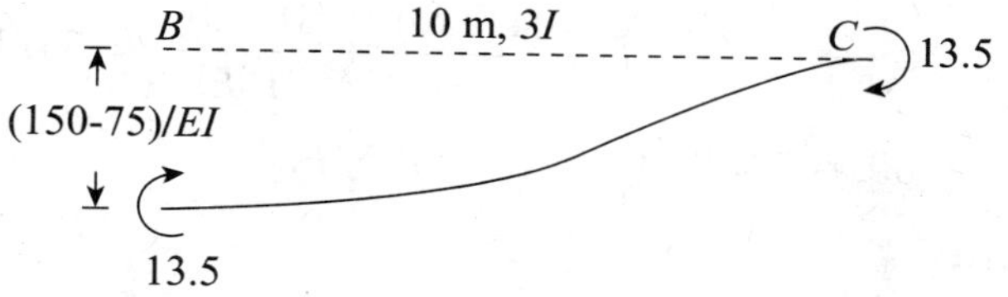

FIG. 4.93 Fixed end moments due to relative settlements between B and C

The moments due to settlement at C

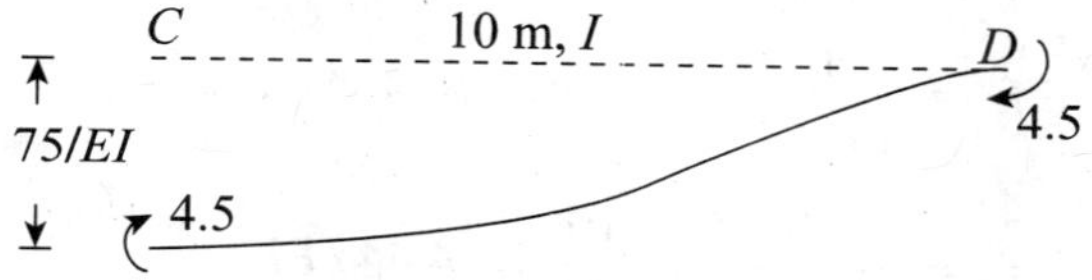

FIG. 4.94 Fixed end moment due to relative settlement between C and D

$$M'_{CD} = M'_{DC} = \frac{6EI(75/EI)}{10^2} = 4.5\ \text{kNm}$$

The moments due to settlement are reversed and the displacements due to loads are computed.

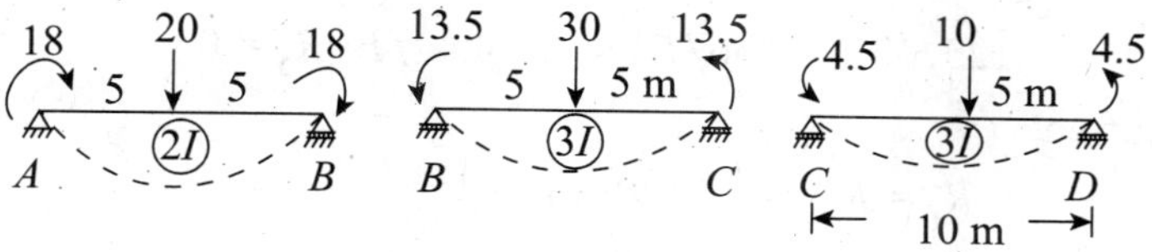

FIG. 4.95

$$\delta L_2 = \theta_{BA} + \theta_{BC}$$

$$\theta_{BA} = \frac{Wl^2}{16(2EI)} - \frac{Ml}{3(2EI)} + \frac{Ml}{6/(2EI)}$$

$$\theta_{BA} = \frac{20(10)^2}{16(2EI)} - \frac{18(10)}{3(2EI)} + \frac{18(10)}{6/(2EI)} = 47.5/EI$$

$$\theta_{BC} = \frac{30(10)^2}{16(3EI)} - \frac{13.5(10)}{3(3EI)} + \frac{13.5(10)}{6/(3EI)} = 55/EI$$

$$\delta L_2 = 47.5/EI + 55/EI = 102.5/EI$$

$$\theta_{CB} = \frac{30(10)^2}{16(3EI)} + \frac{13.5(10)}{3(3EI)} - \frac{13.5(10)}{6/(3EI)} = 70/EI$$

$$\theta_{CD} = \frac{-4.5(10)}{3EI} + \frac{4.5(10)}{6EI} + \frac{10(10)^2}{16EI} = 55/EI$$

$$\delta L_3 = 70/EI + 55/EI = 125/EI$$

Flexibility coefficient

$$f_{22} = \frac{10}{3(2EI)} + \frac{10}{3(3EI)} = 2.78/EI$$

$$f_{32} = \frac{10}{6(3EI)} = 0.56/EI$$

$$f_{23} = \frac{10}{6(3EI)} = 0.56/EI$$

$$f_{33} = \frac{10}{3(3EI)} + \frac{10}{3EI} = 4.44/EI$$

Compatibility equations

$$\delta L_2 + f_{22}R_1 + f_{23}R_2 = 0 \quad (1)$$

$$\delta L_3 + f_{32}R_1 + f_{33}R_2 = 0 \quad (2)$$

$$102.5/EI + (2.78/EI)R_1 + (0.56/EI)R_2 = 0 \quad (3)$$

$$125/EI + (0.56/EI)R_1 + (4.44/EI)R_2 = 0 \quad (4)$$

Solving the above

$R_1 = M_B = -32$ kNm ↺

$R_2 = M_C = -24$ kNm ↺

Support moment $M_B = 32$ kNm

Support moment $M_C = 24$ kNm

Example 4.14 Analyse the continuous beam shown in Fig. 4.96 by force method. Draw the bending moment diagram and shear force diagram.

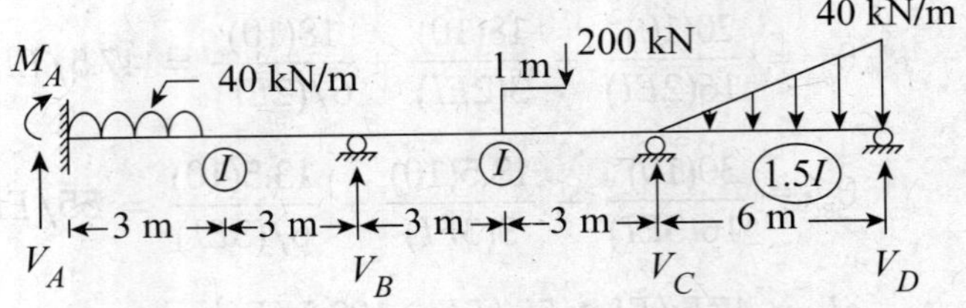

FIG. 4.96 Three-span continuous beam with the reaction components

Solution: The given continuous beam is statically determinate to the third degree. The number of unknowns are M_A, V_A, V_B, V_C and V_D. The equilibrium equations are $\Sigma V = \Sigma M = 0$ and hence the degree of indeterminacy is 5 – 2 = 3. The same degree of indeterminacy can also be obtained logically also. The given continuous beam can be made as a cantilever beam by removing the supports *B*, *C* and *D*. The cantilever beam is statically determinate and the degree of redundancy is zero. The additional three support reactions make a continuous beam. Hence, the number of additional support reactions over and above the cantilever beam gives the degree of indeterminacy. Thus, the degree of indeterminacy is three.

It is not advisable to treat V_B, V_C and V_D, as redundants. The reason being, the computations of deflections at *B*, *C* and *D* of the cantilever beam is tedious and cumbersome. Hence, it is preferable to make the given beam as determinate. The beam is made determinate by removing the external moment M_A and the internal moments at *B* and *C* [*B* and *C* are continuours supports, while *D* is a simple support with zero moment], i.e., $R_1 = M_A$; $R_2 = M_B$ and $R_3 = M_C$ and are assumed as sagging.

The released structure consists of three simply supported beams with the external loads along with BMD for each load to facilitate the usage of the moment area method.

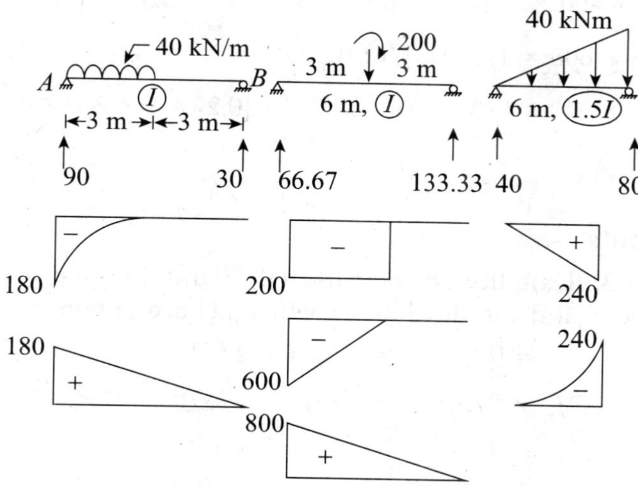

FIG. 4.97 Released structure with BMS in kNm

The rotations at *A*, *B* and *C* due to external loads and moments are calculated and considered +ve, if they are in the same directions of the redundant at *A*, *B* and *C* as indicated in Fig. 4.97.

$$\delta L_1 = \theta A = [\text{Moment of the BMD between } A \text{ and } B \text{ about } B]/\text{span } EI$$

$$\theta_A = \left[-\left(\frac{1}{3}\right)(3)(180)(3+0.75\times 3)+\left(\frac{1}{2}\right)6\,(180)\left(\frac{2}{3}\times 6\right)\right]\Big/6EI$$

$$\boxed{\delta L_1 = \theta A = 202.5/EI}$$

$$\delta L_2 = \theta_B|_{BA} + \theta_B|_{BC} = \{[\text{Moment of the BMD between } A \text{ and } B \text{ about } A]/\text{span} \times EI\}_{BA}$$

$$+\{[\text{Moment of the BMD between } B \text{ and } C \text{ about } C]/\text{span} \times EI\}_{BC}$$

$$\delta L_2 = \{[0.5 \times 6 \times 180 \times 1/3 \times 6]/6EI - (1/3)(3)180(0.25 \times 3)/6EI\}_{BA}$$

$$+\{[0.5 \times 6 \times 800 \times 2/3 \times 6]/6EI - [3 \times 200 \times (3 + 0.5 \times 3)/6EI]$$

$$-[0.5 \times 3 \times 600 \times (3 + 2/3 \times 3)/6EI]\}_{BC}$$

$$\boxed{\delta L_2 = 557.5/EI}$$

$$\delta L_3 = \theta_C|_{CB} + \theta_C|_{CD}$$

$$= \{[\text{Moment of the BMD between } B \text{ and } C \text{ about } B]/\text{span} \times EI\}_{BA}$$

$$+\{[\text{Moment of the BMD between } C \text{ and } D \text{ about } D]/\text{span} \times EI\}_{BC}$$

$$= \{[0.5 \times 6 \times 800 \times 1/3 \times 6]/6EI - [3 \times 200 \times 0.5 \times 3]/6EI$$

$$-[0.1 \times 3 \times 600 \times 1/3 \times 3]/6EI\}_{BC}$$

$$+\{[0.5 \times 6 \times 240 \times 1/3 \times 6/6 \times 1.5EI] - [0.25 \times 6 \times 240 \times 0.2 \times 6/6 \times 1.5EI]\}_{CD}$$

$$\boxed{\delta L_3 = 612/EI}$$

Flexibility coefficients

Since redundancy is 3, flexibility coefficients at 1, 2 and 3 is required. A unit moment is applied at the nodes and the flexibility coefficients are obtained as follows.

$$f_{11} = [l/3EI]_{AB} = 6/3EI = 2/EI$$

$$f_{21} = [l/6EI]_{BA} = 6/6EI = 1/EI$$

$$f_{31} = 0$$

$$f_{12} = [l/6EI]_{AB} = 6/(6 \times EI) = 1/EI$$

$$f_{22} = (l/3EI)_{BA} + (l/3EI)_{BC} = \{6/(3 \times EI)\}_{BA}$$

$$+\{6/(3 \times EI)\}_{BC} = 4/EI$$

$$f_{32} = (l/6EI)_{BC} = 6/6EI = 1/EI$$

$$f_{13} = 0$$

$$f_{23} = (l/6EI)_{BC} = 6/(6EI) = 1/EI$$

$$f_{33} = (l/3EI)_{CB} + (l/3EI)_{CD} = 10/3EI$$

Compatibility equations

$$\delta L_1 + f_{11}R_1 + f_{12}R_2 + f_{13}R_3 = 0$$
$$\delta L_2 + f_{21}R_1 + f_{22}R_2 + f_{23}R_3 = 0$$
$$\delta L_3 + f_{31}R_1 + f_{32}R_2 + f_{33}R_3 = 0$$

Substituting the computed values

$$202.5/EI + (2/EI)R_1 + (1/EI)R_2 = 0$$
$$557.5/EI + (1/EI)R_1 + (4/EI)R_2 + (1/EI)R_3 = 0$$
$$612.0/EI + (1/EI)R_2 + (10/3EI)R_3 = 0$$

The above compatibility conditions can be written in matrix form

$$\begin{Bmatrix} R_1 \\ R_2 \\ R_3 \end{Bmatrix} = -\begin{bmatrix} 2 & 1 & 0 \\ 1 & 4 & 1 \\ 0 & 1 & 10 \end{bmatrix}^{-1} \begin{bmatrix} 202.5 \\ 557.5 \\ 612.0 \end{bmatrix}$$

On solving, we obtain

$$R_1 = -58.65 \text{ kNm} \circlearrowleft$$
$$R_2 = -85.20 \text{ kNm} \circlearrowleft$$
$$R_3 = -158.04 \text{ kNm} \circlearrowleft$$

The negative sign indicates that moments are hogging. Figure 4.98 gives the free body, shear force and bending moment diagram.

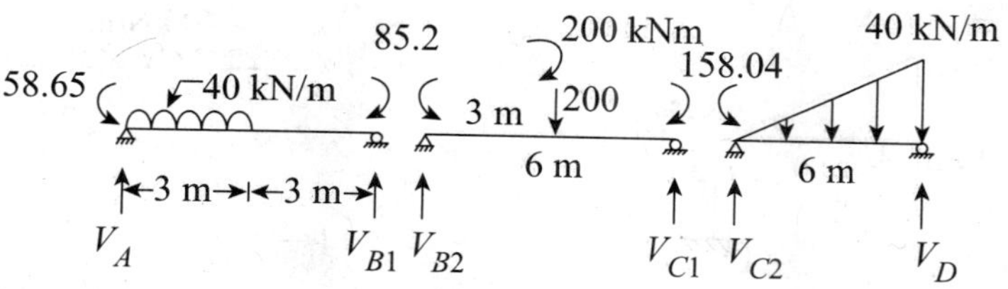

FIG. 4.98 Free body diagram of each span

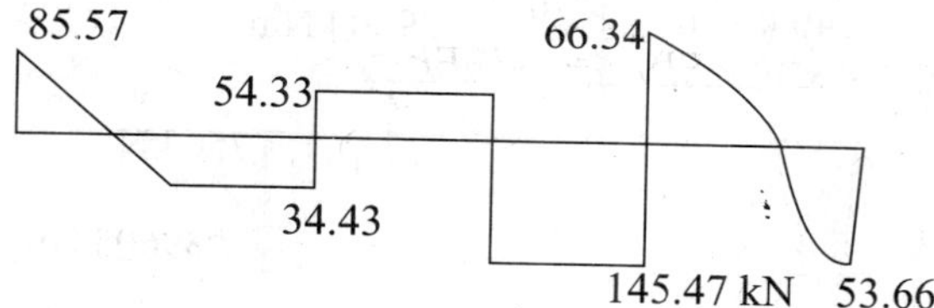

FIG. 4.99 Shear force diagram

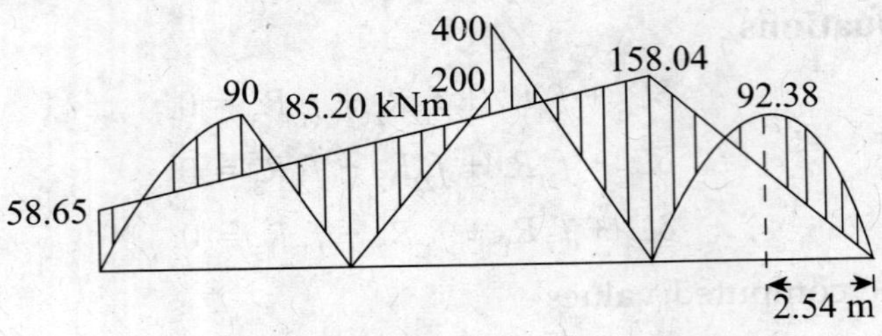

FIG. 4.100 Bending moment diagram

Example 4.15 Analyse the rigid frame shown in by the Flexibility method. Draw BMD, SFD and axial force diagram.

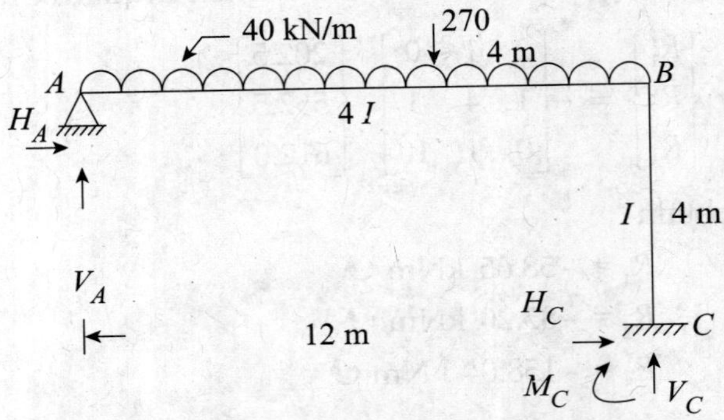

FIG. 4.101 Rigid frame with reaction components

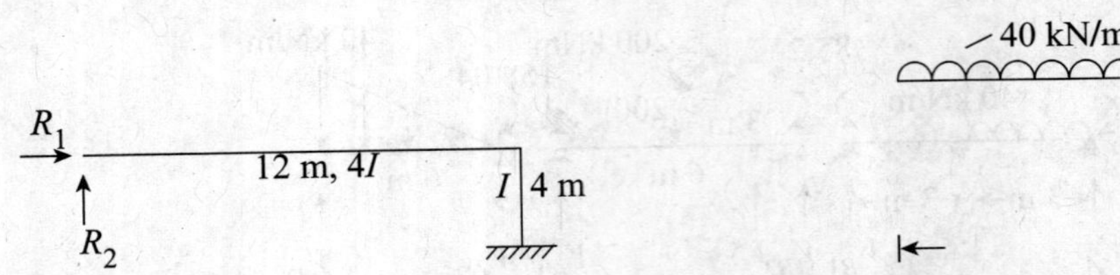

Fig. 4.102 Redundants

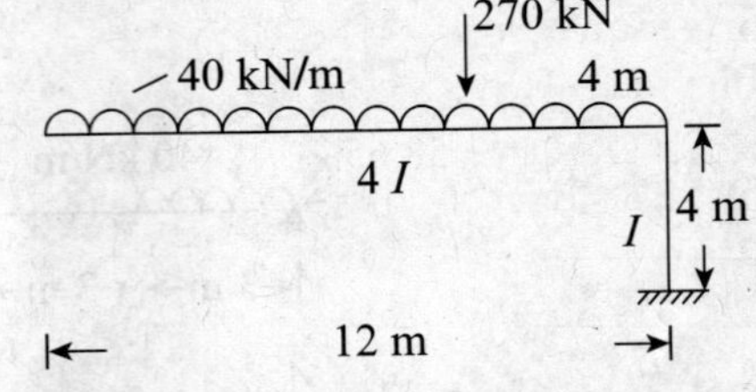

Fig. 4.103 Released structure with external loads

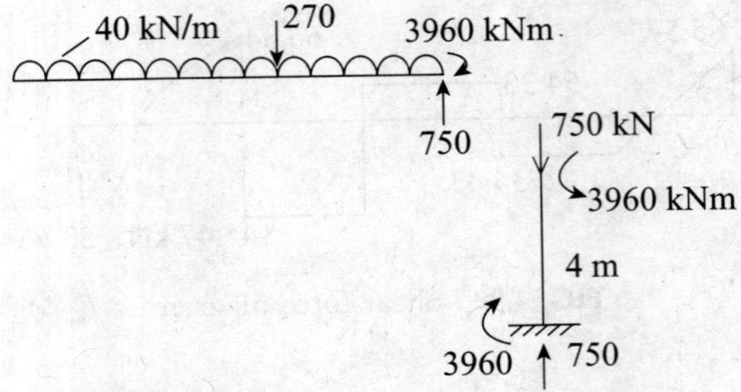

FIG. 4.104 FBD of the released structure

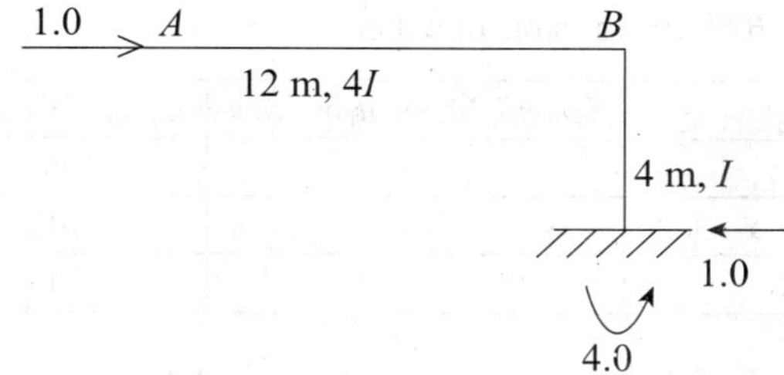

FIG. 4.105 Unit horizontal load at A

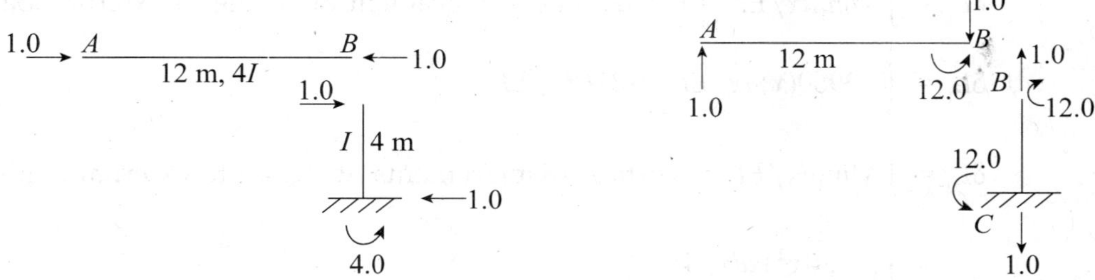

Fig. 4.106 FBD for unit horizontal load at A

Fig. 4.107 FBD for unit vertical load at A

The rigid frame is statically indeterminate to the second degree. H_A and V_A are assumed as redundants R_1 and R_2. Thus, the given rigid frame is a cantilever bent; which is stable and statically determinate. The released structure with external loads is shown in Fig. 4.102. The horizontal and vertical displacements at A due to the external loads are calculated using the unit load method. Using the free body diagram, the expressions for bending moment in each member is worked out and are indicated in Tables 4.2–4.4:

Table 4.2 Expressions for BM, M in

Members	*M*	*Second Moment of Area*	*Limits of Integration*	*Origin*
AD	$-40x^2/2 = -20x^2$	$4I$	0 – 8	*A*
DB	$750x - 20x^2 - 3960$	$4I$	0 – 4	*B*
BC	-3960	I	0 – 4	*B*

Table 4.3 Expressions for BM, m_1 in each member

Members	m_1	*Second Moment of Area*	*Limits of Integration*	*Origin*
AD	0	–	–	–
DB	0	–	–	–
BC	x	I	0 – 4	*B*

Table 4.4 : Expressions for BM, m_2 in each member

Members	m_2	*Second Moment of Area*	*Limits of Integration*	*Origin*
AD	x	$4I$	0 – 8	*A*
DB	$12 - x$	$4I$	0 – 4	*B*
BC	12	I	0 – 4	*B*

In the above Table, 4.2–4.4, M represents the bending moment in the members due to unit horizontal and vertical loads respectively.

$$\delta L_1 = \int Mm_1 dx/EI \text{ = Horizontal displacement at } A \text{ due to external loads}$$

$$\delta L_1 = \int_0^4 -3960(x)dx/EI = -31680/EI$$

$$\delta L_2 = \int Mm_2 dx/EI \text{ = Vertical displacements at } A \text{ due to external loads}$$

$$= \int_0^8 \{(-20x^2)xdx/4EI\}_{AD}$$

$$+ \int_0^4 \{(750x - 20x^2 - 3960)(12 - x)dx/4EI\}_{BD}$$

$$+ \int_0^4 \{(-3960)(12)dx/EI\}_{BC}$$

i.e. $\boxed{\delta L_2 = -221760/EI}$

Flexibility coefficients

$$f_{11} = \int m_1 m_1 dx/EI = \int_0^4 \{x^2 dx/EI\}_{BC} = 64/3EI$$

$$f_{12} = f_{21} = \int m_1 m_1 dx/EI = \int_0^4 \{12xdx/EI\}_{BC} = 96/EI$$

$$f_{22} = \int m_2 m_2 dx/EI = \int_0^8 \{x^2 dx/4EI\}_{AD} + \int_0^4 \{(12 - x)^2 dx/4EI\}_{BD}$$

$$+ \int_0^4 \{(12^2 dx/EI\}_{BC} = 720/EI$$

Compatibility Eqns. in terms of displacements

$$\delta L_1 + f_{11}R_1 + f_{12}R_2 = 0$$

$$\delta L_2 + f_{21}R_1 + f_{22}R_2 = 0$$

Substituting the computed values

$$-31680/EI + (64/3EI)R_1 + (96/EI)R_2 = 0$$

$$-221760/EI + (96/EI)R_1 + (720/EI)R_2 = 0$$

Solving the above equations;

$$R_1 = H_A = 247.5 \text{ kN}$$

$$R_2 = V_A = 275.0 \text{ kN}$$

The positive sign indicates that the assumed direction is correct.

Calculation of other reactions

The free body diagrams of each span are shown with loads and redundant reactions.

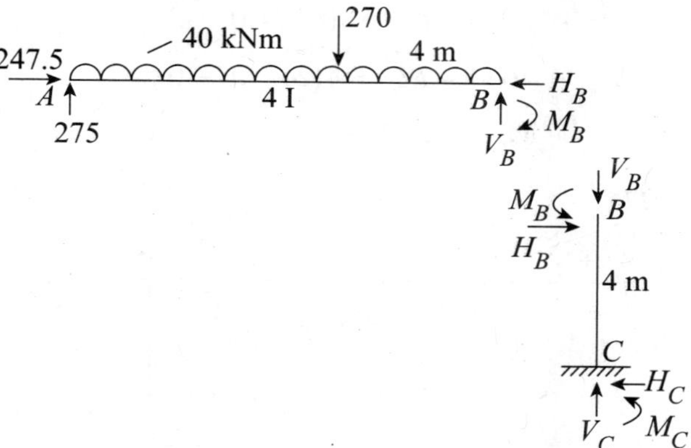

FIG. 4.108 FBD of each element

Consider span *AB*,

$$\Sigma V = 0;$$

$$275 + V_B = 270 + 12(40)$$

$$\boxed{V_B = 475 \text{ kN}}$$

$$\Sigma M_A = 0;$$

$$270(8) + 40(12)^2/2 - 475(12) + M_B = 0$$

$$\boxed{M_B = 660 \text{ kNm}}$$

$$\Sigma H = 0;$$

$$247.5 - H_B = 0$$

$$\boxed{H_B = 247.5 \text{ kN}}$$

Consider span *BC*

H_B, V_B and M_B at joint *B* of the member *BC* will act opposite to the forces at joint *B* of Member *AB*. This is the requirement for joint equilibrium (Ref. Fig. 4.108)

$$\Sigma V = 0;$$

$$V_C - V_B = 0$$

$$V_C = V_B = 475 \text{ kN}$$

$$\Sigma H = 0;$$

$$H_B - H_C = 0$$

$$H_B = H_C = 247.5 \text{ kN}$$

$$\Sigma M_B = 0;$$

$$-M_B - M_C + 4H_B = 0$$

$$-660 - M_C + 4(247.5) = 0$$

$$M_C = 330 \text{ kNm}$$

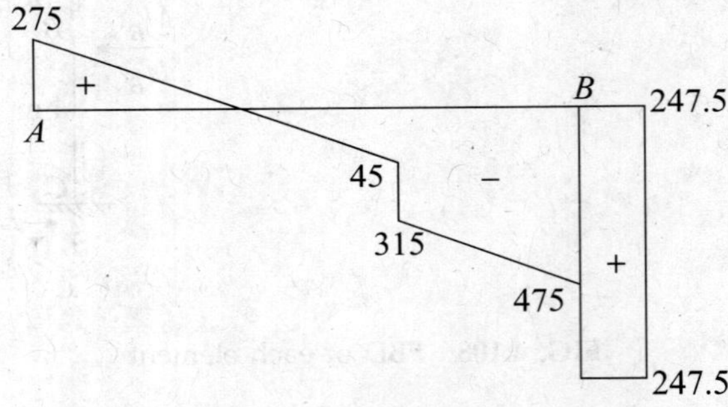

FIG. 4.109 Shear force diagram (kN)

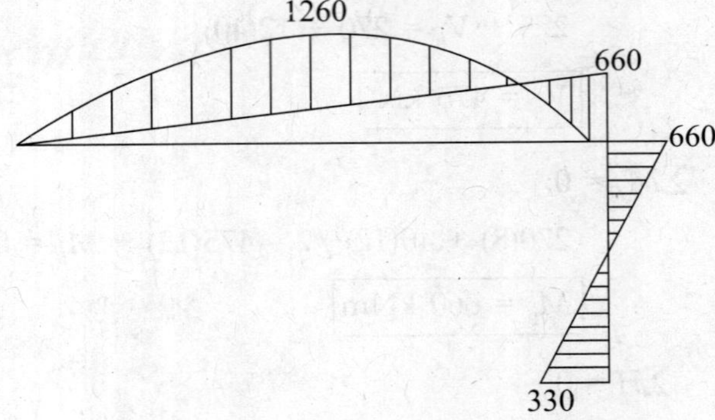

FIG. 4.110 Bending moment diagram

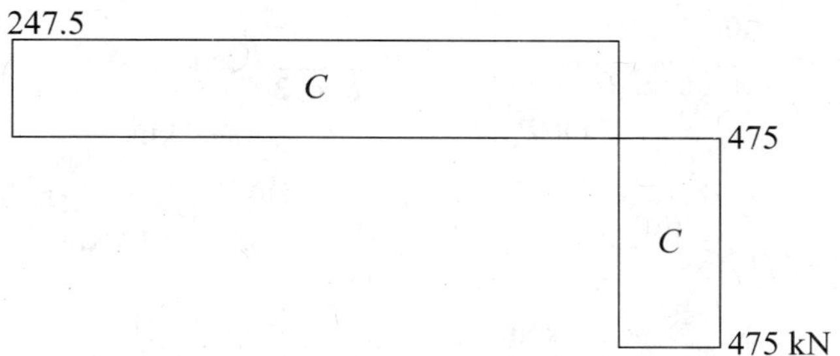

FIG. 4.111 Axial force diagram (kN)

Example 4.16 Analyse the frame shown in Fig. 4.112 and draw the bending moment diagram. (VTU - 2012)

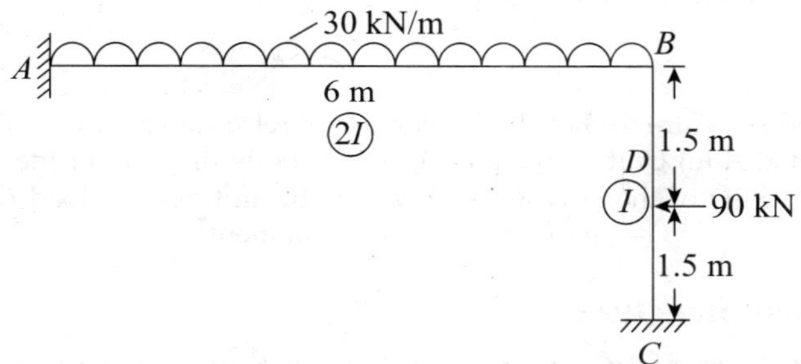

FIG. 4.112

Solution: The given frame is statically indeterminate to the third degree. Three redundants are to be removed to make it a statically determinate structure. Hence, remove the support *A* and in doing so the given structure becomes a cantilever bent *CBA*. This cantilever bent is the released structure which is statically determinate.

δL_1, δL_2 and δL_3 are the horizontal, vertical displacements and rotation at *A* due to the applied loads. They are computed and given below.

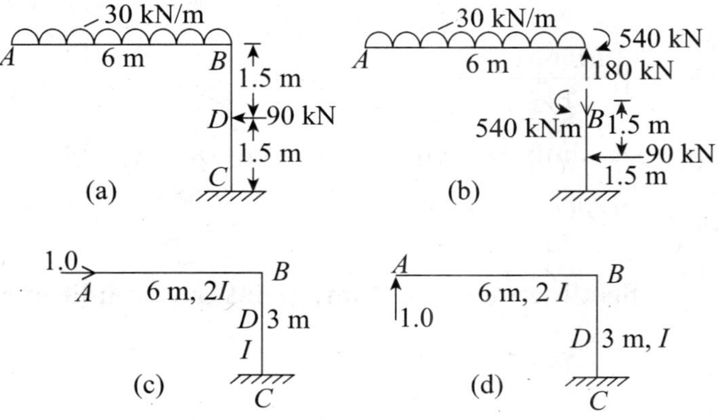

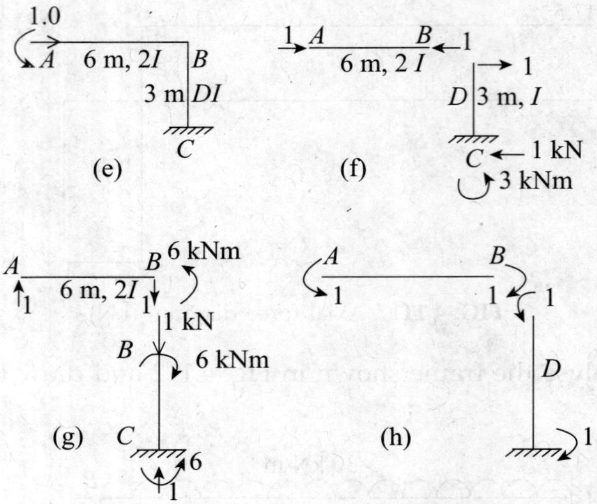

FIG. 4.113 (a) Released structure (b) Free body diagram of released structure (c) Unit horizontal load at A (d) Unit vertical load at A (e) Unit moment at A (f) Free body diagram of the released structure with unit horizontal load (g) FBD of the released structure with unit vertical load (h) FBD of the released structure with unit moment.

Analysis of Released Structure

The displacement at A are δL_1 in the horizontal direction, δL_2 in the vertical direction and δL_3 is the rotation. These displacements were determined using the unit load method.

$$\delta L_1 = \int \frac{Mm_1 dx}{EI} = \text{Horizontal displacement due to external loads at A}$$

where

M = bending moment at any point due to external load

m_1 = bending moment at any point due to unit horizontal load

EI = flexural rigidity of the member

Similarly,

$$\delta L_2 = \int \frac{Mm_2 dx}{EI}$$

and m_2 = Bending moment at any point due to unit vertical load at A

$$\delta L_3 = \int \frac{Mm_3 dx}{EI}$$

and m_3 = Bending moment at any point due to unit moment.

The values of M, m_1, m_2 and m_3 are tabulated for different segments.

Table 4.5 Values of M, m_1, m_2, and m_3

Members	*M*	m_1	m_2	m_3	*I*	*Limits*	*Origin*
AB	$-30x^2/2$	0	x	–1	2*I*	0 – 6	*A*
BD	–540	x	6	–1	*I*	0 – 1.5	*B*
BC	–675	$-(x + 3)$	6	–1	*I*	0 – 1.5	*C*

$$\delta L_1 = \int \frac{Mm_1 dx}{EI} = \int_0^{1.5} -(540)x\frac{dx}{EI} + \int_0^{1.5} -(675)(-x+3)\frac{dx}{EI}$$

$$\boxed{\delta L_1 = -2885.6/EI}$$

$$\delta L_2 = \int \frac{Mm_2 dx}{EI} = \int_0^6 \frac{-15x^2(x)\,dx}{2EI} + \int_0^{1.5} -(540)(6)\frac{dx}{EI}$$

$$+ \int_0^{1.5} -(675)(6)\frac{dx}{EI}$$

$$\boxed{\delta L_2 = -13365/EI}$$

$$\delta L_3 = \int \frac{Mm_3 dx}{EI} = \int_0^6 \frac{-15x^2(-1)\,dx}{2EI} + \int_0^{1.5} (-540)(-1)\frac{dx}{EI}$$

$$+ \int_0^{1.5} (-675)(-1)\frac{dx}{EI}$$

$$\boxed{\delta L_3 = 2362.5/EI}$$

Flexibility coefficients

$$f_{11} = \int m_1 m_1 \frac{dx}{EI} = \int_0^{1.5} \frac{x^2 dx}{EI} + \int_0^{1.5} \frac{(-x+3)^2 dx}{EI} = \frac{9}{EI}$$

$$f_{12} = f_{21} = \int m_1 m_2 \frac{dx}{EI}$$

$$f_{13} = f_{31} = \int m_1 m_3 \frac{dx}{EI}$$

$$f_{13} = f_{31} = \int_0^{1.5} \frac{-x\,dx}{EI} + \int_0^{1.5} (-x+3)(-1)\frac{dx}{EI}$$

$$f_{13} = f_{31} = \frac{-4.5}{EI}$$

$$f_{21} = \int_0^{1.5} \frac{6xdx}{EI} + \int_0^{1.5} 6(-x+3)\frac{dx}{EI} = \frac{27}{EI}$$

$$f_{22} = \int \frac{m_2 m_2 dx}{EI} = \int_0^6 \frac{x^2 dx}{2EI} + \int_0^{1.5} \frac{36dx}{EI} + \int_0^{1.5} \frac{30dx}{EI}$$

$$f_{22} = \frac{144}{EI}$$

$$f_{23} = f_{32} = \int m_2 m_3 dx/EI$$

$$= \int_0^6 \frac{-xdx}{2EI} + \int_0^{1.5} \frac{-6dx}{EI} + \int_0^{1.5} \frac{-6dx}{EI}$$

$$f_{23} = f_{32} = -27/EI$$

$$f_{33} = \int m_3 m_3 \frac{dx}{EI} = \int_0^6 \frac{dx}{2EI} + \int_0^{1.5} \frac{dx}{EI} + \int_0^{1.5} \frac{dx}{EI}$$

$$f_{33} = 6/EI$$

Compatibility equations

$$\delta L_1 + f_{11}R_1 + f_{12}R_2 + f_{13}R_3 = 0$$

$$\delta L_2 + f_{21}R_1 + f_{22}R_2 + f_{23}R_3 = 0$$

$$\delta L_3 + f_{31}R_1 + f_{32}R_2 + f_{33}R_3 = 0$$

Substituting the values

$$-2885.6/EI + (9/EI)R_1 + (27/EI)R_2 + (-4.5/EI)R_3 = 0$$

$$-13365/EI + (27/EI)R_1 + (144/EI)R_2 + (-27/EI)R_3 = 0$$

$$+2362.5/EI + (-4.5/EI)R_1 + (-27/EI)R_2 + (6/EI)R_3 = 0$$

Simplifying the above

$$9R_1 + 27R_2 - 4.5R_3 = 2885.6$$

$$27R_1 + 144R_2 - 27R_3 = 13365.0$$

$$-4.5R_1 - 27R_2 + 6R_3 = -2362.5$$

On solving;

$$R_1 = H_A = 81.6 \text{ kN } (\rightarrow)$$

$$R_2 = V_A = 97.0 \text{ kN } (\uparrow)$$

$$R_3 = M_A = 104.01 \text{ kNm } (\circlearrowleft)$$

Using the free body diagram and equilibrium equations, the final BMD is drawn.

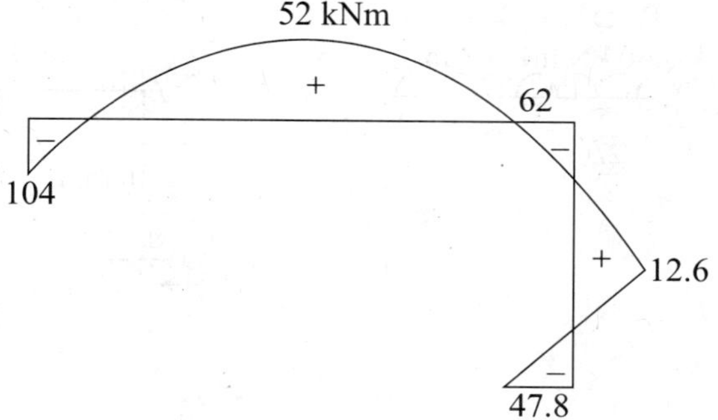

FIG. 4.114 Bending moment diagram

Example 4.17 Analyse the rigid frame shown in Fig. 4.115 by the flexibility method. Draw BMD, SFD and axial force diagram (AFD).

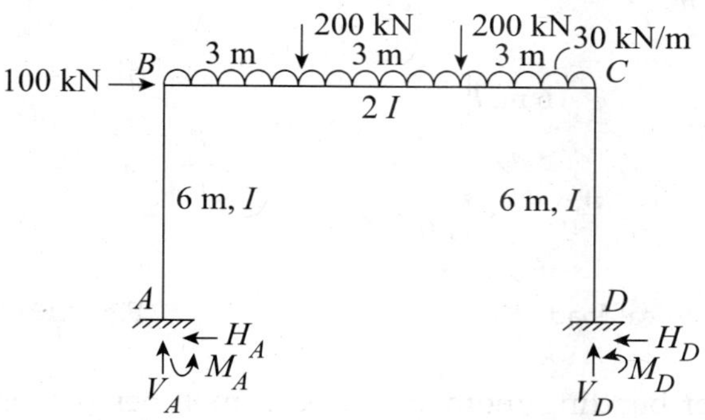

FIG. 4.115 Rigid frame with reaction components

Solution: The above rigid frame is statically indeterminate to the third degree. The frame is made statically determinate by releasing the support D. In doing so, the given frame becomes a cantilever bent. $ABCD$ with a fixed support A. The released structure with external loads is shown in Fig. 4.116.

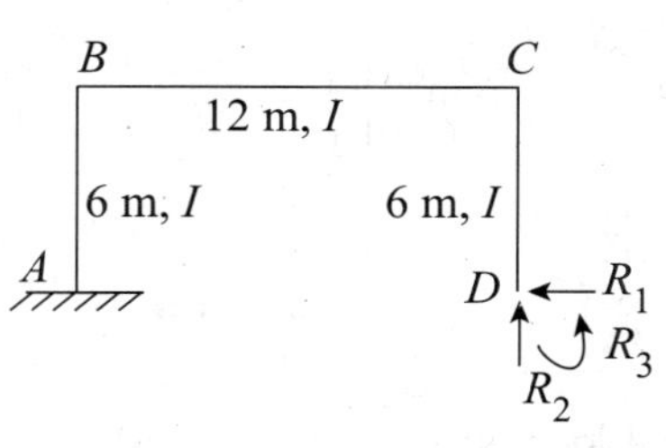

Fig. 4.116 Redundants

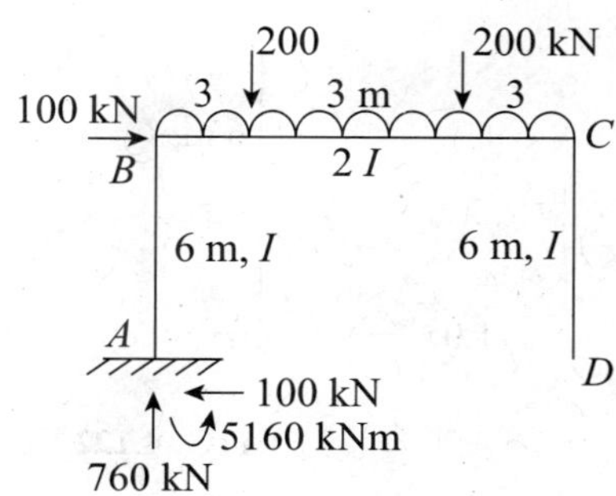

4.117 Released structure with external loads

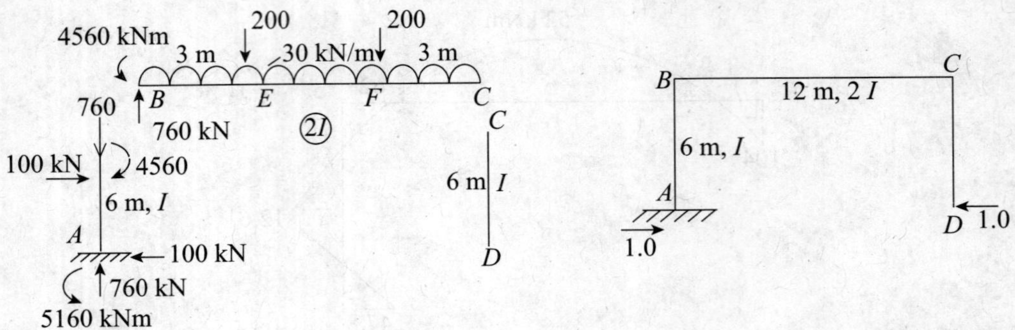

Fig. 4.118 FBD of released structure

Fig. 4.119 Unit horizontal loads

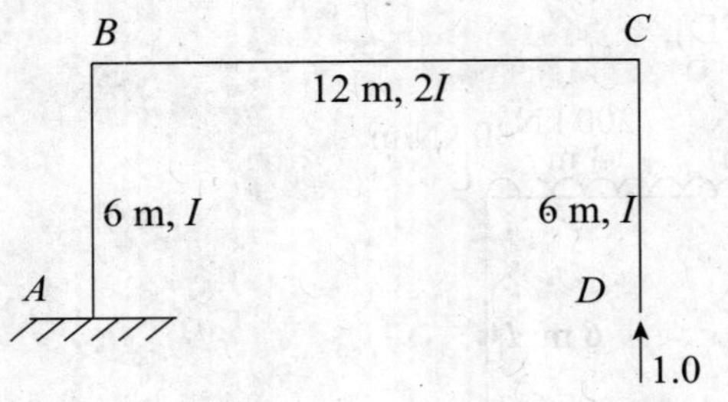

Fig. 4.120 Unit vertical load at *D*

Fig. 4.121 Unit moment at *D*

The expressions for bending moment for each member is worked out based on the free body diagrams (FBD). They are indicated in the following Table 4.6.

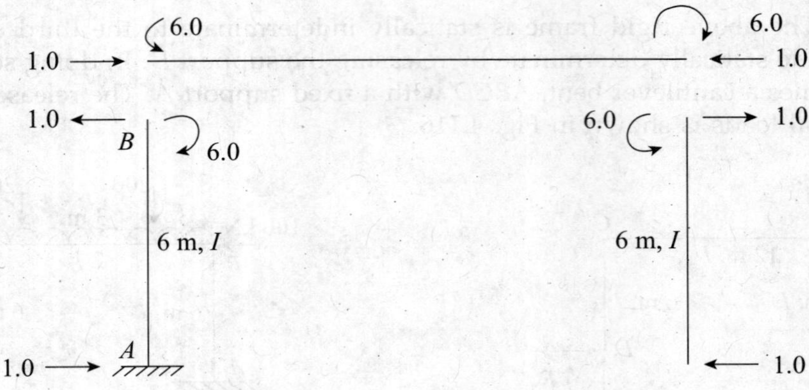

FIG. 4.122 FBD for unit horizontal load at *D*

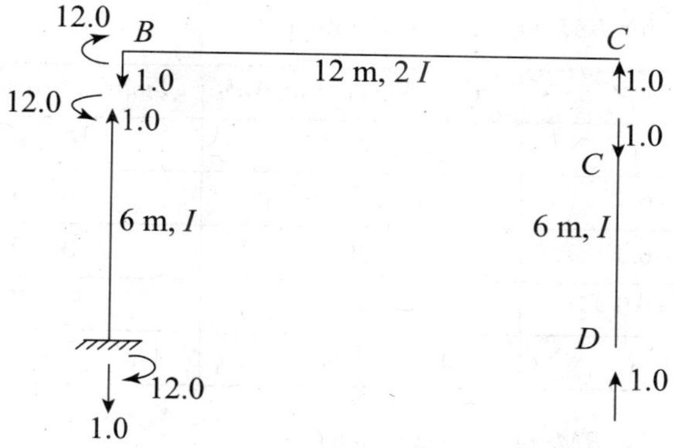

FIG. 4.123 FBD for unit vertical load at D

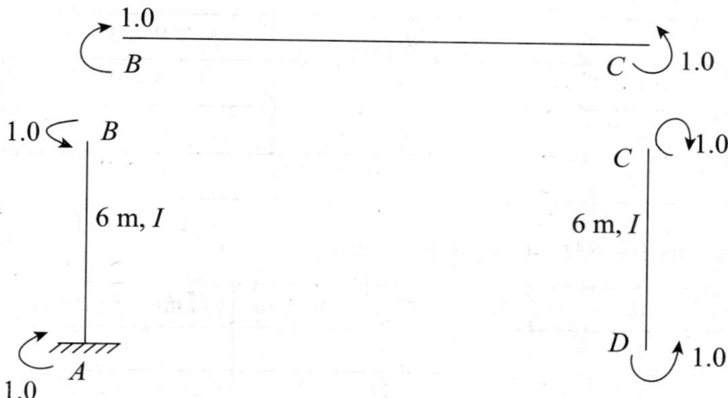

FIG. 4.124 FBD for unit moment at D

Table 4.6 Expressions for BM, M in each member

Members	M	*Second Moment of Area*	*Limits of Integration*	*Origin*
DC	0	–	–	–
CF	$-30x^2/2 = -15x^2$	$2I$	0 – 3	C
FE	$-15x^2 - 200(x - 3)$	$2I$	3 – 9	C
EB	$-15x^2 - 200(x - 3) - 200(x - 9)$	$2I$	9 – 12	C
AB	$100x - 5160$	I	0 – 6	A

Table 4.7 Expressions for *BM*, m_1 in each member

Members	m_1	*Second Moment of Area*	*Limits of Integration*	*Origin*
DC	$-x$	*I*	0 – 6	*D*
CF	–6.0	2*I*	0 – 3	C
FE	–6.0	2*I*	3 – 9	C
EB	–6.0	2*I*	9 – 12	C
AB	$-x$	*I*	0 – 6	*A*

Table 4.8 Expressions for *BM*, m_2 in each member

Members	m_2	*Second Moment of Area*	*Limits of Integration*	*Origin*
DC	0	–	–	–
CF	x	2*I*	0 – 3	C
FE	x	2*I*	3 – 9	C
EB	x	2*I*	9 – 12	C
AB	12	*I*	0 – 6	*A*

Table 4.9 Expressions for *BM*, m_3 in each member

Members	m_2	*Second Moment of Area*	*Limits of Integration*	*Origin*
DC	1.0	*I*	0 – 6	*D*
CF	1.0	2*I*	0 – 3	C
FE	1.0	2*I*	3 – 9	C
EB	1.0	2*I*	9 – 12	C
AB	1.0	*I*	0 – 6	*A*

In the above tables, *M* represents the *BM* in the members due to external loads, while m_1, m_2 and m_3 represent *BM* in the members due to unit horizontal, vertical and moment respectively.

$$\delta L_1 = \int Mm_1 dx/EI = \text{Horizontal displacement at } D \text{ due to external loads}$$

$$\delta L_1 = \int_0^3 \left[(-15x^2)(-6)dx/2EI\right]_{CF} + \int_3^9 \left[(-15x^2 - 200(x - 3))(-6)dx/2EI\right]_{FE}$$

$$+ \int_9^{12} \left[(-15x^2 - 200(x - 3)) - 200(x - 9)(-6)dx/2EI\right]_{EB} +$$

$$+ \int_0^6 \left[(100x - 5160)(-x)dx/EI\right]_{AB}$$

$$\boxed{\delta L_1 = 138600/EI}$$

$$\delta L_2 = \int Mm_2 dx/EI = \text{Vertical displacement at } D \text{ due to external loads}$$

$$= \int_0^3 \{(-15x^2)xdx/2EI\}_{CF} + \int_3^9 [(-15x^2 - 200(x - 3))xdx/2EI]_{FE}$$

$$+ \int_0^{12} [(-15x^2 - 200(x - 3) - 200(x - 9))(x)dx/2EI]_{EB}$$

$$+ \int_0^6 \{[100x - 5160](12)dx/EI\}_{AB}$$

$$\boxed{\delta L_2 = -430200/EI}$$

$$\delta L_3 = \int Mm_3 dx/EI = \text{Rotation at } D \text{ due to external loads}$$

$$= \int_0^3 [(-15x^2)(1)dx/2EI]_{CF} + \int_3^9 \{(-15x^2 - 200(x - 3)(1)dx/2EI\}_{FE}$$

$$+ \int_9^{12} \{(-15x^2 - 200(x - 3) - 200(x - 9))(1)dx/2EI\}_{FB}$$

$$+ \int_0^6 [100x - 5160(1)dx/EI]_{AB}$$

$$\boxed{\delta L_3 = -37980/EI}$$

The flexibility coefficients are obtained from the following equations

$$f_{11} = \int m_1 m_1 dx/EI = \int_0^6 \{-x^2 dx/EI\}_{DC} \times 2 + \int_0^{12} [-6^2 dx/2EI]_{CB} = 360/EI$$

$$f_{12} = f_{21} = \int m_1 m_2 dx/EI = \int_0^{12} \{-6xdx/2EI\}_{CB} + \int_0^6 \{-12xdx / EI\}_{AB} = -432/EI$$

$$f_{13} = f_{31} = \int m_1 m_3 dx/EI = \int_0^6 \{-xdx/EI\}_{DC} \times 2 + \int_0^{12} \{-6dx / 2EI\}_{CB} = -72/EI$$

$$f_{22} = \int m_2 m_2 dx/EI = \int_0^{12} \{x^2 dx/2EI\}_{CB} + \int_0^6 \{12^2 dx / EI\}_{AB} = 1152/EI$$

$$f_{23} = f_{32} = \int m_2 m_3 dx/EI = \int_0^{12} \{xdx/2EI\}_{CB} + \int_0^{6} \{12dx / EI\}_{AB} = 108/EI$$

$$f_{33} = \int m_3 m_3 dx/EI = \int_0^{6} \{1dx/EI\}_{DC} \times 2 + \int_0^{12} \{dx / 2EI\}_{CB} = 18/EI$$

The flexibility coefficients for BC have been evaluated for the whole span.

Compatibility equations in terms of displacement

$$\delta L_1 + f_{11}R_1 + f_{12}R_2 + f_{13}R_3 = 0$$
$$\delta L_2 + f_{21}R_1 + f_{22}R_2 + f_{23}R_3 = 0$$
$$\delta L_3 + f_{31}R_1 + f_{32}R_2 + f_{33}R_3 = 0$$

Substituting the computed values

$$138600/EI + (360/EI)R_1 - (432/EI)R_2 - (72/EI)R_3 = 0$$
$$-430200/EI - (432/EI)R_1 + (1152/EI)R_2 + (108/EI)R_3 = 0$$
$$-37980/EI - (72/EI)R_1 + (108/EI)R_2 + (18/EI)R_3 = 0$$

Solving the above simultaneous equations

$R_1 = H_D$ = 185.0 kNm
$R_2 = V_D$ = 401.4 kNm
$R_3 = M_D$ = 441.4 kNm

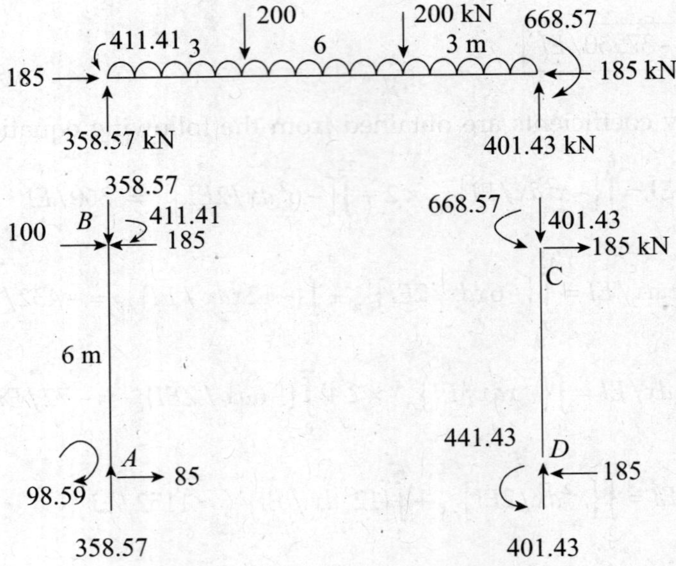

FIG. 4.125 Free body diagram

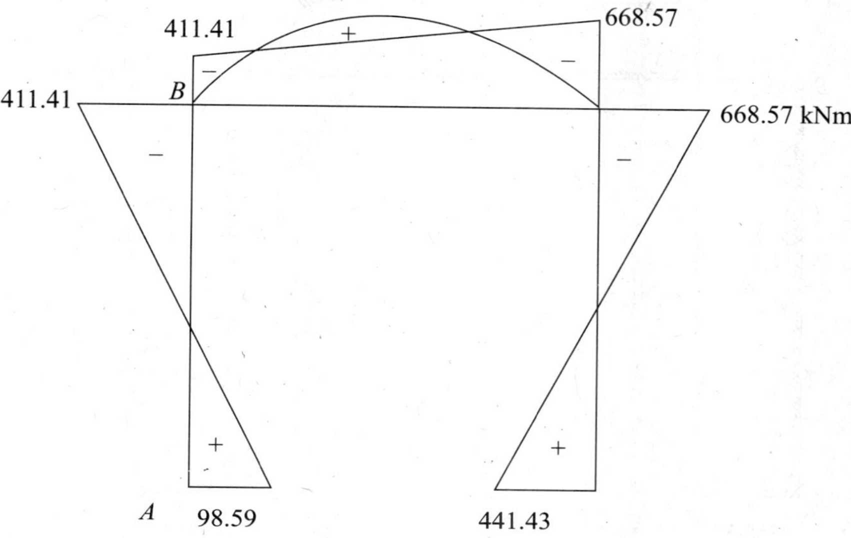

FIG. 4.126 Bending moment diagram

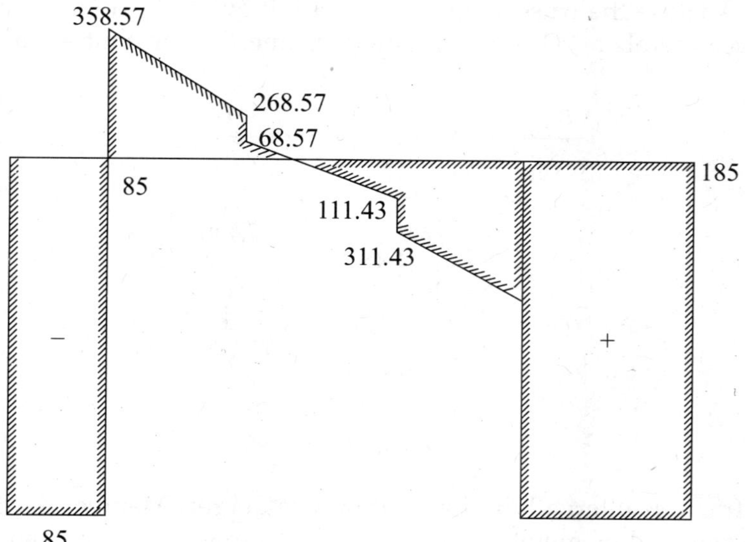

FIG. 4.127 Shear force diagram (kN)

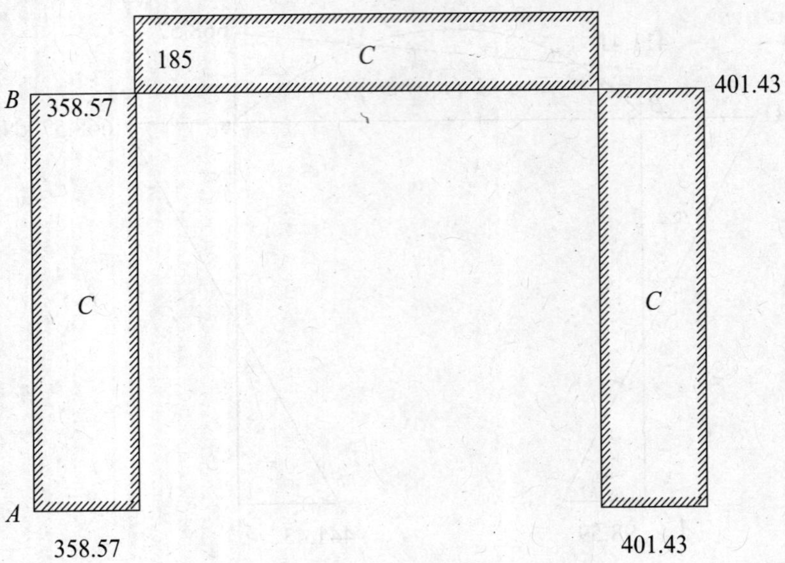

FIG. 4.128 Axial force diagram

Example 4.18 Analyse the truss shown in Fig. 4.129 by the flexibility matrix method; choosing the force member *AD* is redundant. Assume *EA* constant for all members.

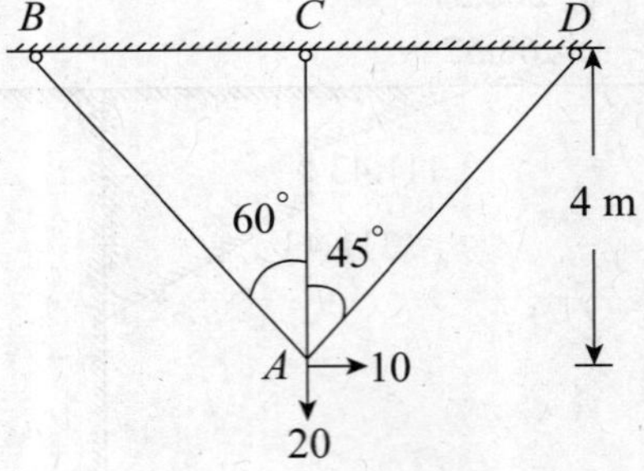

FIG. 4.129

Solution: The frame is internally redundant by one degree. Member *AD* is considered redundant. The removal of member *AD* makes the frame statically determinate.

The displacement in the direction *AD* is determined using unit load method, tensile forces are taken as positive and compression is taken as negative.

Released structure

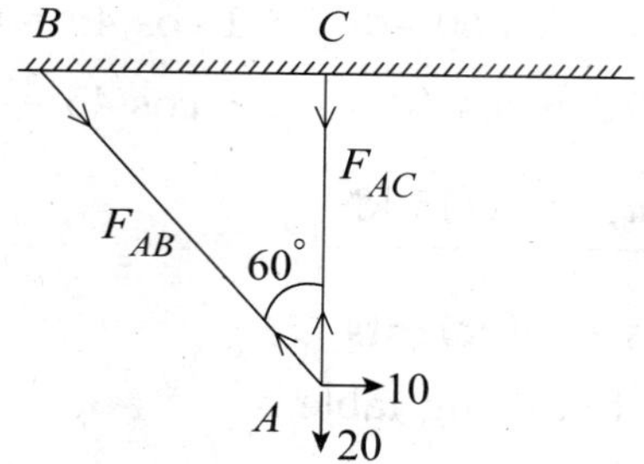

FIG. 4.130 Determinate truss

$\Sigma H = 0$

$$F_{AB} \sin 60 = 10$$

$$\boxed{F_{AB} = 11.55 \text{ kN}}$$

$\Sigma V = 0$

$$F_{AB} \cos 60 + F_{AC} - 20 = 0$$

$$11.55 \cos 60 + F_{AC} - 20 = 0$$

$$\boxed{F_{AC} = 14.23 \text{ kN}}$$

Unit load along *AD*

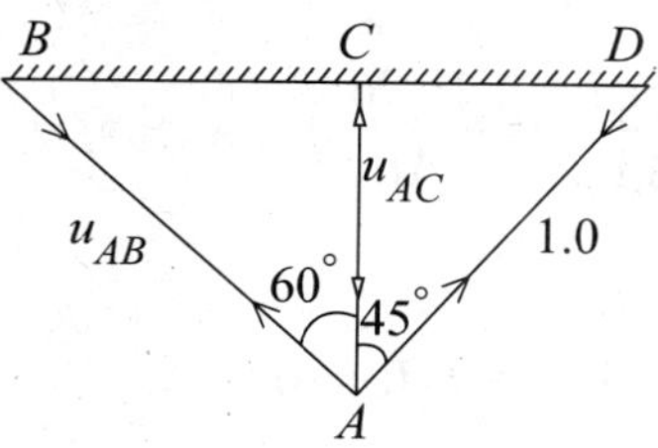

FIG. 4.131 Effect of unit load along *AD*

$\Sigma H = 0;$

$$u_{AB} \sin 60 - 1 \sin 45 = 0$$

$$u_{AB} = 0.816 \text{ kN}$$

$$\Sigma V = 0$$

$$u_{AB} \cos 60 - u_{AC} + 1 \cos 45 = 0$$

$$0.816 \cos 60 - u_{AC} + \cos 45 = 0$$

$$\boxed{u_{AC} = 1.115 \text{ kN}}$$

Displacements and flexibility coefficients

They are computed using the following table

Table 4.10 Computation of δL_1

Members	*P*	*u*	*l*	*Pul/AE*	u^2l/AE	*P + Ru*
AB	11.55	0.816	8	75.40	5.33	+10.94
AC	14.23	−1.115	4	−63.46	4.97	+15.06
AD	–	1.000	$4\sqrt{2}$	–	$4\sqrt{2}$	−0.75
			Σ	11.94	15.96	

$$\delta L_1 = \sum_{i=1}^{3} \frac{P_i u_i l_i}{A_i E_i}$$

$$\boxed{\delta L_1 = 11.94/AE}$$

Flexibility coefficient

$$f_{11} = \frac{\Sigma u_i^2 l_i}{A_i E_i} = \frac{15.96}{AE}$$

Compatibility equation

$$\delta L_1 + f_{11} R = 0$$

$$11.94/AE + (15.96/AE)R = 0$$

$$R = -0.75 \text{ kN (compressive)}$$

$$\therefore \quad F_{AD} = 0.75 \text{ kN (compressive)}$$

Example 4.19 Analyse the pin jointed plane frame shown in Fig. 4.132 by the flexibility method and tabulate member forces. Assume (L/AE) for each member = 0.025 mm/kN.

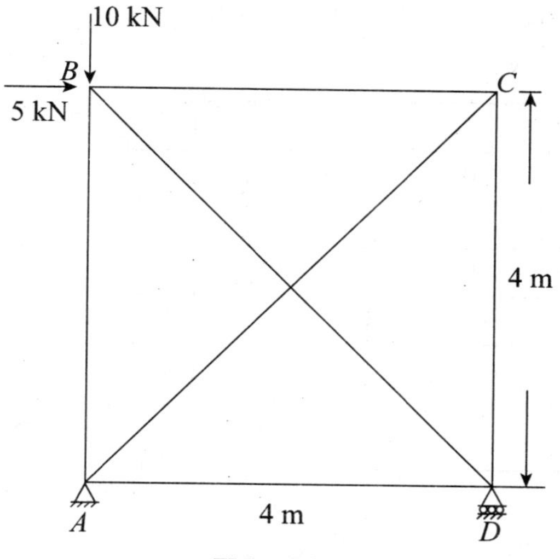

FIG. 4.132

Solution: The number of members are nm = 6 and the number of joints is n_j = 4. The number of members required for statically determinate truss is $(2n_j - 3) = 2(4) - 3 = 5$. The degree of indeterminacy is 6 – 5 = 1, i.e., number of members available–number of members required for a determinate truss. The given truss is statically indeterminate to the first degree. Hence, the member is redundant. The redundant member is taken AC. After removing the member AC, the released structure is analysed.

Released structure

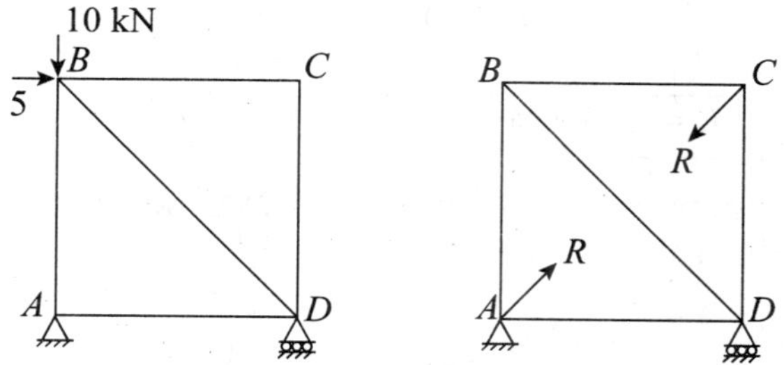

FIG. 4.133 Released structure along with redundants

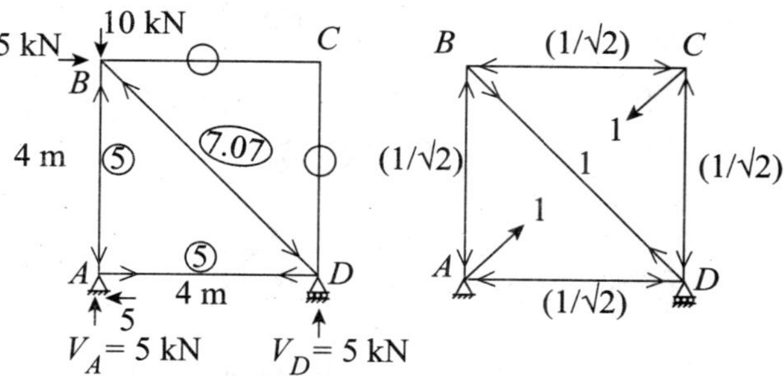

FIG. 4.134 Forces in the released structure for external loads and unit load

Displacements and the flexibility coefficients.

Table 4.11 Computation of δL_1

Members	*P*	*u*	*l*	*Pul/AE*	*u²l/AE*	*P + Ru*
AB	−5	−0.707	4	14.14	2.00	−6.46
BC	0	−0.707	4	0	2.00	−1.46
CD	0	−0.707	4	0	2.00	−1.46
DA	+5	−0.707	4	−14.14	2.00	+3.54
DB	−7.07	+1.000	$4\sqrt{2}$	−40.00	5.66	−5.00
AC	–	+1.000	$4\sqrt{2}$	–	5.66	+2.07
				−40.00	19.32	

Compatibility equation

$$\delta L_1 + f_{11} R = 0$$

and

$$\delta L_1 = \sum_{i=1}^{6} \frac{P_i u_i l_i}{A_1 E_i} = -40/AE$$

$$f_{11} = \sum \frac{u_i^2 l_i}{A_i E_i} = 19.32/AE$$

Substituting the values in the above equation

$$-40/AE + (19.32/AE)R = 0$$

and hence

$$\boxed{R = 2.07 \text{ kN}}$$

The forces in other members are computed using the relation ($P + Ru$) and given in the last column of the Table. 4.11.

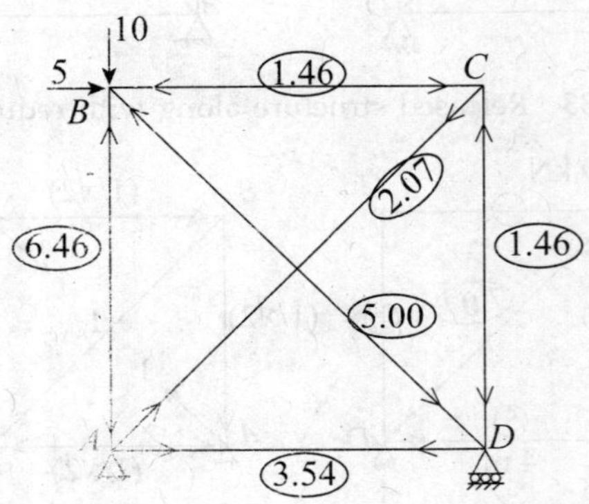

FIG. 4.135 Final forces in the members

Example 4.20 Analyse the pin jointed plane truss shown in Fig. 4.136 by the flexibility method. The numbers given in the parenthesis are the cross-sectional areas of members.

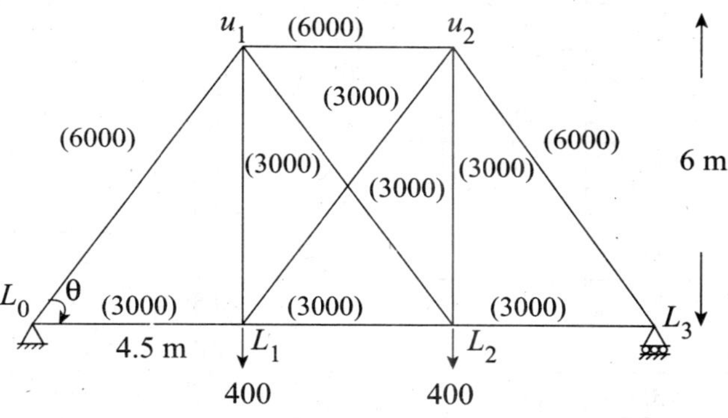

FIG. 4.136

Solution: The given frame is redundant to the first degree. L_1U_2 is treated as a redundant. Removing the member L_1u_2, the determinate truss is analysed by the method of joints.

$$\tan\theta = 6/4.5$$

$$\sin\theta = 0.8$$

$$\cos\theta = 0.6$$

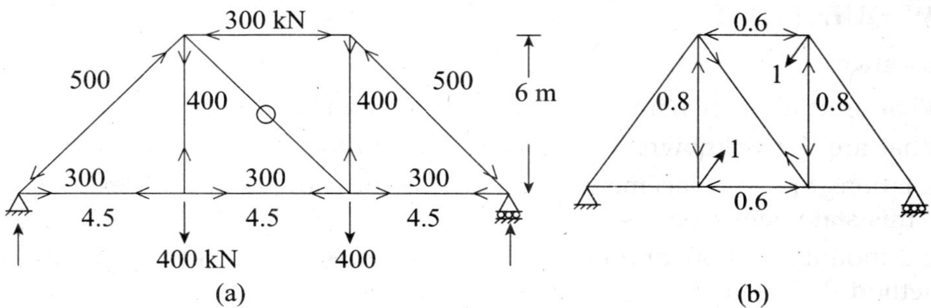

FIG. 4.137 Forces in the released structure due to applied loads and unit load

The forces in the members were calculated using the method of joints. The computed values and their nature is indicated in Fig. 4.137(a). A unit force is applied along L_1u_2 and the resulting member forces are shown in Fig. 4.137(b). The displacements and the redundant forces are computed and given below.

Table 4.12 Computation of δL_1

S.No.	Member Force	l_{mm}	A_{mm^2}	P_{kN}	u	Pul/A	u^2l/AE	$(P + Ru)_{kN}$
1.	L_0u_1	7500	6000	−500	0	0	0.00	−500.0
2.	u_1u_2	4500	6000	−300	−0.6	135	0.27	−445.0
3.	u_2l_3	7500	6000	−500	0	0	0.00	−500.0
4.	L_0L_1	4500	3000	+300	0	0	0.00	+300.0
5.	L_1L_2	4500	3000	+300	−0.6	−270	0.54	−155.4
6.	L_2L_3	3000	3000	+300	0	0	0.00	+300.0
7.	u_1L_1	6000	3000	+400	−0.8	−640	1.28	+207.2
8.	u_1L_2	7500	3000	0	+1.0	0	2.50	+241.0
9.	u_2u_2	6000	3000	+400	−0.8	−640	1.28	+207.2
						−1415	5.87	

$$\delta L_1 + f_{11}R = 0$$

$$-1415 + 5.87R = 0$$

$$R = \frac{-(-1415)}{5.87}$$

$$\boxed{R = 241 \text{ kN}}$$

REVIEW QUESTIONS

Remembrance

4.1. What are the other names of flexibility method?

4.2. What are the unknowns to be determined in the flexibility method?

4.3. Rotations and displacements are taken as unknowns in the flexibility method. Is this statement true?

4.4. Redundant reaction or internal force is taken as unknowns in the flexibility method. Is it true?

4.5. Define flexibility coefficient.

4.6. Write down the flexibility coefficient for a cantilever beam subjected to unit load at the free end.

4.7. What is compatibility condition?

4.8. What is static indeterminacy?

4.9. Name the type of indeterminacy considered in the flexibility method.

4.10. What is the advantage of flexibility coefficients?

4.11. What is primary structure/release structure? Give an example.

4.12. Explain the term consistent deformation.

4.13. Give the choice of redundant in the analysis of propped cantilever beam?

4.14. Is the application of theorem of three moments a flexibility method or stiffness method? If so why?

4.15. How temperature stresses and lack of fit taken care of in the flexibility method?

4.16. If a beam is statically indeterminate to the third degree, how many conditional equations are to be developed?

4.17. What is the maximum slope and vertical displacement of a cantilever subjected to load at the free end?

4.18. What is the maximum slope at the ends of a simply supported uniform beam subjected to a central concentrated load?

4.19. What is the maximum slope at the ends of a simply supported uniform beam subjected to uniformly distributed load?

4.20. A beam is simply supported at the ends. The beam is of uniform section and is subjected to a concentrated load at a distance of $(2l/3)$ firm are end? Determine the slope at the ends.

4.21. A simply supported beam of uniform section is subjected to a clockwise moment M at the ends. Determine the slope at the ends.

4.22. The flexibility coefficient depends on:

(a) actual loading and elastic properties

(b) actual loading and geometry

(c) geometry and elastic properties

4.23. Releases are introduced to render the structure

(a) statically determinate

(b) statically indeterminate

(c) kinematically determinate

4.24. What is the relation between flexibility and stiffness matrix?

4.25. What is the unit for linear flexibility coefficient?

4.26. What is the unit for angular flexibility coefficient?

4.27. List the names of the coefficient of the unknowns in the flexibility method.

Understanding

4.1. Compute the static indeterminacy of the following structures.

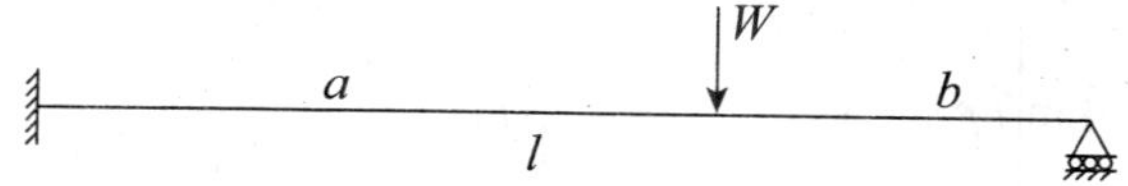

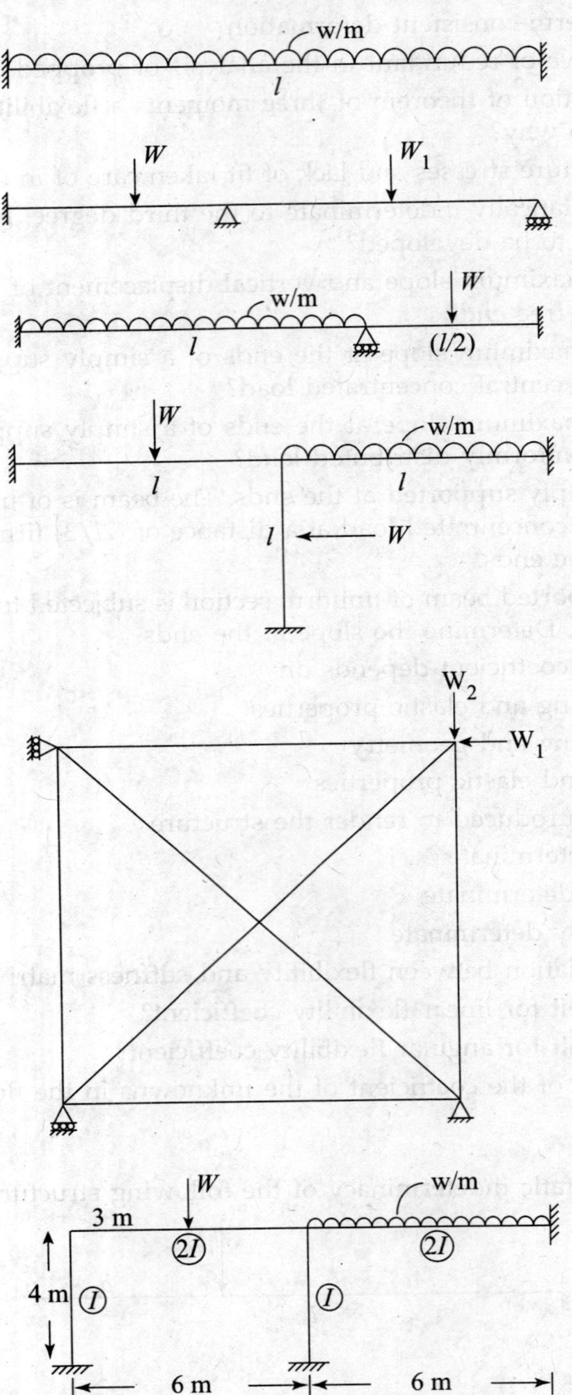

w/m
l
W
W$_1$
w/m
W
l
(l/2)
W
w/m
l
l
l
W
W$_2$
W$_1$
W
w/m
3 m
2I
2I
4 m
I
I
6 m
6 m

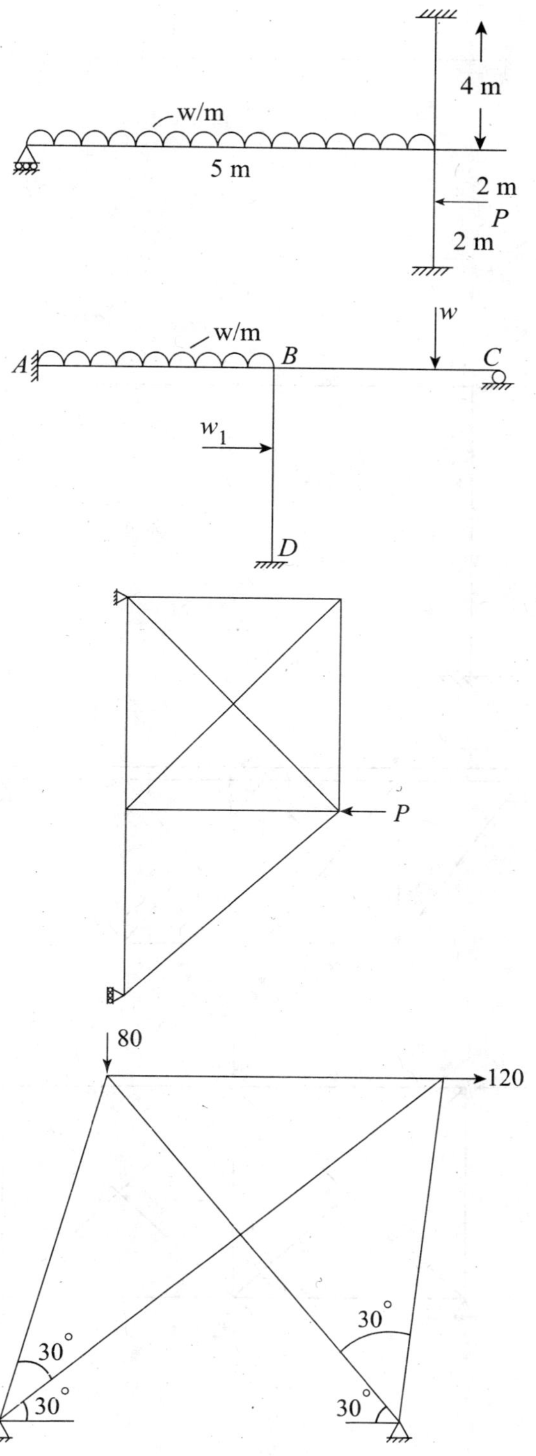
w/m
4 m
5 m
2 m
P
2 m
w
w/m
A
B
C
w_1
D
P
80
120
30°
30°
30°
30°

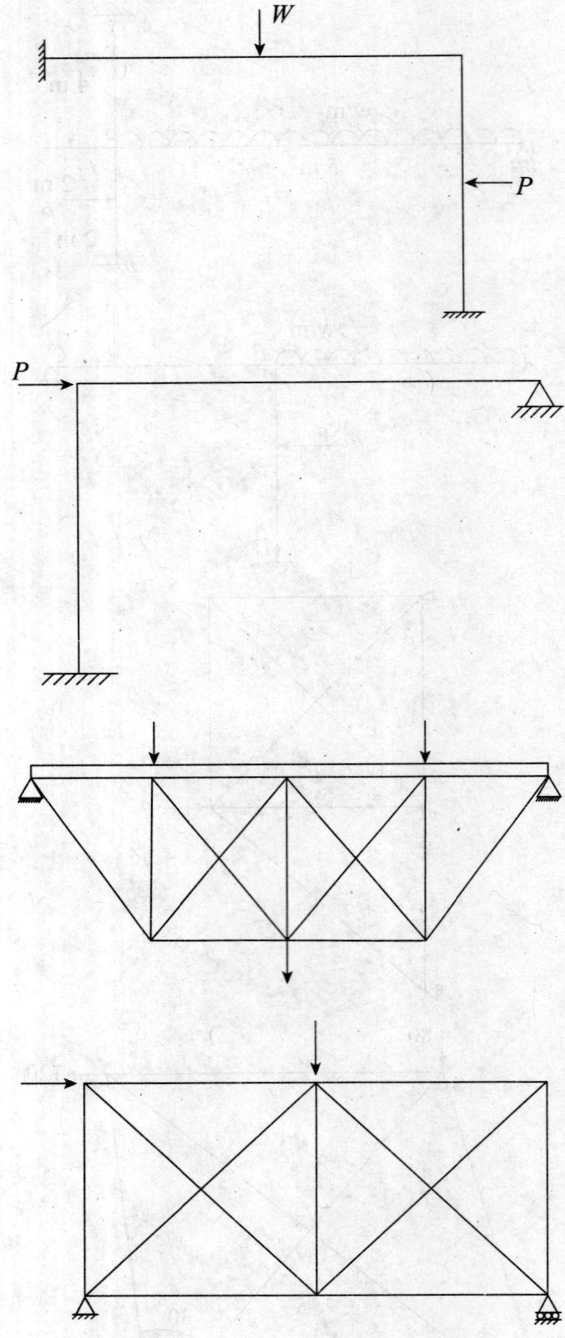
W
P
P

4.2. If the static indeterminacy is 4 then what should be the size of the flexibility coefficient matrix?

4.3. Distinguish between linear flexibility coefficient and angular flexibility coefficient.

4.4. What are the unknowns in the flexibility method?

EXERCISE PROBLEMS

4.1. Using the flexibility method, find the support reaction at B. Find other reactions and draw BMD. EI is constant.

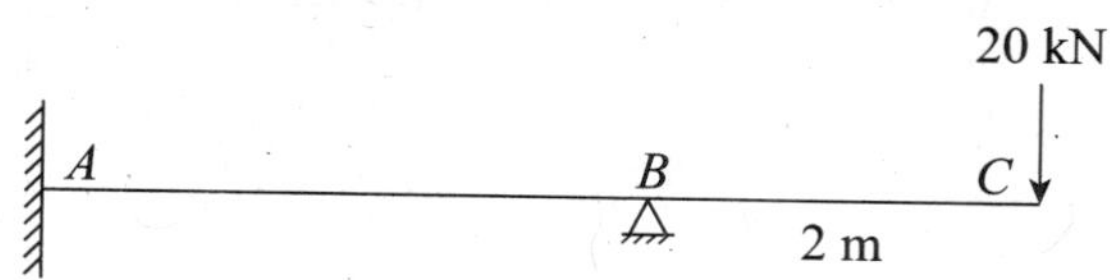

Ans: V_B = 34.98 kN, V_A = 14.98 kN, M_A = 19.92 kNm

4.2. A propped cantilever is loaded as shown below. During loading, the prop sinks by 5 mm. Determine the propped reaction. Draw SFD and BMD. $E = 2(10)^5$ N/mm^2 and $I = 80(10)^6$ mm^4

20 kN/m 10 kNm
A 3 m C 1.5 m 1.5 m B

Ans: V_A = 50 kN, V_B = 10 kN, M_A = 40 kNm

4.3. For the beam shown in figure below, compute the reactions and draw BMD.

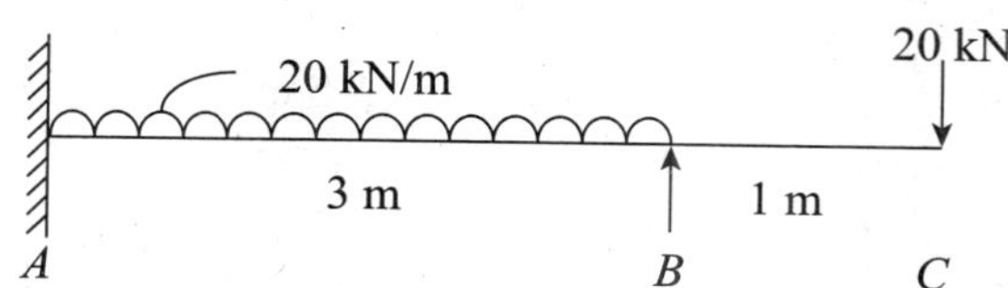

Ans: V_A = 27.5 kN, V_B = 52.5 kN, M_A = 12.5 kNm

4.4. Analyse the beam and determine the end moments by using the flexibility method.

12 kN/m
A 3 m C 3 m

Ans: V_A = 29.3 kN, V_B = 6.8 kN, M_A = 24.8 kNm, M_B = 11.3 kNm

4.5. A fixed beam is shown in figure below. Compute the moments at the supports.

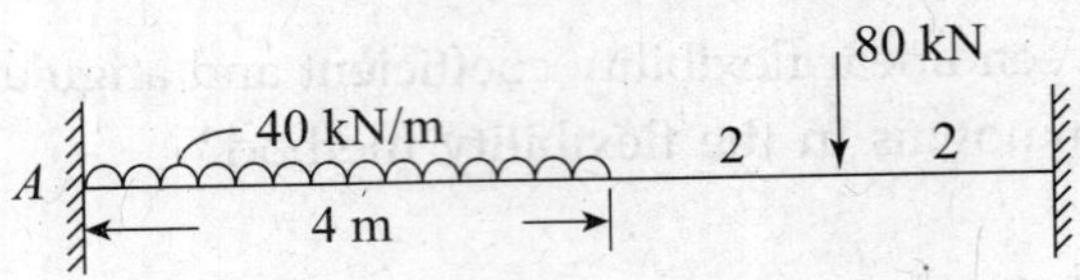

Ans: M_A = 118.7 kNm, M_B = 161.1 kNm

4.6. A vertical column of 8 m height is to carry a crane girder load of 50 kN applied at an eccentricity of 0.2 m. Calculate the end moments assuming both ends are fixed.

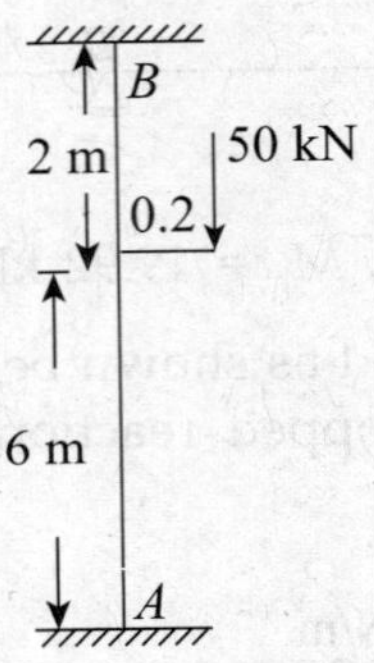

Ans: M_A = 3.14 kNm, M_B = 1.86 kNm

4.7. Compute the value of moment at the intermediate support and draw BMD.

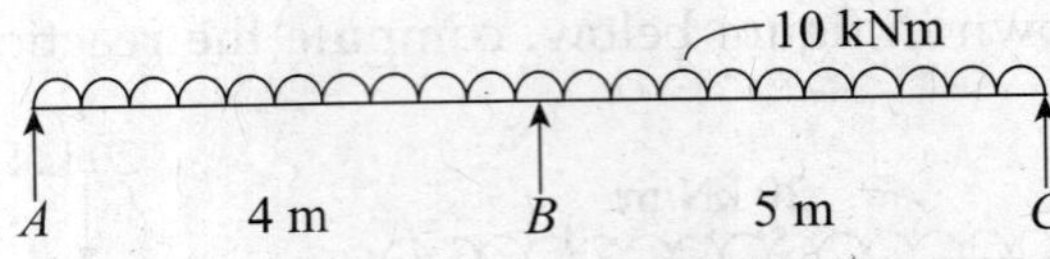

Ans: M_B = 26.3 kNm

4.8. Analyse the continuous beam by the flexibility method. Draw SFD and BMD.

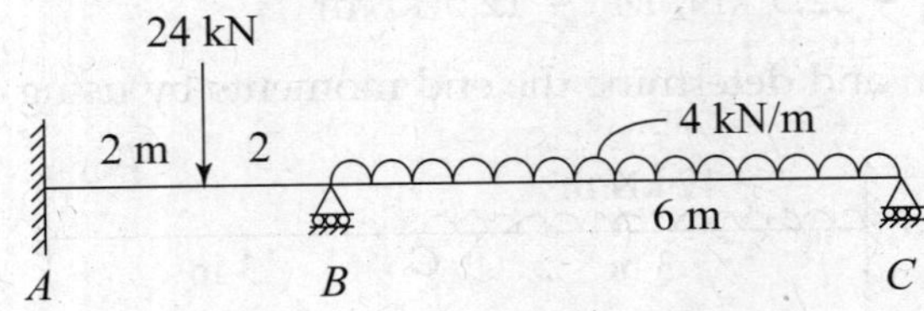

Ans: M_A = 10 kNm, M_B = 16 kNm

4.9. Determine the reactions at A, B and C of the continuous beam shown in figure below.

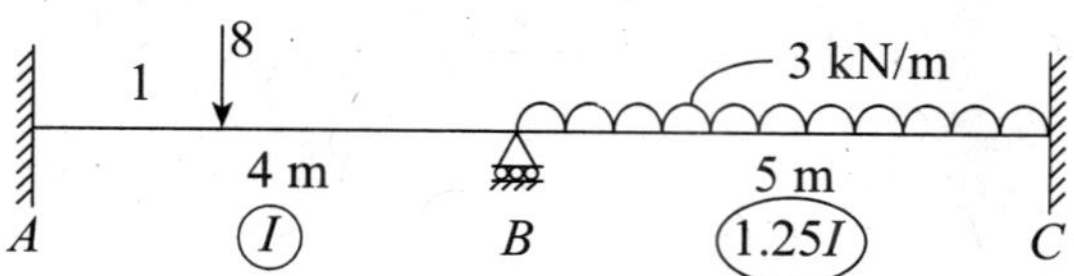

Ans: V_{AB} = 6.75 kN, V_{BA} = 1.25 kN, V_{BC} = 6.31 kN, V_{CB} = 8.69 kN, M_A = –3.31 kNm, M_B = –3.97 kNm, M_C = +7.44 kNm

4.10. Analyse the truss by the flexibility method.

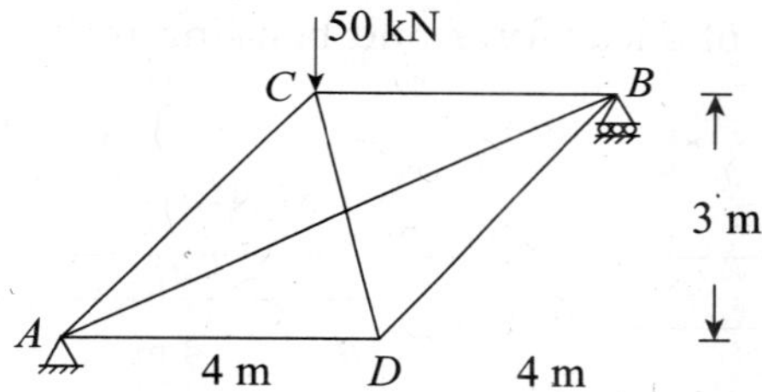

Ans: F_{CD} = 24.46 kN

4.11. Determine the forces in members AC and CE.

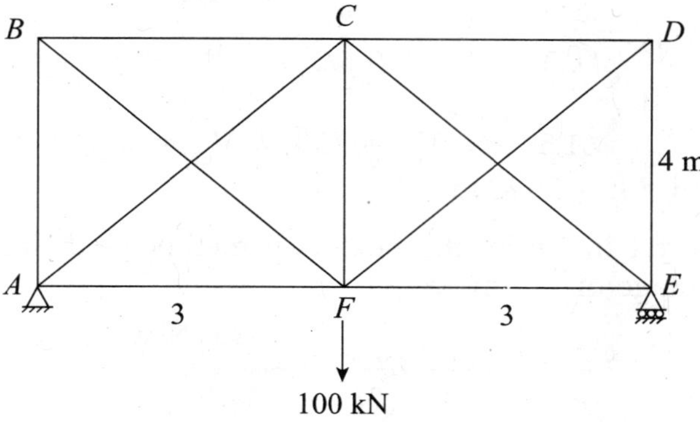

Ans: $F_{BF} = F_{FD}$ = 35.2 kN

4.12. Analyse the frame by the flexibility method.

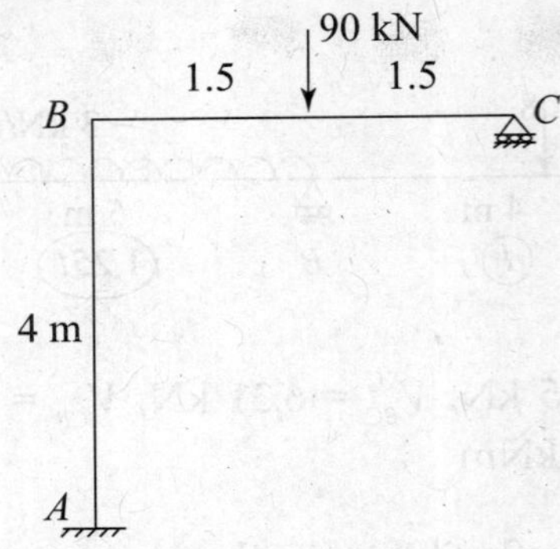

Ans: $M_A = -10.1$ kNm, $M_D = 62.4$ kNm

4.13. Compute the value of shear force and bending moment at B using the flexibility method.

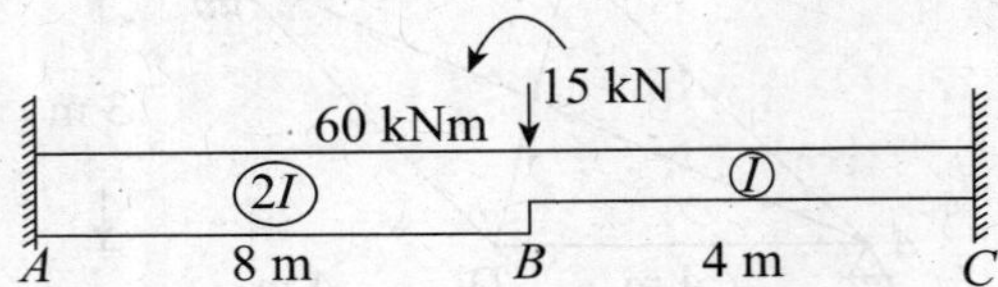

Ans: $(SF)_B = 11.76$ kN; $(BM)_B = 56.75$ kNm

4.14. Determine the support reactions and sketch the BMD using flexibility method.

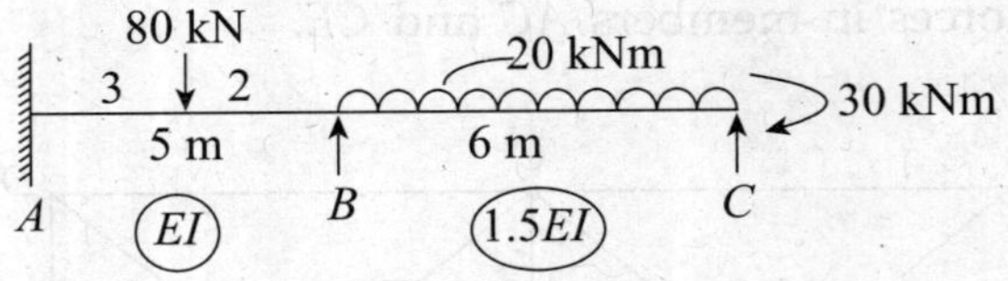

Ans: $V_{AB} = 18.5$ kN, $V_{BA} = 61.50$ kN, $V_{BC} = 70$ kN, $V_{CB} = 50$ kN, $M_A = 22.29$ kNm, $M_B = 89.91$ kNm, $M_C = 30$ kNm

4.15. Analyse the portal frame by the flexibility method and draw the BMD. (Hint: Take the advantage of symmetry)

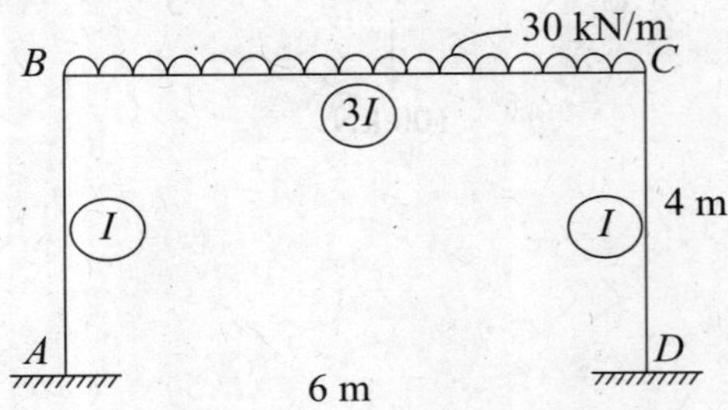

Ans: $M_A = 22.5$ kNm, $M_B = 45$ kNm

Stiffness Method: System Approach 5

Objectives

Concept of stiffness method – Define stiffness coefficient – Development of governing equations – settlement of supports – Analysis of propped cantilever beam, fixed beams, continuous beams, rectilinear frames, rigid frames and pin jointed plane trusses.

5.1 INTRODUCTION

In the flexibility method, the dual nature of force relationships were considered and the relation between flexibility and stiffness matrices were discussed. The analysis of structures can also be carried out by using stiffness method, which is more elegant and simple.

5.2 STIFFNESS COEFFICIENTS

A stiffness coefficient represents a force at a point caused by a unit displacement applied at the same or different point. Units – kN/m, N/mm.

For example, k_{11} is the stiffness coefficient representing a force at point 1 due to unit displacement at the same point 1.

As a second example, k_{23} is the stiffness coefficient representing a force at point 2 due to unit displacement at point 3.

In general, k_{ij} is the stiffness coefficient representing a force at points i due to unit displacement at point j.

Stiffness Matrices are Symmetrical

Consider the external work done equations expressed in matrix format

$$W_E = \left(\frac{1}{2}\right) [\delta]^T [P] = \left(\frac{1}{2}\right) [P]^T \{\delta\} \qquad (1)$$

But $$\{P\} = [k]\{\delta\} \qquad (2)$$

Substituting Eq. (2) in Eq. (1)

$$W_E = \left(\frac{1}{2}\right) [\delta]^T [k]\{\delta\} \qquad (3)$$

Taking transpose of Eq. (2)

$$\{P\}^T = [\delta]^T \{k\}^T \tag{4}$$

Substituting Eq. (4) in Eq. (1)

$$W_E = \left(\frac{1}{2}\right) [\delta]^T \{k\}^T [\delta] \tag{5}$$

From Eqs. (3) and (5)

$$[k] = [k]^T \tag{6}$$

The above equations indicate that the stiffness matrices are symmetrical. Mathematically,

$$k_{ij} = k_{ji}$$

which represents the reciprocal relationship of a stiffness matrix.

5.3 STIFFNESS METHOD

In this method, the analysis is carried out by using.

$$\{P\} = [k] \{D\} \tag{7}$$

with usual notations.

This method considers the independent displacements (kinematic indeterminacy or degrees of freedom) as unknowns and is also called the displacement method. Using the equilibrium equations, simultaneous equations are formulated and the unknown displacements are determined. Stiffness methods are more suitable for general-purpose computer coding since the independent displacements can be chosen more automatically.

Stiffness methods can be carried out by using

- System approach
- Element approach

5.4 SYSTEM APPROACH

Statically indeterminate structures can be analysed by writing the equilibrium equations in terms of the stiffness coefficients and unknown. This method involves calculation of displacement due to external loads. The restrained structure is obtained by fixing all the supports. The analyst has no choice in choosing the unknown displacements.

5.5 BASICS OF THE STIFFNESS METHOD

- The indeterminate structure is first made kinematically determinate by introducing the constraints at the nodes.
- The kinematically determinate structure comprises fixed ended members and hence all the nodal displacements are zeros.

- This results in a state of stress where the clamped joints are not in equilibrium.
- To restore the equilibrium of stress resultants, the joints are subjected to unknown displacements.
- It should be noted that the number of joint equilibrium of forces is equal to kinematic indeterminacy.
- The joint displacements were obtained using equilibrium equations.
- The member end forces are computed using stiffness properties of the member.

As the joint displacements are treated as unknown in this method, this method is aptly called the displacement method. As the stiffness properties of the members are considered in the analysis it is called the stiffness method. As joint equilibrium has been considered in the analysis, it is also called the equilibrium method.

5.6 DEVELOPMENT OF EQUATIONS

Consider a two-span continuous beam as shown in Fig. 5.1(a). The beam is kinematically indeterminate to the second degree. The kinematic redundants are the rotations at B and C respectively. They are denoted as D_1 and D_2 respectively. The kinematic redundants are also called degrees of freedom or independent displacements. Let L_1, L_2 be the span of the beam AB and BC respectively. I_1, I_2 are the moment of the inertia of the beam AB and BC respectively. Let P_{R1}, P_{R2}, P_{R3} and P_{R4} be the reaction at A, B and C respectively. Clockwise rotations and clockwise moments are considered positive in the analysis. The restrained structure is obtained by fixing all the joints and calculated the fixed end moments. The algebraic sums of the fixed end moments at each joint is termed P_1 and P_2 respectively. If the sum of the fixed end moments is clockwise then it is expressed as positive, otherwise negative. The moments are marked clockwise in the following figure and are treated as positive.

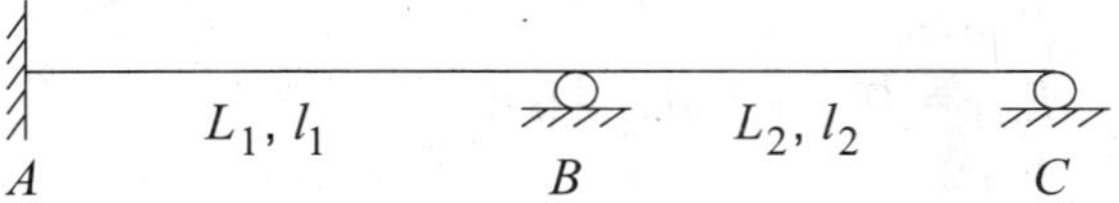

FIG. 5.1(a) Continuous beam

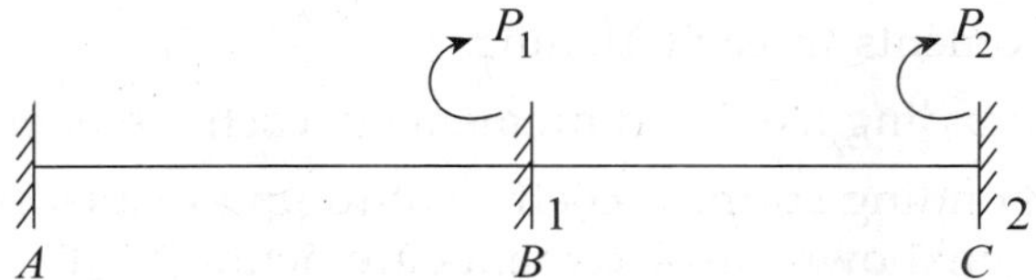

FIG. 5.1(b) Restrained structure with joint moments

Using the equilibrium equations at joints B and C having independent displacements, we have

Joint 1;

$$P_1 + k_{11}D_1 + k_{12}D_2 = 0 \quad (8)$$

Joint 2;

$$P_2 + k_{21}D_1 + k_{22}D_2 = 0 \quad (9)$$

where k_{11}, k_{12}, k_{21} and k_{22} are the stiffness coefficients at 1 and 2 due to unit displacements, which are successively applied at 1 and 2. This can be obtained from Table 5.1. The above equations are expressed in matrix form as

$$\begin{Bmatrix} P_1 \\ P_2 \end{Bmatrix} + \begin{bmatrix} k_{11} & k_{12} \\ k_{21} & k_{22} \end{bmatrix} \begin{Bmatrix} D_1 \\ D_2 \end{Bmatrix} = \begin{bmatrix} 0 \\ 0 \end{bmatrix} \quad (10)$$

In general;

$$\{P\} + [k]\{D\} = [0] \quad (11)$$

$$[D] = -[k]^{-1}\{P\} \quad (12)$$

where

$[D]$ = unknown displacement vector in the given structure

$[k]$ and $[k]^{-1}$ = stiffness matrix and inverse of stiffness matrix

$\{P\}$ = algebraic sum of the fixed end moments at each joint where an unknown displacement exists.

Equations (10) – (12) are the basic equations in system stiffness approach. It is to be understood that solutions can be obtained by using either Eqs. (8, 9) or Eq. (12).

The member end moments are calculated using the slope deflection equations with BMD and SFD drawn based on the FBD.

5.6.1 End Moments and Shear Forces Using Matrix Approach Only

We know

$$\{P_M\} = \{P_{MF}\} + [k_{MD}]\{D\} \quad (13)$$

and

$$\{P_S\} = \{P_{SF}\} + [k_{SD}]\{D\} \quad (14)$$

where

$\{P_M\}$ = end moments in each member

$\{P_{MF}\}$ = corresponding fixed end moments in each member due to external loads

$[k_{MD}]$ = corresponding stiffness coefficient to unit displacement applied at joints where unknown displacements are acting in the direction of moments required.

$\{P_S\}$ = shear forces in each member

$\{P_{SF}\}$ = corresponding fixed end shear force in each member due to external loads

$[k_{SD}]$ = corresponding stiffness coefficient to unit displacements applied at joints where unknown displacements are acting in the direction of SF's required.

$[D]$ = unknown displacement vector

Table 5.1 Member end forces due to end deformations of prismatic members

S. No.	*Beams and deformations*	*End moments*	*End reactions*
(1)	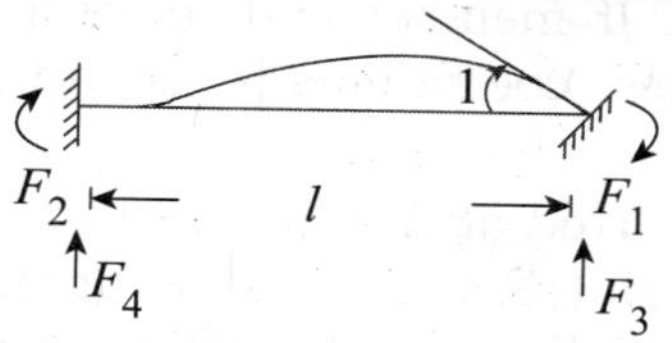	$F_1 = 4EI/l$ $F_2 = 2EI/l$	$F_3 = 6EI/l^2$ $F_4 = -6EI/l^2$
(2)	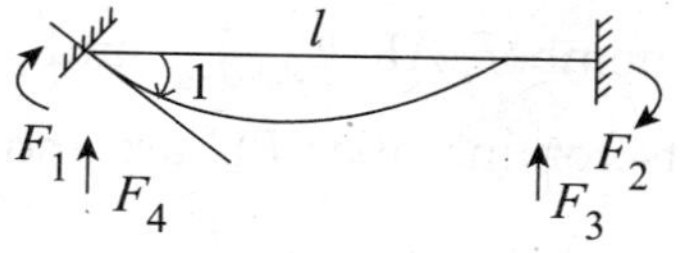	$F_1 = 4EI/l$ $F_2 = 2EI/l$	$F_3 = 6EI/l^3$ $F_4 = -6EI/l^2$
(3)	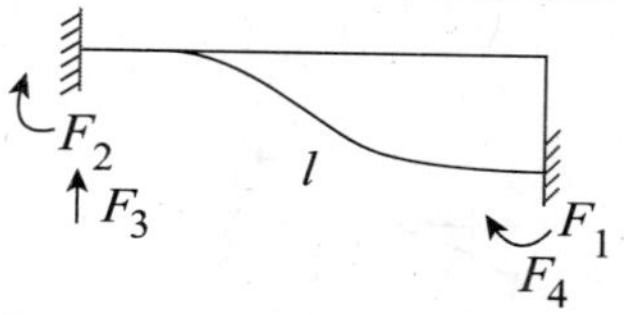	$F_1 = \frac{-6EI}{l^2}$ $F_2 = \frac{-6EI}{l^2}$	$F_3 = 12EI/l^3$ $F_4 = -12EI/l^3$
(4)	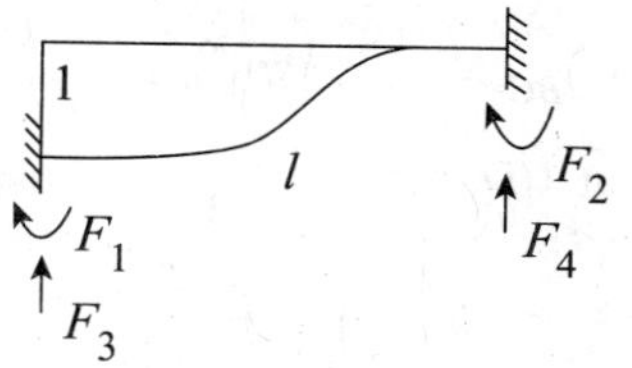	$F_1 = 6EI/l^2$ $F_2 = 6EI/l^2$	$F_3 = -12EI/l^3$ $F_4 = 12EI/l^3$
(5)	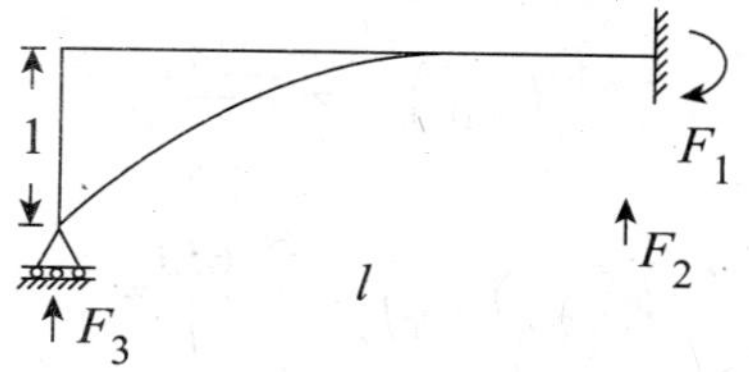	$F_1 = \frac{3EI}{l}$	$F_2 = 3EI/l^3$ $F_3 = -3EI/l^3$
(6)	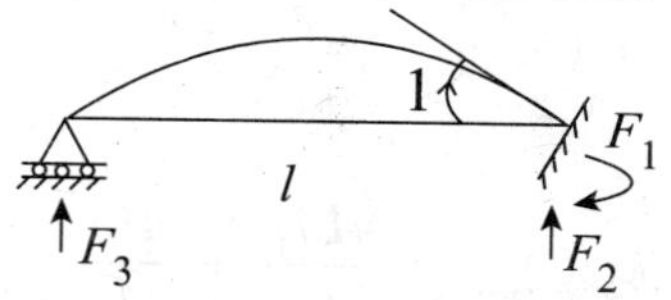	$F_1 = \frac{3EI}{l}$	$F_2 = \frac{3EI}{l^2}$ $F_3 = \frac{-3EI}{l^2}$
(7)	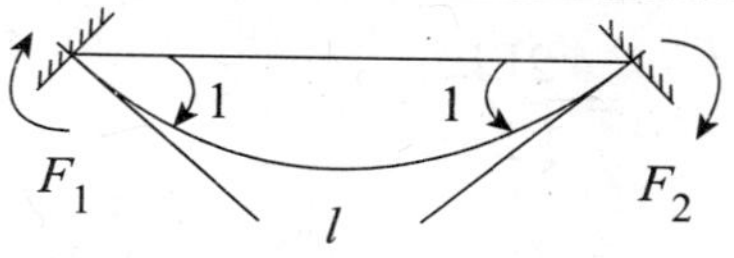	$F_1 = 2EI/l$ $F_2 = -2EI/l$	

Notes:

(1) The positive direction for reactions is upwards.

(2) The positive direction of moments is clockwise.

(3) The effect of axial force and shear force is neglected.

The calculation of the stiffness coefficients is illustrated in Fig. 5.2.

Stiffness coefficients k_{11}, k_{12}, k_{21} and k_{22} are calculated at joints ① and ② by giving unit displacements at ① and ② successively. If member end forces (BM_s and SF_s) are to be calculated through matrices, the stiffness coefficients $[k_{MD}]$ and $[k_{SD}]$ for each member has to be calculated separately.

Consider a two-span continuous beam ABC fixed at A and C. B is an intermediate support. Let the joint B be denoted as 1 and joint C as 2. A unit rotation is applied at joint 1 and the stiffness coefficients due to application of rotation at joint 1 are as follows. $\left.k_{MD_1}\right)_{AB}$ and $\left.k_{SD_1}\right)_{AB}$ are at joint A of member AB; $\left.k_{MD_1}\right)_{BA}$ and $\left.k_{SD_1}\right)_{BA}$ are at joint B of member AB. The stiffness coefficients for member BC are indicated in the Fig. 5.2.

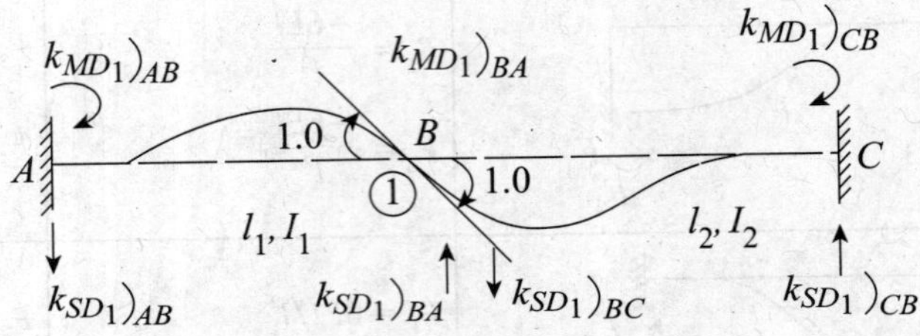

FIG. 5.2 Unit rotation at 1(D_1 = 1)

$$\left.k_{MD_1}\right)_{AB} = \frac{2EI_1}{l_1} \quad ; \quad \left.k_{MD_1}\right)_{BA} = \frac{4EI_1}{l_1}$$

$$\left.k_{MD_1}\right)_{BC} = \frac{4EI_2}{l_2} \quad ; \quad \left.k_{MD_1}\right)_{CB} = \frac{2EI_2}{l_2}$$

$$\left.k_{SD_1}\right)_{AB} = \frac{-6EI_1}{l_1^2} \quad ; \quad \left.k_{SD_1}\right)_{BA} = \frac{6EI_1}{l_1^2}$$

$$\left.k_{SD_1}\right)_{BC} = \frac{-6EI_2}{l_2^2} \quad ; \quad \left.k_{SD_1}\right)_{CB} = \frac{6EI_2}{l_2^2}$$

$$k_{11} = \left.k_{MD_1}\right)_{BA} + \left.k_{MD_1}\right)_{BC} = \frac{4EI_1}{l_1} + \frac{4EI_2}{l_2}$$

$$k_{21} = \left.k_{MD_1}\right)_{CB} = \frac{2EI_2}{l_2}$$

Note:

$\left.k_{SD_1}\right)_{AB}$, $\left.k_{SD_1}\right)_{BA}$, $\left.k_{SD_1}\right)_{BC}$ and $\left.k_{SD_1}\right)_{CB}$ are the reactions due to the moments induced to unit rotation. The upward forces are positive.

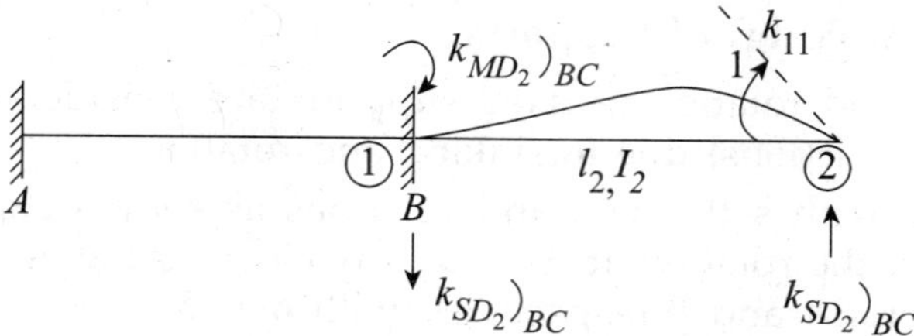

FIG. 5.3 Unit rotation at ② $[D_2 = 1]$

The stiffness coefficients based on unit rotation at ② is entered in column ② of the corresponding matrix.

$$k_{MD_2}\Big)_{BC} = \frac{2EI_2}{l_2} \quad ; \quad k_{MD_2}\Big)_{CB} = \frac{4EI_2}{l_2}$$

$$k_{SD_2}\Big)_{BC} = \frac{-6EI_2}{l_2^2} \quad ; \quad k_{SD_2}\Big)_{CB} = \frac{6EI_2}{l_2^2}$$

$$k_{12} = k_{MD_2}\Big)_{CB} = \frac{4EI_2}{l_2}$$

and

$$k_{22} = k_{MD_2}\Big)_{CB} = \frac{4EI_2}{l_2}$$

Note: $k_{SD_2}\Big)_{BC}$ and $k_{SD_2}\Big)_{CB}$ are the reactions due to the moments induced due to unit rotation. Upward forces are positive.

$$[k] = \begin{bmatrix} k_{11} & k_{12} \\ k_{21} & k_{22} \end{bmatrix} = \begin{bmatrix} \left(\frac{4EI_1}{l_1} + \frac{4EI_2}{l_2}\right) & \frac{2EI_2}{l_2} \\ 2EI_2/l_2 & 4EI_2/l_2 \end{bmatrix}$$

$$[k_{MD}] = \begin{bmatrix} k_{MD_1}\Big)_{AB} & k_{MD_2}\Big)_{AB} \\ k_{MD_1}\Big)_{BA} & k_{MD_2}\Big)_{BA} \\ k_{MD_1}\Big)_{BC} & k_{MD_2}\Big)_{BC} \\ k_{MD_1}\Big)_{CB} & k_{MD_2}\Big)_{CB} \end{bmatrix} = \begin{bmatrix} 2EI_1/l_1 & 0 \\ 4EI_1/l_1 & 0 \\ 4EI_2/l_2 & 2EI_2/l_2 \\ 2EI_2/l_2 & 4EI_2/l_2 \end{bmatrix}$$

$$[k_{SD}] = \begin{bmatrix} k_{SD_1}\Big)_{AB} & k_{SD_2}\Big)_{AB} \\ k_{SD_1}\Big)_{BA} & k_{SD_2}\Big)_{BA} \\ k_{SD_1}\Big)_{BC} & k_{SD_2}\Big)_{BC} \\ k_{SD_1}\Big)_{CB} & k_{SD_2}\Big)_{CB} \end{bmatrix} = \begin{bmatrix} -6EI_1/l_1^2 & 0 \\ 6EI_1/l_1^2 & 0 \\ -6EI_2/l_2^2 & -6EI_2/l_2^2 \\ 6EI_2/l_2^2 & 6EI_2/l_2^2 \end{bmatrix}$$

5.6.2 Settlement (Sinking) of Supports

Sinking of supports and rotation of fixed supports are considered by calculating the fixed end moments (reactions) due to sinking and rotation.

Consider a beam with settlement and rotations as shown below. The beam is of propped type and let the rotation at A be θ_A, say clockwise at A. Let δ_A and δ_B be the settlements at supports A and B respectively with $\delta_B > \delta_A$.

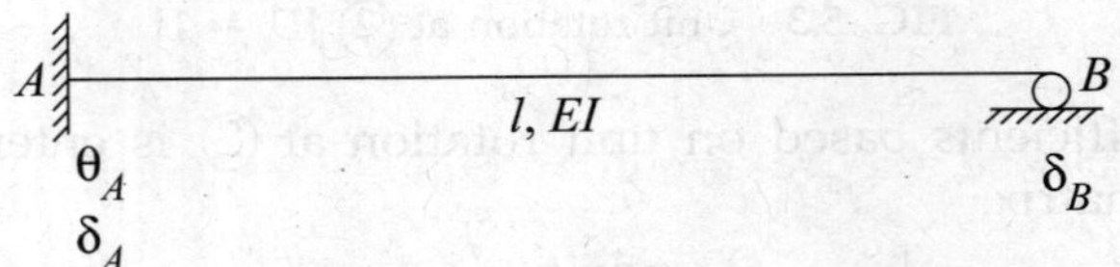

FIG. 5.4

Relative displacements between A and B;

$$\delta = \delta_B - \delta_A$$

Fixed end moments at A and B = $6EI\delta/l^2$ both anticlockwise as indicated in Fig. 5.5. If $\delta_A > \delta_B$; then the moments are clockwise. These moments are due to relative displacements between A and B.

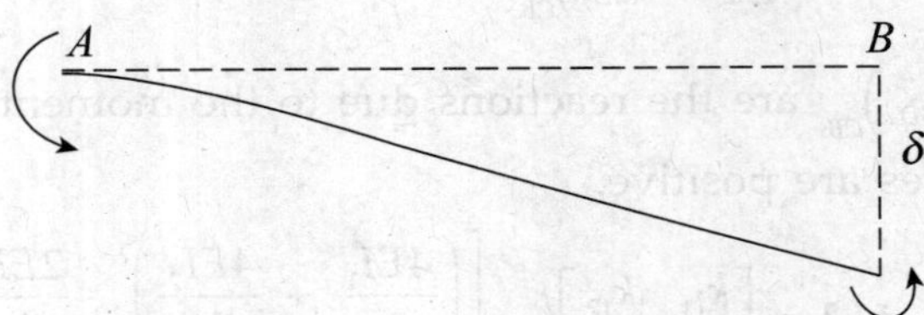

FIG. 5.5 Beam with relative displacement at supports

The rotation at A is in clockwise direction. This causes the fixed end moments and their magnitude being $4EI\theta_A/l$ and $2EI\theta_A/l$ at A and B respectively. If the rotation at A is anticlockwise, then the fixed end moments will also be anticlockwise moments.

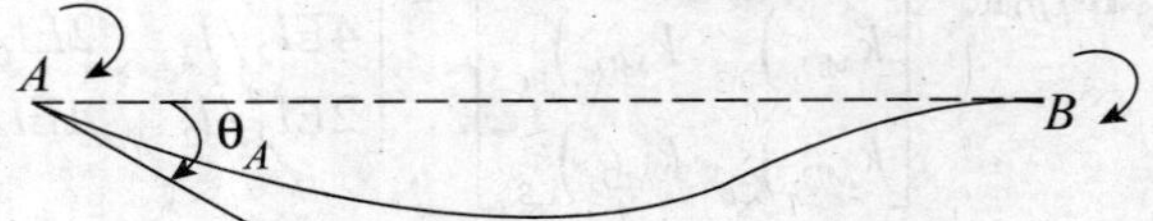

FIG. 5.6 Beam with clockwise rotation at support A

5.7 ANALYSIS OF PROPPED CANTILEVER BEAM

Example 5.1 Analyse the propped cantilever beam shown in Fig. 5.7 using the stiffness method.

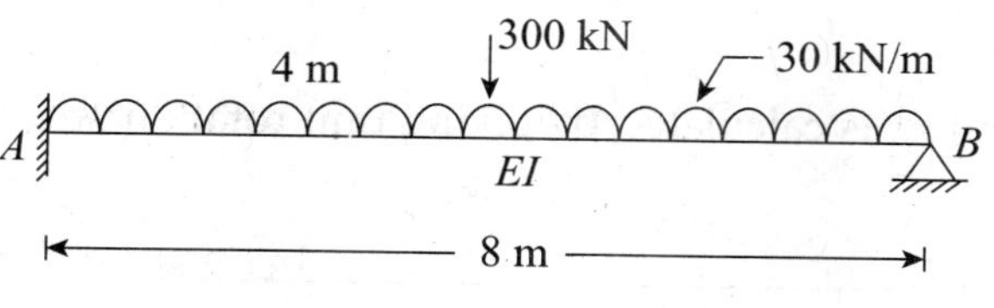

FIG. 5.7

Solution: In the given propped cantilever beam, the slope and deflection at the fixed support *A* are zero. The deflection at the propped support *B* is zero. The rotation at *B* is not equal to zero. Hence, the unknown kinematic redundant is *D*1 (slope at *B*) and is assumed as clockwise.

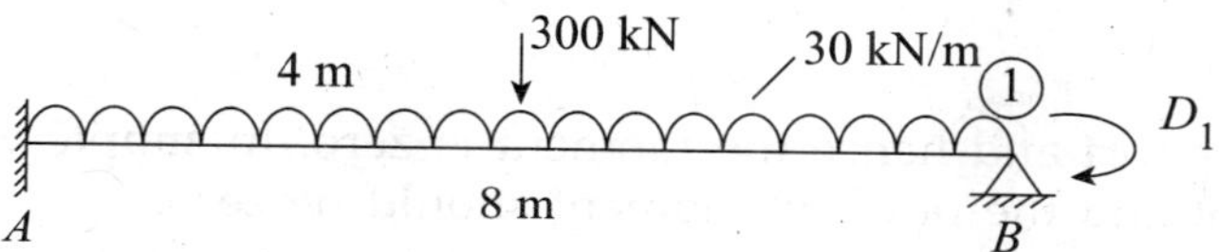

FIG. 5.8 Propped cantilever

1. Fixed End Forces

The fixed end moment and the reactions are

$$M_{FAB} = -M_{FBA} = -\left(\frac{wl^2}{12} + \frac{Wl}{8}\right)$$

$$M_{FAB} = -M_{FBA} = -\left(30 \times \frac{8^2}{12} + \frac{300 \times 8}{8}\right)$$

$$M_{FAB} = -M_{FBA} = 460 \text{ kNm}$$

$$V_A = V_B = \frac{1}{2}(wl + W) = 270 \text{ kN}$$

The above moments and reactions at the ends of the beam *AB* are termed end forces. They are indicated in the Fig. 5.9.

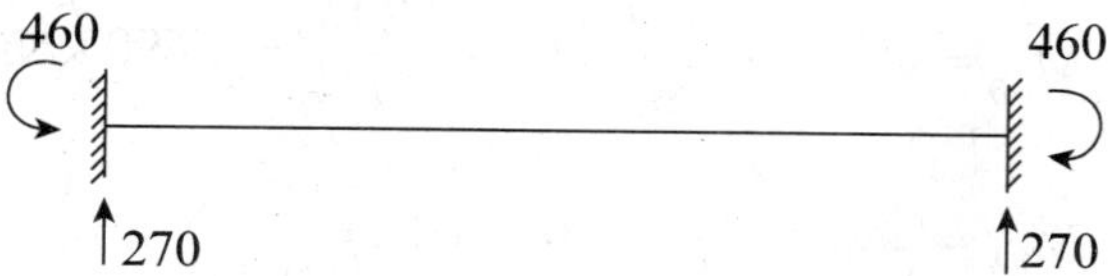

FIG. 5.9 Fixed end forces

Hence, the restraining moment at ① is

$$P_1 = 460 \text{ kNm} \curvearrowright \text{ along } D_1.$$

2. Stiffness Coefficient

The stiffness coefficient k_{11} is calculated by giving unit rotation at B in the direction of D_1.

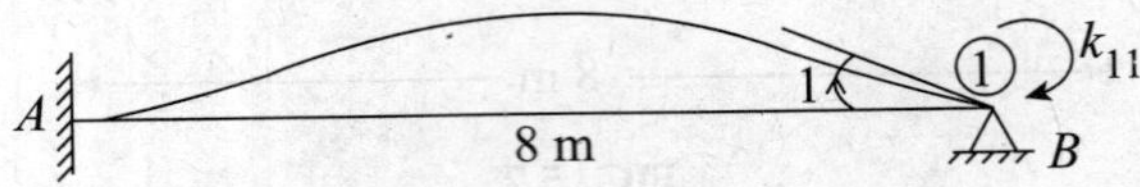

FIG. 5.10

The unit rotation creates a moment at B which is termed k_{11}. From basics,

$$k_{11} = \frac{4EI}{l} = \frac{4EI}{8} = \frac{EI}{2}$$

3. Joint Equilibrium

B is the hinged support and hence the moment is zero. In simple words, the sum of restraining moment and the released moment should be zero.

i.e.

$$P_1 + k_{11}D_1 = 0 \tag{1}$$

$$460 + \left(\frac{EI}{2}\right) D_1 = 0$$

$$\boxed{D_1 = -920/EI}$$

The negative sign indicates that the slope at B is in anticlockwise direction.

4. Determination of Other Reactions

After computing the value of D_1, end moments are obtained using slope deflection equations.

$$M_{AB} = M_{FAB} = \frac{2EI}{l}(2\theta_A + \theta_B) \tag{2}$$

Substituting the values of M_{FAB}, θ_A and θ_B; we get

$$M_{AB} = -460 + \frac{2EI}{8}\left(0 - \frac{920}{EI}\right) = -690 \text{ kNm}$$

$$M_{BA} = 460 + \frac{2EI}{8}(2\theta_B + \theta_A)$$

$$M_{BA} = 460 + \frac{2EI}{8}\left(-2 \times \frac{920}{EI} + 0\right) = 0$$

The reactions/shear forces are obtained from the free body diagram.

5. Free Body Diagram

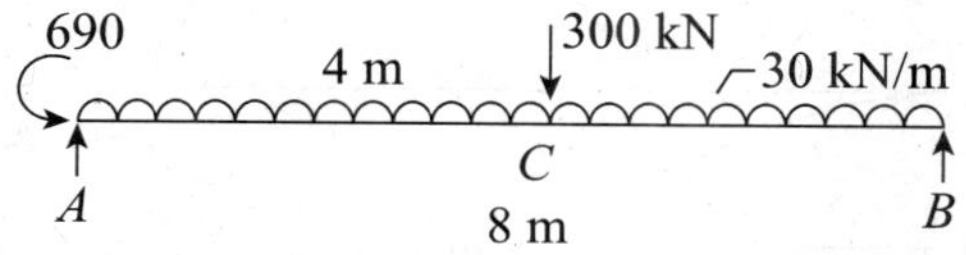

FIG. 5.11

$\Sigma V = 0;$

$$V_A + V_B = 30(8) + 300$$

$$V_A + V_B = 540$$

$\Sigma M_A = 0;$

$$-690 + 300(4) + 30(8)^2/2 - 8V_B = 0$$

$$V_B = 183.75 \text{ kN}$$

$\therefore$ $$V_A = 356.25 \text{ kN}$$

Bending moment at C;

$$M_C = 183.75(4) - 30(4)^2/2 = 495 \text{ kNm}$$

356.25
236.25
+
63.75
−
183.75

FIG. 5.12 Shear force diagram

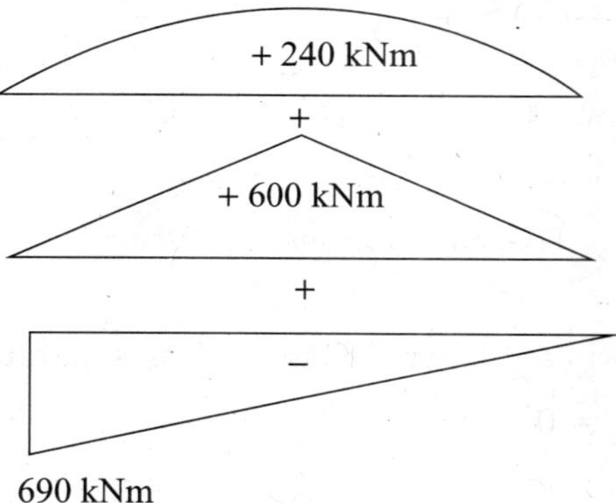

FIG. 5.13 Bending moment diagram

The answers tally with the results obtained by flexibility method. It is to be inferred that the stiffness method is automatic in choosing the unknown displacements. The elastic curve is shown below.

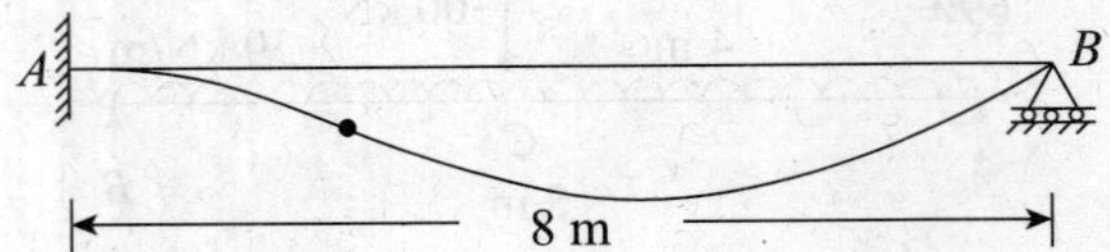

FIG. 5.14 Elastic curve for the propped cantilever beam

5.8 ANALYSIS OF FIXED BEAM WITH VARIABLE MOMENT OF INERTIA

Example 5.2 Analyse the beam by general stiffness approach. Determine the bending moments at salient points.

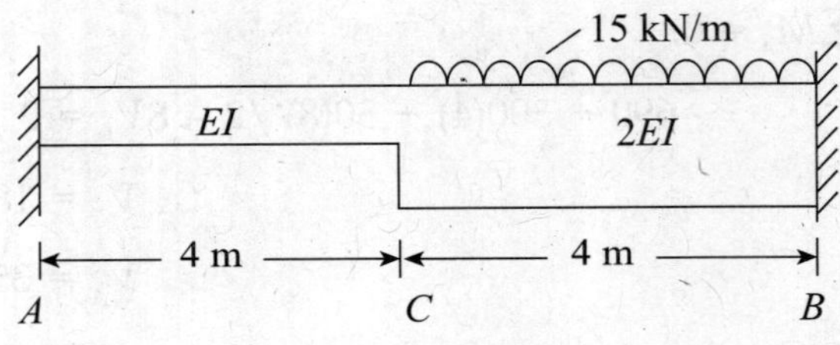

FIG. 5.15

Solution: The fixed beam *AB* has the slopes and deflection at *C*. At the fixed end; the slopes are zeros. i.e., $\theta_A = \theta_B = 0$. Thus, the kinematic redundants are vertical deflection at *C* and slope at *C*. They are designated as $\Delta_C = D_1$ and $\theta_C = D_2$ and are shown below.

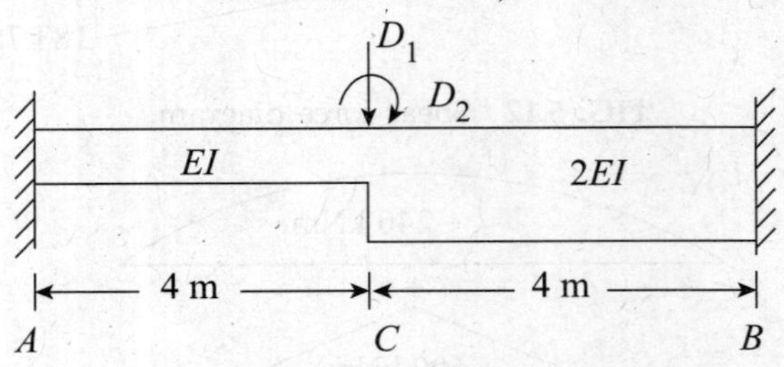

FIG. 5.16 Kinematic redundants

1. Fixed End Moments

Restrain the beam at *C* and cantilever *AC* and *CB* as separate fixed beams. Then;

$$M_{FAC} = 0$$

$$M_{FCA} = 0$$

$$M_{FCB} = \frac{-15(4)^2}{12} = -20 \text{ kNm}$$

$$M_{FBC} = \frac{+15(4)^2}{12} = +20 \text{ kNm}$$

$$V_{BC} = \frac{15(4)}{2} = +30 \text{ kN}$$

$$V_{CB} = \frac{15(4)}{2} = +30 \text{ kN}$$

Joint loads corresponding to coordinates

$$\{P\} = \begin{Bmatrix} P_1 \\ P_2 \end{Bmatrix} = \begin{Bmatrix} -30 \\ -20 \end{Bmatrix}$$

2. Stiffness Matrix

The stiffness coefficients are determined as follows:

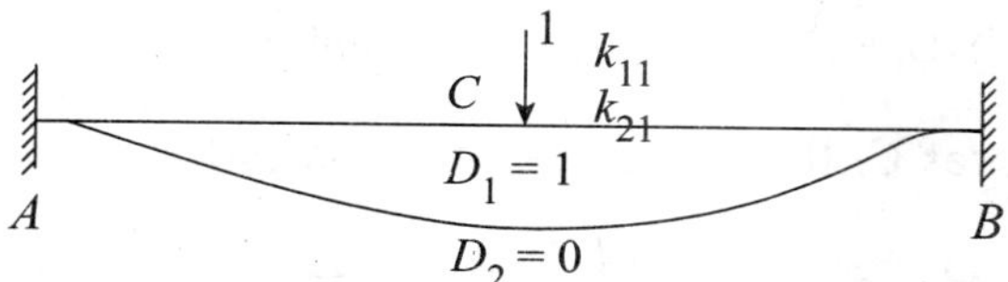

FIG. 5.17 Unit vertical displacement

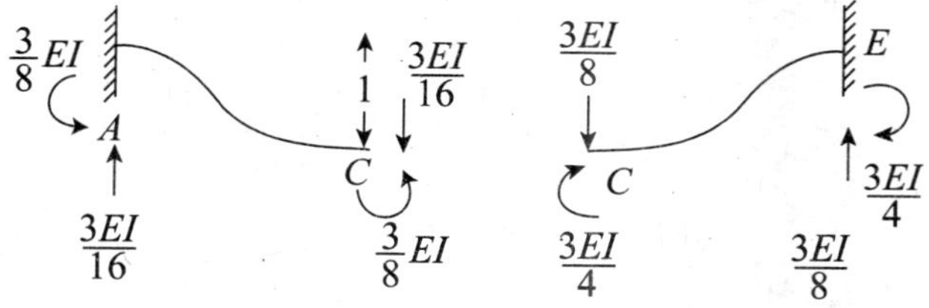

FIG. 5.18 End forces due to unit vertical displacement

$$M_{AB} = -\frac{6EI(1)}{4^2} = \frac{-3EI}{8} \quad ; \quad M_{CB} = \frac{6E(2I)}{4^2} = \frac{3EI}{4}$$

$$M_{CA} = -\frac{6EI(1)}{4^2} = \frac{-3EI}{8} \quad ; \quad M_{BC} = \frac{6E(2I)}{4^2} = \frac{3EI}{4}$$

$$V_{AC} = -\frac{12EI(1)}{4^3} = \frac{-3EI}{16} \quad ; \quad V_{CB} = +\frac{12E(2I)}{4^3} = \frac{3EI}{8}$$

$$V_{CA} = \frac{12EI(1)}{4^3} = \frac{3EI}{16} \quad ; \quad V_{BC} = \frac{12E(2I)}{4^3} = \frac{-3EI}{8}$$

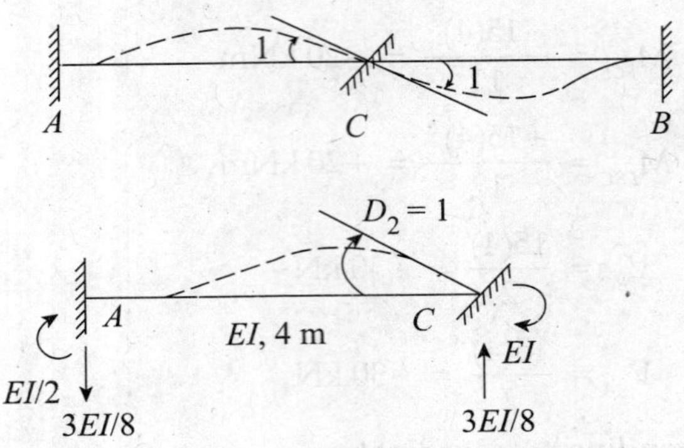

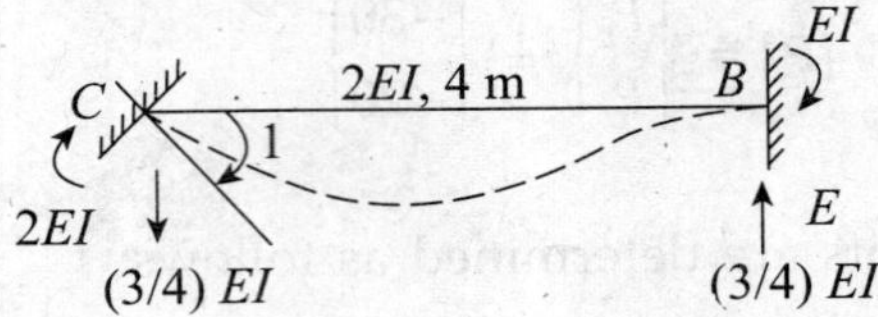

FIG. 5.19 End forces due to unit rotation at 1

Apply unit rotation at C, the end forces are computed as

$$M_{AC} = \frac{2EI}{4} = \frac{EI}{2} \quad ; \quad M_{CB} = \frac{4E(2I)}{4} = 2EI$$

$$V_{AC} = \frac{6EI}{4^2} = -\frac{3EI}{8} \quad ; \quad V_{CB} = \frac{6E(2I)}{4^2} = \frac{3EI}{4}$$

$$M_{CA} = \frac{4EI}{4} = EI \quad ; \quad M_{BC} = \frac{4E(2I)}{4} = 2EI$$

$$V_{CA} = \frac{6EI}{4^2} = -\frac{3EI}{8} \quad ; \quad V_{BC} = \frac{6E(2I)}{4^2} = \frac{3}{4}EI$$

3. Stiffness Coefficients

$$k_{11} = V_{CA} + V_{CB} = \frac{3EI}{16} + \frac{3EI}{8} = \frac{9EI}{16}$$ (Due to unit vertical displacement at C)

$$k_{21} = M_{CA} + M_{CB} = \frac{3EI}{4} + \frac{3EI}{8} = \frac{3EI}{8}$$ (Due to unit vertical displacement at C)

$$k_{12} = V_{CB} + V_{CA} = \frac{3EI}{4} + \frac{3EI}{8} = \frac{3EI}{8}$$ (Due to unit rotation at C)

$$k_{22} = M_{CA} + M_{CB} = EI + 2EI = 3EI$$ (Due to unit rotation at C)

4. Joint Equilibrium

$$P_1 + k_{11}D_1 + k_{12}D_2 = 0$$

$$P_2 + k_{21}D_1 + k_{22}D_2 = 0$$

Substituting the computed values

$$-30 + \frac{9EI}{16} D_1 + \frac{3EI}{8} D_2 = 0$$

$$-20 + \frac{3EI}{8} D_1 + 3EI\ D_2 = 0$$

Solving this

$$D_1 = 53.33/EI$$

$$D_2 = 0$$

5. End Moments

$$M_{AB} = 0 - \left(\frac{3EI}{8}\right)\left(\frac{53.33}{EI}\right) + \left(\frac{EI}{2}\right)(0) = -20 \text{ kNm}$$

$$M_{BA} = 0 - \left(\frac{3EI}{8}\right)\left(\frac{53.33}{EI}\right) + EI\,(0) = -20 \text{ kNm}$$

$$M_{BC} = -20 + \left(\frac{3EI}{4}\right)\left(\frac{53.33}{EI}\right) + 2EI\,(0) = 20 \text{ kNm}$$

$$M_{CB} = +20 + \left(\frac{3EI}{4}\right)\left(\frac{53.33}{EI}\right) + EI\,(0) = +60 \text{ kNm}$$

Example 5.3 Determine the fixed end moments for a fixed beam shown in Fig. 5.20 using system stiffness approach.

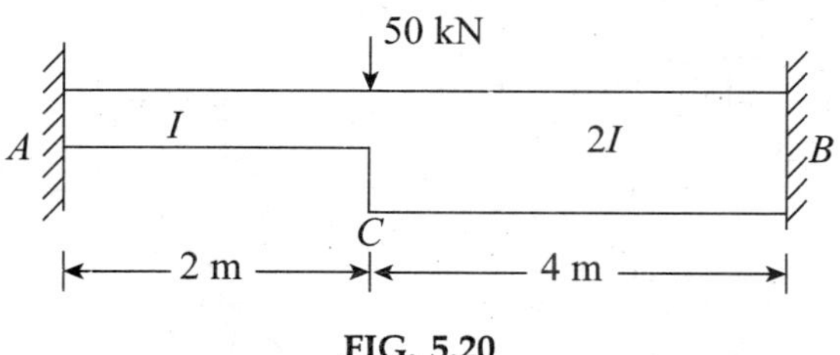

FIG. 5.20

Solution: At the ends of the fixed beam the slope is zero and the vertical deflection is zero. At C there is a change of cross section and a vertical load 50 kN is acting down. This causes a vertical deflection and slope at C. They are treated as kinematic redundants and denoted as D_1 and D_2 respectively.

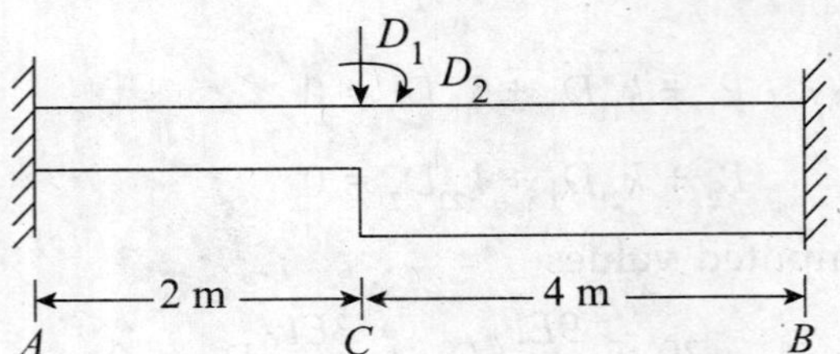

FIG. 5.21 Kinematic redundants

The positive directions of D_1 and D_2 are shown above.

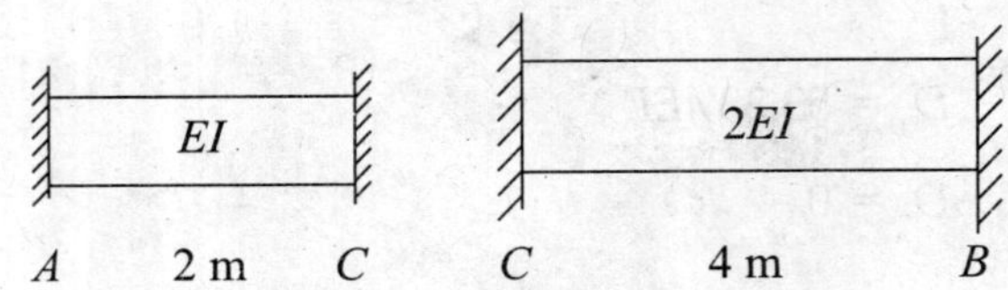

FIG. 5.22 Fixed beams

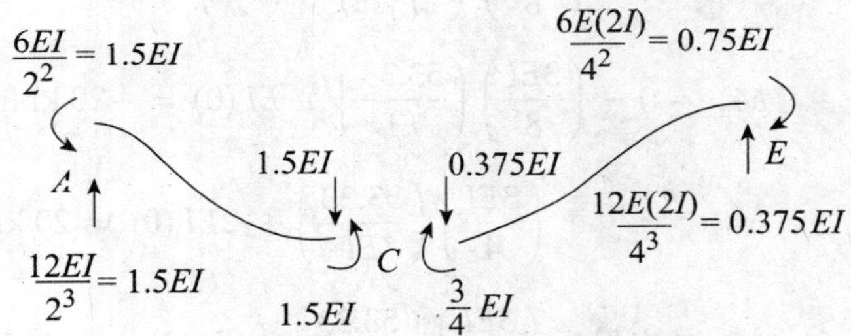

FIG. 5.23 End force due to $D_1 = 1$ and $D_2 = 0$

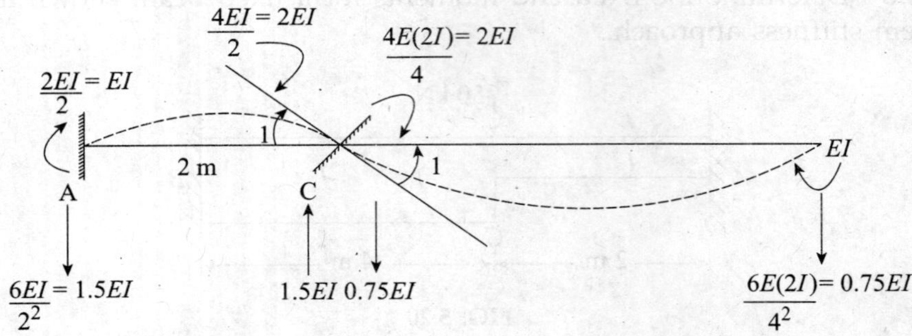

FIG. 5.24 End force due to $D_2 = 1$ and $D_1 = 0$

1. Fixed End Moments

As no loads are acting in span AC and CB respectively;

$$M_{FAC} = 0 \quad ; \quad M_{FCB} = 0$$

$$M_{FCA} = 0 \quad ; \quad M_{FBC} = 0$$

2. Stiffness Coefficients

Referring to the effects of unit vertical displacement and rotation at C:

$$k_{11} = 1.5EI + 0.375EI = 1.875EI$$

$$k_{21} = 0.75EI - 1.5EI = -0.75EI$$

$$k_{12} = 0.75EI - 1.5EI = -0.75EI$$

$$k_{22} = 2EI + 2EI = 4EI$$

3. Joint Loads

$$[P] = \begin{Bmatrix} P_1 \\ P_2 \end{Bmatrix} = \begin{Bmatrix} -50 \\ 0 \end{Bmatrix}$$

4. Joint Equilibrium Equations

We know at joint C;

$$P_1 + k_{11}D_1 + k_{12}D_2 = 0 \tag{1}$$

$$P_2 + k_{21}D_1 + k_{22}D_2 = 0 \tag{2}$$

Substituting the computed values, we obtain

$$-50 + 1.875EID_1 - 0.75EID_2 = 0 \tag{3}$$

$$0 - 0.75EID_1 + 4EID_2 = 0 \tag{4}$$

Solving the above equations

$$D_1 = 28.83/EI$$

$$D_2 = 5.41/EI$$

5. End Moments

$$M_{AB} = 0 - 1.5 \times 28.83 + 1(5.41) = -37.84 \text{ kNm}$$

$$M_{BA} = 0 + 0.75 \times 28.83 + 1(5.41) = +27 \text{ kNm}$$

5.9 ANALYSIS OF CONTINUOUS BEAM WITHOUT SUPPORT SETTLEMENT

Example 5.4 Analyse the continuous beam ABC shown in Fig. 5.25 by stiffness method. Draw BMD. A and C are fixed ends.

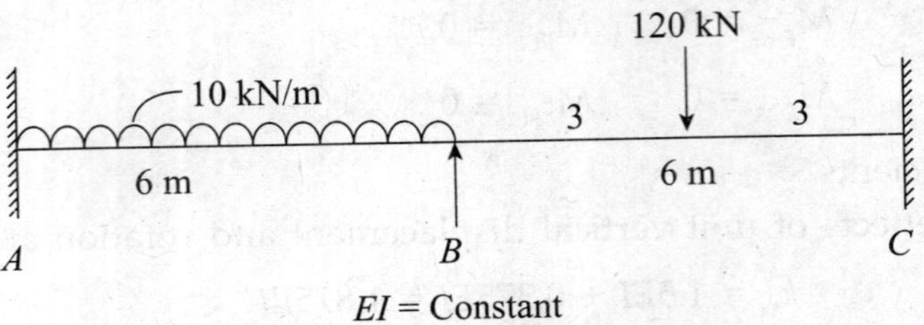

FIG. 5.25

Solution: The ends of the continuous beams ABC is fixed. Hence, the slopes at the fixed support are zeroes. Thus, $\theta_A = \theta_C = 0$. The support B is an intermediate support and that $\theta_B \neq 0$. Let this kinematic redundant is denoted as D_1. This is assumed to act in the clockwise direction.

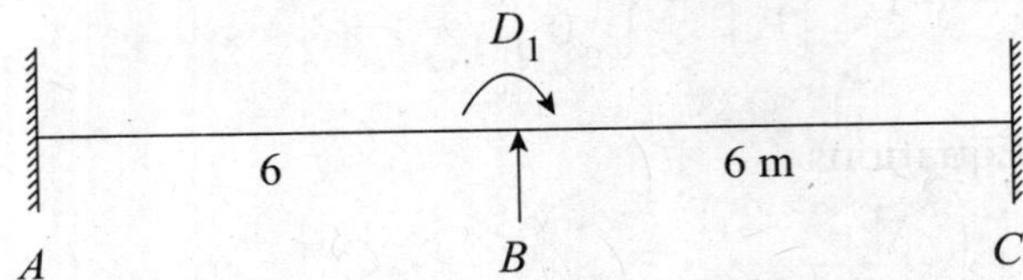

FIG. 5.26 Kinematic redundant

1. Fixed End Moments

Clamp the ends A, B and C and write the fixed end moments

$$M_{FAB} = -10 \times \frac{6^2}{12} = -30 \text{ kNm}$$

$$M_{FBA} = +10 \times \frac{6^2}{12} = +30 \text{ kNm}$$

$$M_{FBC} = -120 \times \frac{6}{8} = -90 \text{ kNm}$$

$$M_{FCB} = +120 \times \frac{6}{8} = +90 \text{ kNm}$$

2. Joint Load

$$P = 30 - 90 = -60 \text{ kNm}$$

3. Stiffness Coefficient/Stiffness Matrix

Apply unit rotation at joint B and compute the moment required. This is denoted as k_{11}.

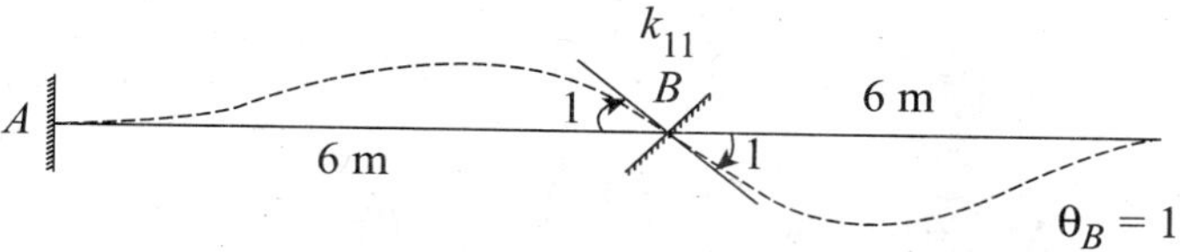

FIG. 5.27

$$k_{11} = \frac{4EI}{6} + \frac{4EI}{6}$$

$$k_{11} = (4/3)EI$$

4. Joint Equilibrium

As no external load is acting at joint B

$$P + k_{11}D_1 = 0$$

$$-60 + \frac{4}{3}EID_1 = 0$$

$$\boxed{D_1 = 45/EI}$$

5. End Moments

The end moments are calculated using the slope deflection equations as

$$M_{AB} = M_{FAB} + \frac{2EI}{l}(2\theta_A + \theta_B)$$

$$M_{AB} = -30 + \frac{2EI}{6} \times \frac{45}{EI} = -15 \text{ kNm}$$

$$M_{BA} = M_{FBA} + \frac{2EI}{l}(2\theta_B + \theta_A)$$

$$M_{BA} = +30 + \frac{2EI}{6} \times \left(2 \times \frac{45}{EI}\right) = +60 \text{ kNm}$$

$$M_{BC} = M_{FBC} + \frac{2EI}{l}(2\theta_B + \theta_C)$$

$$M_{BC} = -90 + \frac{2EI}{6} \times 2 \times \frac{45}{EI} = -60 \text{ kNm}$$

$$M_{CB} = M_{FCB} + \frac{2EI}{l}(2\theta_C + \theta_B)$$

$$M_{CB} = +90 + \frac{2EI}{6}\left(\frac{45}{EI}\right) = +105\,\text{kNm}$$

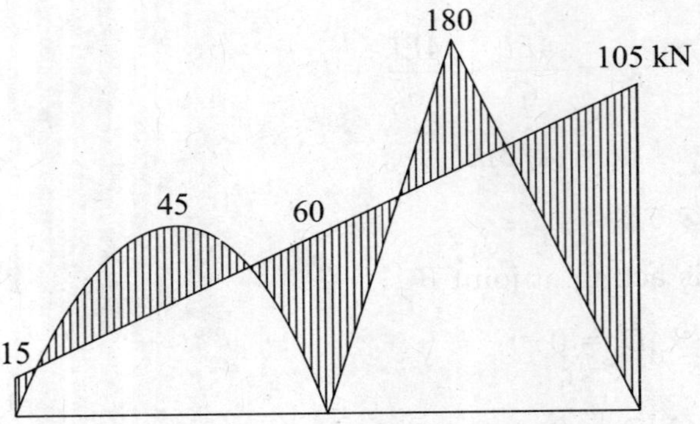

FIG. 5.28 Bending moment diagram

Example 5.5 Analyse the two-span continuous beam shown in Fig. 5.29 by the stiffness method.

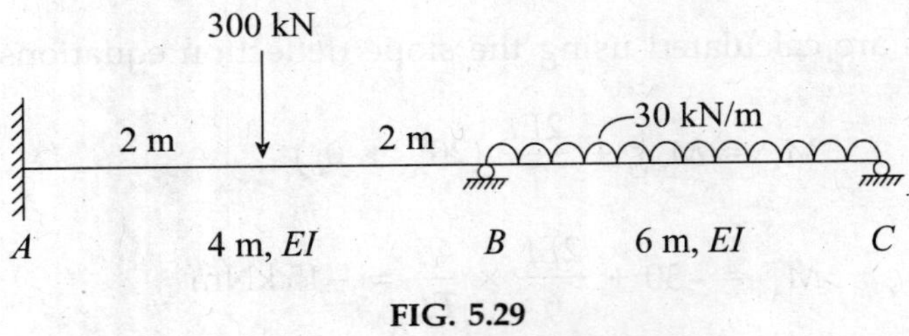

FIG. 5.29

Solution: The kinematic redundants are θ_A, θ_B and θ_C. As the support A is fixed, θ_A is zero. Hence, the two kinematic redundants are $\theta_B = D_1$ and $\theta_C = D_2$ respectively. They are assumed to act in the clockwise direction.

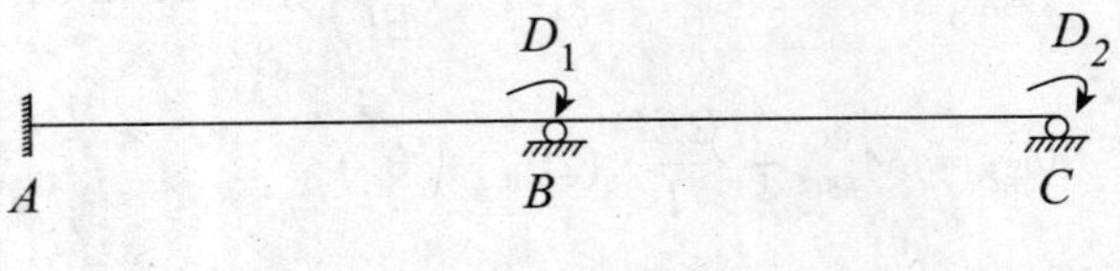

FIG. 5.30

1. Fixed End Forces

The fixed end reactions and moments are calculated by considering spans AB and BC respectively. The calculated values are shown in Fig. 5.30.

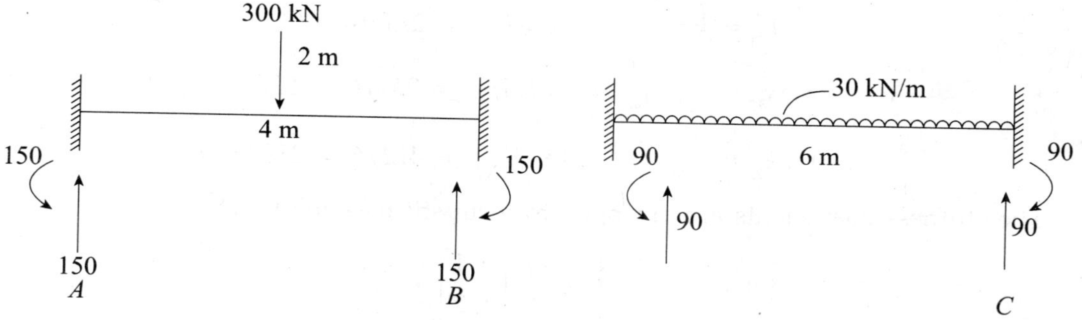

FIG. 5.31 Joint loads

$$P_1 = 150 - 90 = 60 \text{ kNm}$$

$$P_2 = +90 \text{ kNm}$$

$$\{P\} = \begin{bmatrix} 60 \\ 90 \end{bmatrix}$$

2. Stiffness Coefficients and Stiffness Matrix

The stiffness coefficients are calculated by giving unit rotations at ① and ② respectively. The coefficients are shown in Figs. (5.32 and 5.33) below.

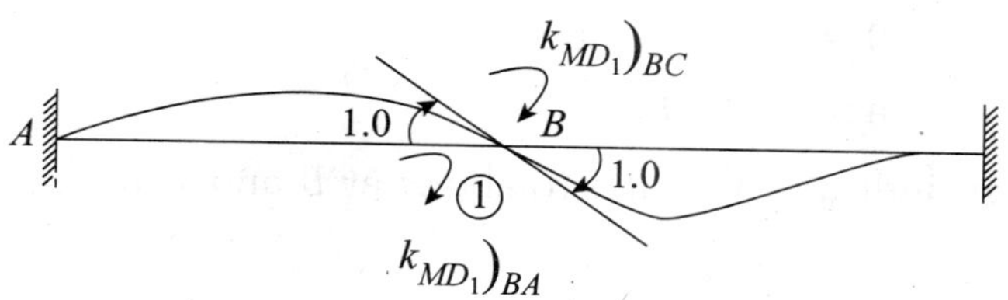

FIG. 5.32 Unit rotation at ①

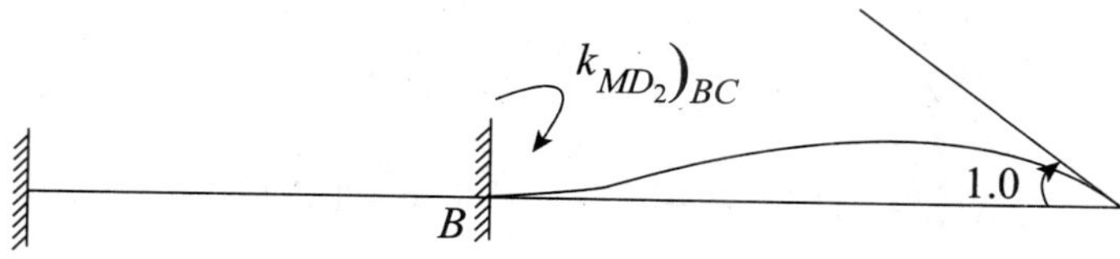

FIG. 5.33 Unit rotation at ②

$$k_{11} = k_{MD_1}\Big)_{BA} + k_{MD_1}\Big)_{BC} = 4EI/l)_{BA} + 4EI/l)_{BC}$$

$$k_{11} = 4EI/4 + 4EI/6 = \left(\frac{5}{3}\right) EI$$

$$k_{21} = k_{MD_1}\Big)_{CB} = 2EI/l)_{CB} = 2EI/6 = EI/3$$

$$k_{12} = k_{MD_2}\Big)_{BC} = 2EI/l)_{BC} = 2EI/6 = EI/3$$

$$k_{22} = k_{MD_2}\Big)_{CB} = 4EI/l)_{CB} = 4EI/6 = 2EI/3$$

The stiffness coefficients are arranged to form stiffness matrix as

$$[k] = \begin{bmatrix} k_{11} & k_{12} \\ k_{21} & k_{22} \end{bmatrix} = \frac{EI}{3}\begin{bmatrix} 5 & 1 \\ 1 & 2 \end{bmatrix}$$

3. Joint Equilibrium

As the supports are unyielding; the unknowns D_1 and D_2 can be found from

$$[D] = [-k]^{-1} \{P\}$$

$$\begin{bmatrix} D_1 \\ D_2 \end{bmatrix} = \left\{\frac{EI}{3}\begin{bmatrix} 5 & 1 \\ 1 & 2 \end{bmatrix}\right\}^{-1} \begin{Bmatrix} 60 \\ 90 \end{Bmatrix}$$

$$\begin{bmatrix} D_1 \\ D_2 \end{bmatrix} = \frac{-1}{EI}\begin{Bmatrix} 10 \\ 130 \end{Bmatrix}$$

Thus,

$$\theta_B = -10/EI \quad \text{and}$$

$$\theta_C = -130/EI$$

The negative sign indicates that the rotations at B and C are in the anticlockwise direction.

4. Determination of Reactions/Moments

The end moments are obtained using the slope deflections as

$$M_{AB} = M_{FAB} + \frac{2EI}{l}(2\theta_A + \theta_B)$$

$$M_{AB} = -150 + \frac{2EI}{4}\left(\frac{-10}{EI}\right) = -155\text{ kNm}$$

$$M_{BA} = M_{FBA} + \frac{2EI}{4}(2\theta_B + \theta_A)$$

$$M_{BA} = +150 + \frac{2EI}{4}\left(2 \times \frac{-10}{EI}\right) = +140\text{ kNm}$$

$$M_{BC} = M_{FBC} + \frac{2EI}{6}\left(2\theta_B + \theta_C\right)$$

$$M_{BC} = -90 + \frac{2EI}{6}\left(2 \times \frac{-10}{EI} - \frac{130}{EI}\right) = -140 \text{ kNm}$$

$$M_{CB} = M_{FCB} + \frac{2EI}{6}\left(2\theta_C + \theta_B\right)$$

$$M_{CB} = +90 + \frac{2EI}{6}\left\{2\left(\frac{-130}{EI}\right) - 10/EI\right\}$$

$$M_{CB} = +90 + \frac{2EI}{6}\left\{\frac{-260}{EI} - \frac{10}{EI}\right\} = 0$$

The above values are used to draw the pure moment diagram.

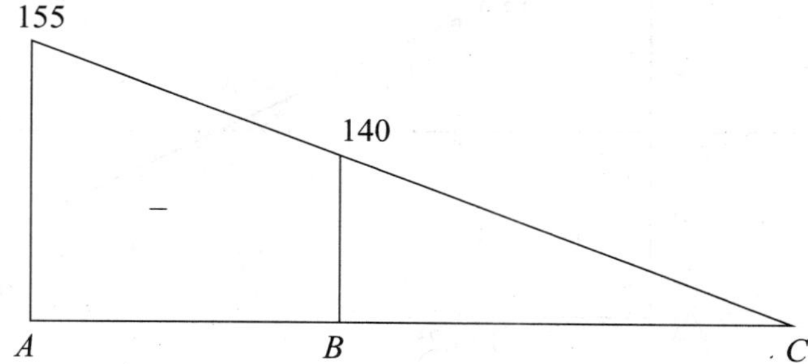

FIG. 5.34 Pure moment diagram

5. Bending Moment and Shear Force Diagrams

The final bending moment diagram is drawn by superimposing the simple beam bending moment diagram over pure moment diagram. It is to be mentioned here that the simple beam BMD is drawn by considering each span as simply supported beam.

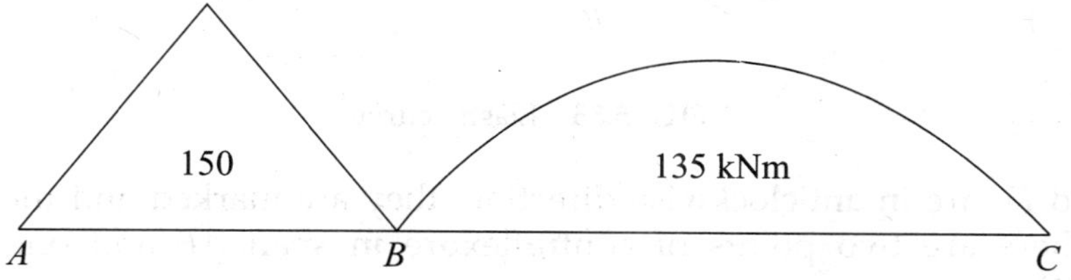

FIG. 5.35 Simple beam BMD

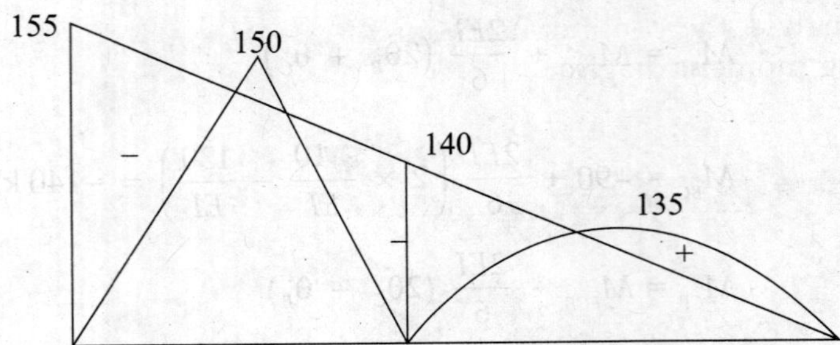

FIG. 5.36 Bending moment diagram

The reactions are found out by drawing the free body diagram. The shear force diagram is drawn by using the applied loads and reactions in the free body diagram.

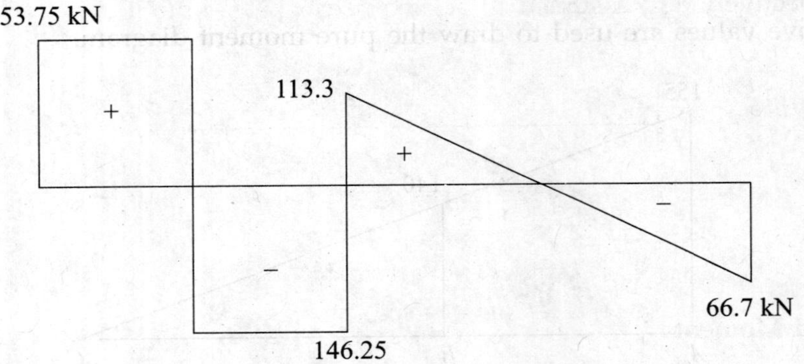

FIG. 5.37 Shear force diagram

The elastic curve is shown below.

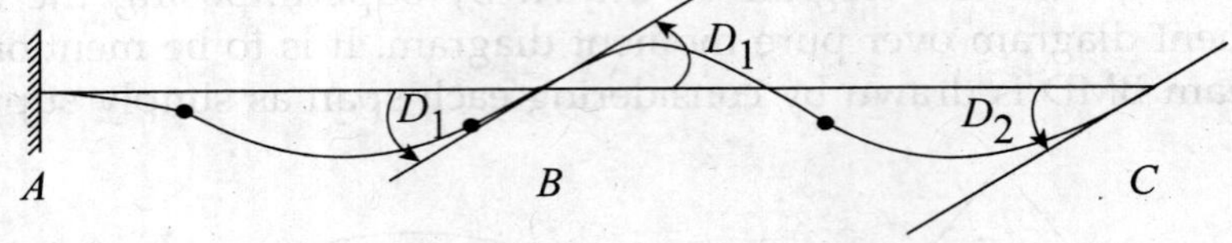

FIG. 5.38 Elastic curve

As D_1 and D_2 are in anticlockwise direction, they are marked and the elastic curve is drawn. There are two points of contraflexure in span *AB* and one in span *BC*. (Ref. Fig. 5.36)

If the support *A* is simply supported, then independent displacements are D_1, D_2 and D_3 at *A*, *B* and *C* respectively. In such case, the stiffness matrix is 3 × 3. The unknowns can be evaluated using calculator.

Example 5.6 Analyse the continuous beam shown in Fig. 5.39 by stiffness method. Draw the bending moment diagram. (VTU – July 2011)

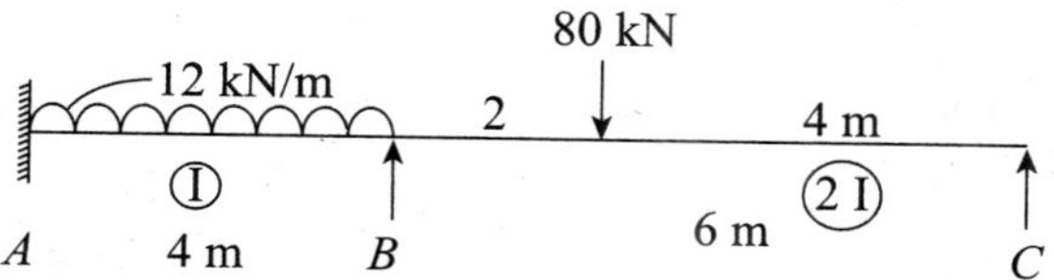

FIG. 5.39

Solution: The slope at A is zero while slopes at B and C are not equal to zero. This is due to the fact that A is a fixed support, B is an intermediate support and C is the end simple support. The kinematic redundants are θ_B and θ_C. They are denoted as D_1 and D_2 respectively and assumed to act in the clockwise direction.

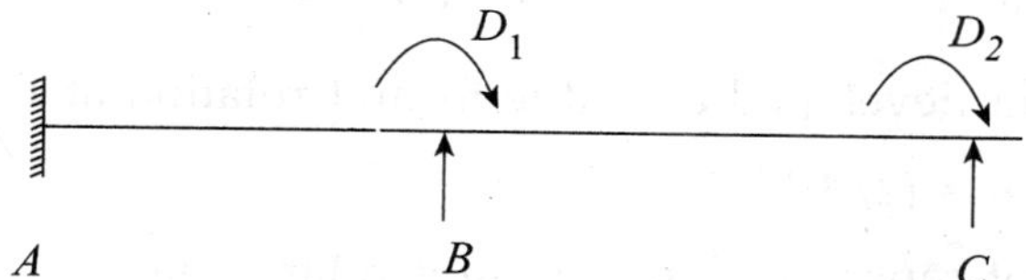

FIG. 5.40 Kinematic redundants

1. Fixed End Moments

$$M_{FAB} = -12(4)^2/12 = -16 \text{ kNm}$$

$$M_{FBA} = +12(4)^2/12 = +16 \text{ kNm}$$

$$M_{FBC} = -2(80)4^2/6^2 = -71.11 \text{ kNm}$$

$$M_{FCB} = +4(80)2^2/6^2 = +35.56 \text{ kNm}$$

2. Joint Loads

$$P_1 = 16 - 71.11 = -55.11 \text{ kNm}$$

$$P_2 = +35.56 \text{ kNm}$$

3. Stiffness Coefficients/Stiffness Matrix

A unit rotation is given at B and C respectively the clockwise direction. The stiffness coefficients are evaluated as follows.

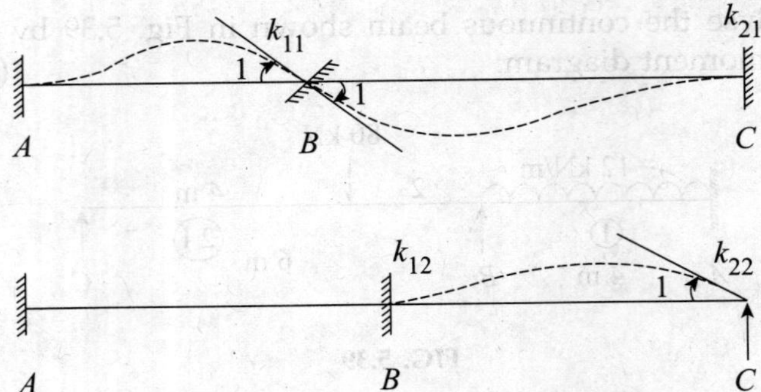

FIG. 5.41 Restrained beam under unit displacement

k_{11} = Moment required at B due to unit rotation of B

$$k_{11} = M_{BA} + M_{BC} = \frac{4EI}{4} + \frac{4E(2I)}{6} = (7/3)\,EI$$

k_{21} = Moment developed at C due to unit rotation at B

$k_{21} = 2(2EI)/6 = (2/3)EI$

k_{12} = Moment caused at B due to unit rotation at C

$k_{12} = 2(2EI)/6 = (2/3)EI$

k_{22} = Moment required at C due to unit rotation at C

$k_{22} = 4(2EI)/6 = (4/3)EI$

Thus, the stiffness matrix is

$$[k] = \begin{bmatrix} 7 & 2 \\ 2 & 4 \end{bmatrix} (EI/3)$$

4. Joint Equilibrium

The joint equilibrium conditions can be written as

$$P_1 + k_{11}D_1 + k_{12}D_2 = 0$$

$$P_2 + k_{12}D_1 + k_{22}D_2 = 0$$

Substituting the values and simplifying

$$7D_1 + 2D_2 = 165.33/EI \qquad (1)$$

$$2D_1 + 4D_2 = -106.68/EI \qquad (2)$$

Solving Eqs. (1) and (2)

$$D_1 = +36.45/EI = \theta_B$$
$$D_2 = -44.89/EI = \theta_C$$

5. End Moments

The end moments are obtained using slope deflection equation as

$$M_{AB} = -16 + \frac{2EI}{4}(2\theta_A + \theta_B)$$

$$M_{AB} = -16 + \left(\frac{EI}{2}\right)\left(\frac{36.45}{EI}\right) = 2.23 \text{ kNm}$$

$$M_{BA} = +16 + \frac{2EI}{4}(2\theta_B + \theta_A) = 52.45 \text{ kNm}$$

$$M_{BC} = -71.11 + \frac{2(2EI)}{6}(2\theta_B + \theta_C) = -52.45 \text{ kNm}$$

$$M_{CB} = +35.56 + \frac{2(2EI)}{6}(2\theta_C + \theta_B) = 0.00$$

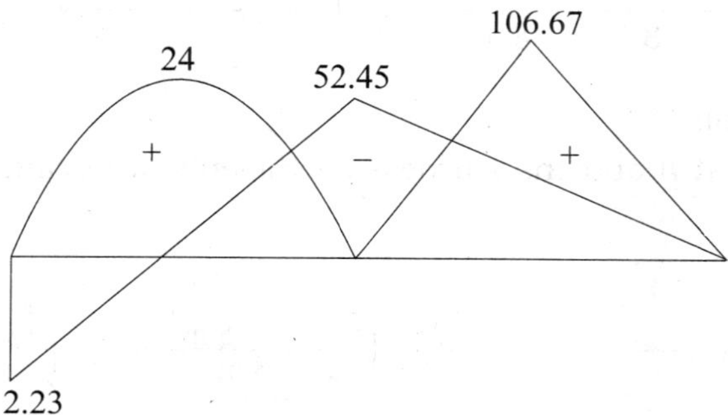

FIG. 5.42 Bending moment diagram

Example 5.7 Analyse the continuous beam shown in Fig. 5.43 by stiffness method. Support *C* is a guided support.

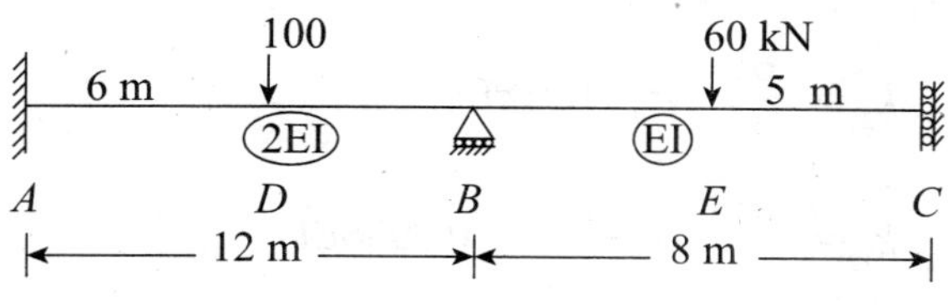

FIG. 5.43

Solution: The continuous beam *ABC* is fixed at *A*, on rollers at *B* and on gliding support at *C*. The kinematic redundants are rotation at *B* and a vertical displacement at *C*. The kinematic redundants are shown below.

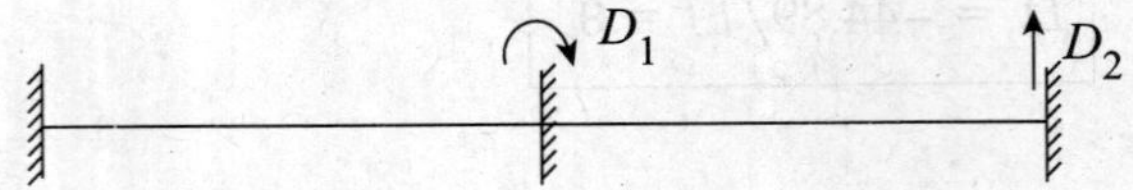

FIG. 5.44 Kinematic redundants

1. Fixed End Forces

The restrained forces due to loading are as follows.

At joint *B*;

$$M_{FBA} = \frac{100 \times 12}{} \quad 150 \text{ kNm}$$

$$M_{FBC} = \frac{-3(60)5^2}{8^2} = -70.31 \text{ kNm}$$

2. Joint Loads

$$P_1 = M_{FBA} + M_{FBC} = 79.69 \text{ kNm} \curvearrowright$$

$$P_2 = \frac{60 \times 3}{8} - \frac{1}{8}\left[\frac{3 \times 60 \times 5^2}{8^2} - \frac{5 \times 60 \times 3^2}{8^2}\right] = 18.8 \text{ kNm}$$

3. Flexibility Coefficients

Apply a unit rotation at *B* and the stiffness coefficients are evaluated.

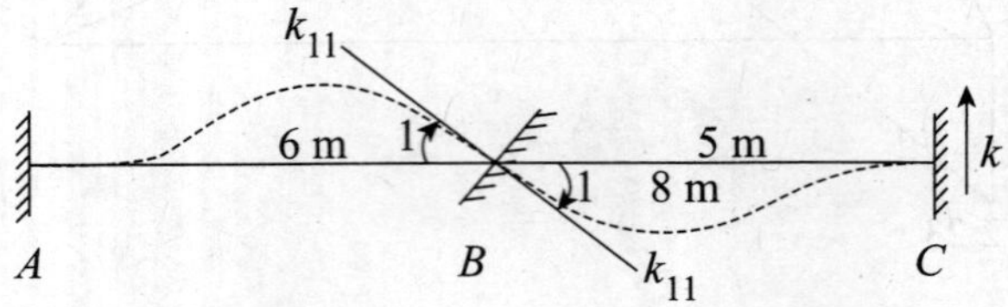

FIG. 5.45 Unit rotation at *B*

$$k_{11} = \frac{4E(2I)}{12} + \frac{4EI}{8} = 1.17EI$$

$$k_{21} = \frac{6EI}{8^2} = 0.0938EI$$

$$k_{12} = \frac{6EI}{l^2} = \frac{6EI}{8^2} = 0.0938EI$$

$$k_{22} = \frac{12EI}{l^3} = \frac{12EI}{8^3} = 0.0234EI$$

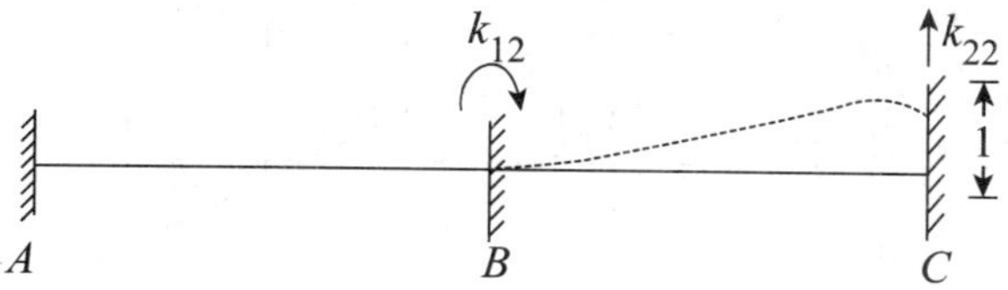

FIG. 5.46 Unit vertical displacement at C

4. Equilibrium Equations

$$P_1 + k_{11}D_1 + k_{12}D_2 = 0$$

$$P_2 + k_{21}D_1 + k_{22}D_2 = 0$$

Substituting the computed values;

$$1.17\ D_1 + 0.0938\ D_2 = -79.69/EI$$

$$0.0938\ D_1 + 0.0234\ D_2 = -18.98/EI$$

Solving

$$D_1 = \theta_B = -4.5438/EI$$

$$D_2 = \Delta_C = -792.897/EI$$

5. End Moments

$$M_{AB} = M_{FAB} + 2\frac{(2EI)}{l}\left(2\theta_A + \theta_B - \frac{3\Delta}{l}\right) = 148.4 \text{ kNm}$$

$$M_{BA} = M_{FBA} + 2\frac{(2EI)}{l}\left(2\theta_B + \theta_A - \frac{3\Delta}{l}\right) = 146.97 \text{ kNm}$$

$$M_{BC} = M_{FBC} + \frac{2EI}{8}\left(2\theta_B + \theta_C - \frac{3\Delta}{l}\right) = -146.97 \text{ kNm}$$

$$M_{CB} = \frac{5(60)3^2}{8^2} + \frac{2EI}{8}\left(\frac{-4.5438}{EI} - \frac{3 \times 792.897}{8EI}\right) = -32.29 \text{ kNm}$$

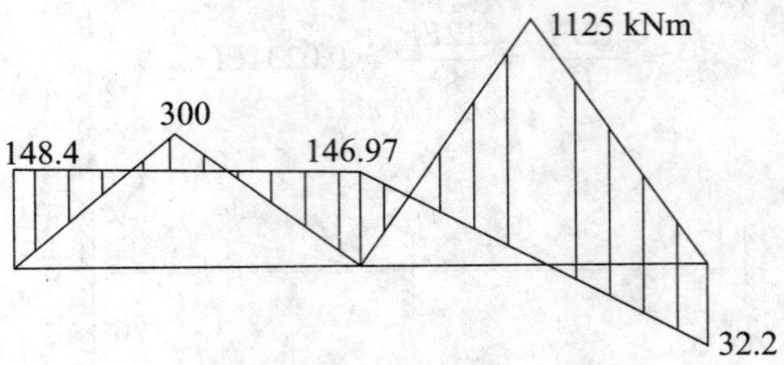

FIG. 5.47 Bending moment diagram

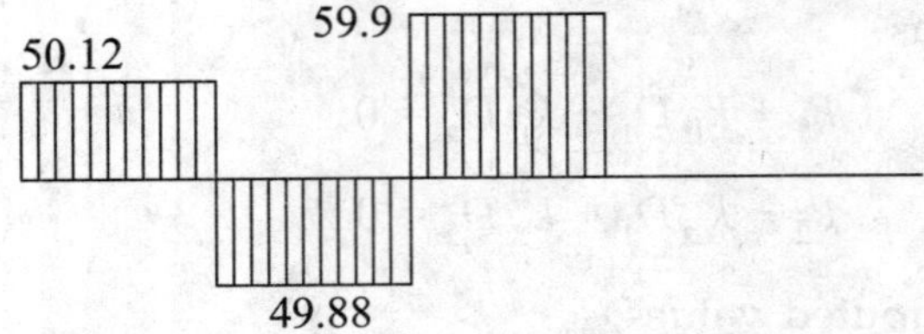

FIG. 5.48 Shear force diagram

Example 5.8 Analyse the beam by stiffness system approach. Draw the bending moment diagram.

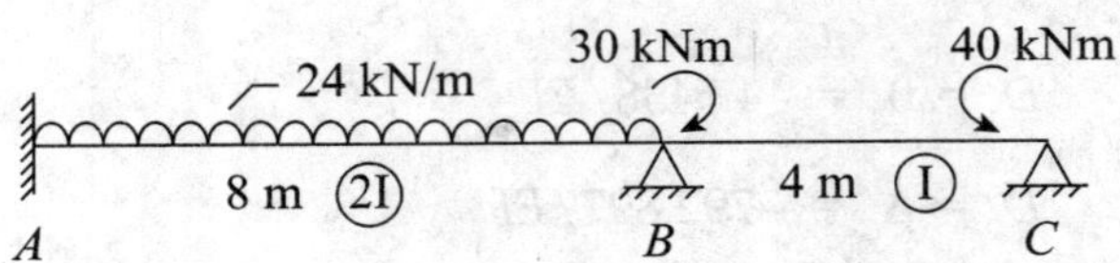

FIG. 5.49

Solution: The continuous beam is kinematically indeterminate to second degree. i.e., $\theta_A = 0$; $\theta_B \neq 0$ and $\theta_C \neq 0$. They are designated as D_1 and D_2 respectively. They are assumed to act in the clockwise direction.

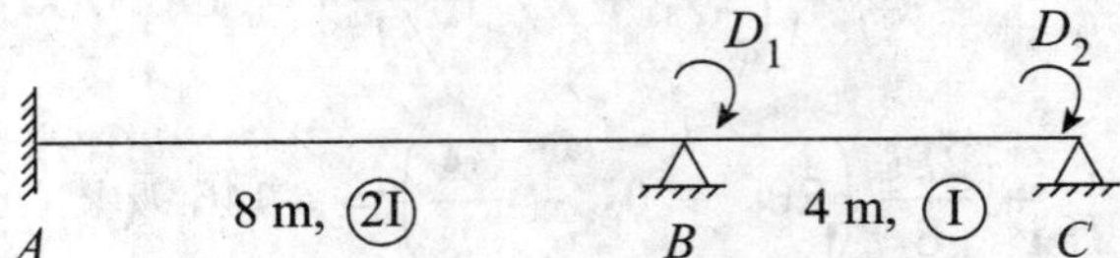

FIG. 5.50 Kinematic redundants

1. Fixed End Moments

$$M_{FAB} = \frac{-wl^2}{12} = -24\frac{(8^2)}{12} = -128 \text{ kNm}$$

$$M_{FBA} = \frac{+wl^2}{12} = +24\frac{(8^2)}{12} = +128\,\text{kNm}$$

2. Joint Loads

$$P_1 = 128 - 30 = 98\ \text{kNm}$$

$$P_2 = +40\ \text{kNm}$$

It is to be noted that at joint B, the applied moment is 30 kNm in the clockwise direction. The resisting moment is 30 kNm in the anticlockwise direction, which is opposite to the direction of assumed direction of D_1. Hence, it is –30 kNm. Similarly, at joint C, the applied moment is 40 kNm in the anticlockwise direction. Hence, the resisting moment is in the clockwise direction. Thus, the joint load is +40 kNm

3. Stiffness Coefficients/Stiffness Matrix

Applying a unit rotation at B and at C; the stiffness coefficients are obtained.

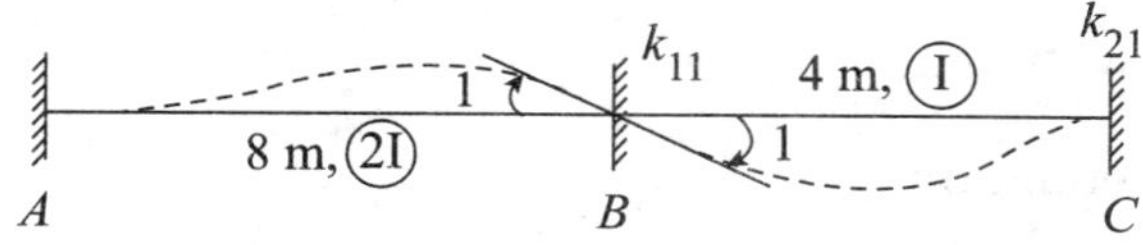

FIG. 5.51

$$k_{11} = \frac{4EI(2I)}{8} + \frac{4EI}{4} = 2EI$$

$$k_{21} = \frac{2EI}{4} = 0.5EI$$

Applying a unit rotation at C and the stiffness coefficients are

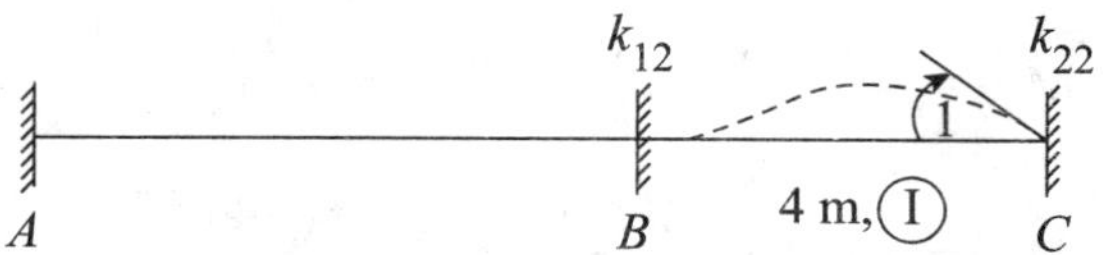

FIG. 5.52 Unit rotation at C

$$k_{12} = \frac{2EI}{4} = 0.5EI$$

$$k_{22} = \frac{4EI}{4} = EI$$

Hence, the stiffness matrix reduces to

$$[k] = \begin{bmatrix} 2EI & 0.5EI \\ 0.5EI & EI \end{bmatrix}$$

Joint equilibrium equations

$$P_1 + k_{11}D_1 + k_{12}D_2 = 0$$

$$P_2 + k_{21}D_1 + k_{22}D_2 = 0$$

Substituting the computed values

$$98 + 2EID_1 + 0.5EID_2 = 0$$

$$40 + 0.5EID_1 + EID_2 = 0$$

i.e.

$$2D_1 + 0.5D_2 = -98/EI \qquad (1)$$

$$0.5D_1 + D_2 = -40/EI \qquad (2)$$

Solving the above equations

$$D_1 = \theta_B = -44.57/EI$$

$$D_2 = \theta_C = -17.71/EI$$

4. End Moments

$$M_{AB} = M_{FAB} + \frac{2E(2I)}{l}(2\theta_A + \theta_B)$$

$$M_{AB} = -128 + \frac{2E(2I)}{8}(0 - 44.57/EI) = -150.28 \text{ kNm}$$

$$M_{BA} = M_{FBA} + \frac{2E(2I)}{l}(2\theta_B + \theta_A)$$

$$M_{BA} = +128 + \frac{2E(2I)}{8} \times 2 \times \frac{(-44.57)}{EI} = +83.43 \text{ kNm}$$

$$M_{BC} = M_{FBC} + \frac{2E(I)}{l}(2\theta_B + \theta_C)$$

$$M_{BC} = 0 + \frac{2EI}{4}\frac{(2(-44.57) - 17.71)}{EI} = -53.43 \text{ kNm}$$

$$M_{CB} = 0 + \frac{2EI}{4}\frac{(2(-17.71) - 44.57)}{EI} = -40 \text{ kNm}$$

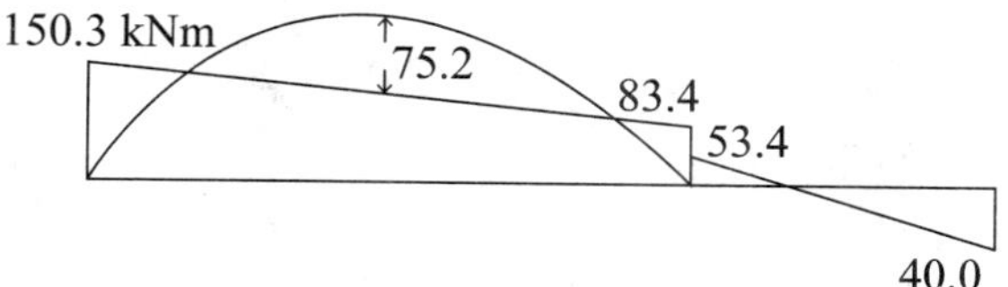

FIG. 5.53 Bending moment diagram

Example 5.9 Analyse the continuous beam shown in Fig. 5.54 by the stiffness method.

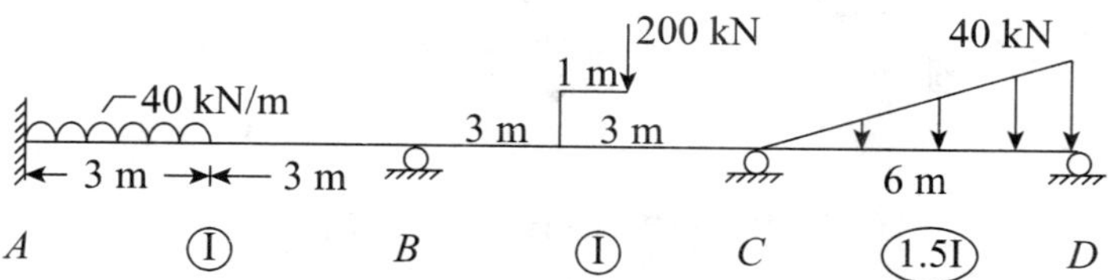

FIG. 5.54 Three-span continuous beam

Solution: The beam has three unknown slopes, viz. θ_B, θ_C and θ_D. These kinematic redundants are assumed to act in the clockwise direction.

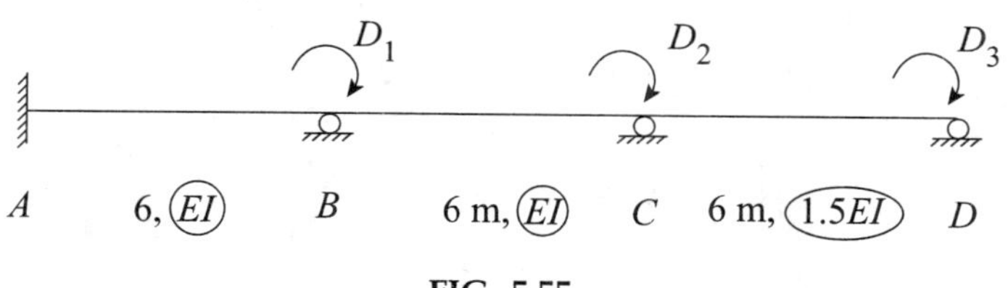

FIG. 5.55

1. Fixed End Forces Due to External Loads

The ends A, B, C and D are restrained. The corresponding fixed end moments and reactions are computed. They are shown in the figure below.

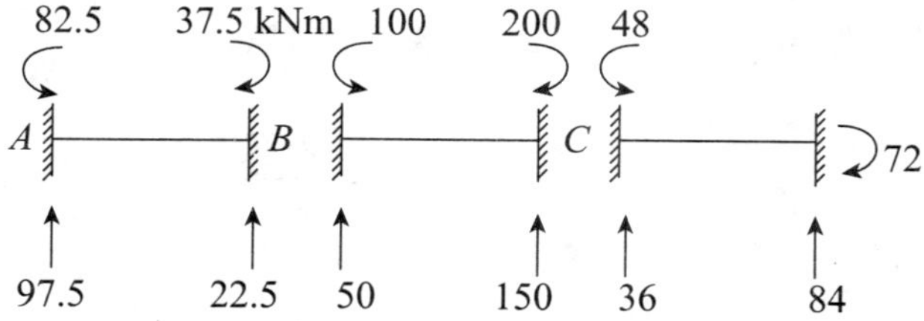

FIG. 5.56 Fixed end forces

2. Joint Loads

$$P_1 = 37.5 - 100 = -62.5 \text{ kNm}$$

$$P_2 = 200 - 48 = +152 \text{ kNm}$$

$$P_3 = 72 \text{ kNm}$$

3. Stiffness Coefficients/Stiffness Matrix

The stiffness matrix is generated by giving unit rotation at ①, ② and ③ respectively. The stiffness coefficients are indicated in the following figures.

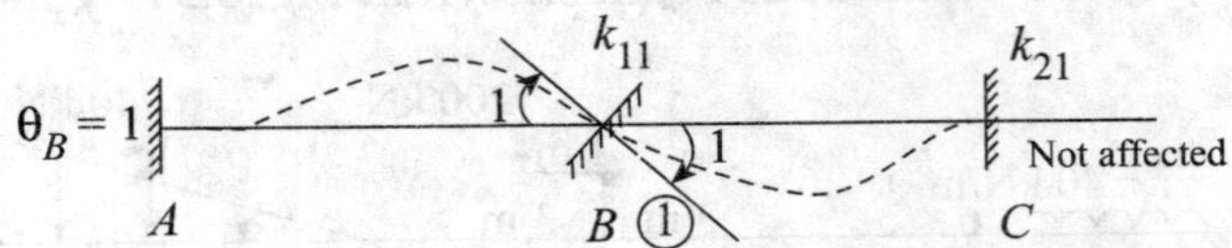

FIG. 5.57 Unit rotation at ①

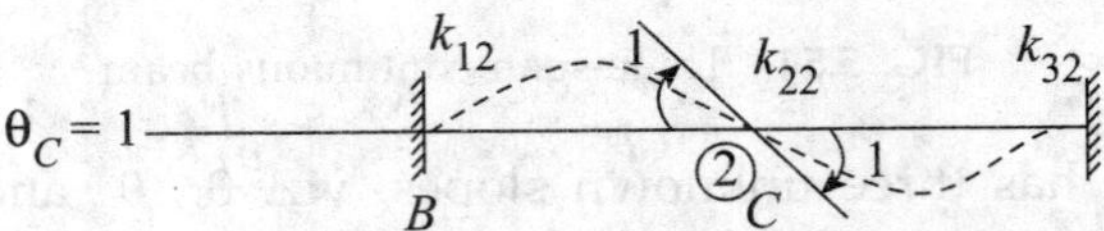

FIG. 5.58 Unit rotation at ②

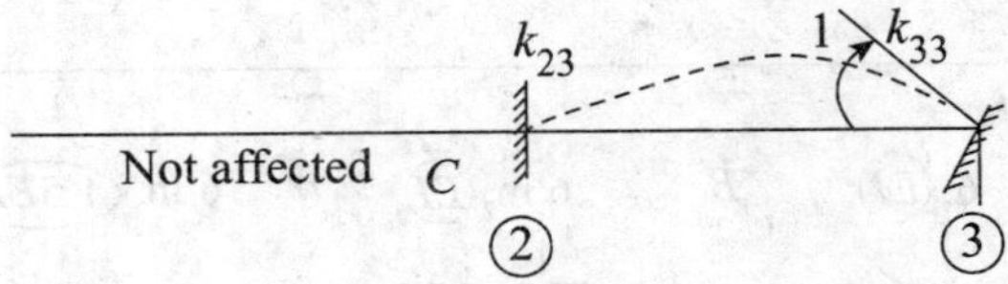

FIG. 5.59 Unit rotation at D

$$k_{11} = \frac{4EI}{6} + \frac{4EI}{6} = \left(\frac{4}{3}\right) EI \quad ; \quad k_{21} = 2EI/6 = \left(\frac{1}{3}\right) EI$$

$$k_{12} = 2EI/6 = \left(\frac{1}{3}\right) EI \quad ; \quad k_{22} = 4EI/6 + \frac{4E(1.5I)}{6} = \left(\frac{5}{3}\right) EI$$

$$k_{13} = 0 \quad ; \quad k_{23} = \frac{2E(1.5I)}{6} = \left(\frac{1}{2}\right) EI$$

$$k_{31} = 0 \quad ;$$

$$k_{32} = \frac{2E(1.5I)}{6} = \left(\frac{1}{2}\right) EI \quad ; \quad [k] = \begin{bmatrix} k_{11} & k_{12} & k_{13} \\ k_{21} & k_{22} & k_{23} \\ k_{31} & k_{32} & k_{33} \end{bmatrix}$$

$$k_{33} = 4E(1.5I)/6 = EI \quad ;$$

Substituting the stiffness coefficients

$$[k] = EI \begin{bmatrix} 4/3 & 1/3 & 0 \\ 1/3 & 5/3 & 0.5 \\ 0 & 0.5 & 1.0 \end{bmatrix}$$

Joint equilibrium equations

$$P_1 + k_{11}D_1 + k_{12}D_2 + k_{13}D_3 = 0 \quad (1)$$

$$P_2 + k_{21}D_1 + k_{22}D_2 + k_{23}D_3 = 0 \quad (2)$$

$$P_3 + k_{31}D_1 + k_{32}D_2 + k_{33}D_3 = 0 \quad (3)$$

Substituting the values

$$-62.5 + EI\,[(4/3)D_1 + (1/3)D_2] = 0 \quad (4)$$

$$+152.0 + EI\,[(1/3)D_1 + (5/3)D_2 + (1/2)D_3] = 0 \quad (5)$$

$$+72.0 + EI\,[(1/2)D_2 + D_3] = 0 \quad (6)$$

Solving the above equations

$$D_1 = 71.56/EI \circlearrowright = \theta_B$$

$$D_2 = -98.72/EI \circlearrowleft = \theta_C$$

$$D_3 = -22.64/EI \circlearrowleft = \theta_D$$

The corresponding elastic curve is shown below. The end moments were obtained using the slope deflection equations. Their values are

$$M_{AB} = -58.65 \text{ kNm}$$

$$M_{BA} = +85.20 \text{ kNm}$$

$$M_{BC} = -85.20 \text{ kNm}$$

$$M_{CB} = +158.04 \text{ kNm}$$

$$M_{CD} = -158.04 \text{ kNm}$$

$$M_{DC} = 0$$

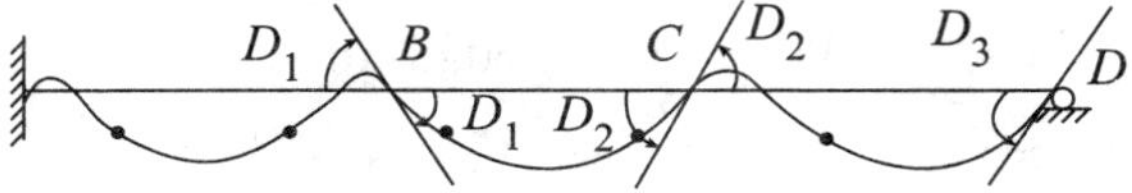

FIG. 5.60 Elastic curve

The reactions are

$V_{AB} = 85.57$ kN, $V_{BA} = 34.43$, $V_{BC} = 54.53$, $V_{CB} = 145.47$, $V_{CD} = 66.34$ kN, $V_{DC} = 53.66$ kN

5.10 ANALYSIS OF CONTINUOUS BEAM WITH SINKING OF SUPPORTS

Example 5.10 Analyse the continuous beam shown in Fig. 5.61 by the stiffness method using system approach. The support B sinks by 5 mm. Sketch the BMD and the elastic curve.

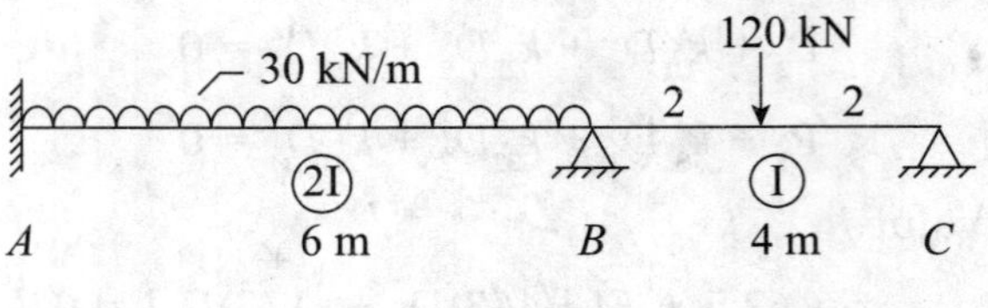

FIG. 5.61

Solution

1. Kinematic Redundants

The kinematic redundants are θ_B and θ_C. They are denoted as D_1 and D_2 respectively.

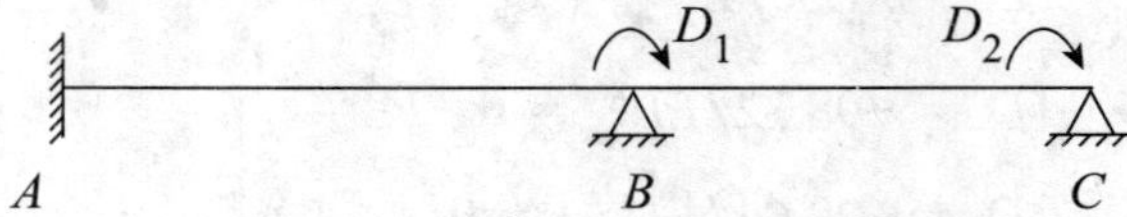

FIG. 5.62 Kinematic redundants

2. Fixed End Moments

Due to external loading

$$M'_{FAB} = -30 \times \frac{6^2}{12} = -90 \text{ kNm}$$

$$M'_{FBA} = +30 \times \frac{6^2}{12} = +90 \text{ kNm}$$

$$M'_{FBC} = -120 \times \frac{4}{8} = -60 \text{ kNm}$$

$$M'_{FCB} = +120 \times \frac{4}{8} = +60 \text{ kNm}$$

Due to settlement of B

$$M''_{FAB} = -6 \times (2 \times 15000) \times \frac{5}{1000} \times \frac{1}{6^2} = -25 \text{ kNm}$$

$$M''_{FBA} = -25 \text{ kNm}$$

$$M''_{BC} = +6 \times 15000 \times \frac{5}{1000} \times \frac{1}{4^2} = +28.13 \text{ kNm}$$

$$M''_{CB} = +28.13 \text{ kNm}$$

Total Moments

$$M_{FAB} = -90 - 25 = -115 \text{ kNm}$$

$$M_{FBA} = +90 - 25 = +65 \text{ kNm}$$

$$M_{FBC} = -60 + 28.13 = -31.87 \text{ kNm}$$

$$M_{FCB} = +60 + 28.13 = +88.13 \text{ kNm}$$

3. Joint Loads

$$P_1 = 65 - 31.87 = 33.13 \text{ kNm}$$

$$P_2 = 88.13 \text{ kNm}$$

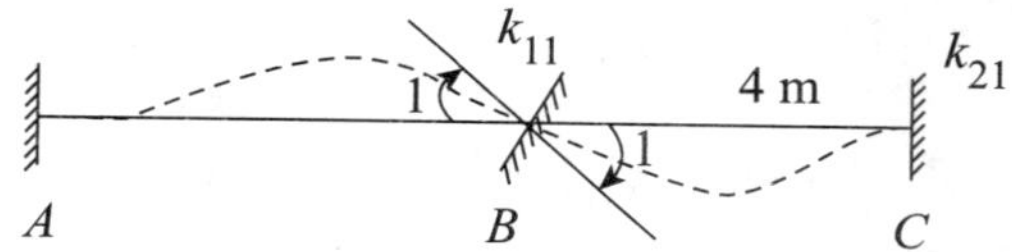

FIG. 5.63 Unit rotation at joint *B*

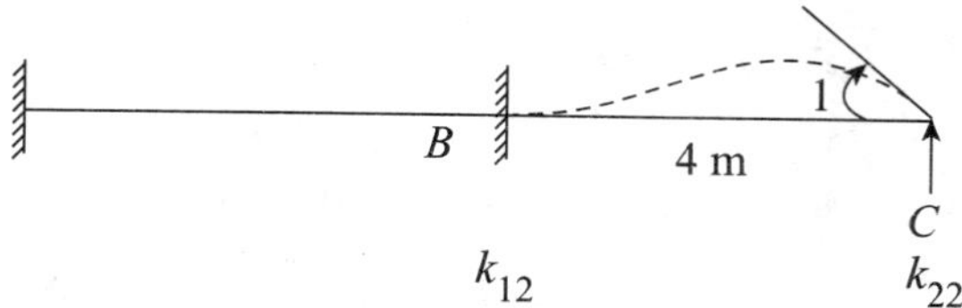

FIG. 5.64 Unit rotation at joint *C*

4. Stiffness Matrix

k_{11} = Moment required at B due to unit rotation B

$$k_{11} = m_{BA} + m_{BC} = \frac{4E(2I)}{6} + \frac{4E(I)}{4} = 7/3EI$$

k_{21} = Moment caused at C due to unit rotation at B

$$k_{21} = 2EI/4 = EI/2$$

k_{22} = Moment required at C due to unit rotation at C

$$k_{22} = 4EI/4 = EI$$

k_{12} = Moment caused at B due to unit rotation at C

$k_{12} = 2EI/4 = EI/2$

Therefore,
stiffness matrix

$$[k] = \left(\frac{1}{EI}\right)\begin{bmatrix}(7/3) & (1/2)\\(1/2) & (1)\end{bmatrix}$$

5. Joint Equilibrium

$$P_1 + k_{11}D_1 + k_{12}D_2 = 0$$

$$33.13 + \left(\frac{7}{3}\right) EI\,D_1 + \left(\frac{1}{2}\right) EI\,D_2 = 0$$

Simplifying

$$14D_1 + 3D_2 = -198.78 \quad (1)$$

$$P_2 + k_{21}D_1 + k_{22}D_2 = 0$$

$$88.13 + \left(\frac{1}{2}\right) EID_1 + (1)EID_2 = 0$$

$$0.5\,D_1 + D_2 = -88.13 \quad (2)$$

Solving Eqs. (1) and (2);

$$D_1 = \theta_B = 5.25/EI$$

$$D_2 = \theta_C = -90.75/EI$$

6. End Moments

The end moments of the members were obtained using the slope deflections as

$$M_{AB} = M_{FAB} + \frac{2(2EI)}{6}\left(2\theta_A + \theta_B - \frac{3\Delta}{l}\right)$$

$$M_{AB} = -90 + \left(\frac{4}{6}\right) \times 15000 \times \left(\frac{5.25}{15000} - \frac{3 \times 5}{1000 \times 6}\right)$$

$$M_{AB} = -90 + 3.5 - 25 = -111.5 \text{ kNm}$$

$$M_{BA} = M_{FBA} + \frac{2(2EI)}{6}\left(2\theta_B + \theta_A - \frac{3\Delta}{l}\right)$$

$$M_{BA} = +90 + \left(\frac{4}{6}\right) \times 15000\left(2 \times \frac{5.25}{15000} - \frac{3 \times 5}{1000 \times 6}\right)$$

$$M_{BA} = 90 + 7 - 25 = +72 \text{ kNm}$$

$$M_{BC} = -60 + \frac{2(EI)}{4}\left(2\theta_B + \theta_C - \frac{3\Delta}{l}\right)$$

$$M_{BC} = -60 + 2 \times \frac{15000}{4}\left(2 \times \frac{5.25}{15000} - \frac{90.75}{15000} + \frac{3 \times 5}{1000 \times 4}\right)$$

$$M_{BC} = -60 + 5.25 - 45.38 + 28.13 = -72 \text{ kNm}$$

$$M_{CB} = +60 + 2 \times \frac{15000}{4}\left(2\theta_C + \theta_B - \frac{3\Delta}{l}\right)$$

$$M_{CB} = 60 + 2 \times \frac{15000}{4}\left(2 \times \frac{-90.75}{15000} + \frac{5.25}{15000} + \frac{3 \times 5}{1000 \times 4}\right)$$

$$M_{CB} = 60 - 90.75 + 2.63 + 28.13 = \text{zero}$$

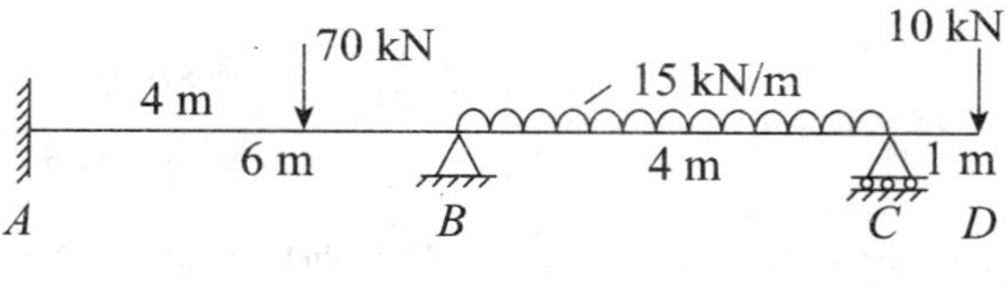

FIG. 5.65 Bending moment diagram

Example 5.11 Analyse the beam shown in Fig. 5.66 by moment distribution method. Support B sinks by 10 mm. $E = 200$ kN/mm^2. $I = 4000(10)^4$ mm^4. Draw BMD and SFD.

FIG. 5.66

Solution

1. Kinematic Redundants

The kinematic redundants are θ_A, θ_B and θ_C. As the end is fixed $\theta_A = 0$. While $\theta_B \neq \theta_C \neq 0$. The slopes θ_B and θ_C are denoted as D_1 and D_2 respectively. They are assumed to act in the clockwise direction.

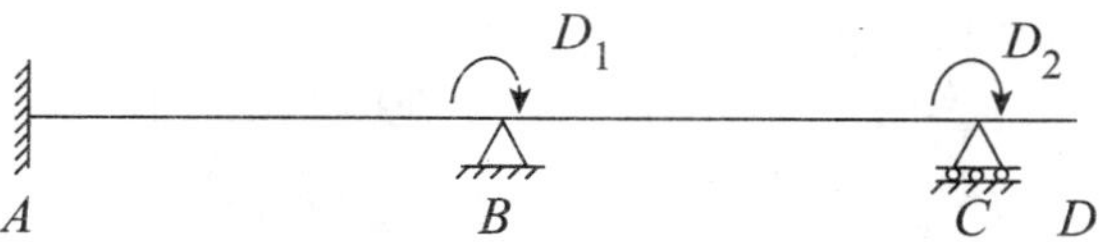

FIG. 5.67 Kinematic redundants

2. Fixed End Moments

The fixed end moments are due to loading as well as due to sinking of supports. They are calculated separately and added. These final moments at the joints are used to determine unrestrained moments.

Due to applied loading

$$M'_{AB} = -W\frac{ab^2}{l^2} = -4(70)\frac{2^2}{6^2} = -31.11\text{ kNm}$$

$$M'_{BA} = +W\frac{a^2b}{l^2} = 70(4)^2\frac{(2)}{6^2} = +62.22\text{ kNm}$$

$$M'_{BC} = -wl^2/12 = -15(4)^2/12 = -20.00\text{ kNm}$$

$$M'_{CB} = +wl^2/12 = +15(4)^2/12 = +20.00\text{ kNm}$$

$$M'_{CD} = -Wa = -10(1) = -10.00\text{ kNm}$$

It is to be mentioned here that the value of *MCD* is the anticlockwise moment which is in opposite nature of the applied clockwise moment due to loading.

Due to sinking of supports.

The settlement/sinking at *B* causes the end moment as follows.

$$M''_{AB} = M''_{BA} = \frac{-6EI\Delta_B}{l^2_{BA}} = \frac{-6\times 8000\times 10}{1000\times 6^2} = -13.33\text{ kNm}$$

The value of settlement at *B* is taken as positive as *B* settles in the clockwise direction with respect to *A*.

$$M''_{BC} = M''_{CB} = \frac{-6EI(-\Delta_B)}{l^2_{CB}} = \frac{6\times 8000\times 10}{1000\times 4^2} = +30\text{ kNm}$$

Thus, the actual fixed end moment is the addition of moment due to applied loading and due to the settlement of supports. Hence,

$$M_{FAB} = -31.11 - 13.33 = -44.44\text{ kNm}$$

$$M_{FBA} = +62.22 - 13.33 = +48.89\text{ kNm}$$

$$M_{FBC} = -20.00 + 30.00 = +10.00\text{ kNm}$$

$$M_{FCB} = +20.00 + 30.00 = +50.00\text{ kNm}$$

$$M_{FCD} = -10.00\text{ kNm}$$

3. Joint Loads

$$P_1 = M_{FBA} + M_{FBC} = 58.89\text{ kNm}$$

$$P_2 = M_{FCB} + M_{FCD} = 40.00\text{ kNm}$$

4. Stiffness Matrix

$$k_{11} = \frac{4EI}{6} + \frac{4EI}{4} = (5/3)EI$$

$$k_{12} = 2EI/4 = (1/2)EI$$

$$k_{21} = 2EI/4 = (1/2)EI$$

$$k_{22} = 4EI/4 = EI$$

Stiffness matrix

$$[k] = EI \begin{bmatrix} (5/3) & (1/2) \\ (1/2) & 1 \end{bmatrix}$$

5. Joint Equilibrium

$$58.89 + \left(\frac{5}{3}\right) EID_1 + \left(\frac{1}{2}\right) EID_2 = 0$$

Simplifying;

$$10D_1 + 3D_2 = -353.34 \qquad (1)$$

$$4000 + \left(\frac{1}{2}\right) EID_1 + EID_2 = 0$$

$$D_1 + 2D_2 = -80.00 \qquad (2)$$

Solving Eqs. (1) and (2), we get the slopes at B and C as

$$D_1 = \theta_B = -27.45/EI$$

$$D_2 = \theta_C = -26.27/EI$$

6. Member End Moments

Substituting the values of θ_B and θ_C in slope deflection equations;

$$M_{AB} = -53.59 \text{ kNm}$$

$$M_{BA} = +30.59 \text{ kNm}$$

$$M_{BC} = -30.59 \text{ kNm}$$

$$M_{CB} = +10.00 \text{ kNm}$$

$$M_{CD} = -10.00 \text{ kNm}$$

Superimposing the above end moments, we sketch the final bending moment diagram.

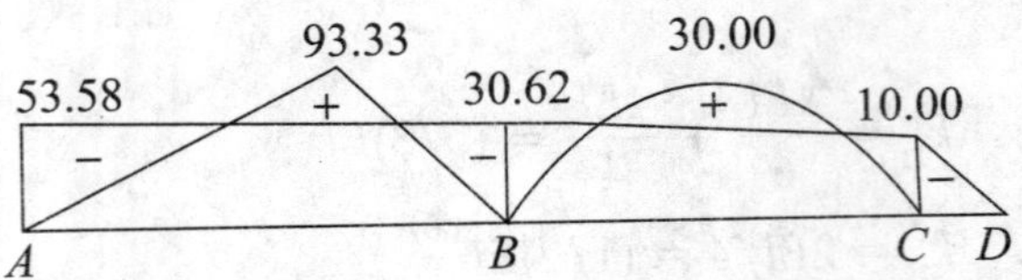

FIG. 5.68 Bending moment diagram

Using free body diagram, the reactions are obtained and SFD is drawn as follows.

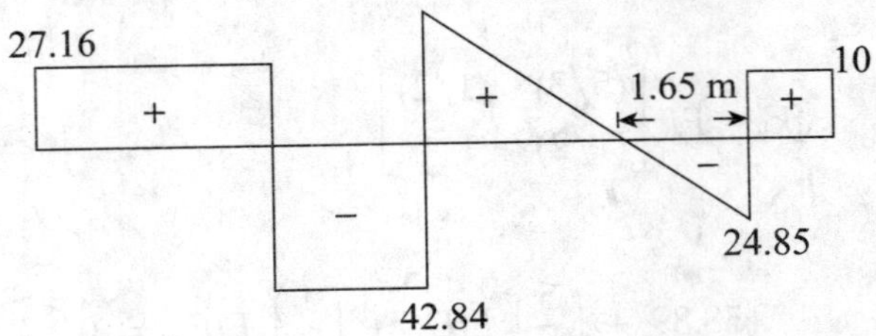

FIG. 5.69 Shear force diagram

Example 5.12 Analyse the continuous beam shown in Fig. 5.70 by stiffness method. The support B and C sink by 10 and 5 mm respectively. E = 200 GPa and I = 80(106) mm^4.

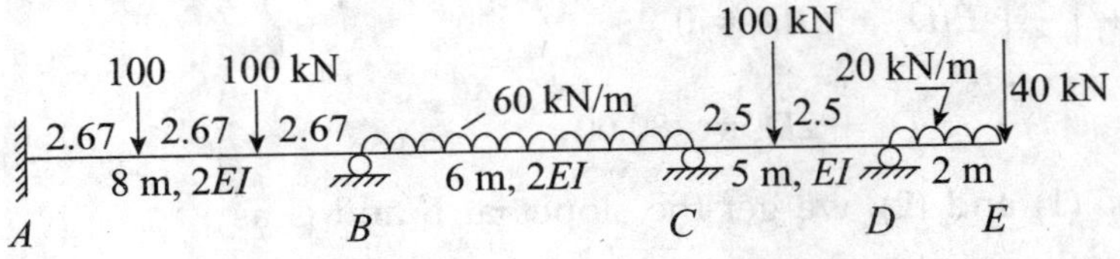

FIG. 5.70 Three-span continuous beam

Solution

1. Kinematic Redundants

The continuous beam has three unknowns, viz. D_1, D_2 and D_3 at B, C and D respectively. They are assumed to act in the clockwise direction.

2. Fixed End Moments Due to External Loads

The fixed end forces due to external loads are given as follows.

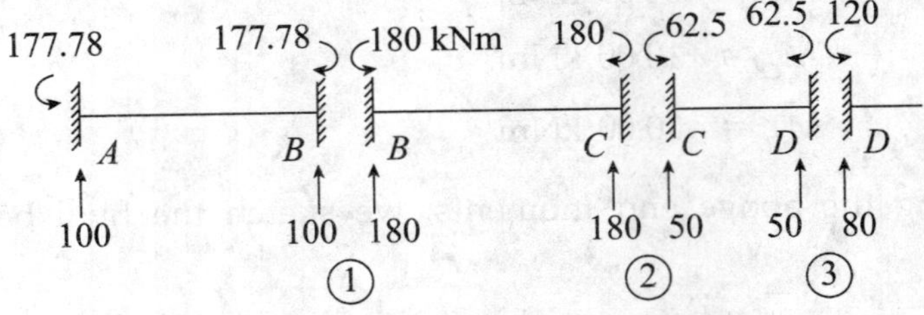

FIG. 5.71 Fixed end forces due to external loads

Fixed end moments due to settlement at B and C

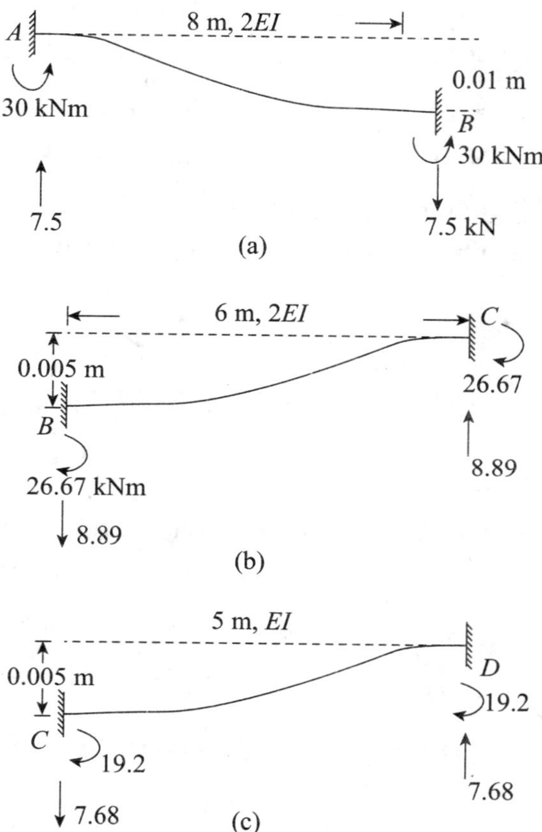

FIG. 5.72 Fixed end forces due to sinking of supports

The end moments and shear force were calculated using

$$\text{Moments at the ends} = \frac{6EI\delta}{l^2}$$

$$\text{Shear force at the ends} = \frac{12EI\delta}{l^3}$$

where δ = relative displacement the between supports.

Table 5.2 Combined effect of external loads and sinking of support

Moments/Member	*BA*	*BC*	*CB*	*CD*	*DC*	*DE*
External Loads	+177.78	−180.00	+180.00	−62.50	+62.50	−120.00
Sinking of Supports	−30.00	+26.67	++26.67	+19.20	+19.20	0.00
Combined Effect	147.78	−153.33	206.67	−43.30	+81.70	−120.00

From the above table;

3. Joint Loads

$$P_1 = 147.78 - 153.33 = -5.55 \text{ kNm}$$ ↺

$$P_2 = 206.67 - 43.30 = +163.37 \text{ kNm}$$ ↻

$$P_3 = 81.70 - 120.00 = -38.30 \text{ kNm}$$ ↺

4. Stiffness Coefficients and Stiffness Coefficients

The stiffness matrix elements are calculated by successively giving unit rotations at joints ①, ②, and ③.

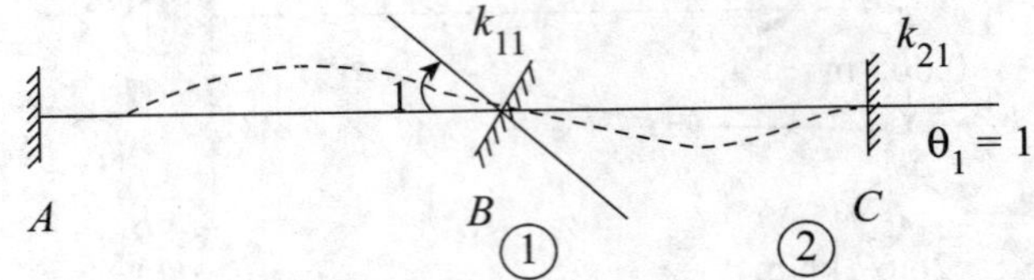

FIG. 5.73 Unit rotation at joint ①

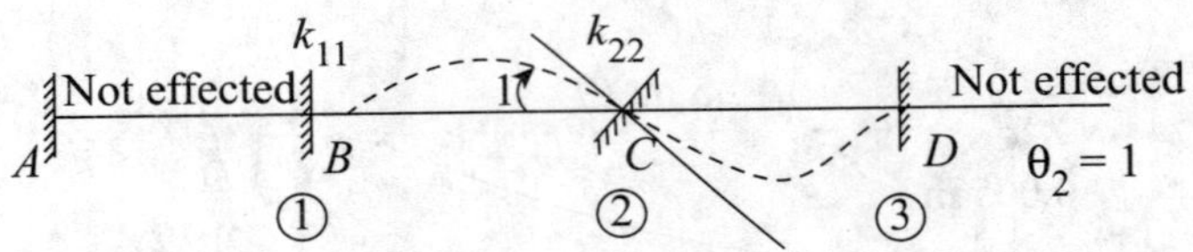

FIG. 5.74 Unit rotation at joint ②

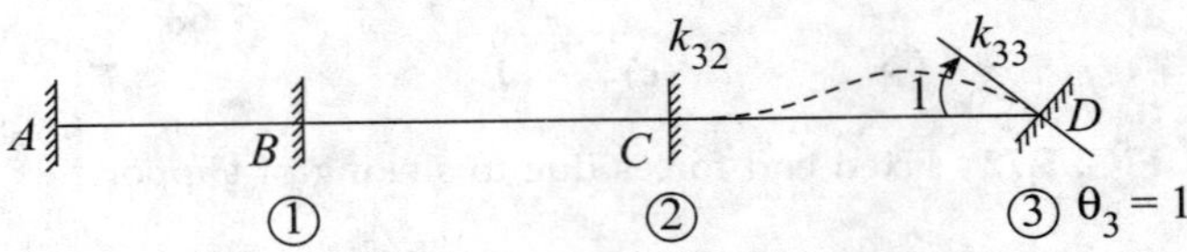

FIG. 5.75 Unit rotation at joint ③

k_{11} = Moment required at B due to unit rotation at B

$$k_{11} = \frac{4(2EI)}{8} + \frac{4(2EI)}{6} = \frac{7EI}{3}$$

k_{21} = Moment caused at C due to unit rotation at C

$$k_{21} = \frac{2(2EI)}{6} = \frac{2EI}{3}$$

k_{22} = Moment required at Ⓒ for unit rotation at Ⓒ

$$k_{22} = \frac{4E(2I)}{6} + \frac{4EI}{5} = \left(\frac{32}{15}\right) EI$$

k_{12} = Moment caused at Ⓑ for unit rotation at Ⓒ

$$k_{12} = \frac{2(2EI)}{6} = \frac{2}{3}EI$$

k_{33} = Moment required at Ⓓ for unit rotation at Ⓓ

$$k_{33} = \frac{4(EI)}{5} = 0.8EI$$

k_{32} = Moment caused at Ⓓ for unit rotation at Ⓒ

$$k_{32} = \frac{2(EI)}{5} = 0.4\,EI$$

5. Joint Equilibrium

$$P_1 + k_{11}D_1 + k_{12}D_2 + k_{13}D_3 = 0 \quad (1)$$

$$P_2 + k_{21}D_1 + k_{22}D_2 + k_{23}D_3 = 0 \quad (2)$$

$$P_3 + k_{31}D_1 + k_{32}D_2 + k_{33}D_3 = 0 \quad (3)$$

Substituting the computed values

$$7D_1/3 + \frac{2}{3}D_2 + 0D_3 = 5.5/EI \quad (4)$$

$$(2/3)D_1 + (32/15)D_2 + 0.4D_3 = -163.37/EI \quad (5)$$

$$0D_1 + 0.4D_2 + 0.8D_3 = 38.30/EI \quad (6)$$

On solving the above simultaneous equations;

$$D_1 = \theta_B = 32.57/EI \curvearrowright$$

$$D_2 = \theta_C = -105.66/EI \curvearrowleft$$

$$D_3 = \theta_D = 100.71/EI \curvearrowright$$

The slopes θ_B and θ_D are positive which means the rotations at B and C are in clockwise direction. θ_C is negative which shows that the slope at C is in the anticlockwise direction. Using the above nature of the slopes the elastic curve is drawn.

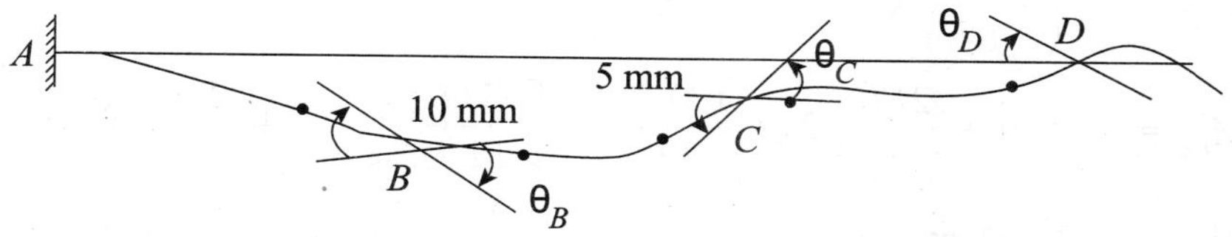

FIG. 5.76 Elastic curve

6. Determination of End Moments of the Members

The values of θ_B, θ_C and θ_D are substituted in the slope deflection equation to obtain the end moments.

$$M_{AB} = M_{FAB} + \frac{2EI}{l}\left(2\theta_A + \theta_B - \frac{3\Delta}{l}\right)$$

$$M_{AB} = -177.78 + \frac{2(2EI)}{8}\left(\frac{32.57}{EI} - \frac{3 \times 10}{1000 \times 8}\right)$$

$$M_{AB} = -177.78 + 16.28 - 30 = -191.50 \text{ kNm}$$

$$M_{BA} = +177.78 + \frac{2E(2I)}{8}\left(2\theta_B + \theta_A - \frac{3\Delta}{l}\right)$$

$$M_{BA} = +177.78 + 32.57 - 30.00 = 180.35 \text{ kNm}$$

$$M_{BC} = -180.00 + \frac{2(2EI)}{6}\left(2 \times \theta_B + \theta_C - \frac{3\Delta}{l}\right)$$

$$M_{BC} = -180.00 + \frac{2(2EI)}{6}\left(\frac{2 \times 32.57}{EI} - \frac{105.66}{EI} + \frac{3 \times 0.005}{6}\right) = -180.35 \text{ kNm}$$

$$M_{CB} = +180.00 + \frac{2(2EI)}{6}\left(2\theta_C + \theta_D - \frac{3\Delta}{l}\right)$$

$$M_{CB} = +180.00 + \frac{2(2EI)}{6}\left(2 \times \frac{-105.66}{EI} + \frac{32.57}{EI} + \frac{3 \times 0.005}{6}\right) = 87.50 \text{ kNm}$$

$$M_{CD} = -62.50 + \frac{2EI}{5}\left(2\theta_C + \theta_D - \frac{3\Delta}{l}\right)$$

$$M_{CD} = -62.50 + \frac{2EI}{5}\left(2 \times \left(\frac{-105.66}{EI}\right) + \frac{100.71}{EI} + \frac{3 \times 0.005}{5}\right) = -87.50 \text{ kNm}$$

$$M_{DC} = +62.50 + \frac{2EI}{5}\left(2\theta_D + \theta_C - \frac{3\Delta}{l}\right)$$

$$M_{DC} = +62.50 + \frac{2EI}{5}\left(2 \times \frac{100.71}{EI} - \frac{105.66}{EI} + \frac{3 \times 0.005}{5}\right) = 120 \text{ kNm}$$

$$M_{DC} = -120 \text{ kNm}$$

The end moments are hogging in nature and the pure moment diagram is drawn. This diagram is superimposed over simple beam diagram to obtain the final BMD.

The free body diagram is drawn along with end moments. The support reactions were computed and the shear force diagram is drawn.

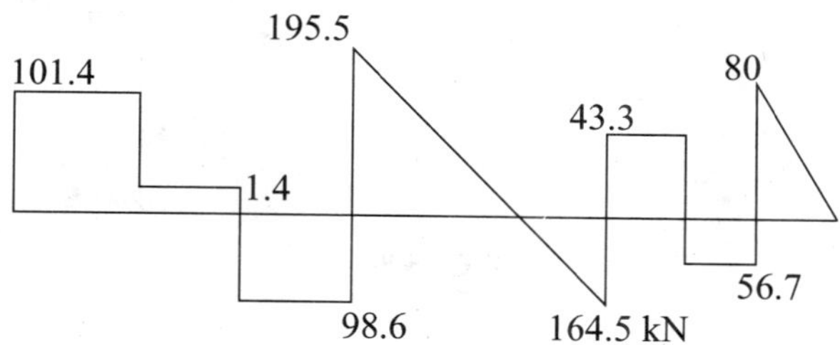

FIG. 5.77 Shear force diagram

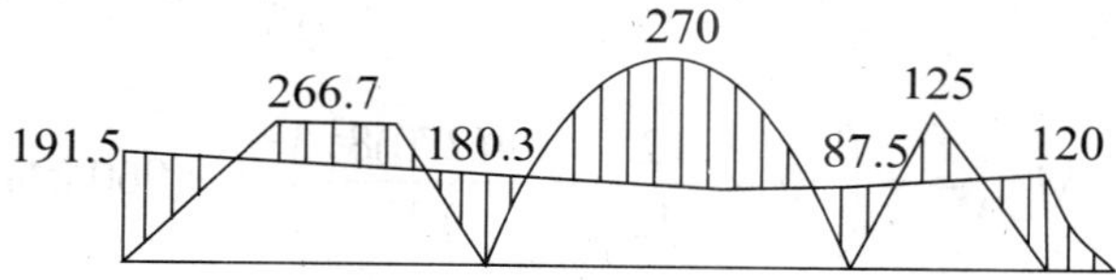

FIG. 5.78 Bending moment diagram

5.11 ANALYSIS OF RECTILINEAR FRAMES

Example 5.13 Analyse the rigid frame shown in Fig. 5.79 by stiffness method.

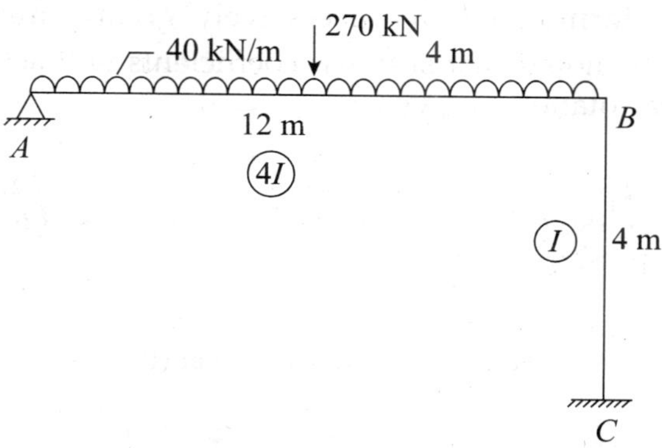

FIG. 5.79

Solution

1. Kinematic Redundants

The frame cannot sway as the hinge at A prevents sway. The beam has two unknown slopes θ_A and θ_B. While at the fixed end $\theta_C = 0$. The kinematic redundants $D_1 = \theta_A$ and $D_2 = \theta_B$ are assumed to act in the clockwise direction.

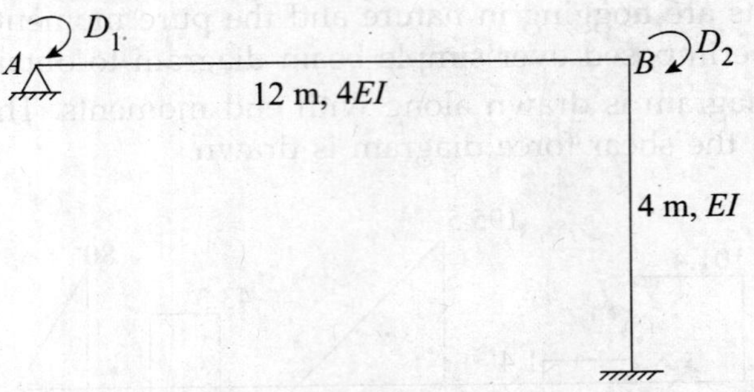

FIG. 5.80

2. Fixed End Moments

$$M_{FAB} = -\left(40 \times \frac{12^2}{12} + \frac{8(270)4^2}{12^2}\right) = -720 \text{ kNm}$$

$$M_{FBA} = +\left(40 \times \frac{12^2}{12} + \frac{4(270)8^2}{12^2}\right) = +960 \text{ kNm}$$

3. Joint Loads

$$P_1 = -720 \text{ kNm}$$

$$P_2 = +960 \text{ kNm}$$

4. Stiffness Coefficients/Stiffness Matrix

The stiffness matrix is formulated by successively giving unit rotations at ① and ② respectively. It is to be noted that stiffness coefficients will act only of two adjacent spans (members) where rotation is given.

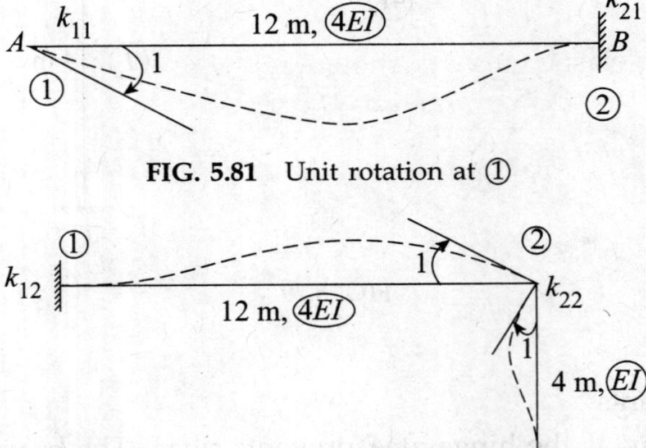

FIG. 5.81 Unit rotation at ①

FIG. 5.82 Unit rotation at ②

$$k_{11} = \frac{4E(4I)}{12} = \left(\frac{4}{3}\right) EI$$

$$k_{21} = \frac{2E(4I)}{12} = \left(\frac{2}{3}\right) EI$$

$$k_{12} = \frac{2E(4I)}{12} = \left(\frac{2}{3}\right) EI$$

$$k_{22} = \frac{4E(4I)}{12} + \frac{4EI}{4} = \left(\frac{7}{3}\right) EI$$

Stiffness matrix

$$[k] = \frac{EI}{} \begin{bmatrix} 4 & 2 \\ 2 & 7 \end{bmatrix}$$

5. Joint Equilibrium

$$P_1 + k_{11}D_1 + k_{12}D_2 = 0 \quad (1)$$

$$P_2 + k_{21}D_1 + k_{22}D_2 = 0 \quad (2)$$

Substituting the values

$$-720 + \left(\frac{4}{2}\right) EID_1 + \left(\frac{2}{3}\right) EID_2 = 0 \quad (3)$$

i.e.

$$2D_1 + D_2 = 1080/EI \quad (4)$$

$$+960 + \left(\frac{2}{3}\right) EID_1 + \left(\frac{7}{3}\right) EID_2 = 0 \quad (5)$$

$$2D_1 + 7D_2 = -2880/EI \quad (6)$$

Solving Eqs. (4) and (6)

$$D_1 = 870/EI, \quad D_2 = -660/EI$$

The corresponding elastic curve is shown below.

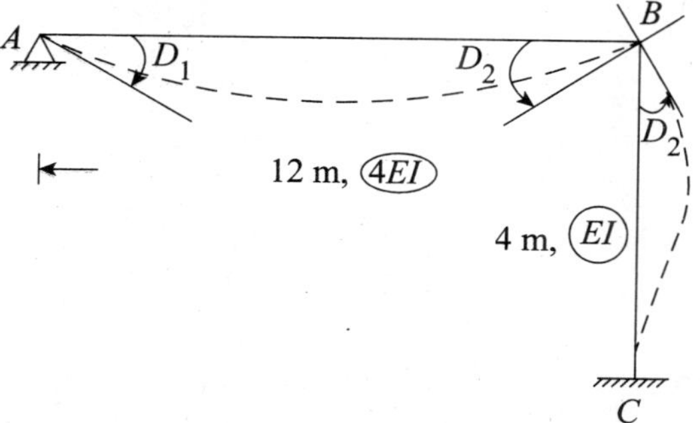

FIG. 5.83 Elastic curve

6. End Moment of the Elements

$$M_{AB} = -720 + \frac{2E(4I)}{12}(2\theta_A + \theta_B)$$

$$M_{AB} = -720 + \frac{2}{3}EI\left(\frac{2 \times 870}{EI} - \frac{660}{EI}\right) = 0$$

$$M_{BA} = +960 + \frac{2}{12}E(4I)(2\theta_B + \theta_A)$$

$$M_{BA} = 960 + \frac{2}{3}EI\left(2\left(\frac{-660}{EI}\right) + \frac{870}{EI}\right) = 660 \text{ kNm}$$

$$M_{BC} = 0 + \frac{2EI}{4}(2\theta_B + \theta_C) = -660 \text{ kNm}$$

$$M_{CB} = 0 + \frac{2EI}{4}(2\theta_C + \theta_B) = -330 \text{ kNm}$$

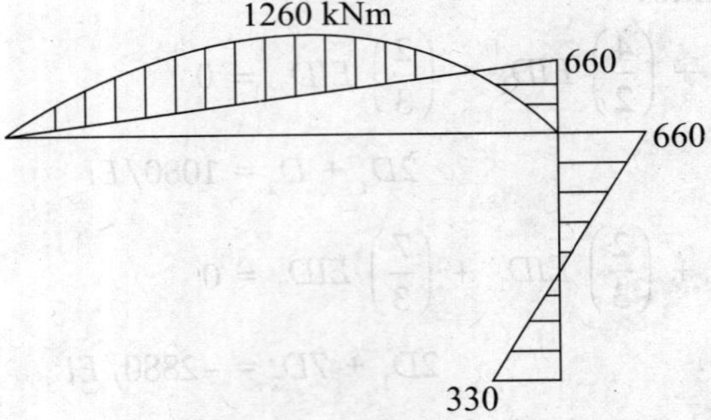

FIG. 5.84 Bending moment diagram

Example 5.14 Analyse the frame by system stiffness approach. Draw the bending moment diagram. (Anna Univ, 2008)

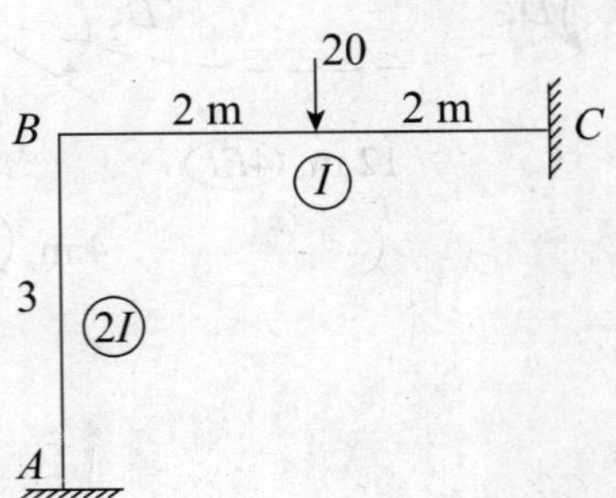

FIG. 5.85

Solution

1. Kinematic Redundants

The supports A and C are fixed. Hence, $\theta_A = \theta_C = 0$. B is a rigid joint and it rotates. Hence, θ_B is the only kinematic redundant to be determined. In other words, it is kinematically indeterminate to the first degree.

2. Fixed End Moments

$$M_{FBC} = -20 \times \frac{4}{8} = -10 \text{ kNm} = P_1$$

$$M_{FCB} = +20 \times \frac{4}{8} = +10 \text{ kNm} = P_2$$

3. Stiffness Coefficient

Apply a unit rotation at $\theta_B = D_1$ and evaluate the only one stiffness coefficient k_{11}.

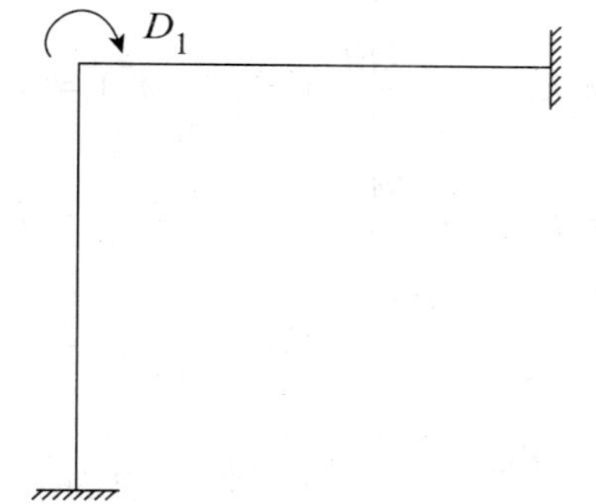

FIG. 5.86 Kinematic redundant

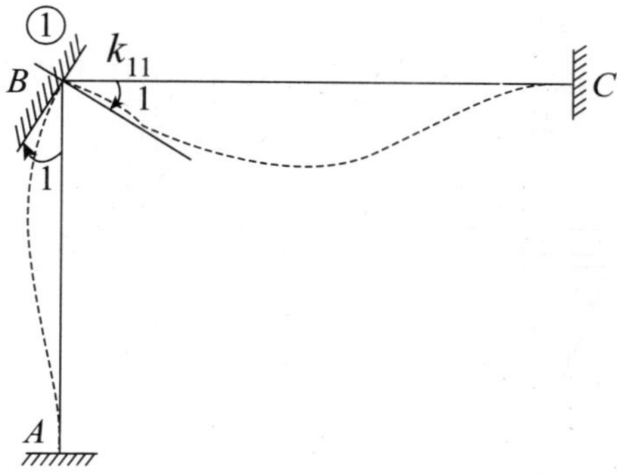

FIG. 5.87 Unit rotation at B

$$k_{11} = \frac{4EI}{4} + \frac{4E(2I)}{3} = \left(\frac{11}{3}\right) EI$$

4. Joint Equilibrium

At Joint B;

$$P_1 + k_{11}D_1 = 0$$

$$-10 + \left(\frac{11}{3}\right) EID_1 = 0$$

$$D_1 = \left(\frac{30}{11}\right)/EI$$

i.e.

$$\theta_B = \left(\frac{30}{11}\right)/EI$$

5. End Moments of the Elements

$$M_{AB} = M_{FAB} + \frac{2EI}{l}(\theta_A + \theta_B) = 3.64 \text{ kNm}$$

$$M_{BA} = M_{FBA} + \frac{2EI}{l}(2\theta_B + \theta_A) = 7.27 \text{ kNm}$$

$$M_{BC} = M_{FBC} + \frac{2EI}{l}(2\theta_B + \theta_C) = 7.27 \text{ kNm}$$

$$M_{CB} = M_{FCB} + \frac{2EI}{l}(2\theta_C + \theta_B) = 11.36 \text{ kNm}$$

$$M_D = \frac{20(4)}{4} - \left[\frac{1}{2}(7.27 + 11.36)\right] = 10.7 \text{ kNm}$$

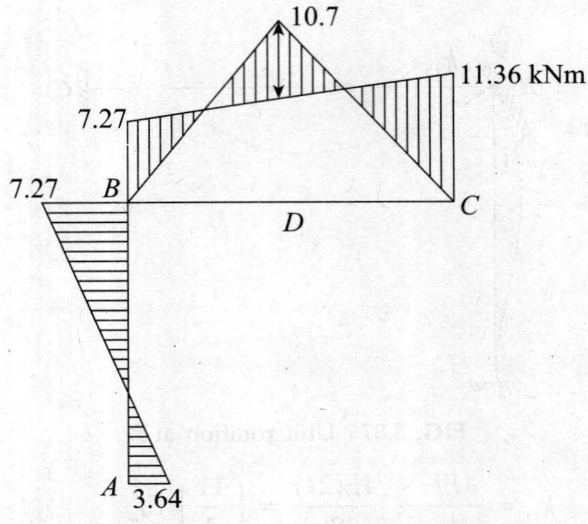

FIG. 5.88 Bending moment diagram

Example 5.15 Draw the bending moment diagram shown in Fig. 5.89. Use system stiffness approach. The frame is built in at *A*, *D* and *C*. It has a stiff joint at *B*. It curves a uniformly distributed load of 50 kN/m on *BC* and is of uniform section throughout.

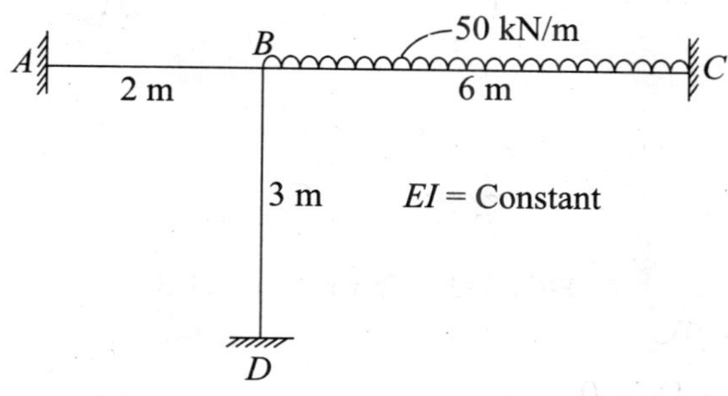

FIG. 5.89

Solution

1. Kinematic Redundants

The supports *A*, *D* and *C* are fixed. Hence, $\theta_A = \theta_C = \theta_D = 0$. The joint *B* is stiff joint. Hence, $\theta_B \neq 0$. Thus, in the above rectilinear frame the only kinematic redundant is θ_B.

2. Fixed End Moments

$$M_{FBC} = -50 \times \frac{6^2}{12} = -150 \text{ kNm}$$

$$M_{FCB} = +50 \times \frac{6^2}{12} = +150 \text{ kNm}$$

3. Joint Load

$$P_1 = -150 \text{ kNm}$$

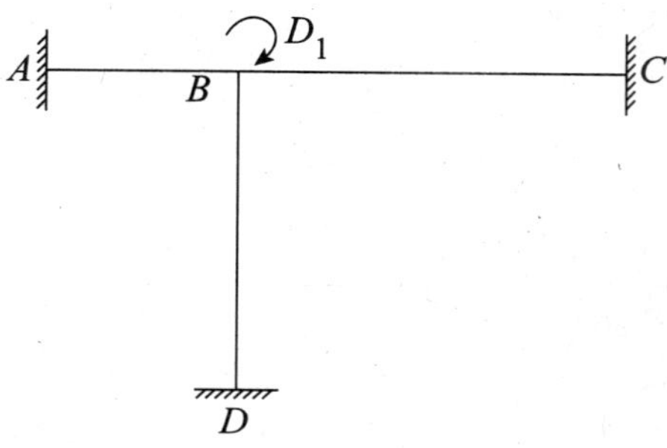

FIG. 5.90 Kinematic redundant

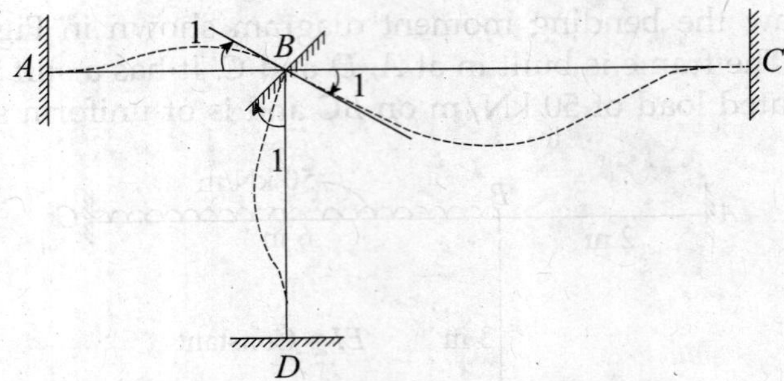

FIG. 5.91 Unit rotation at B

4. Joint Equilibrium at (B)

$$- 150 + 4EI\, D_1 = 0$$

$$D_1 = \theta_B = 37.5/EI$$

5. End Moments of Elements

$$M_{AB} = 0 + \frac{2EI}{l}(2\theta_A + \theta_B) = \frac{2EI}{2} \times \frac{37.5}{EI} = 37.5 \text{ kNm}$$

$$M_{BA} = 0 + \frac{2EI}{l}(2\theta_B + \theta_A) = \frac{4EI}{2} \times \frac{37.5}{EI} = 75 \text{ kNm}$$

$$M_{BC} = -150 + \frac{2EI}{l}(2\theta_B + \theta_C) = 150 + \frac{2EI}{6}\left(\frac{2 \times 37.5}{EI}\right) = 125 \text{ kNm}$$

$$M_{CB} = 150 + \frac{2EI}{6}(2\theta_C + \theta_B) = 150 + \frac{2EI}{6} \times \frac{37.5}{EI} = 162.5 \text{ kNm}$$

$$M_{BD} = 0 + \frac{2EI}{3}(2\theta_B + \theta_D) = \frac{2EI}{3} \times 2 \times \frac{37.5}{EI} = 50 \text{ kNm}$$

$$M_{DB} = 0 + \frac{2EI}{3}(2\theta_D + \theta_B) = \frac{2EI}{3} \times \frac{37.5}{EI} = 25 \text{ kNm}$$

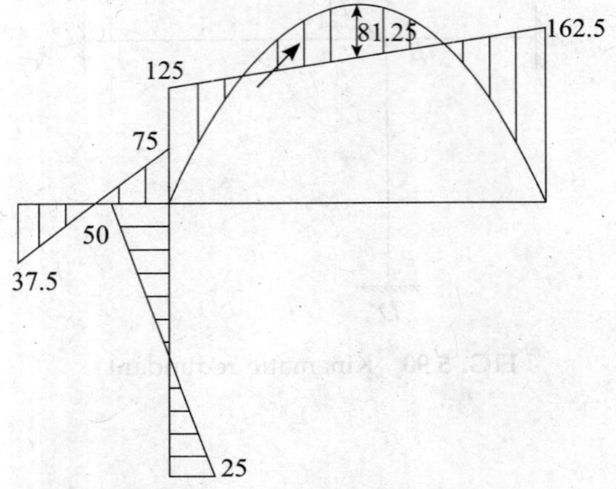

FIG. 5.92 Bending moment diagram

Example 5.16: Compute the end moments of the frame shown in Fig. 5.93 and draw the bending moment diagram.

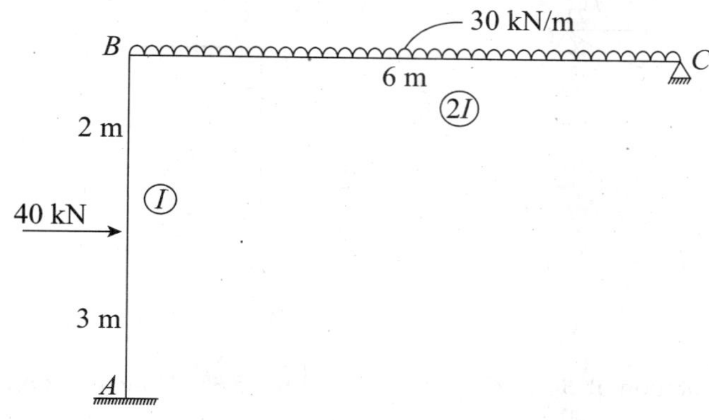

FIG. 5.93

Solution

1. Kinematic Redundants

The kinematic redundants are $\theta_B = D_1$ and $\theta_C = D_2$. They are assumed to act in the clockwise direction.

2. Fixed End Moments

$$M_{FAB} = -\frac{3(40)\,2^2}{5^2} - 19.2 \text{ kNm}$$

$$M_{FBA} = +\frac{2(40)\,3^2}{5^2} = +28.8 \text{ kNm}$$

$$M_{FBC} = -30 \times \frac{6^2}{12} = -90.0 \text{ kNm}$$

$$M_{FCB} = +30 \times \frac{6^2}{12} = +90.0 \text{ kNm}$$

3. Joint Loads

$$P_1 = -90 + 28.8 = -61.2 \text{ kNm}$$

$$P_2 = +90 \text{ kNm}$$

4. Stiffness Coefficients

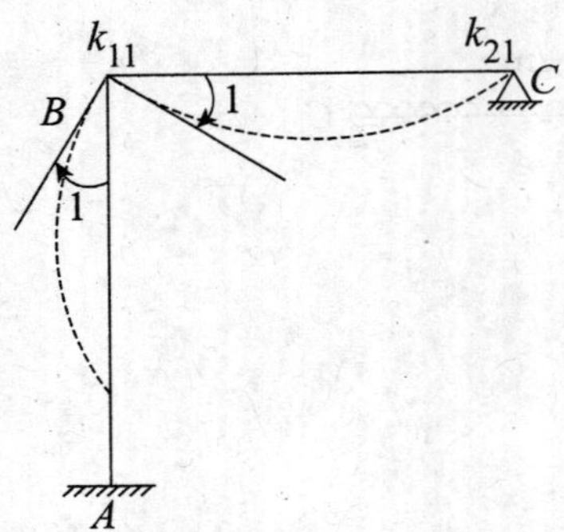

FIG. 5.94 Unit rotation at B

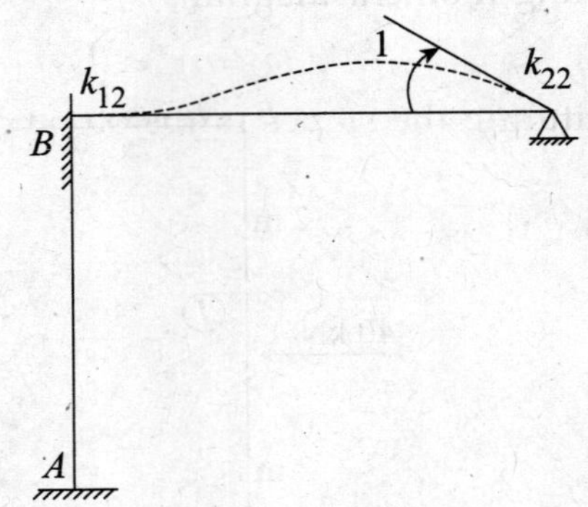

FIG. 5.95 Unit rotation at C

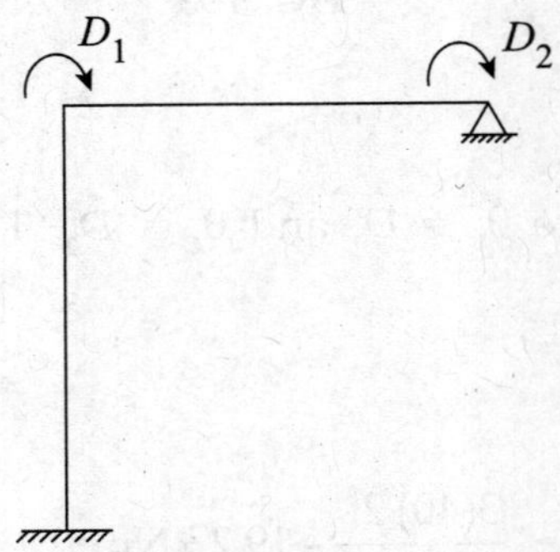

FIG. 5.96 Kinematic redundants

Applying a unit rotation at B;

$$k_{11} = \frac{4EI}{5} + \frac{4E(2I)}{6} = 2.13EI$$

$$k_{21} = 2E(2I)/6 = 0.67EI$$

$$k_{12} = 2E(2I)/6 = 0.67EI$$

$$k_{22} = 4E(2I)/6 = 1.33EI$$

5. Joint Equilibrium Equations

$$P_1 + k_{11}D_1 + k_{12}D_2 = 0$$

$$P_2 + k_{21}D_1 + k_{22}D_2 = 0$$

Substituting the computed values;

$$-\,61.2 + 2.13EID_1 + 0.67EID_2 = 0$$

$$+\,90.0 + 0.67EID_1 + 1.33EID_2 = 0$$

Solving

$$D_1 = \theta_B = 59.436/EI$$

$$D_2 = \theta_c = -97.611/EI$$

Substituting the slopes in the slope deflection equations, we obtain

$$M_{AB} = 4.57 \text{ kNm}$$

$$M_{BA} = 76.35 \text{ kNm}$$

$$M_{BC} = -76.35 \text{ kNm}$$

$$M_{CB} = 0 \text{ kNm}$$

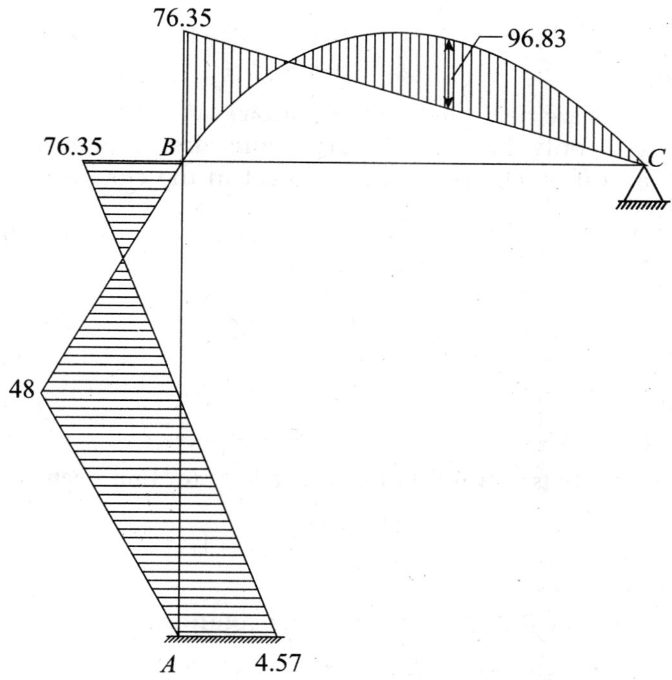

FIG. 5.97 Bending moment diagram

5.12 ANALYSIS OF SYMMETRICAL FRAMES

Example 5.17: Analyse the symmetrical frame by the stiffness method using system approach. Draw the bending moment diagram.

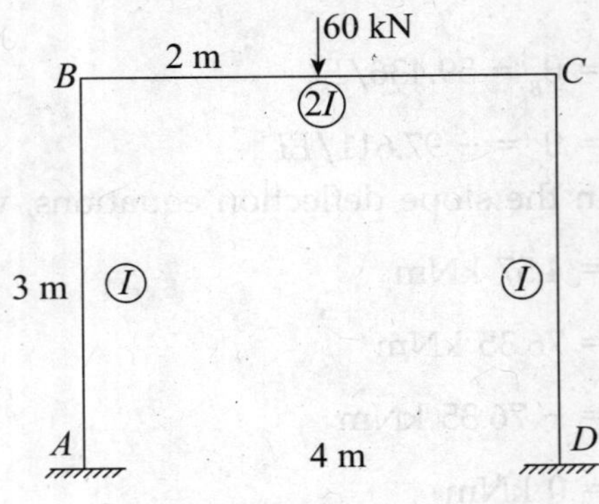

FIG. 5.98

Solution

1. Kinematic Redundants

As the axial deformations are neglected, only rotations can occur at *B* and *C* respectively. Hence, in this analysis only bending deformations are considered. The kinematic redundants $\theta_B = D_1$ and $\theta_C = D_2$ are assumed to act in the clockwise direction.

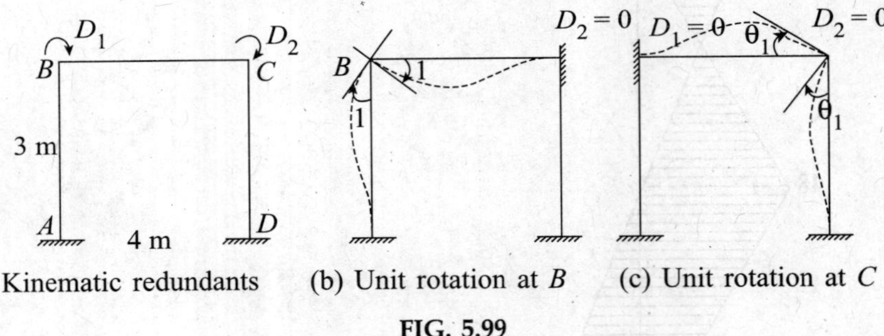

(a) Kinematic redundants (b) Unit rotation at *B* (c) Unit rotation at *C*

FIG. 5.99

2. Fixed End Moments

The restraining moments at *B* and *C* to make the rotations zero are

$$M_{FBC} = -60 \times \frac{4}{8} = -30 \text{ kNm} = P_1$$

$$M_{FCB} = +60 \times \frac{4}{8} = +30 \text{ kNm} = P_2$$

3. Stiffness Coefficients

$$k_{11} = \frac{4EI}{3} + \frac{4E(2I)}{4} = 3.33EI$$

$$k_{12} = \frac{2E(2I)}{4} = EI$$

$$k_{21} = k_{12} = EI$$

$$k_{22} = k_{11} = 3.33EI$$

4. Joint Equilibrium

$$P_1 + k_{11} D_1 + k_{12} D_2 = 0 \quad (1)$$
$$P_2 + k_{21} D_1 + k_{22} D_2 = 0 \quad (2)$$

Substituting the corresponding values

$$-30 + 3.33EID_1 + EID_2 = 0$$
$$3.33D_1 + D_2 = 30/EI \quad (3)$$
$$+30 + EID_1 + 3.33EID_2 = 0$$
$$D_1 + 3.33 D_2 = -30/EI \quad (4)$$

Solving Eqs. (3) and (4);

$$\boxed{\begin{aligned} D_1 &= 12.87/EI \\ D_2 &= -12.87/EI \end{aligned}}$$

The above computed values are correct as the slopes at B and C are equal in magnitude and but opposite in sign (nature) due to symmetry. It is to be noted that for symmetrical single bag frame subjected to symmetrical loading, the slopes at the ends of the beam are of same magnitude with opposite sign.

5. End Moments of the Elements

$$M_{AB} = M_{FAB} + \frac{2EI}{l}(2\theta_A + \theta_B)$$

$$M_{AB} = 0 + \frac{2EI}{l} \times \frac{12.87}{EI} = 8.58 \text{ kNm}$$

$$M_{BA} = M_{FBA} + \frac{2EI}{l}(2\theta_B + \theta_A)$$

$$M_{BA} = 0 + \frac{2EI}{3}\left(2 \times \frac{12.87}{EI}\right) = 17.12 \text{ kNm}$$

$$M_{BC} = M_{FBC} + \frac{2E\,(2I)}{4}\left\{\left(2 \times \frac{12.87}{EI}\right) - \frac{12.87}{EI}\right\}$$

$$M_{BC} = -30 + 12.87 = 17.13 \text{ kNm}$$

$$M_{CB} = +30 + \frac{2E\,(2I)}{4}\left\{\left(2 \times \frac{-12.87}{EI}\right) + \frac{12.87}{EI}\right\}$$

$$M_{CB} = +30 - 12.87 = 17.13 \text{ kNm}$$

$$M_{CD} = 0 + \frac{2E(I)}{3}\left\{\left(2 \times \frac{-12.87}{EI} + 0\right)\right\} = -17.13 \text{ kNm}$$

$$M_{DC} = 0 + \frac{2E(I)}{3}\left\{0 - \frac{12.87}{EI}\right\} = -8.58 \text{ kNm}$$

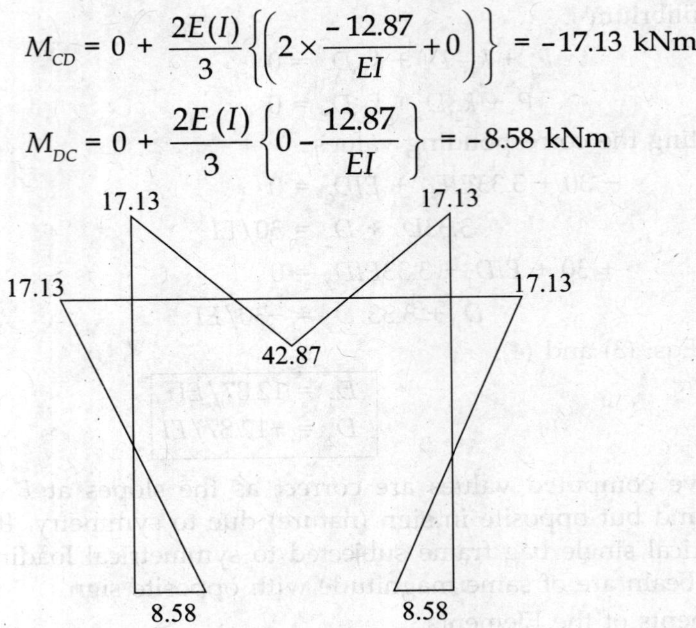

FIG. 5.100 Bending moment diagram

Example 5.18: Analyse the frame by system stiffness approach. Draw the bending moment diagram.

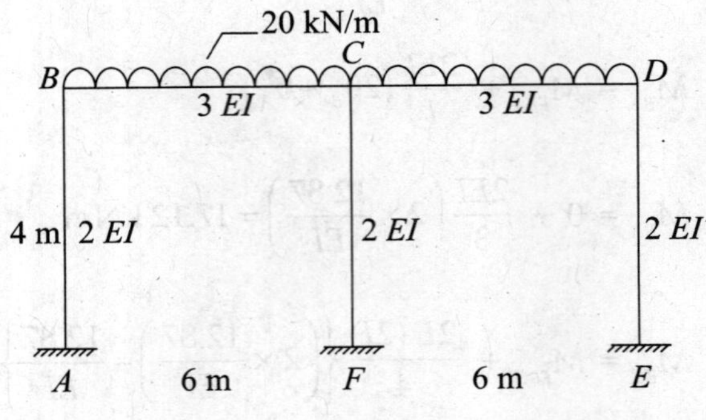

FIG. 5.101

Solution

1. Kinematic Redundants

The frame is symmetrical with respect to loading and geometry. The slopes at the fixed support are zeroes. Hence, $\theta_A = \theta_F = \theta_E = 0$. Due to symmetry $\theta_C = 0$ and also $\theta_D = -\theta_B$. Hence, we can analyse only half frame. The analysis reduces to the estimation of rotation of stiff joint B, i.e., θ_B. This is denoted A as D_1. It is assumed that this kinematic redundant act in the clockwise direction.

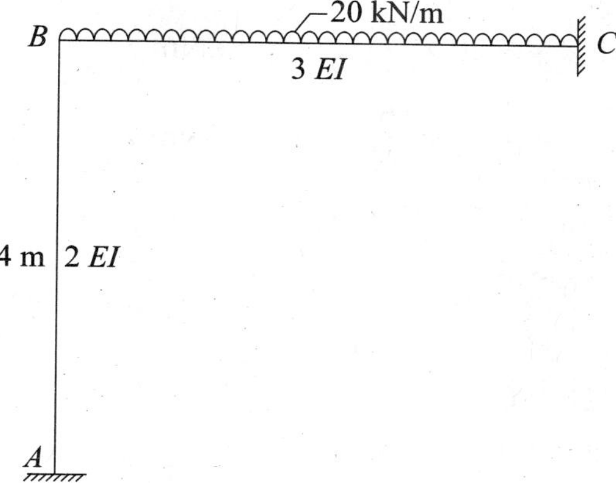

FIG. 5.102

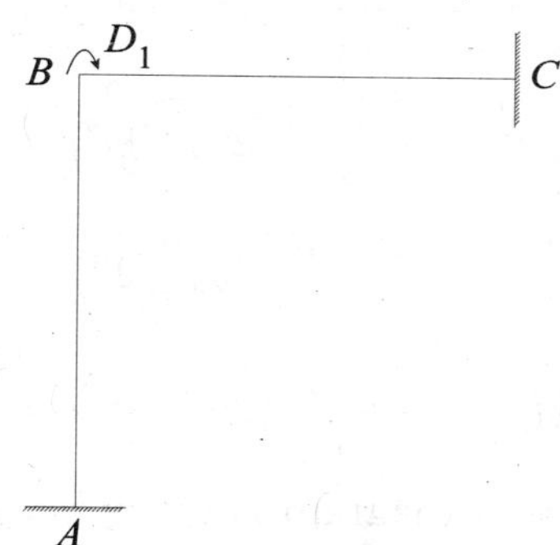

FIG. 5.103 Kinematic redundant

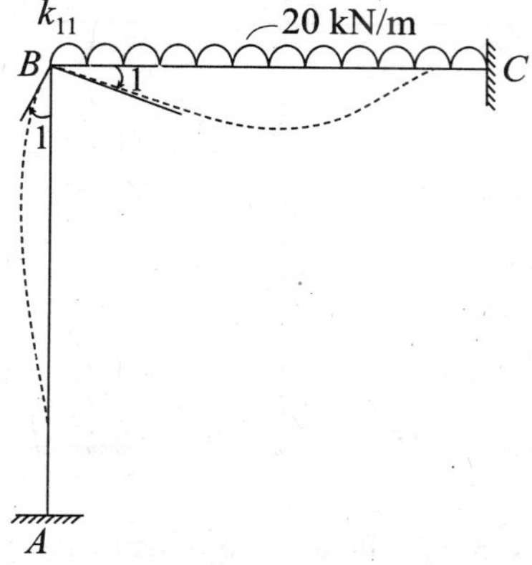

FIG. 5.104 Unit rotation at B

2. Fixed End Moments

$$M_{FBC} = -20 \times \frac{6^2}{12} = -60 \text{ kNm}$$

$$M_{FCB} = +20 \times \frac{6^2}{12} = +60 \text{ kNm}$$

Stiffness coefficient k_{11}

$$k_{11} = \frac{4(3EI)}{6} + \frac{4(2EI)}{4} = 4EI$$

3. Joint Equilibrium at B

$$-60 + 4EID_1 = 0$$

$$D_1 = \theta_B = 15/EI$$

4. End Moments of the Elements

$$M_{AB} = \cancel{M_{FAB}} + \frac{2(2EI)}{4}(2\cancel{\theta_A} + \theta_B) = EI\theta_B = 15 \text{ kNm}$$

$$M_{BA} = \cancel{M_{FBA}} + \frac{2(2EI)}{4}(2\theta_B + \cancel{\theta_A}) = 2EI\theta_B = 30 \text{ kNm}$$

$$M_{BC} = M_{FBC} + \frac{2(3EI)}{6}(2\theta_B + \cancel{\theta_C}) = -30 \text{ kNm}$$

$$M_{CB} = M_{FCB} + \frac{2(3EI)}{6}(2\theta_C + \theta_B) = +75 \text{ kNm}$$

The resulting bending moment diagram is given below.

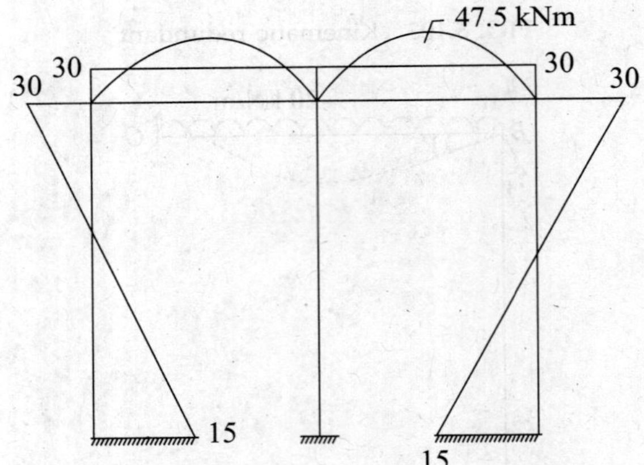

FIG. 5.105 Bending moment diagram

5.13 ANALYSIS OF UNSYMMETRICAL FRAMES

Example 5.19: Draw the bending moment diagram for the frame shown in Fig. 5.106 Use system stiffness approach.

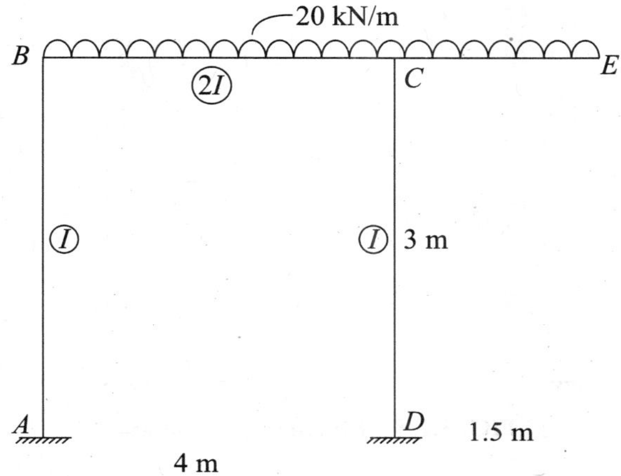

FIG. 5.106

Solution

1. Kinematic Redundants

The frame is asymmetrical with respect to geometry and loading. Hence, the frame sways by an amount Δ. The kinematic redundants are θ_A, θ_B, θ_C, θ_D, θ_E and Δ. As the columns AB and CD are on fixed supports $\theta_A = \theta_D = 0$. Thus, the redundants are θ_B, θ_C and Δ only.

The effect of overhang CE is replaced by a downward ($20 \times 1.5 = 30$ kN) and a moment of $\left(+20 \times \frac{1.5^2}{2} = +22.50 \text{ kNm}\right)$. The analysis of the frame modified loads are.

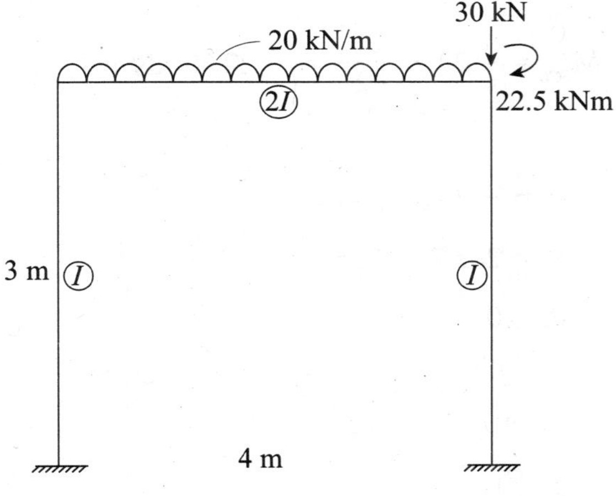

FIG. 5.107

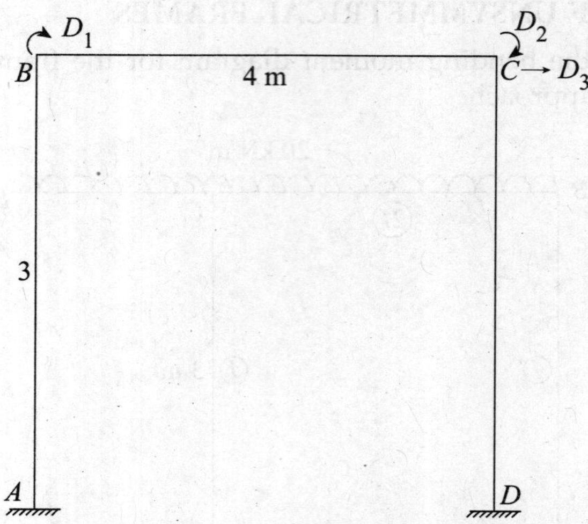

FIG. 5.108 Kinematic redundants

2. Fixed End Moments

$$M_{FAB} = 0$$

$$M_{FBA} = 0$$

$$M_{FBC} = -\,20\,\frac{(4)^2}{12} = -\,26.67 \text{ kNm}$$

$$M_{FCB} = +\,20\,\frac{(4)^2}{12} = +\,26.67 \text{ kNm}$$

$$M_{FCE} = -\,20 \times \frac{1.5^2}{2} = -\,22.50 \text{ kNm}$$

3. Joint Loads

$$P_1 = -\,26.67 \text{ kNm}$$

$$P_2 = 26.67 - 22.50 = 4.17$$

$$P_3 = 0$$

4. Stiffness Coefficients/Stiffness Matrix

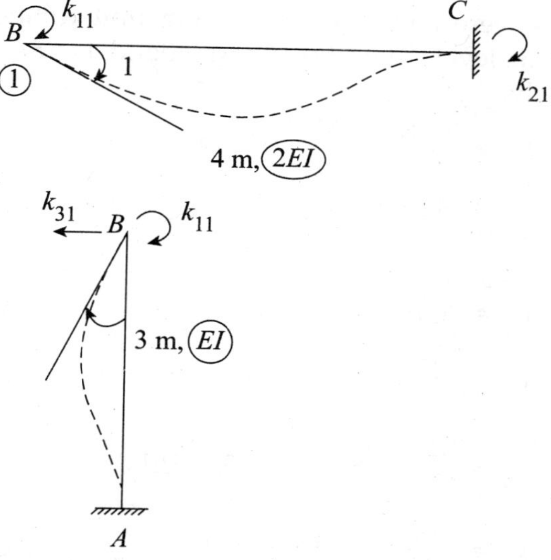

FIG. 5.109 Unit rotation at ①

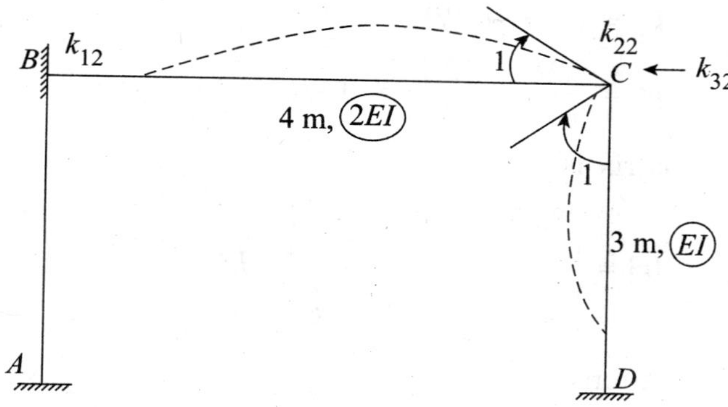

FIG. 5.110 Unit rotation at ②

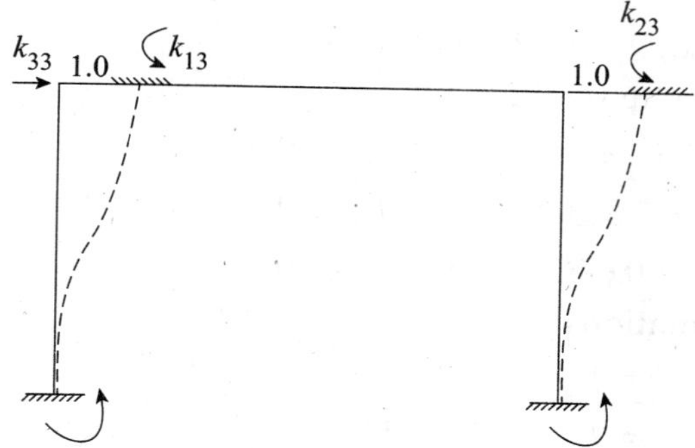

FIG. 5.111 Unit rotation at ③

5. Stiffness Coefficients/Stiffness Matrix

The stiffness coefficients were computed by giving unit rotations at joints 1 and 2; displacement at the beam level. The above correspond to the kinematic redundants D_1, D_2 and D_3 respectively.

$$k_{11} = \frac{4(2EI)}{4} + \frac{4EI}{3} = 3.333EI$$

$$k_{21} = \frac{2(2EI)}{4} = EI$$

$$k_{31} = \frac{-6EI}{3^2} = -0.67EI$$

$$k_{12} = \frac{2(2EI)}{4} = EI$$

$$k_{22} = \frac{4E\,(2I)}{4} + \frac{4EI}{3} = 3.333EI$$

$$k_{32} = \frac{-6EI}{3^2} = -0.67EI$$

$$k_{13} = k_{31} = -0.67EI$$

$$k_{23} = k_{32} = -0.67EI$$

$$k_{33} = \frac{12EI}{3^3} + \frac{12EI}{3^3} = 0.88EI$$

Thus, the stiffness matrix is

$$[k] = \begin{bmatrix} 3.333 & 1 & -0.67 \\ 1 & 3.333 & -0.67 \\ -0.67 & -0.67 & 0.88 \end{bmatrix} EI$$

6. Joint Equilibrium Equations

$$P_1 + k_{11}D_1 + k_{12}D_2 + k_{13}D_3 = 0 \quad (1)$$

$$P_2 + k_{21}D_1 + k_{22}D_2 + k_{23}D_3 = 0 \quad (2)$$

$$P_3 + k_{31}D_1 + k_{32}D_2 + k_{33}D_3 = 0 \quad (3)$$

Substituting the computed values

$$-26.67 + EI(3.333D_1 + 1D_2 - 0.67D_3) = 0 \quad (4)$$

$$+4.17 + EI(D_1 + 3.333D_2 - 0.67D_3) = 0 \quad (5)$$

$$EI(-0.67D_1 - 0.67D_2 + 0.888D_3) = 0 \quad (6)$$

Solving the above equations

$$\theta_B = D_1 = 10/EI$$

$$\theta_C = D_2 = -3.22/EI$$

$$\Delta = D_3 = 5.11/EI$$

7. End Moment of the Elements

The end moments are computed using the slope deflections as

$$M_{AB} = \cancel{M_{FAB}} + \frac{2EI}{3}\left(2\cancel{\theta_A} + \theta_B - \frac{3\Delta}{3}\right) = 3.28 \text{ kNm}$$

$$M_{BA} = \cancel{M_{FBA}} + \frac{2EI}{3}\left(2\theta_B + \cancel{\theta_A} - \frac{3\Delta}{3}\right) = 9.93 \text{ kNm}$$

$$M_{BC} = M_{FBC} + \frac{2E(2I)}{4}(2\theta_B + \theta_C) = -9.93 \text{ kNm}$$

$$M_{CB} = M_{FCB} + \frac{2E(2I)}{4}(2\theta_C + \theta_B) = 30.23 \text{ kNm}$$

$$M_{CD} = M_{FCD} + \frac{2EI}{3}\left(2\theta_C + \cancel{\theta_D} - \frac{3\Delta}{3}\right) = -7.7 \text{ kNm}$$

$$M_{CE} = -20(1.5)^2/2 = -22.5 \text{ kNm}$$

$$M_{DC} = M_{FDC} + \frac{2EI}{3}\left(2\cancel{\theta_D} + \theta_C - \frac{3\Delta}{3}\right) = -5.55 \text{ kNm}$$

Example 5.20: Determine the bending moments of the frame shown in Fig. 5.112. Adopt system stiffness approach.

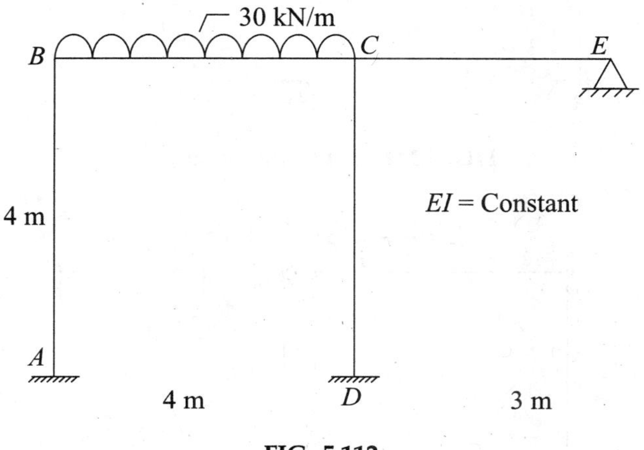

FIG. 5.112

Solution

1. Kinematic Redundants

The kinematic redundants are θ_B, θ_C and θ_E. A and D are fixed supports and hence $\theta_A = \theta_D = 0$. The redundants are assumed to act in the clockwise direction.

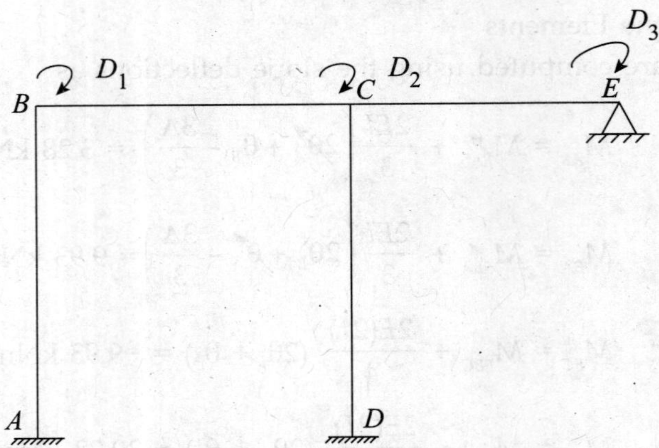

FIG. 5.113 Kinematic redundants

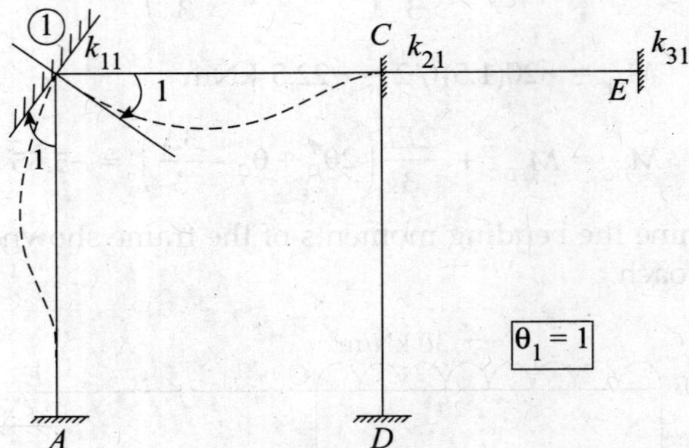

FIG. 5.114 Unit rotation at ①

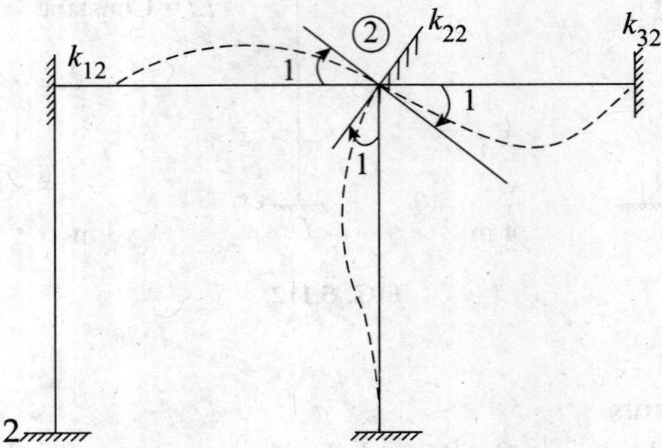

FIG. 5.115 Unit rotation at ②

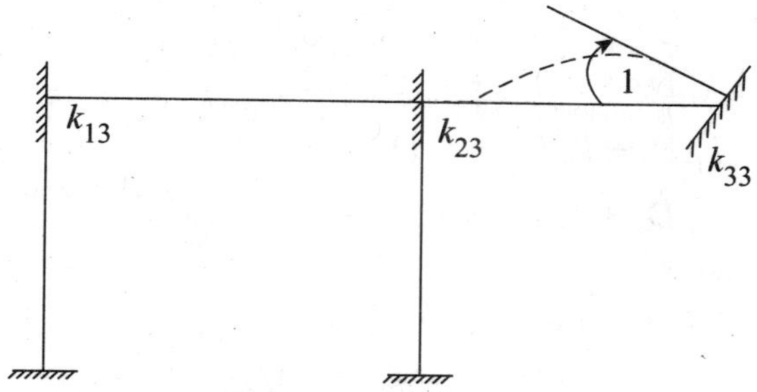

FIG. 5.116 Unit rotation at (3)

2. Fixed End Moments

$$M_{FBC} = -30(4)^2/12 = -40 \text{ kNm} = P_1$$

$$M_{FCB} = +30(4)^2/12 = +40 \text{ kNm} = P_2$$

3. Stiffness Coefficients

$$k_{11} = \frac{4EI}{4} + \frac{4EI}{4} = 2EI$$

$$k_{21} = 2EI/4 = 0.5EI$$

$$k_{31} \Rightarrow \text{No effect} = 0$$

$$k_{12} = k_{21} = 0.5EI$$

$$k_{22} = \frac{4EI}{4} + \frac{4EI}{3} + \frac{4EI}{4} = 3.33EI$$

$$k_{32} = \frac{2}{3}EI = 0.67EI$$

$$k_{13} = k_{31} = 0$$

$$k_{23} = k_{32} = 0.67EI$$

$$k_{33} = 4EI/3 = 1.33EI$$

Hence, the stiffness matrix is

$$[k] = EI \begin{bmatrix} 2 & 0.5 & 0 \\ 0.5 & 3.33 & 0.67 \\ 0 & 0.67 & 1.33 \end{bmatrix}$$

4. Joint Equilibrium

$$P_1 + k_{11}D_1 + k_{21}D_2 + k_{31}D_3 = 0$$
$$P_2 + k_{12}D_1 + k_{22}D_2 + k_{32}D_3 = 0$$
$$P_3 + k_{13}D_1 + k_{23}D_2 + k_{33}D_3 = 0$$

Substituting the values;

$$-40 + EI(2D_1 + 0.5D_2) = 0$$
$$2D_1 + 0.5D_2 = 40/EI \quad (1)$$
$$+40 + EI(0.5D_1 + 3.33D_2 + 0.67D_3) = 0$$
$$0.5D_1 + 3.33D_2 + 0.67D_3 = -40 \quad (2)$$
$$0 + EI(0D_1 + 0.67D_2 + 1.33D_3) = 0$$
$$0D_1 + 0.67D_2 + 1.33D_3 = 0 \quad (3)$$

On solving

$$D_1 = \theta_B = 24.36/EI$$
$$D_2 = \theta_C = -17.44/EI$$
$$D_3 = \theta_E = 8.78/EI$$

5. End Moments

$$M_{AB} = 0 + \frac{2EI}{4}(2\cancel{\theta_A} + \theta_B) = 12.18 \text{ kNm}$$

$$M_{BA} = 0 + \frac{2EI}{4}(2\theta_B + \cancel{\theta_A}) = 24.36 \text{ kNm}$$

$$M_{BC} = -40 + \frac{2EI}{4}(2\theta_B + \theta_C) = -24.36 \text{ kNm}$$

$$M_{CB} = +40 + \frac{2EI}{4}(2\theta_C + \theta_B) = 34.74 \text{ kNm}$$

$$M_{CD} = 0 + \frac{2EI}{4}(2\theta_C + \cancel{\theta_D}) = -17.44 \text{ kNm}$$

$$M_{DC} = 0 + \frac{2EI}{4}(2\cancel{\theta_D} + \theta_C) = -8.72 \text{ kNm}$$

$$M_{CE} = 0 + \frac{2EI}{3}(2\theta_C + \theta_E) = -17.4 \text{ kNm}$$

$$M_{EC} = 0 + \frac{2EI}{3}(2\theta_E + \theta_C) = 0$$

Example 5.21: Draw the bending moment diagram and the elastic curve of the frame shown in Fig. 5.117.

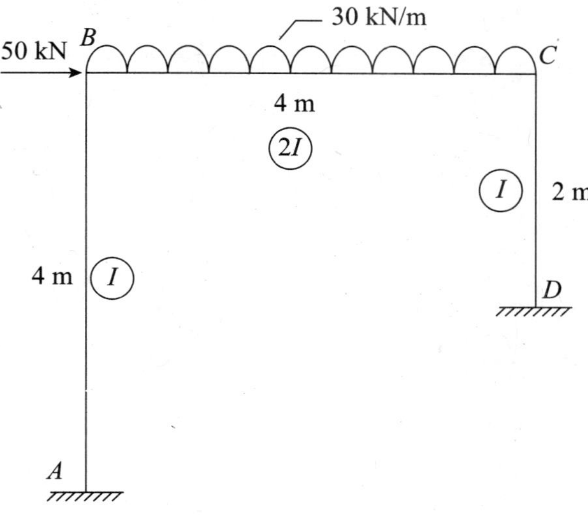

FIG. 5.117

Solution

1. Kinematic Redundants

The kinematic redundants are θ_B, θ_C and Δ where θ_B and θ_C are the slopes at B and C respectively. Δ is the sway of the columns AB and CD. They are denoted as D_1, D_2 and D_3.

2. Fixed End Moments

$$M_{FAB} = M_{FBA} = M_{FCD} = M_{FDC} = 0$$

$$M_{FBC} = -30(4)^2/12 = -40 \text{ kNm}$$

$$M_{FCB} = +30(4)^2/12 = +40 \text{ kNm}$$

3. Joint Loads

$$[P] = \begin{bmatrix} -40 \\ +40 \\ -50 \end{bmatrix}$$

4. Stiffness Coefficients / Stiffness Matrix

A unit rotation at *B* is given. The corresponding stiffness coefficients are as follows.

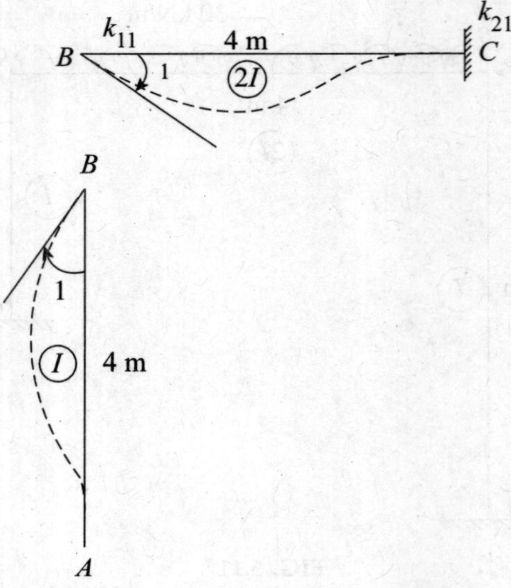

FIG. 5.118 Unit rotation at *B*

$$k_{11} = \frac{4E(2I)}{4} + \frac{4EI}{4} = 3EI$$

$$k_{21} = \frac{2E(2I)}{4} = EI$$

$$k_{31} = \frac{-6EI}{4^2} = -0.375EI$$

A unit rotation at *C* is given, the stiffness coefficients are.

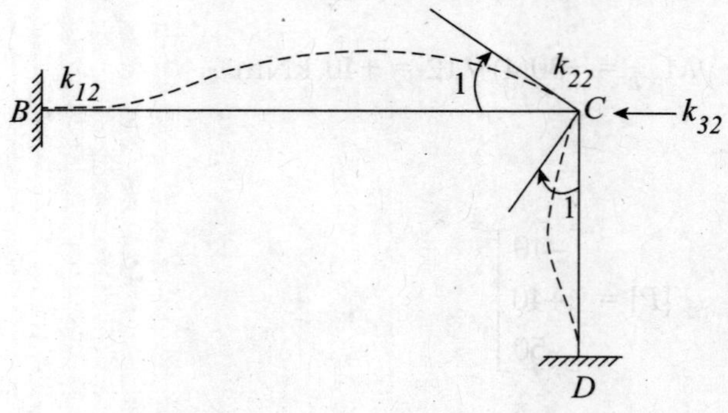

FIG. 5.119 Unit rotation at *C*

$$k_{12} = \frac{2E(2I)}{4} = EI$$

$$k_{22} = \frac{4E(2I)}{4} + \frac{4EI}{2} = 4EI$$

$$k_{32} = \frac{-6EI}{2^2} = -1.5EI$$

A unit horizontal displacement (sway) is applied at the floor level of the beam.

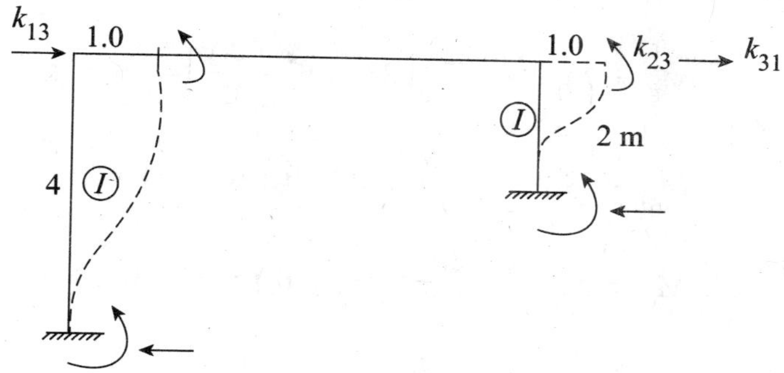

FIG. 5.120

$$k_{13} = \frac{-6EI}{4^2} = -0.375EI$$

$$k_{23} = \frac{-6EI}{2^2} = -1.5EI$$

$$k_{33} = \frac{12EI}{4^3} + \frac{12EI}{2^3} = 1.69EI$$

5. Joint Equilibrium Conditions

$$P_1 + k_{11}D_1 + k_{12}D_2 + k_{13}D_3 = 0$$
$$P_2 + k_{21}D_1 + k_{22}D_2 + k_{23}D_3 = 0$$
$$P_3 + k_{31}D_1 + k_{32}D_2 + k_{33}D_3 = 0$$

Substituting the computed values and simplifying

$$3D_1 + D_2 - 0.375D_3 = 40/EI \quad (1)$$
$$D_1 + 4D_2 - 1.5D_3 = -40/EI \quad (2)$$
$$-0.375D_1 - 1.5D_2 + 1.69D_3 = 50/EI \quad (3)$$

Solving the above equations,

$$D_1 = \theta_B = 18.182$$

$$D_2 = \theta_C = -2.905$$

$$D_3 = \Delta = +31.042$$

6. End Moments

$$M_{AB} = M_{FAB} + \frac{2EI}{l}\left(2\theta_A + \theta_B - \frac{3\Delta}{l}\right) = -2.55 \text{ kNm}$$

$$M_{BA} = M_{FBA} + \frac{2EI}{l}\left(2\theta_B + \theta_A - \frac{3\Delta}{l}\right) = -6.54 \text{ kNm}$$

$$M_{BC} = M_{FBC} + \frac{2EI}{l}(2\theta_B + \theta_C) = -6.54 \text{ kNm}$$

$$M_{CB} = M_{FCB} + \frac{2EI}{l}(2\theta_C + \theta_B) = +52.37 \text{ kNm}$$

$$M_{CD} = M_{FCD} + \frac{2EI}{l}(2\theta_C + \theta_D) = -52.37 \text{ kNm}$$

$$M_{DC} = M_{FDC} + \frac{2EI}{l}(2\theta_D + \theta_C) = -49.5 \text{ kNm}$$

Example 5.22: Analyse the rigid frame by stiffness system approach. Draw the bending moment diagram.

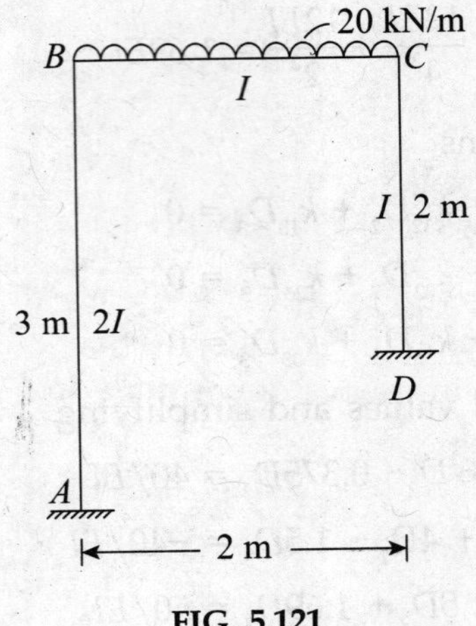

FIG. 5.121

Solution

1. Kinematic Redundants

The kinematic redundants are θ_B, θ_C and Δ. They are denoted as D_1, D_2 and D_3 and the positive directions are given below.

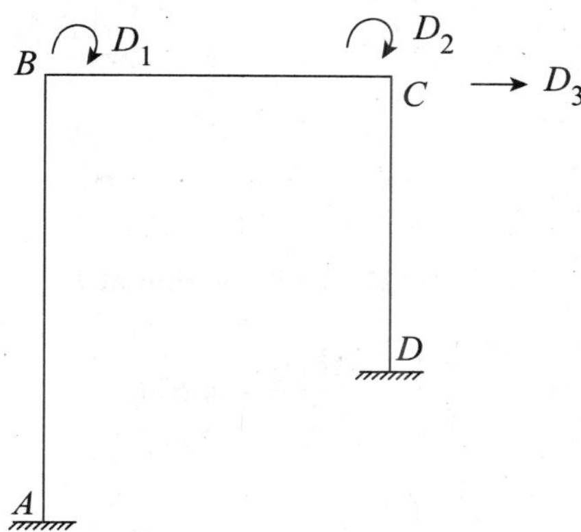

FIG. 5.122 Kinematic redundants

2. Fixed End Moments

$$M_{FBC} = \frac{-20(2)^2}{12} = -6.67 \text{ kNm}$$

$$M_{FCB} = \frac{20(2)^2}{12} = +6.67 \text{ kNm}$$

3. Joint Loads

$$P_1 = -6.67 \text{ kNm}$$

$$P_2 = 6.67 \text{ kNm}$$

$$P_3 = 0$$

As there is no sway load acting at the floor level in this case, $P_3 = 0$.

4. Stiffness Coefficients

They are computed by applying unit rotation at that joint and keeping all other kinematic redundant zero. The following ways of computation can be done.

(i) $D_1 = 1, D_2 = D_3 = 0$

(ii) $D_2 = 1, D_1 = D_3 = 0$

(iii) $D_3 = 1, D_1 = D_2 = 0$

Case (i)

Apply unit rotation at B

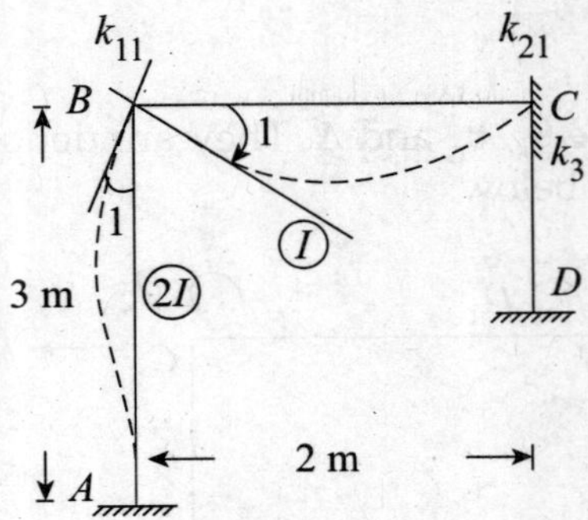

FIG. 5.123 Unit rotation at B

$$k_{11} = \frac{4E(2I)}{3} + \frac{4E(I)}{2} = 4.67EI$$

$$k_{21} = \frac{2EI}{2} = EI$$

$$k_{31} = \frac{-6E(2I)}{3^2} = -1.33EI$$

The negative sign is due to the fact that axial force developed along BC due to unit rotation at B is acting opposite to the positive direction of D_3.

Apply unit rotation at C

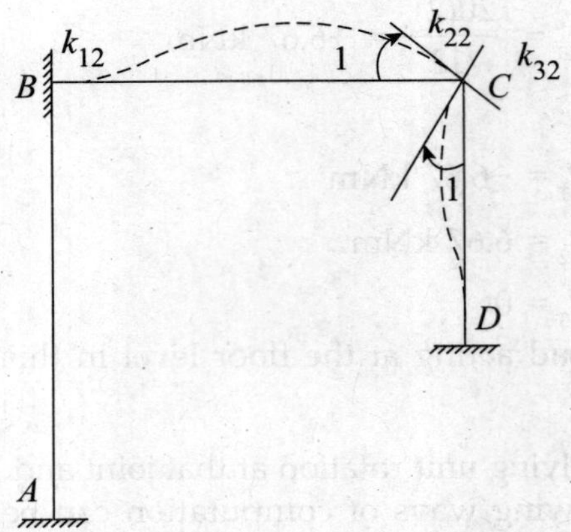

FIG. 5.124 Unit rotation at C

$$k_{12} = \frac{2EI}{2} = EI$$

$$k_{22} = \frac{4EI}{2} + \frac{4EI}{2} = 4EI$$

$$k_{32} = \frac{-6EI}{2^2} = -1.5EI$$

Apply a unit displacement at B to the right

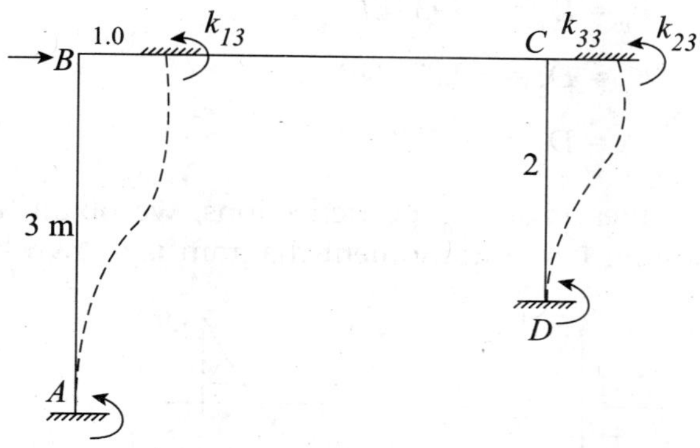

FIG. 5.125 Unit displacement at B

$$k_{13} = k_{31} = -1.33EI$$

$$k_{23} = k_{32} = -1.5EI$$

$$k_{33} = \frac{12E(2I)}{3^3} + \frac{12EI}{2^3} = 2.39EI$$

Using the stiffness coefficients, the stiffness matrix can be written as

$$[k] = \begin{bmatrix} 4.67 & 1.00 & -1.33 \\ 1.00 & 4 & -1.50 \\ -1.33 & -1.50 & 2.39 \end{bmatrix}$$

5. Joint Equilibrium

$$P_1 + k_{11}D_1 + k_{12}D_2 + k_{13}D_3 = 0$$

$$P_2 + k_{21}D_1 + k_{22}D_2 + k_{23}D_3 = 0$$

$$P_3 + k_{31}D_1 + k_{32}D_2 + k_{33}D_3 = 0$$

Substituting the values;

$$4.67D_1 + D_2 - 1.33D_3 = 6.67/EI \quad (1)$$

$$D_1 + 4D_2 - 1.50D_3 = -6.67/EI \quad (2)$$

$$-1.33D_1 - 1.50D_2 + 2.39D_3 = 0 \quad (3)$$

Solving the above equations;

$$\theta_B = D_1 = 1.793/EI$$

$$\theta_C = D_2 = -2.278/EI$$

$$\Delta = D_3 = -0.432/EI$$

Substituting these values in the slope deflections, we obtain the end moments of the members. The resulting bending moment diagram is shown below.

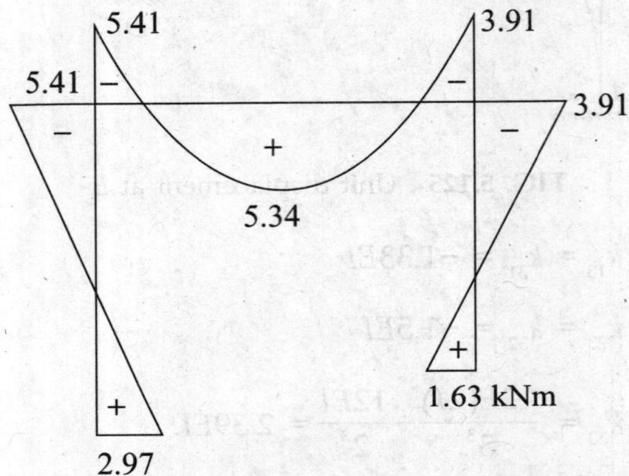

FIG. 5.126

Example 5.23: Analyse the rigid frame shown in Fig. 5.127 by the system stiffness method. Draw the bending moment, shear force and thrust diagram. Also sketch the elastic curve.

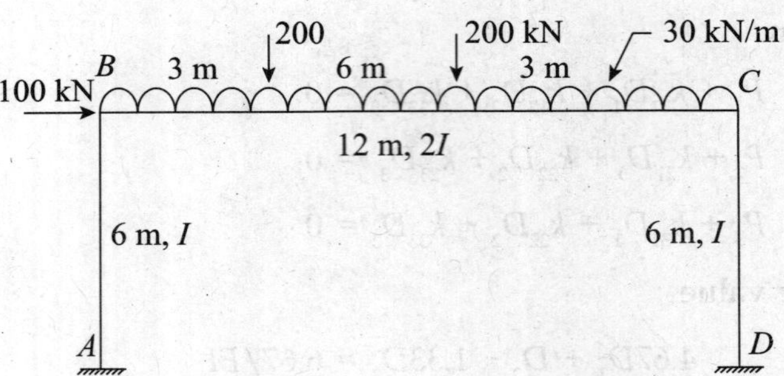

FIG. 5.127 Rigid frame with sway

Solution

1. Kinematic Redundants

The kinematic redundants are three, i.e., rotations at B, C and lateral sway for member BC. They are denoted as D_1, D_2 and D_3. The rotations are assumed in the clockwise direction and the joint displacement at C is towards right.

The reader can recall that a frame is made up of assemblage of beam elements and not restricted to lying along a straight line. Also, in a rigid joint, the angle between members prior to deformation also exists after deformation.

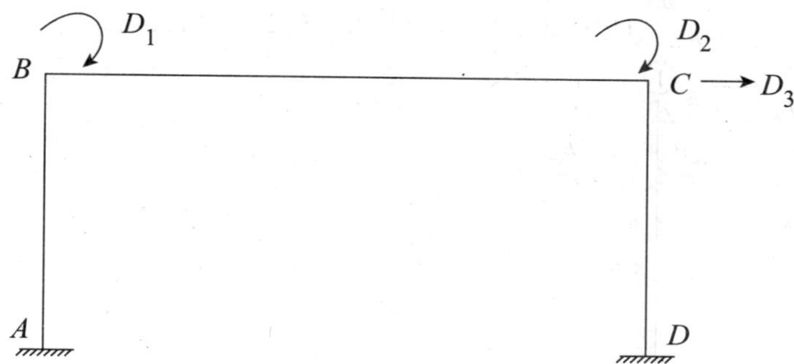

FIG. 5.128 Kinematic redundants

2. Fixed End Forces

All the members are made fixed at ends and the fixed end reactions in the direction of kinematic redundants are calculated at each joint.

$$M_{FBC} = -\left(30 \times \frac{12^2}{12} + \frac{200 \times 3 \times 9}{12}\right) = -\ 810 \text{ kNm} = P_1$$

$$M_{FCB} = +\left(30 \times \frac{12^2}{12} + \frac{200 \times 3 \times 9}{12}\right) = +\ 810 \text{ kNm} = P_2$$

$$P_3 = -100 \text{ kN}$$

It is to be mentioned here that P_3 is negative as it acts opposite to the direction of the redundant at 3. Thus, joint loads are

$$[P_{DL}] = \begin{bmatrix} -810 \\ +810 \\ -100 \end{bmatrix}$$

Negative sign indicates that the fixed and reactions/forces are acting in the opposite direction of that assumed kinematic redundant.

3. Stiffness Coefficient

They are calculated by giving unit displacements at ①, ② and ③ successively, along the kinematic redundants assumed. If the member end forces (bending moment and shear forces) are to be calculated through matrix approaches, then they should be calculated for each member when unit displacement is given.

To begin with, apply unit rotation along ①

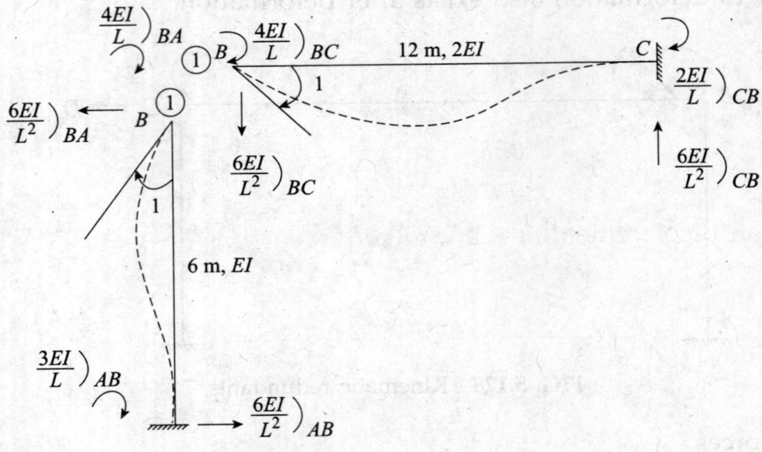

FIG. 5.129 Unit rotation at 1

$$k_{11} = \left.\frac{4EI}{L}\right)_{BA} + \left.\frac{4EI}{L}\right)_{BC} = \frac{4EI}{3}$$

$$k_{21} = \left.\frac{2EI}{L}\right)_{CB} = \frac{EI}{3}$$

$$k_{31} = \left.\frac{6EI}{L}\right)_{BA} = \frac{-EI}{6}$$

Negative sign in the stiffness coefficients indicate that they act in opposite direction to that of assumed kinematic redundant.

Apply a unit rotation at C and the stiffness coefficients are evaluated.

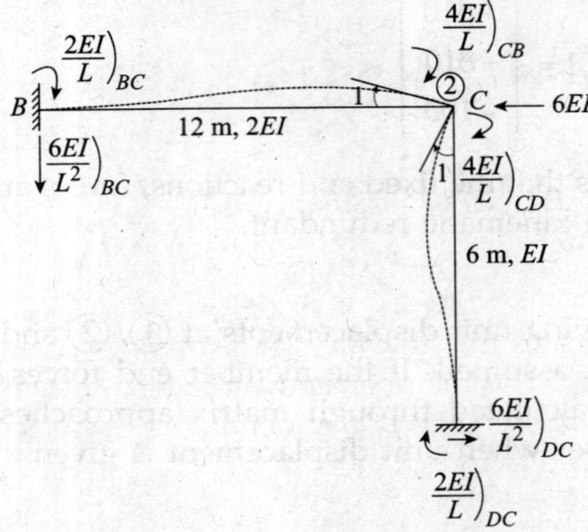

FIG. 5.130 Unit rotation at 2

$$k_{12} = \left.\frac{2EI}{L}\right)_{BC} = \frac{EI}{3}$$

$$k_{22} = \left.\frac{4EI}{L}\right)_{CB} = \left.\frac{4EI}{L}\right)_{CD} = 4\frac{EI}{3}$$

$$k_{32} = -\left.\frac{6EI}{L^2}\right)_{CD} = -\frac{EI}{6}$$

Apply a unit displacement at B and derive the stiffness coefficients

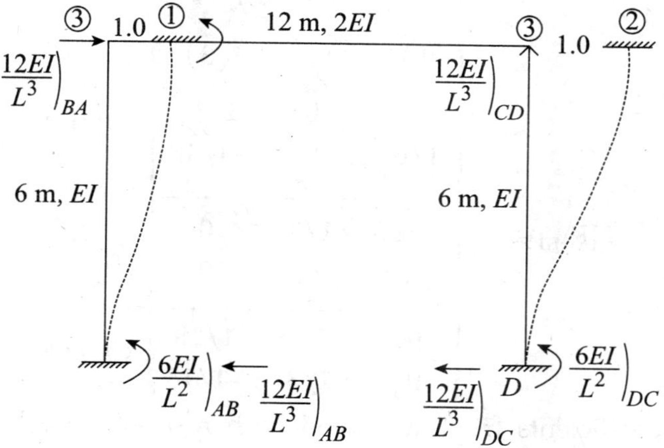

FIG. 5.131 Unit rotation at 3

$$k_{13} = -\left.\frac{6EI}{L^2}\right)_{BA} = -\frac{EI}{6}$$

$$k_{23} = -\left.\frac{6EI}{L^2}\right)_{CD} = -\frac{EI}{6}$$

$$k_{33} = \left.\frac{12EI}{L^3}\right)_{BA} \left.\frac{12EI}{L^3}\right)_{CD} = \frac{EI}{9}$$

The negative sign indicates that they act in the opposite direction to that of assumed direction of displacements. The member end moments and shear force are calculated using stiffness coefficients at each end of the member for each independent displacement. They are represented as $[K_{MD}]$ and $[K_{SD}]$.

$$[K_{MD}] = EI \begin{bmatrix} ① & ② & ③ & \\ 1/3 & 0 & 1/6 & \\ 2/3 & 0 & -1/6 & AB/BA \\ \hline 2/3 & 1/3 & 0 & \\ 1/3 & 2/3 & 0 & BC/CB \\ \hline 0 & 2/3 & -1/6 & \\ 0 & 1/3 & -1/6 & CD/DC \end{bmatrix}$$

$$[K_{SD}] = EI \begin{bmatrix} ① & ② & ③ & \\ -1/6 & 0 & 1/18 & \\ 1/6 & 0 & -1/18 & AB/BA \\ \hline -1/12 & -1/12 & 0 & \\ 1/12 & 1/12 & 0 & BC/CB \\ \hline 0 & -1/6 & 1/18 & \\ 0 & 1/3 & -1/18 & CD/DC \end{bmatrix}$$

The stiffness coefficients $[K_{SD}]$ are noted with respect to each member by seeing from inside the frame.

4. Joint Equilibrium Equations

$$P_1 + k_{11}D_1 + k_{12}D_2 + k_{13}D_3 = 0$$
$$P_2 + k_{21}D_1 + k_{22}D_2 + k_{23}D_3 = 0$$
$$P_3 + k_{31}D_1 + k_{32}D_2 + k_{33}D_3 = 0$$

Substituting the values

$$-810 + \left[\frac{4}{3}D_1 + \frac{1}{3}D_2 - \frac{1}{6}D_3\right] EI = 0$$

$$810 + \left[\frac{1}{3}D_1 + \frac{4}{3}D_2 - \frac{D_3}{6}\right] EI = 0$$

$$-100 + \left[-\frac{1}{6}D_1 - \frac{1}{6}D_2 + \frac{D_3}{9}\right] EI = 0$$

Solving the above equations

$$D_1 = 938.57/EI \curvearrowright = \theta_B$$
$$D_2 = -681.43/EI \circlearrowleft = \theta_C$$
$$D_3 = 1285.71/EI \rightarrow = \delta$$

5. End Moments and Shear Forces

These are calculated using matrices as

$$\{P_M\} = \{P_{MF}\} + [K_{MD}]\,[D]$$

$$\{P_S\} = \{P_{SF}\} + [K_{SD}]\,[D]$$

$$[P_M] = \begin{bmatrix} 0 \\ 0 \\ -810 \\ +810 \\ 0 \\ 0 \end{bmatrix} + \begin{bmatrix} 1/3 & 0 & -1/6 \\ 2/3 & 0 & -1/6 \\ 2/3 & 1/3 & 0 \\ 1/3 & 2/3 & 0 \\ 0 & 2/3 & -1/6 \\ 0 & 1/3 & -1/6 \end{bmatrix} \frac{1}{EI} \begin{Bmatrix} 938.571 \\ -681.429 \\ 1285.714 \end{Bmatrix}$$

$$\begin{bmatrix} M_{AB} \\ M_{BA} \\ M_{BC} \\ M_{CB} \\ M_{CD} \\ M_{DC} \end{bmatrix} = \begin{bmatrix} 98.57 \\ 411.43 \\ -411.43 \\ 668.57 \\ -668.57 \\ -441.43 \end{bmatrix}$$

$$[P_S] = \begin{Bmatrix} 0 \\ 100 \\ 380 \\ 380 \\ 0 \\ 0 \end{Bmatrix} + EI \begin{bmatrix} -1/6 & 0 & 1/18 \\ 1/6 & 0 & -1/18 \\ -1/12 & -1/12 & 0 \\ 1/12 & 1/12 & 0 \\ 0 & -1/6 & 1/18 \\ 0 & -1/6 & -1/18 \end{bmatrix} \frac{1}{EI} \begin{Bmatrix} 938.571 \\ -681.429 \\ 1285.714 \end{Bmatrix}$$

$$= \begin{Bmatrix} \rightarrow & \leftarrow & \uparrow & \uparrow & \rightarrow & \leftarrow \\ -85.0 & 185.00 & 358.57 & 401.43 & 185.0 & -185.0 \\ V_{AB} & V_{BA} & V_{BC} & V_{CB} & V_{CD} & V_{DC} \end{Bmatrix}_{\text{kN}}$$

The answers are same as in the flexibility method. Elastic curve is shown in Fig. 5.132.

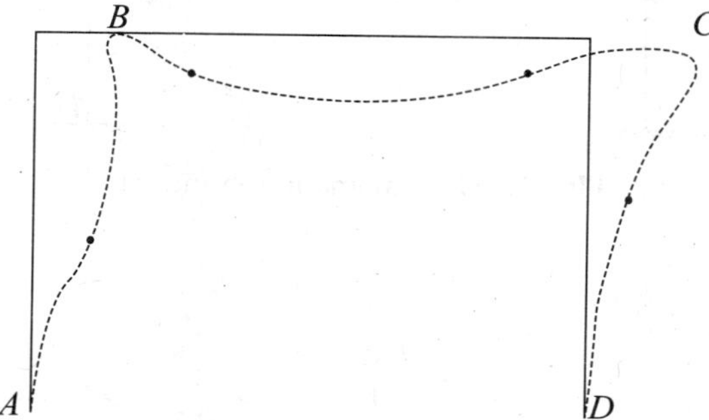

FIG. 5.132 Elastic curve

Example 5.24: Generate the stiffness matrix and determine the kinematic redundants for the frame shown in Fig. 5.133. Use system approach.

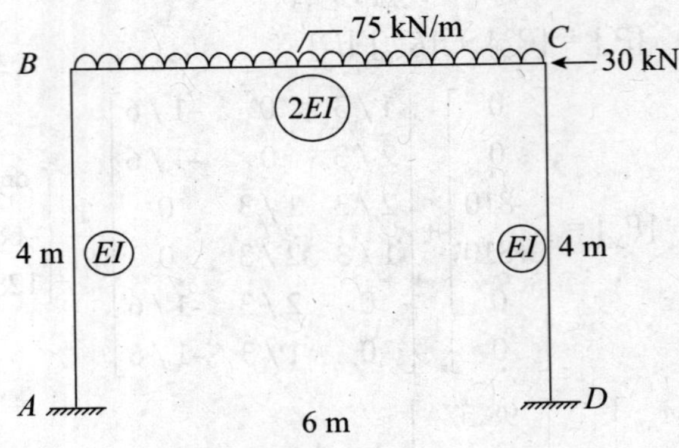

FIG. 5.133

Solution

1. Kinematic Redundants

The kinematic redundants are θ_B, θ_C and Δ. The sway Δ is towards left. The positive direction of Δ is assumed to be towards left. The positive directions for slopes at B and C are assumed to act in the clockwise direction. The kinematic redundants and the restrained structure are shown below.

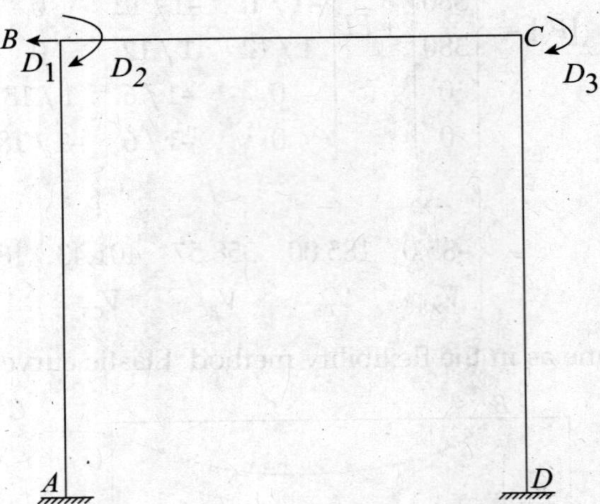

FIG. 5.134 Kinematic redundants

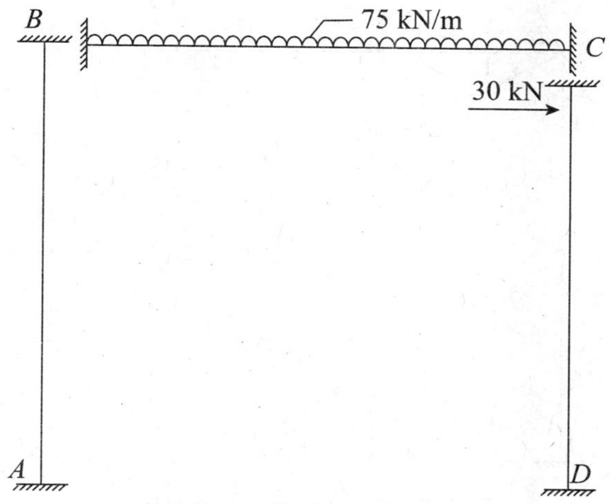

FIG. 5.135 Restrained structure

2. Stiffness Coefficients

Apply a unit displacement to act towards left and the end forces are computed as follows:

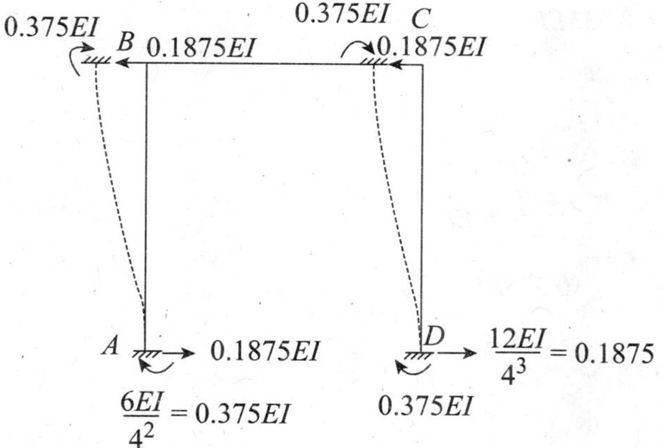

FIG. 5.136 Unit displacement at B ($D_1 = 1$, $D_2 = D_3$)

Referring to the above;

$$k_{11} = 0.1875EI + 0.1875EI = 0.375EI$$

$$k_{21} = 0.375EI$$

$$k_{31} = 0.375EI$$

Apply a unit rotation at B and compute the end forces.

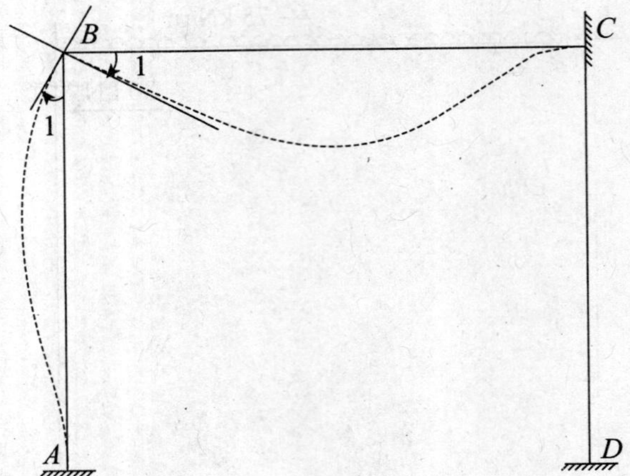

FIG. 5.137 Unit rotation at B

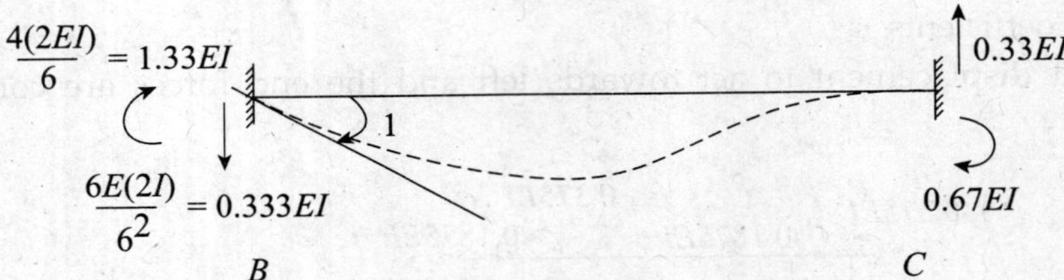

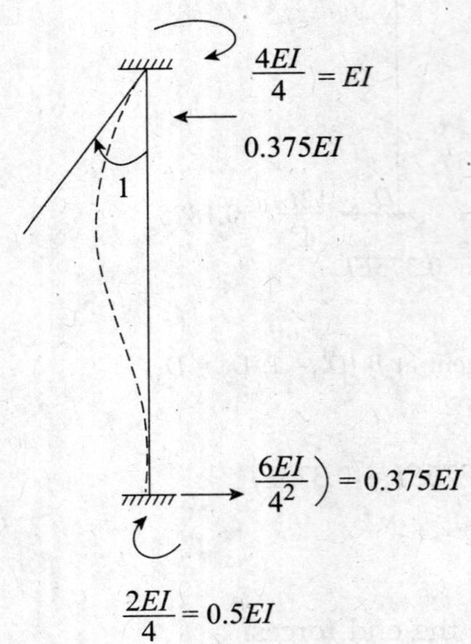

FIG. 5.138 End forces

Referring to the above figure

$$k_{12} = 0.375EI$$

$$k_{22} = 1.33EI + EI = 2.33EI$$

$$k_{32} = 0.67EI$$

Apply a unit rotation at C and compute the end forces

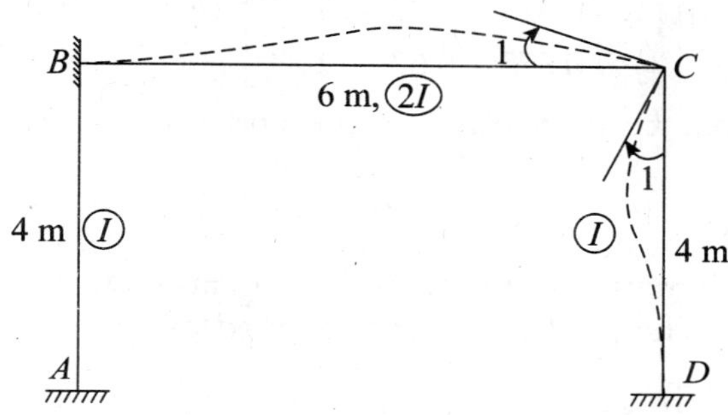

FIG. 5.139 Unit rotation at C

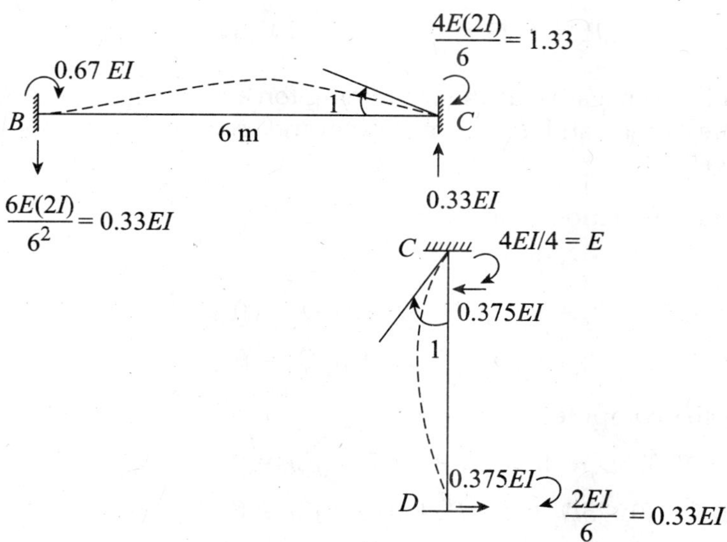

FIG. 5.140 End forces due to unit rotation at C

From the above figure;

$$k_{13} = 0.375EI$$

$$k_{23} = 0.67EI$$

$$k_{33} = 1.33EI + EI = 2.33EI$$

The stiffness matrix can be written as

$$[k] = \begin{bmatrix} k_{11} & k_{12} & k_{13} \\ k_{21} & k_{22} & k_{23} \\ k_{31} & k_{32} & k_{33} \end{bmatrix}$$

$$[k] = \begin{bmatrix} 0.375 & 0.375 & 0.375 \\ 0.375 & 2.33 & 0.67 \\ 0.375 & 0.67 & 2.33 \end{bmatrix} (EI)$$

It is to be noted that k_{13} is positive as the end force k_{13} is acting in the positive direction of D_1.

Kinematic Redundants

The values of the kinematic redundants D_1, D_2 and D_3 are obtained using joint equilibrium conditions. The joint loads are obtained as

$$P_1 = -30 \text{ kN} \rightarrow$$

$$P_2 = -75 \times \frac{6^2}{12} = -225 \text{ kNm} \circlearrowleft$$

$$P_3 = +75 \times \frac{6^2}{12} = +225 \text{ kNm} \circlearrowright$$

The joint load P_1 is negative as the restraining force is acting in the opposite direction of D_1. The values of P_2 and P_3 are negative and positive as clockwise direction is assumed positive.

The equilibrium equations can be written as

$$P_1 + k_{11}D_1 + k_{12}D_2 + k_{13}D_3 = 0$$

$$P_2 + k_{21}D_1 + k_{22}D_2 + k_{23}D_3 = 0$$

$$P_3 + k_{31}D_1 + k_{32}D_2 + k_{33}D_3 = 0$$

Substituting the computed values

$$-30 + (0.375D_1 + 0.375D_2 + 0.375D_3)EI = 0$$

$$-225 + (0.375D_1 + 2.33D_2 + 0.67D_3)EI = 0$$

$$225 + (0.375D_1 + 0.67D_2 + 2.33D_3)EI = 0$$

On solving

$$D_1 = \Delta = 106.67/EI$$

$$D_2 = \theta_B = 122.21/EI$$

$$D_3 = \theta_C = -148.87/EI$$

5.14 ANALYSIS OF PIN JOINTED TRUSSES

Example 5.25: Analyse the pin jointed frame using the stiffness method.

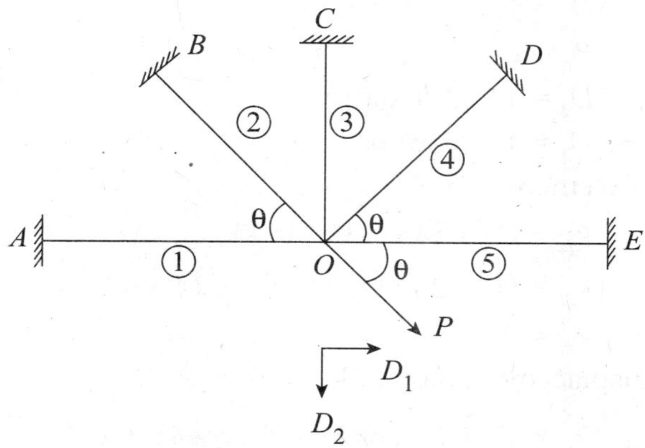

FIG. 5.141

Axial load P = 200 kN and θ = 45°. Axial stiffness, (AE/L) is same for all members.

Solution: The above system has two independent displacements D_1 and D_2 at O as indicated.

$$P_1 = P \cos \theta = 200/\sqrt{2} \text{ kN}$$
$$P_2 = P \sin \theta = 200/\sqrt{2} \text{ kN}$$

1. Stiffness Matrix

The stiffness matrix is based on the unit displacement applied along D_1 and D_2 successively at O. The displacements along each member is calculated. If the displacement induces tension in the member then it is positive and if it is compression in the member then it is negative. The displacements are shown in Fig. 5.142.

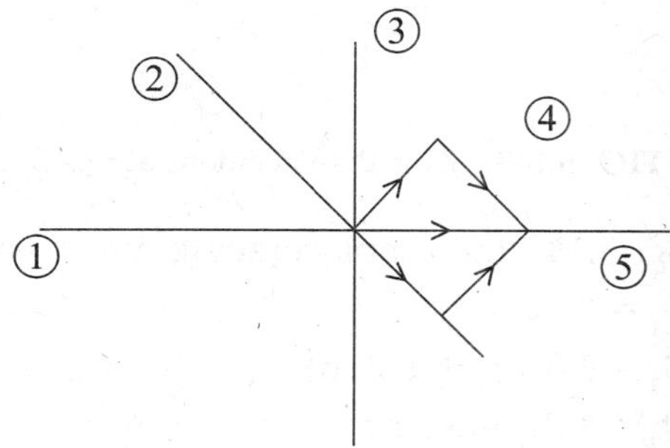

FIG. 5.142 Unit displacement along D_1

Based on the above Fig. 5.142 the displacements along each member

$D_1 = 1.0$ (tension)

$D_2 = 1/\sqrt{2}$ (tension)

$D_3 = 0$

$D_4 = 1/\sqrt{2}$ (comp)

$D_5 = 1.0$ (comp)

Total vertical displacement along D_2

$$k_{21} = D_2 \sin 45 - D_4 \sin 45$$

$$k_{21} = (1/\sqrt{2}) \sin 45 - (1/\sqrt{2}) \sin 45$$

$$k_{21} = 0$$

Total horizontal displacement along D_1

$$k_{11} = D_1 + D_2 \cos 45 + D_4 \cos 45 + D_5$$

$$k_{11} = 1 + \left(\frac{1}{\sqrt{2}}\right)\left(\frac{1}{\sqrt{2}}\right) + \left(\frac{1}{\sqrt{2}}\right)\left(\frac{1}{\sqrt{2}}\right) + 1$$

$$k_{11} = 3AE/l$$

The displacements along D_1 and D_2 are calculated based on the assumed directions. If (AE/l) is different for each member, the corresponding (AE/l) shall be multiplied with corresponding displacement to get stiffness coefficients.

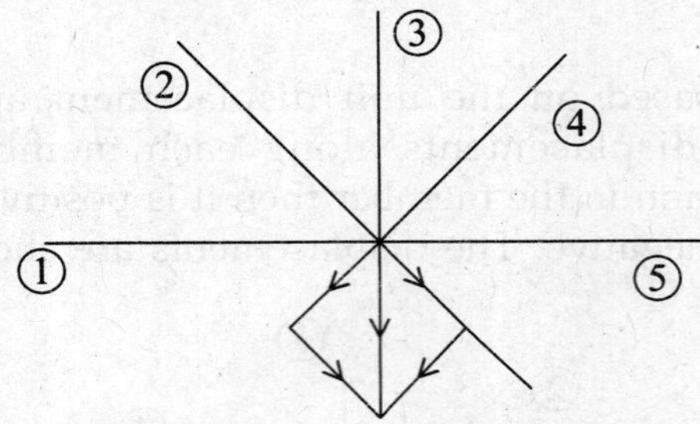

FIG. 5.143 Unit displacement along D_2

Based on the above Fig. 5.143 the displacements along each member is

$D_1 = 0$

$D_2 = 1/\sqrt{2}$ (tension)

$D_3 = 1.0$ (tension)

$D_4 = 1/\sqrt{2}$ (tension)

$D_5 = 0$

Total vertical displacement along D_2

$$k_{22} = D_2 \sin 45 + D_3 + D_4 \sin 45 = 2.0$$

$$k_{22} = \frac{2AE}{l}$$

Total horizontal displacement along D_1

$$k_{12} = D_2 \cos 45 - D_4 \cos 45 = 0$$

Thus, the stiffness matrix is written as

$$[k] = \frac{AE}{l}\begin{bmatrix} 3 & 0 \\ 0 & 2 \end{bmatrix}$$

2. Joint Equilibrium Conditions

$$-P_1 + k_{11}D_1 + k_{12}D_2 = 0$$
$$-P_2 + k_{21}D_1 + k_{22}D_2 = 0$$

$$\begin{bmatrix} k_{11} & k_{12} \\ k_{21} & k_{22} \end{bmatrix}\begin{bmatrix} D_1 \\ D_2 \end{bmatrix} = \begin{bmatrix} P_1 \\ P_2 \end{bmatrix}$$

$$\left\{\frac{AE}{l}\right\}\begin{bmatrix} 3 & 0 \\ 0 & 2 \end{bmatrix}\begin{bmatrix} D_1 \\ D_2 \end{bmatrix} = \begin{bmatrix} 200/\sqrt{2} \\ 200/\sqrt{2} \end{bmatrix}$$

Solving

$$D_1 = 47.14l/AE$$
$$D_2 = 70.71l/AE$$

In pin jointed frames, the loads act at the joints and hence [P] is negative. The member forces are obtained using the matrix $[k_{MD}]$. $[k_{MD}]$ is obtained based on displacements in each member due to D_1 and D_2.

$$[K_{MD}] = \begin{bmatrix} \textcircled{D_1} & \textcircled{D_2} \\ 1.0 & 0 \\ 1/\sqrt{2} & 1/\sqrt{2} \\ 0 & 0 \\ -1/\sqrt{2} & 1/\sqrt{2} \\ -1.0 & 0 \\ 0 & 1/3 \end{bmatrix}$$

The member forces are obtained using the equation.

$$[P_M] = [k_{MD}]\{D\}$$

$$= \frac{AE}{l}\begin{bmatrix} 1 & 0 \\ 1/\sqrt{2} & 1/\sqrt{2} \\ 0 & 1 \\ -1/\sqrt{2} & 1/\sqrt{2} \\ -1 & 0 \end{bmatrix}\frac{1}{AE}\begin{bmatrix} 47.14 \\ 70.71 \end{bmatrix}$$

$$\{P_M\} = \begin{bmatrix} 47.14 \\ 83.33 \\ 70.71 \\ 16.67 \\ -47.14 \end{bmatrix}$$

They are as shown in Fig. 5.144

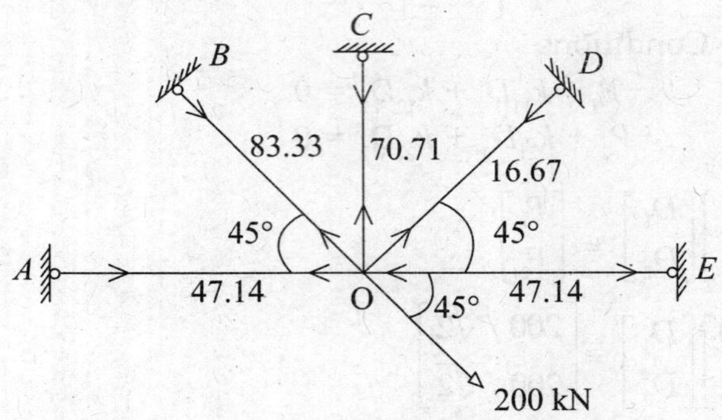

FIG. 5.144

Example 5.26: Analyse the pin jointed frame shown in Fig. 5.145 by the stiffness method. Axial stiffness members are $AE/l)_{OA}$ = 20 kN/mm; $AE/l)_{OB}$ = 60 kN/mm and $AE/l)_{OC}$ = 30 kN/mm.

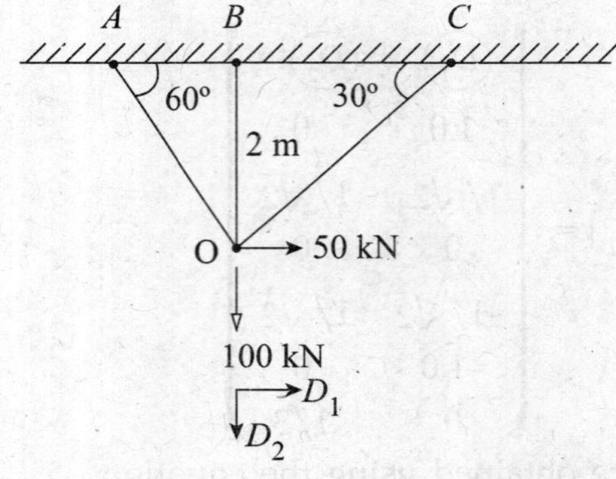

FIG. 5.145

Solution The pin jointed frame has two independent displacements D_1 and D_2 at O.

$$[P] = \begin{bmatrix} 50 \\ 100 \end{bmatrix}$$

1. Stiffness Matrix

The stiffness matrix is based on unit displacement along D_1 and D_2 at O. The displacement along each member is calculated based on the displacement diagram shown below.

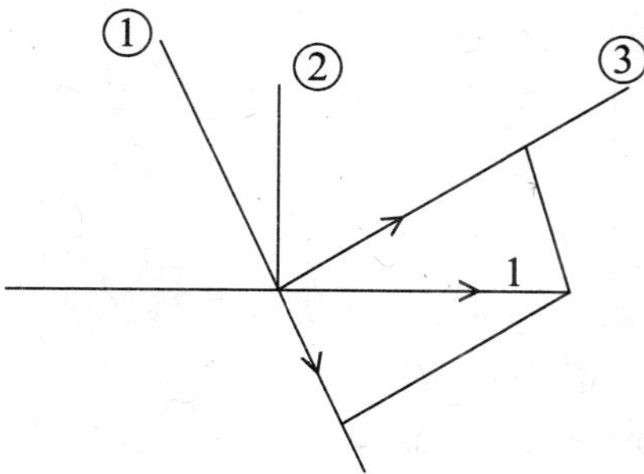

FIG. 5.146 Unit displacement along D_1

The displacement along each member is

$$D_1 = 0.5 \text{ (tension)}$$

$$D_2 = 0.0$$

$$D_3 = \frac{\sqrt{3}}{2} \text{ (compression)}$$

$$k_{11} = \left.\frac{AE}{l}\right)_{OA} \times 0.5 \times \cos 60 + \left.\frac{AE}{l}\right)_{OC} \times \frac{\sqrt{3}}{2} \times \cos 30 = 27.5 \text{ kN}$$

$$k_{21} = \left.\frac{AE}{l}\right)_{OA} \times 0.5 \times \sin 60 - \left.\frac{AE}{l}\right)_{OC} \times \frac{\sqrt{3}}{2} \times \sin 30 = -4.33 \text{ kN}$$

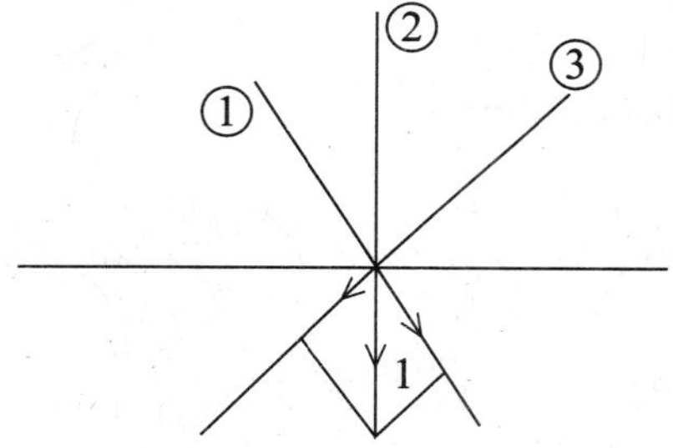

FIG. 5.147 Unit displacement along D_2

Using the above Fig. 5.147 the displacements along each member is

$$D_1 = \frac{\sqrt{3}}{2} \text{ (tension)}$$

$$D_2 = 1.0 \text{ (tension)}$$

$$D_3 = 0.5 \text{ (tension)}$$

Thus,

$$k_{12} = \left.\frac{AE}{l}\right)_{OA} \frac{\sqrt{3}}{2} \cos 60 - \left.\frac{AE}{l}\right)_{OC} \times 0.5 \cos 30 = 27.5 \text{ kN/m}$$

$$k_{22} = \left.\frac{AE}{l}\right)_{OA} \frac{\sqrt{3}}{2} \sin 60 + \left.\frac{AE}{l}\right)_{OB} (1) + \left.\frac{AE}{l}\right)_{OC} \times 0.5 \times \sin 30 = 82.5 \text{ kN/m}$$

The stiffness matrix can be written as

$$[k] = \begin{bmatrix} 27.5 & -4.33 \\ -4.33 & 82.5 \end{bmatrix}$$

2. Joint Equilibrium

$$-P_1 + k_{11}D_1 + k_{12}D_2 = 0$$
$$-P_2 + k_{21}D_1 + k_{22}D_2 = 0$$

Substituting the above computed values

$$-50 + 27.5D_1 - 4.33D_2 = 0$$
$$-100 - 4.33D_1 + 82.5D_2 = 0$$

Solving the above equations;

$$D_1 = 2.026 \text{ mm}$$
$$D_2 = 1.318 \text{ mm}$$

The member forces are obtained as

$$\{P_M\} = [k_{MD}]\,[\delta] = \begin{bmatrix} 10.0 & 17.32 \\ 0 & 60 \\ -25.98 & 15.0 \end{bmatrix} \begin{bmatrix} 2.026 \\ 1.318 \end{bmatrix}$$

$$[P_M] = \begin{bmatrix} 43.09 \\ 79.08 \\ -32.87 \end{bmatrix}$$

The member forces are shown below.

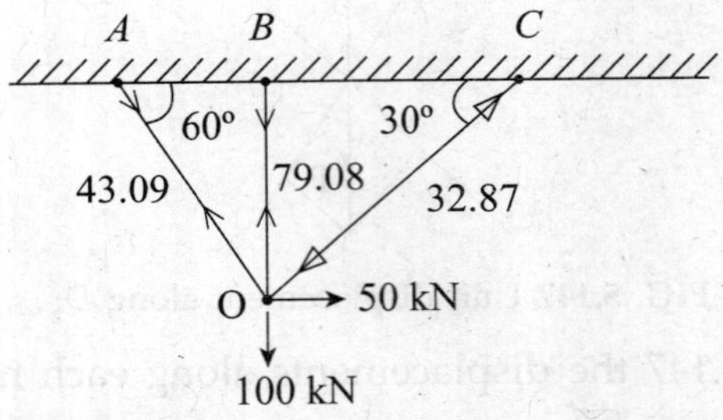

FIG. 5.148

5.15 COMPARISON OF SYSTEM STIFFNESS APPROACH WITH FLEXIBILITY APPROACH

In the flexibility method, the analyst has different choice of redundants. The proper selection of redundants should be able to generate the well conditioned flexibility matrix. The released structure is obtained by releasing the redundants. The number of equations to be solved is equal to the number of redundants. These equations involve the flexibility coefficients. They represent the displacements due to unit value of the redundants. The flexibility coefficients in the flexibility matrix need not be dimensionally homogeneous. This method is preferable for the analysis of trusses, lattice girders and transmission towers.

In stiffness method, the analyst does not have choice of redundants. The restrained structure is obtained by fixing all the supports. The stiffness method has all similarities in approach with the flexibility method. However, the stiffness method has an edge over the flexibility method with respect to simplicity, computational effort and accuracy. While applying the direct stiffness method, we infer that the stiffness matrix is a banded matrix; on the other hand, the elements in the flexibility matrix is sparsely populated. The hand calculations do not give an appreciation for the stiffness method. However, application of computer oriented direct stiffness method for complex structures reveals the choice of the same over the flexibility method.

The basic fact is that matrix method of structural analysis has been developed for large structures which are complex in nature. As the analysis of such structures require more time, computer oriented direct stiffness method is sought.

REVIEW QUESTIONS

Remembrance

5.1. Mention other names of the stiffness method?

5.2. Name the unknowns to be determined in the stiffness method?

5.3. What is kinematic indeterminacy?

5.4. Define stiffness coefficient?

5.5. Generate the stiffness matrix for a cantilever beam subjected to unit load at the free end?

5.6. What is direct stiffness coefficient and cross stiffness coefficient?

5.7. What is the unit for linear stiffness coefficient?

5.8. What is the unit for angularity stiffness coefficient?

5.9. Does the principle of superposition used in the stiffness method, if so justify.

5.10. How temperature variation, prestrain are taken into account in stiffness method.

5.11. How many degrees of freedom are there in a truss joint?

5.12. What is degrees of freedom and write the relationship between degrees of freedom and number of independent displacements?

5.13. The unknowns to be determined in stiffness method are: (a) forces, (b) stresses, (c) strains and (d) deformations.

5.14. The stiffness matrix is banded and symmetrical. Is it true?

5.15. What type of deformations de we consider in the analysis of trusses? Is it flexural /torsional or axial?

5.16. Stiffness method is preferable for
(i) low degree of static indeterminacy and low degree of kinematic indeterminacy
(ii) low degree of static indeterminacy and high degree of kinematic indeterminacy
(iii) high degree of static indeterminacy and low degree of kinematic indeterminacy
(iv) high degree of static indeterminacy and high degree of kinematic indeterminacy

5.17. The joint equilibrium equations in the stiffness method is in terms of
(i) stiffness coefficients and joint actions
(ii) stiffness coefficients and equivalent joint loads
(iii) stiffness coefficients and joint displacements
(iv) stiffness coefficients and flexibility coefficients

5.18. Name a kinematic determinate beam.

5.19. List the factors which make the stiffness method preferable?

5.20. How many types of restrained structure possible in stiffness method?

5.21. Write the end forces when a unit rotation is given in clockwise direction of a fixed beam.

5.22. Write the end forces developed when a unit displacement given at the far end of a fixed beam?

5.23. Is the moment distribution method, a stiffness method or flexibility method?

Understanding

5.1. Why stiffness method is suitable for computer programming?

5.2. Why stiffness method is considered to be better than the flexibility method?

5.3. What does the element of stiffness matrix represent?

5.4. A well conditioned matrix is preferable in flexibility method for greater accuracy. Does this statement true for stiffness method?

5.5. Write the equilibrium condition for the beam shown in figure.

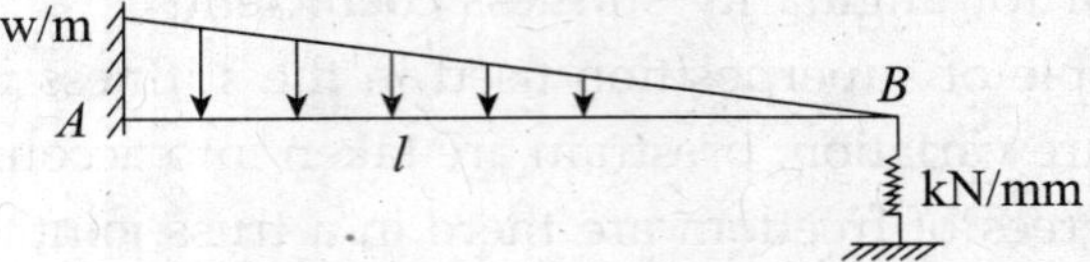

5.6. Identify the kinematic redundants.

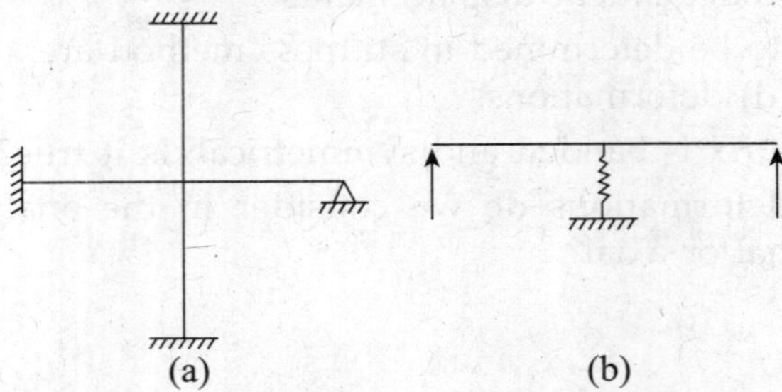

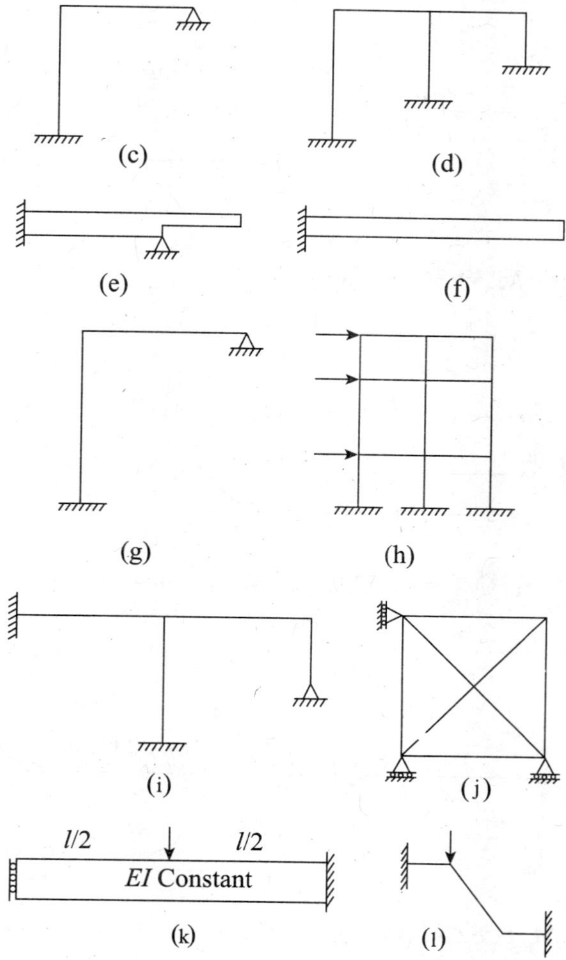

EXERCISE PROBLEMS

5.1 Analyse the given problem and draw the bending moment diagram.

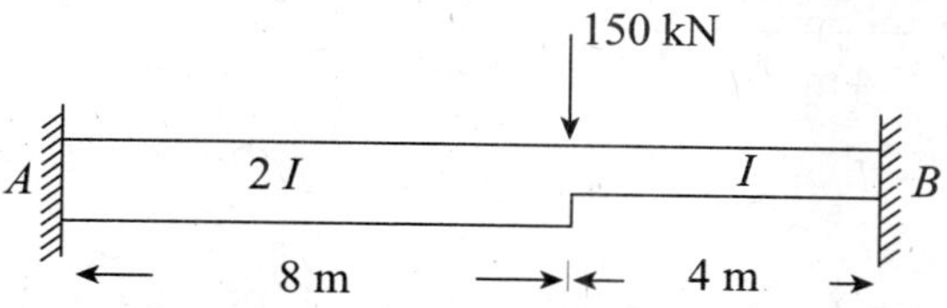

Ans: $M_A = -97.3$ kNm $M_B = +64.9$ kNm

5.2

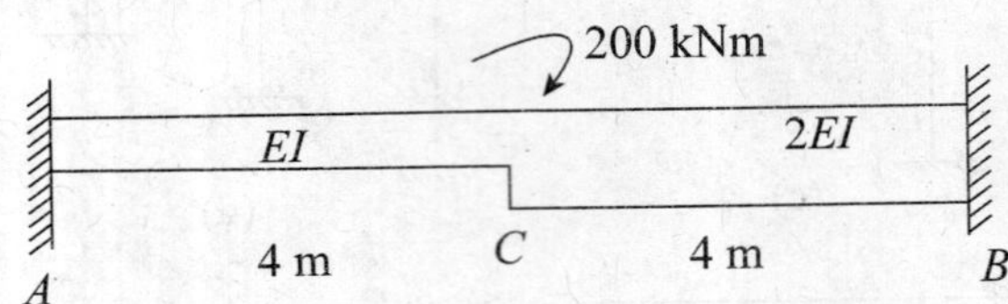

Ans: M_A = 54.54 kNm M_B = 36.36 kNm

5.3

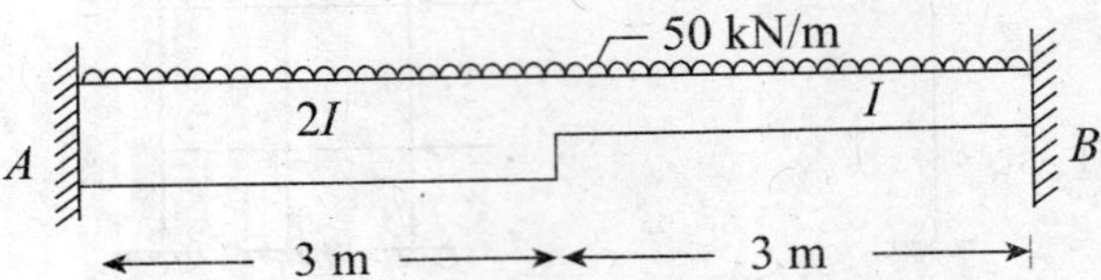

Ans: M_A = −173.86 kNm M_B = +132.95 kNm

5.4

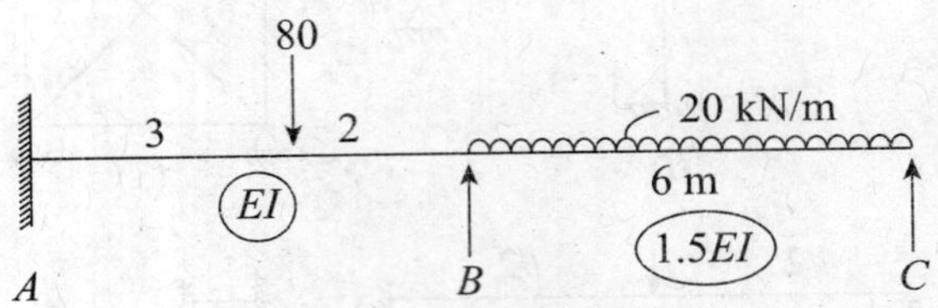

Ans: M_A = 26 kNm, M_B = 82 kNm, M_C = 30 kNm

5.5

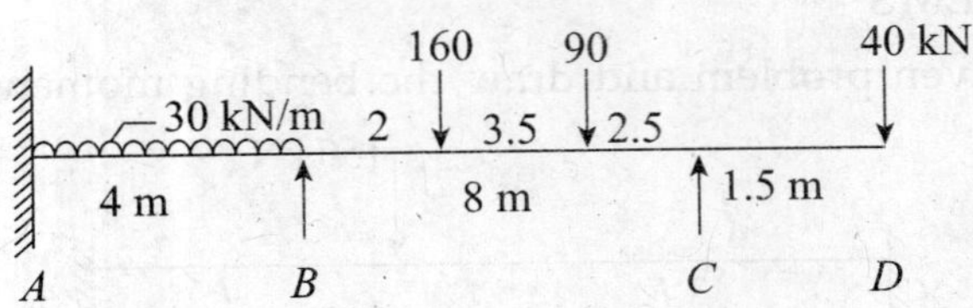

I_{AB} = 170(10^6) mm^4; $I_{BC} = I_{CD}$ = 340 $(10)^6$ mm^4, E = 200 kN/mm^2; support C sinks by 20 mm

Ans: M_A = 48.54 kNm, M_B = 217 kNm, M_C = −60 kNm

5.6

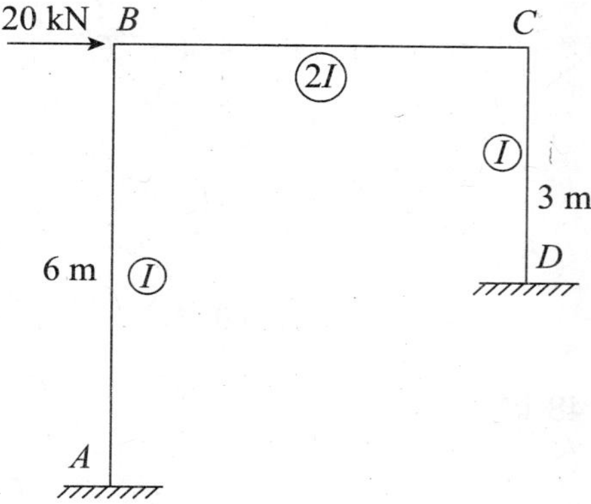

Ans: $M_A = -10$ kNm, $M_B = -10$ kNm, $M_C = -20$ kNm, $M_D = -30$ kNm

5.7

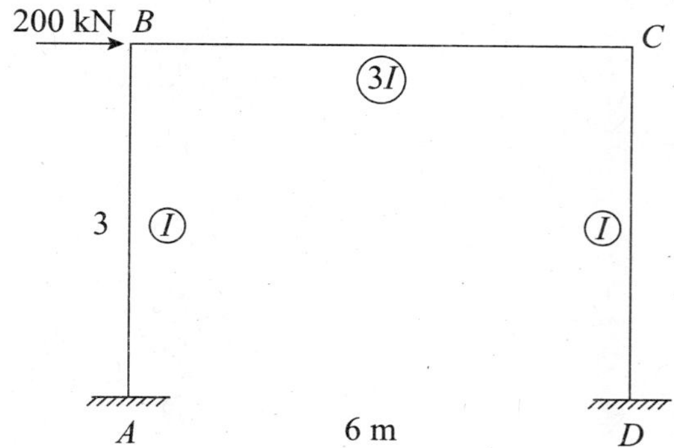

Ans: $M_A = 165$ kNm, $M_B = 135$ kNm, $M_C = 135$ kNm, $M_D = 165$ kNm

5.8

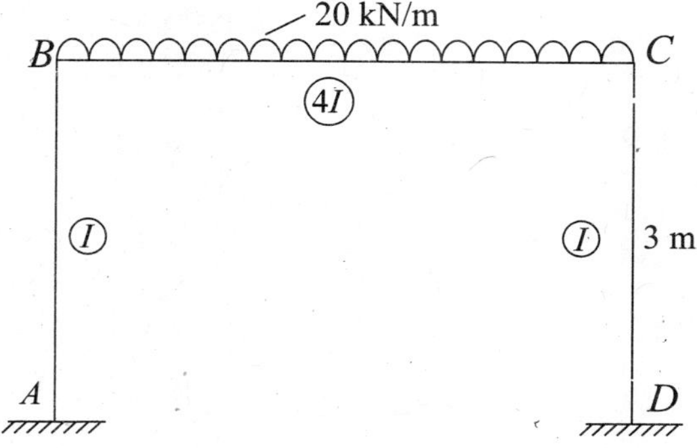

Ans: $M_A = 15$ kNm, $M_B = 30$ kNm, $M_C = -30$ kNm, $M_D = 15$ kNm

5.9

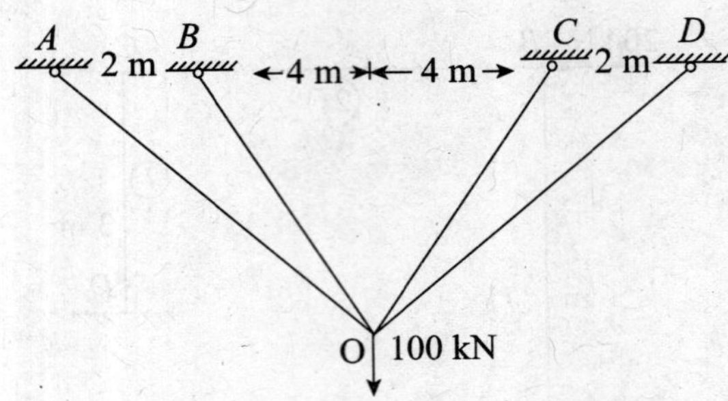

Ans: F_{OA} = 29 kN, F_{OB} = 48 kN

6 Approximate Method of Analysis

Objectives:

Define approximate method – Analysis of propped cantilever – Fixed and continuous beams – Analysis of portal frames subjected to lateral loads – Portal method – Cantilever method and factor method – Analysis of Multistorey frames subjected to vertical loads – Analysis of mill bents – Analysis of indeterminate trusses.

6.1 INTRODUCTION

The finite element method is being widely used design and is a part of the curriculum for graduate studies. However, the practising engineer requires an important tool so that he can readily check the results of the computer packages.

Several investigators—Sutherland and Bowman (1950), Norris (1976), White et al. (1976), Dewolf (1981), Epstein (1988), Behr et al. (1989) have contributed to the development of the approximate method of the analysis of frames subjected to vertical loads.

In practice, we come across multibay, multistoreys, and frames which are highly indeterminate. The structural analysis of such frames is based on the geometry and the material properties. In the exact analysis of indeterminate structures, we satisfy the equilibrium, boundary and compatibility conditions. In the approximate method of analysis, we do not satisfy the compatibility conditions. The relative values of moment of inertia, material properties are not used in the approximate analysis. The statically indeterminate structure is made statically determinate by using independent assumptions. One of the techniques is to introduce the internal hinges.

Consider a propped cantilever beam. The initial step is to determine the statical indeterminacy. The number of unknown reactions in case of propped cantilever is four, i.e., three at the fixed end and one is the prop reaction. The equilibrium equations are three. Hence, the static degree of indeterminacy is $(4 - 3) = 1$. Thus, by introducing a hinge, we can convert it to a statically determinate beam. The location of internal hinges is the next step. This location depends on the type of loading and type of supports.

The type of loading could be concentrated/point load, uniformly distributed load and moments. The type of support includes simple support and fixed support. The internal hinges/inflexion points can also be located by drawing the deflected profile. The location of inflection points/internal hinges for standard cases are given by Epstein (1988).

Table 6.1 Location of interior hinges (k)

S. No.	*Beam Types*	*Loading*		
1	kl, kl, l	0.25	0.21	0.33
2	kl	0.27	0.25	0.33
3	l, l, kl, kl	0.15	0.12	0.22
4	kl, kl	0.16	0.15	0.22

Location of inflection points

Consider a beam AB simply supported at the ends.

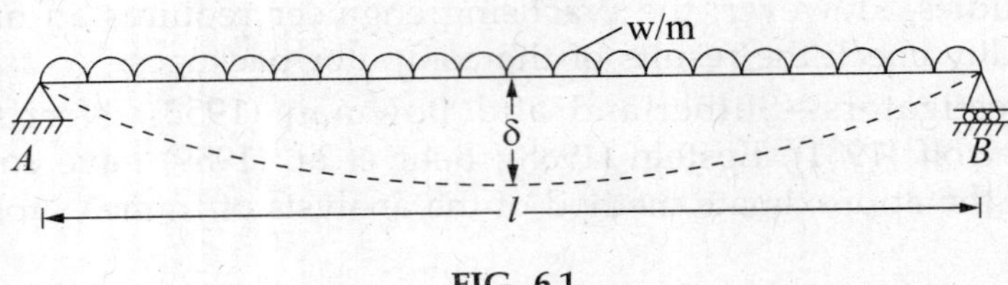

FIG. 6.1

As the ends are simply supported, $M_A = M_B = 0$, i.e., zero bending moments at the simply supported ends. The deflected profile is shown in dotted line. Now, fix the ends A and B respectively. Then the slopes at the ends A and B are zero. The points of zero bending moments are shifted at a suitable distance from the fixed end.

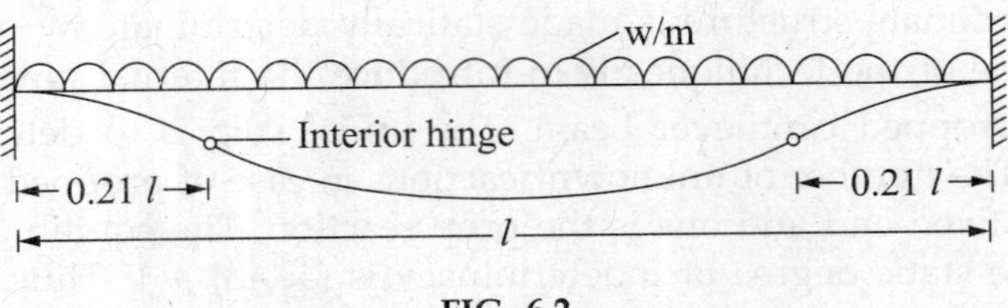

FIG. 6.2

Its location $0.21l$ can be computed from the moment area method. In real situation, in case of multistorey buildings, the ends of the beam is neither simply supported nor restrained.

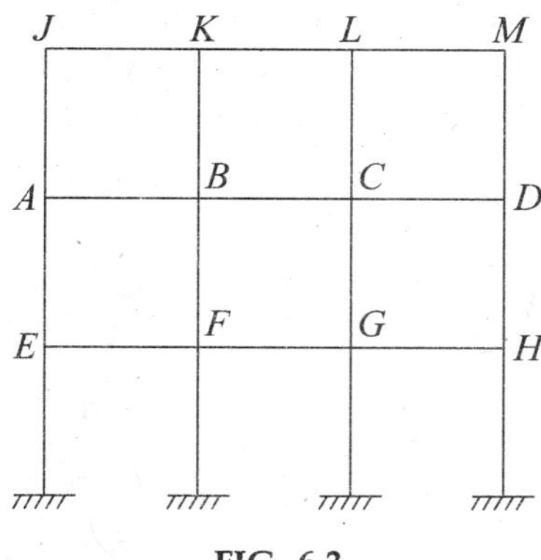

FIG. 6.3

The beams are elastically restrained. Consider the beam *ABCD* at the second level of the multistorey building shown above. The ends of the beam *AB*, i.e., *A* is restrained against rotation by the columns *AE* and *AJ* respectively. Hence, for such beams which are elastically restrained/partially restrained the location of inflection points varies. The location of interior hinges/inflection points is greater than zero and less than 0.21 times the span of the beam. Many textbooks specify the distance as one-tenth of the span ($l/10$) for partially restrained beams subjected to uniformly distributed load. A few examples are presented in this chapter, to illustrate the application of the above methods. It should be noted that the location of inflection point depends on the type of boundary conditions and kind of loadings. Certain guidelines were given by Epstein.

6.2 INDETERMINATE BEAMS

6.2.1 Propped Cantilever Beams

Example 6.1 Compute the end moments of a propped cantilever beam shown in Fig. 6.1. Use the approximate method.

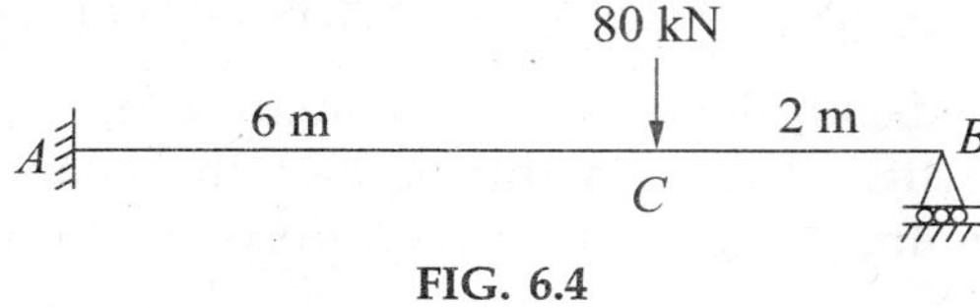

FIG. 6.4

Solution: The inflection point for an eccentrically loaded propped cantilever is given by the formula as

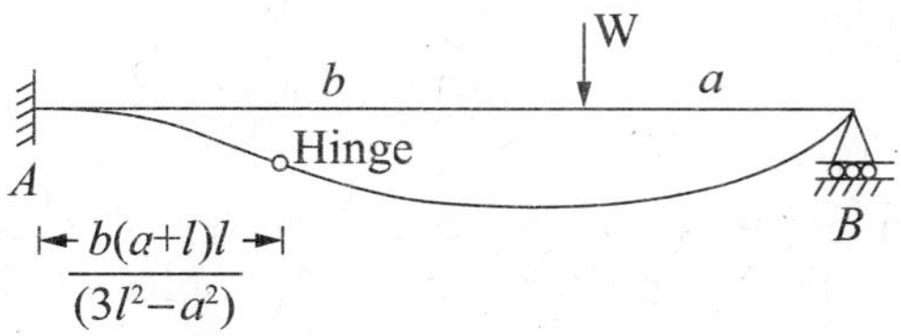

FIG. 6.5 Location of inflection point

Hence, the inflection point is

$$kl = \frac{6(2 + 8)8}{3 \times 8^2 - 2^2}$$

$$\boxed{kl = 2.55 \text{ m}}$$

Thus, the free body diagram is as follows:

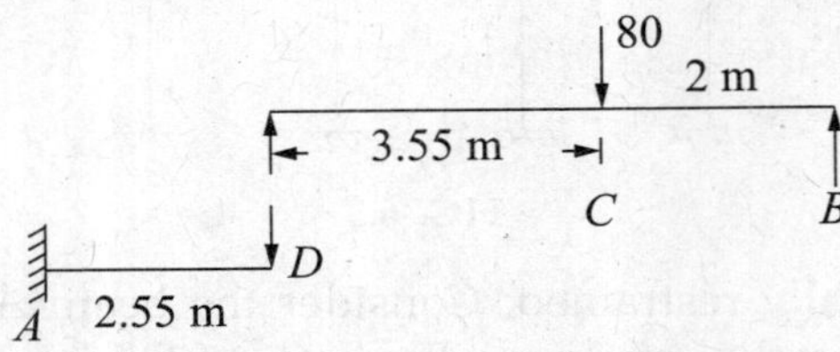

FIG. 6.6 Free body diagram of beam *AB*

Considering the free body diagram of *DCB*

$$V_D = \frac{80 \times 2}{5.55}$$

$$V_D = 28.83 \text{ kN}$$

Using the free body diagram of *AD*:

$$M_A = 2.55 \times V_D$$

$$\boxed{M_A = 73.52 \text{ kNm}}$$

The exact value of M_A from consistent deformation method is 74.4 kNm. Hence, the approximate method is able to predict about 99 per cent of the exact value.

Example 6.2 A propped cantilever beam of span 4 m is subjected to uniformly distributed load of intensity 40 kN/m throughout the span. Compute the propped reaction and the indeterminate moment by the approximate method.

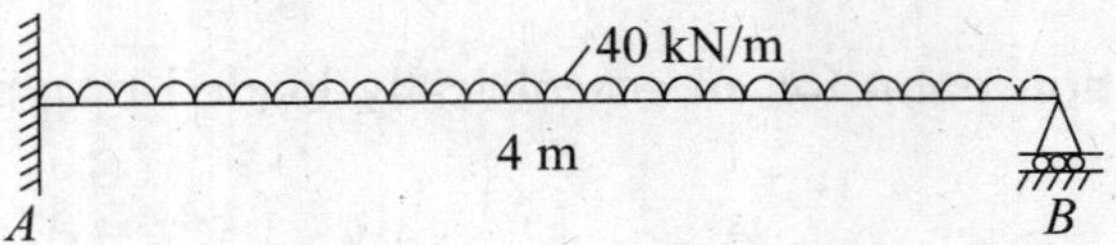

FIG. 6.7

Solution: The inflection point for a propped cantilever beam subjected to uniformly distributed is 0.25 times of the span (Ref. Table 6.1, S. No. 2). By introducing an internal hinge 0.25 × 4 = 1.0 m from the fixed end the resulting free body diagram is shown below;

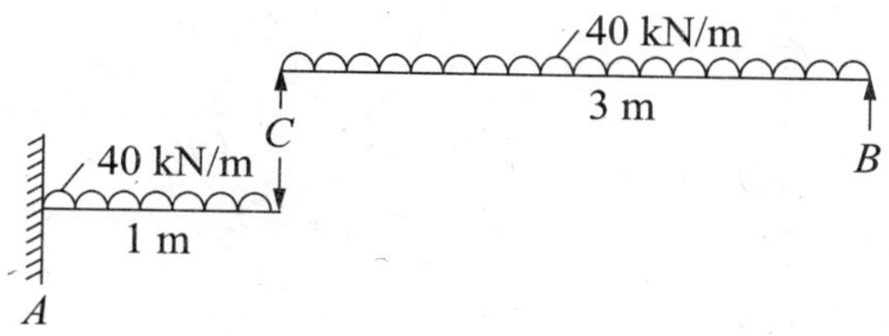

FIG. 6.8 Free body diagram

Considering the free body diagram of *CB*:

$$V_C = V_B = \frac{40 \times 3}{2} = 60 \text{ kN}$$

The indeterminate moment M_A is found out by considering the free body diagram of cantilever *AC*.

$$M_A = 60(1) + \frac{40\,(1)^2}{2} = 80 \text{ kN}$$

The propped reaction fairly agrees with those values obtained from consistent deformation method (Ref. Ex 9.4, Basic Structural Analysis)

Reactions	*Approx. Method*	*Consistent Deformation Method*
V_B (kN)	60	59.9
M_A (kNm)	80	80.4

6.2.2 Fixed Beams

Example 6.3 Determine the end moments of the fixed beam shown in Fig. 6.9 by the approximate method.

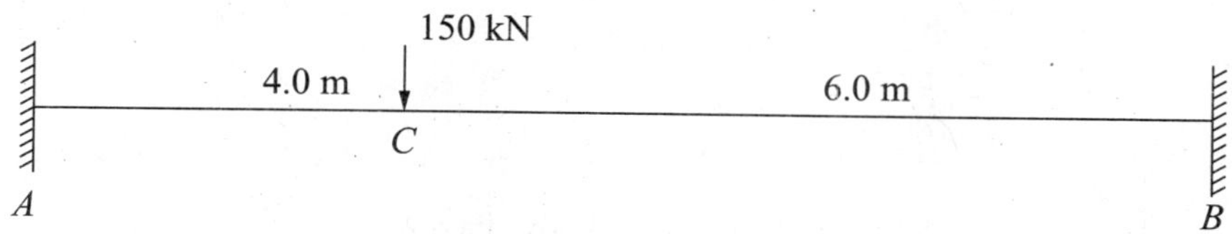

FIG. 6.9

Solution: The fixed beam under vertical loads has two degrees of statical indeterminacy. Hence, two releases in the form of internal hinges are to be provided. The inflection points/interior hinges are computed using the formula given below.

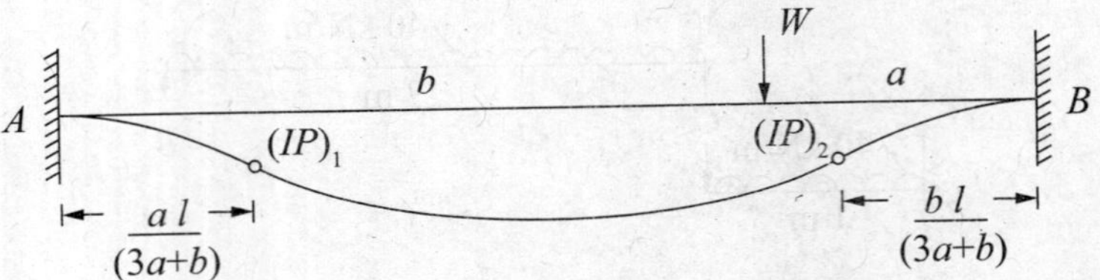

FIG. 6.10 Location of inflection points in a fixed beam

The inflection points $(IP)_1$ and $(IP)_2$ are computed as

$$(IP)_1 = \left(\frac{a}{3a + b}\right) l$$

$$(IP)_1 = \left(\frac{6}{3(6) + 4}\right) l$$

$$(IP)_1 = 0.27l = 2.7 \text{ m}$$

$$(IP)_2 = \frac{bl}{(3a + b)}$$

$$(IP)_2 = \left(\frac{4}{3(6) + 4}\right) l$$

$$(IP)_2 = 0.18l = 1.8 \text{ m}$$

Hence, the free body diagram with inflection points is drawn as follows.

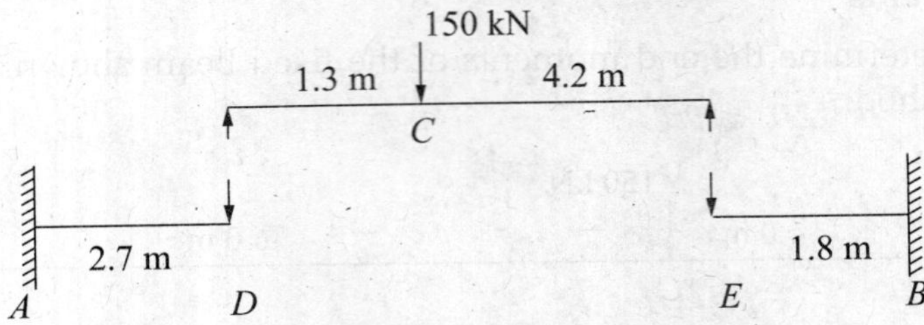

FIG. 6.11 Free body diagram

Considering the free body diagram of *DCE*:

$$V_D = \frac{150 \times 4.2}{5.5}$$

$$V_D = 114.55 \text{ kN}$$

$$\therefore \quad V_E = 35.45 \text{ kN}$$

Considering the cantilevers AD and EB,

$$M_A = 2.7\ V_D = 309 \text{ kNm}$$

$$M_B = 1.8\ V_E = 64 \text{ kNm}$$

The standard beam formulae using elastic analysis give

$$M_A = \frac{Wab^2}{l^2} = 216 \text{ kNm}$$

$$M_B = \frac{Wba^2}{l^2} = 144 \text{ kNm}$$

The above comparison shows that for a fixed beam subjected to eccentric loading requires a good approximation of inflection points. The reader gets better results if an initial guess at a distance of $0.25l$ from the fixed end.

Example 6.4 For a rigidly fixed beam of span 5 metres carrying a udl of 10 kN/m over the entire span, locate the point of contraflexure (inflection points) and analyse the beam.

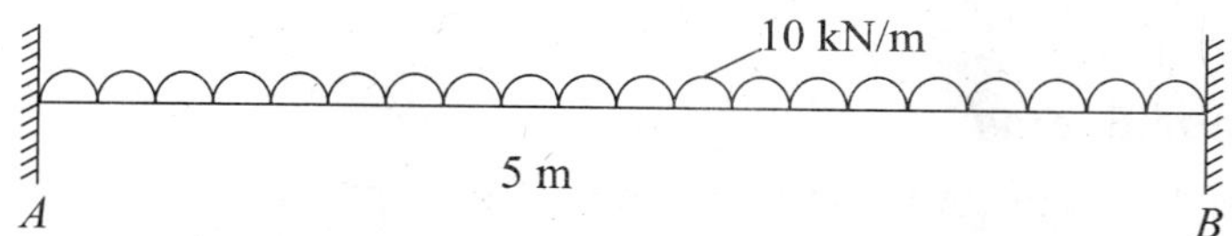

FIG. 6.12

Solution: The inflection points for a fixed beam subjected to uniformly distributed load is at a distance of 0.21 times the span of the beam from the fixed end. That is, $0.21 \times 5 = 1.05$ metres from the ends.

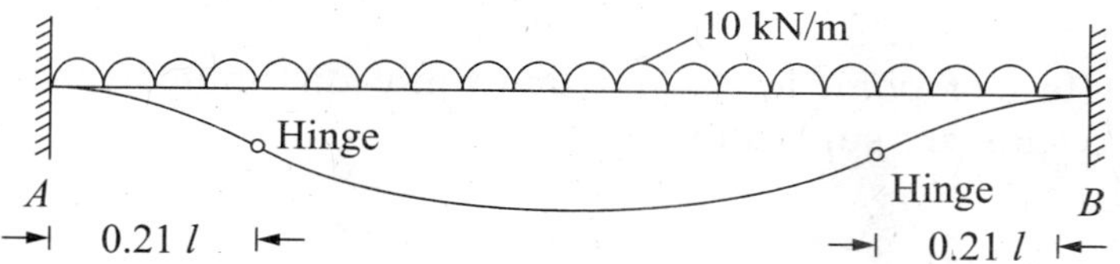

FIG. 6.13 Location of interior hinges in a fixed beam

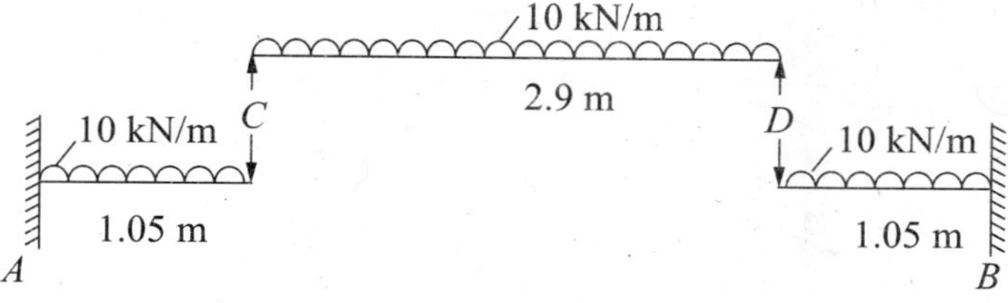

FIG. 6.14 Free body diagram

Considering the simply supported beam CD

$$V_C = V_D = \frac{10(2.9)}{2}$$

$$= 14.5 \text{ kN}$$

This vertical reaction acts down on the cantilever AC. Hence,

$$M_A = 14.5(1.05) + \frac{10(1.05)^2}{2}$$

$$\boxed{M_A = 20.74 \text{ kNm}}$$

The exact moment is

$$M_A = \frac{10(5)^2}{12}$$

$$M_A = 20.83 \text{ kNm}$$

Thus, the approximate method gives the moment values very close to the exact value.

6.2.3 Continuous Beams

Example 6.5 Analyse the continuous beam by the approximate method. EI is constant.

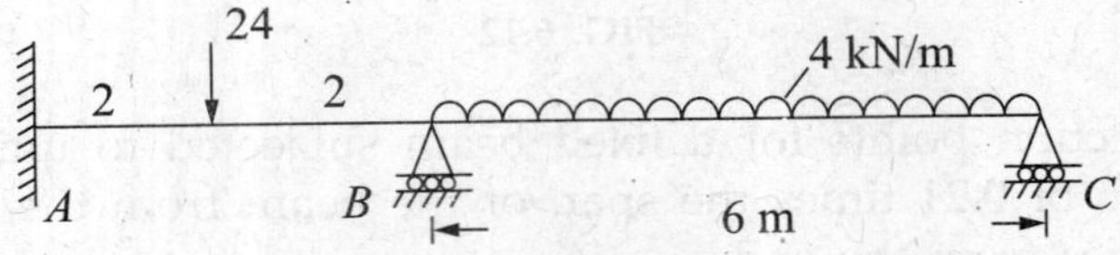

FIG. 6.15

Solution: The effect of point loads and uniformly distributed loads are considered separately. The results are superimposed.

Part I

Consider the beam ABC with point loads only.

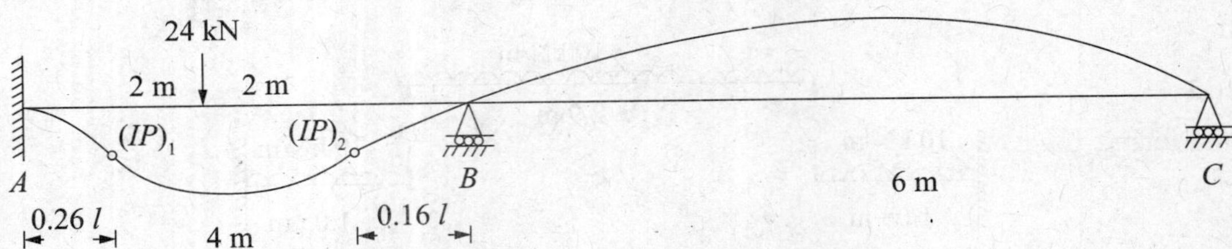

FIG. 6.16

In the beam *ABC*, *A* is a fixed support, *B* is an intermediate support and *C* is on roller.

It is obvious from the basics that there is a fixed end moment at *A*, a hogging bending moment over the intermediate support at *B* and zero moment at *C*. The inflection points in span *AB* can be determined using Table 6.1. The support *B* can be considered partially fixed, i.e., it neither corresponds to Beam Type 1 nor Beam Type 2 in Table 6.1. Therefore, the inflection point can be taken as an average value of $0.25l$ and $0.27l$. Hence, kl for the span *AB* is $0.26l$ from the end *A*. The inflection point from *B* is $0.15l$ from *B*. Finally, the inflection points are $0.26l$ from *A* and $0.15l$ from *B*.

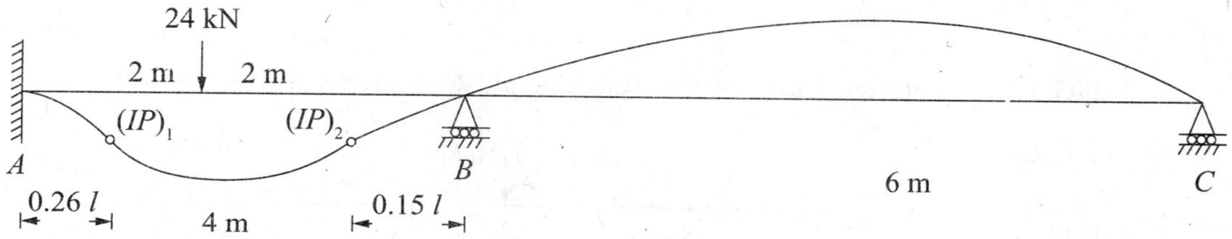

FIG. 6.17 Inflection points for span *AB* due to point load

The reactions at the inflection points can be calculated as follows.

FIG. 6.18

Hence,
$$V_D = \frac{24(0.35l)}{0.59l} = 14.2 \text{ kN}$$

$\therefore$
$$V_E = 24 - 14.2 = 9.8 \text{ kN}$$

The fixed end moments at *A* due to point load and the intermediate moment at *B* are computed from

$$(M_A)_1 = 14.2(0.26)4 = 14.8 \text{ kNm}$$

and
$$(M_B)_1 = -9.8(0.15)4 = -5.9 \text{ kNm}$$

Part 2

Considering the span *DBE*, taking moment about *D*

$$2.96V_B = 10(3.86) + 4(0.9)\left(2.96 + \frac{0.9}{2}\right)$$

$$V_B = 17.19 \text{ kN}$$

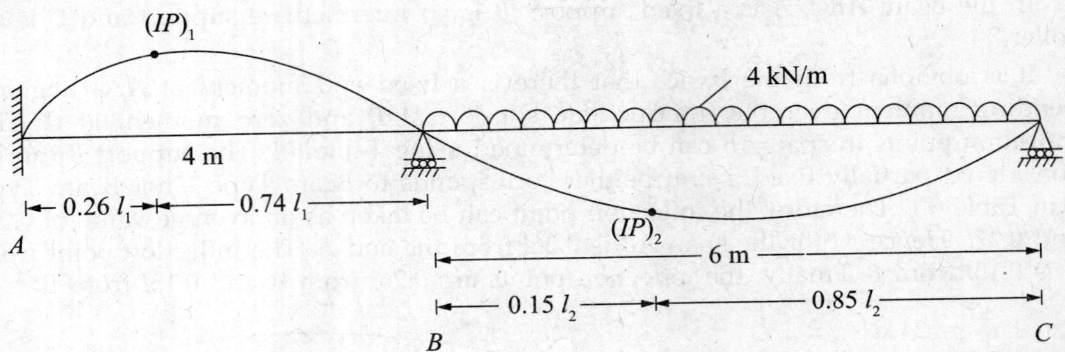

FIG. 6.19 Inflection points for *ABC* due to applied uniformly distributed load

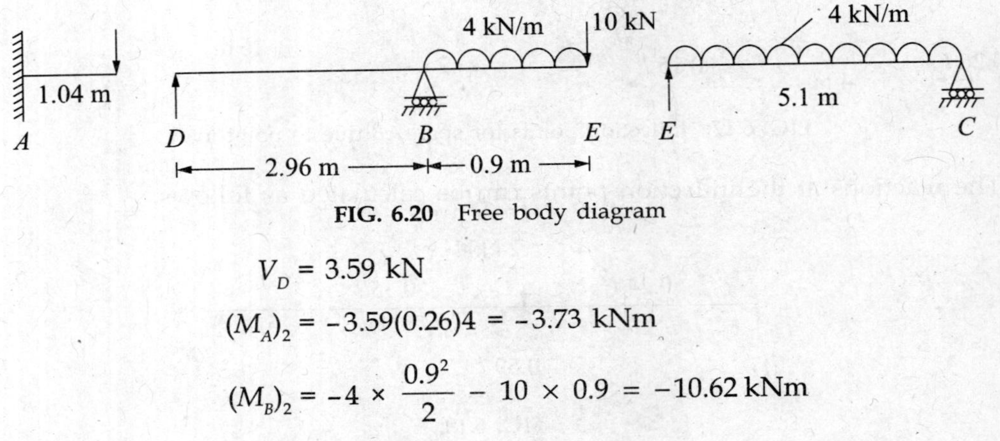

FIG. 6.20 Free body diagram

$$V_D = 3.59 \text{ kN}$$

$$(M_A)_2 = -3.59(0.26)4 = -3.73 \text{ kNm}$$

$$(M_B)_2 = -4 \times \frac{0.9^2}{2} - 10 \times 0.9 = -10.62 \text{ kNm}$$

Superimposing with the results of Part 1

$$M_A = (M_A)_1 + (M_A)_2 = 11.07 \text{ kNm}$$

$$M_B = (M_B)_1 + (M_B)_2 = 16.52 \text{ kNm}$$

Example 6.6 A continuous beam *ABCD* is of uniform section as shown in Fig. 6.21. *EI* is constant. Draw the shear force diagram and the bending moment diagram.

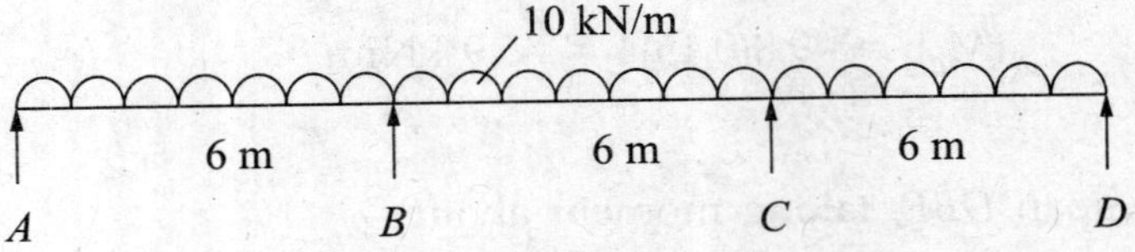

FIG. 6.21

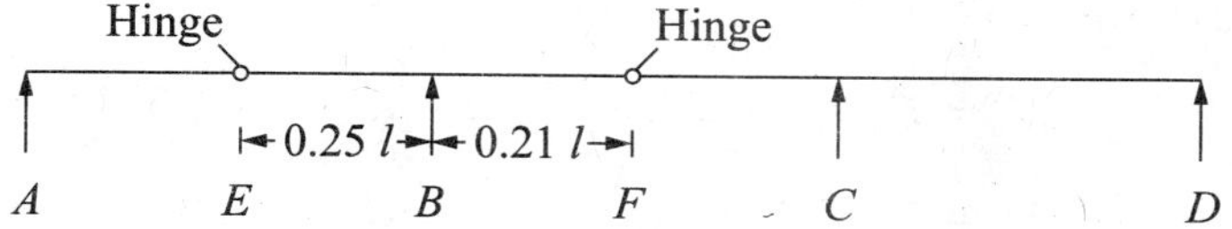

FIG. 6.22 Location of inflection points

Solution: The end span *AB* can be considered a propped cantilever subjected to udl. The prop is at *A* and the fixed support being *B*. Hence, the inflection point is located at 0.25*l* from *B*. (Ref. Table 6.1, Beam Type 2).

The interior span *BC* is considered a fixed beam due to symmetry. The hinge in span *BC* is located at a distance of 0.21*l*, i.e., 1.26 metres from the support *B*.

The free body diagram is drawn with the introduction of internal hinges as

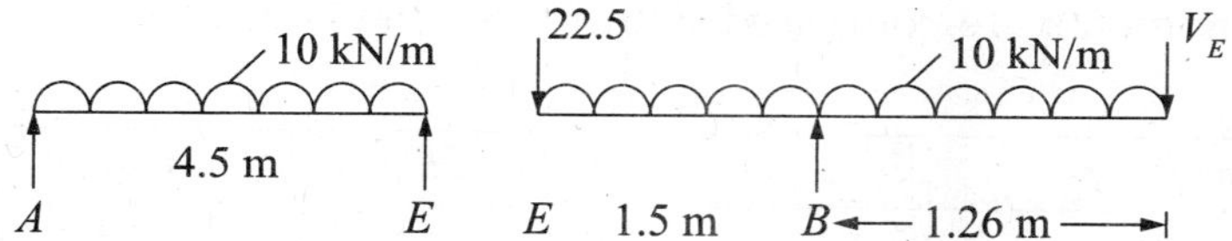

FIG. 6.23 Free body diagram of the continuous beam

Considering the beam *AE*

$$V_A = V_E = \frac{\text{Total load}}{2}$$

$$V_A = V_E = \frac{10 \times 4.5}{2}$$

$$= 22.5 \text{ kN}$$

Consider the span *EBF*

$$\Sigma V = 0;$$

$$22.5 + V_F + 10(2.76) = V_B$$

$$V_B - V_F = 50.1 \qquad (1)$$

$$\Sigma M_E = 0$$

$$1.5V_B = 10\frac{(1.5 + 1.26)^2}{2} + 2.76V_F$$

$$1.5V_B = 38.09 + 2.76VF$$

$$1.5V_B - 2.76V_F = 38.09 \qquad (2)$$

Solving Eqs. (1) and (2)

$$V_B = 79.50 \text{ kN}$$

$$V_E = 29.40 \text{ kN}$$

$$M_B = 22.5(1.5) + 10\frac{(1.5)^2}{2} = 45 \text{ kNm}$$

The positive bending moment is span AB is calculated as

$$M_G = 22.5(3) - \frac{10 \times 3^2}{2}$$

$$M_G = 22.5 \text{ kNm}$$

The above moments are compared with those obtained from the theorem of three moments as

Reactions	*Approximate Method*	*Theorem of Three Moments*
M_B (kNm)	45.00	36.00
M_G (kNm)	22.50	28.80

Example 6.7 Compute the moments at the intermediate support B and C by the approximate method.

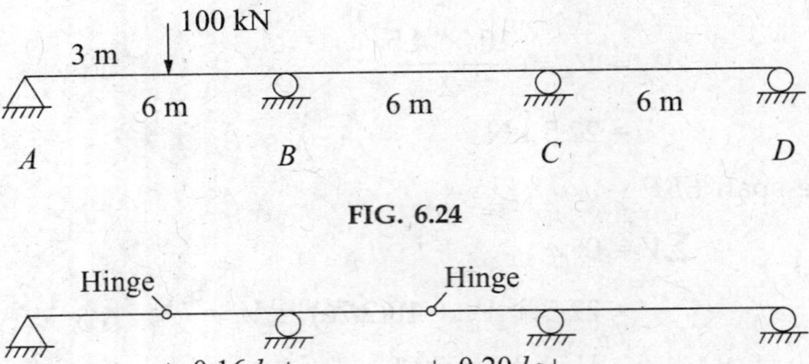

FIG. 6.24

FIG. 6.25

Solution: The location of the hinge in span AB is found out by referring to Table 6.1 (Beam Type 4). Similarly, the location of the hinge in span BC is obtained by taking the average value of Beam Type 2 and Beam Type 3, i.e., $(0.27l + 0.15l)/2 \simeq 0.20l$. Consider the beam AEB and $BE = 0.16(6) \simeq 1$ m. Hence, $AE = 5$ m. Now, consider AE.

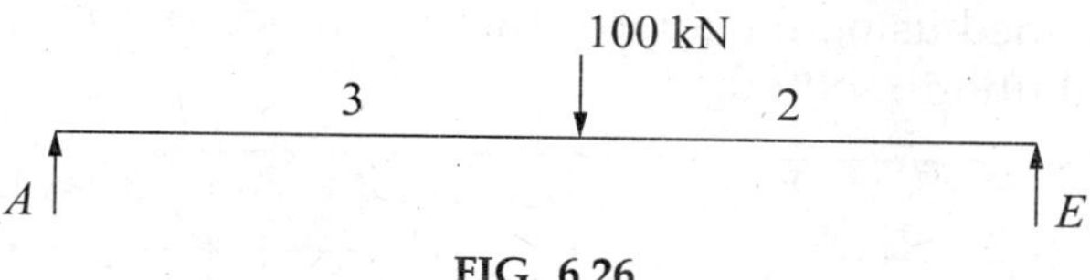

FIG. 6.26

Taking moment about A,

$$5V_E = 100 \times 3$$

$$V_E = 60 \text{ kN}$$

Knowing the value of V_E, consider the free body diagram of EBF

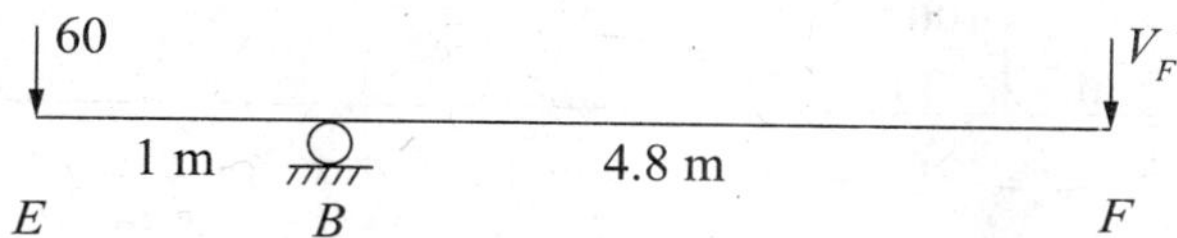

FIG. 6.27

$$\Sigma M_E = 0;$$

$$5.8V_F = 1V_B$$

$$5.8V_F - V_B = 0$$

$$\Sigma V = 0;$$

$$-V_F + V_B = 60$$

Solving the above equations,

$$V_F = 12.5 \text{ kN} \quad \text{and}$$

$$V_B = 72.5 \text{ kN}$$

Now the value of M_B is obtained as

$$M_B = V_E \times E_B$$

$$M_B = 60 \times 1 = 60 \text{ kNm}$$

and

$$M_C = V_F \times F_C$$

$$= -12.5 \times 1.2$$

$$= -15 \text{ kNm}$$

The moments obtained using the approximate method are compared with the results of the moment distribution method.

Reactions	*Approximate Method*	*Moment Distribution Method*
MB	+60.00	+59.82
MC	–15.00	–15.00

The above results illustrate that if the locations of the beam inflection points are accurate, the moments in the beam will also be accurate. The deflected profile of the beam is shown below.

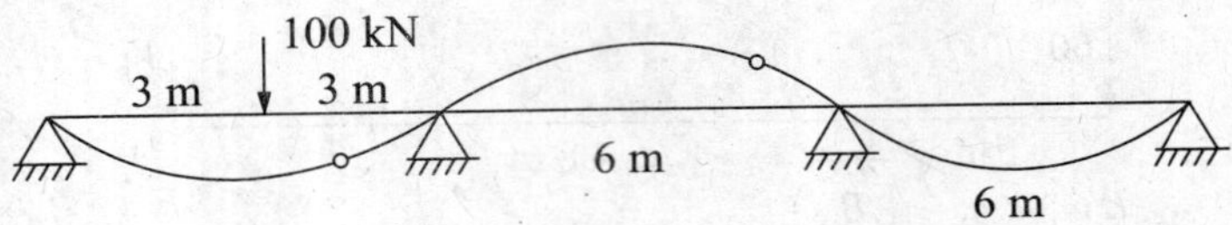

FIG. 6.28 Deflected profile of the loaded beam

6.3 PORTAL FRAMES AND RECTILINEAR FRAMES

The approximate method developed in earlier sections for continuous beams can be extended to frames also. These frames consist of beams and columns rigidly connected by monolithic joints. They are indeterminate structures in general and the moments, shear force and axial force can be estimated only if the material properties (E = modulus of elasticity of the material) and sectional properties (I = second moment of the area of the section) are known. The following assumptions are made in analysing the frames.

(1) Axial deformations of the members are neglected.

(2) After loading, the rigid joints (where beams and columns are connected) maintain the angles between the members. This is essential to draw the deflected shape of the structure.

(3) While drawing the deflected shape of the structure, one point of inflection/notional hinge/point of contraflexure is to be provided in an unloaded member.

Thus, following the above guidelines the indeterminate frame is made as determinate frame by suitably providing the internal hinges. It is to be noted that once the location of internal hinges/inflection points are located, the moments in beams and columns without sway can be calculated using statics. The unsymmetrical frames/frames subjected to sway require an additional assumption on shear or axial force distribution. A few examples with appropriate location of hinges are given below.

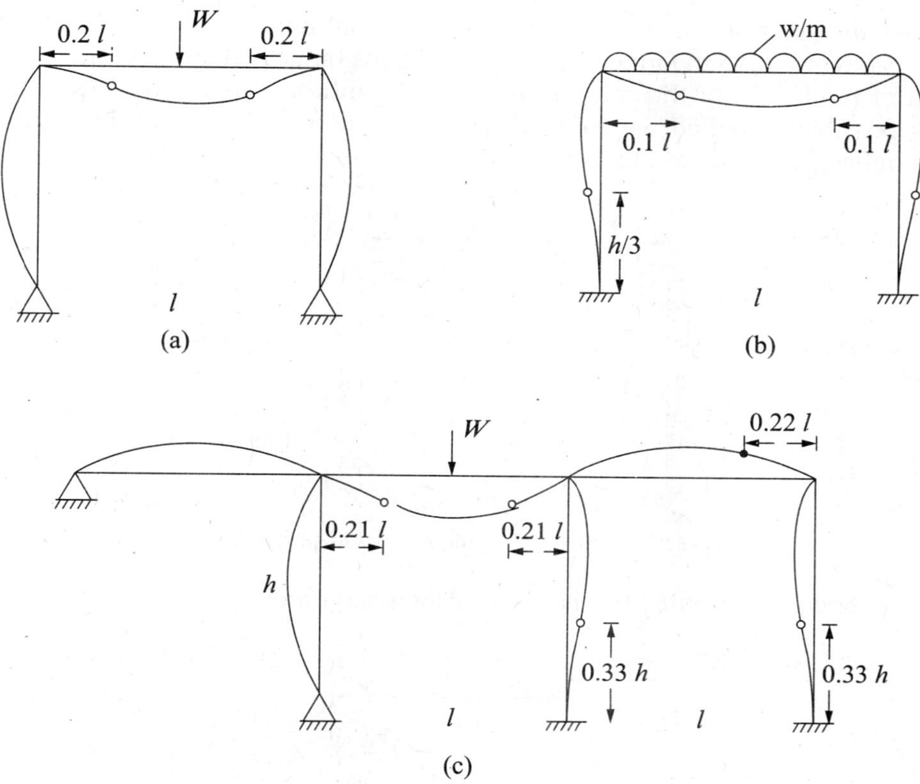

FIG. 6.29 Hinges in rigid frames

Example 6.8 Analyse the given frame by the approximate method.

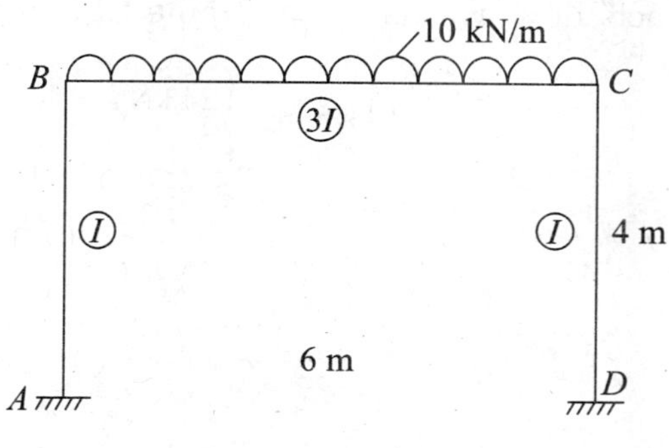

FIG. 6.30

Solution: A single bay, single storey rigid jointed portal frame having one axis of symmetry is shown above. The columns *AB* and *CD* are rigidly connected by the beam *BC*. The base of the columns are fixed. The inflection points/interior hinges are located at 0.1 times the span on the beam *BC*, i.e., the notional hinges are at 0.6 metres from *B* and *C* respectively. Also the inflection points/interior hinges on the columns are located one-third of the height of the column. It is to be mentioned here the above location of hinges is possible provided we have stiffer beam when compared to the column. The probable deflected profile of the frame along with the internal hinges are shown below.

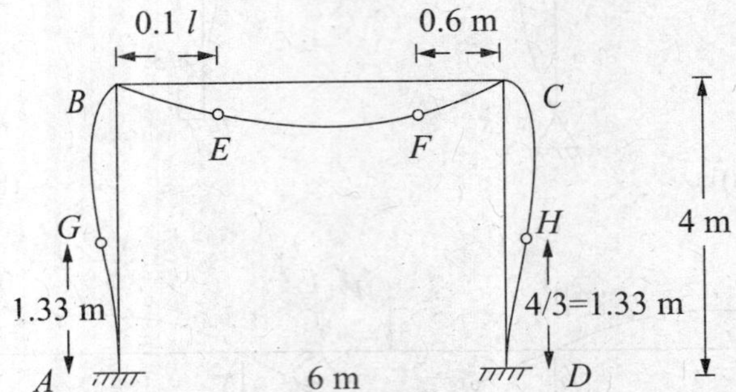

FIG. 6.31 Deflected profile of a fixed portal frame

The free body diagram of the beam is drawn separately.

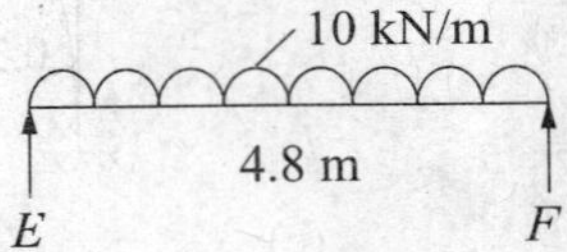

FIG. 6.32 Free body diagram of *EF*

$$V_F = V_E = \frac{10(4.8)}{2} = 24 \text{ kN}$$

The vertical reaction acts downwards on the frame *GBE* and hence

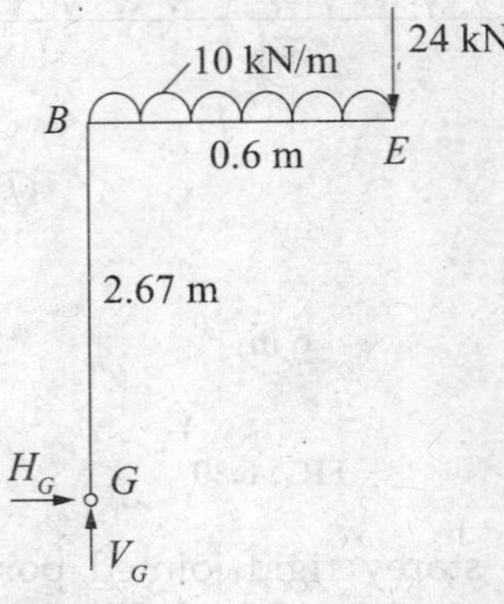

FIG. 6.33 Free body diagram of *GBE*

The horizontal reaction *HG* is evaluated by taking moment about *B*,

$$\Sigma M_B = 0$$

$$2.67\,H_G = 24 \times 0.6 + 10 \times \frac{0.6^2}{2}$$

$$\boxed{HG = 6.07 \text{ kN}}$$

$$\therefore \quad M_B = 6.07 \times GB$$

$$= 16.2 \text{ kNm}$$

$$M_A = 6.07 \times GA$$

$$= 8.0 \text{ kNm}$$

$$\text{Moment at midspan of beam} = 10 \times \frac{4.8^2}{8} = 28.8 \text{ kNm.}$$

The bending moment diagram is drawn using the above values as

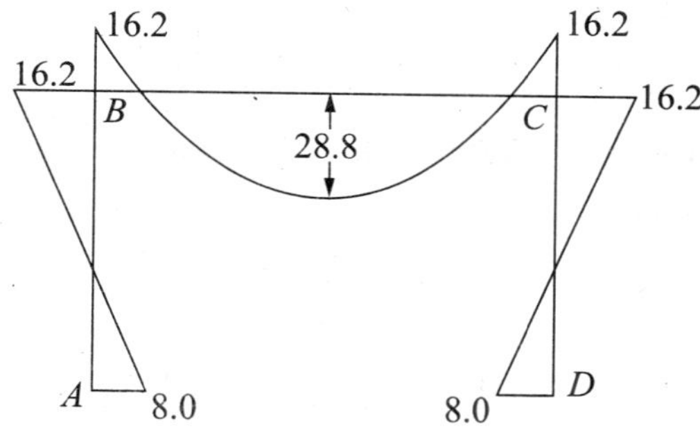

FIG. 6.34 Bending moment diagram

Note: The above bending moment values closely agree with the moments obtained using the moment distribution method. The readers should note that all such frames analysed by the approximate method will not give satisfactory results with the traditional methods, viz., the moment distribution method. The reason being the moment of inertia of beam and columns are not taken into account in approximate method. Therefore, the approximate method can be considered a time saver and the results can only be used for preliminary designs.

6.3.1 Single Fixed Portal Frame Subjected to Lateral Load

Consider a single portal frame which is fixed at the base and subjected to horizontal load (lateral load) at the top. The degree of indeterminacy of such fixed portal frame is three. Hence, to make it statically determinate equal number of releases are to be made. Thus, the three assumptions include the introduction of two hinges in the columns and equal resistance to shear by the columns.

In practice, the column hinges are located at mid height of the column. This assumption is valid if the beam is very rigid when compared to the column. If the beam in the portal frame is very flexible, then the column hinge is to be located near to the top of column. However, it is to be mentioned here that the maximum beam moment and column moment are influenced by the ratio of the beam stiffness to the column stiffness.

The deflected profile of a fixed portal frame having stiff beam under lateral load is shown below. The corresponding bending diagram is also shown. The maximum bending moment on the column and beam is $(Ph/4)$ where P is the horizontal force and h is the height of the column.

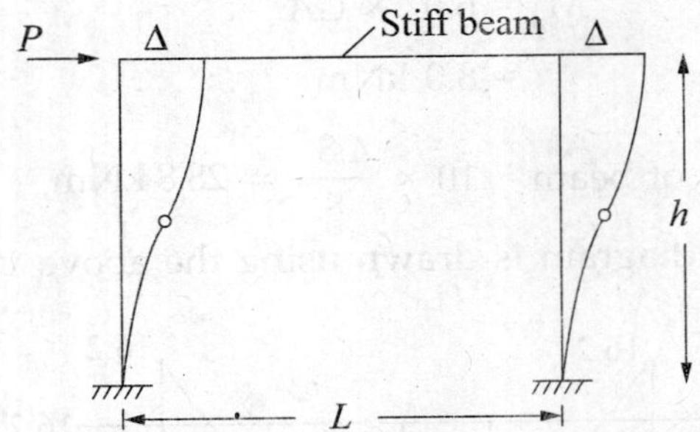

FIG. 6.35 Deflected profile of the fixed portal frame (having a stiff beam)

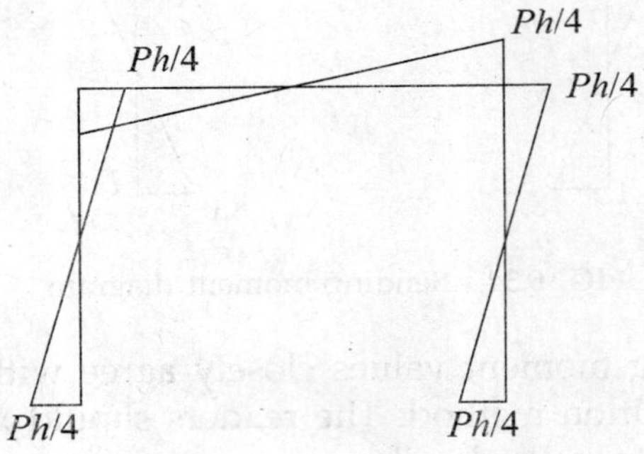

FIG. 6.36 Bending moment diagram

Now consider another fixed portal frame where the beam is highly flexible. It means the beam stiffness is very much less than the column stiffness. The corresponding deflected shape is antisymmetric. The bending moment diagram shows that the maximum bending moment is at the base of the column. Its value being $(Ph/2)$. As the beam is highly flexible, it cannot resist any bending moment.

Thus, the beam flexibility influences the bending moment in the columns as well as the lateral deflection at the top. Denoting the ratio of the beam stiffness to the column stiffness as 'α', i.e.

$$\alpha = \frac{(EI_b/l)}{(EI_c/h)}$$

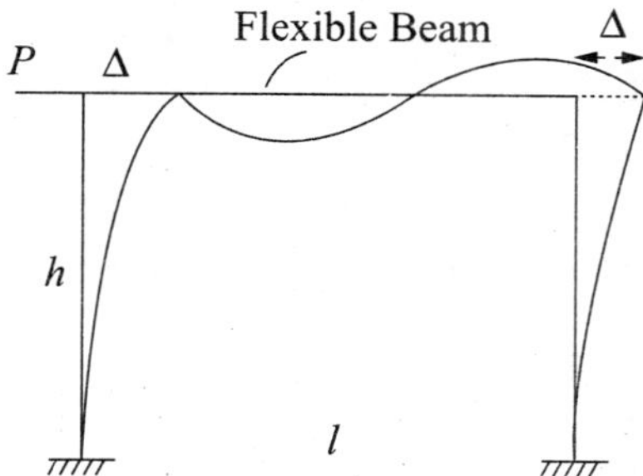

FIG. 6.37 Deflected profile of the fixed portal frame (having a flexible beam)

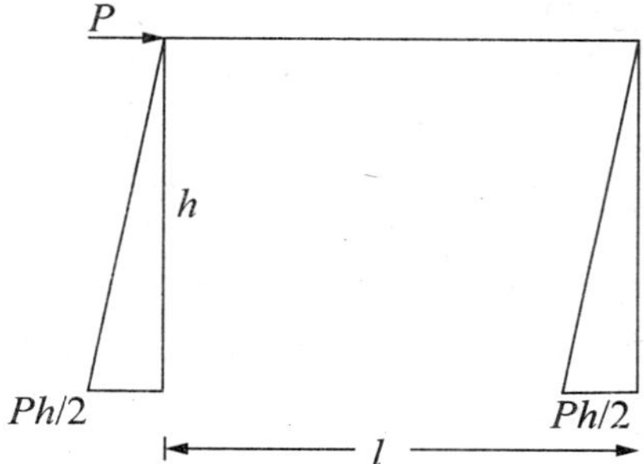

FIG. 6.38 Bending moment diagram

The maximum bending moment of the column and the beam are obtained as follows.

$$M_{c\max} = \frac{Ph}{4}\left(1 \pm \frac{1}{1+6\alpha}\right)$$

$$M_{b\max} = \frac{Ph}{4}\left(1 - \frac{1}{1+6\alpha}\right)$$

Example 6.9 Analyse the portal frame shown in Fig. 6.39 by the approximate method.

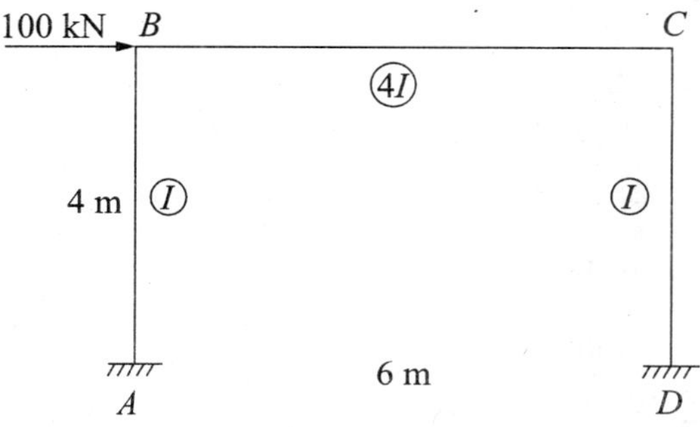

FIG. 6.39

Solution: Assume the hinges at the mid height of the column. The horizontal shear is resisted equally by individual columns. With the above assumptions draw the free body diagram and then analysis made.

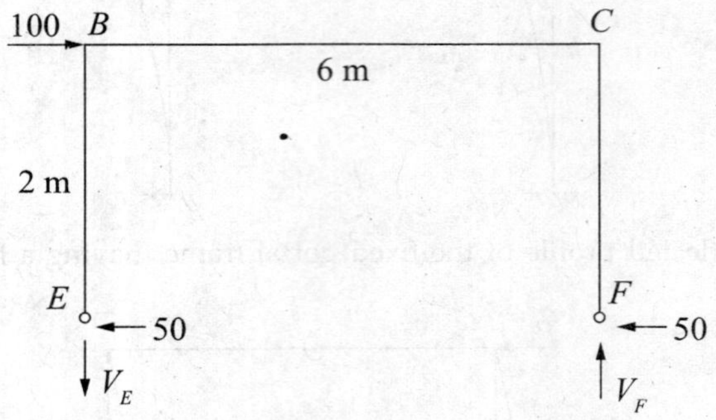

FIG. 6.40 Free body diagram of a determinate frame

Taking moment about E,

$$6V_F = 100 \times 2$$

$$V_F = 33.33 \text{ kN}$$

$$\Sigma V = 0, \quad -V_E + V_F = 0$$

$$\therefore \quad V_E = 33.33 \text{ kN}$$

Now, consider the lower half of the frame,

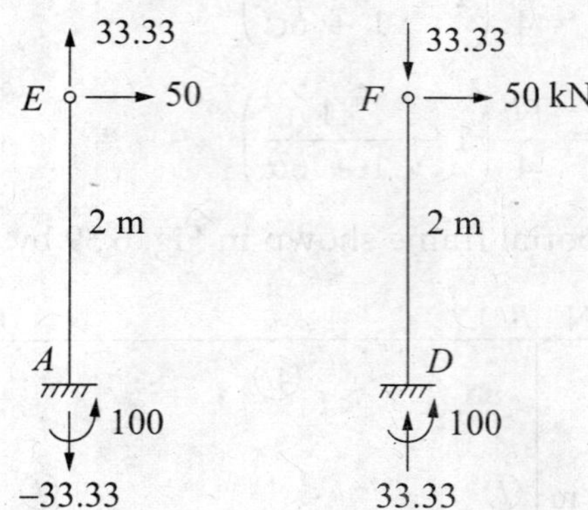

FIG. 6.41 Free body diagram of the lower half of the columns

Using the equilibrium equations

$$\Sigma V = 0;$$

$$+33.33 - V_A = 0$$

$$V_A = 33.33 \text{ kN}$$

$$\Sigma M = 0;$$

$$50(2) - M_A = 0$$

$$M_A = 100 \text{ kNm}$$

The final bending moment diagram is as follows.

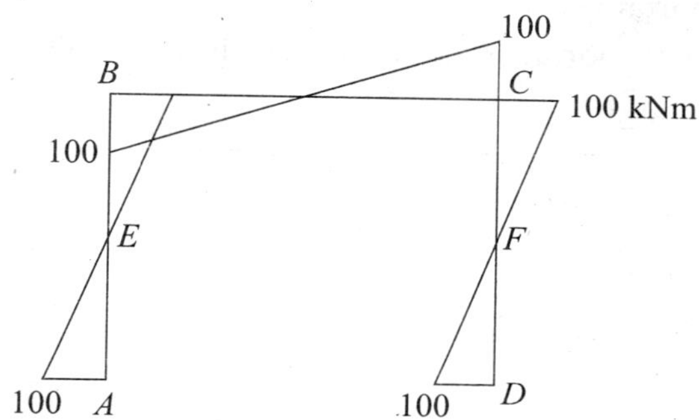

FIG. 6.42 Bending moment diagram

6.4 ANALYSIS OF MULTISTOREY FRAMES FOR LATERAL LOADS

The multistorey frames/building frames are subjected to gravity loads/vertical loads as well as lateral loads/horizontal loads due to wind loads. In the following section, analysis of multistorey frames subjected to horizontal loads/lateral loads are presented. The following approximate methods are presented which are in practice.

(1) Portal method
(2). Cantilever method
(3) Factor method

Portal method was developed by Smith in 1915. Cantilever method was proposed by Wilson in 1908. Factor method was formulated byWilbur in 1934. The above three methods are available in several textbooks. However, the other refreshing and alternative approaches were made by Assudani and Varyani (1967), Arya (1972) and Manicka Selvam (1987). With the development of finite element method, the interest in improving the approximate methods of analysis started diminishing. However, the above methods are proven and time tested. The following sections give the development and applications of the above methods.

6.4.1 Portal Method

The Portal method is simple and can be easily applied. The following assumptions are made:

(1) The internal hinges in each girder are located at midspan of the girder.
(2) The internal hinges in each column are located at mid height of the column.

(3) The total horizontal shear in each storey level is distributed amongst the columns such that interior column resists twice as much the exterior column shear.

The first two assumptions make the frame statically determinate. This leads to the estimation of shears and moments in the columns and at the base of the structure. The third assumption refers to the condition of proportional column shear. That is, each storey is made up of a number of single portal frames, in other words, each bay is treated as a separate portal frame.

Consider the three-bay portal frame as shown in Fig. 6.43.

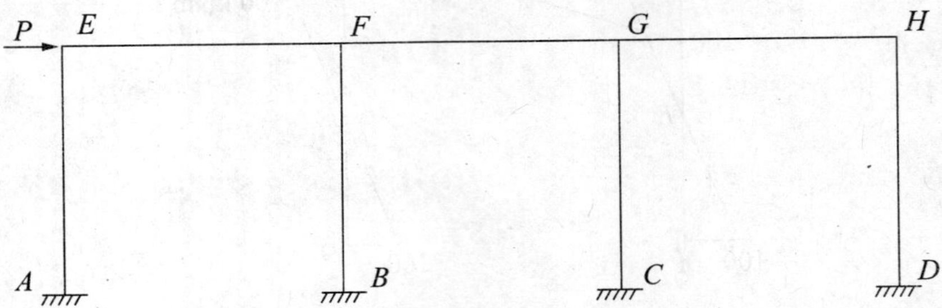

FIG. 6.43 Three bay single storey frame

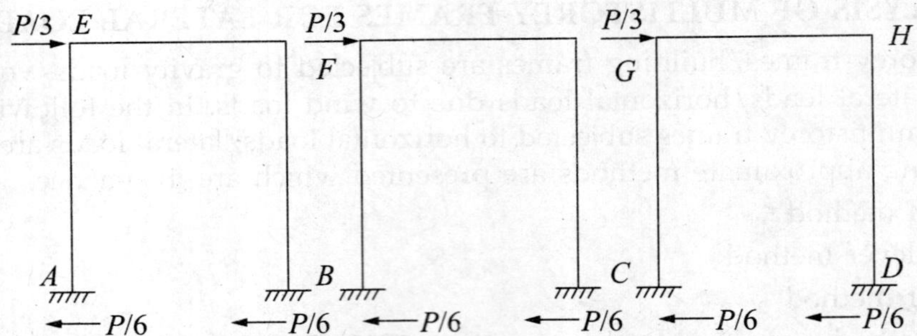

FIG. 6.44 Three-single bay single storey frame

The horizontal shear in the exterior column *AE* is $(P/6)$. While for the interior columns *BF* and *CG* the horizontal shear is $(P/6) + (P/6) = P/3$. Hence, the horizontal shear of interior columns is twice of that of the horizontal shear of exterior columns.

6.4.1.1 *Limitations*

(1) The distribution of horizontal shear is that the interior column takes twice as much shear as the exterior column. Here, the width of the bay is not considered. That is, whether the width of the bay is smaller or larger, the horizontal shear in the interior columns is double that of the column shear in the exterior column. In simple words, the distribution of horizontal shear does not depend on the width of bay. Of late, the width of bay has been included in proportioning horizontal shear and the method which includes the proportioning of horizontal shear according to the width of bay is termed the modified Portal method.

(2) The Portal method does not consider the stiffness of the columns and beams. In other words, the moment of inertia of columns and beams are not considered in the distribution of shears and estimation of moments.

(3) The interior hinges are located at mid height of the columns and at midspan of the beams. It should be remembered that a caution is to be exercised when we use the Portal method for checker board loading patterns. Also experts advise that the hinges at topmost column and the bottom column can be located at 0.55 times the respective height from the top and bottom respectively. The above assumption may provide good results.

(4) The Portal method is found to be satisfactory for 25-storey high buildings with moderate height width ratio.

The application of the Portal method is illustrated in the following example.

Example 6.10 Analyse the frame shown in Fig. 6.45 by the Portal method. Draw the bending moment diagram.

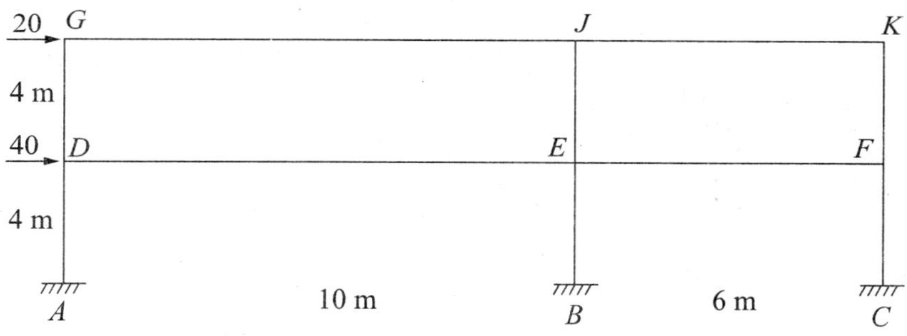

FIG. 6.45

Solution

Step 1: Assume hinges at the mid height of columns and at midspan of the beam. The loading along with internal hinges are shown. We start the analysis from the topmost storey.

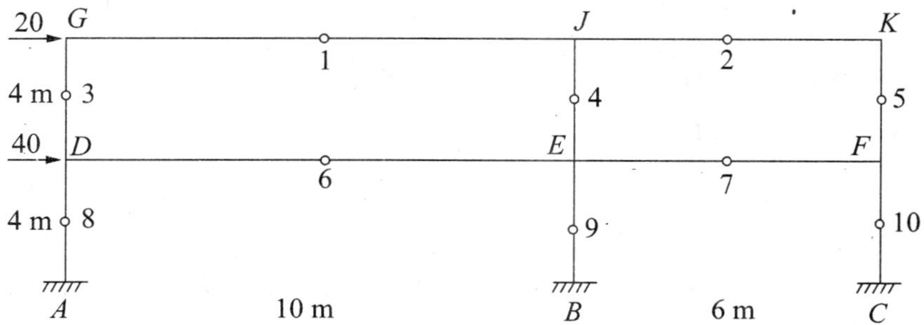

FIG. 6.46 Portal frame with assumed internal hinges

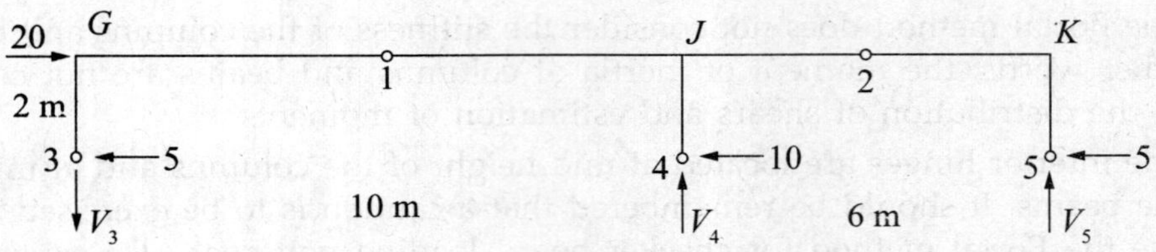

FIG. 6.47 Free body diagram of the half frame (upper half)

Consider the free body diagram, 3G1

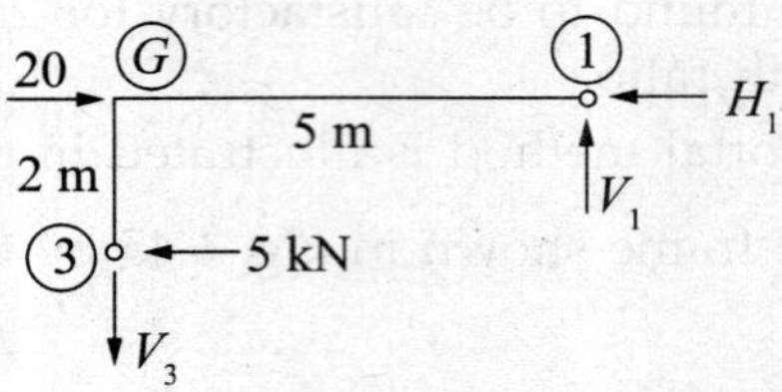

FIG. 6.48

$$\Sigma H = 0$$

$$20 = 5 + H_1$$

$$\boxed{H_1 = 15 \text{ kN}}$$

$$\Sigma V = 0$$

$$-V_3 + V_1 = 0$$

$$-V_3 + 2 = 0$$

$$\boxed{V_3 = 2 \text{ kN}}$$

$$\Sigma M_G = 0$$

$$5V_1 = 2(5)$$

$$\boxed{V_1 = 2 \text{ kN}}$$

Consider the free body diagram of 41J2

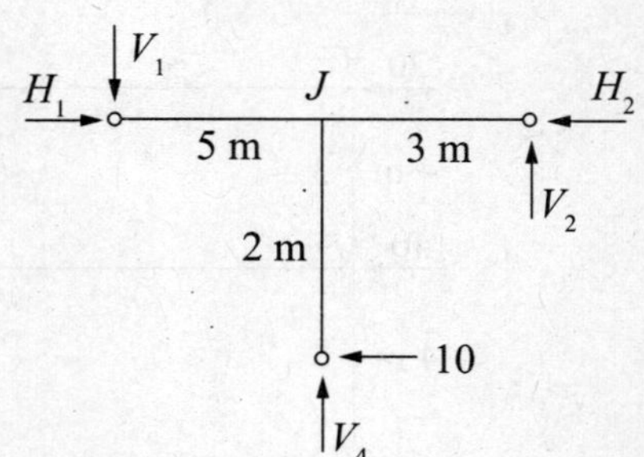

FIG. 6.49

$$\Sigma H = 0;$$

$$15 - H_2 - 10 = 0$$

$$\boxed{H_2 = 5 \text{ kN}}$$

$$\Sigma M_4 = 0$$

$$2H_1 - 5V_1 - 3V_2 - 2H_2 = 0$$

$$2(15) - 5(2) - 3V_2 - 2(5) = 0$$

$$\boxed{V_2 = 3.3 \text{ kN}}$$

$\Sigma V = 0$

$$-V_1 + V_2 + V_4 = 0$$

$$-V_1 + 3.3 + V_4 = 0$$

$$\boxed{V_4 = -1.3 \text{ kN}}$$

Consider the free body diagram 2k5

$\Sigma V = 0,$

$$-V_2 + V_5 = 0$$

$$-3.3 + V_5 = 0$$

$$\boxed{V_5 = 3.3 \text{ kN}}$$

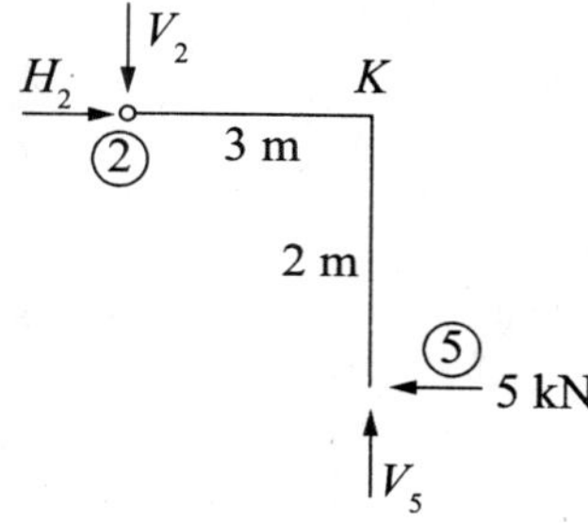

FIG. 6.50

After evaluating the unknowns (in the internal hinges) of the upper half, we analyse the middle portion of the frame, i.e., 3 – 4 – 5 – 8 – 9 – 10. The free body diagram is shown below.

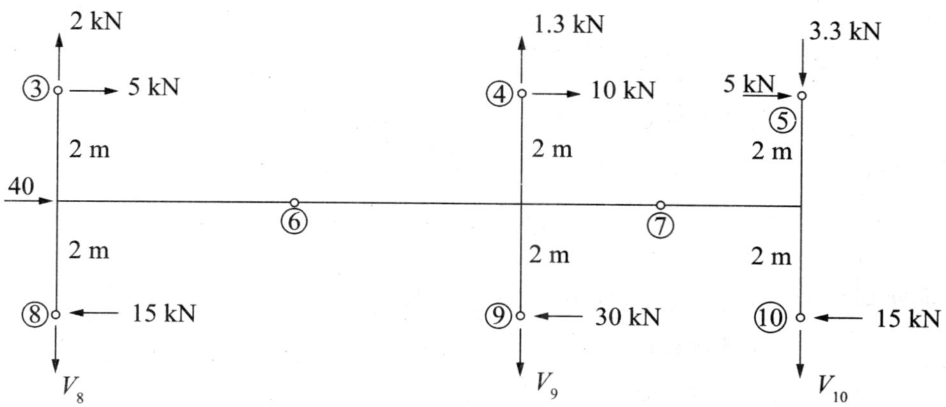

FIG. 6.51 Free body diagram of middle portion of the frame

Consider the free body diagram 3 – 6 – 8

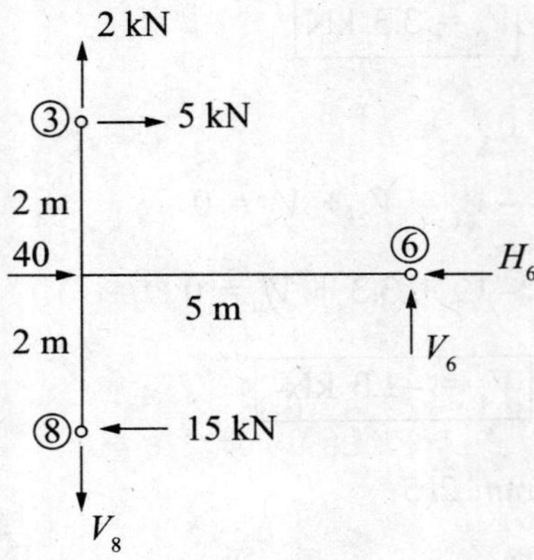

FIG. 6.52

$\Sigma H = 0,$

$$40 + 5 - 15 - H_6 = 0$$

$$\boxed{H_6 = 30 \text{ kN}}$$

$\Sigma M_8 = 0,$

$$40(2) + 5(4) - 2H_6 - 5V_6 = 0$$

$$40(2) + 5(4) - 2(30) = 5V_6$$

$$\boxed{V_6 = 8 \text{ kN}}$$

$\Sigma V = 0,$

$$2 + 8 - V_8 = 0$$

$$\boxed{V_8 = 10 \text{ kN}}$$

Consider the free body diagram of 6 – 4 – 7 – 9

$\Sigma H = 0,$

$$+30 + 10 - 30 - H_7 = 0$$

$$\boxed{H_7 = 10 \text{ kN}}$$

$\Sigma M_9 = 0,$

$$+30(2) + 10(4) - 2H_7 - 8(5) = 0$$

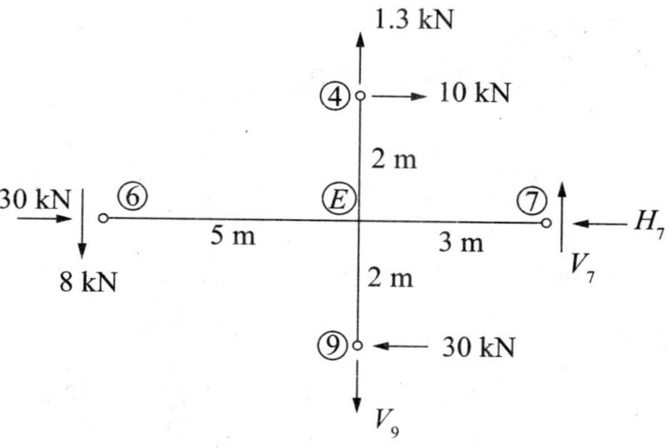

FIG. 6.53

$$-3V_7$$

$$60 + 40 - 2(10) - 40 = 3V_7$$

$$\boxed{V_7 = 13.3 \text{ kN}}$$

$\Sigma V = 0,$

$$-8 + 1.3 + 13.3 - V_9 = 0$$

$$\boxed{V_9 = 6.6 \text{ kN}}$$

Consider the free body diagram 7 – 5 – 10

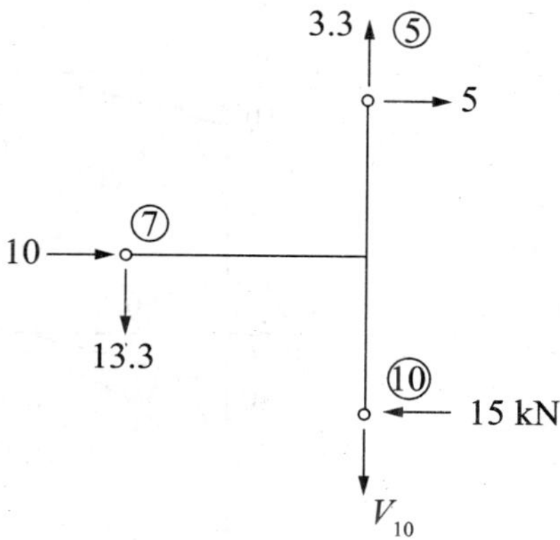

FIG. 6.54

$$\Sigma V = 0,$$

$$-13.3 + V_{10} = 0$$

$$-3.3$$

$$\boxed{V_{10} = 16.6 \text{ kN}}$$

Now consider the lower half of the frame 8 – 9 – 10

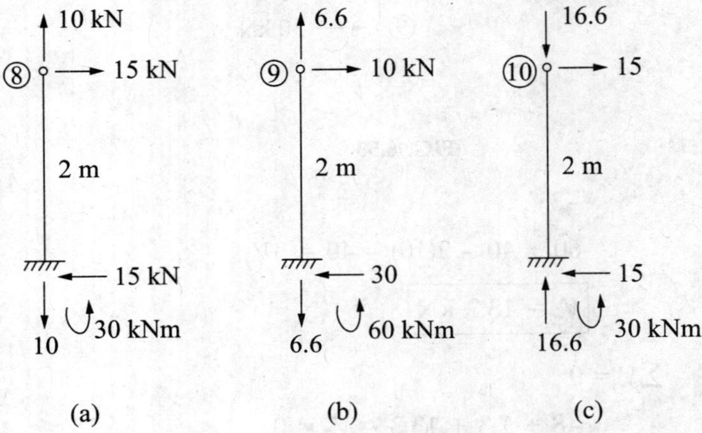

FIG. 6.55 Free body diagram of the lower half of the frame

The final bending moment diagram is shown below.

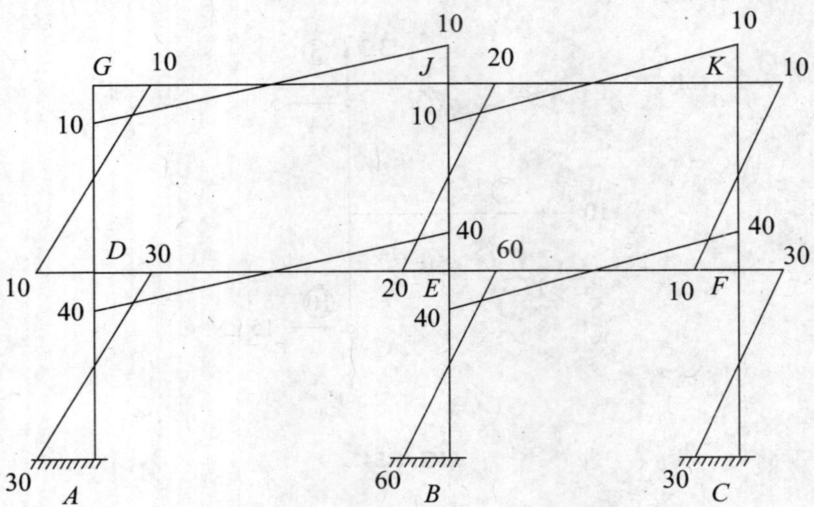

FIG. 6.56 Bending moment diagram

The following points are to be noted at the end of the analysis as:

(1) If the lateral load (horizontal load) is acting from left to right; all the column moments are clockwise in direction and the beam moments are acting in the anticlockwise direction.

(2) Σ column moments = Σ beam moments at each joint of the multistorey frame.

6.4.2 Cantilever Method

A tall building frame under lateral loading is considered a vertical cantilever. The equilibrium of the cantilever is examined using the free body diagram. We know that a cantilever beam is one which is fixed at one end and the other end is free. The flexural stress varies linearly changing from tension to compression through the centroid of the cross section. The axial force in each column is assumed to be proportional to the distance from the centroid of the column areas.

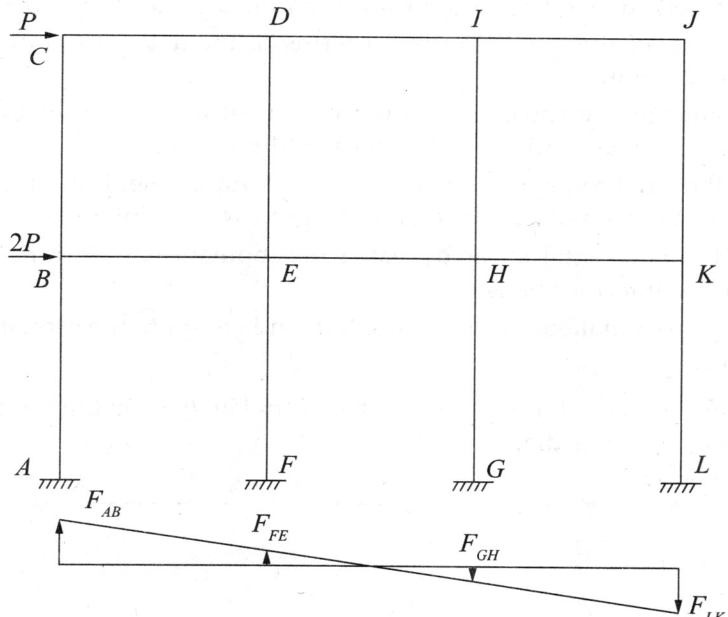

FIG. 6.57 Distribution of axial stresses

6.4.2.1 Assumptions made in the Cantilever Method

(1) Wind loads act at the floor levels only.

(2) Internal hinge is located at the mid height of the column in each storey.

(3) Internal hinge is also located at the centre of each span of the beam.

(4) The intensity of axial stress in each column of a storey is proportional to the horizontal distance of that column from the vertical centroidal axis of all columns of that storey.

6.4.2.2 *Limitations of the Cantilever Method*

(1) It is not justified to assume irrespective of relative stiffness of columns and beams, the internal hinges at midspan of beam and columns.

(2) It is preferable for tall slender building with 25 to 35 storeys. It may not be preferable for short and stocky buildings.

(3) This method is not a self corrective method like Kani's method. If an error happens in any step is carried forward if unnoticed. We have to redo the analysis to eliminate the error. It is the same setback for Portal method as well as the Factor method.

6.4.2.3 *Application of the Cantilever Method*

(1) Locate the interior hinges for the columns at mid height and at midspan for the beams.

(2) Locate the centroid of the column areas and draw the axial stress diagram.

(3) Proportionate all the axial stresses in terms of the axial stress in the windward side of the column.

(4) Start the analysis by considering the uppermost half storey of the frame. Mark the horizontal shear and the axial force in the free body diagram.

(5) Evaluate the axial force in the columns by taking moment of all the forces about the internal hinges which is preferably right extreme interior hinge.

(6) Evaluate the horizontal shear by using the moment equilibrium of the internal hinges in the interior spans.

(7) Apply the horizontal equilibrium equation and compute the remaining horizontal shear force.

Example 6.11 Analyse the building frame shown in Fig. 6.58 by the Cantilever method. Draw the bending moment diagram.

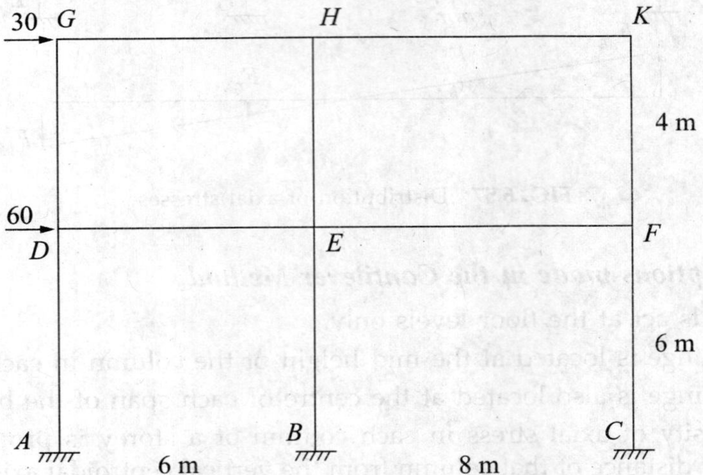

FIG. 6.58

Solution: The internal hinges are located at midspan and mid height of beams and columns respectively. The joints of the frame are marked with letters, viz., *A*, *B*, . . . *K*. The interior hinges are located with numerals, viz., 1, 2, . . . All the column areas are equal. Hence, the centroid of the column areas, i.e., the vertical axis is located by taking moments of areas about the windward side of the column *ADG*.

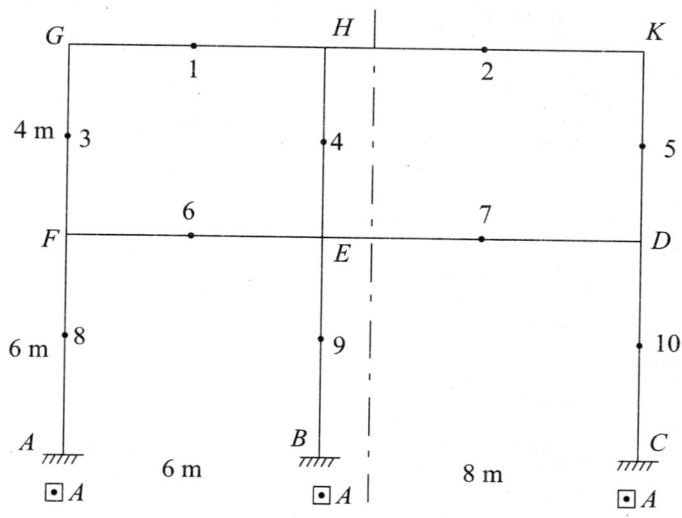

FIG. 6.59 Determinate structure

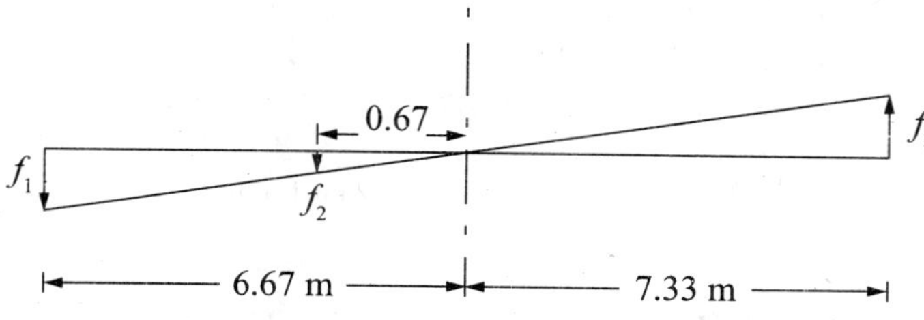

FIG. 6.60 Axial stress diagram

Step 1

Location of the Centroidal Axis

All the column areas are equal in this case. Taking moment of all the areas about the left end column,

Total area of the columns = $A + A + A = 3A$

$$\Sigma A_i x_i = A_1 x_1 + A_2 x_2 + A_3 x_3$$

$$3A\bar{x} = A(0) + A(6) + A(14)$$

$$3A\bar{x} = 20A$$

$$\therefore \quad \boxed{\bar{x} = 6.67 \text{ m}}$$

Step 2

Determination of Axial Forces

Let f_1, f_2 and f_3 be the stresses in the columns *AFG*, *BEH* and *CDK* respectively. We can express the stresses f_2 and f_3 acting in the interior column and the right extreme column in terms of f_1 by similar triangles.

Referring to the axial stress diagram,

$$\frac{0.67}{6.67} = \frac{f_2}{f_1}$$

$$\boxed{f_2 = 0.100\,f_1}$$

$$\frac{f_1}{f_3} = \frac{6.67}{7.33}$$

$$\boxed{f_3 = 1.099\,f_1}$$

Hence, the axial forces are

$$V_1 = f_1 A$$

$$V_2 = 0.100\,f_1 A$$

$$V_3 = 1.099\,f_1 A$$

The direction of axial forces are shown below.

↓ 6 m ↓ 8 m ↓

V_1 $V_2 = 0.100\,V_1$ $V_3 = 1.099\,V_1$

FIG. 6.61 Direction of axial forces

Step 3

Values of Axial Forces and Horizontal Shear of the Top Storey

Consider the top half of the storey, let axial forces in the columns be V_1, V_2 and V_3 acting at hinges 3, 4 and 5 respectively. The corresponding horizontal shears are S_1, S_2 and S_3 at the above hinges 3, 4 and 5. The corresponding free body diagram is shown below.

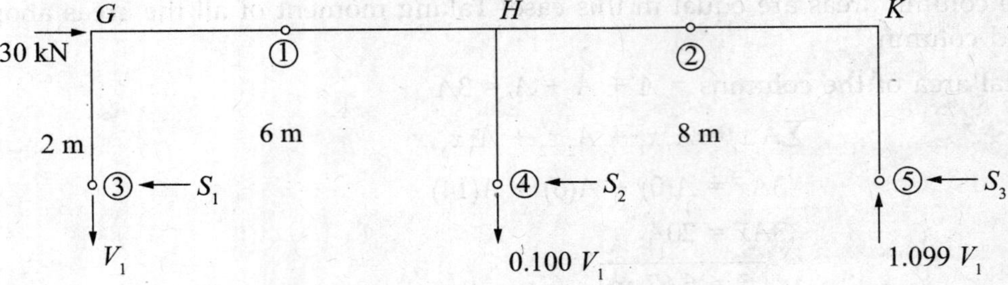

FIG. 6.62

The procedure is followed in a systematic manner.

$$\Sigma M_5 = 0,$$

$$30(2) - 14V_1 - 0.100V_1(8) = 0$$

$$\boxed{\begin{aligned} V_1(\text{Axial Force}) &= 4.05 \text{ kN} \\ V_2 = 0.100(4.05) &= 0.41 \text{ kN} \\ V_3 = 1.099(4.05) &= 4.45 \text{ kN} \end{aligned}}$$

$$\Sigma M_1 = 0,$$

$$2S_1 - 3V_1 = 0$$

$$2S_1 - 3(4.05) = 0$$

$$\boxed{S_1 = 6.08 \text{ kN}}$$

$$\Sigma M_2 = 0,$$

$$2S_3 - (1.099)4 = 0$$

$$2S_3 - (1.099 \times 4.05)4 = 0$$

$$\boxed{S_3 = 8.90 \text{ kN}}$$

$$\Sigma H = 0,$$

$$S_1 + S_2 + S_3 = 30$$

$$6.08 + S_2 + 8.90 = 30$$

$$\boxed{S_2 = 15.02 \text{ kN}}$$

Step 4

Axial Forces and Horizontal Shear in the Midstorey

Consider the midstorey height of 3 – 4 – 5 – 8 – 9 – 10. The free body diagram is drawn below. The axial forces and the horizontal shear in the bottom storey are obtained using the equilibrium equations. The axial forces in the lowermost columns are denoted as V_1', $V_2' = 0.100V_1'$ and $V_3' = 1.099V_1'$. They are in same proportion as that of upper half of the storey. They are marked at the locations 8, 9, 10 respectively. Also the horizontal shears S_1', S_2' and S_3' are marked in the location of internal hinges 8, 9 and 10 respectively.

The procedure is as follows. Take moments of all the forces about the right extreme hinge and obtain the axial force in the left side column as V_1'. Then in proportion obtain all other axial forces in the column. Take moment about the internal hinges of the girder and obtain the horizontal shears. Use the horizontal equilibrium and obtain the horizontal shear in the interior column. This procedure can be used for all the tall structures subjected to lateral loads. Applying the above procedure in a sequential manner.

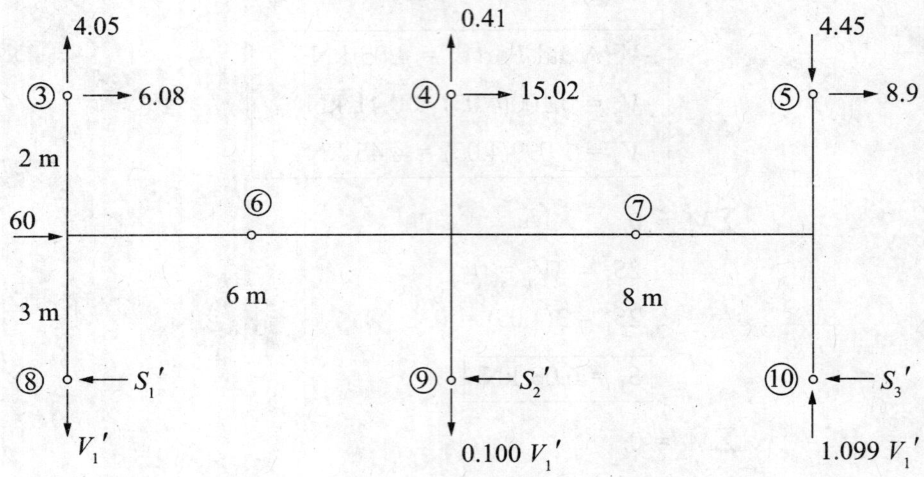

FIG. 6.63

$\Sigma M_{10} = 0,$

$$(6.08 + 15.02 + 8.9)5 + 60(3) + 4.05 \times 14 + 0.41 \times 8$$

$$-14V_1' - (0.1\ V_1')8 = 0$$

$$\boxed{\begin{aligned} V_1' &= 26.35 \text{ kN} \\ V_2' &= 2.64 \text{ kN} \\ V_3' &= 28.96 \text{ kN} \end{aligned}}$$

$\Sigma M_6 = 0,$

$$3S_1' + 6.08(2) + 4.05(3) - 26.35(3) = 0$$

$$\boxed{S_1' = 18.25 \text{ kN}}$$

$\Sigma M_7 = 0,$

$$4.45(4) + 8.9(2) + 3S_3' - 28.96(4) = 0$$

$$\boxed{S_3' = 26.75 \text{ kN}}$$

$\Sigma H = 0,$

$$60 + 6.08 + 15.02 + 8.9 - 18.25 - S_2' - 26.75 = 0$$

$$\boxed{S_2' = 45.00 \text{ kN}}$$

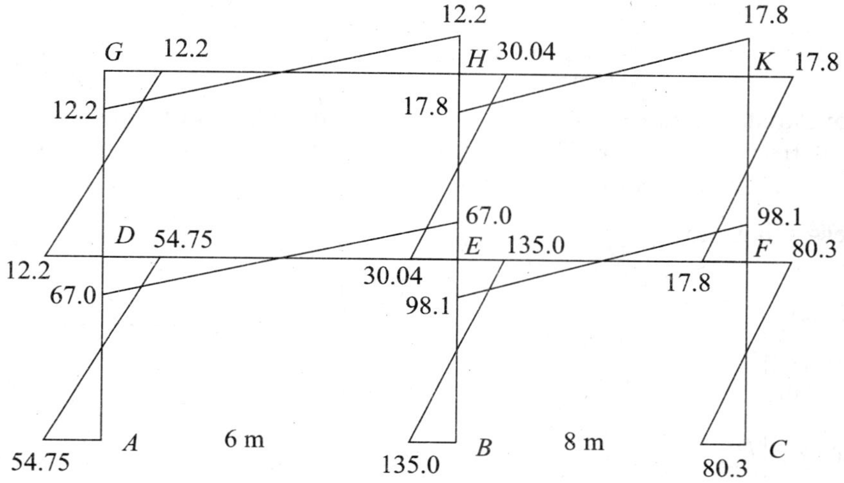

FIG. 6.64 Bending moment diagram

While drawing the bending moment diagram, we should remember that the sum of the beam moments is equal to the column moments.

Example 6.12 Analyse the frame by Cantilever method. Take the area of the central column double the area of the outer column.

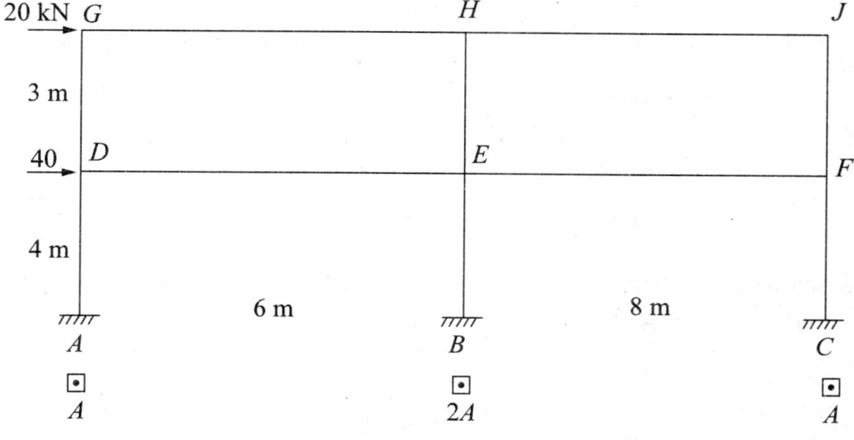

FIG. 6.65

Step 1

Location of Centroidal Axis

Taking moment of all areas about the windward side column *ADG*,

Total area of columns $A + 2A + A = 4A$

$$4A\bar{x} = 2A(6) + A(14)$$

$$\boxed{\bar{x} = 6.5 \text{ m}}$$

The above distance is measured from the column *ADG* and hence $\bar{x} = 6.5$ m from *A* and 0.5 m from column *B*. The location of interior hinges along with the vertical centroidal axis is shown below.

Step 2 Stress Diagram

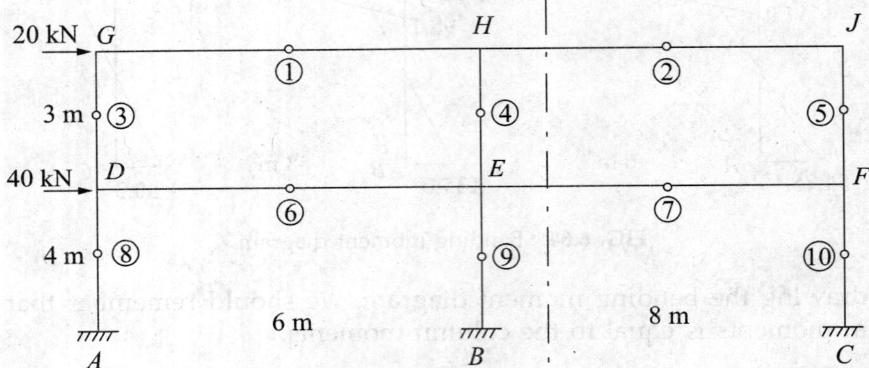

FIG. 6.66 Determinate structure

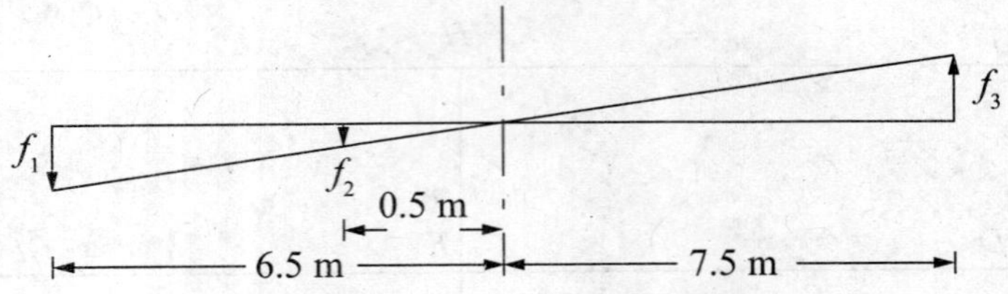

FIG. 6.67 Distribution of axial stress

The stresses in the columns are f_1, f_2 and f_3 which are acting as shown in the figure below. From similar triangles

$$\frac{f_2}{f_1} = \frac{0.5}{6.5}$$

$$\boxed{f_2 = 0.077\, f_1}$$

$$\frac{f_3}{f_1} = \frac{7.5}{6.5}$$

$$\boxed{f_3 = 1.154\ f_1}$$

The corresponding axial forces are

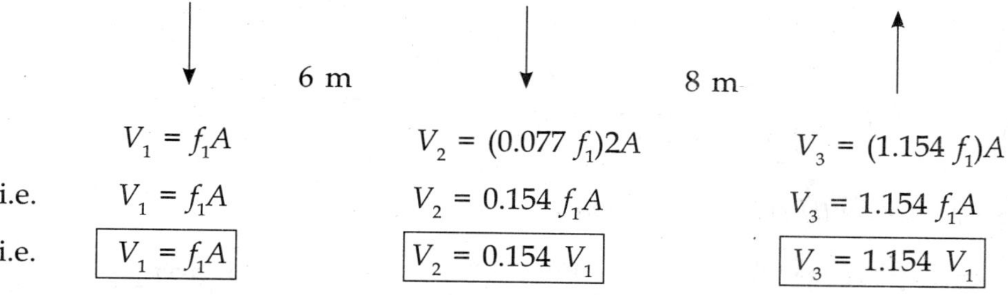

Axial forces and horizontal shear in top storey

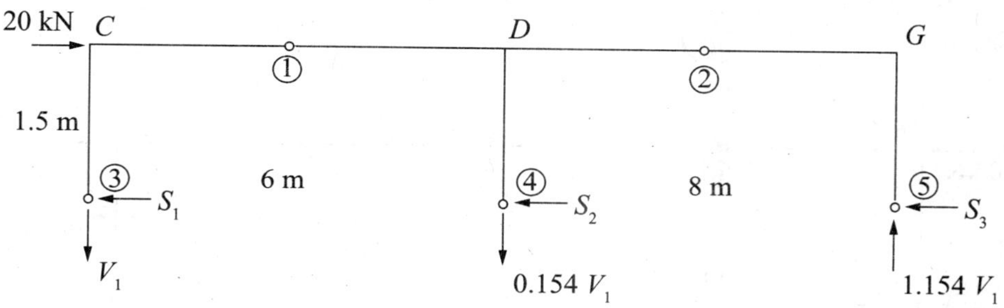

FIG. 6.68

Taking moment about the right extreme hinge

$$\Sigma M_5 = 0$$

$$20(1.5) - 14V_1 - 8V_2 = 0$$

$$20(1.5) - 14V_1 - 8(0.154V_1) = 0$$

$$\therefore \quad \boxed{\begin{aligned} V_1 &= 1.97 \text{ kN} \\ V_2 &= 0.30 \\ V_3 &= 2.27 \text{ kN} \end{aligned}}$$

$$\Sigma M_1 = 0;$$

$$1.5S_1 - 3(1.97) = 0$$

$$\boxed{S_1 = 3.94 \text{ kN}}$$

$\Sigma M_2 = 0;$

$$1.5S_3 - 2.272(4) = 0$$

$$\boxed{S_3 = 6.10 \text{ kN}}$$

$\Sigma H = 0;$

$$S_1 + S_2 + S_3 = 20$$

$$3.94 + S_2 + 6.10 = 20$$

$$\boxed{S_2 = 10 \text{ kN}}$$

Axial forces and horizontal shear in midstorey

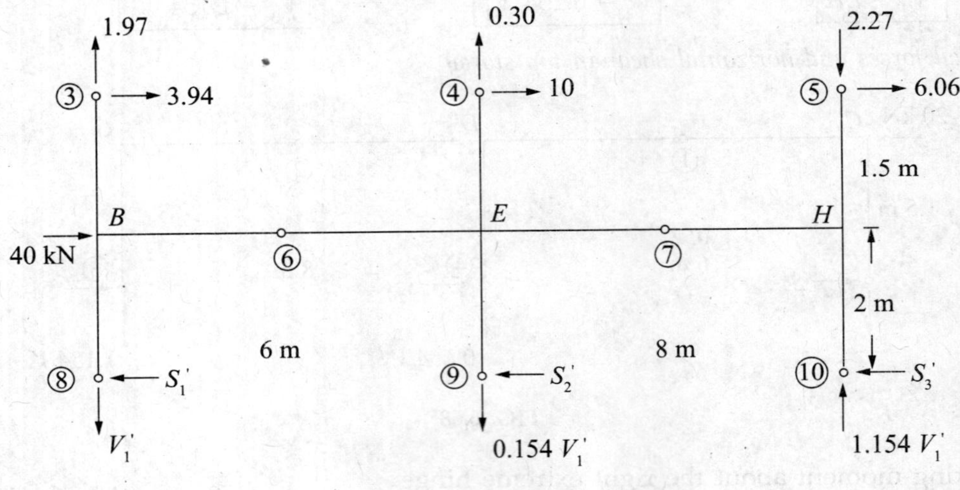

FIG. 6.69

Taking moment about the right extreme hinge

$\Sigma M_{10} = 0$

$$40(2) + (3.94 + 10 + 6.06)3.5 + 1.97(14) + 0.3(8)$$

$$- V_1'(14) - 0.154V_1'(8) = 0$$

$$\boxed{V_1' = 11.82 \text{ kN}}$$

$\Sigma M_6 = 0$

$$2S_1' + 3.94(1.5) + 1.97(3) - 3(11.82) = 0$$

$$\boxed{S_1' = 11.82 \text{ kN}}$$

$$\Sigma M_7 = 0$$

$$2S_3' + 6.06(1.5) - 1.154V_1'(4) + 2.27(4) = 0$$

$$2S_3' + 6.06(1.5) - 1.154(11.82)4 + 2.27(4) = 0$$

$$\boxed{S_3' = 18.20 \text{ kN}}$$

$$\Sigma H = 0$$

$$S_1' + S_2' + S_3' = 40 + 3.94 + 10 + 6.06$$

$$11.82 + S_2' + 18.20 = 60$$

$$\boxed{S_2' = 30.0 \text{ kN}}$$

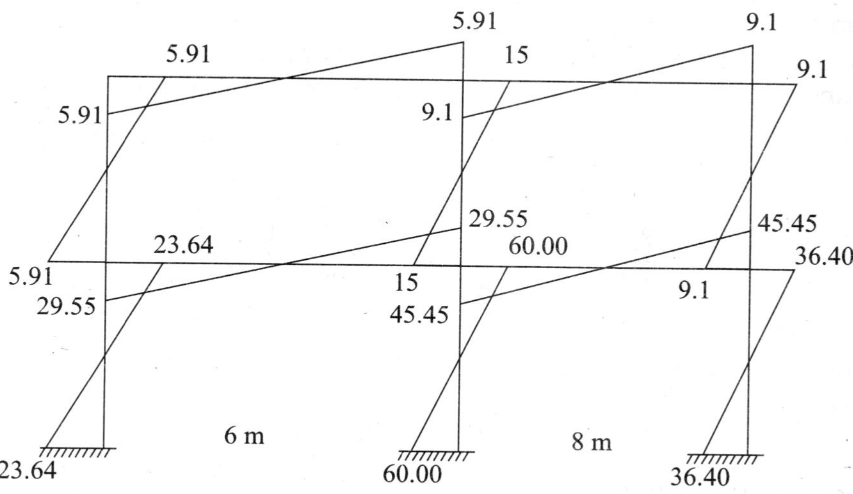

FIG. 6.70 Bending moment diagram

6.4.3 Factor Method

Wilbur formulated this method in 1934. This method is more rational than the other methods, viz., the Portal method and Cantilever method. It takes into the elastic action of the frame. This method relies on the stiffness values k, i.e., (I/l). The procedure is systematic and heavily depends on girder factors and column factors. The axial force, shear force and bending moments obtained by this factor method closely agrees with the results from the slope deflection method. The factor method is based on two factors, viz., Girder factor (g) and Column factor (c).

The Column factor is defined as

$$c = \frac{\text{Sum of girder of stiffness at a joint}}{\text{Sum of stiffness of all the memebers meeting at a joint}} \quad (1)$$

$$g = \frac{\text{Sum of column of stiffness at a joint}}{\text{Sum of stiffness of all the memebers meeting at a joint}} \quad (2)$$

Adding Eqs. (1) and (2),

$$\boxed{c + g = 1.0} \quad (3)$$

In other words,

$$\boxed{c = 1 - g} \quad (4)$$

The above equation implies that the Girder factor can be readily used to determine the Column factor.

In the analysis of the frames, we assume the columns to be fixed at the bottom. For such columns, we use Eq. (4) as follows. Applying Eq. (2) for the fixed base columns,

$$g = \frac{\Sigma k}{\infty} \quad (5)$$

i.e. $g = 0$

and $\boxed{c = 1}$ for columns with end fixed

A step-by-step procedure is given below.

(1) Compute the Girder factor

$$g = \frac{\Sigma k_c}{\Sigma k}$$

(2) Compute the Column factor

$$c = 1 - g$$

The Column factor for the column fixed at the base is 1.00

(3) In the frame the relative stiffness values ($k = I/l$) are written at the centre of the beam as well as at the centre of the column respectively.

(4) The frame is drawn and for the girder, (in a particular floor level) write the Girder factor at the ends on either side of the girder element. Consider a joint G where a girder GH and a column GD meet. Let the moment of inertia of the girder GH be I and the span 6 m. Then the stiffness of the girder $(I/6)$. The girder is connected to the column whose moment of inertia is I and the height of the column is 3 m. Thus, the stiffness of the column 'GD' is $(I/3)$. Then the

total stiffness of the beam and the column meeting at the joint is $\sum k = (I/6) + (I/3) = (I/2)$. The Girder factor is $(I/3)/(I/2) = 0.67$. The Column factor is 0.33. Let the stiffness of girder $(I/6)$ be normalised by multiplying by $(12/I)$, i.e., $I/6 \times 12/I = 2$.

Similarly, the girder is computed at the other hand H. Let it be 0.53. They are entered as follows:

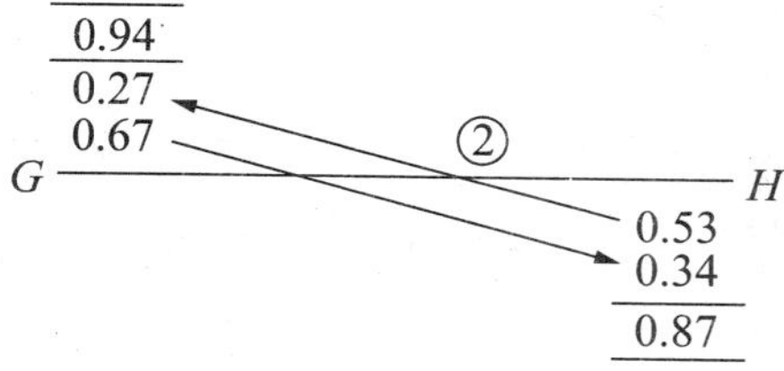

Half of the Girder factor is added to the other end and then added.

(5) The Girder moment factor 'G' is calculated by multiplying the value obtained in the above step by the stiffness value of the beam. 'G' of the girder GH is obtained as (0.94×2) and for the girder HG for the other end is (0.87×2).

$G = 1.88$

G —————————— H

$G = 1.74$

The above girder moment factors are entered as shown above.

(6) In a similar manner the Column moment factors 'C' are computed for each column using the stiffness of the column 'k' and the Column factor 'C'. These column factors are entered on either side of the column element.

(7) The Column factors 'C' represent the approximate relative value of moments at the end of each column. These column factors are used to obtain the column end moments.

(8) The Girder factors 'G' represent the approximate relative value of moments at the ends of each girder. These Girder factors are used to determine the girder moments.

(9) Applying the principles of statics, the sum of the girder moments at each joint is equal to the sum of the column end moments at that joint. The moments of the girders in the intermediate column locations are obtained using the girder moment factors in proportion.

The application of the method is illustrated with the following example.

Example 6.13 Analyse the frame by the Factor method. Draw the bending moment diagram.

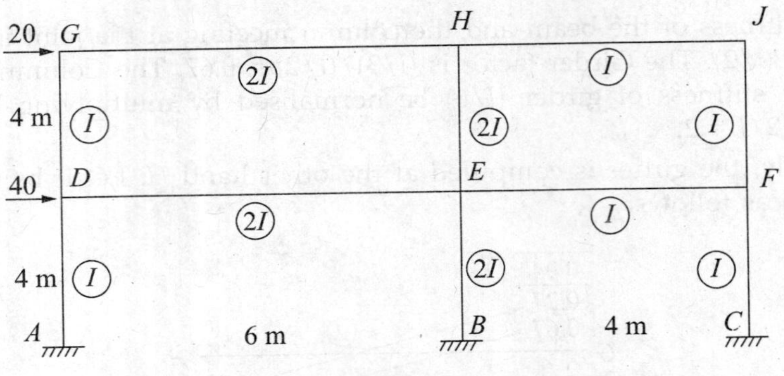

FIG. 6.71

Solution

Stiffness values: k

The stiffness values are calculated for the columns and girders. These values are multiplied by a common factor to obtain the same as relative values.

$$k_{gn} = \frac{2I}{6} \times \frac{12}{I} = 4 \qquad k_{DE} = \frac{2I}{6} \times \frac{12}{I} = 4$$

$$k_{HJ} = \frac{I}{4} \times \frac{12}{I} = 3 \qquad k_{EF} = \frac{I}{4} \times \frac{12}{I} = 3$$

$$k_{GD} = \frac{I}{4} \times \frac{12}{I} = 3 \qquad k_{DA} = \frac{I}{4} \times \frac{12}{I} = 3$$

$$k_{HE} = \frac{2I}{4} \times \frac{12}{I} = 6 \qquad k_{EB} = \frac{2I}{4} \times \frac{12}{I} = 6$$

$$k_{JF} = \frac{I}{4} \times \frac{12}{I} = 3 \qquad k_{FC} = \frac{I}{4} \times \frac{12}{I} = 3$$

These values are entered and used subsequently.

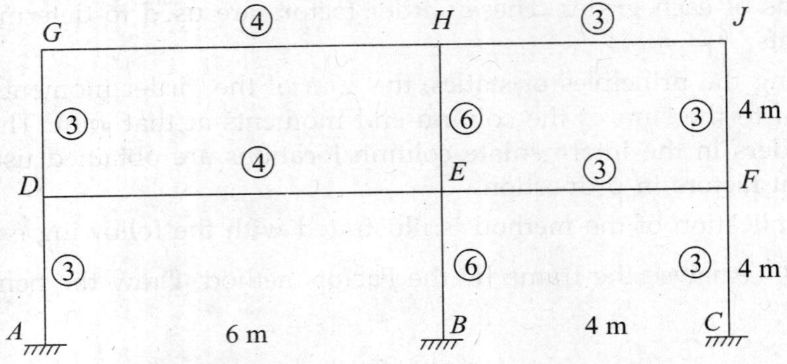

FIG. 6.72 Stiffness values of columns and girders

Girder factors (g):

$$g_{GH} = \frac{3}{3+4} = 0.43$$

$$g_{HG} = g_{HJ} = \frac{6}{6+4+3} = 0.46$$

$$g_{JH} = \frac{3}{3+3} = 0.50$$

$$g_{DE} = \frac{3+3}{3+4+3} = 0.60$$

$$g_{ED} = g_{EF} = \frac{6+6}{6+4+6+3} = 0.63$$

$$g_{FE} = \frac{3+3}{3+3+3} = 0.67$$

Column factors (c):

$$\therefore \qquad c = 1 - g$$

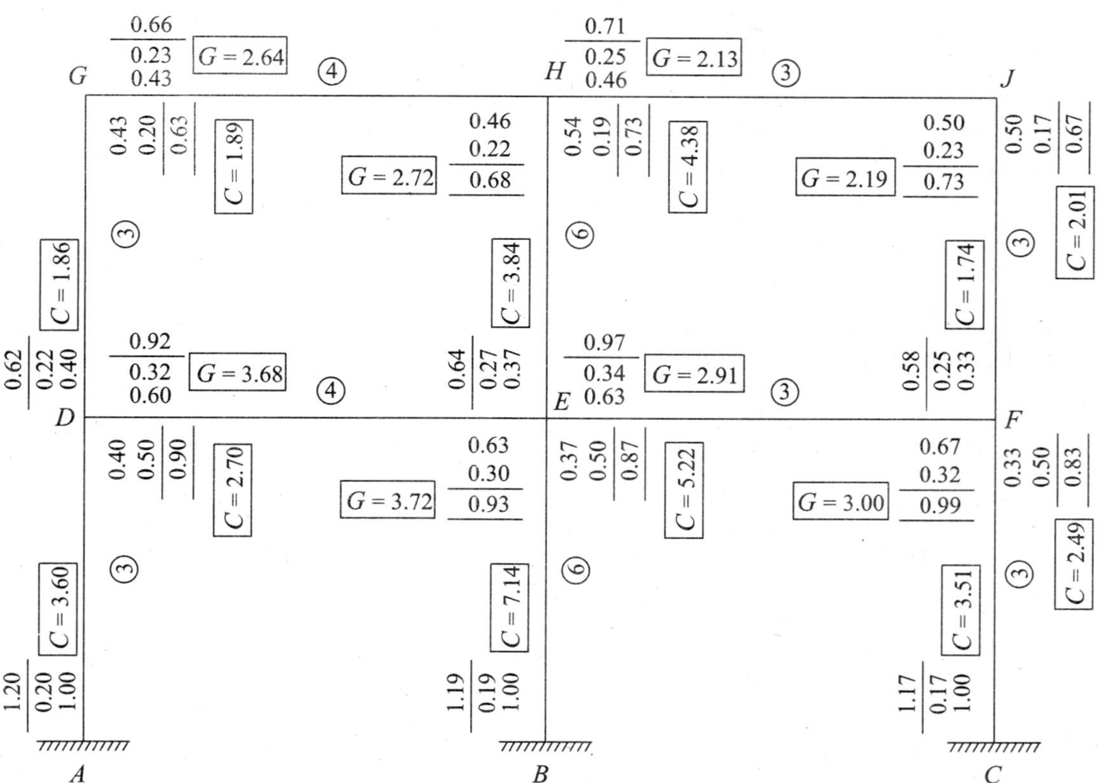

FIG. 6.73 Girder moment factor (*G*) and column moment factor (*C*)

The column factors are given in the following.

Determination of column moments (M_{cr}) using correction factors (A_r)

$$A_r = \frac{Q_r h_r}{\Sigma C_r}$$

For the top storey

$$Q_1 = 20 \text{ kN}; \quad h_1 = 4 \text{ m},$$

$$\Sigma C_r = 1.86 + 1.89 + 3.84 + 4.38 + 1.74 + 2.01 = 15.72$$

$$A_r = \frac{20 \times 4}{15.72} = 5.09$$

Column moments

$$M_C(GD) = 1.86 \times 5.09 = 9.47 \text{ kNm}$$

$$M_C(DG) = 1.89 \times 5.09 = 9.62 \text{ kNm}$$

$$M_C(HE) = 4.38 \times 5.09 = 22.29 \text{ kNm}$$

$$M_C(EH) = 3.84 \times 5.09 = 19.55 \text{ kNm}$$

$$M_C(JF) = 2.01 \times 5.09 = 10.23 \text{ kNm}$$

$$M_C(FJ) = 1.74 \times 5.09 = 8.86 \text{ kNm}$$

For the bottom storey

$$Q_2 = 60 \text{ kN}; \quad h_2 = 4 \text{ m}$$

$$\Sigma C_r = 3.60 + 2.70 + 7.14 + 5.22 + 3.51 + 2.49 = 24.66$$

Correction factor $A_r = \dfrac{60 \times 4}{24.66} = 9.73$

$$M_C(DA) = 2.70 \times 9.73 = 26.27 \text{ kNm}$$

$$M_C(AD) = 3.60 \times 9.73 = 35.03 \text{ kNm}$$

$$M_C(EB) = 5.22 \times 9.73 = 50.80 \text{ kNm}$$

$$M_C(BE) = 7.14 \times 9.73 = 69.47 \text{ kNm}$$

$$M_C(FC) = 2.49 \times 9.73 = 24.23 \text{ kNm}$$

$$M_C(CF) = 3.51 \times 9.73 = 34.15 \text{ kNm}$$

Determination of girder moments (M_{Gr}) using column moments (M_{cr})

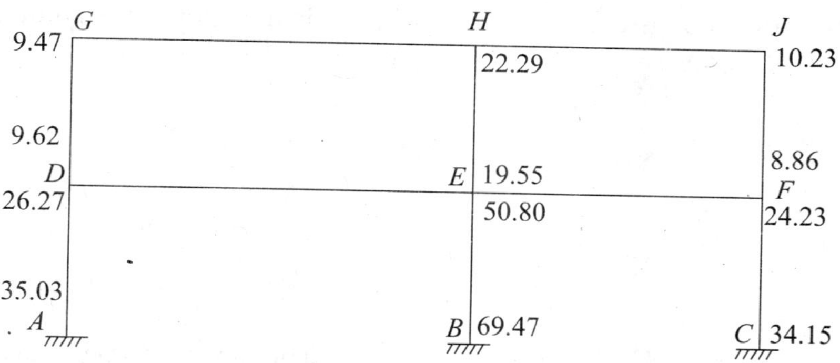

FIG. 6.74

Moments at the end joints

Joint G

$$M_{GH} = M_{GD}$$

∴ $$M_{GH} = 9.47 \text{ kNm}$$

Joint D

$$M_{DE} = M_{DG} + M_{DA}$$

∴ $$M_{DE} = 9.62 + 26.27$$

$$M_{DE} = 35.89 \text{ kNm}$$

Joint J

$$M_{JH} = M_{JF}$$

$$M_{JH} = 10.23 \text{ kNm}$$

Joint F

$$M_{FE} = M_{FJ} + M_{FC}$$

$$M_{FE} = 33.09 \text{ kNm}$$

Moments at the intermediate joints

Joint E

Sum of the girder moments = Sum of the column moments

i.e. $$\Sigma M_G = \Sigma M_C$$

where ΣM_G is the sum of girder moments and ΣM_C is the sum of the column moments.

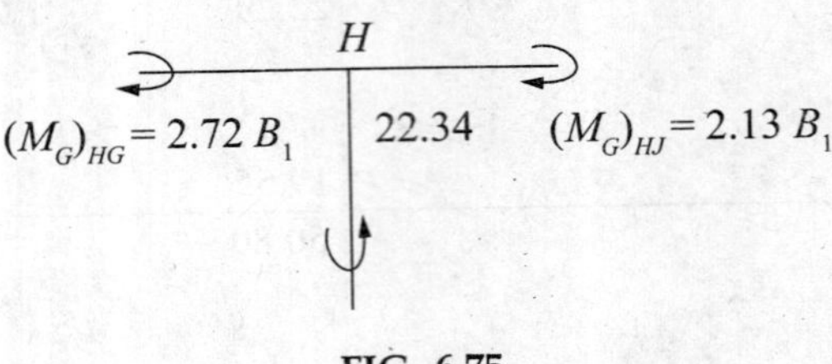

FIG. 6.75

The girder moment in HG is denoted as $(M_G)_{HG}$. The girder moment factor for HG is 2.72. The girder moment is proportional to the girder moment factor. The proportionality constant is taken as B_1. Hence,

$$2.72B_1 + 2.13B_1 = 22.34$$

$$\boxed{B_1 = 4.61}$$

$$\therefore \quad \boxed{\begin{aligned} M_{HG} &= 2.72B_1 = 12.54 \text{ kNm} \\ M_{HJ} &= 2.13B_1 = 9.82 \text{ kNm} \end{aligned}}$$

Joint E

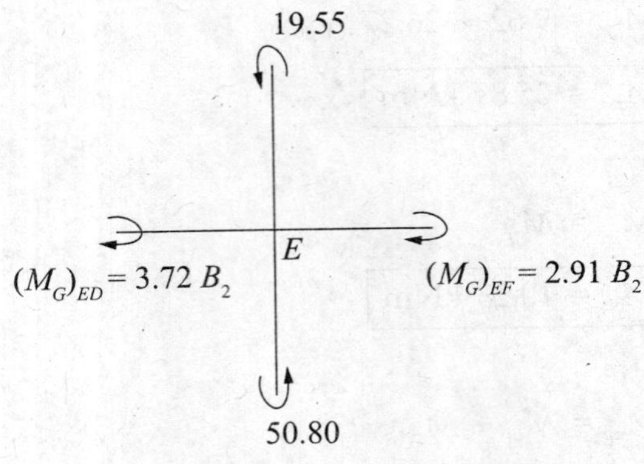

FIG. 6.76

$$3.72B_2 + 2.91B_2 = 19.55 + 50.80$$

$$6.63B_2 = 70.35$$

$$\boxed{B_2 = 10.61}$$

$$\boxed{\begin{aligned} (M_G)_{ED} &= 3.72 \times 10.61 = 39.47 \text{ kNm} \\ (M_G)_{EF} &= 2.91 \times 10.61 = 30.88 \text{ kNm} \end{aligned}}$$

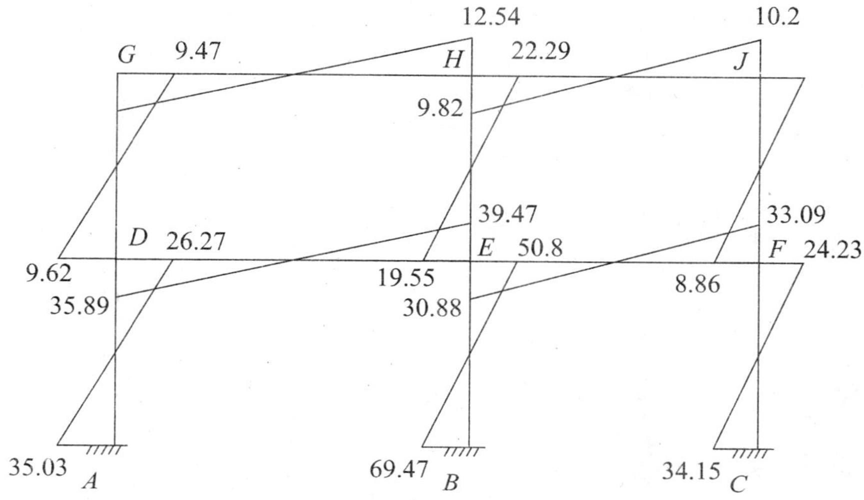

FIG. 6.77 Bending moment diagram

6.5 ANALYSIS OF MULTISTOREY FRAMES FOR VERTICAL LOADS

6.5.1 Method of Inflexion

Consider a rigid frame shown in Fig. 6.78. The frame has a static indeterminacy 12. This can be easily calculated by three times, the number of girders in the frame as the base is fixed. There are four girders and hence the static indeterminacy is 3 × 4 = 12.

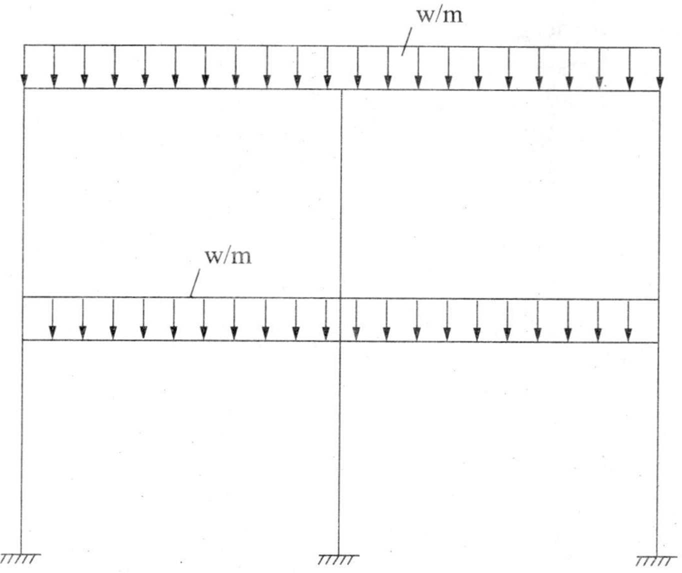

FIG. 6.78 Two-bay storey frame

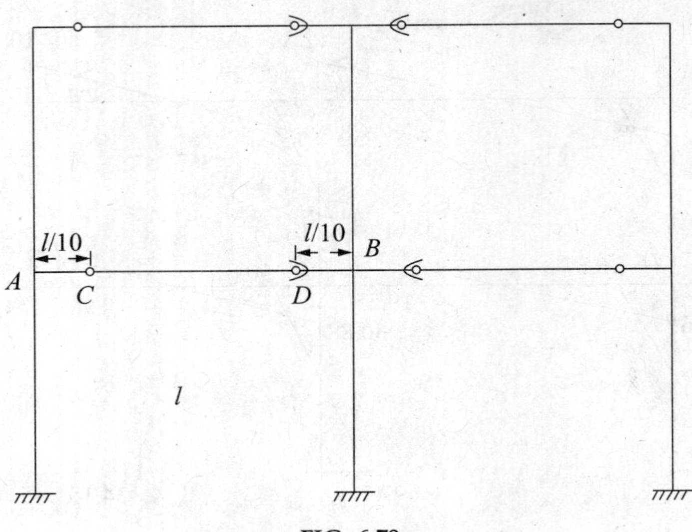

FIG. 6.79

Alternatively, the static indeterminacy for a fixed portal frame can be calculated as three times the number of bays multiplied by the number of storeys. Hence, 3 × 2 × 2 = 12. Therefore, the number of releases to be 12 to make it statically determinate. Hence, the following assumptions are made.

(1) A point of inflection occurs in the girder at one-tenth of the span, i.e., ($l/10$) from each end. In other words, internal hinges are introduced at ($l/10$) from either end of the girder.

(2) The axial force in the girder is zero.

The above assumptions are used to determine the positive bending moment in the span and the negative bending moment at the ends of the beam. This is computed by analysing the free body diagram of the beam at each floor level.

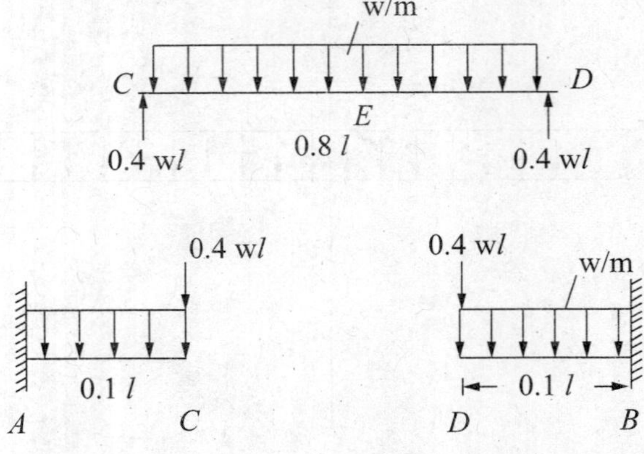

FIG. 6.80 Free body diagram of the girder

The maximum positive bending at the centre of the beam is

$$M_E = \frac{w(0.8l)^2}{8} = 0.08wl^2$$

Then $$V_C = \frac{w(0.8l)}{2} = 0.4wl$$

The maximum negative bending moment for the cantilever AC and BD are

$$M_A = 0.4wl(0.1l) + \frac{w(0.1l)(0.1l)}{2}$$

i.e. $$\boxed{M_A = 0.045wl^2}$$

$$R_A = 0.4wl + (0.1l)w$$

$$\boxed{R_A = 0.5wl}$$

The reaction R_A is transmitted to the column as axial load. The end moment M_A is shared by the upper column and the lower column in proportion to the column stiffnesses.

Though the above assumptions have been specified in many textbooks, the codes of practice and researchers suggested modifications. IS 456 : 2008 specifies in *cl* 23.12 that for continuous beams and frames 'l_0' may be assumed as 0.7 times the effective span where l_0 is the distance between points of zero moments in the beam. Also Behr et al. (1989) suggested a 'Revised Approximate Method' with the following assumptions.

(1) Locate the inflection points at approximately $0.10l$ from the girder ends. More inflection points clear to the columns when the beams are stiff relative to columns and more inflection points out as $0.21l$ when columns are very stiff to beams.

(2) Locate the points of inflection in each column at their mid heights. This assumption is used in lieu of the aforementioned assumption regarding negligible axial forces in the girders. When the frames have hinged supports at the ground level, locate the point of inflection in the first storey columns at their bases.

Epstein (1989) pointed out that the first assumption leads to thumb rules and the second assumption of locating inflection points at mid heights of upper level column in general holds good for many frames. However, the second assumption may not hold good for checker-board or similar loading patterns. It is expected that more alternative approaches to the approximate analysis of vertically loaded rectangular rigid frames would be available in future too.

The following example illustrates the application of the method.

Example 6.14 Analyse the frame by the approximate method. Compute the moments (1) assuming the inflection points at one-tenth of span from either end of the girder (2) IS 456 : 2000 method (3) Kani's method.

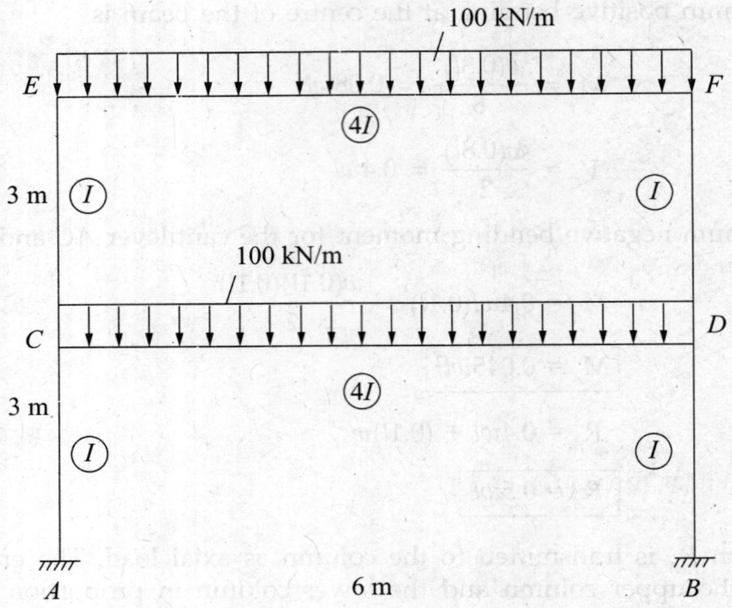

FIG. 6.81

Solution

Inflection point at (l/10) from either end of girder EF

Maximum positive *BM* at centre of *EF* and *CD* are

$$= 0.08\ wl^2 = 0.08 \times 100 \times 6^2 = 288 \text{ kNm}$$

Maximum shear force at *E* and *C*

$$= 0.5\ wl = 0.5 \times 100 \times 6 = 300 \text{ kN}$$

Maximum negative *BM* at *E* and *C*

$$M_E = M_C = 0.045\ wl^2$$

$$M_E = M_C = 0.045 \times 100 \times 6^2$$

$$\boxed{M_C = 162 \text{ kNm}}$$

At joint E

$$M_{EC} = M_{EF} = 162 \text{ kNm}$$

At joint C

$$M_{CE} = 162 \times \frac{I/3}{(I/3 + I/3)} = 81 \text{ kNm}$$

and

$$M_{CA} = 162 - 81 = 81 \text{ kNm}$$

Base moment

$$M_{AC} = 81/2 = 40.5 \text{ kNm}$$

Inflection point at (l/15) from either end of the girder (IS 456 – 2000)
Maximum positive *BM* at the centre of *EF* and *CD* are

$$= \frac{w(0.7l)^2}{8} = 0.06\ wl^2 = 0.06 \times 100 \times 6^2$$

$$= 216 \text{ kNm}$$

Maximum shear force at *E* and *C*

$$= 0.5\ wl$$

$$= 300 \text{ kN}$$

Maximum negative *BM* at *E* and *C*

$$M_E = M_C = 0.04 \times 100 \times 6^2$$

$$\boxed{M_C = 144 \text{ kNm}}$$

At joint E

$$M_{EC} = M_{EF} = 144 \text{ kNm}$$

At joint C

$$M_{CE} = \frac{144 \times (I/3)}{(I/3 + I/3)} = 72 \text{ kNm}$$

$$M_{CA} = 144 - 72 = 72 \text{ kNm}$$

Base moment

$$M_{AC} = 72/2 = 36 \text{ kNm}$$

Kani's method
The frame has been analysed using Kani's method. The results are shown below.

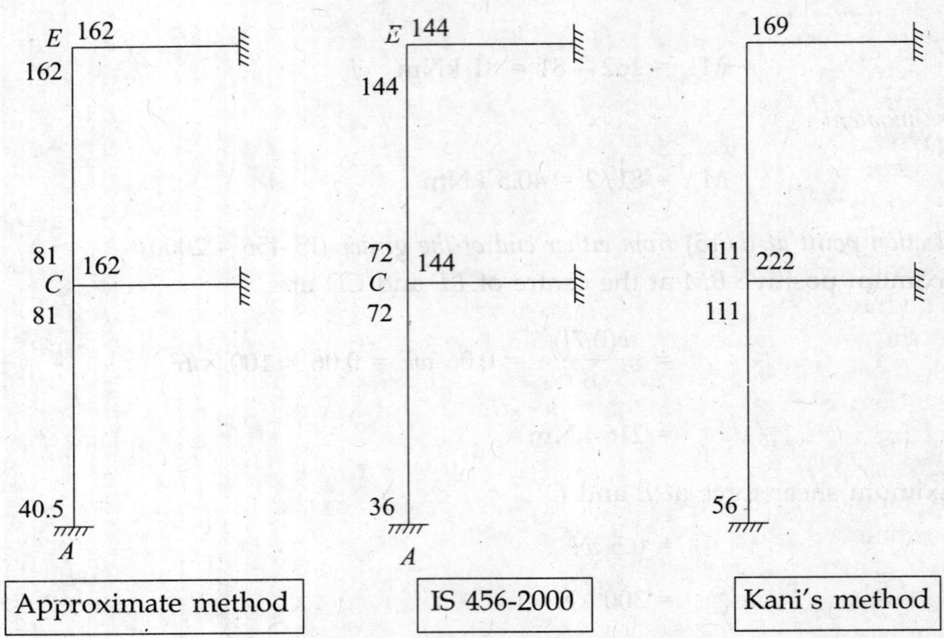

FIG. 6.82 Values of end moments of the frame

6.5.2 Trussed Portals and Mill Bents

A fixed, single bay, single storey portal frame consists of two vertical columns and a girder. The trussed portal is constructed using two vertical columns and a truss. The truss is connected to the column through pins. The truss is considered to be stiff and it behaves like a rigid girder. Figure 6.83 shows a typical Trussed portal.

FIG. 6.83 Trussed portal

6.5.3 Substitute Frame Method

All the framed structures are three dimensional. The three-dimensional analysis requires computer application of software packages. For convenience, the three-dimensional frames are reduced to plane frames which are two dimensional. These plane frames are subjected to gravity loading and lateral loads. The analysis for gravity loading as well as for lateral loads can be done separately and the results can be superimposed. In analysing the plane frames subjected to gravity loading (vertical loads), the frame is subdivided and the analysis is performed. They can be analysed by (i) using inflection points and (ii) the substitute frame method.

The substitute frames are an assemblage of the beams and columns at the floor level under consideration; with the columns above and below with their far ends fixed. The substitute frame method of analysis is also known as the two cycle moment distribution method. In general, when we apply moment distribution method, we use about five cycles so that the results converge. Two cycle moment distribution is carried out as usual. The resulting moment often two cycles of moment distribution gives the approximate moment for the floor beam design. In reality the dead load is permanent. Hence, in the analysis, all the spans are loaded with dead load. The live load is variable. The bending moments change with the loading position of the live load. The concept of influence line helps us in arranging the live load on different spans simultaneously. The particular arrangement of live load would be useful in estimating maximum hogging bending moment at the supports and maximum positive bending moment in the span. In simple words, each floor is analysed separately in a multistorey frame subjected to vertical loads. It is to be mentioned here that the substitute the frame method/two cycle moment distribution method is a simplified moment distribution method. The method considers the loads on neighbouring two nearest spans. The substitute frame method is based on the following assumption, that the moments carried from floor to floor through columns can be neglected when compared with the beam moments.

Example 6.15 Analyse the frame shown in Fig. 6.84 by the substitute frame method. Compute the positive moment at midspan of *BC*. Dead load = 15 kN/m and Live load = 20 kN/m.

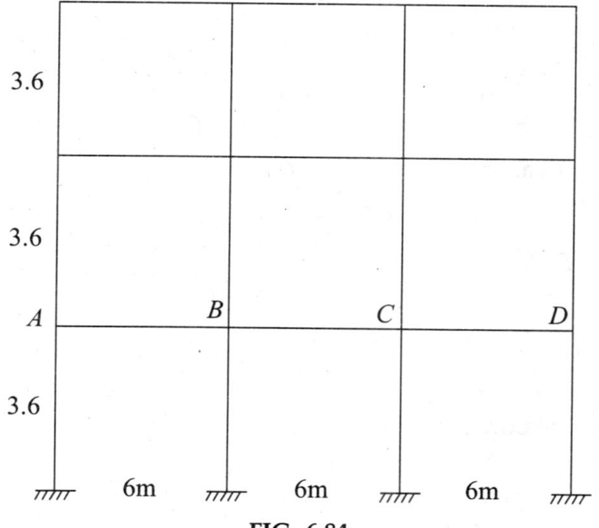

FIG. 6.84

Solution: To obtain the maximum positive moment in midspan *BC*, all spans in that floor level are loaded with dead load. The span *BC* is loaded with *DL+LL* to obtain the maximum positive moment in span *BC*. The far ends of the column are fixed and the corresponding substitute frame is shown below.

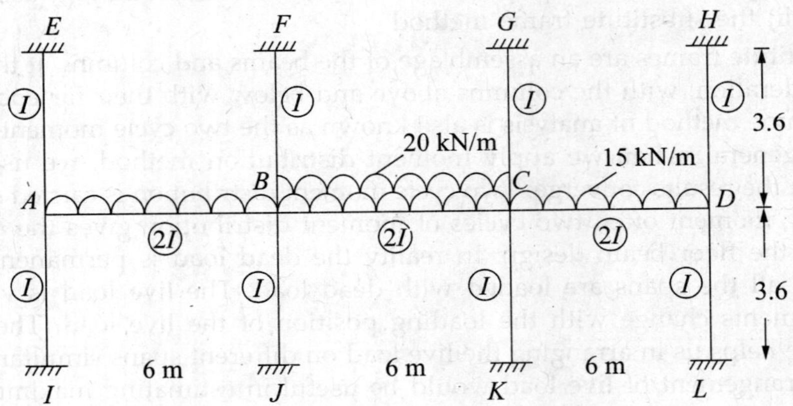

FIG. 6.85 Substitute frame

	A	B		C		D
Member	*AB*	*BA*	*BC*	*CB*	*CD*	*DC*
DF	0.37	0.27	0.27	0.27	0.27	0.37
FEMS	−45.00	+45.00	−105.00	+105.00	−45.00	+45.00
Bal	+16.65	+16.20	+16.20	−16.20	−16.20	−16.65
CO	+8.10	+8.33	−8.10	+8.10	−8.33	−8.10
Bal	−3.00	−0.06	−0.06	+0.06	+0.06	+3.00
Total	−23.25	+69.47	−96.96	+96.96	−69.47	+23.25

The *BM* at midspan can be obtained from

35 kN/m

96.96 96.96

6 m

FIG. 6.86 Substitute frame

$$\text{Maximum positive moment} = 35 \times \frac{6^2}{8} - 96.96$$

$$\text{Maximum positive moment} = 60.54 \text{ kNm}$$

Fixed end moments

$$M_{FAB} = -15 \times \frac{6^2}{12} = -45 \text{ kNm}; \quad M_{FBC} = -\frac{35 \times 6^2}{12} = -105 \text{ kNm}$$

$$M_{FBA} = +15 \times \frac{6^2}{12} = +45 \text{ kNm}; \quad M_{FCB} = +\frac{35 \times 6^2}{12} = +105 \text{ kNm}$$

$$M_{FCD} = -45 \text{ kNm}; \qquad M_{FDC} = +45 \text{ kNm}$$

Distribution factors

Joint	*Members*	*Relative Stiffness*	Σk	*DF* = $k/\Sigma k$
A	*AF* *AB* *AI*	0.28*I* 0.33*I* 0.28*I*	0.89*I*	0.31 0.37 0.31
B	*BA* *B* *BC* *BJ*	0.33 0.28 0.33 0.28	1.22	0.27 0.23 0.27 0.23

Example 6.16 Calculate the moment at midspan *BC* if it is loaded with live load on span *AB* and *CD* in addition to the dead load.

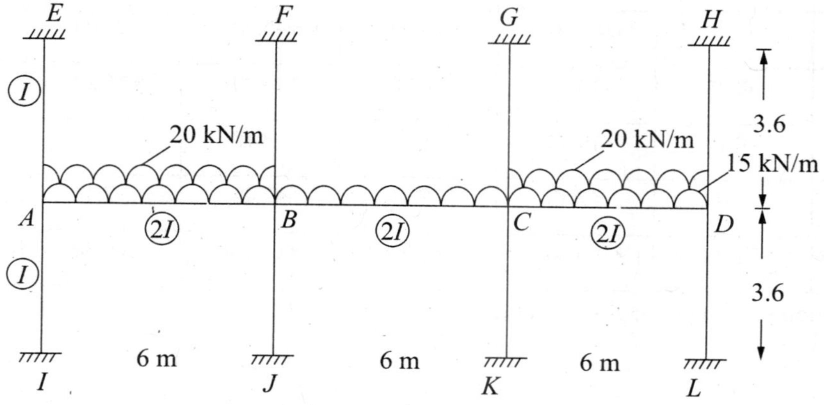

FIG. 6.87

Solution

Fixed end moments

$$M_{FAB} = -35 \times \frac{6^2}{12} = -105 \text{ kNm}; \quad M_{FBC} = -15 \times \frac{6^2}{12} = -45 \text{ kNm}$$

$$M_{FBA} = +35 \times \frac{6^2}{12} = +105 \text{ kNm}; \quad M_{FCB} = +15 \times \frac{6^2}{12} = +45 \text{ kNm}$$

$$M_{FCD} = -105 \text{ kNm}; \qquad M_{FDC} = +105 \text{ kNm}$$

Distribution factors

Joint	*Members*	*Relative Stiffness*	Σk	$DF = k/\Sigma k$
	AE	0.28		0.31
A	*AB*	0.33	0.89	0.37
	AI	0.28		0.32

Joint	*A*	*B*		*C*		*D*
Members	*AB*	*BA*	*BC*	*CB*	*CD*	*DC*
DF	0.37	0.27	0.27	0.27	0.27	0.37
FEMS	-105.00	+105.00	-45.00	+45.00	-105.00	+105.00
Bal	+38.85	-16.20	-16.20	+16.20	+16.20	-38.85
Co	-8.10	+19.43	+8.10	-8.10	-19.43	+8.10
Bal	+3.00	-7.43	-7.43	+7.43	+7.43	-3.00
Total	-71.25	+100.80	-60.53	+60.53	-100.80	+71.25

Maximum Bending Moment at B = 100.80 kNm

Analyse the frame shown in Fig. 6.88 for moments in columns.

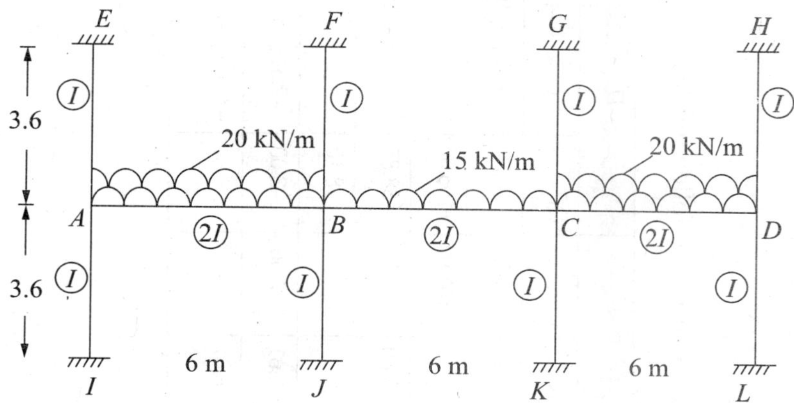

FIG. 6.88

Distribution factors

Joint	*Members*	*Relative Stiffness*	Σk	$DF = k/\Sigma k$
A	AE	0.28	0.89	0.31
	AB	0.33		0.37
	AI	0.28		0.31
B	BA	0.33	1.22	0.27
	BF	0.28		0.23
	BC	0.33		0.27
	BJ	0.28		0.23
C	CB	0.33	1.22	0.27
	CG	0.28		0.23
	CD	0.33		0.27
	CK	0.28		0.23
D	DC	0.33	0.89	0.37
	DH	0.28		0.32
	DL	0.28		0.31

The two cycle moment distribution/substitute frame method of analysis is shown in the following table.

	A			B				C				D		
Members	*AE*	*AI*	*AB*	*BA*	*BF*	*BJ*	*BC*	*CB*	*CG*	*CK*	*CD*	*DC*	*DH*	*DL*
DF	0.31	0.31	0.37	0.27	0.23	0.23	0.27	0.27	0.23	0.23	0.27	0.37	0.31	0.31
FEMS			-105.00	+105.00			-45.00	+45.00			- 105.00	+105.00		
Bals	+32.55	+32.55	+38.85	-16.20	-13.80	-13.80	-16.20	+16.20	+13.80	+13.80	+16.20	-38.85	-32.55	-32.55
Co			-8.10	+19.43			-8.10	+8.10			-19.43	+8.10		
Bal	+2.51	+2.51	+3.00	-3.06	-2.61	-2.61	-3.06	+3.06	2.61	2.61	3.06	-3.00	-2.51	-2.51
Final Moments	35.06	35.06	-	-	-16.41	-16.41	-	-	16.41	16.41	-	-	-35.06	-35.06

Example 6.17 Calculate the maximum bending moment at B for the loading shown below.

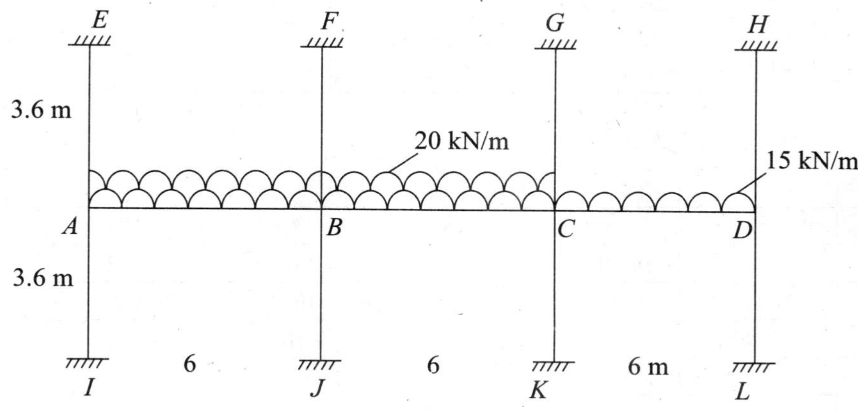

FIG. 6.89

Solution

FEMS

$$M_{FAB} = -35 \times \frac{6^2}{12} = -105 \text{ kNm}; \quad M_{FBC} = -105 \text{ kNm}$$

$$M_{FBA} = +35 \times \frac{6^2}{12} = +105 \text{ kNm}; \quad M_{FCB} = +105 \text{ kNm}$$

$$M_{FCD} = -15 \times \frac{6^2}{12} = -45 \text{ kNm}; \quad M_{FDC} = +45 \text{ kNm}$$

Joint	*Members*	*Relative Stiffness*	Σk	$DF = k/\Sigma k$
	AE	0.28		0.38
A	AB	0.17	0.73	0.23
	AI	0.28		0.38
	BF	0.28		0.31
B	BC	0.17		0.19
			0.90	
	BJ	0.28		0.31
	BA	0.17		0.19

Joint	Members	Relative Stiffness	$\sum k$	$DF = k/\sum k$
	BF	0.28		0.23
	BC	0.33		0.27
B	BJ	0.28	1.22	0.23
	BA	0.33		0.27

Two Cycle Moment Distribution Method

	A	B		C		D
Members	*AB*	*BA*	*BC*	*CB*	*CD*	*DC*
DF	0.37	0.27	0.27	0.27	0.27	0.37
FEMS	-105.00	+105.00	-105.00	+105.00	-45.00	+45.00
Bal	+38.85	0.00	0.00	-16.20	-16.20	-16.65
Co	0.00	+19.43	-8.10	0.00	-8.33	-8.10
Bal		-3.06	-3.06	+2.25	+2.25	+3.00
Total	-66.15	121.37	-116.16	+91.05	-67.28	+23.25

$$\text{Simply supported } BM = \frac{wl^2}{8} = 15 \times \frac{6^2}{8} = 67.5 \text{ kNm}$$

$$\text{Net moment} = 67.5 - 116.16 = -48.7 \text{ kNm}$$

The hinges are introduced at the mid height of the column.

Assumption is made in the analysis of mill bents

(1) The horizontal reactions are equal.

(2) The point of inflection is at half height from $(h/2)$ the fixed base. If the columns are partially restrained, then the inflection points are $(h/3)$ from the base.

Example 6.18 Analyse the Mill bent frame shown in Fig. 6.90 by the approximate method. Determine the horizontal shear, axial load and the moment at the base of the column.

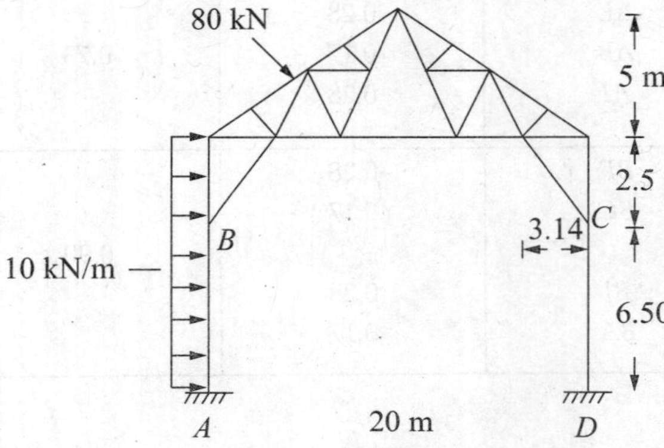

FIG. 6.90

Solution: Mill bents a-re commonly used in industrial structures. The components of the Mill bent include (1) vertical columns (2) roof truss (3) side girts and (4) knee brace. The knee brace gives lateral stability of the Mill bent. The lateral loads are the wind loads. The vertical faces to the Mill bent is covered by sheeting. These side sheets are supported on girts which are regularly spaced on the sides of the column.

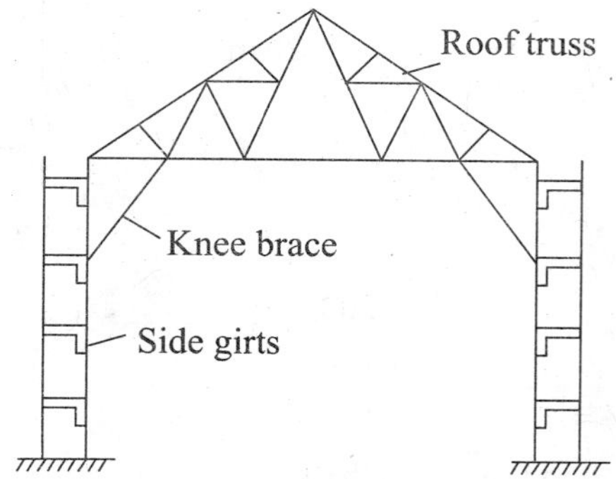

FIG. 6.91 Mill bent details

Assume the inflection point at half the height of the column, analyse the frame above the point of inflection.

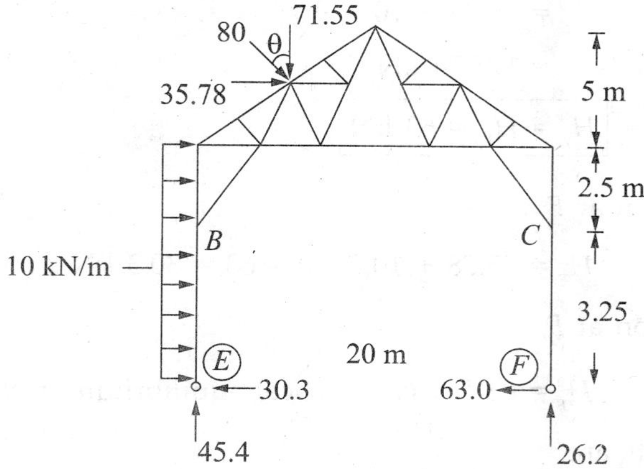

FIG. 6.92

Taking moment E,

$$\Sigma M_E = 0;$$

$$10(5.75)\left(\frac{5.75}{2}\right) = 71.55(5) - 20V_F = 0$$

$$\boxed{V_F = 26.2 \text{ kN}}$$

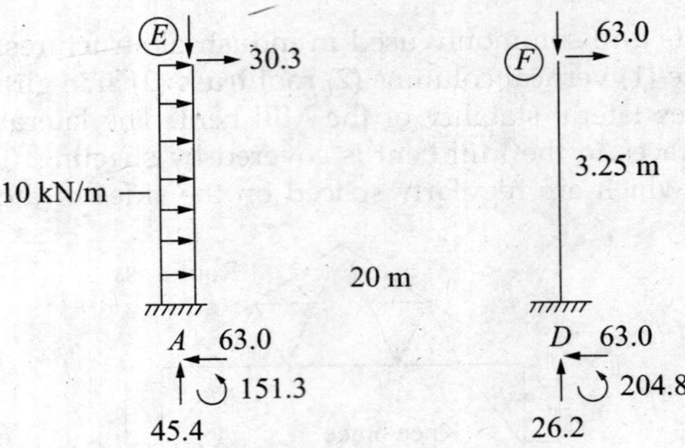

FIG. 6.93

$$\Sigma V = 0;$$

$$V_F + V_E = 71.55$$

$\therefore$ $$\boxed{V_E = 45.4 \text{ kN}}$$

Total horizontal load

$$= 35.78 + 10 \times 9$$

$$= 125.78 \text{ kN}$$

$$\boxed{H_A = H_D = 63 \text{ kN}}$$

Horizontal reaction at E,

$$H_E = 35.78 + 10(5.75) - 63 = 30.3 \text{ kN}$$

Horizontal reaction at F,

$$H_F = 63 \text{ kN } (\because \text{ vertical equilibrium of column } FD)$$

The base moments are

$$M_A = 30.3(3.25) + 10\left(\frac{3.25}{2}\right)^2 = 151.3 \text{ kNm}$$

$$M_F = 63 \times 3.25 = 204.8 \text{ kNm}$$

6.6 ANALYSIS OF INDETERMINATE TRUSSES

Consider a parallel chord indeterminate truss shown in Fig. 6.94.

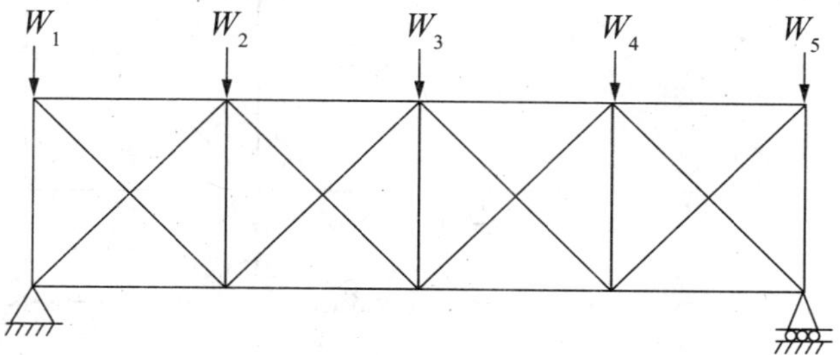

FIG. 6.94 Indeterminate truss

Number of members in the truss = 21

Number of joints in the truss = 10

Number of members > $(2nJ - 3)$

$$nm > (2nJ - 3)$$

$$21 > 17$$

Hence, the static indeterminacy is (21 – 17) = 4. Therefore, four assumptions are required to make the above indeterminate truss to statically determinate truss. This is possible by adopting any of the following methods.

Method 1

Diagonal carry equal forces

The braced structure shown above is made statically determinate by assuming that the diagonal members are equally loaded. This assumption is justified because the extension of one diagonal under the loads is equal to the compression of another diagonal in the same panel. As the extension or compression is same within the panel, force in each diagonal member is same provided the cross-sectional areas are same.

Method 2

Diagonals resist tension only

In case of transmission towers, long slender diagonal members cannot carry any compressive stress. In such cases, the diagonals are able to take up tension only.

Method 3

The given truss is split into two or more trusses. In principle, this method of splitting the given truss seems to be erroneous. The result will lead to over design. Hence, this can be used for minor structures.

Example 6.19 Analyse the truss shown in Fig. 6.95 by

(i) assuming the diagonals are equally strong in compression and tension.

(ii) diagonal resist tension only.

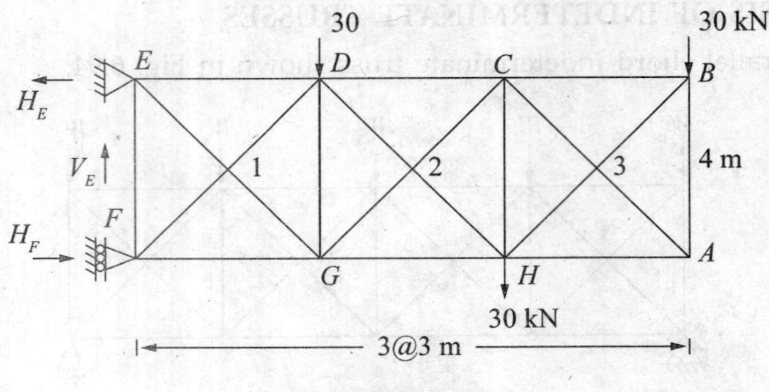

FIG. 6.95

Solution

(i) *Diagonals are equally strong in compression and tension*

$$\Sigma M_E = 0;$$

$$30(3) + 30(6) + 30(9) - 4H_F = 0$$

$$\boxed{H_F = 135 \text{ kN}}$$

$$\Sigma H = 0;$$

$$H_F - H_E = 0$$

$$\therefore \quad H_E = 135 \text{ kN}$$

$$\Sigma V = 0;$$

$$V_E - 30 - 30 - 30 = 0$$

$$\boxed{V_E = 90 \text{ kN}}$$

Panel *ABCH*

The panel shear = 30 kN

Each diagonal resists half the panel shear, i.e., 15 kN. Thus, resolving the diagonal force of F_{HB} vertically,

$$F_{HB} \sin \theta = 30$$

$$F_{HB}\ (4/5) = 30$$

$$\boxed{F_{HB} = 18.75 \text{ kN (compressive)}}$$

In this panel, the other diagonal is AC. As it is equally strong in tension, as per the assumption in Method 1,

$$\boxed{F_{CA} = 18.75 \text{ kN (tensile)}}$$

Joint B

Knowing the force of HB, the forces F_{BC} and F_{BA} are evaluated by the method of joints.

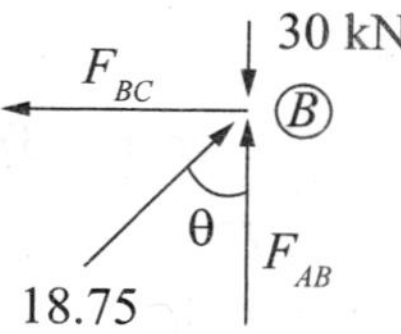

FIG. 6.96

$$\Sigma H = 0$$

$$18.75 \sin\theta - F_{BC} = 0$$

i.e.

$$\boxed{F_{BC} = 11.25 \text{ kN}}$$

$$\Sigma V = 0$$

$$18.75 \cos\theta + F_{AB} - 30 = 0$$

$$18.75(0.8) + F_{AB} - 30 = 0$$

$$\boxed{F_{AB} = 15 \text{ kN}}$$

Panel *DCHG*

The panel shear force is the sum of all the vertical loads acting to the right of the section passing through the diagonal member 2. Hence, the panel shear force is 30 + 30 = 60 kN. Half of this panel shear is resisted by diagonal GC and DH respectively.

$$F_{GC} \sin\theta = 0$$

$$F_{GC} (4/5) = 30$$

$$\boxed{F_{GC} = 37.5 \text{ kN}}$$

The other diagonal in that panel is DH. Hence,

$$\boxed{F_{DH} = 37.5 \text{ kN}}$$

The nature of this force is tensile while FGC is compressive.

Consider joint C

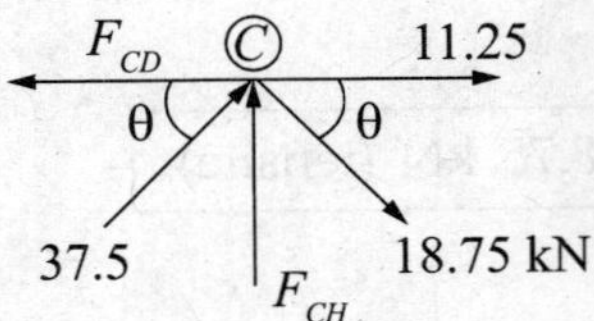

FIG. 6.97

$\Sigma V = 0;$

$$F_{CH} - 18.75 \sin\theta + 37.5 \sin\theta = 0$$

$$F_{CH} = -18.75 \sin\theta$$

$$\boxed{F_{CH} = -15 \text{ kN}}$$

$\Sigma H = 0;$

$$-F_{CD} + 37.5 \cos\theta + 18.75 \cos\theta + 11.25 = 0$$

$$\boxed{F_{CD} = 45 \text{ kN}}$$

$\Sigma M_{02} = 0;$

$$-45(2) + 30(1.5) + 30(4.5) - 2F_{GH} = 0$$

$$\boxed{FGH = 45 \text{ kN}}$$

Panel *DEFG*

Half the panel shear $= \frac{1}{2}$ (30 + 30 + 30) = 45. Resolving the forces vertically,

$$F_{GE} \sin\theta = 45$$

$$F_{GE} (0.8) = 45$$

$$\boxed{F_{GE} = 56.25 \text{ kN}}$$

$\therefore$ $$\boxed{F_{DF} = 56.25 \text{ kN}}$$

Joint D

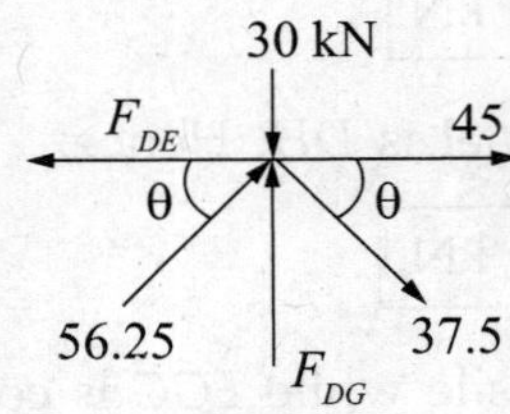

FIG. 6.98

$\Sigma H = 0;$

$$45 - F_{DE} + 37.5 \cos\theta + 56.25 \cos\theta = 0$$

$$\boxed{F_{DE} = 101.25 \text{ kN}}$$

$\Sigma V = 0;$

$$F_{DG} - 30 - 37.5 \sin\theta + 56.25 \sin\theta = 0$$

$$\boxed{F_{DG} = 15 \text{ kN}}$$

$\Sigma M_1 = 0;$

$$2F_{GF} - 101.25 \times 2 + 30(1.5) + 30(4.5) + 30(7.5) = 0$$

$$\boxed{F_{GF} = 101.25 \text{ kN}}$$

Joint F

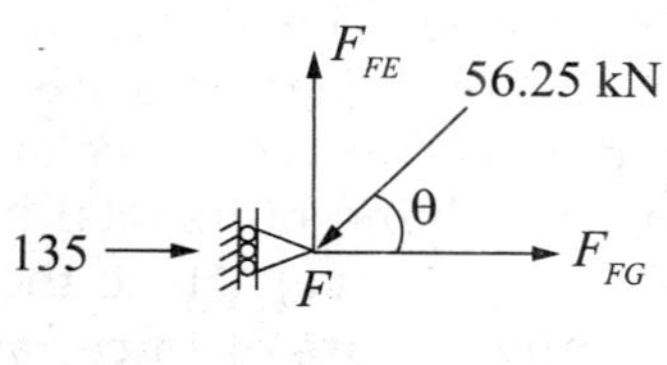

FIG. 6.99

$$F_{FE} - 56.25 \sin\theta = 0$$

$$F_{FE} = 56.25 \times 0.8$$

$$\boxed{F_{FE} = 45 \text{ kN}}$$

$\Sigma H = 0;$

$$135 - 56.25 \cos\theta - F_{FG} = 0$$

$$\boxed{F_{FG} = 101.3 \text{ kN}}$$

(ii) *Diagonals carry tension only*

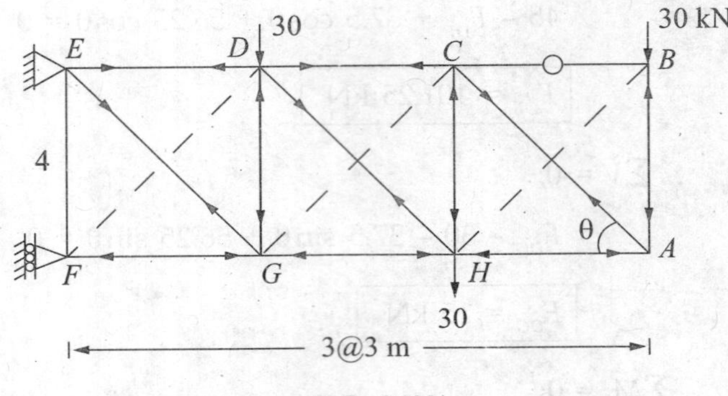

FIG. 6.100

The nature of forces is marked by the method of inspection. For example, consider joint *B*. At joint *B*, *CB*, *HB* and *AB* are meeting. A vertical force of 30 kN is acting down. This force can be balanced at joint *B* by an upward force. Hence, F_{BA} should be upwards at joint *B*. As there is no horizontal force at *B*, F_{BC} = 0. The force in the diagonals *HB*, *GC* and *FD* are taken as zero. The reason is that they are able to resist compressive force and hence neglected. Therefore, F_{BH} = 0. After marking the nature of force at *B*, move on to joint *A*. Three members *AB*, *AC* and *AH* are meeting at *A*. The vertical force F_{AB} is acting down. This can be balanced by the vertical component of F_{AC}. Hence, F_{AC} acts upwards. Resolving horizontal it gives a component to the left. Hence, the horizontal force F_{AH} (compressive) acts to the right to balance the horizontal component of F_{AC}. After marking the nature of forces at joint '*B*' and '*A*', we consider joint *C*. At joint *C*, F_{CA} gives a horizontal component towards right. This will be balanced by the force F_{CD} (tensile) which acts to the left from *C*. Resolving vertically at *C*; F_{CA} gives downward component. This is balanced by vertical upward force of F_{CH}. This procedure is continued till all the forces are marked. The method of joints has been used to evaluate all the forces. The computed forces are given in Table 6.2.

This table shows the member forces do vary between Method 1 and Method 2. The results of the approximate methods do not match as the methods are based on different assumptions. However, the literature points out that the results of Method 1 agree satisfactorily with the results of classical methods.

Table 6.2 Values of the member forces (kN)

Members	*Method 1*	*Method 2*	*Nature*
ED	101.25	67.50	>—<
DC	45.00	22.50	>—<
CB	11.25	0.00	>—<

Contd...

Contd...

BA	15.00	30.00	⟷
AH	11.25	22.50	⟷
HG	45.00	67.50	⟷
GF	101.25	135.00	⟷
FE	45.00	0.00	≻—≺
DG	15.00	90.00	⟷
CH	15.00	30.00	⟷
BA	15.00	30.00	⟷
AC	18.75	37.50	≻—≺
GC	37.5	0.00	⟷
DH	37.50	75.00	≻—≺
FD	56.25	0.00	⟷
EG	56.25	112.50	≻—≺

6.7 NOTE ON THE APPROXIMATE METHODS

Approximate methods can be used to determine the member sizes for preliminary design only. The accuracy of the analysis depends on the location of interior hinges. This in turn depends on the structural behaviour and complexity of the structure. Hence, the approximate methods are to be used with caution.

REVIEW QUESTIONS

Remembrance

6.1. Does the equilibrium condition is satisfied in the approximate method of analysis?
6.2. Does the boundary conditions are satisfied in the approximate method of analysis?
6.3. Does the compatibility condition is satisfied in the approximate method of analysis?
6.4. What are the advantages of the approximate method of analysis?
6.5. Is the design based on the approximate method is conservative or economical?
6.6. List the type of structures that can be analysed by the approximate method of analysis?

6.7. Name the methods used to analyse the multistorey frames subjected to lateral loads.

6.8. How many inflection points are there in a propped cantilever?

6.9. How many inflection points are there in a fixed beam?

6.10. Name the lateral loads that act in a rigid frame.

6.11. What is the value of axial force in the girder of a frame when we apply approximate method of analysis?

6.12. Under what type of boundary conditions of a portal frame, we assume the interior hinge at half the height of the column?

6.13. Under what type of boundary conditions of a portal frame, we assume the inter hinge at one-third height of the column?

6.14. List the assumptions made in the Portal method.

6.15. List the assumptions made in Cantilever method.

6.16. What are the assumptions made in the approximate method of analysis of indeterminate truss?

6.17. Does the Portal method require flexural rigidities and axial stiffness of the members?

Understanding

6.1. In a rigid frame (fixed at their bases), the number of girders is '*n*'. What is the degree of static indeterminacy?

6.2. In a rigid frame fixed at their bases, the number of bays is five and the number of storeys is three, compute the static indeterminacy.

6.3. Locate the inflection point of a propped cantilever of span 8 m subjected to a point load of 100 kN at midspan.

6.4. Locate the inflection point of a fixed beam of span 10 metres subjected to a concentrated load of 100 kN.

6.5. Is bending deformation considered in the Cantilever method or shear deformations?

6.6. What kind of force is evaluated first in the Cantilever method? Is it axial, shear or moment?

6.7. A space frame is to be analysed. How the approximate method of analysis can be used?

6.8. In the Portal method, is the shear force or axial force evaluated first?

6.9. What girder moment factor and column moment factor indicate?

6.10. Why the factor method is considered to be accurate when compared to the Portal method and the Cantilever method?

6.11. Locate the point of inflection of a propped cantilever subjected to udl and of span l.

6.12. Locate the point of inflection of a propped cantilever subjected to a moment at the prop.

6.13. Convert the statically indeterminate truss shown in the figure below to a statically determinate truss such that shear in each panel is carried by the diagonal in tension.

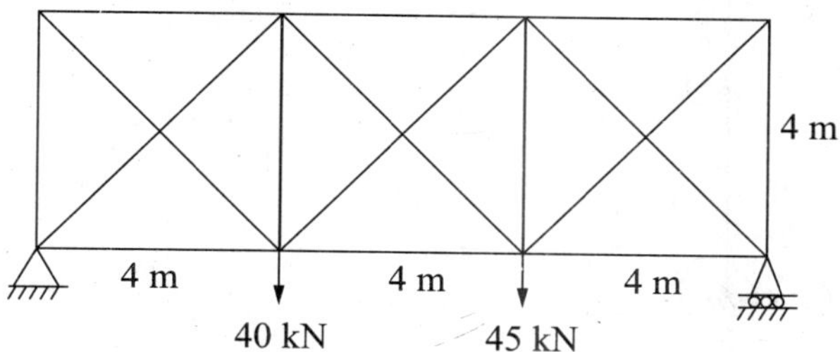

6.14. Locate the point of inflection for the trussed portal.

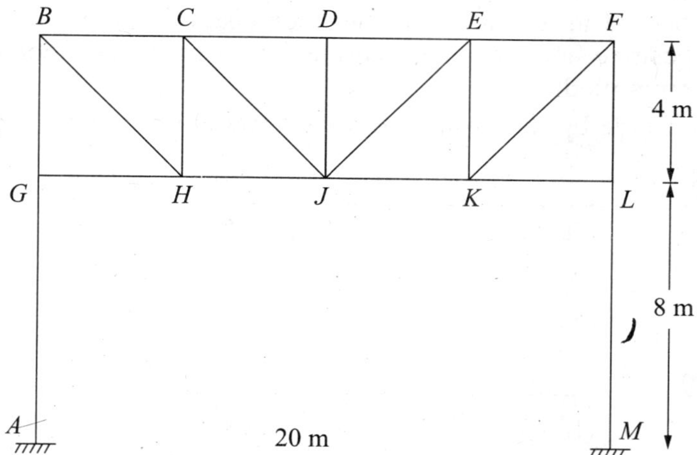

6.15. Why the separate approximate method of analysis is given for gravity loads and lateral loads?

6.16. Does the Portal method or the Cantilever method or the factor method give lateral deflections of a tall multistorey frame?

EXERCISE PROBLEMS

6.1. Analyse the portal frame shown in the figure below by the approximate method. Assume the inflection point at a distance of one-tenth of the span of the beam.

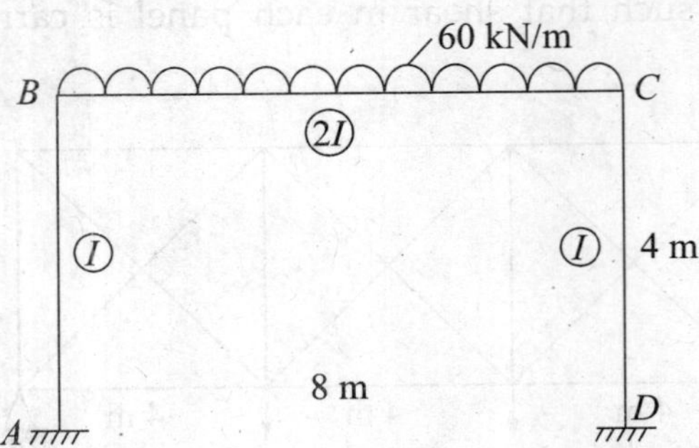

Ans: $M_A = 86.4$ kNm; $M_B = 172.8$ kNm

6.2. A six-storey has frames spaced at 4 metres centre to centre. Each frame has four bays of 6 metres span. In an intermediate floor, the floor beam is 200 × 600 mm and the columns are 200 × 600 mm. The floor thickness is 150 mm. The storey height is 3.6 m. The live load on the floor beam is 30 kN/m. Determine the maximum positive and negative bending moment in the floor beam. Use the substitute frame method.

6.3. Analyse the frame by the Portal method. Draw the bending moment diagram.

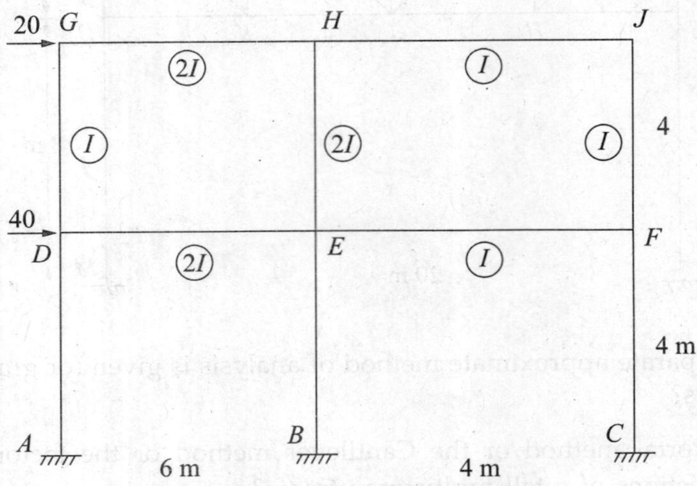

Ans: $M_{AD} = 30$ kNm; $M_{BE} = 60$ kNm; $M_{CF} = 30$ kNm;

$M_{DG} = 10$ kNm; $M_{EH} = 20$ kNm; $M_{FJ} = 10$ kNm;

$M_{DE} = 40$ kNm; $M_{ED} = 20$ kNm; $M_{FE} = 40$ kNm;

$M_{HG} = 10$ kNm; $M_{HE} = 10$ kNm; $M_{HJ} = 10$ kNm

6.4. Analyse the frame by the Cantilever method. Compute the base moments.

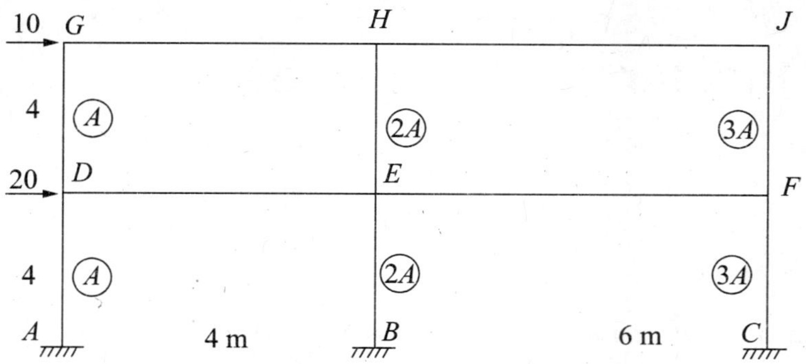

Ans: M_{AD} = 2.76 kNm; M_{BE} = 50.0 kNm M_{CF} = 21.7 kNm

6.5. Analyse the frame by the Factor method. *EI* is constant for all the elements.

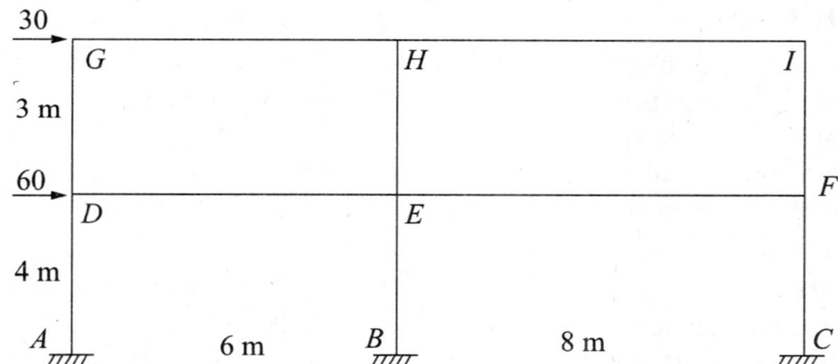

Ans: M_{AD} = 71.36 kNm; M_{BE} = 75.22 kNm; M_{CF} = 43.72 kNm;
M_{DE} = 59.19 kNm; M_{DA} = 46.29 kNm; M_{DG} = 12.90 kNm;
M_{GD} = 14.55 kNm; M_{HG} = 12.21 kNm; M_{EH} = 21.17 kNm;
M_{HE} = 9.48 kNm; M_{IH} = 11.90 kNm; M_{FC} = 65.15 kNm

7 Indeterminate Trusses

Objectives:

Degree of redundancy (internal and external redundancy. Analysis of externally indeterminate truss by strain energy method. Analysis of internally indeterminate truss by strain energy method. Analysis of truss due to lack of fit/fabrication error. Analysis of truss due to temperature effects. Effect of variation in cross-sectional area. Trussed beams.

7.1 INTRODUCTION

A structural member is in equilibrium under the action of a known type of force system. The unknown reactions and member forces are evaluated using equations of equilibrium in statically determinate structure. In the analysis of such structures, we do not use the size of the members or the elastic properties of the material (viz., modulus of elasticity) for the determination of reactions or member forces.

On the other hand, a statically indeterminate structure is one having more supports or more members than are necessary to maintain equilibrium of the structure. In such kind of indeterminate structures, equilibrium equations alone are insufficient for the analysis. It should be supplemented by additional equations. The additional equations necessary to solve statically indeterminate structures from the prescribed conditions of deflection or rotation are the requirements of consistent deformation. The following are the example of statically indeterminate structures.

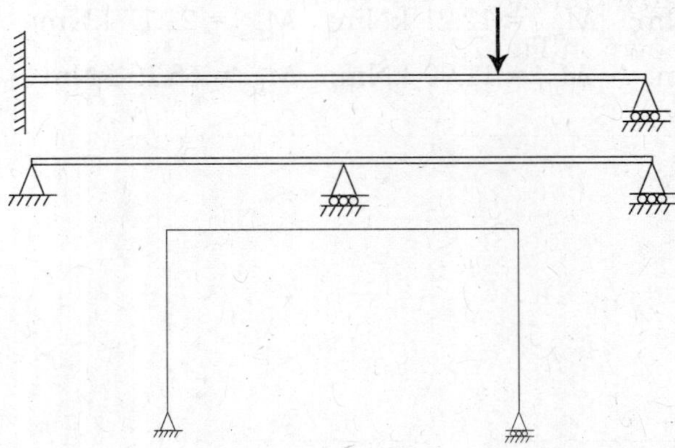

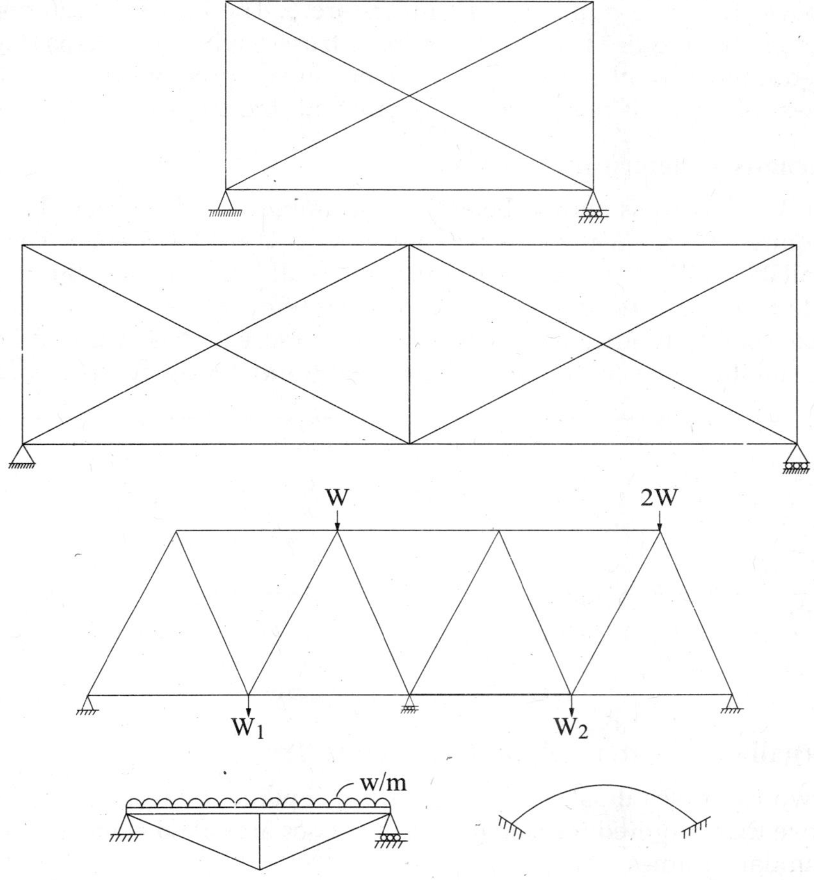

FIG. 7.0 Indeterminate structures

7.1.1 Internally Indeterminate Truss

Consider the truss shown in Fig. 7.1.

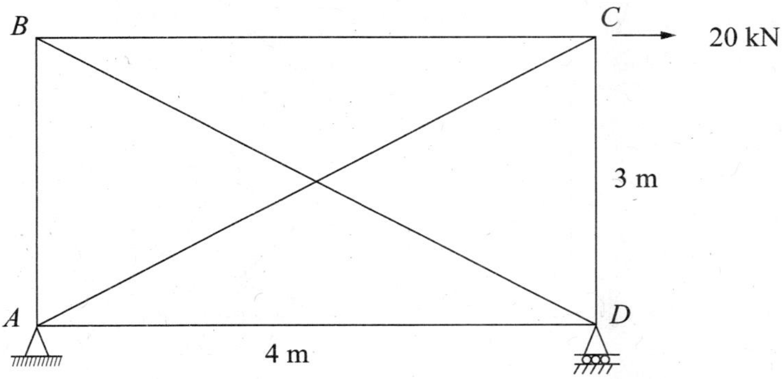

FIG. 7.1 Indeterminate truss

In the above truss, the number of members are 6. The number of joints are 4. The number of members required for a determinate truss having 4 joints is $2nj - 3 = 5$. It means the redundant member is $6 - 5 = 1$. Hence, the degree of statical indeterminacy is 1. It implies that even if one member is removed, the truss is stable.

7.1.2 Externally Indeterminate Truss

Consider the Warren truss shown below. The number of joints is 13. The number of members required for statically determinate truss is $2 \times 13 - 3 = 23$. The number of members available is 23. Therefore, the truss is internally determinate. But it is externally indeterminate to the second degree. That is, if we remove the supports C and *D*; still it is stable externally. In other words, it has two more reactions at C and *D* more than required for stability. Hence, the truss is externally indeterminate to second degree.

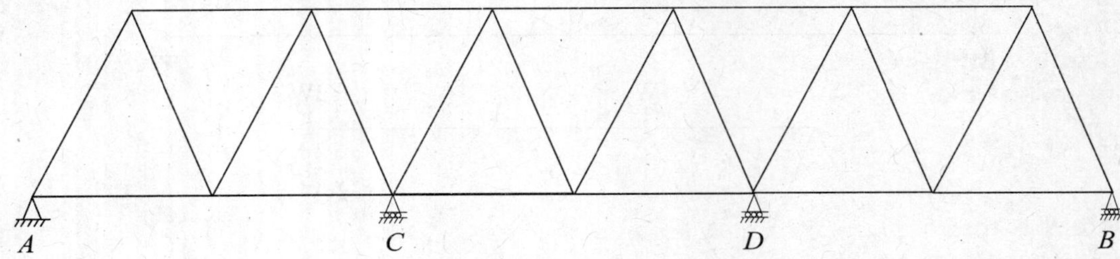

FIG. 7.2 Indeterminate truss

7.1.3 Internally and Externally Indeterminate Truss

The above two examples illustrate that if a truss contains more members or reactions over and above than required for stability, such trusses are called statically indeterminate truss or redundant frames.

Some trusses may be statically indeterminate both internally as well as externally. In these kind of trusses $n_m > (2n_j - 3)$ and the number of external reactions are more than three. A typical truss of such kind is given below.

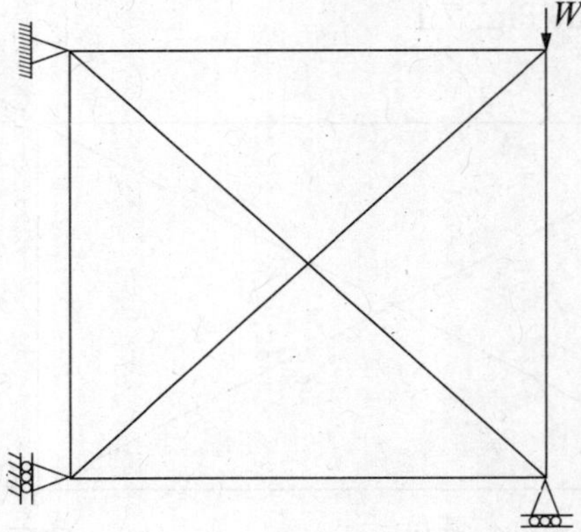

FIG. 7.3 Indeterminate truss

It is to be mentioned here that the member forces in a statically indeterminate truss cannot be found out from statics alone. The forces in the members can be computed by the application of Castigliano's second theorem.

7.2 CASTIGLIANO'S SECOND THEOREM

It states that the redundant reaction of a statically indeterminate structure is low as to make the total strain energy stored up a minimum. This is also called theorem of least work or theorem of minimum strain energy or theorem of minimum resilience.

Statement

The work done in stressing a structure under a given system of loads is the least possible consistent with the maintenance of equilibrium. The differential coefficient of the work done, with respect to one of the forces is thus equal to lack of fit. In other words, the first partial derivative of the total internal strain energy in an indeterminate structure with respect to the force in a redundant member is equal to the initial lack of fit in that member

$$\frac{\partial U}{\partial R} = \lambda$$

where R is the force in any redundant member and U is the total strain energy of the whole frame and λ is the lack of it (initial misfit) of the particular member.

7.2.1 Application of Castigliano's Theorem

Consider the frame shown if Fig. 7.4. This frame is statically indeterminate to one degree. Let AD is a redundant member. Even after removing the member AD, the truss is statically determinate frame. This statically determinate frame can be analysed using static equilibrium conditions.

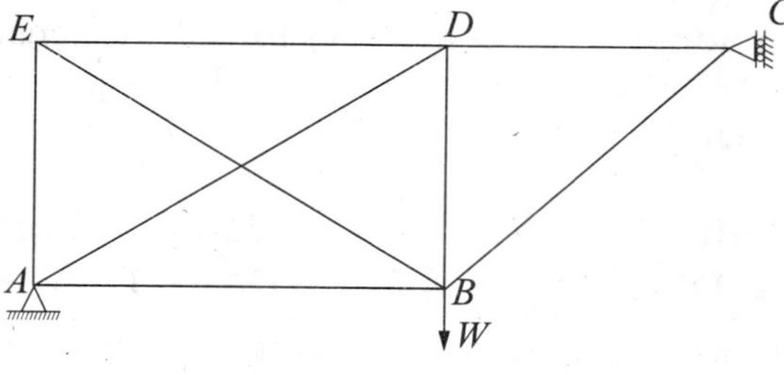

FIG. 7.4

Let P_i be the force in any member 'i' of the statically determinate frame, due to the applied loading. Now apply unit load in the direction of the redundant member at the joint location where the member was connected. Let u_i be axial force in the member 'i' due to unit load. Then the axial force in the member 'i' is Ru_i.

The strain energy stored in the member due to total axial force $(P + Ru_i)$ is

$$\frac{(P + Ru_i)^2 \, l_i}{2A_iE}$$

where l_i is the length of the member and A_i is the cross-sectional area.

The total strain energy of the frame is

$$U = \sum \frac{(P_i + Ru_i)^2 \, l_i}{2A_iE} \tag{1}$$

Applying Castigliano's second theorem

$$\frac{\partial U}{\partial R} = \sum \frac{(P_i + Ru_i) \, l_i u_i}{A_iE} = \lambda \tag{2}$$

Expanding the above expression

$$\sum \frac{P_i u_i l_i}{A_iE} + R \sum \frac{u_i^2 l_i}{A_iE} = \lambda \tag{3}$$

$$R = \frac{\lambda - \Sigma P_i u_i l_i / A_iE}{\Sigma u_i l_i^2 / A_iE} \tag{4}$$

If the redundant member is an exact fit in the unstrained condition, then $\lambda = 0$, the above equation reduces to

$$\boxed{R = \frac{-\Sigma P_i u_i l_i / A_iE}{\Sigma u_i l_i^2 / A_iE}} \tag{5}$$

It is to be noted that the summation of the quantity in the numerator is for all the members excluding the redundant member. The summation of the quantity in the denominator is done for all the members including the redundant member.

The sign convention adopted is positive (+) for all the tensile member forces and negative (–) for all the compressive member forces. This sign convention is adopted in tabulation of the computed values.

7.3 PROCEDURE FOR THE ANALYSIS OF INDETERMINATE TRUSS—SINGLE DEGREE OF REDUNDANCY

(1) Remove the redundant member, and determine the member forces (P) in the remaining members due to external loads. The member forces of the statically determinate frame (with redundant member being removed) can be found by the method of joints/method of sections.

(2) Apply a unit tensile force in the redundant member and determine the member forces (u) in all the members including the redundant member.

(3) It is to be noted that in step (1) the member forces are evaluated for the truss by excluding the redundant member and in step (2), the member forces are computed for all the members of the truss including redundant member.

(4) Tabulate the length of the member l, axial rigidity of the member AE and member forces P, u respectively.

(5) Determine the redundant force R in the member using Eq. (5)

(6) The other member forces are computed using

$$\text{Member force} = (P + Ru)$$

Example 7.1 Determine the forces in the members of the cantilever truss shown in Fig. 7.5 AE is constant.

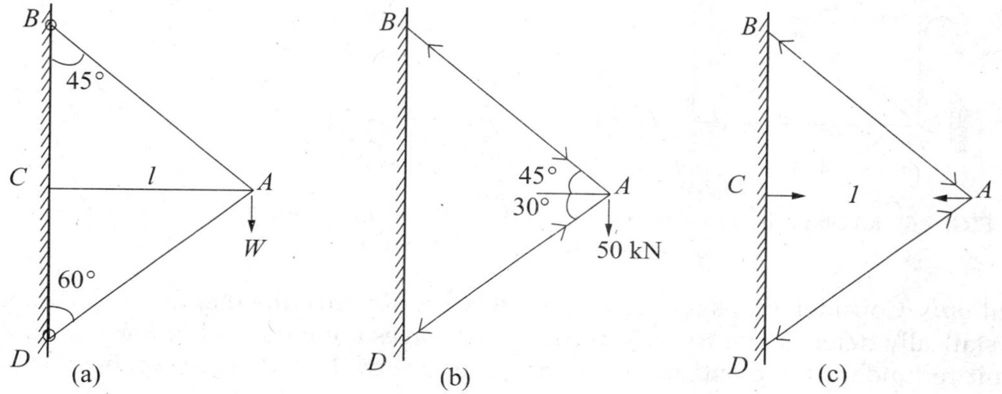

FIG. 7.5 (a) Cantilever indeterminate truss (b) determinate truss (c) determinate truss with a redundant force.

Solution: Three members AB, AC and AD are meeting at joint A. Three independent equations are required to determine the member forces. Two equilibrium forces are formulated by summing up the forces in the horizontal and vertical directions respectively. Third equation is formulated as follows. AC is taken as a redundant member. The determinate truss (after removing AC) member forces were obtained using the method of joints. They are listed in the Table as P. A unit force is applied through A in the direction of AC, the resulting member forces u are listed in the Table. As AE is constant, it is not mentioned in the Table. The redundant force R is obtained and the final member forces $(P + Ru)$ are computed in the last column of the Table.

Members	P	u	l	$\frac{Pul}{AE}$	$\frac{u^2l}{AE}$	$P + Ru$
AB	44.84	−0.518	$1.414l$	$-32.84l$	$0.38l$	44.35
AC	0	+1.00	l	0	1.00	0.95
AD	−36.60	−0.732	$1.155l$	$+30.94l$	$0.62l$	−37.30
				$-1.90l$	$2l$	

$$R = \frac{\Sigma Pul/AE}{\Sigma u^2 l/AE}$$

$$R = 1.90l/2l = 0.95 \text{ kN}$$

(+) sign indicates tensile force.

Example 7.2 In the pin jointed work, all the members have the same cross-sectional area. Determine the axial force in the member AC.

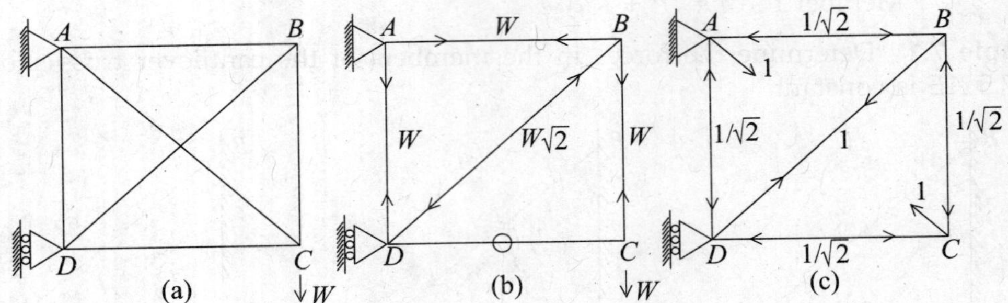

FIG. 7.6 (a) Statically indeterminate truss (b) statically indeterminate truss (c) determinate truss with unit redundant force.

Solution: Consider AC as the redundant member. Remove the member AC and analyse the statically determinate truss. The computed forces are indicated in Fig. 7.6(b). Apply a unit redundant force and the member forces are evaluated. The member forces due to the applied unit redundant force are indicated in Fig. 7.6(c). The redundant force in member Ac is obtained using the computed values. The computed member forces are tabulated.

Members	p (kN)	u (kN)	l (m)	$\frac{Pul}{AE}$	$\frac{u^2l}{AE}$	$P + RU$
AB	W	$-(1/\sqrt{2})$	l	$-\frac{Wl}{\sqrt{2}}$	$\left(\frac{l}{2}\right)$	$0.396W$
BC	W	$-(1/\sqrt{2})$	l	$\frac{Wl}{\sqrt{}}$	$\left(\frac{l}{2}\right)$	$0.396W$
CD	0	$-(1/\sqrt{2})$	l	0	$\left(\frac{l}{2}\right)$	$-0.604W$
DA	W	$-(1/\sqrt{2})$	l	$-\frac{Wl}{\sqrt{2}}$	$\left(\frac{l}{2}\right)$	$0.396W$
DB	$-W\sqrt{2}$	+1.00	$l\sqrt{2}$	$-2Wl$	$l\sqrt{2}$	$-0.56W$
AC	0	+1.00	$l\sqrt{2}$	0	$l\sqrt{2}$	$0.854W$
				$-4.121Wl$	$4.828l$	

$$R = -\left(-\frac{4.121Wl}{4.828l}\right)$$

$$\boxed{R = 0.854W}$$

The (+) sign indicates R is tensile force. The common factor AE is not entered in the table.

Example 7.3 Determine the force in the tie bar BC. The members in tension have a cross-sectional area 'a' and the members in compression have a cross-sectional area '$2a$'.

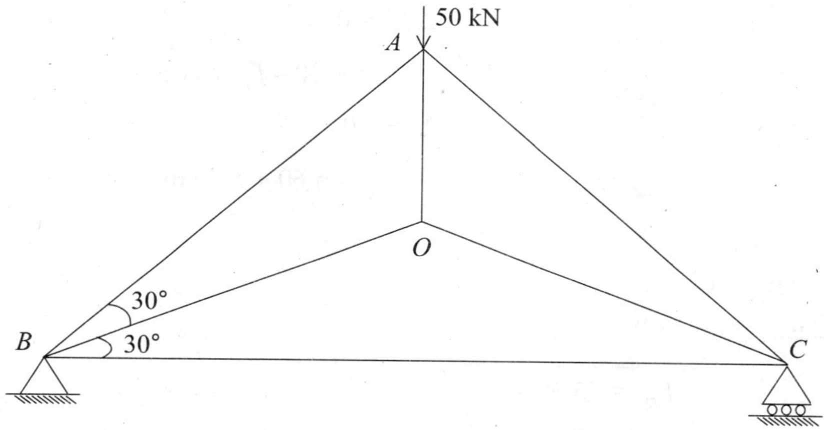

FIG. 7.7

Solution: The total number of members are six. The number of joints are 4. Hence, the required numbers of members for a statically determinate truss is 2 × 4 – 3 = 5. Hence, its statical indeterminacy is (6 – 5) = 1. Thus, remove the redundant member BC and analyse the statically determinate frame.

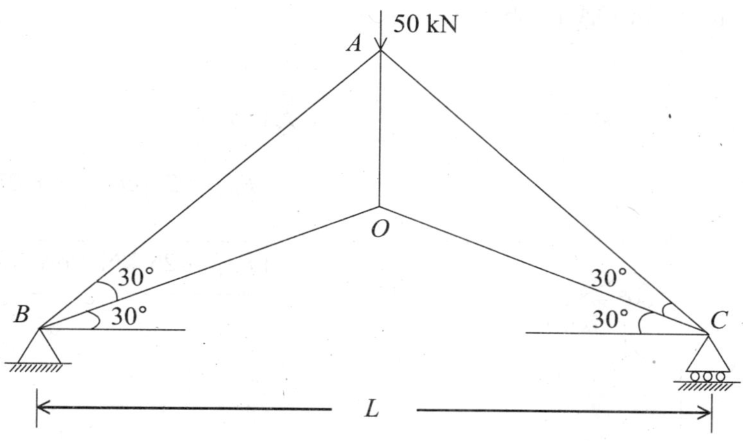

FIG. 7.8

The lengths of the members are found out using geometry as

$$\cos 60 = \frac{(L/2)}{AB} \qquad \cos 30 = \frac{(L/2)}{OB}$$

$$AB = \frac{(L/2)}{\cos 60} \qquad OB = \frac{(L/2)}{\cos 30}$$

$$\therefore \boxed{AC = AB = L} \qquad \boxed{OB = OC = OA = 0.577L}$$

Consider joint B and using method of joints

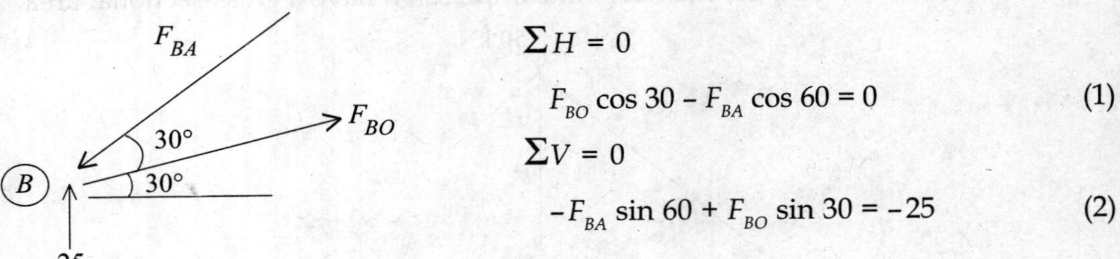

$$\Sigma H = 0$$

$$F_{BO} \cos 30 - F_{BA} \cos 60 = 0 \tag{1}$$

$$\Sigma V = 0$$

$$-F_{BA} \sin 60 + F_{BO} \sin 30 = -25 \tag{2}$$

FIG. 7.9a

Solving Eqs. (1) and (2)

$$F_{BA} = 43.30 \text{ kN}$$
$$F_{BO} = 25.00 \text{ kN}$$

Applying the principle of symmetry

$$F_{CA} = 43.30 \text{ kN (compressive)}$$
$$F_{CO} = 25.00 \text{ kN (tensile)}$$

The member force in OA is obtained as

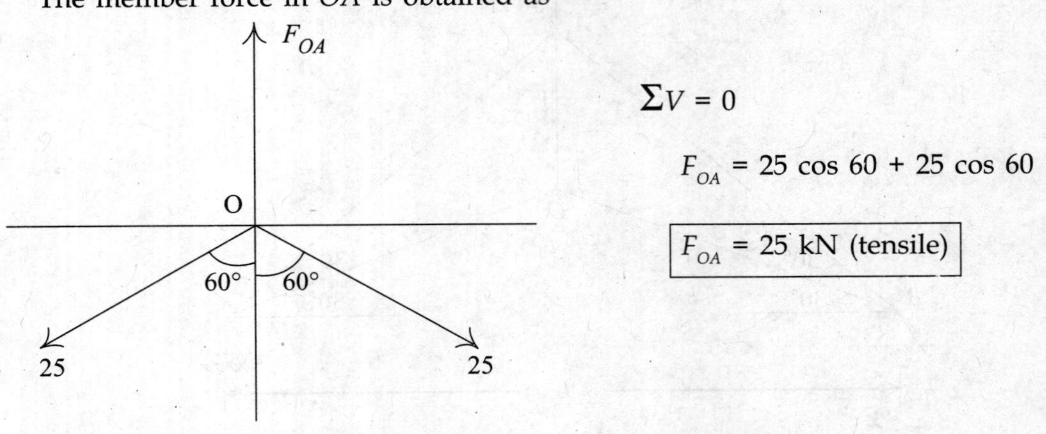

$$\Sigma V = 0$$

$$F_{OA} = 25 \cos 60 + 25 \cos 60$$

$$\boxed{F_{OA} = 25 \text{ kN (tensile)}}$$

FIG. 7.9b

The computed member forces are indicated in Fig. 7.10.

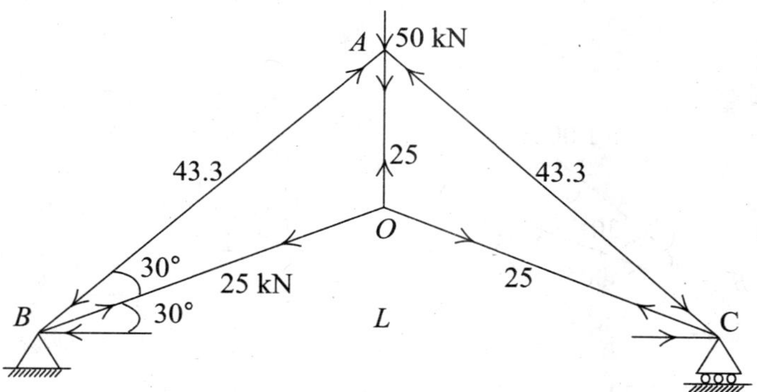

FIG. 7.10 Member forces in statically determinate truss

Now apply unit tensile force in the tie member and once again analyse the truss (after removing the externally applied load of 50 kN). Consider the joint *B*.

Assume initially F_{AB} and F_{BO} as tensile (the force should act away from the joint *B*).

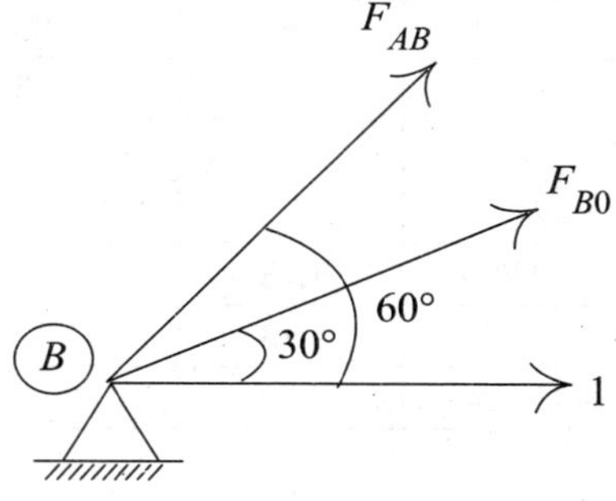

FIG. 7.11

$$\Sigma H = 0$$

$$F_{BO} \cos 30 + F_{AB} \cos 60 + 1 = 0$$

$$F_{BO} \cos 30 + F_{AB} \cos 60 = -1 \qquad (3)$$

$$\Sigma V = 0$$

$$F_{BO} \sin 30 + F_{AB} \sin 60 = 0 \qquad (4)$$

Solving Eqs. (3) and (4)

$$F_{BO} = -1.732 \text{ (compressive)}$$

$$F_{AB} = +1.00 \text{ (tensile)}$$

Using symmetry;

$$F_{CO} = -1.732 \text{ (comp)}$$

$$F_{CA} = +1.00 \text{ (tensile)}$$

These forces are marked and shown below.

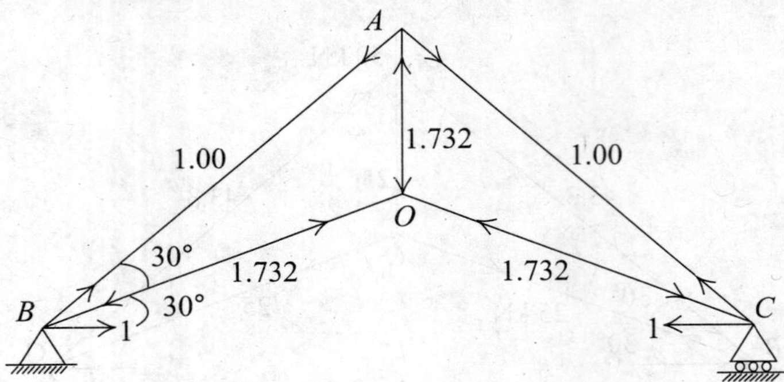

FIG. 7.12 Member forces due to unit redundant force.

The member forces due to the applied load (excluding the redundant member BC) and due to unit force (including the redundant member BC) are tabulated. For convenience (l/a) is omitted in the Table.

Member	*P*	*u*	*l*	*A*	*Pul/AE*	u^2l/AE	*P + Ru*
BA	−43.30	+1.00	*l*	2*a*	−21.65	0.5	−26.85
AC	−43.30	+1.00	*l*	2*a*	−21.65	0.5	−26.85
CB	0	+1.00	*l*	*a*	0	1.0	+16.45
AO	+25.00	−1.732	0.577*l*	*a*	−24.98	1.73	−3.50
OB	+25.00	−1.732	0.577*l*	*a*	−24.98	1.73	−3.50
OC	+25.00	−1.732	0.577*l*	*a*	−24.98	1.73	−3.50
					−118.24	7.19	

$$R = -\frac{\Sigma Pul/AE}{\Sigma u^2 l/AE} = 16.45 \text{ kN}$$

Example 7.4 Analyse the frame shown in Fig. 7.13 and compute the member forces.

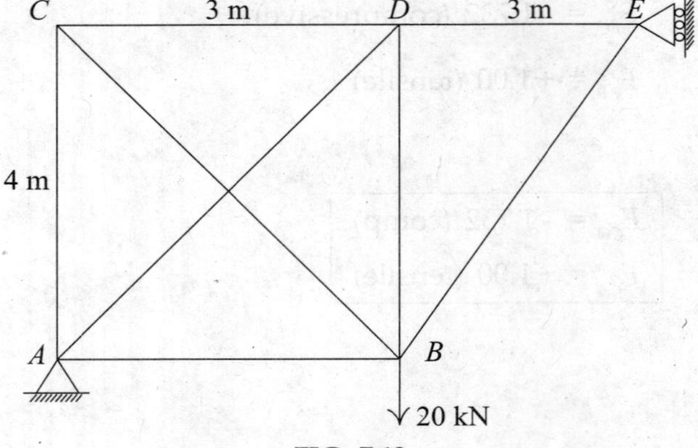

FIG. 7.13

Solution: The given frame is to be analysed for statical indeterminacy. The total number of members is 8. The number of joints is 5. The number of members for a statically determinate truss is $(2n_J - 3)$, i.e., $(2 \times 5 - 3) = 7$. As the number of members exceeds $(8 - 7) = 1$, the truss is internally indeterminate by one degree. As the number of unknown reactions is 3 and the equilibrium equations $(\sum V = \sum H = \sum M = 0)$ is three, the truss is externally determinate.

Thus, the given truss is externally determinate but internally indeterminate to the first degree. The member BC is treated as a redundant member.

Analysis of statically Determinate Truss

The member BC is removed and analyse the statically determinate truss by method of joints.

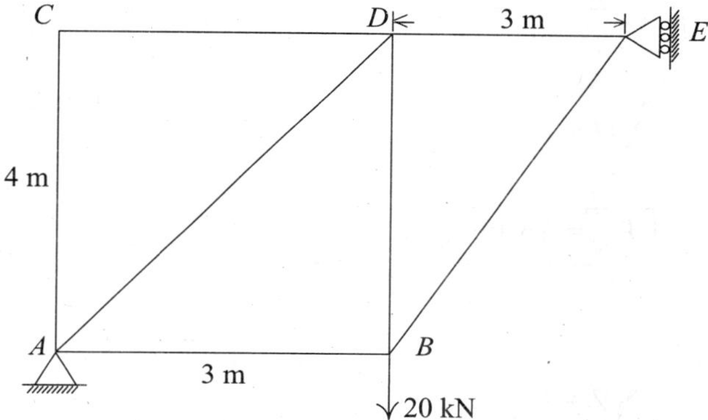

FIG. 7.14 Statically determinate truss

The reactions are found out as

$$\sum V = 0;$$

$$V_A + \cancel{V_E} = 20$$

$$V_A = 20 \text{ kN}$$

There cannot be V_E on the gliding roller. Only H_E is present, which acts perpendicular to the base of the roller.

$$\sum M = 0$$

Taking moment about the hinge support A,

$$\sum M_A = 0;$$

$$20 \times 3 = 4H_E$$

$$\boxed{H_E = 15 \text{ kN}}$$

$$\sum H = 0$$

$$H_A + H_E = 0$$

$$-15 + H_A = 0$$

$$\boxed{H_A = 15 \text{ kN}}$$

By the method of inspection at joint C,

$$\boxed{F_{AC} = F_{CD} = 0}$$

Similarly, from the joint B,

$$\boxed{F_{AB} = F_{BE} = 0}$$

Joint E

FIG. 7.15

$$\Sigma H = 0;$$

$$\boxed{F_{DE} = 15 \text{ kN}}$$

Joint B

$$\Sigma V = 0$$

$$\boxed{F_{BD} = 20 \text{ kN}}$$

FIG. 7.16

The computed member forces are marked.

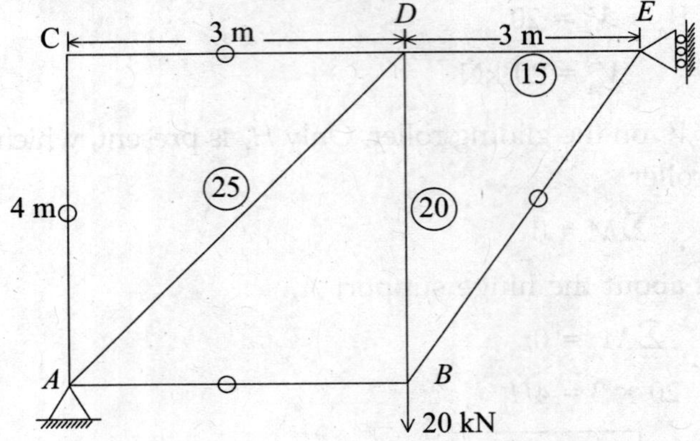

FIG. 7.17 Statically determinate truss

Statically determinate truss with unit redundant force

A unit redundant force is applied in the direction of BC. The redundant unit force can be tensile or compressive. As a matter of choice, We apply unit tensile force always as the redundant unit force.

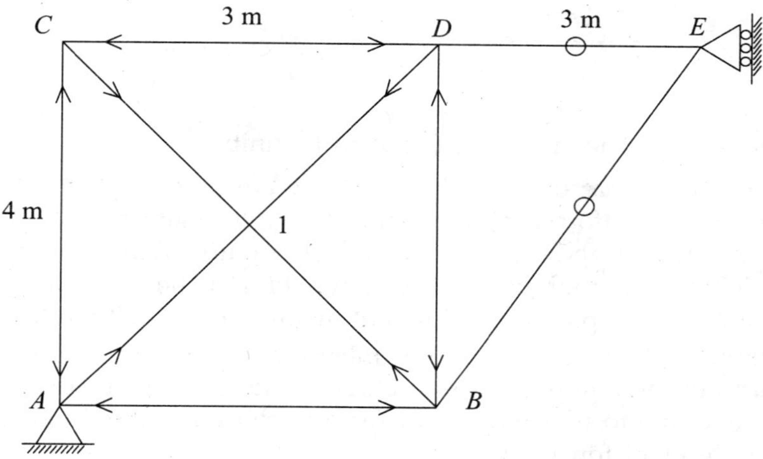

FIG. 7.18

At Joint C

$$\Sigma H = 0$$

$$F_{CD} = 1 \cos \theta$$

$$\boxed{F_{CD} = 0.6}$$

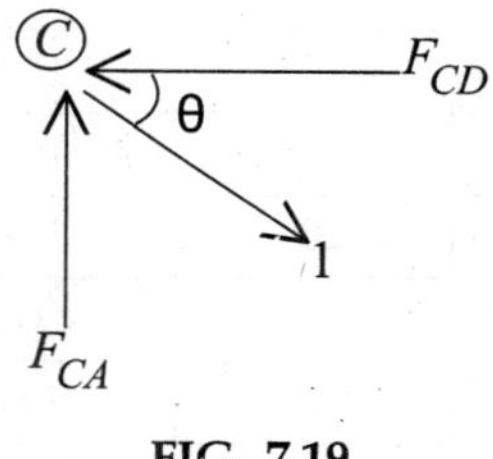

FIG. 7.19

$$\Sigma V = 0$$

$$F_{CA} = 1 \sin \theta$$

$$\boxed{F_{CA} = 0.8}$$

$$\Sigma H = 0$$

$$F_{AB} = 1 \cos \theta$$

$$\boxed{F_{AB} = 0.6}$$

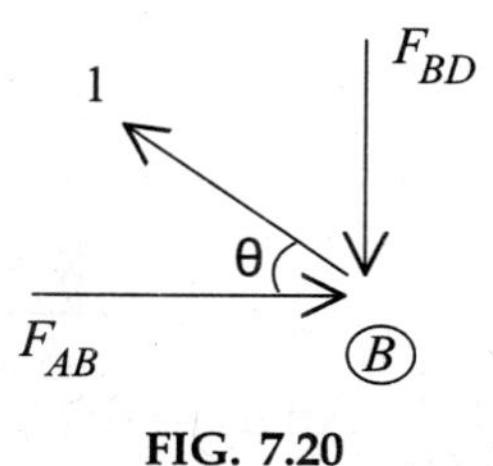

FIG. 7.20

$$\Sigma V = 0$$

$$F_{BD} = 1 \sin \theta$$

$$\boxed{F_{BD} = 0.8}$$

Joint E

As there are no external loads, the horizontal reaction at E is zero,

$$\therefore \boxed{\begin{aligned} F_{DE} &= 0 \\ F_{BE} &= 0 \end{aligned}}$$

Joint D

As F_{CD} and F_{BD} are known, F_{AD} is found to be unity.

It should be noted that the unit applied force affects the panel $ABCD$ only. After computing the member forces of the statically determinate truss (by excluding the redundant member BC) and the member forces for unit redundant force (including the member BC), they are tabulated. The summation of the values of $\sum(Pul/AE)$ and $\sum u^2l/AE$ are used to compute the redundant member force R. Then the other member forces are computed. Using the relation member force = $P + Ru$; where P is the member force of a statically determinate truss excluding the redundant member BC and u is the member force due to unit redundant force (including the redundant member BC) and R is the redundant force.

Members	*P*	*u*	*l*	*Pul/AE*	u^2	u^2l/AE	*(P + Ru)*
AC	0	−0.8	4	0	0.64	2.56	−8.75
CD	0	−0.6	3	0	0.36	1.08	−6.56
DE	−15	0	3	0	0	0	−15.00
EB	0	0	0	0	0	0	0
BD	20	−0.8	4	−64	0.64	2.56	11.25
BA	0	−0.6	3	0	0.36	1.08	−6.56
AD	−25	+1	5	−125	1	5.00	−14.06
BC	0	+1	5	0	1	5.00	10.94
				−189		17.28	

$$R = -\frac{\sum Pul/AE}{\sum u^2l/AE}$$

$$R = -\left(\frac{-189}{17.28}\right)$$

$$\boxed{R = 10.94 \text{ kN}}$$

Example 7.5 Analyse the truss shown in Fig. 7.21 and determine the forces in all the members AE = constant.

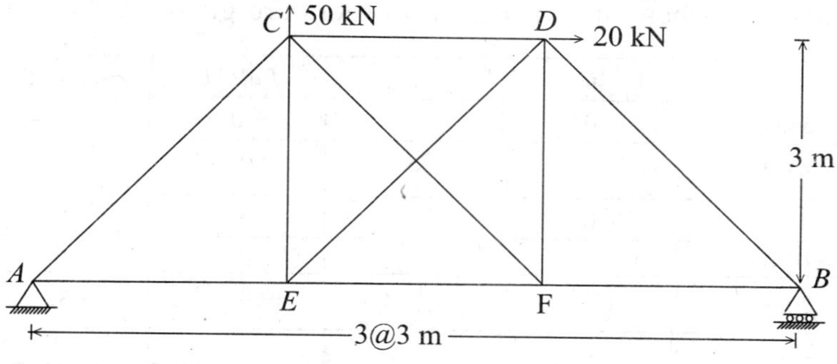

FIG. 7.21

Solution: Consider *ED* as a redundant member. Remove the member *ED* and the determinate truss is analysed by method of joints. The nature of member forces are shown in Fig. 7.22 and their magnitude are given in the table.

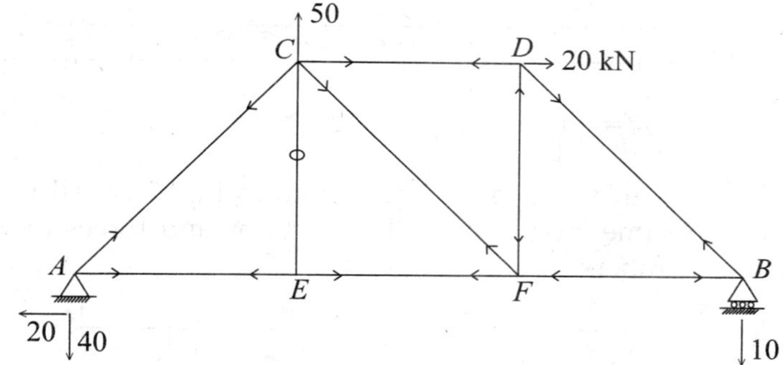

FIG. 7.22 Determinate truss

Apply a unit force along *ED* and the member forces are computed.

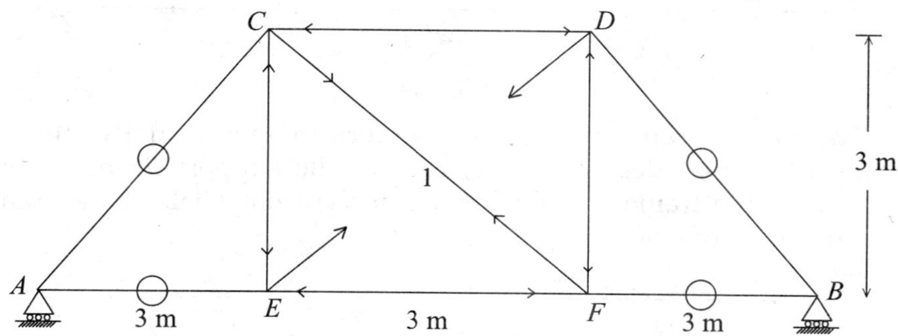

FIG. 7.23 Determinate truss with unit redundant force

Contd...

The redundant member force in *ED* and others are given in the following table.

Members	*P*	*u*	*l*	*Pul/AE*	*u2l/AE*	*P + Ru*
AC	56.58	0	$3\sqrt{2}$	0	0	56.58
CD	30.00	$-(1\sqrt{2})$	3	$-90/\sqrt{2}$	3/2	28.78
DB	14.14	0	$3\sqrt{2}$	0	0	14.14
BF	-10.00	0	3	0	0	-10.00
FE	20.00	$-(1\sqrt{2})$	3	$-60/\sqrt{2}$	+3/2	+18.78
EA	20.00	0	3	0	0	20.00
EC	0	$-(1\sqrt{2})$	3	0	3/2	-1.22
FD	-10.00	$-(1\sqrt{2})$	3	$30/\sqrt{2}$	3/2	-11.22
FC	14.14	+1	$3\sqrt{2}$	60	$3\sqrt{2}$	+15.86
ED	0	+1	$3\sqrt{2}$	0	$3\sqrt{2}$	+1.72
				-24.85	+14.49	

$$R = -\left[\frac{-24.85}{14.49}\right] = 1.72 \text{ kN}$$

Example 7.6 A Warren girder is loaded as shown in Fig. 7.24 All members have the same cross section and same length. Find the reactions and forces in all the members. *E* is same for all the members.

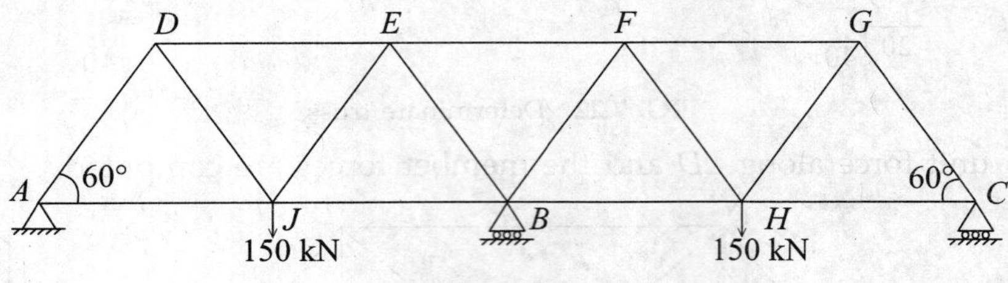

FIG. 7.24

Solution: The above frame is statically determinate internally but externally indeterminate to the first degree. Hence, remove the support *B* and determine the member forces. As the frame is symmetrical, analyse only half of the frame by the method of joints/methods of sections.

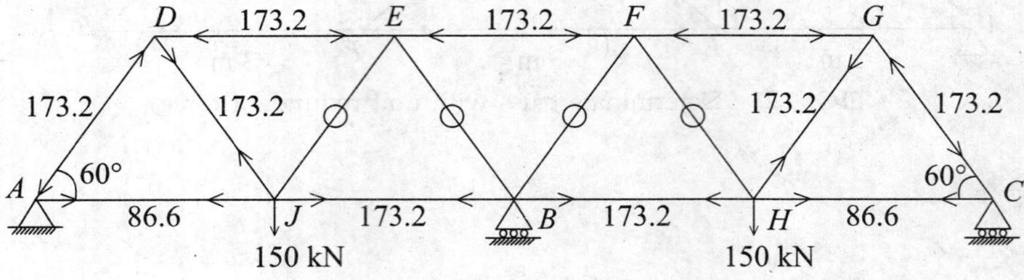

FIG. 7.25 Statically determinate truss

Due to symmetry

$$V_A = V_C = 150 \text{ kN}$$

Use Methods of Joints

Joint A

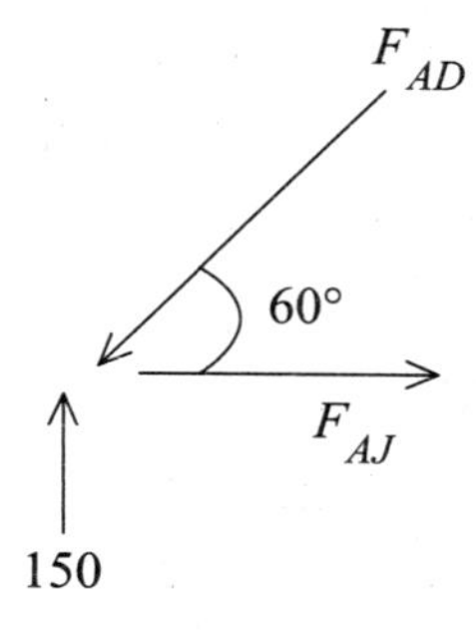

FIG. 7.26

$\Sigma V = 0$

$$F_{AD} \sin 60 = 150$$

$$\boxed{F_{AD} = 173.2 \text{ kN}}$$

$\Sigma H = 0$

$$F_{AJ} = F_{AD} \cos 60$$

$$\boxed{F_{AJ} = 86.6 \text{ kN}}$$

Joint D

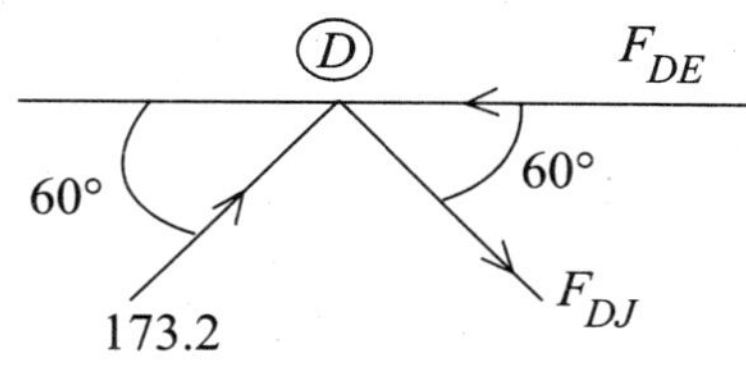

FIG. 7.27

$\Sigma V = 0$

$$173.2 \sin 60 = F_{DJ} \sin 60$$

$$\boxed{F_{DJ} = 173.2 \text{ kN}}$$

$\Sigma H = 0$

$$F_{DE} = 173.2 \cos 60 + 173.2 \cos 60$$

$$\boxed{F_{DE} = 173.2 \text{ kN}}$$

Joint J

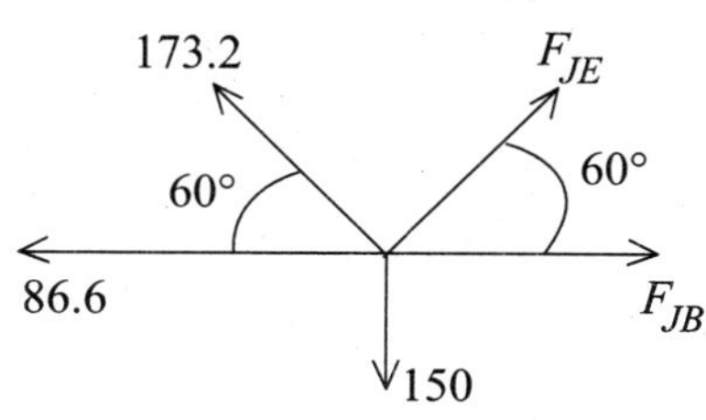

FIG. 7.28

$\Sigma V = 0$

$$173.2 \sin 60 + F_{JE} \sin 60 = 150$$

$$\boxed{F_{JE} = 0}$$

$\Sigma H = 0$

$$F_{JB} = -173.2 \cos 60 - 86.6 + F_{JB} + F_{JE} \cos 60$$

$$\boxed{F_{JB} = 173.2 \text{ kN}}$$

Now apply a unit force in the direction of *VB* and analyse the truss.

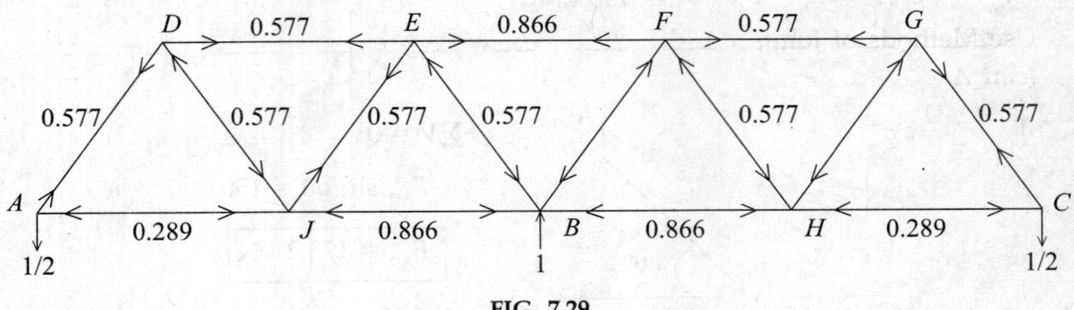

FIG. 7.29

Analyse joint *A* and the remaining member forces can be written by the method of inspection.

Joint A

$$\Sigma V = 0$$

$$F_{AD} \sin 60 = 1/2$$

$$\boxed{F_{AD} = 0.577 \text{ kN}}$$

$$\Sigma H = 0$$

$$F_{AJ} = F_{AD} \cos 60$$

$$\boxed{F_{AJ} = 0.289 \text{ kN}}$$

As *l* and *A* are constant they are omitted in the table.

Member	*P*(kN)	*u*(kN)	*Pu*	*u*²	*P*+*Ru*
AD	−173.2	0.577	−99.94	0.333	−62.8
DE	−173.2	0.577	−99.94	0.333	−62.8
EF	−173.2	0.866	−150.00	0.750	−7.5
FG	−173.2	0.577	−99.94	0.333	−62.8
GC	−173.2	0.577	−99.94	0.333	−62.8
GH	+173.2	0.577	−99.94	0.333	+62.8
HF	0	0.577	0	0.333	+110.4
FB	0	−0.577	0	0.333	−110.4
BE	0	−0.577	0	0.333	−110.4
EJ	0	−0.577	0	0.333	+110.4

Contd...

Contd...

JD	+173.2	-0.577	-99.94	0.333	+62.8
AJ	86.6	-0.289	-25.03	0.084	+31.3
JB	+173.2	-0.866	-150.00	0.750	+7.5
BH	+173.2	-0.866	-150.00	0.750	+7.5
HC	86.6	-0.289	-25.03	0.084	+31.3
			-1099.7	5.748	

$$R = -\frac{\Sigma Pul/AE}{\Sigma u^2 l/AE} = 191.32 \text{ kN}$$

Example 7.7 Analyse the truss shown in Fig. 7.30. Axial rigidity AE is constant. Determine the horizontal reactions at the supports A and B and the member forces respectively.

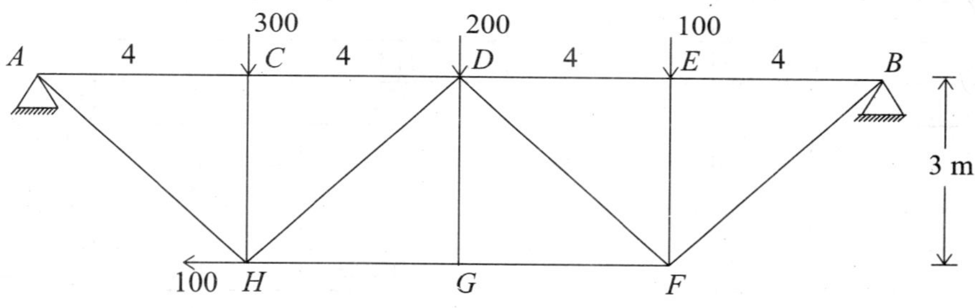

FIG. 7.30

Solution: The truss is supported on hinges at A and B. Hence, the number of reactions are four. The equilibrium equations are three. Therefore, the degree of redundancy is (4 – 3) = 1. Thus, the truss is externally indeterminate to the first degree. The horizontal reaction at B is treated as a redundant. Hence, the horizontal redundant is removed and the hinge B is replaced as roller at B. Thus, the truss is made statically determinate. Analyse the truss by method of joints. The computed forces are shown below.

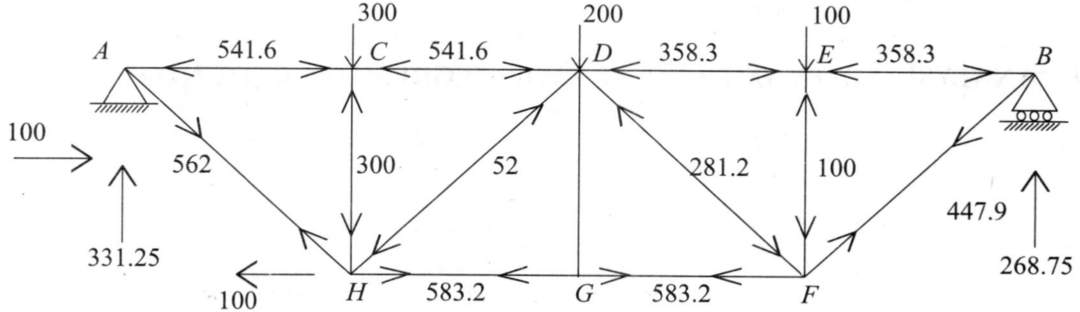

FIG. 7.31 Statically determinate truss

Apply a unit horizontal force in the direction of the redundant. Again, analyse the truss and mark the forces.

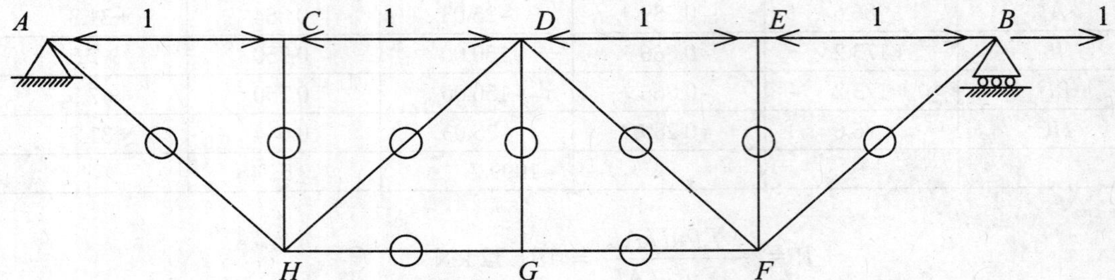

FIG. 7.32 Statically determinate truss (unit redundant force)

The computed member forces using above two stages of analysis are tabulated.

Member	*P(kN)*	*u(kN)*	*Pu*	u^2	*P + Ru*
AC	-541.6	1	-541.6	1	-91.6
CD	-541.6	1	-541.6	1	-91.6
DE	-358.3	1	-358.3	1	91.7
EB	-358.3	1	-358.3	1	91.7
BF	+447.9	0	0	0	+447.9
FE	-100.0	0	0	0	-100.0
FD	-281.2	0	0	0	-281.2
DG	0	0	0	0	0
DH	-52.0	0	0	0	-52.0
HC	-300.0	0	0	0	-300.0
HA	+552.0	0	0	0	+552.0
HG	583.2	0	0	0	+583.2
GF	583.2	0	0	0	+583.2
			-1799.8	4	

$$R = -\left(\frac{-1799.8}{4}\right) = 450 \text{ kN}$$

7.4 ANALYSIS OF FRAMES WITH TWO DEGREES OF REDUNDANCY

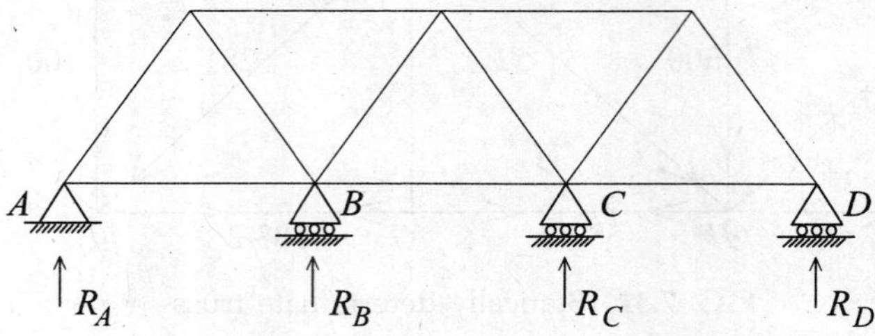

FIG. 7.33

Solution: The truss is supported on a hinge at A and roller at D. They form the end supports. The intermediate supports B and C are on rollers. The truss is stable even after the removal of B and C.

There are two redundants R_B and R_C externally. It means that the structure can be made stable with a hinge support at A and a roller support at D. The member forces can be found if the redundants R_B and R_C are known. They are determined as follows. Remove the redundants R_B and R_C. Let u be the load produced in the truss by a unit vertical load at B with the redundants removed. Let v be the loads under the same condition due to unit load at C. Let P be the loads in the members due to the given loads. Then R_{Bu} will be the loads in the members due to R_B. R_Cv is the member force due to RC.

The deflection at B produced by P equals $\Sigma(Pul/AE)$. The deflection at B produced by R_B equals $\Sigma(R_Bu^2l/AE)$. The deflection at B produced by R_C equals $\Sigma R_C \dfrac{vul}{AE}$.

As the support at B is not yielding (not settled/not subsiding), the resultant deflection at B is zero.

Hence,

$$R_B\Sigma\frac{u^2l}{AE} + R_C\Sigma\frac{vul}{AE} + \Sigma\frac{Pul}{AE} = 0 \quad (1)$$

By extending the same logic at the joint C,

$$R_C\Sigma\frac{v^2l}{AE} + R_B\Sigma\frac{uvl}{AE} + \Sigma\frac{Pvl}{AE} = 0 \quad (2)$$

The above equations are solved simultaneously to determine R_B and R_C. Thus, the force in the member is

$$F = P + R_Bu + R_cv \quad (3)$$

If the support at B and C settle by λ and λ_1 respectively, then the above equation change to

$$R_B\Sigma\frac{u^2l}{AE} + R_C\Sigma\frac{uvl}{AE} + \Sigma\frac{Pul}{AE} = \lambda \quad (4)$$

$$R_C\Sigma\frac{v^2l}{AE} + R_B\Sigma\frac{uvl}{AE} + \Sigma\frac{Pvl}{AE} = \lambda_1 \quad (5)$$

The application of the equation is illustrated in the following example.

Example 7.8 Analyse the truss shown in Fig. 7.34. AE is constant.

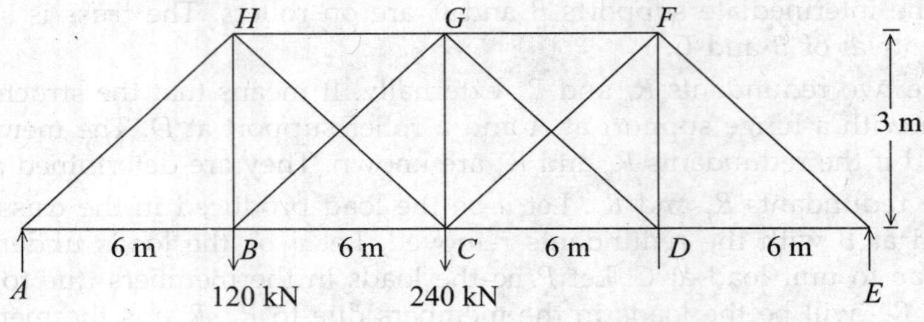

FIG. 7.34

Solution: The total number of joints n_j is eight. The number of member n_m is fifteen. Hence, the statical indeterminacy is $n_m - (2n_j - 3)$ i.e. $15 - (2 \times 8 - 3) = 2$. The redundant members considered are HC and CF respectively. The analysis is done as follows:

(1) Remove the redundant members HC and CF. Then analyse the statically determinate frame and determine the member forces. They are denoted as P.

(2) Apply a unit tensile forces along the direction of redundant member HC. Determine the member forces due to the above unit load. Let these forces are denoted as u.

(3) Apply a unit tensile force along the direction of redundant member CF. Let the member forces be noted as v.

(4) The redundant force $X_1 = F_{HC}$ and $X_2 = F_{CF}$ are determined using the following equations

$$X_1 \Sigma \frac{v^2 l}{AE} + X_2 \Sigma \frac{uvl}{AE} + \Sigma \frac{Pul}{AE} = 0 \qquad (6)$$

$$X_2 \Sigma \frac{v^2 l}{AE} + X_1 \Sigma \frac{uvl}{AE} + \Sigma \frac{Pvl}{AE} = 0 \qquad (7)$$

(5) Then member forces are determined from

$$F = P + X_1 u + X_2 v \qquad (8)$$

Statically determinate truss analysis

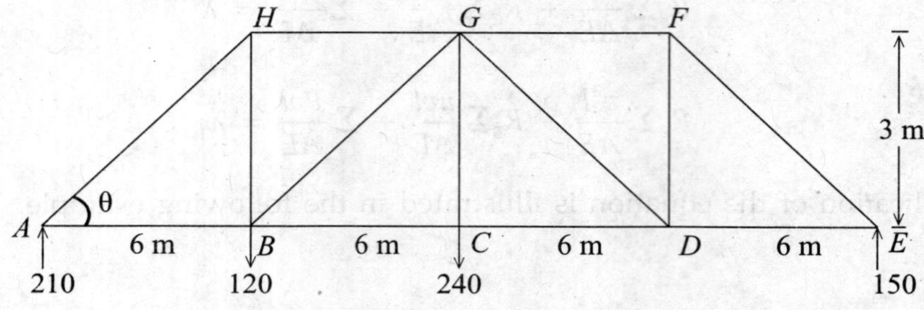

FIG. 7.35

The reactions are determined as

$$\Sigma V = 0$$

$$V_A + V_E = 120 + 240$$

$$V_A + V_E = 360 \qquad (1)$$

$$\Sigma M_A = 0$$

$$120(6) + 240(12) - 24V_E = 0$$

$$\boxed{\begin{array}{l} V_E = 150 \text{ kN} \\ V_A = 210 \text{ kN} \end{array}}$$

Consider joint A

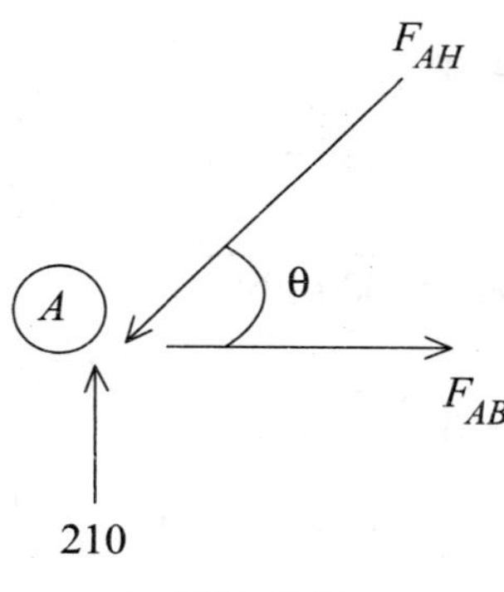

FIG. 7.36

$$\Sigma H = 0$$

$$F_{AB} - F_{AH} \cos \theta = 0$$

$$F_{AB} - 0.8942\ F_{AH} = 0$$

$$\Sigma V = 0$$

$$F_{AH} \sin \theta = 210$$

$$F_{AH}\ (0.4471) = 210$$

$$\boxed{\begin{array}{l} F_{AH} = 469.7 \text{ kN} \\ F_{AB} = 420.0 \text{ kN} \end{array}}$$

Joint E

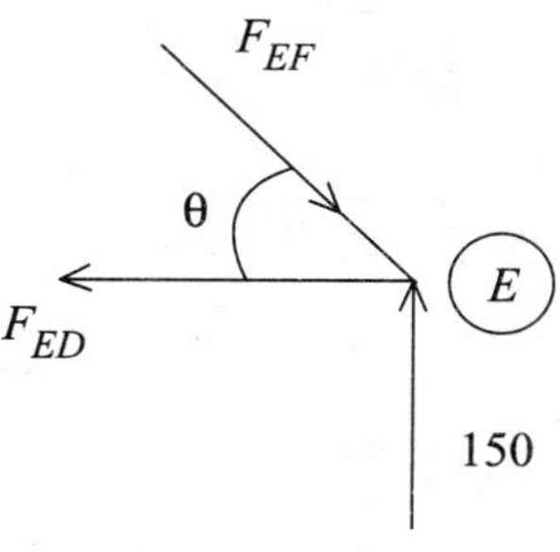

FIG. 7.37

$$\Sigma V = 0$$

$$F_{EF} \sin \theta = 150$$

$$F_{EF} = 150/0.4471$$

$$\boxed{F_{EF} = 335.5 \text{ kN}}$$

$$\Sigma H = 0$$

$$F_{EF} \cos \theta - F_{ED} = 0$$

$$F_{ED} = 0.8942 F_{EF}$$

$$\boxed{F_{ED} = 300.0 \text{ kN}}$$

Joint H

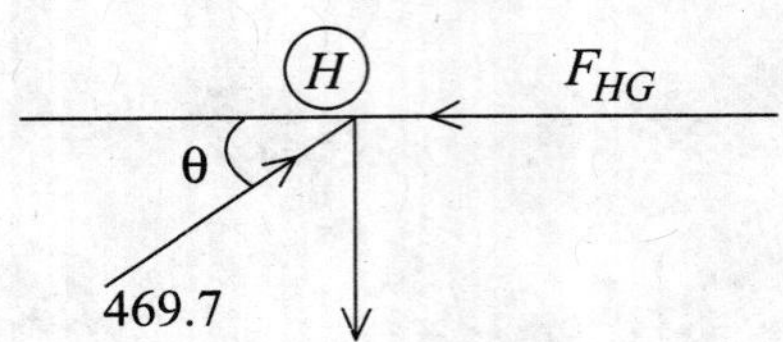

FIG. 7.38

$$\Sigma V = 0$$

$$F_{HB} = 469.7 \sin \theta$$

$$\boxed{F_{HB} = 210.0 \text{ kN}}$$

$$\Sigma H = 0$$

$$F_{HG} = 469.7 \cos \theta$$

$$\boxed{F_{HG} = 420 \text{ kN}}$$

Joint F

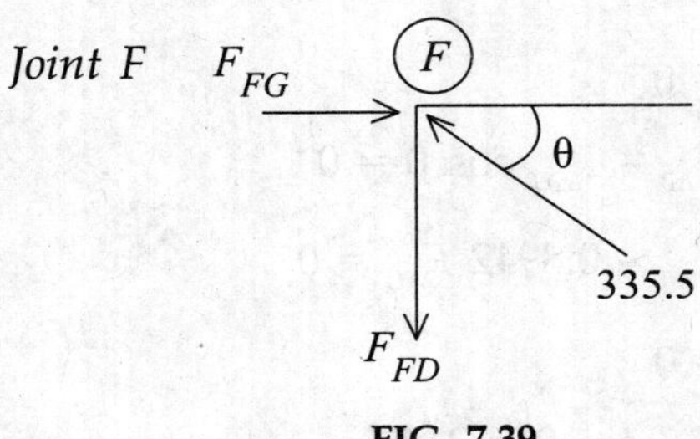

FIG. 7.39

$$\Sigma H = 0$$

$$F_{FG} = 335.5 \cos \theta$$

$$\boxed{F_{FG} = 300 \text{ kN}}$$

$$\Sigma V = 0$$

$$F_{FD} = 335.5 \sin \theta$$

$$\boxed{F_{FD} = 150 \text{ kN}}$$

Joint D

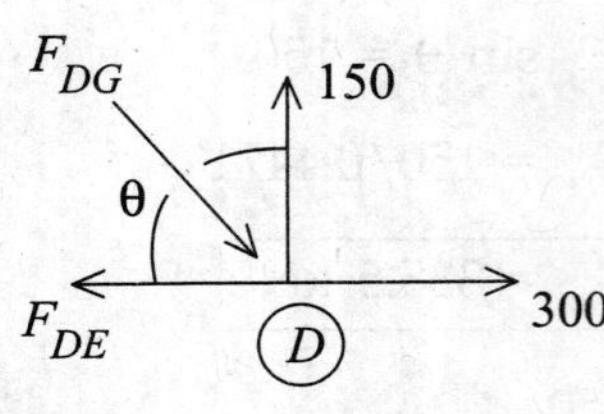

FIG. 7.40

$$\Sigma V = 0$$

$$F_{DG} \sin \theta = 150$$

$$\boxed{F_{DG} = 335 \text{ kN}}$$

$$\Sigma H = 0$$

$$-F_{DC} + 335.5 \cos \theta + 30 = 0$$

$$\boxed{F_{DC} = 600 \text{ kN}}$$

Joint C

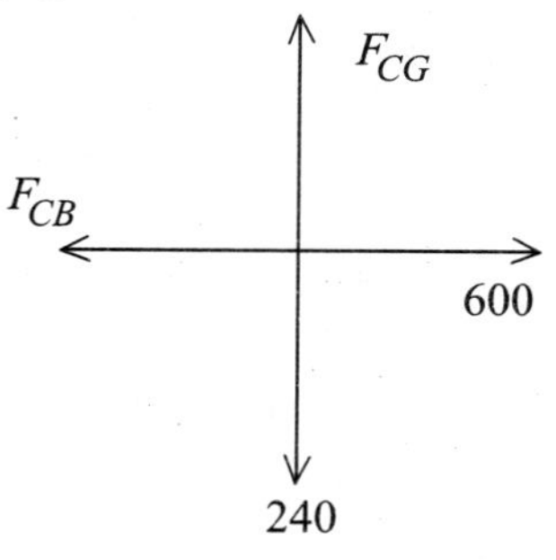

FIG. 7.41

$$\Sigma V = 0$$

$$\boxed{F_{CG} = 240 \text{ kN}}$$

$$\Sigma H = 0$$

$$\boxed{F_{CB} = 600 \text{ kN}}$$

Joint B

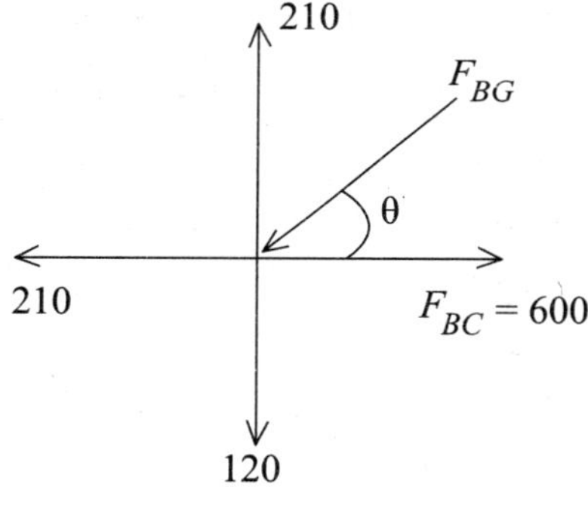

FIG. 7.42

$$\Sigma H = 0$$

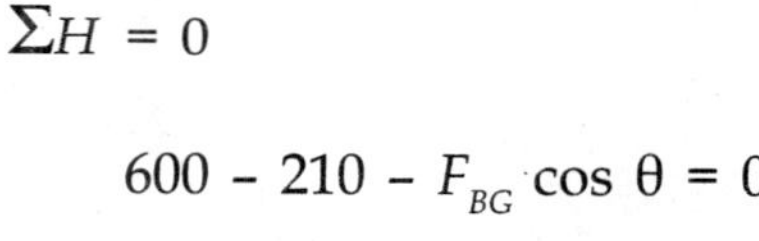

$$600 - 210 - F_{BG} \cos \theta = 0$$

$$\boxed{F_{BG} = 201 \text{ kN}}$$

The above member forces P are tabulated.

Analysis of determinate truss with redundant X_1

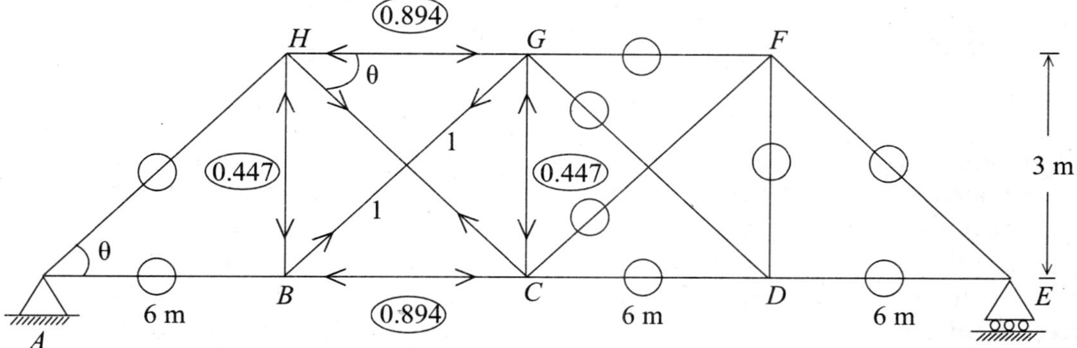

FIG. 7.43 Determinate truss with unit redundant along *HC*

Consider joint H

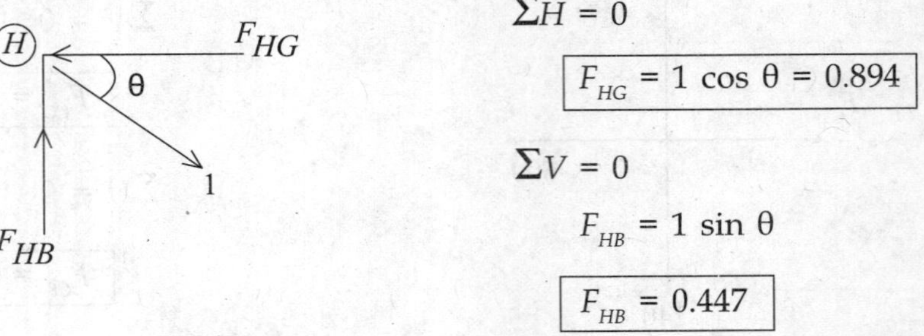

$\Sigma H = 0$

$$\boxed{F_{HG} = 1 \cos \theta = 0.894}$$

$\Sigma V = 0$

$$F_{HB} = 1 \sin \theta$$

$$\boxed{F_{HB} = 0.447}$$

The other member forces are obtained and shown as above.

Analysis of determinate truss with redundant X_2

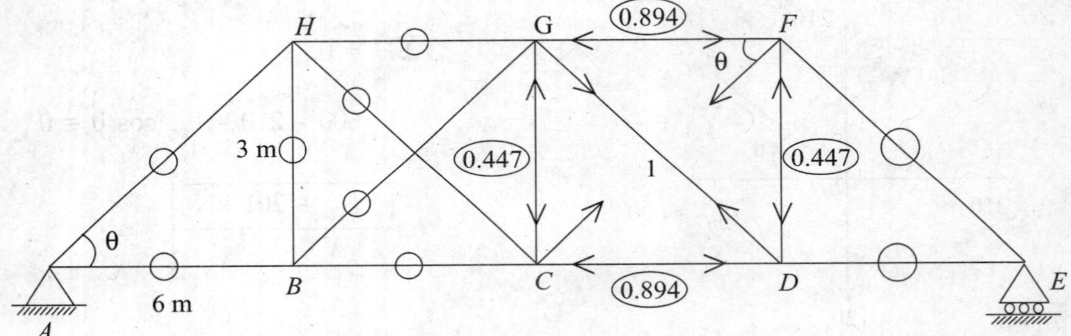

FIG. 7.44 Determinate truss with redundant force along *CF*

Consider joint F

$\Sigma H = 0$

$$F_{GF} = 1 \cos \theta = 0.894$$

$\Sigma V = 0$

$$F_{FD} = 1 \sin \theta = 0.447$$

FIG. 7.45

The other member forces are computed and marked in the above Fig. 7.44 It is to be mentioned here that the effect of the redundant force (unit tension) is in that panel only. This important clue helps in analysing the truss quickly without loss of accuracy.

Table 7.1 gives the values of the computed forces for the estimation of the redundant forces X_1 and X_2 respectively. The reader should note that they are internal redundant forces. In this example X_1 and X_2 are the external redundants. There is a possibility that X_1 and X_2 could be one internally redundant force and the other externally redundant force or vice versa. The basic equations are the same for the determination of two degrees of redundancy.

Table 7.1 Computations of member forces of a truss with 2 degree redundancy

Member	*P* (kN)	*u*	*l* (m)	*Pul*	*v*	*Pvl*	u^2l	v^2l	*uvl*	*F* (kN)
AH	-469.7	0	6.71	0	0	0	0	0	0	-469.7
HG	-420.0	-0.894	6.00	2252.9	0	0	4.795	0	0	-523.8
GF	-300.0	0	6.00	0	-0.894	1609.2	0	4.795	0	-459.3
FE	-335.5	0	6.71	0	0	0	0	0	0	-335.5
ED	300	0	6.00	0	0	0	0	0	0	+300.0
DC	600	0	6.00	0	-0.894	-3218.4	0	4.795	0	+440.68
CB	600	-0.894	6.00	-3218.4	0	0	4.795	0	0	+496.21
AB	420	0	6.00	0	0	0	0	0	0	+420.0
BH	210	-0.447	3.00	-281.6	0	0	0.60	0	0	+158.1
GC	240	-0.447	3.00	-321.8	-.447	-321.8	0.60	0.60	0.60	+108.5
FD	150	0	3.00	0	-.447	-201.2	0	0.60	0	70.3
BG	-201	+1	6.71	-1348.7	0	0	6.71	0	0	-84.9
GD	-335.5	0	6.71	0	+1	-2251.2	0	6.71	0	-157.3
HC	0	+1	6.71	0	0	0	6.71	0	0	+116.1
CF	0	0	6.71	0	+1	0	0	6.71	0	+178.2
				-2917.6		-4383.4	24.2	24.21	0.60	

The governing equations are (6) and (7); on substitution,

$$24.21X_1 + 0.6X_2 = 2917.6$$

$$0.6X_1 + 24.21X_2 = 4383.4$$

On solving $X_1 = 116.1$ kN and $X_2 = 178.2$ kN.

The member forces are computed using the relation

$$F = P + X_1u + X_2v$$

For example,

$$F_{GC} = 240 - 0.447 \times 116.1 - 0.447 \times 178.2$$

i.e. $$\boxed{F_{GC} = 108.5 \text{ kN}}$$

Similarly, other member forces were calculated and tabulated in the last column of the above table.

7.5 LACK OF FIT (MISFIT) IN MEMBERS OF INDETERMINATE TRUSS

In an indeterminate truss, a member may be slightly too long or too short. This difference in length of the member has no effect in the various members. Consider an indeterminate truss *ABCD* shown in Fig. 7.46. The member *AB* is shorter than its exact length by an amount λ as shown. When this member is forced into position, the member is subjected to tension. This generates forces in other members of the truss. The force in the member *R* due to lack of fit is calculated from Eq. (1) as

$$R = \frac{\lambda - \Sigma P_i u_i l_i / A_i E}{\Sigma u_i l_i^2 / A_i E} \tag{1}$$

where λ is the lack of fit.

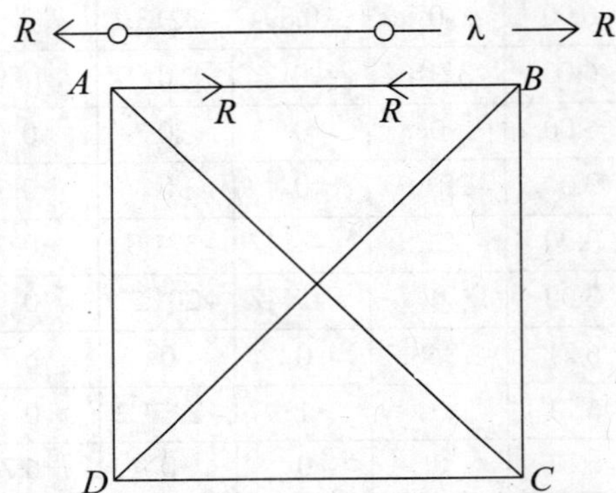

FIG. 7.46 Determinate truss by removal of member with lack of fit

If *AE* is constant the above equation is modified as

$$R = \frac{\lambda AE - \Sigma P_i u_i l_i}{\Sigma u_i^2 l_i} \tag{2}$$

The Eq. (1) can be rewritten as

$$R = \frac{\lambda}{\Sigma u_i l_i^2 / A_i E} - \frac{\Sigma P_i u_i l_i / A_i E}{\Sigma u_i^2 l_i / A_i E} \tag{3}$$

The denominator in the first part of the equation represents the relative displacement of *B* with respect to *A* due to unit force. In other words, the relative displacement *B* and *A* is $R\Sigma u_i l_i^2 / A_i E$.

If the member is short, then *R* is tensile and if the member is long then it is compressive.

Example 7.9 Determine the force in the member *BC* of the given pin jointed truss shown in Fig. 7.47. The bar *BC* was last to be added and was initially 1 mm long. Assume the cross-sectional area of member as 1000 mm² and E = 200 kN/mm².

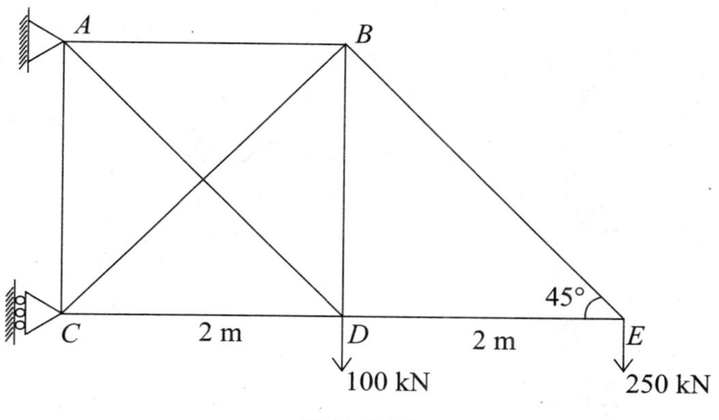

FIG. 7.47

Solution: Remove the redundant member BC and analyse the determinate truss by method of joints.

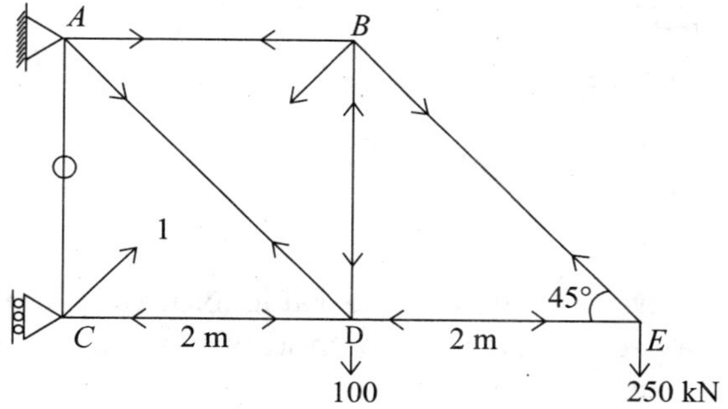

FIG. 7.48 Statically determinate truss

Joint E

$$\Sigma V = 0$$

$$F_{EB} \sin 45 = 250$$

$$F_{EB} = 353.6 \text{ kN}$$

$$\Sigma H = 0$$

$$F_{ED} = 353.6 \cos 45$$

$$F_{ED} = 250 \text{ kN}$$

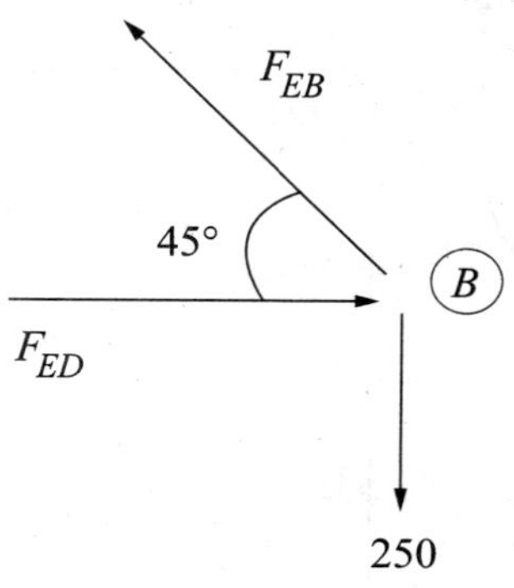

FIG. 7.49

Joint B

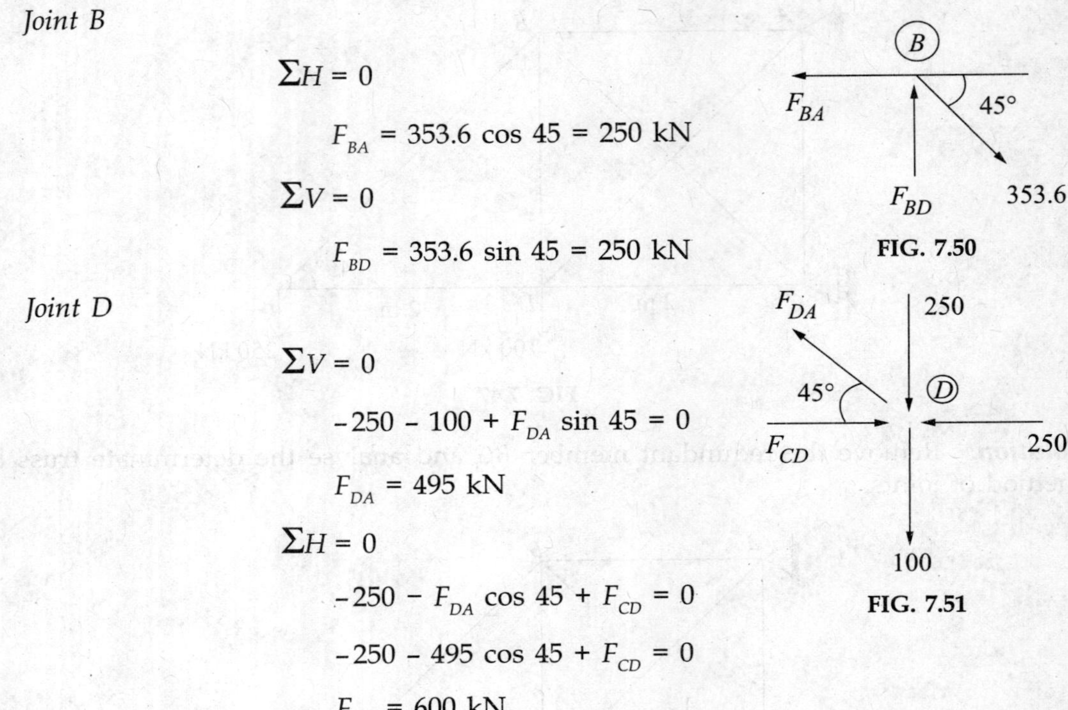

$$\Sigma H = 0$$

$$F_{BA} = 353.6 \cos 45 = 250 \text{ kN}$$

$$\Sigma V = 0$$

$$F_{BD} = 353.6 \sin 45 = 250 \text{ kN}$$

FIG. 7.50

Joint D

$$\Sigma V = 0$$

$$-250 - 100 + F_{DA} \sin 45 = 0$$

$$F_{DA} = 495 \text{ kN}$$

$$\Sigma H = 0$$

$$-250 - F_{DA} \cos 45 + F_{CD} = 0$$

$$-250 - 495 \cos 45 + F_{CD} = 0$$

$$F_{CD} = 600 \text{ kN}$$

FIG. 7.51

The computed forces are tabulated in the Table. Now apply unit and tensile force in the direction of *BC*. Analyse the truss by method of joints again for the unit redundant force and remove the external loads. Thus, the analysis is done for member forces for unit redundant force including the redundant member. As *AE* is constant for all members it is omitted in the table.

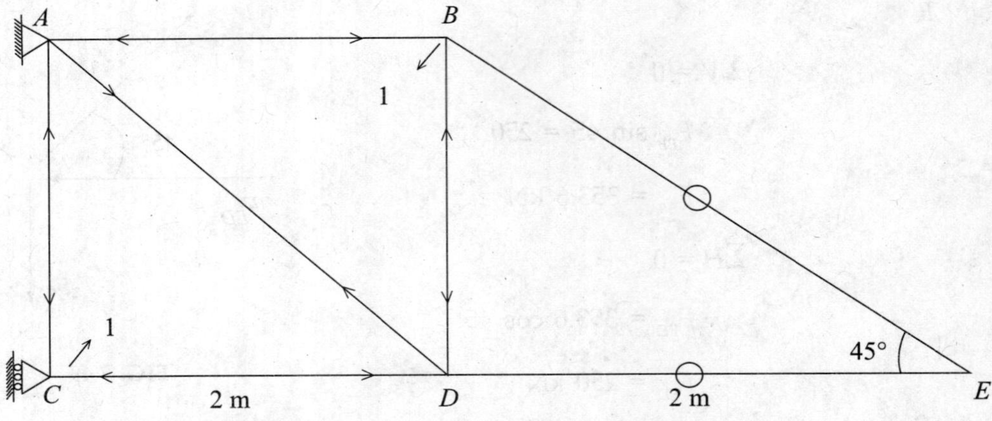

FIG. 7.52 Determinate truss with unit redundant force

Member	P	*u*	*l*	*Pul*	u^2l	*P + Ru*
AB	+250	-0.707	2	-353.5	1	312.14
BE	+353.6	0	$2\sqrt{2}$	0	0	353.60
ED	-250	0	2	0	0	-250.00
DC	-600	-0.707	2	+848.4	1	-537.85
CA	0	-0.707	2	0	1	+62.15
AD	+495	-0.707	$2\sqrt{2}$	-989.7	1.414	+557.14
				-494.8	4.414	

$$R = \left[AE\delta - \Sigma Pul/\Sigma u^2 l\right]$$
$$R = [-1000 \times 200 \times 10^{-3} - (-494.8/4.414)$$
$$R = -87.9 \text{ kN}$$

−ve sign indicates the redundant force is compressing. As the member is long the resulting force is compressive. The units used are area is mm², $\dot{E}$ is in kN/mm² and δ is in metres. The above units have been used such that the computed member forces are to be in kN.

7.6 STRESSES DUE TO CHANGE IN TEMPERATURE

If a redundant frame is made up of the same material and when it is subjected to uniform change in temperature, every member of the truss is shortened or lengthened in the same proportion and no stresses are developed/induced.

If the redundant frame is made up of more than one material and those materials have different coefficient of linear expansion then distortion takes place. Hence, the tendency to distort may induce considerable magnitude of stresses in a redundant frame.

Consider a frame *ABCD* in which all the members are of the same material except *AC*. All the members except *AC* have a coefficient of linear expansion of α. AC is made up of the material whose coefficient expansion is β.

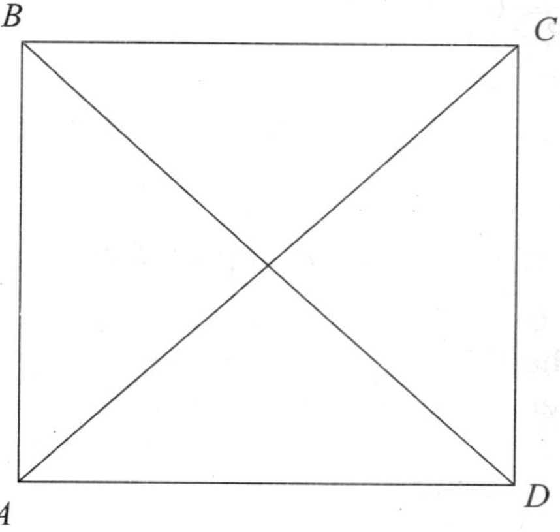

FIG. 7.53

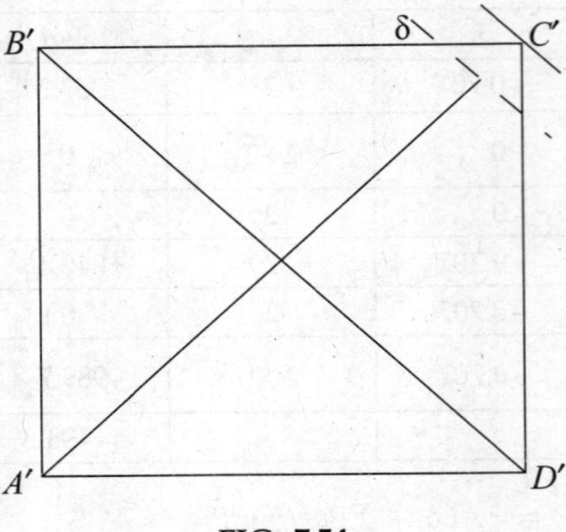

FIG. 7.54

If AC is removed and all the members are subjected to an increase temperature of $t°$, then each member length increases. The new length becomes $(1 + \alpha t)$ times the original length of the member. The frame distorts and the new dimensions are as above. The new shape is similar to the original configuration with increased length. For example, new length of AB is $A'B'$ which is $(1 + \alpha t)$ times the length of AB.

The removed member AC is heated to the temperature t, its new length is $(1 + \beta t)$ times of AC. This is due to the coefficient of linear expansion of AC is β. Essentially, it is short of the correct length by

$$\boxed{\delta = l(\alpha - \beta)t}$$

If this is forced into position, the stresses in the structure will be that of those which would exist if the temperature of the complete redundant frame heated through $t°$.

Let the tensile load in AC after heating be denoted by R, the forces in all bars of the same frame are found interms of R. If U is the strain energy, then

$$\boxed{dU/dR = l(\alpha - \beta)t}$$

Instead of heating, if there is a fall in temperature then the value of t is to be negative.

Example 7.10 The frame given below is supported as shown in Fig. 7.55. All members have cross-sectional area of 1200 mm². If there is a rise in temperature of member BD by 30° C, determine the forces due to change of temperature. Coefficient of linear expansion α = 0.000015 per degree centigrade per unit length. E = 200 kN/mm².

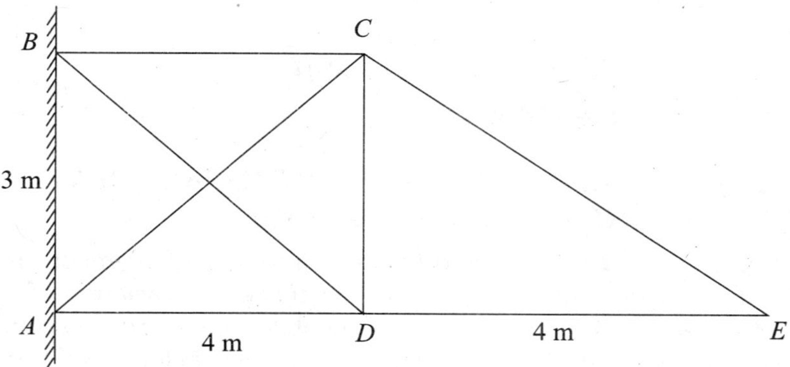

FIG. 7.55

Solution: Remove the member *BD* and apply unit a tensile force in the direction of *BD*.

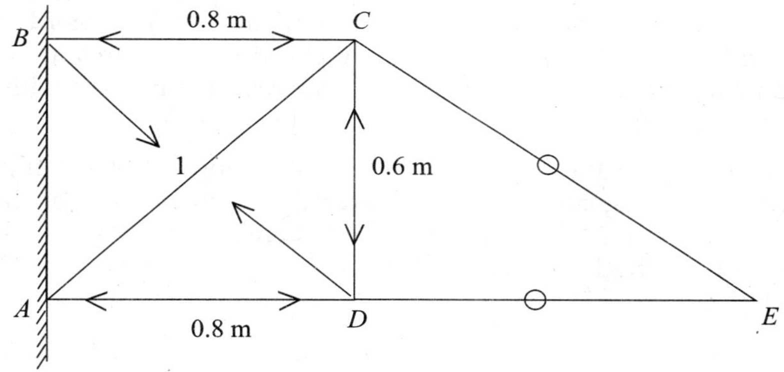

FIG. 7.56

Using the method of joints, the member forces are evaluated and are shown above. The values are tabulated for computational purposes.

Member	*Length*	*Area*	*u*	u^2l/A	*F = Ru (kN)*
BC	4000	1200	−0.8	2.13	30.93
CE	5000	1200	0	0	–
ED	4000	1200	0	0	–
DA	4000	1200	−0.8	2.67	30.93
AC	5000	1200	+1.0	4.17	−38.66
CD	3000	1200	−0.6	−1.50	+23.20
BD	5000	1200	+1.0	+4.17	−38.66
				11.64	

$$R = \frac{-0.000015 \times 5000 \times 30}{(11.64/2 \times 10^5)}$$

$$R = -38.66 \text{ kN}$$

7.7 EFFECT OF VARIATION IN CROSS-SECTIONAL AREA ON THE REACTION OF INDETERMINATE FRAME

In case of determinate trusses wherein the total number of members is equal to the two times the number of joints minus three, the change in variation of cross-sectional area do not effect the member forces. The member forces are determined by using graphical method/method of joints/method of sections/tension coefficient method. In simple words, the member forces are evaluated using statics.

In case of redundant pin jointed frames, $n_m > (2n_j - 3)$ the cross-sectional areas play an important role in computing member forces. There are three possibilities, viz., (i) top chord member areas are greater that bottom chord members; (ii) top chord member areas are equal to bottom chord members; and (iii) top chord member areas are less than bottom chord members. If the top chord member area are different from bottom chord members, it effects the reaction of the redundant frame. This is illustrated in the following example.

Example 7.11 Compute the reactions at the centre hinge D of the frame shown in Fig. 7.57. Vertical and diagonal members have area of a mm² (ii) Upper chord members have an area of ma mm² and (iii) Lower chord members have a cross-sectional area of na mm². Compute the horizontal reaction for (m = 2, n = 1); (m = 2, n = 2) and (m = 1, n = 2) at D.

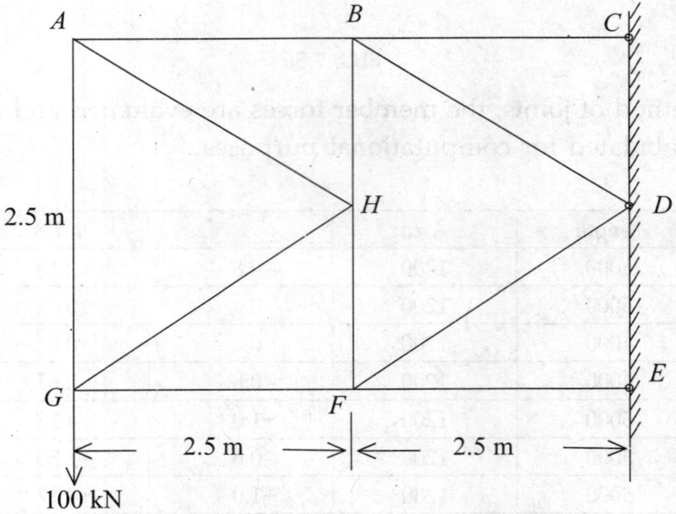

FIG. 7.57

Solution: The member BC is treated as a redundant member. Remove the member BC and it reduces to a statically determinate frame. The member forces are computed using method of joints.

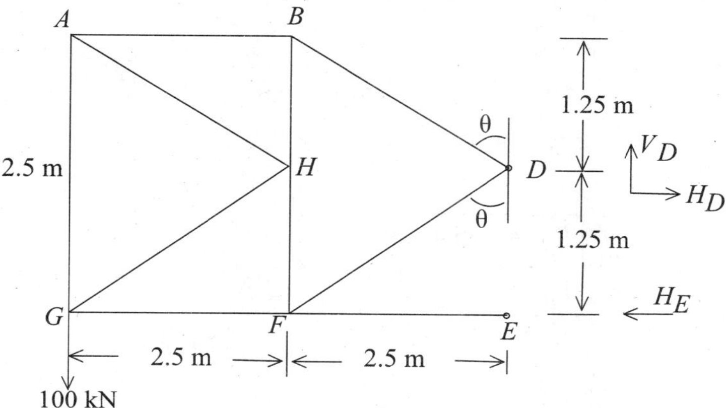

FIG. 7.58 Statically determinate frame

In the above frame, the reactions are found out first. Take moment about E, H_D can be found out

$$\Sigma M_E = 0$$

$$100(5) = 1.25 H_D$$

$$\boxed{H_D = 400 \text{ kN}}$$

$$\Sigma V = 0$$

$$V_D - 100 = 0$$

$$\boxed{V_D = 100 \text{ kN}}$$

$$\Sigma H = 0$$

$$400 - H_E = 0$$

$$\boxed{H_E = 400 \text{ kN}}$$

Joint D

From geometry:

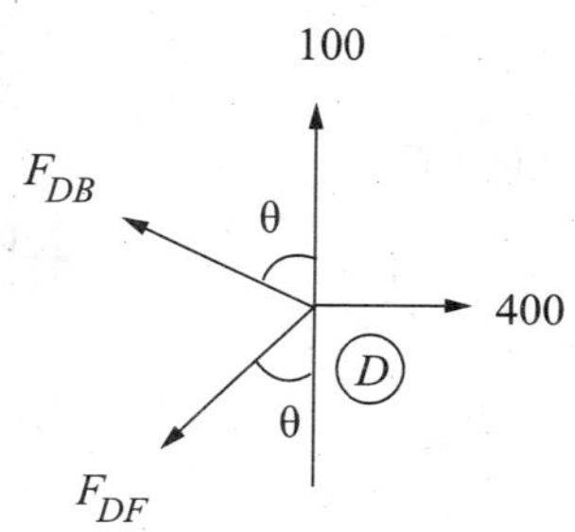

FIG. 7.59

$$\tan\theta = 2.5/1.25$$

$$\sin\theta = 0.8944$$

$$\cos\theta = 0.4472$$

$\sum H = 0;$

$$400 - F_{DB} \sin\theta - F_{DF} \sin\theta = 0$$

$$F_{DB} + F_{DF} = 447.23 \tag{1}$$

$\sum V = 0;$

$$100 + F_{DB} \cos\theta - F_{DF} \cos\theta = 0$$

$$-F_{DF} + F_{DB} = -223.61 \tag{2}$$

Solving

$$\boxed{\begin{aligned} F_{DF} &= 335.4 \text{ kN} \\ F_{DB} &= 111.8 \text{ kN} \end{aligned}}$$

Joint B

$\sum V = 0$

$$F_{BH} = 111.8 \cos\theta$$

i.e. $F_{BH} = 50$ kN

$\sum H = 0$

$$-F_{BA} + 111.8 \sin\theta = 0$$

$$\boxed{F_{BA} = 100 \text{ kN}}$$

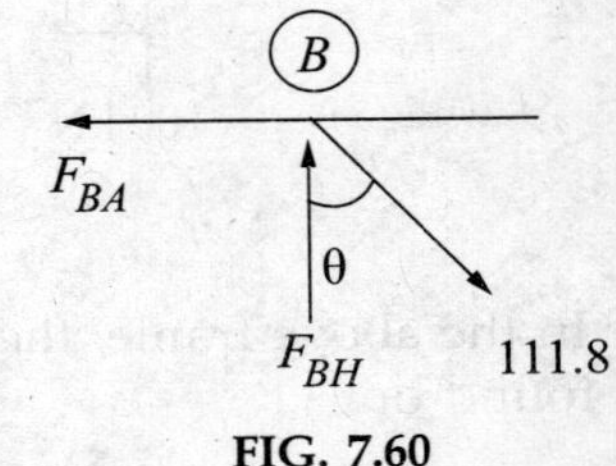

FIG. 7.60

Joint A

$\sum H = 0$

$$F_{AH} \sin\theta = 100$$

$$\boxed{F_{AH} = 111.8 \text{ kN}}$$

$\sum V = 0$

$$F_{AG} = 111.8 \cos\theta$$

$$\boxed{F_{AG} = 50 \text{ kN}}$$

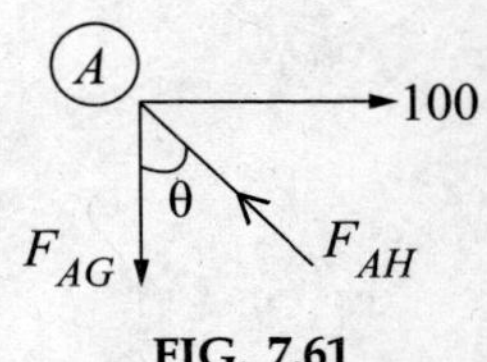

FIG. 7.61

Joint G

$\sum V = 0$

$$50 - F_{GH} \cos\theta = 0 - 100$$

$$F_{GF} \cos\theta = -50$$

$$0.4472\ F_{GH} = -50$$

$$\therefore \quad \boxed{FGH = -111.8 \text{ kN}}$$

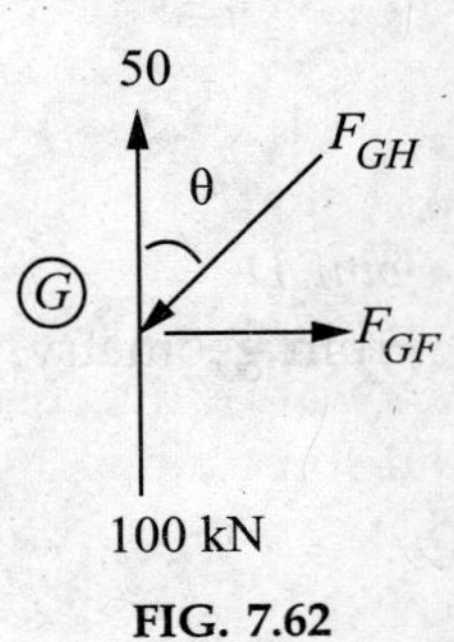

FIG. 7.62

$\Sigma H = 0$

$$F_{GF} - F_{GH} \sin\theta = 0$$

$$F_{GF} = -111.8 \times 0.8944$$

$$\boxed{F_{GF} = -100 \text{ kN}}$$

Analysis of determinate frame due to unit redundant force.

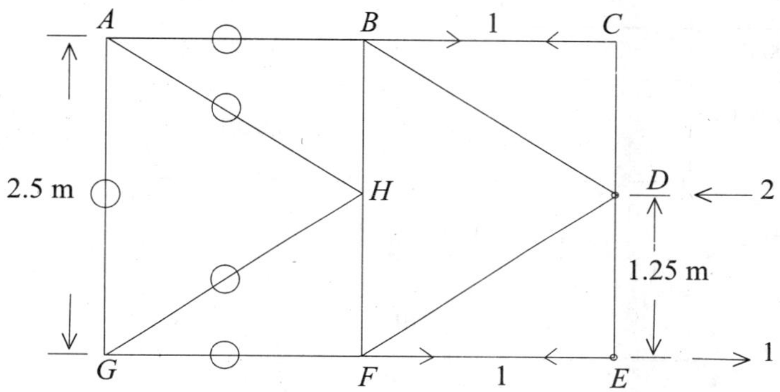

FIG. 7.63 Member forces due to unit redundant force

As unit force is applied along *BC*, by the method of inspection, we can say that the forces in the panel *ABFG* are zeros.

$\Sigma M_E = 0;$

$$1.25\, H_D = 1(2.5)$$

$$\boxed{H_D = 2 \text{ kN}}$$

$\Sigma H = 0;$

$$H_B + H_E = H_D$$

i.e. $$H_B + H_E = 2 \text{ kN}$$

$\therefore$ $\boxed{H_E = 1 \text{ kN}}$ $(\because\ HB = 1)$

As $H_E = 1;\ F_{FE} = 1.0$ kN

Analyse the truss *FHBD*, by the method of joints and the member forces due to unit force along *BC* are tabulated.

Members	P	u	l	A	$\frac{Pul}{AE} \times aE$	$\frac{u^2 l}{AE} \times aE$
BH	−50.0	+0.5	1.25	a	−31.25	0.31
BC	0	+1.0	2.50	ma	0	$2.5/m$
BD	+111.8	−1.12	2.80	a	−350.6	3.51
FH	−150.0	+0.5	1.25	a	−93.75	0.31
FD	+335.4	−1.12	2.80	a	−1051.81	3.51
FE	−400.0	+1.00	2.50	na	$-1000/n$	$2.5/n$
					$-\left(1527.41 + \frac{1000}{n}\right)$	$7.64 + 2.5\left(\frac{1}{m} + \frac{1}{n}\right)$

$$\therefore \quad F_{BC} = \frac{1527.41 + 1000/n}{7.64 + 2.5\left(\frac{1}{m} + \frac{1}{n}\right)}$$

i.e.

$$\boxed{F_{BC} = \frac{(610.96 + 400/n)}{3.06 + \left(\frac{1}{m} + \frac{1}{n}\right)}}$$

The horizontal reaction at C is F_{BC}. But we require horizontal reaction at D. Let us denote the horizontal reaction at D as X. The value of X can be found out by taking moment about the bottom hinge at E.

Hence,

$$\Sigma ME = 0;$$

$$2.5\ F_{BC} + 1.25X = 500$$

i.e.

$$\boxed{X = 400 - 2F_{BC}}$$

Substituting the value of F_{BC}, the horizontal reaction at D is given as

$$X = 400 - 2\left[\frac{(610.96) + (400/n)}{3.06 + \left(\frac{1}{m} + \frac{1}{n}\right)}\right]$$

Case 1 Top boom areas are twice that of bottom boom area

i.e. $m = 2$ and $n = 1$

$$\boxed{X = -43.4 \text{ kN}}$$

Case 2 Top boom areas are equal to bottom boom areas

$$m = 2, \quad n = 2$$

$$\boxed{X = 0}$$

Case 3 Bottom boom areas are twice that of top boom areas.

$$m = 1, \quad n = 2$$

$$\boxed{X = 44.3 \text{ kN}}$$

The above three cases illustrate that the reaction direction and magnitude gets changed by proportioning the cross-sectional areas of top boom and bottom boom respectively.

7.8 TRUSSED BEAMS

This is a combination of beam and truss. In case of beams they are subjected to bending stresses. In the truss members, they are subjected to either compressive force in struts and tensile force in case of ties. As both kinds of action are predominant, they are called trussed beam or beam trusses. They are used in the construction of railway passenger coaches. The analysis of such structures is possible through energy principles.

Figure 7.64 shows a king post trussed beam. It is pinned at *C*. The point *D* is braced by steel ties to the supports *A* and *B*. The beam carries a *udl* of 30 kN/m run. The area of strut is 4000 mm^2 and the ties have a cross-sectional area of 1500 mm^2. The beam is 500 mm deep and has second moment of area of $500(10)^6$ mm^4.

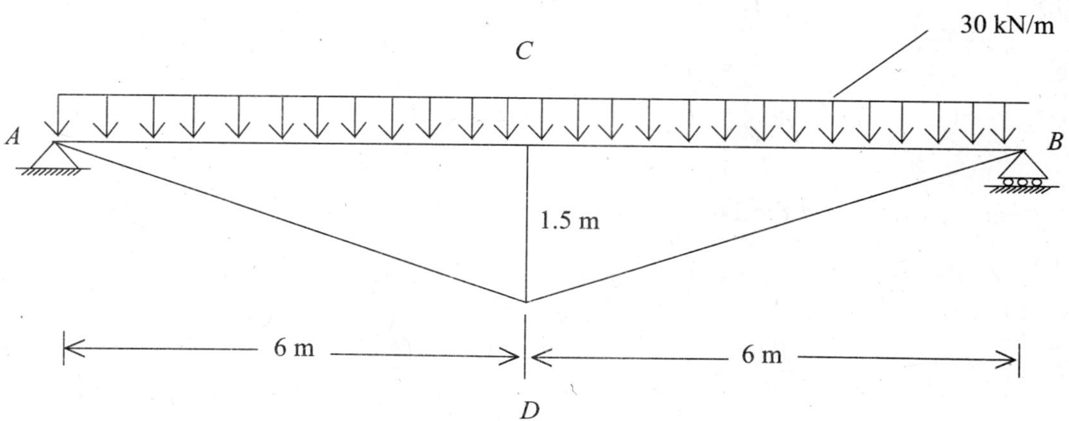

FIG. 7.64 Trussed beam

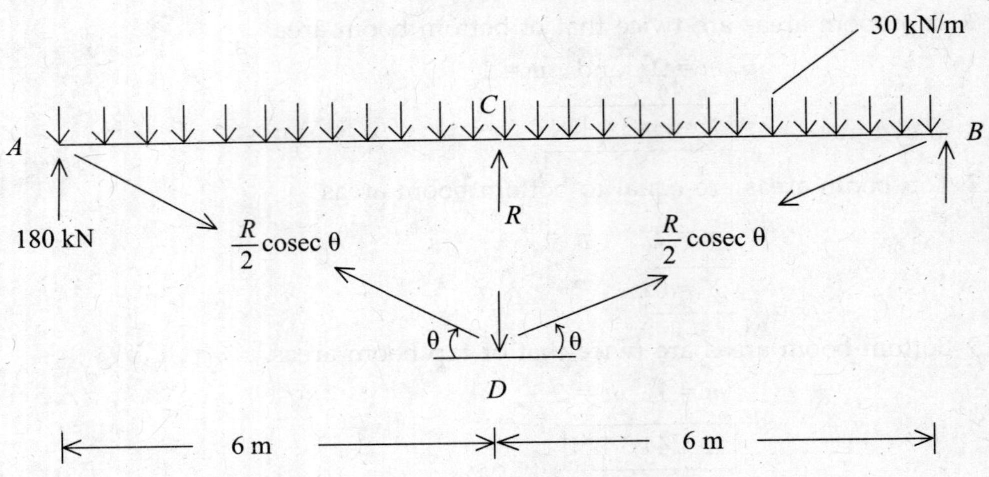

FIG. 7.65 Free body diagram of trussed beam

Let R be the compressive force in the strut CD. The forces meeting at the joint are F_{DA}, F_{DC} and F_{DB}. As F_{DC} is 'R' and due to symmetry $F_{DA} = F_{DB}$. The force in F_{DC} is compressive whereas F_{DA} and F_{DB} are tensile.

Consider joint D and resolving vertically;

$$F_{DA} \sin\theta + F_{DB} \sin\theta = R \quad (1)$$

$$2F_{DA} \sin\theta = R$$

$$\therefore \quad \boxed{F_{DA} = F_{DB} = R \operatorname{cosec} \theta/2}$$

FIG. 7.66

The reaction at A;

$$R_A = 180 - R/2$$

and

$$M_X = R_{Ax} - 30x^2/2$$

i.e.

$$M_x = (180 - R/2)\, x - 15x^2 \quad (2)$$

$$\frac{\partial M_x}{\partial R} = -\frac{x}{2} \quad (3)$$

Bending energy of the beam

$$U_B = \int \frac{M^2}{2EI}\, ds$$

$$\therefore \quad \frac{\partial U_B}{\partial R} = \int \left(\frac{M}{EI}\right)\left(\frac{\partial M}{\partial R}\right) ds$$

$$\frac{\partial U_B}{\partial R} = \int_0^6 \left(\frac{M}{EI}\right)\left(\frac{\partial M}{\partial R}\right) ds + \int_6^{12} \left(\frac{M}{EI}\right)\left(\frac{\partial M}{\partial R}\right) ds$$

Due to symmetry

$$\frac{\partial U_B}{\partial R} = 2\int_0^6 \left(\frac{M}{EI}\right)\left(\frac{\partial M}{\partial R}\right) ds$$

$$\frac{\partial U_B}{\partial R} = \frac{2}{EI}\int_0^6 \{(180 - 0.5R)\,x - 15x^2\}\,(-x/2)\,dx$$

$$\frac{\partial U_B}{\partial R} = \frac{(36R - 8100)}{EI}$$

$$\boxed{\frac{\partial U_B}{\partial R} = \frac{(36R - 8100)}{E \times 500 \times 10^{-6}}}$$

The strain energy of the strut is

$$U_s = \frac{R^2 l}{2AE}$$

Then $$\frac{\partial U_s}{\partial R} = \frac{R(12)}{AE}$$

i.e $$\boxed{\frac{\partial U_s}{\partial R} = \frac{12R}{E \times 400 \times 10^{-6}}}$$

The strain energy in the tie is

$$U_t = 2\left(\frac{R}{2}\operatorname{cosec}\theta\right)^2 \times 6\,\frac{\sec\theta}{2AE}$$

$$U_t = 26.29\,\frac{R^2}{AE}$$

$$\boxed{\frac{\partial U_t}{\partial R} = \frac{52.58R}{1500 \times 10^{-6} E}}$$

Applying Castigliano's second theorem

$$\frac{\partial U}{\partial R} = \frac{\partial U_B}{\partial R} + \frac{\partial U_s}{\partial R} + \frac{\partial U_t}{\partial R}$$

$$\frac{(36R - 8100)}{E \times 500 \times 10^{-6}} + \frac{12R}{E \times 4000 \times 10^{-6}} + \frac{52.58R}{1500 \times 10^{-6} E} = 0$$

Solving this $\boxed{R = 147.2 \text{ kN}}$

Substituting the value of R in the bending moment expression;

$$M_x = 147.2x - 15x^2$$

The above expression will be maximum, when

$$\frac{dM_x}{dx} = 0$$

$$147.2 - 30x = 0$$

i.e. $\boxed{x = 4.9 \text{ m}}$

$$\therefore \quad M_{max} = 178(4.9) - 15(4.9)^2$$

$$M_{max} = 512.05 \text{ kNm}$$

Using the flexural formula

$$f_{max} = \frac{M_{max}}{I} y_{max}$$

$$f_{max} = \frac{512.05(10)^6 \times 250}{500 \times 10^6}$$

$$\boxed{f_{max} = 256 \text{ N/mm}^2}$$

REVIEW QUESTIONS

Remembrance

7.1. List some common types of indeterminate structures.

7.2. In the analysis of indeterminate structures, we require additional equations to be satisfied in addition to equilibrium equations. Give a suitable example.

7.3. What is a redundant member?

7.4. What is a redundant support?

7.5. Name the theorem which is applied to determine the force in redundant members having lack of fit.

7.6. Give an example of externally indeterminate structure.

7.7. Give an example of internally indeterminate structure.

7.8. State Castigliano's second theorem.

7.9. Give an example of externally and internally indeterminate structure.

7.10. Where are trussed beams used in practice?

Understanding

7.1. Is the size of the member required for the analysis of statically indeterminate structure?

7.2. Is Castigliano's theorem valid for inelastic systems?

7.3. What are the unknowns considered in Engessor theorem of compatibility.

7.4. What is axial rigidity and where it is being used?

EXERCISE PROBLEMS

7.1.

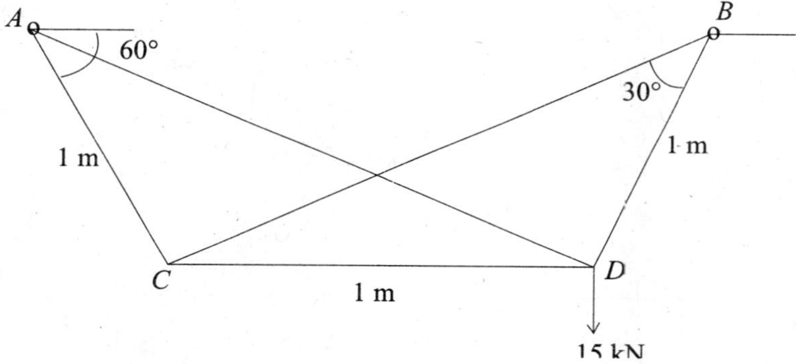

F_{CD} = 0.587 kN

7.2.

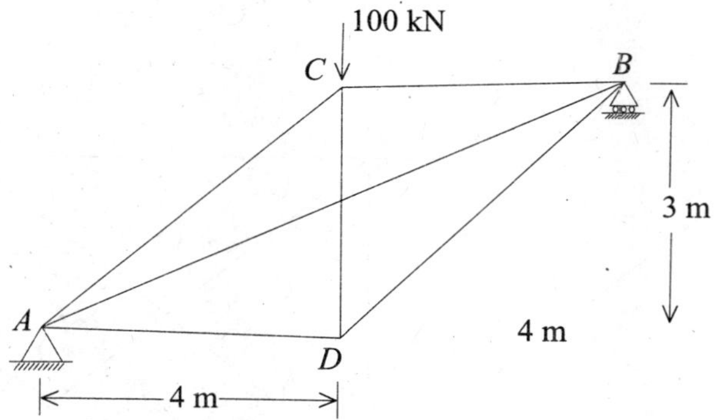

F_{CD} = 48.86 kN (comp)

7.3.

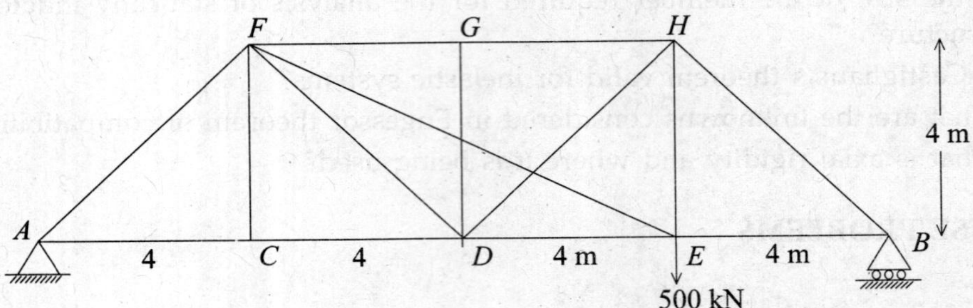

F_{FE} = 134.73 kN

7.4.

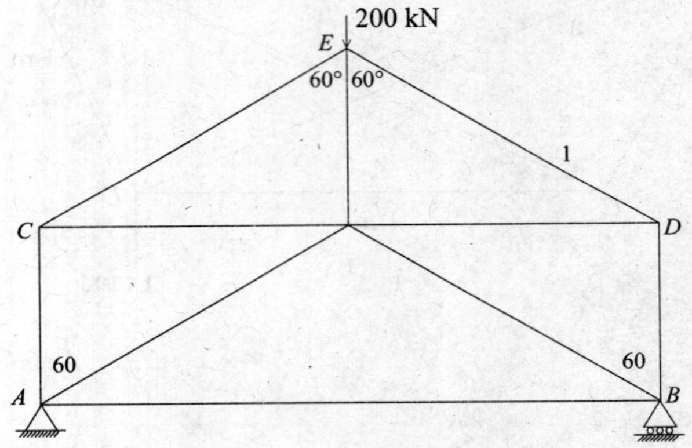

F_{AB} = 83.65 kN

7.5.

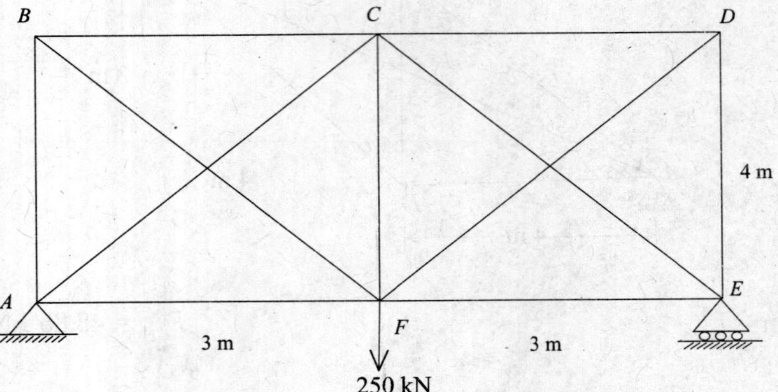

$F_{BF} = F_{FD}$ = 8.8 kN

7.6.

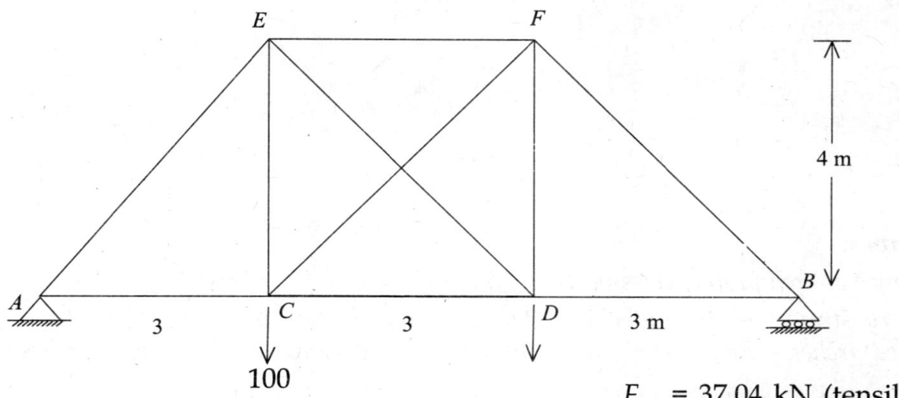

F_{CF} = 37.04 kN (tensile)

7.7.

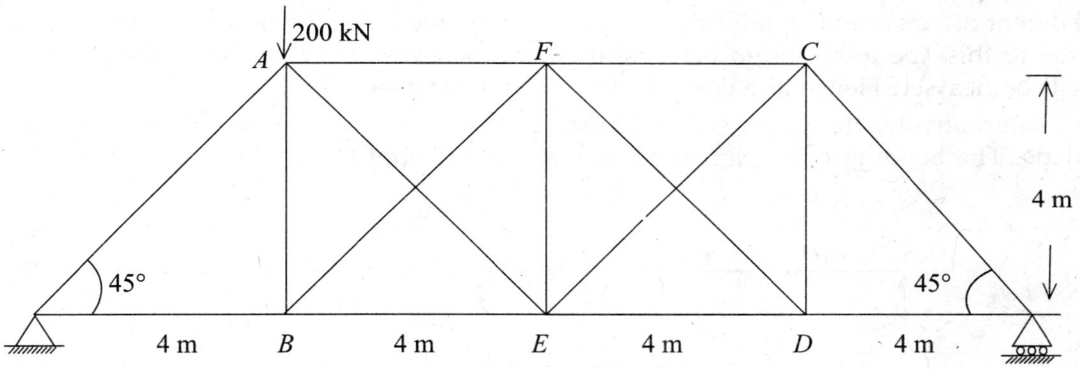

F_{AB} = 16.42 (comp)
F_{CD} = 17.05 (tensile)

7.8.

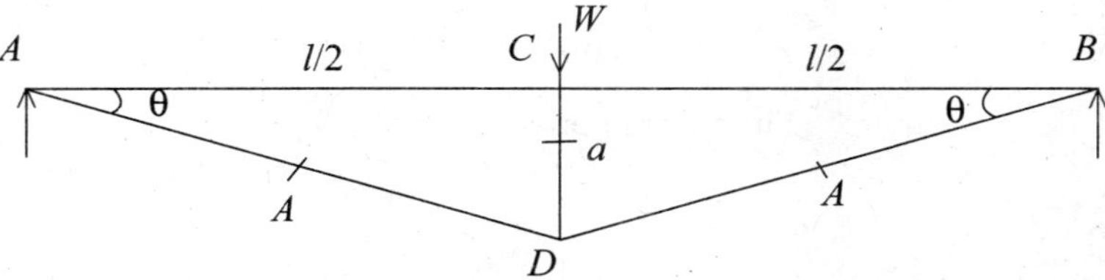

Determine F_{CD}.

$$F_{CD} = \frac{WL^3/48E_BI}{\frac{\operatorname{cosec}^2\theta\sec\theta}{4AE_T} + \frac{\tan\theta}{2aE_s} + \frac{L^2}{48E_BI}}$$

8

Two Hinged Arches

Objectives:

Importance of arch profile, classification, Analysis of circular arches, analysis of parabolic arches, Secondary stresses – effect of rise in temperature, yielding of supports, elastic shortening of ribs, Tied arches – Bowstring girder, Influence line diagram of two hinged arches.

8.1 INTRODUCTION

In a simply supported beam subjected to symmetrical loading, the maximum bending moment occurs at midspan. If the span is large, the dead load bending moment is large. Due to this, the total design bending moment increases and that the section designed will be heavier. Hence, the cost of construction becomes more.

Alternatively, the given straight beam is bent into a curved shape in the vertical plane. The bending of a beam is caused by a horizontal thrust 'H' at the ends.

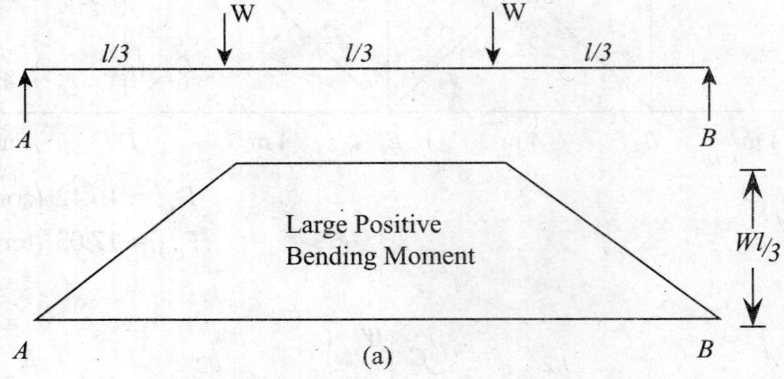

Simply supported beam BMD

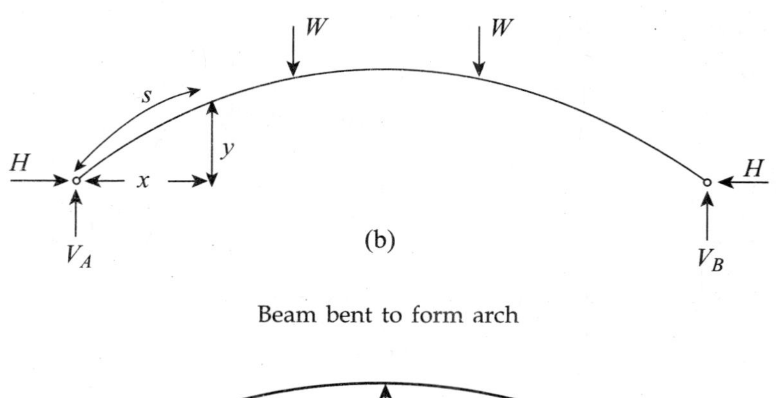

(b)

Beam bent to form arch

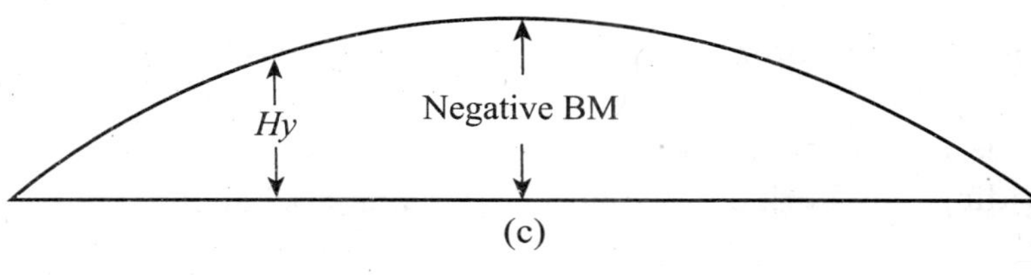

(c)

BMD due to H

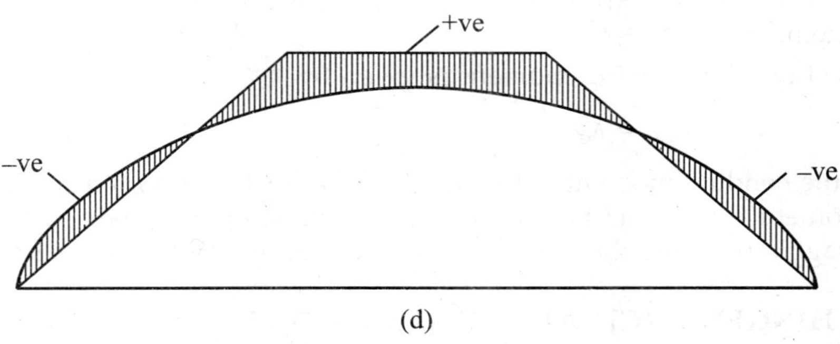

(d)

Final bending moment diagram

FIG. 8.1

The above diagrams illustrate how the bending moment can be reduced by bending the given beam in the form of an arch.

An arch can be defined as a curved structure in which both supports are restrained in direction but not necessarily held in position. Thus, the two hinged arch is hinged at the springings at A and B. The unknown reactions are four, viz., V_A, H_A, V_B and H_B. But the equilibrium equations are three. Hence, one more equation is necessary to determine the horizontal thrust H. This can be determined from elastic deformations.

Consider a two hinged arch shown in Fig. 8.2.

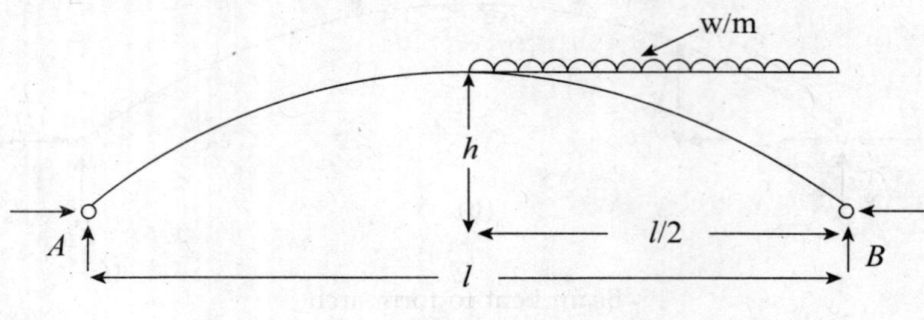

FIG. 8.2

An equation for horizontal thrust 'H' can be derived based on the following assumptions

(i) The arch is free from stresses before loading

(ii) After loading, the horizontal span AB remains unchanged.

Arch action

The action of the horizontal thrust upon the structure is known as arch action. As it is evident from the above expression that its effect is very important as it reduces the size of the member.

At any section, the bending moment is given by

$$M_x = M_s - Hy$$

where M_s is the bending moment at the section for a freely supported straight beam and Hy is the moment due to the horizontal thrust. The simply supported beam moment is positive (sagging). While the moment due to horizontal thrust is negative (hogging.)

8.2 TWO HINGED ARCH AT THE SAME LEVEL

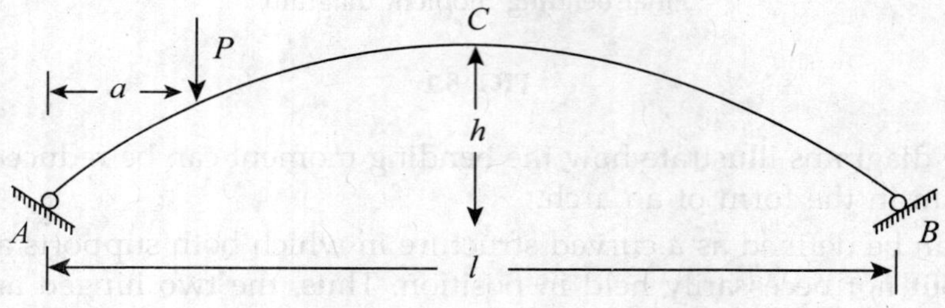

FIG. 8.3

The vertical reactions are found out as

$$V_A = P(l - a)/l$$

$$V_B = Pa/l$$

The above reactions are found out by treating it as a simply supported beam and due to horizontal thrust, the bending moment will be $-Hy$, where 'y' is the vertical height of the section from the springings.

Hence,

$$M_x = M_s - Hy \qquad (1)$$

The total strain energy is given by

$$U = \int_0^l \frac{(M_s - Hy)^2\, ds}{2EI} \qquad (2)$$

If the supports do not yield then

$$\frac{\partial U}{\partial H} = 0 \qquad (3)$$

i.e. $$\int_0^l (M_s - Hy)(-y)\frac{ds}{EI} = 0 \qquad (4)$$

$$-\int_0^l \frac{M_s y ds}{EI} + \int_0^l Hy^2 \frac{ds}{EI} = 0 \qquad (5)$$

$$\boxed{H = \frac{\int_0^l \frac{M_s y ds}{EI}}{\int_0^l \frac{y^2\, ds}{EI}}} \qquad (6)$$

where M_s is the bending moment at any point of a corresponding simply supported beam under the given loading

y is the height of arch at any point

I is the moment of inertia of the arch section at any point

ds is the elemental length measured along the curve of the arch rib

In case moment of inertia at any section of arch varies

$$I = I_c \sec\theta$$

where I_c = moment of inertia of the arch at the crown

and $ds = dx \sec\theta$.

Substituting the value of I and ds in Eq. (6) (EI_c gets cancelled as it appears in numerator as well as in the denominator) the equation reduces to

$$H = \frac{\int_0^l M_s y dx}{\int_0^l y^2 dx} \tag{7}$$

If the support yields by an amount 'δ' then

$$H = \frac{\int M_s y dx - EI_c \delta}{\int y^2 dx} \tag{8}$$

The horizontal thrust for the arches (parabolic and segmental) can be determined using the Eqs. (7) and (8). Alternatively the analysis can be done by the summation as

$$H = \frac{\Sigma My \frac{\delta s}{I}}{\Sigma y^2 \frac{ds}{I}} \tag{9}$$

In applying Eq. (9), the arch rib is divided into a number of parts of equal length δs. The values of y and M are determined at the midordinates along the length δs. Then the value of H is found out by evaluating the numerator and denominator separately.

8.3 CLASSIFICATION OF TWO HINGED ARCHES

8.3.1 Type I

The arches are of simple geometrical form such as (i) segmental arch and (ii) parabolic arch. The horizontal thrust for such arches is obtained by carrying out the integration over the length of the arch. The expressions are given in Eqs. (7) and (8) respectively.

8.3.2 Type II

The centre line of the arch is constructed using portions of several curves. In simple words, the shape of the arch departs from a simple geometrical shape.

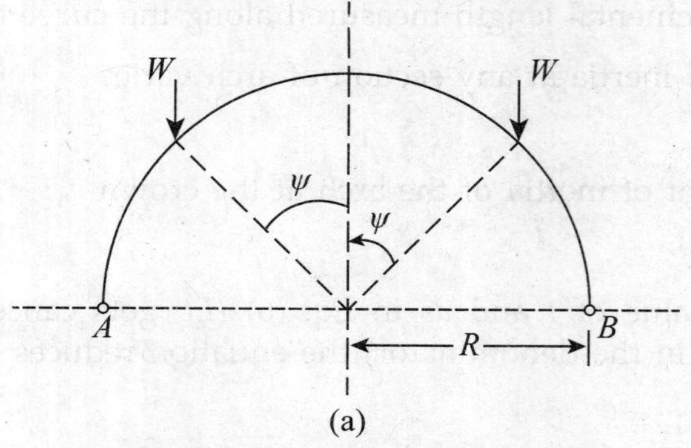

(a)

Semicircular arch (Type I)

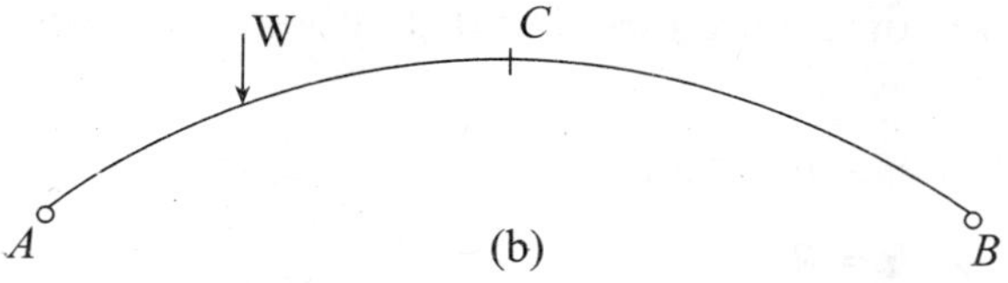

(b)

Parabolic arch (Type I)

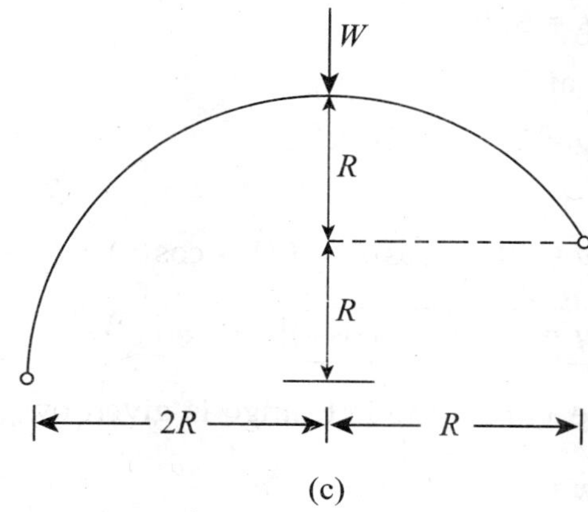

(c)

Two pinned arch (Type II)

FIG. 8.4

8.4 TWO HINGED SEGMENTAL ARCHES

Example 8.1 Derive an expression for horizontal thrust of a symmetrical circular arch subjected to uniformly distributed load.

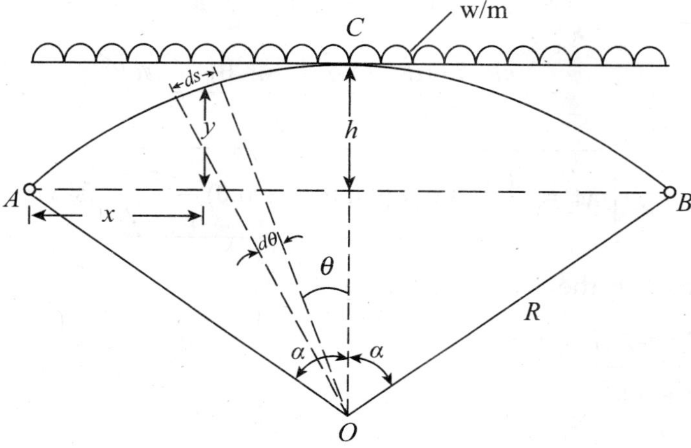

FIG. 8.5

Solution: Let the radius of the arch be R and it subtends an angle of 2α at the centre. The rise of arch is

$$h = R - R\cos\alpha$$

$$h = R(1 - \cos\alpha) \qquad (1)$$

The elemental length is

$$ds = Rd\theta \qquad (2)$$

The ordinate 'y' of the arch is

$$y = h - (R - R\cos\theta)$$

Using Eq. (1);

$$y = R(1 - \cos\alpha) - R(1 - \cos\theta)$$

$$\boxed{y = R(\cos\theta - \cos\alpha)}$$

The horizontal distance x from the left hinge is given by

$$x = \frac{l}{2} - R\sin\theta$$

$$x = R\sin\alpha - R\sin\theta$$

$$x = R(\sin\alpha - \sin\theta)$$

The simply supported beam moment is given by

$$M_s = \frac{wl}{2}x - \frac{wx^2}{2}$$

$$= \frac{wl}{2}R(\sin\alpha - \sin\theta) - \frac{w}{2}R^2(\sin\alpha - \sin\theta)^2$$

$$= \frac{wR\sin\alpha R(\sin\alpha - \sin\theta)}{2} - \frac{wR^2}{2}(\sin\alpha - \sin\theta)^2$$

$$\boxed{M_s = \frac{wR^2}{2}\sin\alpha(\sin\alpha - \sin\theta) - \frac{wR^2}{2}(\sin\alpha - \sin\theta)^2}$$

Horizontal thrust in the arch is given by

$$H = \frac{\int M_s y\,ds/EI}{\int y^2 ds/EI}$$

Substituting the values

$$H = \frac{wR \int_0^{2\alpha} [\sin\alpha(\sin\alpha - \sin\theta) - (\sin\alpha - \sin\theta)^2 (\cos\theta - \cos\alpha)]\, d\theta}{\int_0^{2\alpha} (\cos\theta - \cos\alpha)^2\, d\theta}$$

8.4.1 Integration Method

Example 8.2 A segmental arch of uniform moment of inertia has a span of 20 metres and subtends 90° at the centre. One half of the arch is loaded with a uniformly distributed load of 20 kN/m. Determine the horizontal thrust.

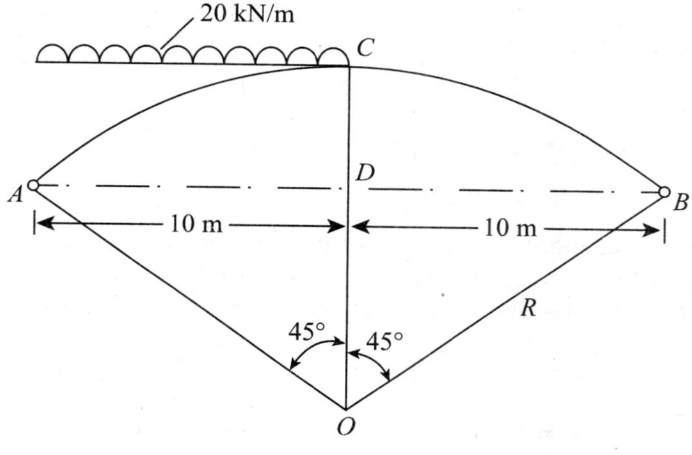

FIG. 8.6

Solution

Radius of the arch

As the arch subtends 90° at '*O*';

$$\angle DOB = 45^\circ$$

and hence

$$DB = DO$$

i.e. $$\boxed{DO = 10 \text{ m}}$$

$$R = \sqrt{DB^2 + DO^2} = 14.14 \text{ m}$$

Hence, rise $$CD = 14.14 - 10 = 4.14 \text{ m}$$

Simply supported beam expressions

The reactions and bending moment expressions are written by considering the arch as a simply supported beam as

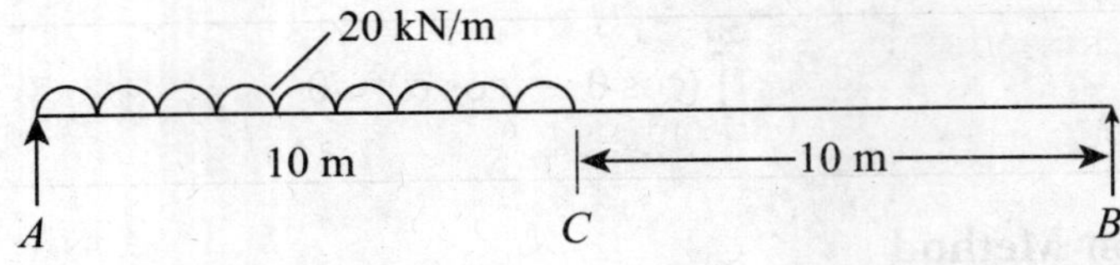

FIG. 8.7

$$\Sigma V = 0;$$

$$V_A + V_B = 200 \tag{1}$$

$$\Sigma M_A = 0;$$

$$20 \times \frac{10^2}{2} - 20V_B = 0$$

$$V_B = 50 \text{ kN} \tag{2}$$

$$\therefore \quad V_A = 150 \text{ kN} \tag{3}$$

It is to be mentioned here that the bending moment expressions are to be written for each portion of the span separately. Here we have to write expressions for *AC* and *BC*. Hence, two expressions are necessary one for the left side and the other for the right side.

Zone AC

$$M = 150x - 20\frac{x^2}{2} \quad \text{for} \quad 0 < x \le 10 \tag{4}$$

and x is measured from A.

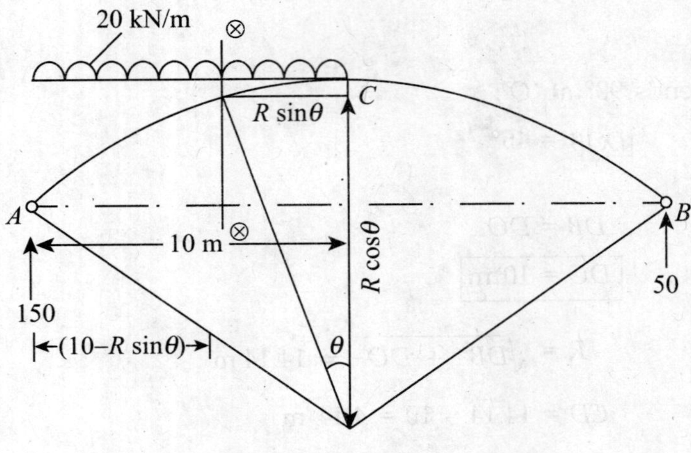

FIG. 8.8

and $$x = 10 - R\sin\theta \quad (5)$$

Substituting (5) in (4);

$$M = 150(10 - R\sin\theta) - 10(10 - R\sin\theta)^2 \quad (6)$$

Expanding and simplifying

$$M = 500 - 10R^2\sin^2\theta + 50R\sin\theta \quad (7)$$

Zone *BC*

$$M = 50x \quad (8)$$

$$M = 50(10 - R\sin\theta) \quad (9)$$

The expressions (7) and (9) are used to evaluate the horizontal thrust as

$$H = \frac{\int Myds}{\int y^2 ds}$$

as EI is constant.

$$y = R\cos\theta - 10$$

and $$ds = Rd\theta$$

The numerator is evaluated separately for the whole span.

Integration of the numerator

Zone AC

$$\int Myds = \int_0^{\pi/4} (500 - 10R^2\sin^2\theta + 50R\sin\theta)(R\cos\theta - 10)Rd\theta$$

$$= 500R^2\int_0^{\pi/4}\cos\theta d\theta - 5000R\int_0^{\pi/4}d\theta - 10R^4\int_0^{\pi/4}\sin^2\theta\cos\theta d\theta$$

$$+100R^3\int_0^{\pi/4}\sin^2\theta d\theta + 50R^3\int_0^{\pi/4}\sin\theta\cos\theta d\theta - 500R^2\int_0^{\pi/4}\sin\theta d\theta$$

$$= 500R^2\left[\sin\theta\right]_0^{\pi/4} - 5000R\left[\theta\right]_0^{\pi/4} - 10R^4\left[\sin\theta\right]_0^{\pi/4} + 10R^4\left[\sin\theta - \frac{\sin^3\theta}{3}\right]_0^{\pi/4}$$

$$+ 100R^3\left[\frac{\theta}{2} - \frac{\sin 2\theta}{4}\right]_0^{\pi/4} + 50R^3\left[\frac{-\cos 2\theta}{4}\right]_0^{\pi/4} - 500R^2\left[-\cos\theta\right]_0^{\pi/4}$$

$$\boxed{\int Myds = 14470.86}$$

Zone BC

$$\int M_s y ds = \int_0^{\pi/4} 50(10 - R\sin\theta)(R\cos\theta - 10)ds$$

$$= \int_0^{\pi/4} 50(10 - R\sin\theta)(R\cos\theta - 10)Rd\theta$$

$$= 50R\left[10R\int_0^{\pi/4}\cos\theta d\theta - R^2\int_0^{\pi/4}\sin\theta\cos\theta d\theta - 100\int_0^{\pi/4}d\theta + 10R\int_0^{\pi/4}\sin\theta d\theta\right]$$

$$= 50R\left[-\frac{R^2}{4} - 78.54 + 10R\right]$$

$$\boxed{\int M_s y ds = 9102.69}$$

Integration of denominator

$$\int y^2 ds = 2\int_0^{\pi/4}(R\cos\theta - 10)^2 Rd\theta$$

$$= 2R\int_0^{\pi/4}\left(\frac{R^2}{2} + \frac{R^2}{2}\cos 2\theta + 100 - 20R\cos\theta\right)d\theta$$

$$= 2R\left[\frac{R^2}{2}\theta + \frac{R^2}{4}\sin 2\theta + 100\theta - 20R\sin\theta\right]_0^{\pi/4}$$

$$\boxed{\int y^2 ds = 200.8}$$

$$H = \frac{14470.86 + 9102.69}{200.8} = 117.4\text{ kN}$$

8.4.2 Graphical Summation Method

The above method requires the evaluation of integrals. This process becomes cumbersome if the bending moment expressions are many for the given arch. However, a time saving method, viz., graphical summation method can be used. The graphical summation method is approximate when compared to the exact integral solution.

In this method, the arch is divided into number of equal parts each of a length δs. The values of the ordinate y and bending moment M are found at the midordinates of each of these lengths. This involves the approximation of dividing the span rather than the length of the arch rib. This is valid provided the arch is of 'flat' type and the number of divisions is large say between 8 and 12. The computation of horizontal thrust requires the value of the numerator of the horizontal thrust expression, i.e., $\Sigma M_s y ds$ and the denominator value of $\Sigma y^2 ds$.

Though this method is approximate; it is readily useful for practical constructions.

Example 8.3 Determine the horizontal thrust of the arch problem in Example 8.1 by using the graphical summation method.

Solution: The given arch is of flat type and is divided into ten equal parts.

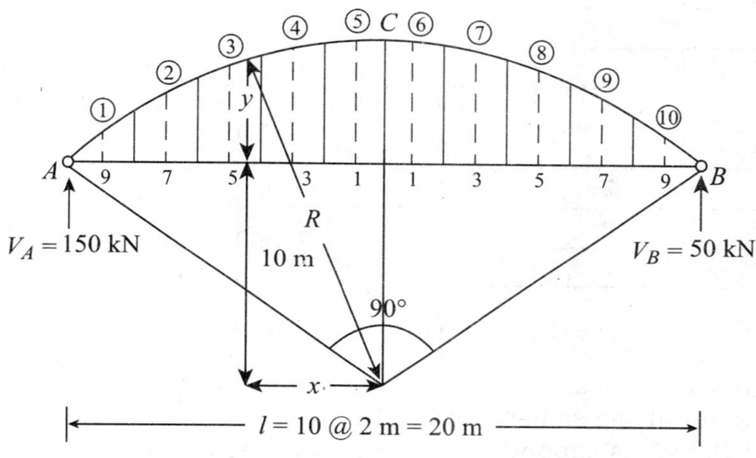

FIG. 8.9 Segmental arch

From geometry, we can write

$$(10 + y)^2 + x^2 = R^2$$

$$(10 + y)^2 + x^2 = 14.14^2$$

Simplifying,

$$y = \sqrt{14.14^2 - x^2} - 10$$

It is to be noted that 'x' is measured from the crown C on either side. Thus, knowing the value of x, y and y^2 can be computed. The computed values are given in the Table.

The bending moment expression from A to C is

$$M = 150(10 - x) - 20(10 - x)^2/2$$

and the bending moment expression from B to C is

$$M = 50(10 - x)$$

The values of M and y for each of the section are given below.

Section	*x (m)*	*y (m)*	y^2	*M (kNm)*	*My(kN-m²)*
1	9	0.906	0.821	140	126.84
2	7	2.286	5.225	360	822.96
3	5	3.226	10.410	500	1613.00
4	3	3.818	14.578	560	2138.08
5	1	4.105	16.847	540	2216.70
6	1	4.105	16.847	450	1847.25
7	3	3.818	14.578	350	1336.30
8	5	3.226	10.410	250	806.50
9	7	2.286	5.225	150	342.90
10	9	0.906	0.821	50	45.30
			95.762		11295.83

$$H = \Sigma Myds / \Sigma y^2 ds = 117.96 \text{ kN}$$

Example 8.4 A segmental arch rib, span 36 m and rise 6 m is hinged at the springings. The two hinges are at the same level and the cross section of the rib is uniform. A vertical load of 100 kN is supported by the arch at the crown.

Determine the following:

(i) the horizontal thrust in the arch rib

(ii) the maximum bending moment in the rib

(iii) maximum change in the horizontal thrust due to a change in temperature of 30°C.

Assume $\alpha = 15(10)^{-6}$ per°C, $E = 200$ kN/mm^2; and $I = 15(10)^9$ mm^4

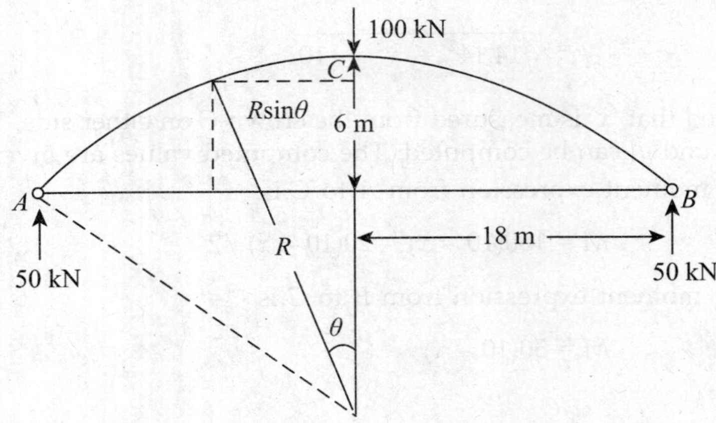

FIG. 8.10

Solution: (i) The radius of the arch can be determined by using geometry of the circle;

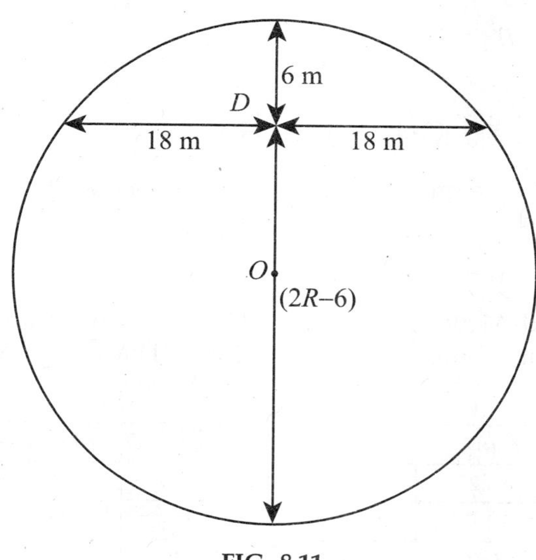

FIG. 8.11

$$(2R - 6)6 = 18 \times 18$$

and hence $$R = 30 \text{ m}$$

$$\therefore \quad OD = 30 - 6 = 24 \text{ m}$$

As there is only crown load acting; we can use graphical summation which gives the solution faster. The given arch is divided into 10 equal parts. Hence, each division is (36/10) = 3.6 metres.

$$H = \frac{\Sigma \frac{M_s y}{EI} ds}{\Sigma \frac{y^2}{EI} ds}$$

where M_s is the simply supported beam moment.

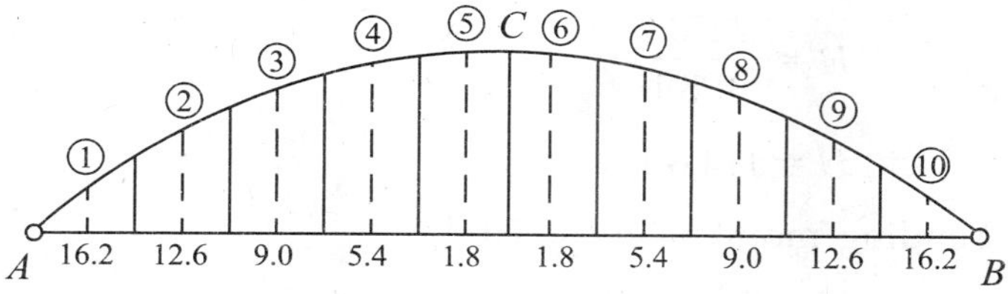

FIG. 8.12

The ordinate y is computed from the geometry as

$$x^2 + (24 + y)^2 = 30^2$$

and on simplification

$$y = \sqrt{30^2 - x^2} - 24$$

The bending moment expression from A to C as well as from B to C (due to symmetry) is written as

$$M = 50(18 - x)$$

The values of y and M were computed for different values of x from 1.8 to 16.2 which are the midpoint of the divided segments. The computed values are given in the table.

Section	*x (m)*	*y (m)*	y^2	*M (kNm)*	*My*
1	16.2	1.25	1.56	90	112.50
2	12.6	3.23	10.43	270	872.10
3	9.0	4.62	21.34	450	2079.00
4	5.4	5.51	30.36	630	3471.30
5	1.8	5.95	35.40	810	4819.50
6	1.8	5.95	35.40	810	4819.50
7	5.4	5.51	30.36	630	3471.30
8	9.0	4.62	21.34	450	2079.00
9	12.6	3.23	10.43	270	872.10
10	16.2	1.25	1.56	90	112.50
			198.18		22708.8

$$H = \frac{\Sigma \dfrac{M_s y ds}{EI}}{\Sigma \dfrac{y^2}{EI} ds}$$

$$H = \frac{22708.8}{198.18}$$

$$H = 114.6 \text{ kN}$$

(ii) Maximum bending moment in the rib

$$M = 50(18 - 30 \sin\theta) - 114.6(30 \cos\theta - 24)$$

$$M = 3650.4 - 1500 \sin\theta - 3438 \cos\theta$$

$$\frac{dM}{d\theta} = 0$$

i.e. $\quad -1500\cos\theta + 3438\sin\theta = 0$

$$\tan\theta = 0.4363$$

$$\theta = 23°\ 34'$$

and $\quad \sin\theta = 0.3998 \quad$ and $\quad \cos\theta = 0.9166$

Maximum Hogging Moment = 3650.4 − 1500 × 0.3998 − 3438 × 0.9166

Max. Hogging Moment = −99.08 kNm

(iii) Maximum Sagging Moment (occurs when θ = 0; at the crown)

Max. Sagging Moment = 3650.4 − 1500(0) − 3438(1)

i.e. Max. Sagging Moment = 212.4 kNm

(iv) Change in H due to rise in temperature $t°$

$$\delta H = \frac{\alpha t l}{\Sigma(y^2 ds/EI)}$$

As EI is constant

$$\delta H = \frac{EI\alpha t l}{\Sigma y^2 ds}$$

$$= \frac{200000 \times 15 \times 10^9 \times 15 \times 10^{-6} \times 30 \times 36}{198.18 \times 3.6 \times 10^9}$$

$\delta H = 6.8$ kN

8.5 TWO HINGED PARABOLIC ARCHES

Example 8.5 A two hinged parabolic arch of span 'l' and rise 'h' has the second moment of area varying as a secant of the slope of the rib axis. The arch carries a concentrated load of W at a distance of 'a' from the left hand support. Estimate the horizontal thrust at the supports.

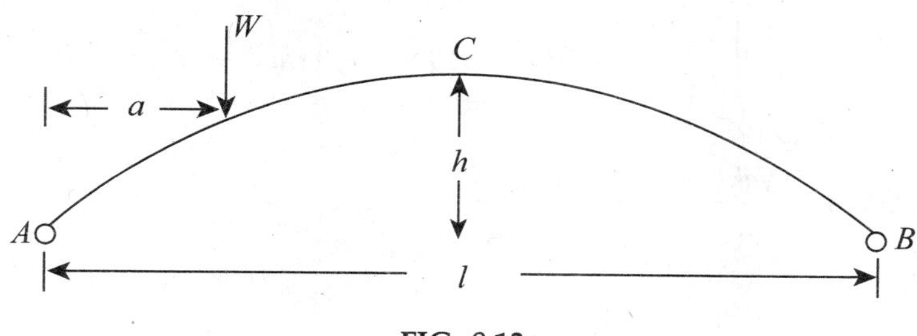

FIG. 8.13

Solution: Taking A as the origin, the centre line of the arch rib is defined by the equation

$$y = \frac{4h}{l^2} x(l - x) \tag{1}$$

and

$$ds = dx \sec\theta \tag{1a}$$

$$I = I_c \sec\theta \tag{1b}$$

where I_c is the second moment of area at the crown.

The horizontal thrust is estimated using

$$H = \frac{\int \frac{M_s y ds}{EI}}{\int y^2 \frac{ds}{EI}} = \frac{\int \frac{M_s y dx \sec\theta}{EI_c \sec\theta}}{\int \frac{y^2 dx \sec\theta}{EI_c \sec\theta}} \tag{2}$$

$$\therefore \quad \boxed{H = \frac{\int M_s y dx}{\int y^2 dx}} \tag{3}$$

where M_s is the simply supported beam bending moment.

A general expression for M_s can be written as

$$M_s = Wa \frac{(l - x)}{l} - W[a - x] \tag{4}$$

Thus, the numerator of H can be computed as

$$\int M_s y dx = \int_0^l \left\{ Wa \frac{(l - x)}{l} - W[a - x] \right\} y dx \tag{5}$$

Substituting 'y' expression from Eq. (1) in (5)

$$\int M_s y dx = \int_0^l \frac{Wa(l - x)}{l} \left(\frac{4hx(l - x)}{l^2} \right) dx - \int_0^l W[a - x] \left(\frac{4hx(l - x)}{l^2} \right) dx$$

$$= \frac{4h}{l^2} \left\{ \frac{a}{l} \int_0^l (l^2 x - 2lx^2 + x^3)\, dx - \int_0^a \left(alx - x^2(a + l) + x^3 \right) dx \right\}$$

$$\int M_s y dx = \frac{Wahl}{} \left(\quad - \frac{a^2}{l^2} + \frac{a^3}{l^3} \right) \tag{6}$$

and

$$\int y^2 dx = \int_0^l \left[\frac{4h}{l^2} x(l - x) \right]^{1/2} dx$$

i.e. $$\int y^2 dx = \frac{8}{15} h^2 l \qquad (7)$$

Substituting Eqs. (6) and (7) in Eq. (3);

$$\boxed{H = \left(\frac{5}{8}\right) W \left(\frac{a}{h}\right)\left(1 - 2\left(\frac{a}{l}\right)^2 + \left(\frac{a}{l}\right)^3\right)} \qquad (8)$$

The above equation can further be simplified if $a = \alpha l$, then

$$H = \left(\frac{5}{8}\right) W \left(\frac{\alpha l}{h}\right)\left[1 + 2\alpha^2 + \alpha^3\right]$$

On factorisation

$$\boxed{H = \left(\frac{5}{8}\right) W \left(\frac{l}{h}\right) \alpha\,(1 - \alpha)\left(1 + \alpha - \alpha^2\right)} \qquad (9)$$

The above expression can be readily used, for example, if W is acting at a distance '$l/3$' from A, then $\alpha = 1/3$. Then for such a case

$$H = \left(\frac{5}{8}\right) W \left(\frac{l}{h}\right)\left(\frac{1}{3}\right)\left(1 - \frac{1}{3}\right)\left(1 + \frac{1}{3} - \left(\frac{1}{3}\right)^2\right)$$

$$= 0.509\,\frac{Wl}{h}$$

Example 8.6 Derive an expression for horizontal thrust, when a point load 'L' acts at 'k' distance from the left support and also act at 'k' distance from right support (two loads totally) a symmetrical two hinged parabolic arch of rise 'R' and of span 'S'.

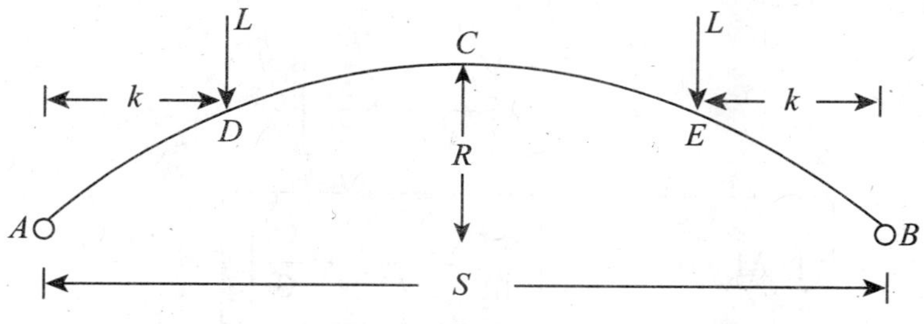

FIG. 8.14

Solution: Equation of the parabolic arch

$$y = \frac{4Rx}{S^2}(S - x) \qquad (1)$$

Simply supported bending moment expressions

Due to symmetry; $V_A = V_B = L$ (2)

The BM expressions for zone *AD* and *DC* are written as follows:

$$\boxed{M_s = V_A x = Lx} \quad \text{for} \quad 0 < x \leq k \tag{3}$$

$$M_s = V_A x - L(x - k)$$

i.e.

$$\boxed{M_s = Lk} \quad \text{for} \quad k < x \leq S/2 \tag{4}$$

The value of the integral $\int M_s y dx$ is obtained by taking advantage of symmetry

$$\int M_s y dx = \int_0^k M_s y dx + \int_k^{S/2} M_s y dx \tag{5}$$

Region AD

$$\int_0^k Lx\left(\frac{4Rx}{S^2}\right)(S - x)\,dx = \frac{4RL}{S^2}\int_0^k \left(Sx^2 - x^3\right)dx$$

$$= \frac{4RL}{S^2}\left[S\frac{x^3}{3} - \frac{x^4}{4}\right]_0^k$$

$$\boxed{\int_0^k M_s y dx = \frac{4RL}{S^2}\left[S\frac{k^3}{3} - \frac{k^4}{4}\right]} \tag{6}$$

Region DC

$$\int_k^{S/2} Lk\left(\frac{4Rx}{S^2}\right)(S - x)\,dx = \frac{4RL}{S^2}\int_k^{S/2} (kxS - kx^2)\,dx$$

$$= \frac{4RL}{S^2}\left[kS\frac{x^2}{2} - k\frac{x^3}{3}\right]_k^{S/2}$$

$$\boxed{\int_k^{S/2} M_s y dx = \frac{4RL}{S^2}\left[\frac{kS^3}{12} - \frac{Sk^3}{2} + \frac{k^4}{3}\right]} \tag{7}$$

Adding the above results

$$\int_0^S M_s y dx = \frac{2}{3}\frac{RL}{S^2}\left(k^4 - 2Sk^3 + S^3k\right) \tag{8}$$

$$H = \frac{\int_0^S M_s y dx}{\int_0^S y^2 dx} = \frac{10}{8}\left(\frac{L}{RS^3}\right)\left(k^4 - 2Sk^3 + S^3k\right) \tag{9}$$

Example 8.7 A two hinged arch loaded with uniformly distributed load over half of the span. Determine the horizontal thrust.

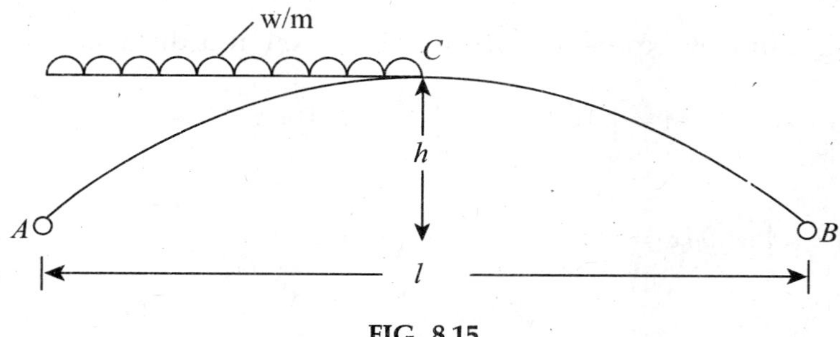

FIG. 8.15

Solution

Equation of the parabolic arch

$$y = \frac{4hx}{l^2}(l - x) \tag{1}$$

For the parabolic arch

$$\int_0^l y^2 dx = \frac{8}{15} lh^2 \tag{2}$$

Simply supported bending moment expression

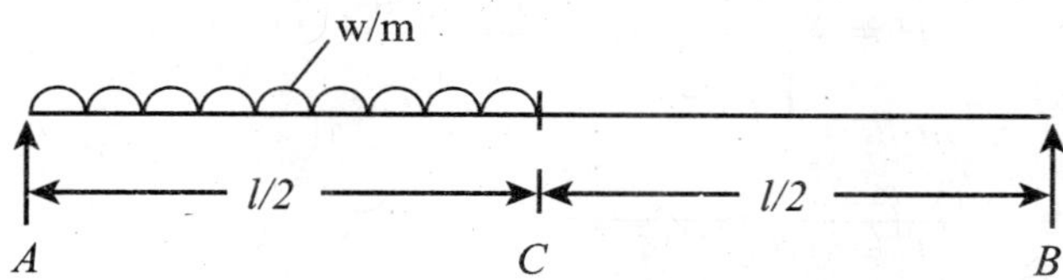

The reactions of the above simply supported beam are

$$\Sigma V = 0;$$

$$V_A + V_B = \frac{wl}{2} \tag{3}$$

$$\Sigma M = 0;$$

$$w\left(\frac{l}{2}\right)\left(\frac{l}{4}\right) - V_B l = 0 \qquad (4)$$

$$V_B = \frac{wl}{8}$$

and hence

$$V_A = \frac{3wl}{8}$$

The bending moment expression M can be written at a distance x from A;

$$M = \left(\frac{3wl}{8}x - \frac{wx^2}{2}\right) \quad \text{for } 0 < x \leq l/2 \qquad (5)$$

$$\therefore \quad \int_0^l M_s y dx = \int_0^{l/2}\left(\frac{3wl}{8}x - \frac{wx^2}{2}\right) y dx + \int_0^{l/2} \frac{wl}{8} xydx \qquad (6)$$

(Zone AC) (Zone BC)

Substituting the expression for y

$$\int_0^l M_s y dx = \int_0^{l/2}\left(\frac{3wl}{8}x - \frac{wx^2}{2}\right)\left(\frac{4hx}{l^2}(l - x)\right) dx + \int_0^{l/2} \frac{wl}{8} x\left(\frac{4hx(l - x)}{l^2}\right) dx$$

$$\boxed{\int_0^l M_s y dx = 0.0332whl^3}$$

The horizontal thrust is determined as

$$H = \frac{\int_0^l M_s y dx}{\int_0^l y^2 dx} = \frac{0.0332whl^3}{\left(\frac{8}{15}\right) lh^2}$$

$$\boxed{H = 0.0623\, wl^2/h}$$

The numerator corresponds to the BM at midspan. In simple words,

$$\boxed{\text{Horizontal thrust} = \frac{\text{BM at midspan}}{\text{Central rise of arch}}}$$

Example 8.8 A symmetrically loaded arch with uniformly distributed load throughout the span. Calculate the horizontal thrust.

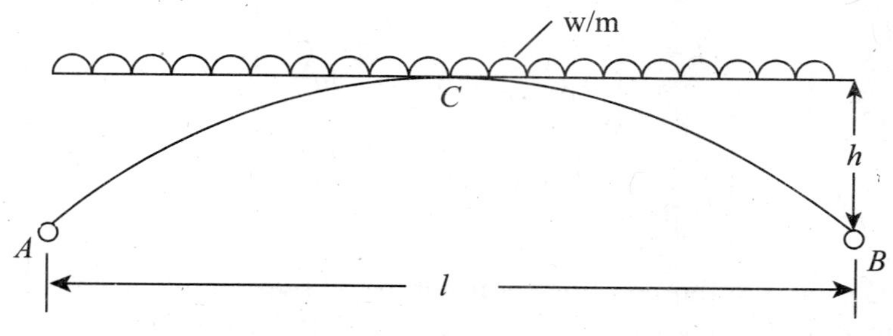

FIG. 8.16

Solution

Equation of the parabolic arch

$$y = \frac{4hx}{l^2}(l - x) \tag{1}$$

For the parabolic arch

$$\int y^2 dx = \frac{8}{15} lh^2 \tag{2}$$

Simply supported bending moment expression

w/m

A B

l

$\frac{wl}{2}$ $\frac{wl}{2}$

$$M_s = \frac{wl}{2}x - \frac{wx^2}{2} \quad \text{for BM at } x \text{ from } A \tag{3}$$

$$\therefore \quad \int_0^l M_s y dx = \int_0^l \left(\frac{wl}{2}x - \frac{wx^2}{2}\right)\left(\frac{4hx}{l^2}(l - x)\right) dx$$

$$= \int_0^l \left(\frac{2wh}{l^2}\right)(lx - x^2)^2 dx$$

$$\boxed{\int_0^l M_s y dx = \frac{whl^3}{15}}$$

Horizontal thrust

$$H = \frac{\int_0^l M_s y\,dx}{\int_0^l y^2 dx} = \frac{whl^3}{15 \times \frac{8}{15} lh^2}$$

$$\therefore \quad \boxed{H = \frac{wl^2}{8h}}$$

Example 8.9 Prove that the bending moment in the two hinged arch is zero everywhere if it is carrying a uniformly distributed load over its entire span.

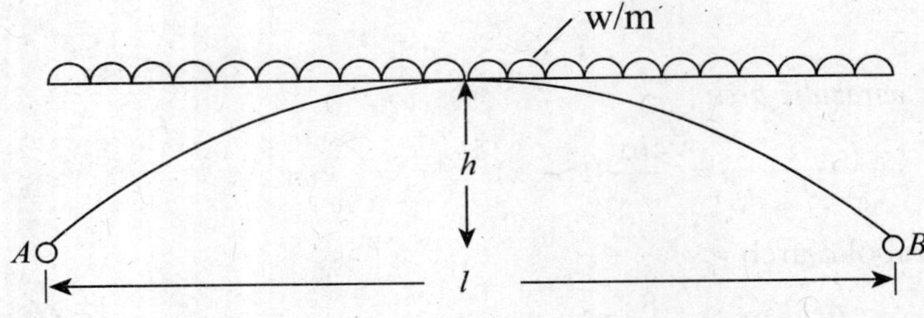

FIG. 8.17

Solution: The equation of parabolic arch with A as the origin

$$y = \frac{4x}{l^2}(l - x) \tag{1}$$

Due to symmetry

$$V_A = \frac{\text{Total load}}{2}$$

$$V_A = \frac{wl}{2} \tag{2}$$

The horizontal thrust is given by (Ref. Example no. 8.6)

$$H_A = \frac{wl^2}{8h} \tag{3}$$

The bending moment at a distance x is given by

$$M_x = V_A x - Hy - \frac{wx^2}{2}$$

$$M_x = \frac{wl}{2}x - \frac{wl^2}{8h}y - \frac{wx^2}{2}$$

$$M_x = \frac{wl}{2}x - \frac{wl^2}{8h}\left(\frac{4h}{l^2}x(l - x)\right) - \frac{wx^2}{2}$$

$$M_x = \frac{wl}{2}x - \frac{wl}{2}x(l - x) - \frac{wx^2}{2}$$

$$\boxed{M_x = 0}$$

This shows the bending moment is zero everywhere.

Example 8.10 Find the horizontal thrust for two hinged parabolic arch shown in Fig. 8.18 The moment of inertia at any section is $I_c \sec\theta$ where θ is the slope at the section and I_c is the moment of inertia at the crown. Neglect the effect of rib shortening.

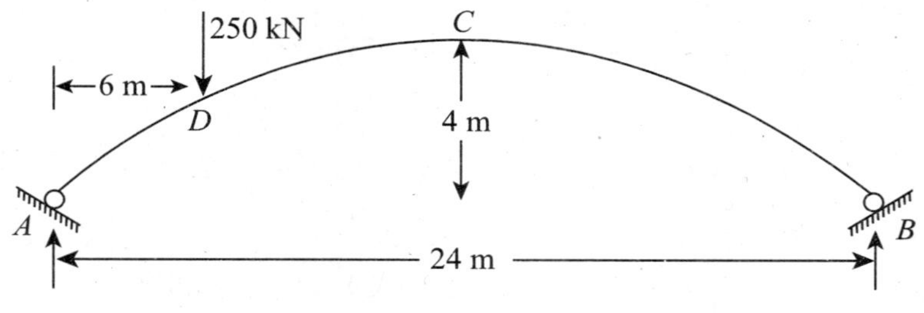

FIG. 8.18

Solution: The simply supported beam moments are obtained by knowing the reactions as follows. Considering *AB* as simply supported beam,

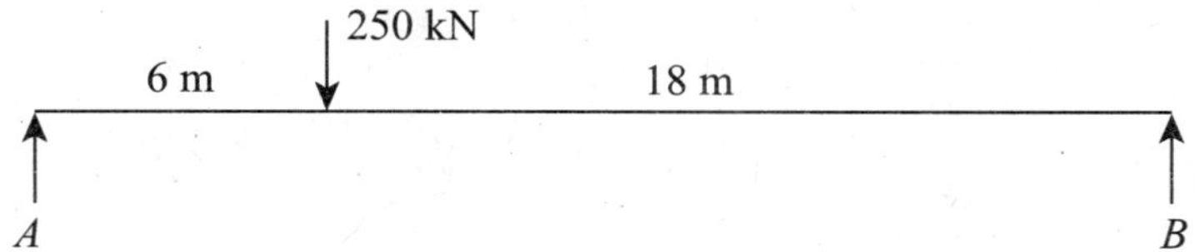

FIG. 8.19

$$\Sigma V = 0;$$

$$V_A + V_B = 250 \qquad (1)$$

$$\Sigma M = 0;$$

$$250(6) - 24V_B = 0 \qquad (2)$$

$$\boxed{V_B = 62.5 \text{ kN}}$$

$$\therefore \qquad \boxed{V_A = 187.5 \text{ kN}}$$

$$\frac{\int M_s y dx}{\int y^2 dx} = H \qquad (3)$$

The equation of the parabola is given by

$$y = \frac{4hx(l - x)}{l^2}$$

$$y = \frac{4(4)x(24 - x)}{24^2}$$

i.e.
$$\boxed{y = \frac{x(24 - x)}{36}} \quad (4)$$

$$\int_0^{24} M_s y dx = \int_0^{6} 187.5xydx + \int_0^{18} 62.5xydx$$

$$= \int_0^{6} 187.5x\left\{\frac{x(24 - x)}{36}\right\}dx + \int_0^{18} 62.5x\left\{\frac{x(24 - x)}{36}\right\}dx$$

$$= \int_0^{6}\left(125x^2 - 5.208x^3\right)dx + \int_0^{18}\left(41.67x^2 - 1.736x^3\right)dx$$

$$= 7312.61 + 35446.89 = 42759.5$$

For the parabolic arch

$$\int_0^{l} y^2 dx = \frac{8}{15}h^2 l \quad (5)$$

$$= \frac{8}{15}(4)^2(24)$$

$$= 204.8$$

Hence,
$$H = \int_0^{24} M_s y dx \Big/ \int_0^{24} y^2 dx$$

$$H = \frac{24759.5}{204.8}$$

$$\boxed{H = 208.8 \text{ kN}}$$

Bending moments at D and C

The ordinate at D is obtained from

$$y_D = \frac{6(24 - 6)}{36}$$

$$\boxed{y_D = 3 \text{ m}}$$

Hence, $M_D = 187.5(6) - 208.8(3) = 498.6$ kNm

$M_C = 62.5(12) - 208.8(4) = -85.2$ kNm

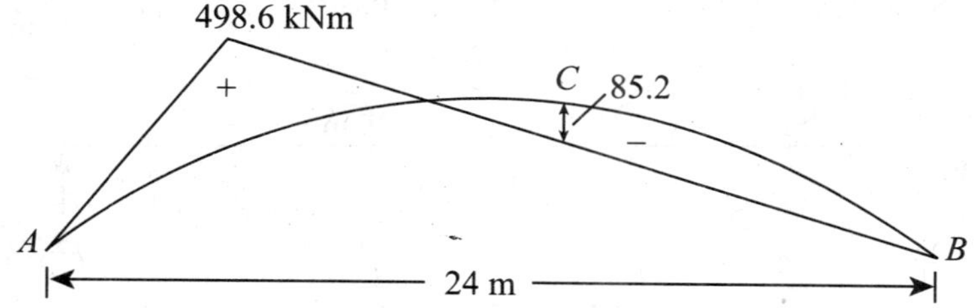

FIG. 8.20 Bending moment diagram

Example 8.11 A uniformly distributed load 40 kN/m covers the left half of the span of a parabolic arch span 36 metres and central rise of 8 m. Determine the maximum bending moment. Assume that moment of inertia at a section varies as secant of the slope at the section. Neglect the effect of rib shortening. (Oct 1998, BU)

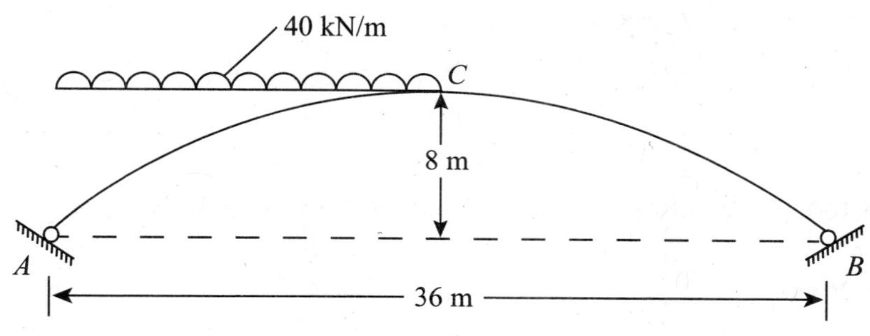

FIG. 8.21

Solution

Equation of arch:

$$y = \frac{4hx(l - x)}{l^2}$$

$$y = \frac{4(8)x(36 - x)}{36^2}$$

$$y = 0.0247x(36 - x)$$

For the parabolic arch, i.e. $\boxed{y = 0.8892x - 0.0247x^2}$

$$\int y^2 dx = \frac{8}{15} h^2 l$$

$$\int y^2 dx = \frac{8}{15} (8)^2 (36)$$

$$\boxed{\int y^2 dx = 1228.8}$$

Simply supported beam bending moment expression

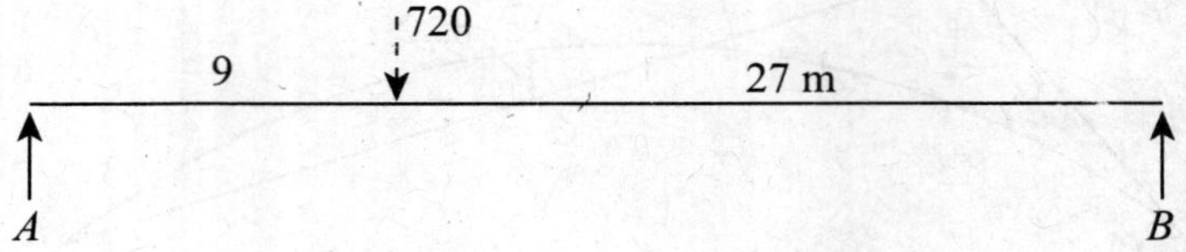

FIG. 8.21(a)

The total load of magnitude 40 × 18 = 720 kN acts at 9 metres from A. The reactions are

$$V_A + V_B = 720 \tag{1}$$

$$\Sigma M_A = 0;$$

$$720(9) - 36V_B = 0$$

$$V_B = 180 \text{ kN}$$

$$\therefore \quad V_A = 540 \text{ kN}$$

The expression for bending moment at any section can be written for the entire arch.

Zone AC (A as origin)

$$M_x = 540x - 40\frac{x^2}{2}$$

i.e. $$M_x = 540x - 20x^2 \quad \text{for} \quad 0 < x < 18$$

Zone BC (B as origin)

$$Mx = 180x \quad \text{for} \quad 0 < x < 18$$

Thus,

$$\int_0^{36} M_s y dx = \int_0^{18} M_s y dx + \int_0^{18} M_s y dx$$

(Zone AC) (Zone BC)

$$= \int_0^{18} (540x - 20x^2)(0.8892x - 0.0247x^2)\,dx + \int_0^{18} 180x\,(0.8892x - 0.0247x^2)\,dx$$

$$= \int_0^{18} (0.494x^4 - 31.12x^3 + 480.17x^2)\,dx + \int_0^{18} (160x^2 - 4.45x^3)\,dx$$

$$= 303430 + 194254.2 = 497684.2$$

$$\therefore \qquad H = \frac{\int_0^{36} M_s y dx}{\int y^2 dx} = \frac{497684.2}{1228.2} = 405.2 \text{ kN}$$

Maximum –ve BM in span BC

$$M_x = 180x - 404.7y$$

$$= 180x - 404.7\,(0.8892x - 0.0247x^2)$$

$$M_x = 10x^2 - 180x$$

$$\frac{dMx}{dx} = 20x - 180 = 0$$

$$\boxed{x = 9 \text{ m}}$$

Substituting this value in the above expression

$$\text{Max – ve BM} = 10(9)^2 - 180(9)$$

$$\boxed{\text{Max – ve BM} = 2810 \text{ kNm}}$$

Maximum +ve BM in span AC

$$M_x = 540x - 20x^2 - 405.2y$$

$$M_x = 540x - 20x^2 - 405.2\,(0.8892x - 0.0247x^2)$$

$$M_x = 540x - 20x^2 - 360.3x + 10x^2$$

$$M_x = -10x^2 + 180x$$

$$\frac{dM_x}{dx} = 0; \quad \text{i.e.} \quad x = 9 \text{ m}$$

Hence, max +ve BM

$$M_x = -10(9)^2 + 180(9)$$

$$\boxed{\text{Max + ve BM} = 810 \text{ kNm}}$$

Shear force and normal thrust at quarter span

$$y = 0.8892x - 0.0247x^2$$

$$\frac{dy}{dx} = 0.8892 - 0.0494x$$

At $x = 9$ m; $\tan\theta = 0.8892 - 0.0494(9)$

$$\tan\theta = 0.4446$$

$$\sin\theta = 0.4062; \quad \cos\theta = 0.9137$$

Shear force at 9 m;

$$SF = H\sin\theta - V\cos\theta$$

$$SF = 405.2(0.4062) - (540 - 360)0.9137$$

$$\boxed{SF = 0}$$

Normal thrust

$$NT = H\cos\theta + V\sin\theta$$

$$NT = 405.2 \times 0.9137 + (540 - 360)0.4062$$

$$\boxed{NT = 443.35 \text{ kN}}$$

Example 8.12 A two hinged parabolic arch of span 24 m with a central rise of 4 m uniformly loaded over the right half of the span with 30 kN/m and a concentrated load of 100 kN at the crown. Determine the horizontal thrust assuming secant variation for the moment of inertia.

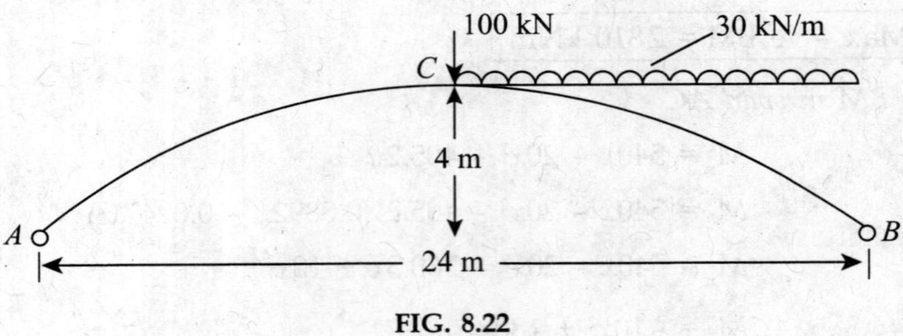

FIG. 8.22

Solution

Equation of the parabolic arch

$$y = \frac{4hx(l - x)}{l^2}$$

$$y = \frac{4(4)x(24 - x)}{24^2}$$

$$y = 0.0278x\ (24 - x)$$

i.e. $$\boxed{y = -0.0278x^2 + 0.6672x}$$

For the parabolic arch

$$\int y^2 dx = \frac{8}{15}h^2 l = \left(\frac{8}{15}\right)4^2(24)$$

$$\boxed{\int y^2 dx = 204.8}$$

Simply supported bending moment equation

100 kN, 30 kN/m, A, 12 m, C, 12 m, B

$$\Sigma V = 0;$$

$$V_A + V_B = 100 + 30(12)$$

$$V_A + V_B = 460 \quad (1)$$

$$\Sigma M_A = 0;$$

$$100(12) + 30(12)(12 + 6) = 24V_B \quad (2)$$

$$\boxed{\begin{array}{l} V_B = 320 \text{ kN} \\ V_A = 140 \text{ kN} \end{array}}$$

$$\int_0^l M_s y dx = \underbrace{\int_0^{12} (140x) y dx}_{\text{Region } AC} + \underbrace{\int_0^{12}\left(320x - \frac{30x^2}{2}\right) y dx}_{\text{Region } BC}$$

$$= \int_0^{12} (140x)\left(-0.0278x^2 + 0.6672x\right) + \int_0^{12} \left(320x - 15x^2\right)\left(-0.0278x^2 + 0.6672x\right) dx$$

On integration

$$\int_0^l M_s y dx = 120394.2$$

$$H = \frac{\int_0^l M_s y dx}{\int y^2 dx} = \frac{120394.2}{204.8} = 587.9 \text{ kN}$$

Example 8.13 A two hinged parabolic arch of span 24 m is subjected to 100 kN and 150 kN acting left and right quarter span respectively. Calculate the horizontal thrust and bending moment at the loaded points. The rise of the arch is 4 m. Determine the horizontal thrust assuming secant variation for the moment of inertia.

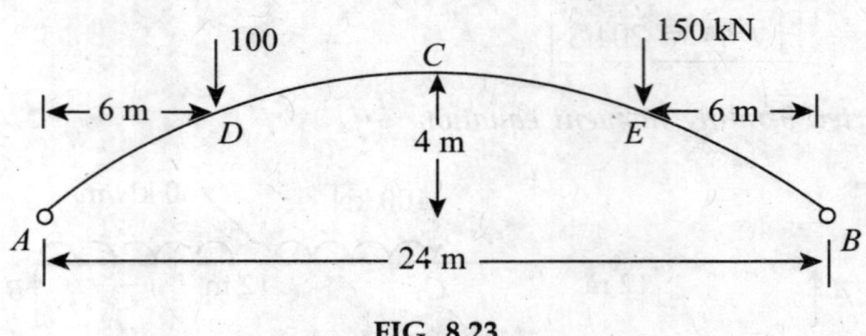

FIG. 8.23

Equation of the parabolic arch

$$y = \frac{4h}{l^2}x(l - x)$$

$$y = \frac{4(4)}{24^2}x(24 - x)$$

$$\boxed{y = -0.0278x^2 + 0.6667x}$$

Simply supported beam bending moment expression

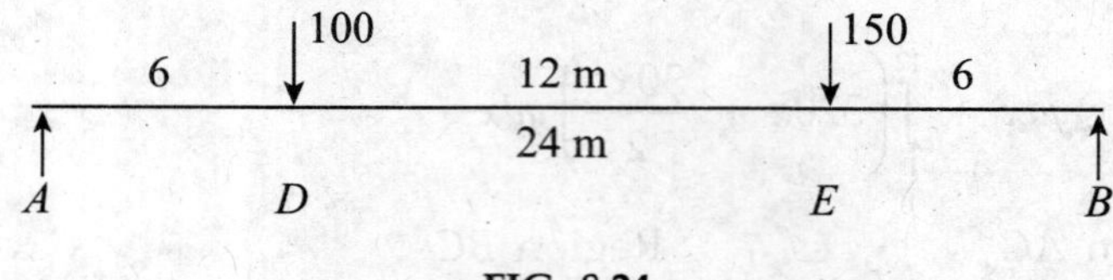

FIG. 8.24

The reactions are

$$\Sigma V = 0;$$

$$V_A + V_B = 250 \qquad (1)$$

$$\Sigma M = 0;$$

$$100(6) + 150(18) - 24V_B = 0$$

$$V_B = 137.5 \text{ kN}$$

$$\therefore \qquad V_A = 112.5 \text{ kN}$$

$$\int_0^{24} M_s y dx = \int_0^6 (112.5x) y dx + \int_6^{18} (112.5x - 100(x - 6)) y dx$$

(Region AD) (Region DE)

$$+ \int_0^6 137.5\, xy\, dx$$

(Region EB) (2)

The above expression is rewritten as

$$\int_0^{24} M_s y dx = \int_0^6 (112.5x) y dx + \int_6^{18} (12.5x + 600) y dx + \int_0^6 137.5 xy\, dx$$

Substituting the equation of the arch in the above expression gives

$$\int_0^6 (112.5x)(-0.0278x^2 + 0.6667x)\, dx = 4391$$

$$\int_0^{18} (12.5x + 600)(-0.0278x^2 + 0.6667x)\, dx = 33034.18$$

$$\int_0^6 137.5xy dx = 5366.79$$

$$\int_0^{24} M_s y dx = 42791.97$$

$$\int y^2 dx = \frac{8}{15} h^2 l$$

$$= \frac{8}{15} \times 4^2 \times 24 = 204.8$$

$$\therefore \quad \boxed{H = \frac{\int_0^{24} M_s y\, dx}{\int_0^{24} y^2\, dx}}$$

Bending moment at D and E

$$y_D = -0.0278(6)^2 + 0.6672(6) = 3.00 \text{ m}$$

also $$y_E = y_D$$

Thus,

$$M_D = 112.5(6) - 208.95(3) = 48.15 \text{ kNm}$$

$$M_E = 137.5(6) - 208.95(3) = 198.15 \text{ kNm}$$

Example 8.14 A two hinged parabolic arch has an effective span of 30 metres and a central rise of 5 m. Supports are at the same level. The arch carries a uniformly distributed load of 20 kN/m for a length of 10 m symmetrically placed on the arch. In addition, there is a point load of 400 kN at the left quarter point. If the moment of inertia of the cross section of the arch varies as the secant of the slope of the arch axis, determine the horizontal thrust.

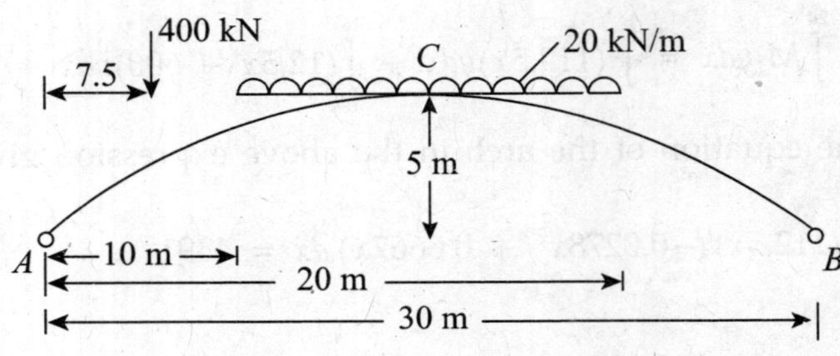

FIG. 8.25

Solution: The given arch subjected to loading is considered as follows.

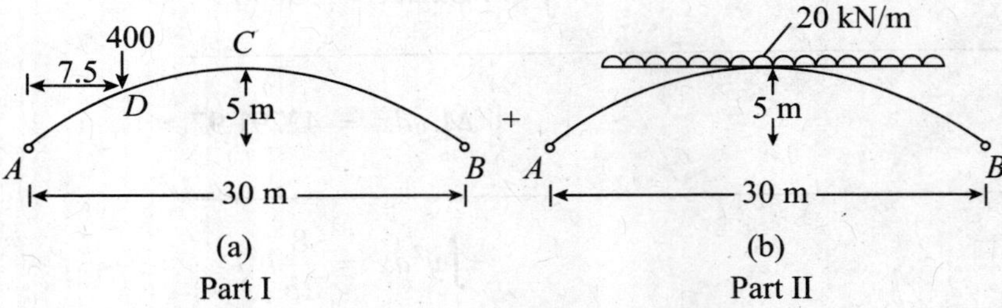

FIG. 8.26

Part I Solution

The horizontal thrust due to point load 'W' at a distance 'αl' from the left support is given by

$$H_1 = \left(\frac{5}{8}\right) W \left(\frac{l}{h}\right) (\alpha)(1 - \alpha)(1 + \alpha - \alpha^2)$$

In the given problem $\alpha = 7.5/30 = 0.25$, $l/h = 6$.

Substituting the above values

$$H_1 = \left(\frac{5}{8}\right) 400(6)(0.25)(0.75)(1 + 0.25 - 0.25^2)$$

$$H_1 = 334 \text{ kN}$$

Part 2 Solution

Consider the uniformly distributed load $w\delta x$ over an elemental length δx at a distance x from A. Treating this elemental load as a concentrated load, the incremental horizontal reaction δH_2 is given by

$$\delta H_2 = \frac{5l}{8h}(w\delta x)\left[\frac{x}{l} + \left(\frac{x}{l}\right)^4 - 2\left(\frac{x}{l}\right)^3\right]$$

Therefore, the horizontal reaction 'H_2' due to partially uniformly distributed load is

$$H_2 = \frac{5l}{8h}(wdx)\int_{x_1}^{x_2}\left[\frac{x}{l} + \left(\frac{x}{l}\right)^4 - 2\left(\frac{x}{l}\right)^3\right] dx$$

where x_1 and x_2 are the distances of the head of udl and the tail end of the udl. Thus,

$$H_2 = \left(\frac{5}{8}\right) w \left(\frac{l}{h}\right)\left[\frac{x^2}{2l} + \frac{x^5}{5l^4} - \frac{x^4}{4l^3}\right]_{x_1}^{x_2}$$

Denoting $x_1 = \alpha_1 l$ and $x_2 = \alpha_2 l$, the above equation is modified as

$$H_2 = \left(\frac{5}{8}\right)(wl)\left(\frac{l}{h}\right)\left[\frac{1}{2}\left(\frac{x}{l}\right)^2 + \frac{1}{5}\left(\frac{x}{l}\right)^5 - \frac{1}{2}\left(\frac{x}{l}\right)^4\right]_{x_1}^{x_2}$$

$$H_2 = \left(\frac{5}{8}\right)(wl)\left(\frac{l}{h}\right)\left[\frac{1}{2}\left(\alpha_2^2 - \alpha_1^2\right) + \frac{1}{5}\left(\alpha_2^5 - \alpha_1^5\right) - \frac{1}{2}\left(\alpha_2^4 - \alpha_1^4\right)\right]$$

In this problem, $\alpha_1 = (10/30)$ and $\alpha_2 = (20/30)$ and thus

$$H_2 = \left(\frac{5}{8}\right)(20 \times 30)6\left[\frac{1}{2}\left(0.67^2 - 0.33^2\right) + \frac{1}{5}\left(0.67^5 - 0.33^5\right) - \frac{1}{2}\left(0.67^4 - 0.33^4\right)\right]$$

$$\boxed{H_2 = 225 \text{ kN}}$$

The horizontal thrust due to applied loading is

$$H = H_1 + H_2$$

$$= 334 + 225$$

$$\boxed{H = 559 \text{ kN}}$$

Example 8.15 A parabolic two hinged arch is subjected to uniformly varying load as shown in Fig. 8.27. Derive an expression for horizontal thrust.

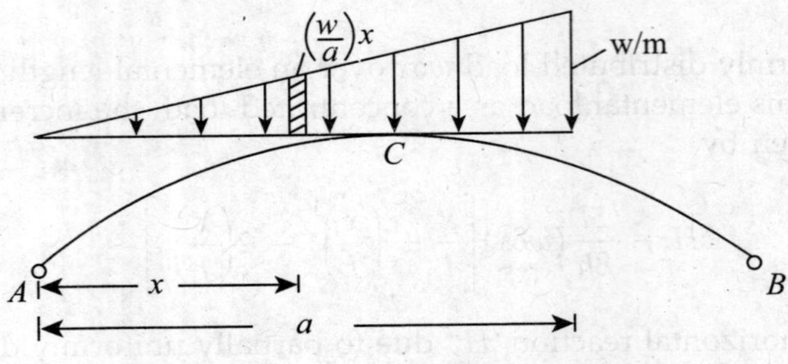

FIG. 8.27

Solution: Consider a section 'x' from A. The load that is acting at δx is given by $\left(\frac{w}{a}\right)x$. This load is acting on the arch as shown below.

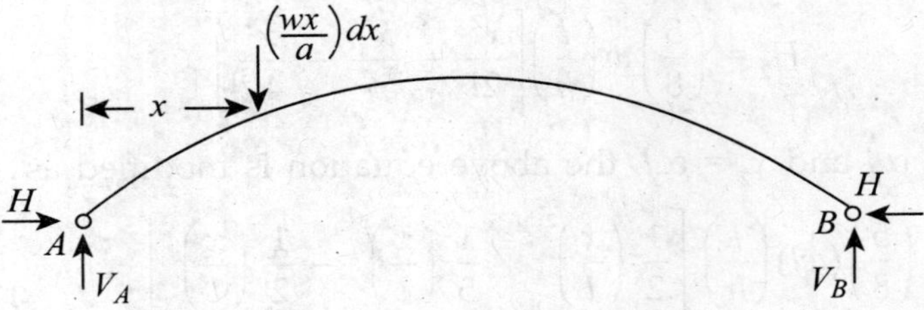

FIG. 8.28

Using the first principles;

$$dH = \left(\frac{5}{8}\right) w \left(\frac{x}{a}\right) dx \frac{x(l-x)\left(l^2 + lx - x^2\right)}{hl^3}$$

Rearranging the terms,

$$dH = \frac{5}{8}\left(\frac{w}{ahl^3}\right) x^2 (l-x)\left(l^2 + lx - x^2\right) dx$$

The total horizontal thrust is obtained by

$$H = \frac{5}{8}\left(\frac{w}{ahl^3}\right)\int_0^a \left(l^3x^2 - 2lx^4 + x^5\right) dx$$

$$H = \frac{5}{8}\left(\frac{w}{ahl^3}\right)\left[l^3\frac{x^3}{3} - 2l\frac{x^5}{5} + \frac{x^6}{6}\right]_0^a$$

$$\boxed{H = \frac{wa^2}{48hl^3}\left(10l^3 - 12la^2 + 5a^3\right)}$$

The horizontal thrust can be calculated for various values of 'a' as follows.

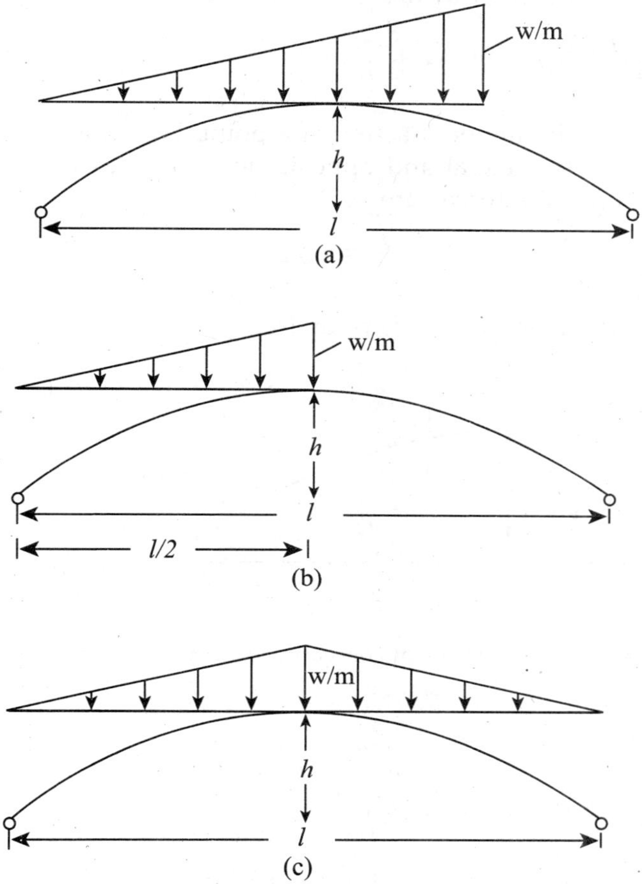

w/m

h

l

(a)

$$H = \frac{wl^2}{16h}$$

w/m

h

l

l/2

(b)

$$H = \frac{61}{1536}\frac{wl^2}{h}$$

w/m

h

l

(c)

$$H = \frac{61}{768}\frac{wl^2}{h}$$

FIG. 8.29

Example 8.16 Derive an expression for horizontal thrust of a two hinged arch subjected to a moment M acting at a point βL from the left support.

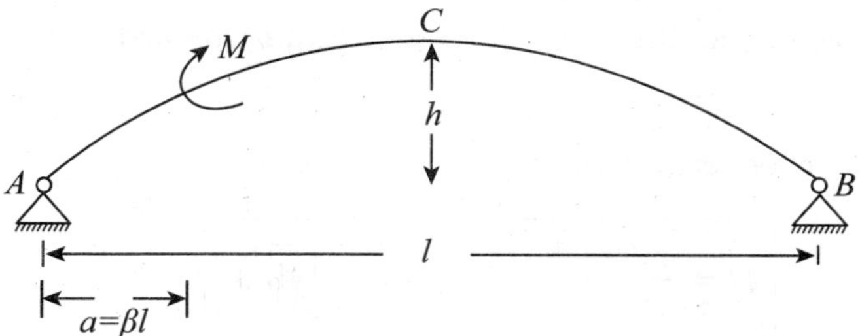

FIG. 8.30

Solution: When a two hinged parabolic arch is subjected to a concentrated load W at a distance βl from left support, the horizontal thrust is derived as

$$H = \frac{5W}{8}\left(\frac{l}{h}\right)(\beta - 2\beta^3 + \beta^4) \tag{1}$$

This result can be made use of for the moment acting at a point. We know that the effect of moment can be obtained by two equal and opposite loads at a distance $\delta\beta l$. The value of such two equal and opposite forces are

$$W = \frac{M}{\delta\beta l} \tag{2}$$

where $\delta\beta l$ is a small distance which the load W is kept apart.

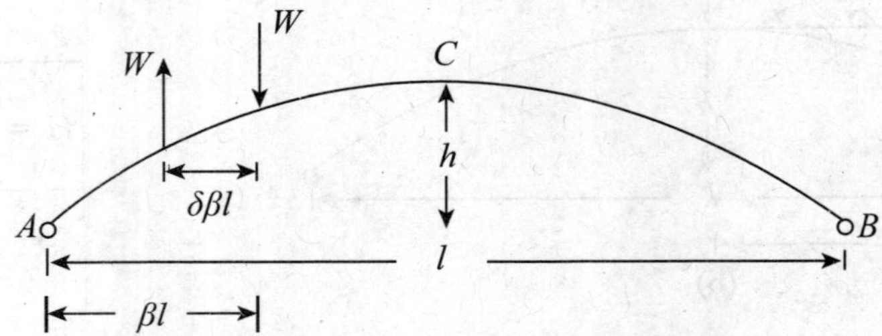

FIG. 8.31 Effect of moment in terms of concentrated loads

Due to downward load W

Using Eq. (1)

$$H_1 = \frac{5W}{8}\left(\frac{l}{h}\right)(\beta - 2\beta^3 + \beta^4) \tag{3}$$

where H_1 is the horizontal thrust developed due to downward load W.

Due to upward load W

Applying Eq. (1) [W with a negative sign and β is replaced as $(\beta - \beta l)$].

$$H_2 = -\frac{5W}{8}\left(\frac{l}{h}\right)\left((\beta - \delta\beta) - 2(\beta - \delta\beta)^3 + (\beta - \delta\beta)^4\right) \tag{4}$$

Thus, superimposing the effects of downward and upward 'W'

$$H = H_1 + H_2 \tag{5}$$

and taking the first order terms only

$$\boxed{H = \frac{5W}{8}\left(\frac{l}{h}\right)(1 - 6\beta^2 + 4\beta^3)} \tag{6}$$

The above equation can be used to determine the horizontal thrust due to the application of moment M at various points on the arch.

Moment acting at the left support $\beta = 0$

$$\boxed{H = \frac{5M}{8}\left(\frac{l}{h}\right)}$$

The positive value indicates that the arch is subjected to sagging bending moment with a tendency to move outwards.

Moment acting at the right support $\beta = 1$

$$\boxed{H = -\frac{5M}{8}\left(\frac{l}{h}\right)}$$

The negative value indicates that the arch is subjected to hogging bending moment with a tendency to move inwards at the supports.

Moment acting at the crown $\beta = 0.5$

$$H = \frac{5}{8}M\left(1 - 6(0.5)^2 + 4(0.5)^3\right)$$

$$\boxed{H = 0}$$

The above results show that when a moment is applied at the crown, left half of the arch is subjected to hogging bending moment and the right half is subjected to sagging bending moment.

Example 8.17 A two pinned arch of constant section is shown in Fig. 8.32. Determine the horizontal thrust at the hinges and find the *BM* at '*F*'.

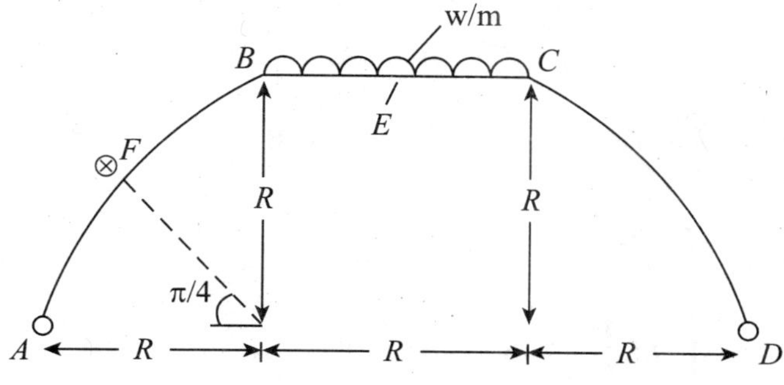

FIG. 8.32

Solution: Let H be the horizontal thrust A and D respectively. Due to symmetry in loading and geometry;

$$V_A = V_D = \frac{\text{Total load}}{2}$$

i.e.

$$V_A = \frac{wR}{2}$$

The horizontal thrust is obtained by using the condition that the partial derivative of strain energy with respect to H is zero as the supports are not yielding.

$$\frac{\partial U}{\partial H} = \frac{1}{EI}\int_0^R M\frac{\partial M}{\partial H}\,Rd\theta + \frac{1}{EI}\int_R^{3/2R} M\frac{\partial M}{\partial H}\,dx$$

The above equation is written by taking the advantage of symmetry.

Curved portion AB

The moment at any section can be written with reference to Fig. 8.33.

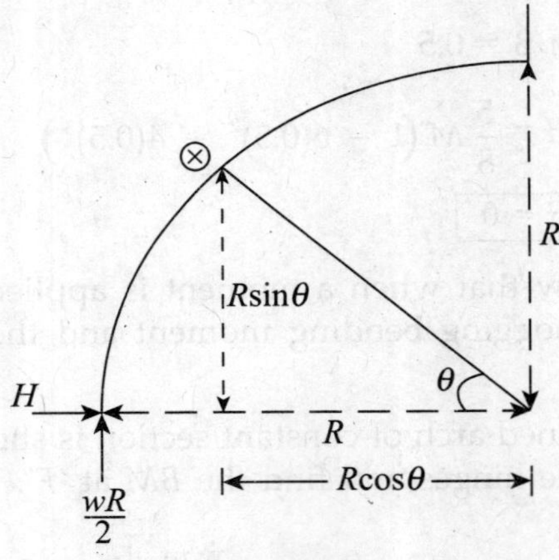

FIG. 8.33

$$M = \frac{wR}{2}(R - R\cos\theta) - HR\sin\theta$$

$$\frac{\partial M}{\partial H} = -R\sin\theta$$

$$\frac{\partial U}{\partial H} = \int_0^{\pi/2} M\frac{\partial M}{\partial H}\,Rd\theta$$

$$= \int_0^{\pi/2}\left(\frac{wR^2}{2}(1 - \cos\theta) - HR\sin\theta\right)(-R\sin\theta)\,Rd\theta$$

$$= -\frac{wR^4}{2}\int_0^{\pi/2}\sin\theta\,d\theta + \frac{wR^4}{2}\int_0^{\pi/2}\sin\theta\cos\theta\,d\theta + HR^3\int_0^{\pi/2}\sin^2\theta\,d\theta$$

$$= -\frac{wR^4}{2}[-\cos\theta]_0^{\pi/2} + \frac{wR^4}{2}\left[\frac{-\cos 2\theta}{4}\right]_0^{\pi/2} + \frac{HR^3}{2}\left[\theta - \frac{\sin 2\theta}{2}\right]_0^{\pi/2}$$

$$= -\frac{wR^4}{2}[1] + \frac{wR^4}{2}\left[\frac{1}{4} + \frac{1}{4}\right] + \frac{HR^3}{2}[1.5708]$$

$$= -0.5wR^4 + 0.25wR^4 + 0.7854HR^3$$

$$\boxed{\frac{\partial U}{\partial H} = -0.25wR^4 + 0.7854HR^3} \qquad (1)$$

Straight beam portion BC

The moment at any section in this region *BC* can be written with reference to Fig. 8.34.

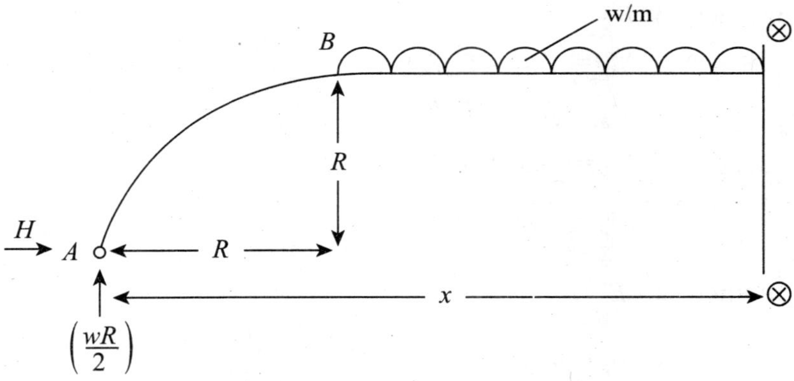

FIG. 8.34

$$M_{xx} = \frac{wR}{2}x - HR - w\frac{(x - R)^2}{2}$$

$$\frac{dM}{dH} = -R$$

$$\frac{\partial U}{\partial H} = \int_R^{\frac{3}{2}R}\left[\left(\frac{wR}{2}\right)x - HR - \frac{w(x - R)^2}{2}\right](-R)dx$$

$$\frac{\partial U}{\partial H} = \int_R^{3/2R}\left[\frac{w}{2}\left(Rx - x^2 + R^2 + 2\,Rx\right) - HR\right](-R)dx$$

$$= \int_R^{3/2R}\left[\frac{w}{2}\left(R^2 - x^2 + 3Rx\right) - HR\right](-R)\,dx$$

$$\boxed{\frac{\partial U}{\partial H} = 0.292wR^4 + 0.5HR^3} \quad (2)$$

Adding (1) and (2);

$$\frac{\partial U}{\partial H} = -0.542wR^4 + 1.2854HR^3 \quad (3)$$

As

$$\frac{\partial U}{\partial H} = 0$$

We get

$$0.542wH^4 = 1.2854HR^3$$

$$\therefore \quad \boxed{H = 0.422wR}$$

Bending moment at B

$$M_B = V_A R - H_A R$$

$$M_B = (V_A - H_A)R$$

$$M_B = (0.5wR - 0.422wR)R$$

$$\boxed{M_B = 0.078wR^2}$$

Bending moment at 'E'

$$M_E = V_A(R + R/2) - HR - \frac{w(R/2)^2}{2}$$

$$M_E = \frac{3V_A R}{2} - HR - \frac{wR^2}{8}$$

$$M_E = \frac{3}{2}\left(\frac{wR}{2}\right) R - 0.422wR^2 - \frac{wR^2}{8}$$

$$\boxed{M_E = 0.203wR^2}$$

Bending moment at 'F'

$$M_F = V_A(R - R\cos(\pi/4)) - H_A R \sin(\pi/4)$$

$$M_F = \frac{wR}{2}(R - 0.7071R) - 0.422wR \times R \times 0.7071$$

$$M_F = 0.1465wR^2 - 0.2984wR^2$$

$$\boxed{M_F = -0.152wR^2}$$

Example 8.18 Calculate the horizontal thrust and draw the bending moment diagram for the arch shown in Fig. 8.35.

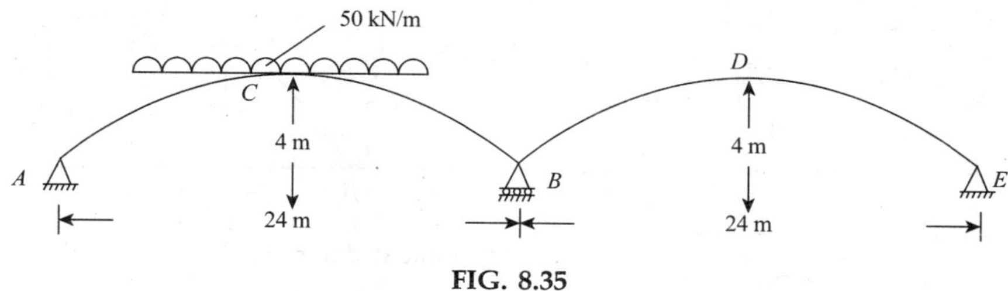

FIG. 8.35

Solution: The applied loading is unsymmetrical. As the arch *AB* is loaded the roller support at *B* moves to the right by δ. On the other hand, if the arch *BE* is loaded it will move to the left by δ. It follows the logic, that if both the arches *AB* and *BE* are loaded simultaneously and the results can be superimposed. This results in the horizontal thrust 2*H* and the movement to the left and to the right get cancelled.

Thus, the horizontal thrust can be found out from

$$2H = \frac{\text{Maximum BM of simply supported beam}}{\text{Rise of the arch}}$$

$$2H = \frac{wl^2}{8h}$$

$$H = \frac{wl^2}{16h}$$

i.e.
$$H = \frac{50 \times 24^2}{16 \times 4}$$

$$\boxed{H = 450 \text{ kN}}$$

Bending moment diagram

$$M_c = \frac{wl^2}{8} - \frac{wl^2}{16h} \times h$$

$$M_c = \frac{wl^2}{16}$$

$$M_c = \frac{50 \times 24^2}{16}$$

$$\boxed{M_c = 1800 \text{ kNm}}$$

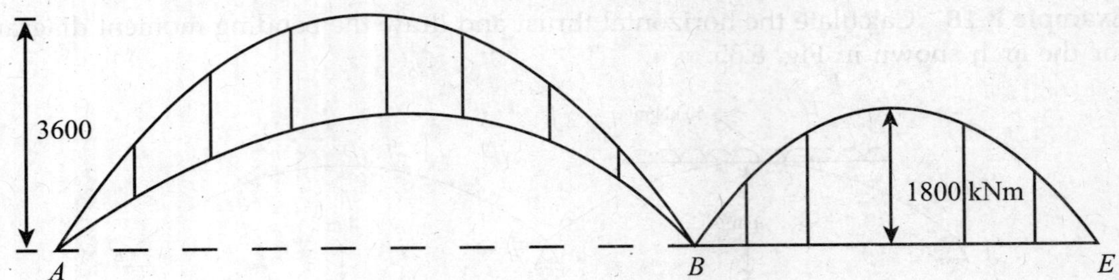

FIG. 8.36 Bending moment diagram

8.6 LINEAR ARCH

Example 8.19 Determine the horizontal thrust for the linear arch shown in Fig. 8.37 and draw the bending moment diagram by neglecting the rib shortening and shear.

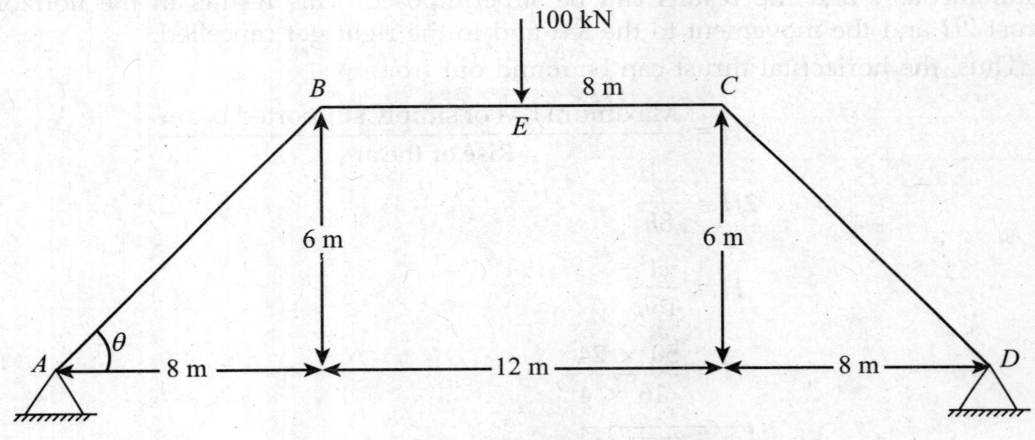

FIG. 8.37

Solution: The simply supported beam reactions are found as follows:

$$\Sigma V = 0;$$

$$V_A + V_D = 100 \tag{1}$$

$$\Sigma M_A = 0;$$

$$100(12) - 28V_D = 0 \tag{2}$$

$$V_D = 42.86 \text{ kN}$$

$$\therefore \quad V_A = 57.14 \text{ kN}$$

From the geometry

$$\tan\theta = \frac{6}{8} = 0.75$$

$$\sin\theta = 0.6$$

$$\cos\theta = 0.8$$

The horizontal thrust is given by

$$H = \frac{\Sigma M_s yds}{\Sigma y^2 ds} \qquad (\because \; EI \text{ is constant})$$

It is known from the above Fig. 8.37 $ds = dx/\cos\theta$

Portion	*Origin*	*Limits*	M_s	y	*ds*
AB	*A*	0 – 8	57.14*x*	0.75*x*	1.25*dx*
BE	*B*	0 – 4	57.14(*x* + 8)	6	*dx*
EC	*C*	0 – 8	42.86(*x* + 8)	6	*dx*
CD	*D*	0 – 8	42.86*x*	0.75*x*	1.25*dx*

$$\Sigma M_s yds = \int_0^8 57.14x(0.75x)1.25dx + \int_0^4 57.14(x+8)6dx + \int_0^8 42.86(x+8)6dx$$

$$+ \int_0^8 42.86x(0.75x)1.25dx$$

$$= 53.57 \int_0^8 x^2 dx + 342.84 \int_0^4 xdx + 2742.72 \int_0^4 dx + 257.16 \int_0^8 xdx + 2057.28 \int_0^8 dx$$

$$+ 40.18 \int x^3 dx$$

$$\boxed{\Sigma M_s yds = 88688}$$

$$\Sigma y^2 ds = \int_0^8 (0.75x)^2 (1.25x)dx + \int_0^4 6^2 dx + \int_0^8 6^2 dx + \int_0^8 0.75x(1.25x)dx$$

$$= 0.7031 \int_0^8 x^3 dx + 36 \int_0^4 dx + 36 \int_0^8 dx + 0.9375 \int_0^8 x^2 dx$$

$$\boxed{\Sigma y^2 ds = 1311.98}$$

Thus, the horizontal thrust is given by

$$H = \frac{\Sigma M_s y ds}{\Sigma y^2 ds}$$

$$H = \frac{88688}{1311.98}$$

$$\boxed{H = 67.6 \text{ kN}}$$

The bending moment diagram is drawn by drawing the BMD of the corresponding simply supported beam. Hence, for a point load 100 kN at a distance of 12 m is 57.14 × 12 = 685.68. As the beam *BC* is horizontal, the maximum negative BM is 67.6 × 6 = 405.6 kNm. Thus, by superimposing the positive BMD and negative BMD, the final BMD is obtained.

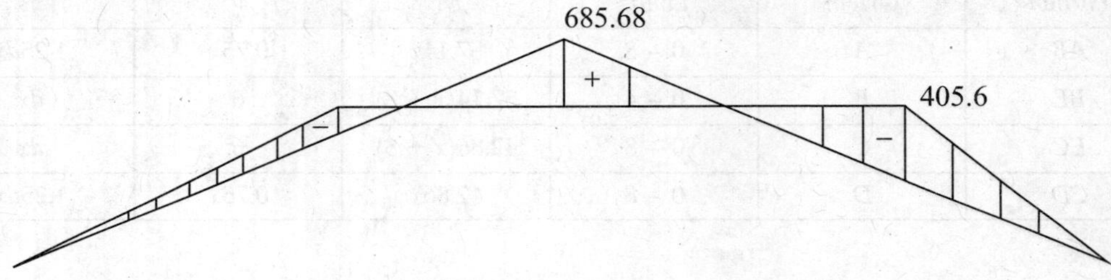

FIG. 8.38 Final BMD

8.7 SECONDARY STRESSES IN TWO HINGED ARCH

8.7.1 Temperature Thrust in a Two Hinged Arch

A two hinged arch is fixed in position at the ends and lateral movement of the abutments is prevented. Any rise or fall in temperature causes a change in the stresses in the rib. These stresses are called temperature stresses. The estimation of temperature stresses plays an important role in the design of reinforced concrete and steel arches.

Consider a two hinged arch as shown in Fig. 8.39.

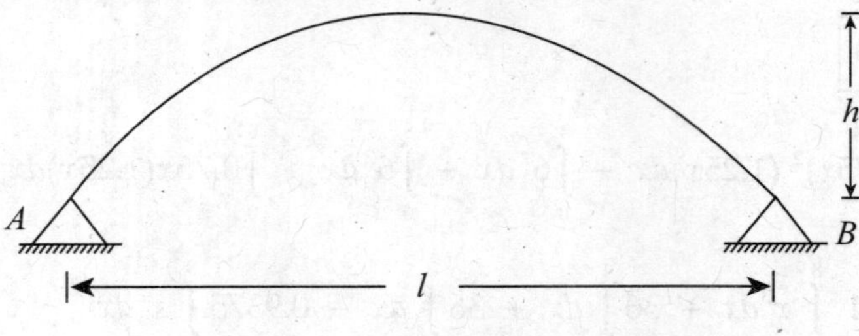

FIG. 8.39

Let the rise in temperature be '*t*'. Had there been a roller in *B* in place of hinge at *B*, the dimensions of the member increases. As it is resting on a frictionless roller, the deformed arch is as shown below.

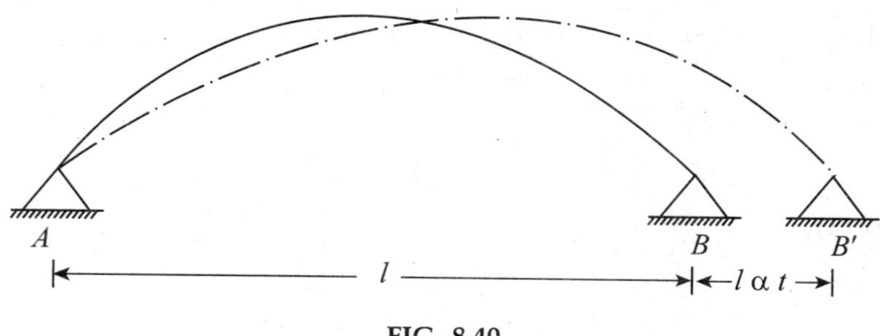

FIG. 8.40

As the arch is free to expand, the new span AB' is $l + l\alpha t = l(1 + \alpha t)$ where α is the coefficient of expansion of the material.

Now, assume a horizontal thrust H is applied at B' to bring back from B' to B.

Applying Castigliano's theorem;

$$\text{the horizontal movement of } H = \frac{dU}{dH} = l\alpha t \tag{1}$$

$$\therefore \quad \int \left(\frac{M}{EI}\right)\left(\frac{dM}{dH}\right) ds = l\alpha t \tag{2}$$

But
$$M = Hy \tag{3}$$

Substituting (3) in (2)

$$\int \left(\frac{Hy}{EI}\right) y ds = l\alpha t \tag{4}$$

$$\int \frac{Hy^2}{EI} ds = l\alpha t \tag{5}$$

i.e.
$$\boxed{H = \frac{l\alpha t}{\int y^2 \frac{dS}{EI}}}$$

Example 8.20 Derive the expression for horizontal thrust of a two pinned parabolic arch whose moment of area at any section is $I_c \sec\theta$ where I_c is the value of the crown and θ is the slope of the rib at that section. The arch rib is subjected to a temperature rise is 't' degrees.

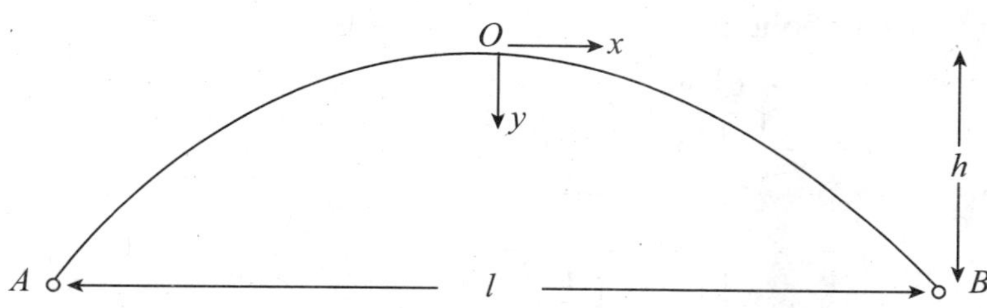

FIG. 8.41

Solution: Taking the origin at the centre of the arch

$$y = h\left(1 - \frac{4x^2}{l^2}\right) \quad (1)$$

and

$$M_x = Hy \quad (2)$$

Using (2) in (1)

$$M_x = Hh\left(1 - \frac{4x^2}{l^2}\right) \quad (3)$$

$$\therefore \quad \frac{\partial M_x}{\partial H} = h\left(1 - \frac{4x^2}{l^2}\right) \quad (4)$$

$$\therefore \quad \frac{\partial M_x}{\partial H} = 2\int_0^{l/2} M\,\frac{\partial M_x}{\partial H}\left(\frac{1}{EI}\right) \quad (5)$$

$$\frac{\partial U}{\partial H} = 2\int_0^{l/2} \frac{Hh^2}{I_x}\left(1 - \frac{4x^2}{l^2}\right) ds = l\alpha t \quad (6)$$

We know

$$I_x = I_c\,\frac{ds}{dx} \quad (7)$$

Substituting Eq. (7) in Eq. (6)

$$\frac{2Hh^2}{EI_c}\int_0^{l/2}\left(1 - \frac{4x^2}{l^2}\right)^2 dx = l\alpha t$$

$$\boxed{H = \frac{15EI_c l\alpha t}{8h^2}}$$

The above expression can be written in the form

$$H = \frac{l\ t}{\overline{15}\ \overline{EI}}$$

and we know for a parabolic arch

$$\int \frac{y^2 ds}{EI} = \left(\frac{8}{15}\right)\frac{h^2}{EI}$$

i.e.

$$\boxed{H = \frac{l\alpha t}{\int \frac{y^2 ds}{EI}}}$$

where α = coefficient of expansion
t = change of temperature
l = span of the arch

8.7.2 Elastic Shortening of Rib

The two hinged arch is made up of materials like steel or concrete. These materials are compressible under loading. The above sections show that the horizontal thrust is induced by loading and temperature changes. The elastic shortening causes deformation which causes the relieving a small axial thrust in H. This logic leads to the final value of H less than that calculated assuming that the arch is incompressible. The effect of rib shortening increases the numerical value of the expression in the denominator. Thus, the value of H for external loading, temperature change and elastic shortening is given by

$$H = \frac{\int \frac{Myds}{EI} + l\alpha t}{\int \frac{y^2 ds}{EI} + \int \frac{ds}{AE}}$$

where A is the cross-sectional area of the rib at any point in the rib.

8.7.3 Yielding of Supports

The yielding of supports of the two hinged arch causes spread. This is given by

$$\Delta = H\left(\frac{1}{k_A} + \frac{1}{k_B}\right)$$

where k_A and k_B are the stiffnesses of the elastic supports at A and B respectively. Thus, incorporating the effect of yielding

$$H = \frac{\int \frac{Myds}{EI} + l\alpha t}{\int \frac{y^2 ds}{EI} + \int \frac{ds}{AE} + \left(\frac{1}{k_A} + \frac{1}{k_B}\right)}$$

The above expression is more in general and we can infer that the change in temperature, elastic shortening of ribs and the spread alters the horizontal thrust.

For a two hinged parabolic arch with secant variation of moment of inertia of the cross section, the above equation reduces to

$$H = \frac{\int_0^l Mydx + (l\alpha t)EI}{(lI/A) + \Delta EI + \int_0^l y^2 dx}$$

Example 8.21 A two hinged parabolic arch of 30 metres span is of arch rib with $I = 10(10)^9$ mm^4. E = 12000 N/mm^2, the temperature variation is 40°C, area of the cross section $12(10)^4$ mm^2, the coefficient of expansion per degree centigrade as $10(10)^{-6}$. Each support yield horizontally by 0.01 mm/kN. The cross section of the arch rib follows the secant law. i.e. $I = I_c \sec\theta$.

Determine the horizontal thrust due to

(i) External loads only

(ii) External loads and temperature variation only

(iii) External loads, temperature rise, and rib shortening

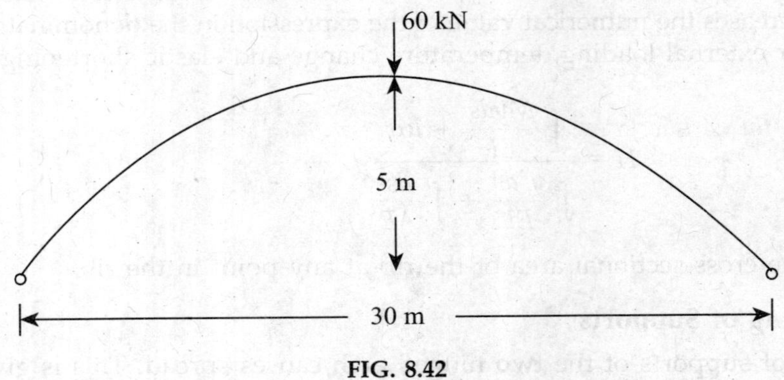

FIG. 8.42

Solution: The horizontal thrust due to loads, temperature rise and elastic shortening of ribs is determined using the derived expression as

$$\int_0^l Mydx = \frac{Wh\alpha l^2}{3}\left(1 - 2\alpha^2 + \alpha^3\right)$$

where αl is the distance from the left support where a single point load W acts.

$$\int_0^l Mydx = \frac{Whl^2}{3}\left(\alpha - 2\alpha^3 + \alpha^4\right)$$

$$= \frac{60 \times 5 \times 30^2}{3}\left(0.5 - 2(0.5)^3 + 0.5^4\right)$$

i.e.

$$\int_0^l Mydx = 28125 \text{ kNm}^3$$

$$\int_0^l y^2 dx = \frac{8}{15}h^2 l$$

$$= \frac{8}{5} \times 5^2 \times 30 = 400 \text{ m}^3$$

$$EI = (10(10)^9 \times 12000)\ 10^{-9} = 12(10)^4 \text{ kNm}^2$$

$$LI/A = \frac{30(10 \times 10^9)\ 10^{-6}}{12 \times 10^4} = 2.5 \text{ m}^3$$

$$\delta_H = 2 \times 0.01 \times 10^{-3} = 20(10)^{-6} \text{ m/kN}$$

The above values are substituted in the expression as

Horizontal thrust due to external loads

$$H = \frac{\int_0^l Mydx}{\int_0^l y^2 dx} = \frac{28125}{400} = 70.3 \text{ kN}$$

Horizontal thrust due to external load and temperature rise

$$H = \frac{\int_0^l Mydx + (l\alpha t)\ EI}{\int_0^l y^2 dx}$$

$$= \frac{28125 + 0.012(12)10^4}{400}$$

$$H = 73.91 \text{ kN}$$

Horizontal thrust due to external loads, temperature rise, rib shortening of the arch

$$H = \frac{\int_0^l Mydx + (l\alpha t)\ EI}{(lI/A) + \Delta EI + \int_0^l y^2 dx}$$

$$H = \frac{28125 + 0.012 + 12 \times 10^4}{2.5 + 20 \times 10^{-6} \times 12 \times 10^4 + 400}$$

$$H = 73.02 \text{ kN}$$

8.8 TIED ARCH

In this structural system, the arch is connected by a tie beam. This tied arch is supported on a hinge at one end and the other end is on the roller. Hence, the reactions can be found out by three equations of equilibrium.

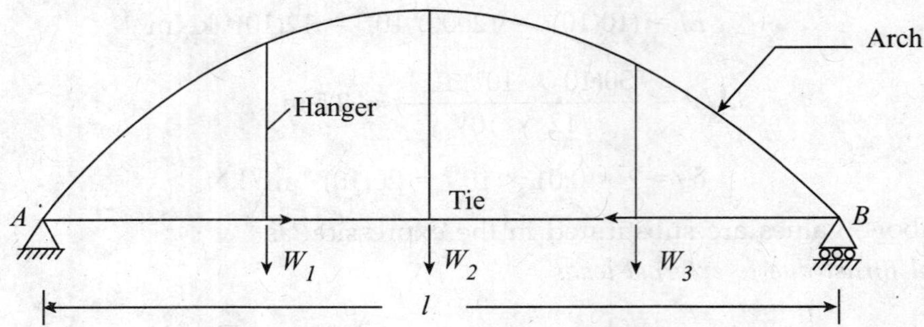

FIG. 8.43 Bowstring girder

But it is internally indeterminate. The redundant force is the tension in the tie. Using Castigliano's theorem;

$$\frac{dU}{dT} = \frac{dU^T}{dT} + \frac{dU^A}{dT} = 0$$

where U = Total strain energy, U^T = Strain energy of the tie, and U^A = Strain energy of the arch.

(or)

$$\frac{Tl}{A_t E_t} + \int M \frac{dM}{dT} \frac{dS}{EI} = 0$$

where $A_t E_t$ = axial rigidity of the tie and we know,

$$M = M_s - Ty$$

and that
$$\frac{dM}{dT} = -y$$

Substituting the above in the governing equation,

$$\frac{Tl}{A_t E_t} + T \int y^2 \frac{ds}{EI} = \int M_s y \frac{ds}{EI}$$

$$\boxed{T = \frac{\int M_s y ds/EI}{\int y^2 \frac{ds}{EI} + \frac{l}{A_t E_t}}}$$

Example 8.22 Determine the horizontal thrust in a parabolic arch rib of a bowstring girder shown in Fig. 8.44. The moment of inertia of the cross section of the arch varies as the secant of the slope of the arch axis. A point load of 60 kN acts at the left quarter point. The area of the cross section of the tie is 30000 mm². The value of $I_c = 10^{10}$ mm⁴. E is constant for all the members. Draw the bending moment diagram.

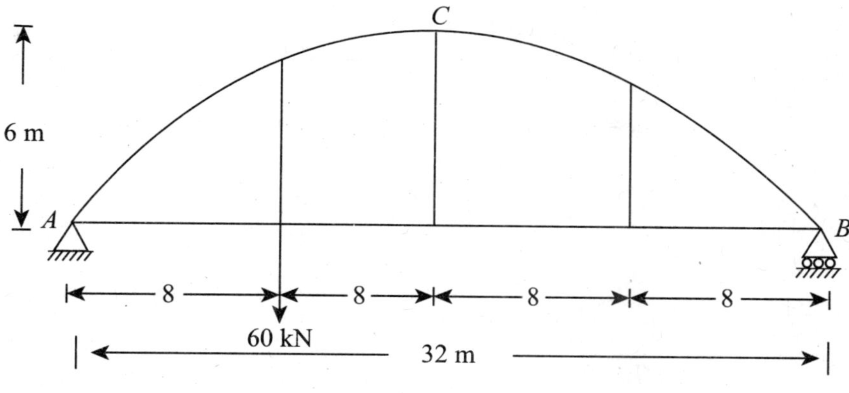

FIG. 8.44

Solution: The horizontal force is given by

$$T = \frac{\int M_s y ds / EI}{\int \frac{y^2}{EI} ds + \frac{l}{A_t E_t}}$$

It is known that

$$\frac{ds}{EI} = \frac{dx}{EI_c}$$

and for the tie

$$\frac{l}{A_t E_t} = \frac{lI_c}{A_t (E_t I_c)} = \frac{32 \times 10 \times 10^{10} \times 10^{-12}}{30000 \times 10^{-6} \times E_t I_c}$$

$$\frac{l}{A_t E_t} = \frac{320 \times 10^{-2}}{0.03\, E_t I_c} = \frac{100.67}{E_t I_c}$$

Substituting in the equation above and cancelling the term $E_t I_c = EI_c$ in the numerator and the denominator,

$$T = \frac{\int_0^l M_s y dx}{100.67 + \int_0^l y^2 dx}$$

The values of the numerator and the denominator are

$$\int_0^l M_s y dx = \frac{Whl^2}{3} \left(\alpha - 2\alpha^3 + \alpha^4\right)$$

$$= \frac{60(6)32^2}{3}\,(0.25 - 2(0.25)^3 + 0.25^4)$$

$$= 27360$$

$$\int_0^l y^2 dx = \frac{8h^2 l}{15} = \left(\frac{8}{15}\right) 6^2 \times 32 = 614.4$$

$$\therefore \quad T = \frac{27360}{100.67 + 614.4}$$

$$\boxed{T = 38.26 \text{ kN}}$$

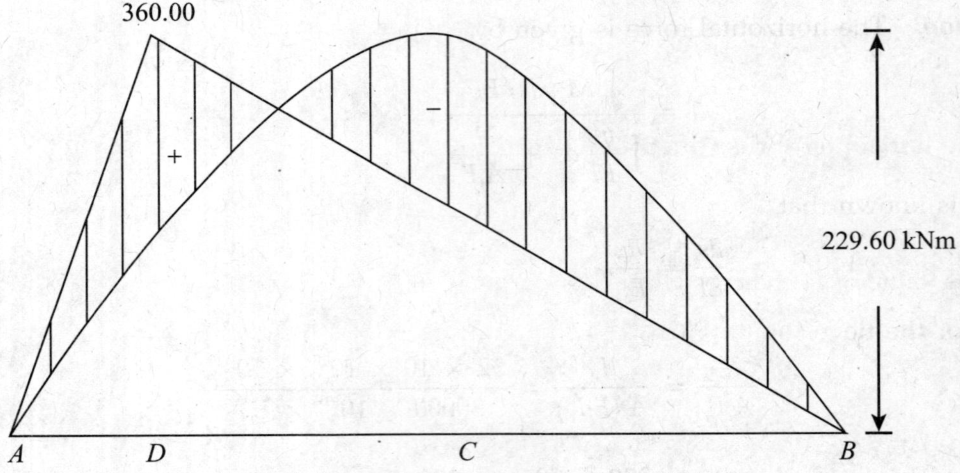

FIG. 8.45 Bending moment diagram

The bending moment diagram is drawn as

(i) Draw the positive bending moment diagram caused by the load 60 kN. $M_D = \frac{8 \times 60 \times 24}{32}$ i.e. $MD = 360$ kNm.

(ii) Determine the maximum negative bending moment at C as $Th = 38.26(6) = 229.60$ kNm. The negative bending moment caused by T at any point on the arch is $-Ty$, and since this term varies directly with y, the negative bending moment diagram is similar to the shape of the arch itself.

(iii) The combined bending moment diagram is drawn by superimposing the data of step (i) and (ii).

8.9 ILD FOR TWO HINGED PARABOLIC ARCHES

8.9.1 Influence Line for Horizontal Thrust

Consider a parabolic arch of span of 'l' as shown in Fig. 8.46. Let the central rise be 'h'. It is assumed that the moment of inertia of the arch section varies as the secant of the slope of the arch axis. Let a unit load act at a distance αl from the left support.

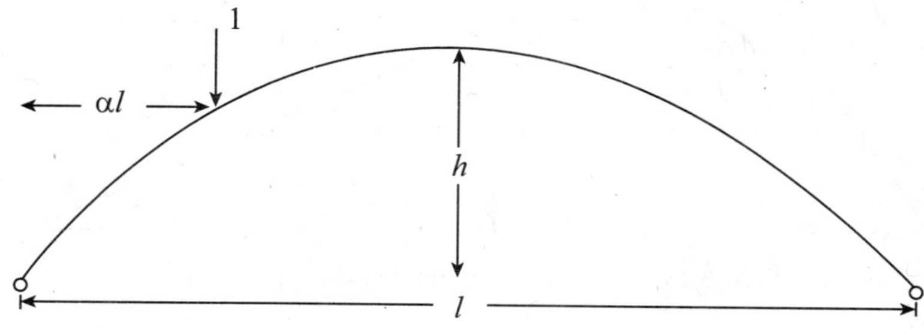

FIG. 8.46

The horizontal thrust due to unit load is given by

$$H = \frac{5W}{8}\left(\frac{l}{h}\right) \alpha\,(1 - \alpha)\,(1 + \alpha - \alpha^2)$$

The value of H changes with 'α' for a given arch. The maximum value of H is when $\alpha = 0.5$, i.e. when the unit load is at the crown. Thus, substituting the value of $\alpha = \frac{l}{2}$,

$$\boxed{H = \frac{25Wl}{128h}}$$

Thus, the influence line diagram for horizontal thrust is given by

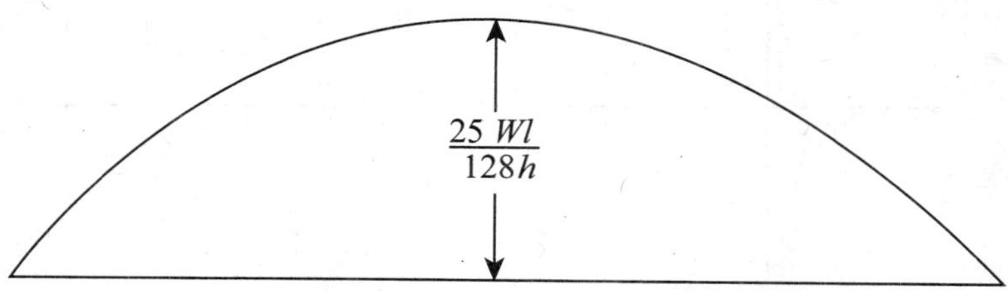

FIG. 8.47 Influence line for H

8.9.2 Influence Line for Bending Moment

The bending moment at a section is given by

$$M = M_s - Hy$$

where M_s is the simply supported beam moment, H is the horizontal thrust and y is the ordinate of the arch axis. Thus, the influence line for bending moment is obtained by superimposing the Hy on the ILD of M_s. Thus,

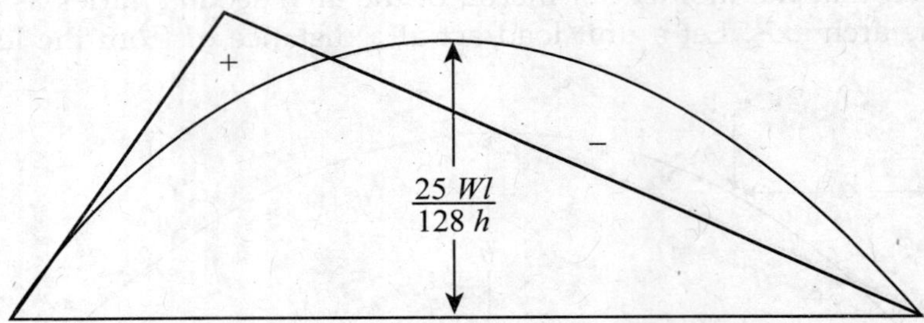

FIG. 8.48 Influence line for bending moment

8.9.3 Influence Line for Normal Thrust

The normal thrust is given by

$$N = H\cos\theta + V\sin\theta$$

At a section, the slope is constant, i.e., θ is constant. Hence, $\sin\theta$ and $\cos\theta$ are constant for a section. The normal thrust can be obtained by superposition of the effect of H and V. The ILD for normal thrust is shown below.

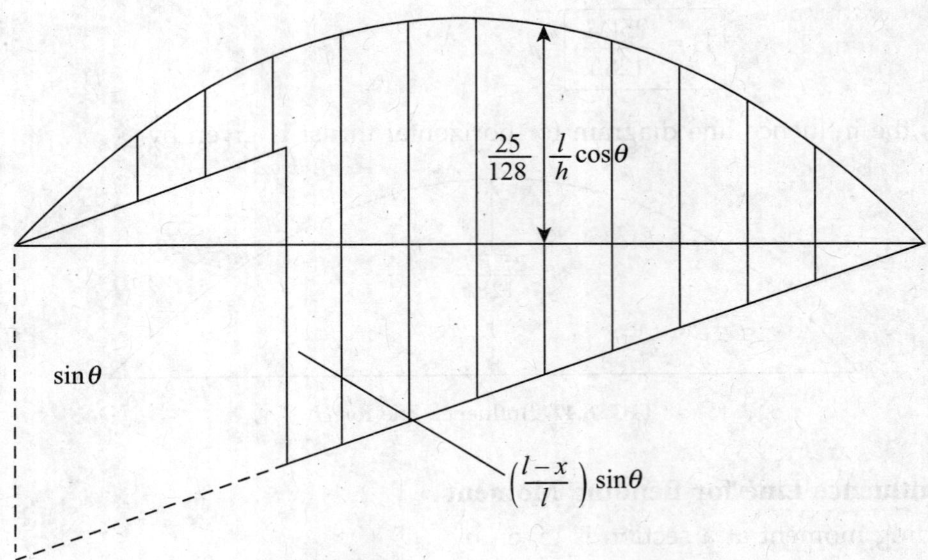

FIG. 8.49 Influence line for normal thrust

8.9.4 Influence Line for Shear Force

The shear force at any section is given by

$$R = V\cos\theta - H\sin\theta$$

The influence line diagram is drawn using the superposition as

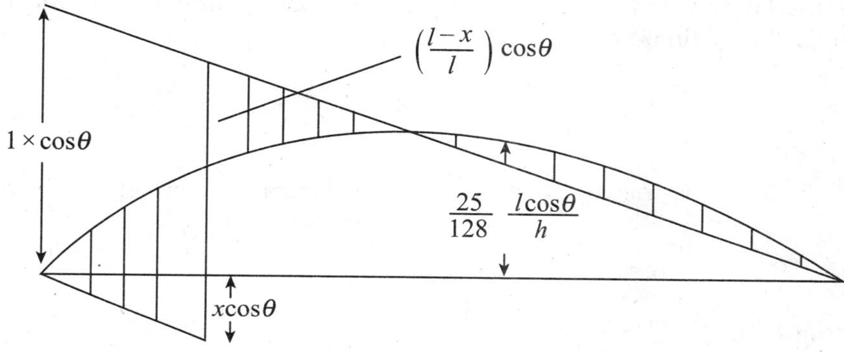

FIG. 8.50 ILD for shear force

Example 8.23 A symmetrical two hinged parabolic arch of span 40 m and rise 8 m is traversed by a single point load of 100 kN rolling horizontally from left to right. The moment of inertia of arch section varies as the secant of the slope of the arch axis. Sketch the influence line diagram for bending moment at a section 12 metres horizontally from the left hinge.

Determine the maximum positive and negative bending moment values at a section due to the rolling load.

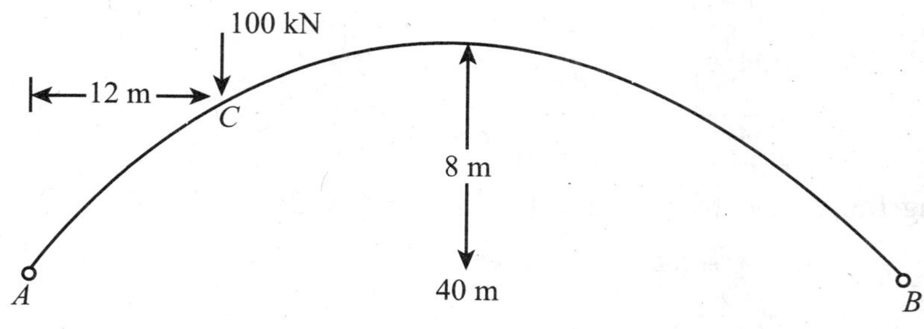

FIG. 8.51

Solution: Maximum positive bending moment

This occurs at the loading point C. The equation of the arch is given by

$$y = \frac{4h}{l^2}x(l - x) \qquad (1)$$

i.e. $$y_c = \frac{4(8)}{40^2}(12)(40-12) = 6.72\text{ m}$$

The moment at C is given by

$$M_c = M_s - Hy_c \tag{2}$$

where M_s is the corresponding simply supported beam moment, H is the horizontal thrust and y_c is the ordinate of the arch

$$H = \frac{5W}{8}\left(\frac{l}{h}\right)(\alpha - 2\alpha^3 + \alpha^4) \tag{3}$$

where $\alpha = \left(\frac{x}{l}\right)$ and x is the distance of the point load from the left support hinge.

$$\alpha = \frac{12}{40} = 0.3$$

Substituting α in the H expression

$$H = \frac{5}{8}W\left(\frac{40}{8}\right)(0.3 - 2(0.3)^3 + 0.3^4)$$

$$H = 0.794W$$

i.e. $$\boxed{H = 79.4\text{ kN}}$$

and

$$M_s = \frac{Wab}{l} \tag{4}$$

$$M_s = \frac{100 \times 12 \times 28}{40}$$

$$\boxed{M_s = 840\text{ kNm}}$$

Substituting the values of M_s, H and y_C in the Eq. (2),

$$M_c = 840 - 79.4(6.72)$$

$$\boxed{M_c = 306.4\text{ kNm}}$$

Maximum negative bending moment

This occurs at a distance x with the right hinge,

$$M_x = M_s - H_x y_x \tag{5}$$

where

$$M_s = \frac{Wx(l-x)}{l}$$

$$H_x = \frac{5W}{8}\left(\frac{l}{h}\right)\left\{\left(\frac{x}{l}\right) - 2\left(\frac{x}{l}\right)^3 + \left(\frac{x}{l}\right)^4\right\}$$

$$y_x = \frac{4hx(l - x)}{l^2}$$

Substituting the above expressions in Eq. (5) and simplifying

$$M_x = 2.5x(4 - x) - 312.5\left\{\left(\frac{x}{l}\right) - 2\left(\frac{x}{l}\right)^3 + \left(\frac{x}{l}\right)^4\right\}\{0.02x(40 - x)\} \quad (6)$$

Using the value of 'l'

$$M_x = (40x - x^2)\,2.5 - 6.25\,(0.025x - 3.125(10)^{-5}\,x^3 + 3.9063(10)^{-7}\,x^4) \quad (7)$$

M_x will be maximum when

$$\frac{dM_x}{dx} = 0 \quad (8)$$

Differentiating M_x expression with respect to x and equating to zero, is quite involved. Alternatively M_x was calculated for various values of x and given in the following form.

S. No.	x (m)	M_x (kNm)
1.	0	0
2.	10	332.5
3.	15	259.5
4.	20	218.8

It is noted from the above computations, that maximum value of negative bending moment is 332.5 kNm at quarter span from the right hinge. i.e. 10 m from B. The maximum value of positive bending moment is beneath the point load.

Example 8.24 Draw the influence line diagram for horizontal thrust H, bending moment M and thrust at D of the parabolic arch shown in Fig. 8.52. Consider $I_X = I_C \sec\theta$.

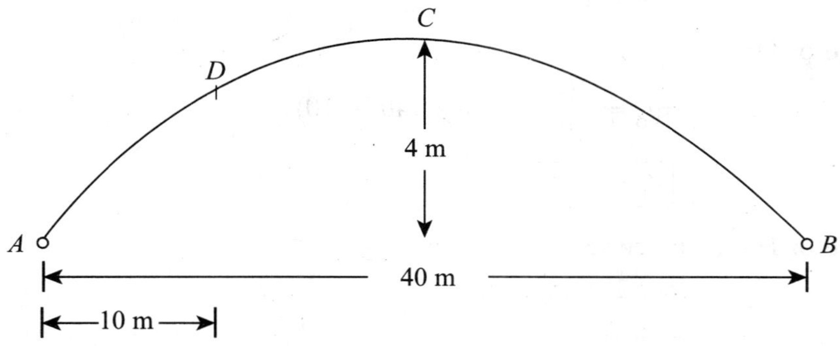

FIG. 8.52

Solution: When a unit load travels, the horizontal thrust is given by

$$H = \frac{5W}{8}\left(\frac{l}{h}\right)\alpha(1-\alpha)\left(1+\alpha-\alpha^2\right)$$

where $$\alpha = \frac{x}{l}$$

The values of H are calculated for different values of α as follows.

$\alpha = 0.25$

$$H = \frac{5}{8}(1)\left(\frac{40}{4}\right)0.25\,(1-0.25)\,(1+0.25-0.25^2) = 1.392$$

$\alpha = 0.50$

$$H = \frac{5}{8}(1)\left(\frac{40}{4}\right)0.50\,(1-0.50)\,(1+0.5-0.5^2) = 1.953$$

$\alpha = 0.75$

$$H = \frac{5}{8}(1)\left(\frac{40}{4}\right)0.75\,(1-0.75)\,(1+0.75-0.75^2) = 1.392$$

The above ordinates are used to draw the influence line for H.

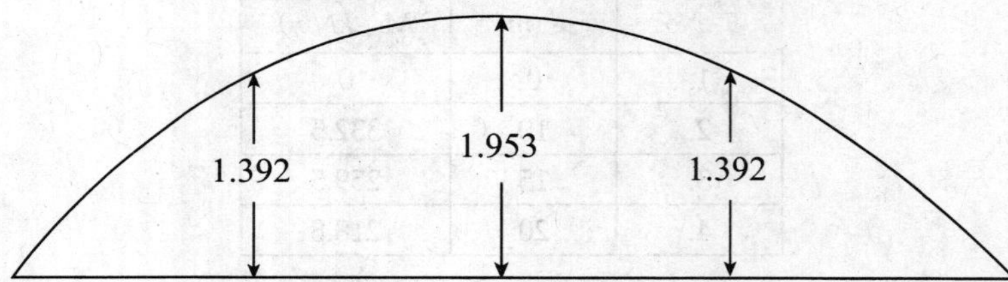

FIG. 8.53 Influence line for horizontal thrust

Influence line for D

The influence line diagram for horizontal thrust is superimposed on the influence line diagram of bending moment

$$M = M_s - Hy$$

Ordinate of D,

$$y_D = \frac{4(4)}{40^2} \times 10 \times (40-10)$$

$$\boxed{y_D = 3\text{ m}}$$

Ordinate of Hy at the centre,

$$Hy_c = 1.953 \times 3$$

$$\boxed{Hy_c = 5.86}$$

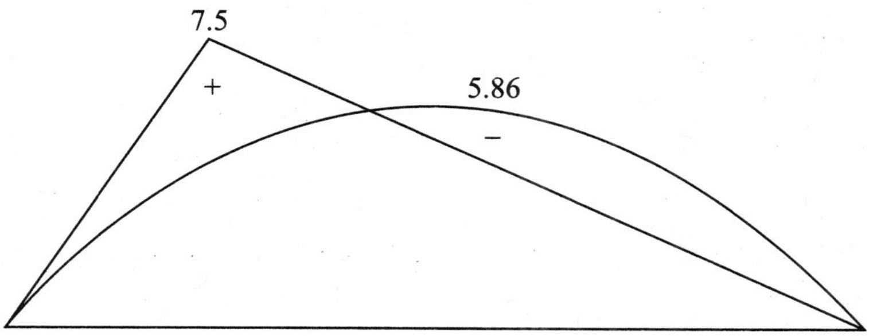

FIG. 8.54 Influence line for moment at D

Influence line for radial shear

Radial shear at a section is given by = $V \cos\theta - H \sin\theta$. The trigonometric terms are to be evaluated using the equation of the arch,

$$y = \frac{4hx}{l^2}(l - x)$$

$$\frac{dy}{dx} = \frac{4h}{l^2}(l - 2x)$$

0.74

0.25

FIG. 8.55 ILD for shear

Substituting the values of h = 3 m, l = 40 m,

$$\frac{dy}{dx} = \frac{4(3)}{40^2}(40 - 2 \times 10)$$

$$\tan\theta = 0.15$$

$$\therefore \qquad \sin\theta = 0.1483, \cos\theta = 0.9889$$

Influence line for normal thrust

Normal thrust at D is

$$N_D = V \sin\theta + H \cos\theta$$

The expression shows that the influence line for normal thrust can be obtained by superposition as follows. The ordinates of influence line for shear force and horizontal thrust is multiplied by $\sin\theta$ and $\cos\theta$ respectively. The sum of the above values give the normal thrust.

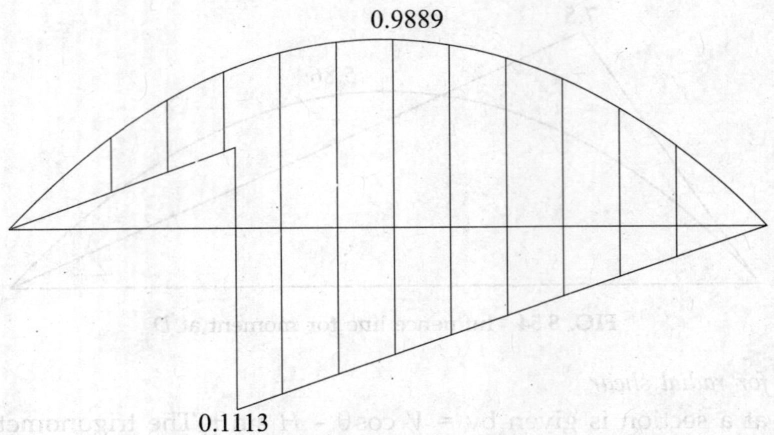

FIG. 8.56 Influence line for normal thrust

REVIEW QUESTIONS

Remembrance

8.1. Define an arch.
8.2. What is a two hinged arch?
8.3. How are arches classified?
8.4. What is a segmental arch?
8.5. State the section in a two hinged arch where I is maximum.
8.6. State the section in a two hinged arch where I is minimum.
8.7. Name the theorem which can be used to evaluate the horizontal thrust.
8.8. Write the expression for H of a parabolic two hinged arch.
8.9. What is the value of BM at any section of a two hinged parabolic arch subjected to udl?
8.10. Name the analytical methods to determine the horizontal thrust in a two hinged arch?
8.11. Name the structural components in a bowstring girder.
8.12. List the causes for secondary stresses in an arch.
8.13. What is the value of the horizontal thrust at each support of a two hinged semicircular arch of radius R which carries a concentrated load.
8.14. For a semicircular two hinged arch shown in the figure below, a moment of 10 kNm applied at B produces a displacement of 8 mm at A. If a concentrated load of 80 kNm is applied at A. What will be the rotation at B.

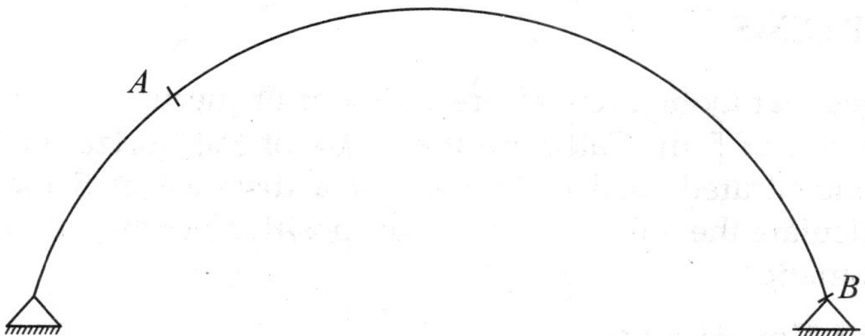

Understanding

8.1. Name the reaction components in a two hinged arch.

8.2. What is the significance of secant variation of moment of inertia in a two hinged arch?

8.3. How do you construct the bending moment diagram in a two hinged arch?

8.4. If the span and rise of a two hinged segmental arch is known, how do you find the radius of arch.

8.5. Is two hinged arch statically determinate or indeterminate?

8.6. The equation of the two hinged parabolic arch is given by $y = kx(l - x)$ where x is measured from the left hinge. If the span of the arch is 20 metres and the rise is 2 metres. What is the value of k?

8.7. Distinguish between the structural behaviour of a beam and an arch.

8.8. How radial shear and normal thrust can be found from horizontal thrust?

8.9. A semicircular arch is of radius 'r' subjected to a crown load. Another semicircular arch is of radius '$2r$' and subjected to a load at the crown of same magnitude. Will the horizontal thrust differ?

8.10. What type of arch can be designed if the arch bridge abutments is expected to move little laterally?

8.11. How horizontal thrust is provided in a tied arch?

8.12. What kinds of support provided in a bowstring girder?

8.13. What is the effect of horizontal thrust on bending moment in a two hinged arch?

8.14. The bending moment due to any load at all sections of an arch is less than the bending moment at all sections of a simple beam having the same span of arch under the same load. Why?

8.15. In a two hinged arch, an increase in temperature induces the bending moment. Is it true or false?

8.16. Define linear arch.

EXERCISE PROBLEMS

8.1. A two hinged parabolic arch where moment of inertia $I = I_c \sec\theta$ has a span of 25 m and a rise of 5 m. Calculate the value of the horizontal thrust if the arch carries a concentrated load of 100 kN at a distance of 5 metres from the left support. Calculate the value of maximum positive bending moment and negative bending moment.

Ans: 58 kN, 214.4 kNm, 94 kNm

8.2. Determine the horizontal thrust in the symmetrical two hinged parabolic arch, the moment of inertia varies as the secant of the slope of the arch axis.

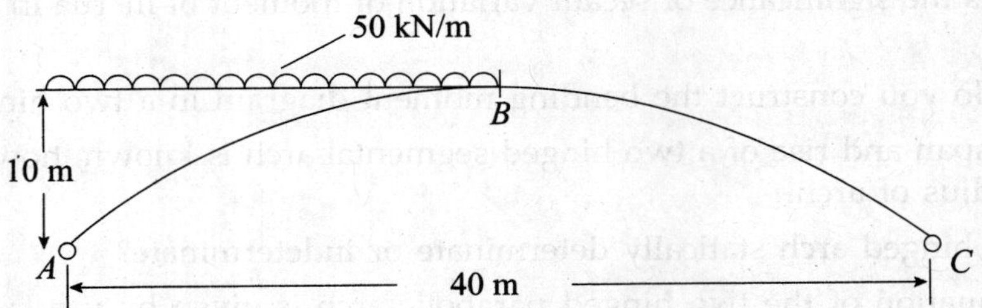

Ans: 500 kN

8.3. Derive an expression for horizontal thrust in the symmetrical parabolic two hinged arch when a point load Pacts at 'αl' distance from the left support. The span of the arch is 'l' and its rise is 'r'.

Ans: $H = \left(\frac{5}{8}\right)(W)\left(\frac{\alpha l}{r}\right)(1 - 2\alpha^2 + \alpha^3)$

8.4. Determine the horizontal thrust in the symmetrical two hinged parabolic arch, the moment of inertia varies as the secant slope of the arch axis.

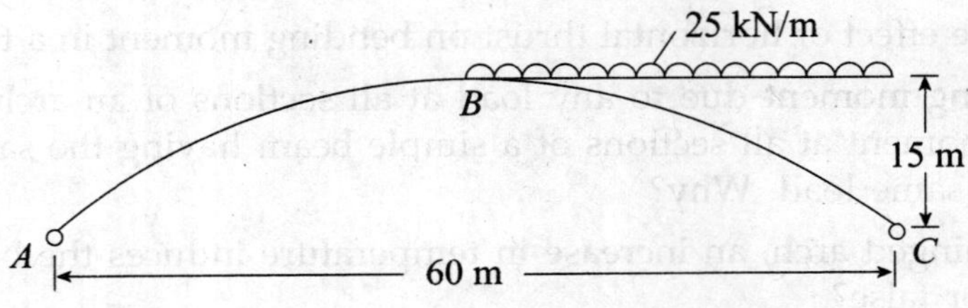

Ans: 375 kN

8.5. A two hinged parabolic arch is loaded as shown in the figure below. Compute the horizontal thrust from the first principles.

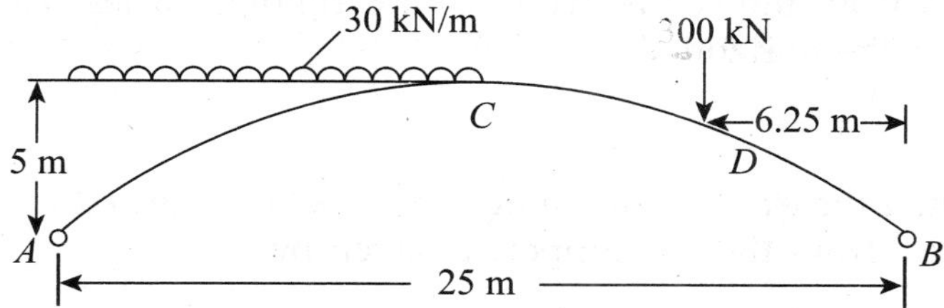

Ans: 443 kN

8.6. Calculate the maximum negative bending moment for the above data and also compute the radial shear and normal thrust at 10 metres from the right support.

Ans: 112.5 kNm at 11.8 metres from *B*; –51.5 kN, 440.5 kN

8.7. A two hinged parabolic arch is subjected to loads as shown in the figure below. Determine the horizontal thrust assuming secant variation for the moment of inertia. Compute the maximum positive and negative bending moment and draw the BMD. Also find the normal thrust and radial shear at quarter span from left support. Neglect the rib shortening (use fundamental principles).

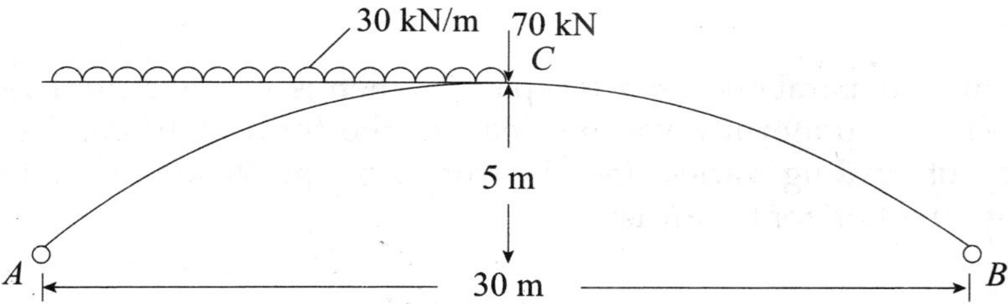

Ans: 421.95 kN, 363.5 kNm @ 7.97 m from *A*, –482.4 kNm @ 7.21 m from *B*, 446.96 kN, 6.61 kN.

8.8. A two hinged arch of variable cross section is shown in the figure below. The relevant second moment of the area at any section is proportional to the secant of the slope of the centre line at the section. Loads of 250 kN are applied at points 10 metres from the ends. Calculate the horizontal thrust, bending moment under the loads, and at midspan and sketch the bending moment diagram.

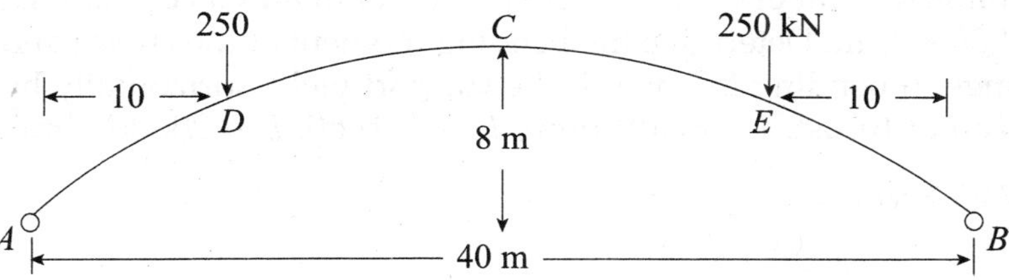

Ans: 348 kN, 412 kNm, 284 kNm

8.9. A two hinged parabolic arch has a span of 40 metres and a central rise of 10 m. Calculate the horizontal thrust when a load of 100 kN acts at 45° to the vertical at 10 metres from the centre. The flexural rigidity varies with secant angle of inclination to the horizontal.

Ans: 62.2 kN

8.10. The horizontal thrust for a two hinged parabolic subjected to a point load W at a distance 'αl' from the left support is given by

$$H = \left(\frac{5}{8}\right) W \left(\frac{l}{h}\right) \alpha\ (1 - \alpha)\ (1 + \alpha - \alpha^2)$$

where 'l' is the span and 'h' is the rise of the arch. Compute the value of for (l/h) = 3(1)6 and α = 0.3(0.2)0.8. Use the above formula and check the results of Exercise Problem no. 8.

8.11. A two hinged parabolic arch of span 'l' has central rise one-fourth of the span. Find the horizontal thrust in the arch when it is subjected to a central concentrated load W at the quarter span $I = I_c \sec\theta$ where I_c is the second moment of the area at the crown.

Ans: 0.556 W

8.12. A two hinged parabolic arch of span 30 metres has a central rise of 5 m. It is subjected to uniformly varying load in the form of triangular loading. The intensity of loading varies linearly from zero to 50 kN/m at the other end. Compute the horizontal thrust.

Ans: 562.5 kN

8.13. A two hinged parabolic arch of horizontal span 40 m and central rise 8 m is carrying uniformly distributed load of 20 kN/m for the entire span length. The temperature of the arch rib has been increased by 10°C. Calculate the horizontal and vertical components of the reaction at the hinges. Given $I = I_C \sec\theta$, $I_C = 10^{10}$ mm^4, $\alpha = 12(10)^{-6}$ per degree centigrade, $E = 2(10)^5$ N/mm^2.

Ans: 781.25 kN, 400 kN

8.14. A two hinged parabolic arch rib of span of 40 m and a central rise of 5 m carries a udl of 50 kN/m. Determine the bending moment at the crown and also calculate the change in bending moment if the support yield horizontally by 0.06 mm/kN of horizontal thrust. $I = 8(10)^8$ mm^4, $I_X = I_C \sec\theta$, $E = 200$ kN/mm^2.

Ans: Zero, 273.5 kNm

8.15. A segmental arch has a span of 40 metres and rise 8 m. It is loaded with 200 kN at the crown. Calculate the horizontal thrust by using graphical summation method. Take 10 equal divisions. If the temperature rises by 20°C, calculate the increase in horizontal thrust and change in bending stress. Take the coefficient of expansion $\alpha = 15(10)^{-6}$ per °C. $E = 2(10)^5$ N/mm². $I = 10^{10}$ mm⁴. The section is 700 mm deep throughout the length.

Ans: 190 kN, 16.8 kN, 4.7 N/mm²

8.16. Two parabolic arches have hinges at their outer ends and rest on roller at *B*. Arch *AB* is loaded with a load of 250 kN at 6 m from *A*. If EI_C is same for both arches and $I = I_C \sec\theta$, determine the horizontal thrust at hinges and draw bending moment diagram.

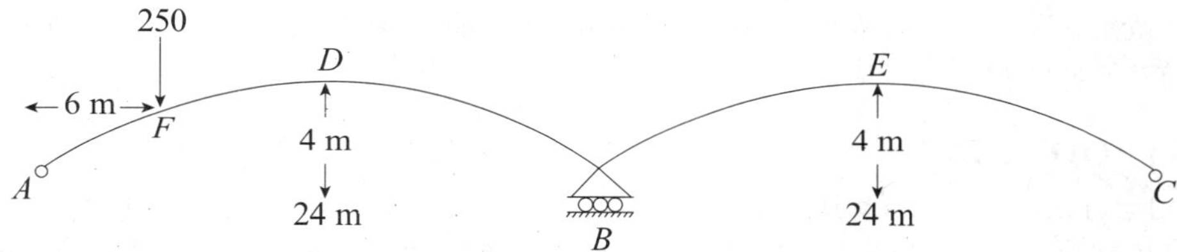

Ans: 104.4 kN

8.17. Determine the horizontal thrust in the two hinged arch shown in the figure below.

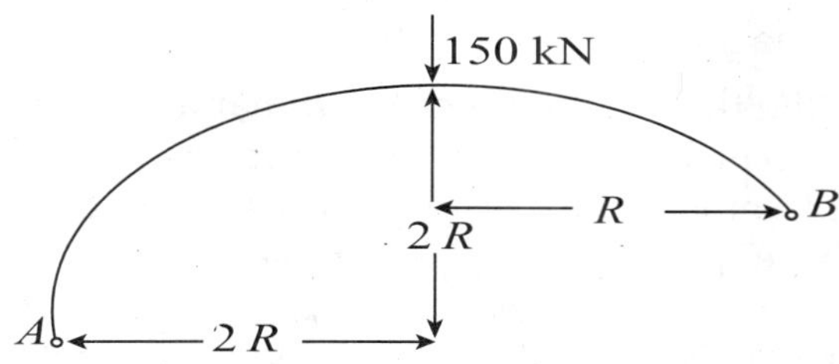

Ans: 42 kN

9 Basic Principles of Structural Dynamics

Objectives:

Deterministic and non-deterministic analysis, Types of dynamic loading, damping element, spring element, equivalent spring stiffnesses, degrees of freedom, equations of motion, Single degree freedom of system, Undamped free vibration analysis, Free vibration of damped structural dynamic force system, Logarithmic decrement, Forced vibrations.

9.1 INTRODUCTION

Structural dynamics deals with the analysis of structural systems subjected to dynamic loads. Dynamic loads can be defined as time varying loads, where the magnitude, position and direction of the loads vary with time. The structural response to a dynamic load, namely, the resulting displacements, bending moments, shear and axial forces, stresses, etc., will also be dynamic or time varying. The analysis in structural dynamics can be deterministic and non-deterministic analysis.

9.1.1 Deterministic Analysis

In this type of analysis, the time variation of the dynamic loading is completely known and such loadings are termed prescribed dynamic loadings. This type of dynamic analysis leads to a displacement-time history corresponding to a prescribed dynamic loading and other response quantities such as stresses, strains, bending moments, shear and axial forces can be obtained as a secondary phase of analysis from the previously obtained displacements. Mathematical expressions can be written to determine any response at any time, t. Deterministic analysis is relatively simpler and can be used for all types of structures. Availability of instruments to measure any type of dynamic loads including the complex seismic and wind gusts make this method more reliable.

Examples Harmonic loadings due to rotating and reciprocating machines on structures, blast loadings, seismic and wind loadings.

9.1.2 Non-deterministic Analysis

In this type of analysis, the time variation of the dynamic loading is completely unknown and defined by statistical variation. These loadings are called random dynamic loading. A non-deterministic analysis provides statistical information about the displacements and other response quantities has to be evaluated separately by independent analysis. It will be assumed that unpredictability or randomness will exist in both input and output data. Future values cannot be easily predicted. Non-deterministic analysis is relatively more mathematically involved and requires thorough knowledge of theory of probability and statistics.

Examples Seismic and wind loadings, waves in a rough sea and gust pressures on aeroplanes.

9.2 TYPES OF DYNAMIC LOADINGS

A structural system can be subjected to different types of dynamic loadings during its lifetime. Prescribed dynamic loadings used in deterministic analysis are described in terms of time, t using mathematical expressions. These loadings are described below:

Periodic Loadings: They are repetitive loadings which exhibit the same time variation successively over a large number of cycles.

Examples

Simple harmonic loadings: These loadings occur due to unbalanced mass in rotating and reciprocating machineries and can be expressed as $F(t) = F_0 \sin\omega t$ or $F_0 \cos\omega t$, where F_0 is the amplitude of the force and ω = forcing frequency.

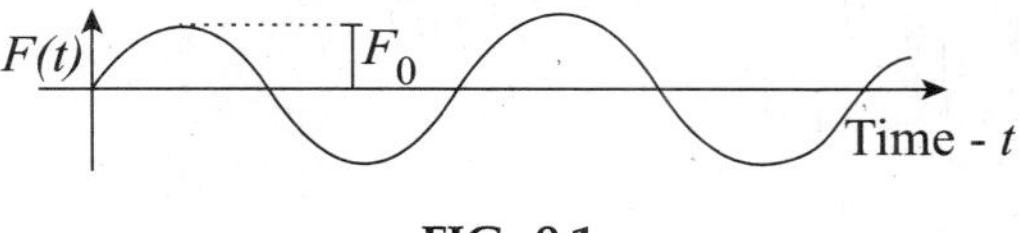

FIG. 9.1

Non-harmonic periodic loadings: General forms of periodic loading occur at different situations. Hydrodynamic pressures generated by the propeller of a ship and wave loadings on off-shore structures are typical examples of this class. Any periodic loading can be represented by Fourier series as the sum of a series of simple harmonic components and the response procedure will be common for any periodic loadings.

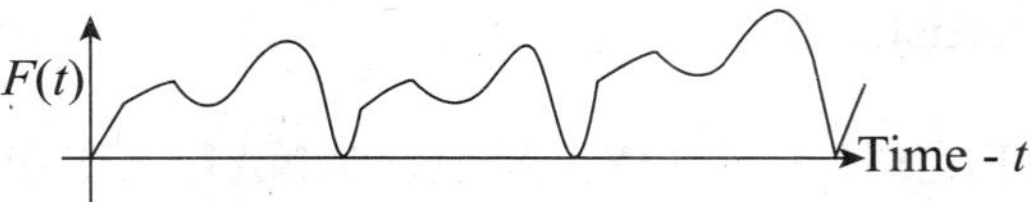

FIG. 9.2

Non-periodic Loadings: This type of loading is not repetitive and can be short duration impulsive loadings or long duration loads.

Examples

Blast loadings on structures can be represented by a triangular impulse and is a typical example of short duration impulsive loadings.

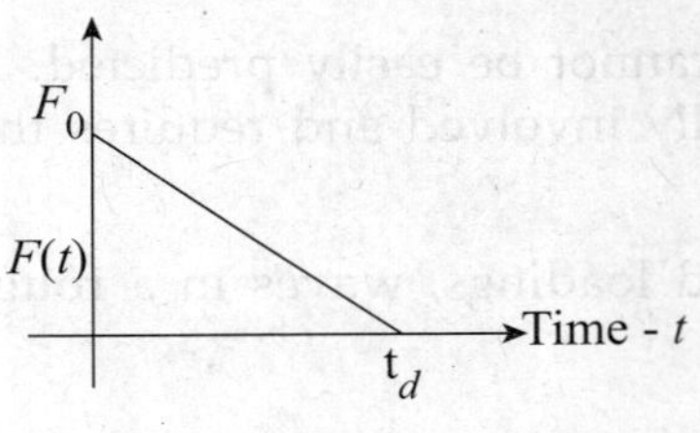

FIG. 9.3

Seismic and wind loadings on structures are typical examples of long duration loads.

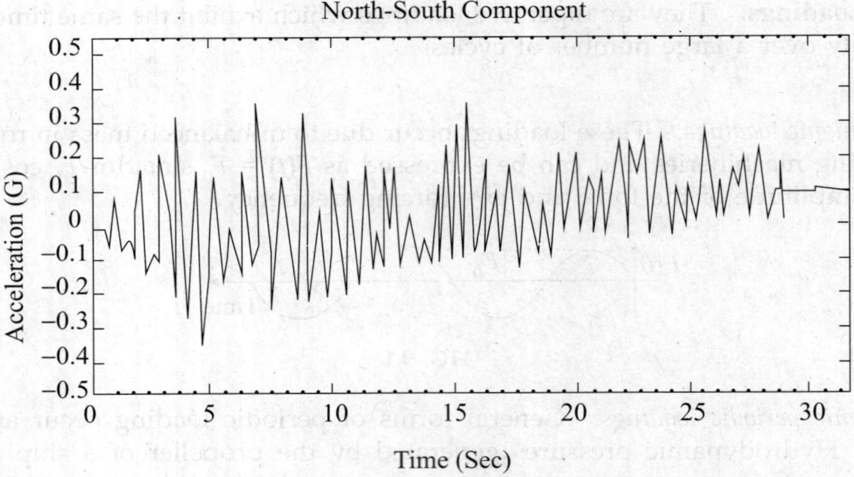

FIG. 9.4 Elcentro earthquake, May 18-1940

Random-dynamic Loadings: The relation between the load and time is random with loading being defined statistically.

9.3 CHARACTERISTICS OF A DYNAMIC ANALYSIS PROBLEM

The problems in structural dynamics differ from a static problem in the following aspects:

(i) In a dynamic problem, the load and the responses vary with time and hence, there will be multiple solutions available as compared to a single solution in a static problem. A dynamic analysis establishes a series of solutions corresponding to a displacement-time history. A dynamic analysis is more complex and time-consuming than a static analysis.

(ii) A simple beam subjected to a static load, P as shown in Fig. 9.5, can be analyzed by the force-equilibrium relation with displacements, reactions, shear forces and bending moments obtained directly. Indeterminate beams require compatibility condition along with force-equilibrium relation.

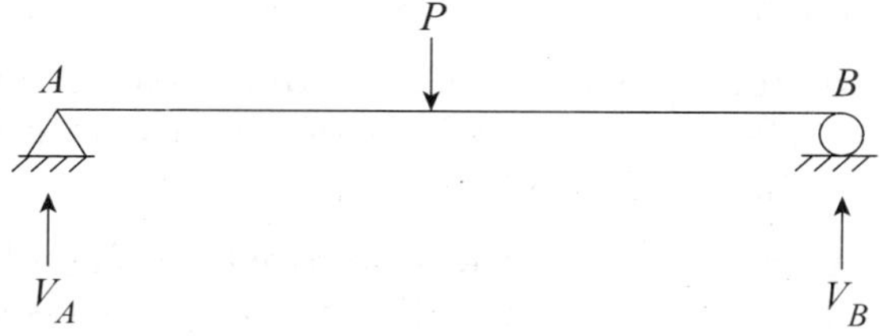

FIG. 9.5 Simply supported beam with static loading

On the other hand, if the beam is subjected to a dynamic loading, $P(t)$ or $F(t)$ as shown in Fig. 9.6, it is associated with accelerations causing inertia forces. In analyzing dynamic problems, the equilibrium equations involve inertia forces along with elastic forces (spring or stiffness force) and damping forces (energy dissipating force). If the dynamic forces cause large motions, so that the inertia forces are significant, then the problem requires a dynamic analysis. If the dynamic forces cause small motions, so that the inertia forces are not significant, then static analysis is sufficient.

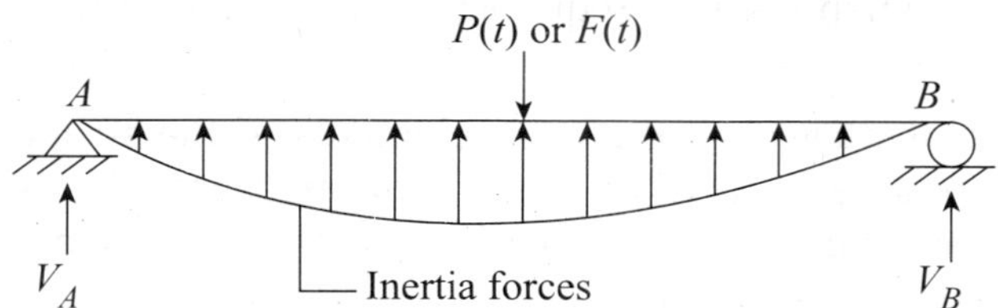

FIG. 9.6 Simply supported beam with dynamic loading

9.4 DEFINITIONS

Vibration or oscillation: Any motion repeating over a period of time about an equilibrium position is called vibration or oscillation.

Though both 'vibration' and 'oscillation' indicate to and fro motion of a body about an equilibrium position, vibration involves deformation of the body whereas oscillation only rigid body movement.

Mass: It is the amount of matter present in the system and defines the inertial characteristics of the system. Mathematically,

Mass, m = Weight/acceleration due to gravity = W/g

Units = kg or tonnes, T (If W is in newtons, N then m is in kg; if W is in kN, then m is in tonnes, T)

Stiffness: It is defined as the force required to cause unit deformation, denoted by k.

Units = N/m, kN/m, N/mm and kN/mm

Periodic motion: Any motion repeating itself after equal intervals of time is called periodic motion. Most of the vibrations are of periodic motion.

Free vibrations: A structural system vibrating on its own due to initial disturbance or initial conditions is called free vibration. No external dynamic force or excitation force will be acting on the system.

Forced vibrations: A structural system vibrating due to an external dynamic force is called forced vibrations. The dynamic analysis of the system depends on the type of dynamic force acting on the system.

Damping: It is the property of the system due to which the vibrations are gradually reduced over a period of time. In most of the dynamic analysis, viscous damping is used to model damping.

Undamped vibrations: Most of the structural systems possess low damping and can be neglected for finding basic properties like natural frequency and natural period. If damping is ignored in the analysis, then the resulting vibrations are said to be undamped (undamped free and forced vibrations).

Damped vibrations: If damping present in the system is considered, then the resulting vibrations are said to be damped (damped free and forced vibrations). Damping becomes important at resonance.

(Resonance occurs when natural frequency of the system = forcing frequency of the excitation force.)

Linear and non-linear vibrations: If the basic elements of the structural system (spring, damper and mass) behave linearly during vibration, then the system is said to have linear vibrations. The governing differential equations are linear and the principle of superposition is valid.

On the other hand, if any of the basic elements of the system behaves non-linearly, then the system is said to have non-linear vibrations. The governing differential equations are non-linear and the principle of superposition is not valid.

Usually, stiffness is considered non-linear, while mass and damping is considered linear. With increasing amplitudes, (displacements) the vibrations tend to become non-linear. A non-linear analysis is more complex than a linear analysis.

Cycle: The vibration of a structural system starting from its equilibrium position and returning back to the same position is called a cycle or one cycle of vibration. An angular displacement of 2π radians constitutes a cycle of vibration.

Amplitude: It is the maximum displacement or velocity or acceleration of a vibrating system from its equilibrium (mean or static equilibrium) position and is usually expressed in mm, mm/sec or mm/sec^2 (m, m/sec, m/sec^2).

Period: It is the time taken to complete one cycle of vibration and is expressed in seconds (s). The natural period is a property of the structural system when subjected to free vibration and is denoted by T_n. Forcing period is associated with the external dynamic force and denoted by T.

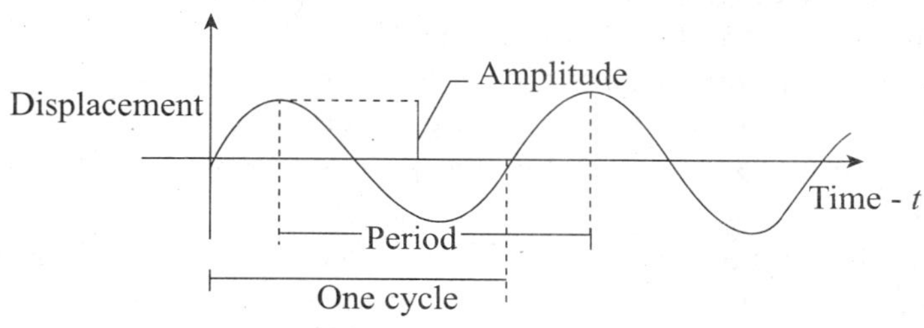

FIG. 9.7

Frequency: Linear or cyclic frequency (f) is defined as the number of cycles/unit time and is expressed as cycles/sec or Hertz (Hz), while angular frequency, ω is defined as the angular displacement/unit time and is expressed as radians/sec.

Natural frequency is a property of the structural system under free vibration and is denoted by ω_n.

Natural frequency depends on the distribution of mass and stiffness of the system and it is observed that damping does not affect the natural frequency of the system appreciably. Forcing frequency is associated with the external dynamic force and denoted by ω.

A structural system with N degrees of freedom will have N natural frequencies and the lowest natural frequency is called the *fundamental natural frequency*. Under each frequency, the structure vibrates with a particular shape called 'mode shape' and the shape corresponding to the fundamental natural frequency is called the fundamental mode of vibration. The fundamental natural frequency and the fundamental mode of vibration is very important in the analysis of multi degrees of freedom systems.

Period and frequency is related as $f = 1 / T$; $T = 2\pi/\omega$ and $\omega = 2\pi f$

Resonance: When the frequency of the external dynamic force is equal to or matches with one of the natural frequencies of the vibrating system, the amplitude of the motion becomes very large and we say that the structural system is under resonance. Many failures have taken place due to this phenomenon. The design shall be such that the natural frequencies and forcing frequencies do not coincide.

9.5 IMPORTANCE OF THE SUBJECT

Vibrations of structures are undesirable and can cause the following problems:

(i) Overstressing due to additional stresses because of dynamic load causing collapse of structures.

(ii) Causes cracking requiring costly rehabilitation.

(iii) Damage to safety related equipment and delicate instruments.

(iv) Irritation to the occupants.

(v) Causes fatigue failures under repeated loadings such as periodic loadings.

9.6 SOLUTION PROCEDURE IN DYNAMIC PROBLEMS

In a structural dynamics problem, the variables (input and output) are time-dependent with input being the external excitation force and the output being the response. A structural system in vibration requires means for storing potential energy, *PE* represented by spring (elastic force), a means for storing kinetic energy, *KE* represented by mass (inertia force) and a means by which the energy is dissipated represented by a damper (damping force).

Vibrations involve transfer of *PE* to *KE* and vice versa. If the system is undamped, there is no loss of energy and the vibrations continue forever, which may not be logical. The system is always damped and some energy is dissipated in each cycle of vibration and the vibration always decays after some time (free vibration). If an external excitation is maintained, a steady state of vibration persists (forced vibration). The actual response depends on the initial conditions as well as on external excitation sources.

The analysis in a structural dynamics problem involves the following steps:

9.6.1 Mathematical Modelling

Practical structural systems are very complex with distributed masses and stiffness and are very difficult to consider all the details correctly for a mathematical analysis. Usually, the important features required to predict the behaviour of the system, are considered. A system which involves only the important features that are required to predict the behaviour of the system is termed a mathematical model. It is an idealized system which serves as a link between the actual structural system and the mathematically feasible system. A mathematical model should have enough details so that the actual behaviour of the structural system is predicted within the required accuracy and the model should not be very complex. A mathematical model provides an approximate solution which is sufficient from safety and economical considerations.

A mathematical model can be linear or non-linear depending upon the accuracy required and the importance of the structure. Engineering judgement is essential to come up with a suitable mathematical model. If required, the mathematical model can be improvised as the analysis progresses.

A mathematical model for a structural system subjected to dynamic loads shall include the mass of the structural system, elastic properties (stiffness), energy dissipating element and an external source of excitation. The elements associated with the mathematical model are called pure elements as each element represents a single property of the system. The pure elements associated with the mathematical model are:

(a) *Mass element, m:* The entire mass of the system is included in the rigid block and represents the inertial characteristics of the system.

(b) *Spring element, k:* The elastic restoring force and the potential energy of the system is represented by a weightless spring (stiffness).

(c) *Damping element, c:* The energy dissipating capacity of the system is represented by the damping element.

(d) *Excitation force, F(t):* The external dynamic force acting on the system is represented by this and depends on the type of dynamic force acting.

For an SDF system, the mathematical model is shown in Fig. 9.8 with m, k and c representing the mass, stiffness and damping characteristics of the system. Rollers provided represents a single displacement co-ordinate.

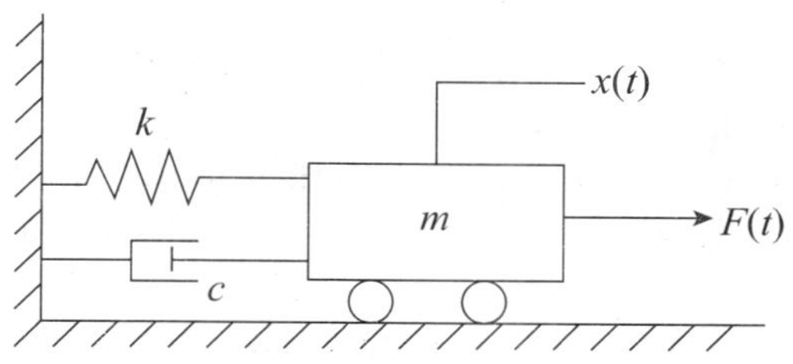

FIG. 9.8 Idealisation of SDF system

9.6.2 Governing Equations of Motion

These are the mathematical expressions defining the dynamic displacements and the solution of these equations provides the required displacement-time histories. These equations are ordinary differential equations for discrete systems and partial differential equations for continuous systems.

These equations are obtained by considering the free body diagram (FBD) of each mass. The FBD of a mass can be obtained by isolating the mass and indicating all the forces, namely, elastic forces, damping forces, inertia forces and the external excitation forces on the mass. The equations of motion may be linear or non-linear depending upon the type of analysis required. The governing equations of motion can be formulated using d'Alembert's principle, virtual work principle and Hamilton's principle. For structural engineering problems, d'Alembert's principle is commonly used, while for rigid body assemblages virtual work principle is adopted.

9.6.3 Solution of the Equations of Motion

The solution of the governing equation of motion depends on the type of structural system and dynamic force. For simple problems, closed form solutions can be obtained, while for complex problems, numerical approaches are adopted.

9.6.4 Interpretation of the Results

The solution of the governing equations of motion yields displacement, velocity and acceleration of each mass, from which the desired forces are obtained for design.

9.7 DAMPING ELEMENT

A mass when displaced from its equilibrium position comes back to its original position with certain velocity and gets displaced to the other side of the equilibrium position and thus starts vibrating. This vibration will continue infinitely, if there is no loss of energy from the structural system. However, in practice, vibration gradually decays with time due to loss of energy and the vibrating system comes to rest. These forces causing loss of energy are called damping forces. In actual structural systems, the energy dissipation can be due to:

(i) Friction with air
(ii) Friction between the particles of the structural members
(iii) Friction at connections
(iv) Friction with partition walls
(v) Yielding of the material
(vi) Opening and closing of cracks (concrete)

Therefore, it is very difficult to mathematically include all the parameters in a dynamic analysis. Practically, damping is modelled using any of the following damping elements:

1. *Viscous damping element*: This damping element is the most commonly used damping element since the mathematical formulations become simple. It is assumed that the structural systems when vibrating in a fluid medium such as air, water or soil, resistance offered by the fluid medium to the motion results in loss of energy. The energy loss depends on the size and shape of the vibrating system, viscosity of fluid, frequency of vibration and velocity of the vibrating system. The damping force is proportional to the velocity of the vibrating system.

   If F_D is the damping force and x^o is the velocity of the system, then $F_D = cx^o$, where c is a constant called the damping coefficient or damping constant. Using viscous damping, the governing differential equation becomes linear. The concept of equivalent viscous damping is often used, when other types of damping elements are considered in modelling.
2. *Coulomb or dry friction damping element:* The damping force is constant and acts opposite to the motion of the vibrating body. Mathematically, $F_D = \mu N$, where F_D is the damping force, μ is the coefficient of dynamic friction and N is the normal force acting to the direction of motion. The damping force is due to the friction between two rubbing surfaces and is influenced by the materials, surface conditions and weight of the system. It is independent of the velocity. The analysis has to be carried out over half-cycle intervals by applying the damping force opposite to the motion.
3. *Structural or hysteresis damping:* When materials are deformed, energy is absorbed and dissipated through heat by the materials.

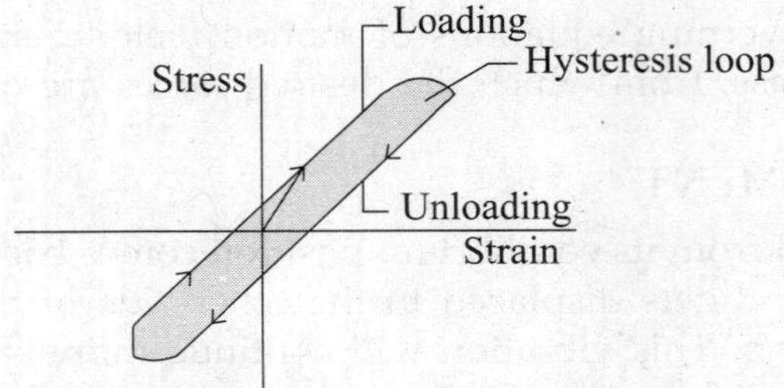

FIG. 9.9 Typical hysteretic curve

The damping is due to the internal friction between the molecules. A member under vibration shows a stress-strain curves, which forms a hysteresis loop when subjected to loading and unloading as shown in Fig. 9.9. The area of the loop represents the energy dissipated due to molecular friction/cycle/unit volume. The damping force depends on the size of the loop, which depends on the material types, frequency of the system and excitation force and the magnitude of the dynamic stress. The formation of the hysteresis loop with loading and unloading paths is indicated in Fig. 9.9.

4. *Negative Damping:* The structural system draws energy from external source continuously instead of dissipating the energy. With this, the amplitudes of motion tend to increase leading to instability. The supply of external energy to the system is called negative damping.

   The building up of amplitudes in transmission lines and suspension bridges under the action of wind at critical speeds are examples of negative damping. The failure of Tacoma's Narrow bridge is a classical example of negative damping. Providing helical spoilers for chimneys, Stockbridge dampers for bridges reduces instability and breaks the wind vortex shedding.

   In general, damping in structural systems is determined experimentally by free and forced vibration approach with many methods available.

9.8 SPRING ELEMENT

Springs are always related to the stiffness of the system. The characteristics of a spring are described based on the relation between the force and displacement. This relationship is as shown below.

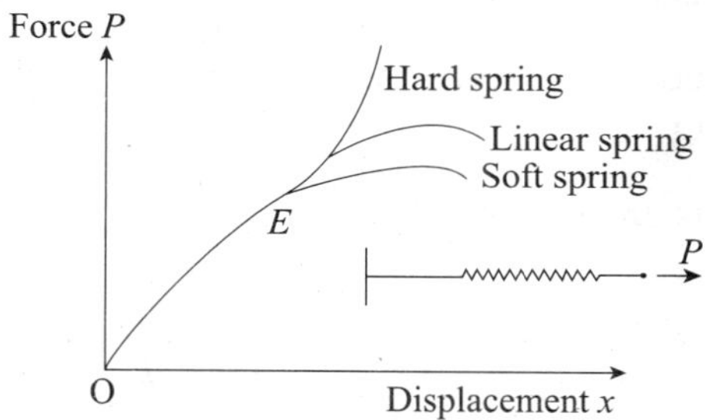

FIG. 9.10 Behaviour of springs

The behaviour of soft spring, linear spring and hard spring are shown as above. It is noted that beyond the linear range E, for the given displacement–hard spring carries more force while soft spring requires less force. The linear spring always represents an average value. In normal situations, the deformations are very small, it is usual to assume that the behaviour of the system will be linear. It is expressed mathematically as

$$F \alpha\ x \quad (1)$$

i.e.

$$F = kx \quad (2)$$

where k = stiffness of the string

In mathematical modelling, a linear spring is always adopted as it is the simplest. However, if non-linear analysis is to be carried out it is usual to assume a soft spring. Since the materials used in construction tend to lose the stiffness after proportionality limit.

9.8.1 Equivalent Spring Stiffness: k_e

The springs can be arranged in series or in parallel. The equivalent stiffness is computed for the above cases and used in analysis.

9.8.1.1 *Case (i) Springs in Parallel*

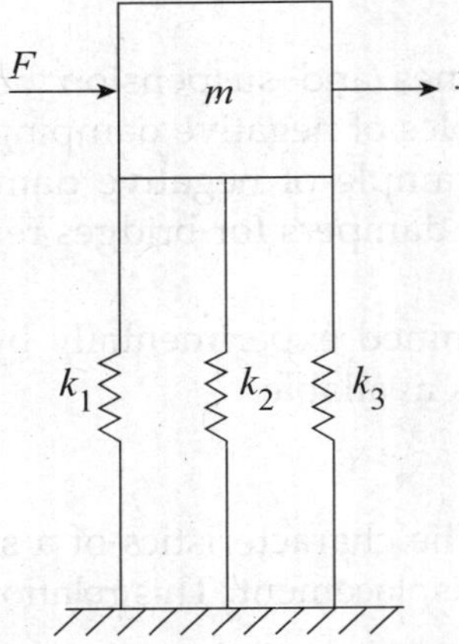

Fig. 9.11 Springs in parallel

In this case, the displacement of each spring will be same but the force in the springs will be different depending upon the stiffness of the spring.

$$\text{Total force} = F_1 + F_2 + F_3 + \ldots\ldots + F_n \tag{3}$$

$$k_e x = k_1 x + k_2 x + k_3 x + .. + k_n x \tag{4}$$

$$k_e = k_1 + k_2 + k_3 + \ldots. + k_n \tag{5}$$

$$k_e = \sum_{i=1}^{n} k_i \tag{6}$$

In simple words, the equivalent stiffness of springs in parallel will be the sum of the stiffness of each spring.

9.8.1.2 *Case (ii) Springs in Series*

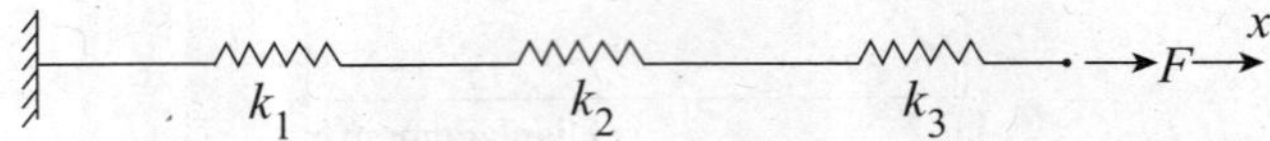

FIG. 9.12 Springs in series

In this case, the force in each spring will be same but the displacements will be different.

$$x = x_1 + x_2 + x_3 + x_4 + \ldots. + x_n \tag{7}$$

$$\frac{F}{k_e} = \frac{F}{k_1} + \frac{F}{k_2} + \frac{F}{k_3} + \frac{F}{k_4} + \ldots + \frac{F}{k_n} \tag{8}$$

$$\frac{1}{k_e} = \frac{1}{k_1} + \frac{1}{k_2} + \frac{1}{k_3} + \frac{1}{k_4} + \ldots + \frac{1}{k_n} \tag{9}$$

$$\frac{1}{k_e} = \sum_{i=1}^{n} \frac{1}{k_i} \tag{10}$$

The reciprocal of the equivalent stiffness for the springs in series is the sum of the reciprocal of each of stiffness.

9.9 DEGREES OF FREEDOM

The minimum number of independent co-ordinates required to completely define the displacements of the system are called degrees of freedom (DOF). The number of independent displacements required to completely represent the effects of all significant inertial forces of structure are called 'Dynamic degrees of freedom' of a structure. Structural systems usually have infinite dynamic degrees of freedom and such systems are analyzed as continuous systems. The structural systems analyzed assuming finite degrees of freedom are called discrete systems. Better accuracy can be obtained by increasing the degrees of freedom.

From analysis point of view, structural systems are classified as single degree freedom (SDF) system, multi-degree freedom (MDF) systems and continuous system. *Single degree freedom system:* A single co-ordinate (x or y) is sufficient to describe the dynamic motion of the system. Typical examples of SDF system are shown in Fig. 9.13.

9.10 EQUATIONS OF MOTION

These are mathematical expressions which define the dynamic displacement and the solution of this equation provides the required displacement, time histories.

The differential equations of motion will be ordinary differential equations for discreet systems. While these will be partial differential equations for continuous system.

9.10.1 Methods of Formulation—Single Degree Freedom Systems

Direct equilibrium using d'Alembert's principle

The equation of motion of a dynamic system can be represented based on Newton's second law of motion.

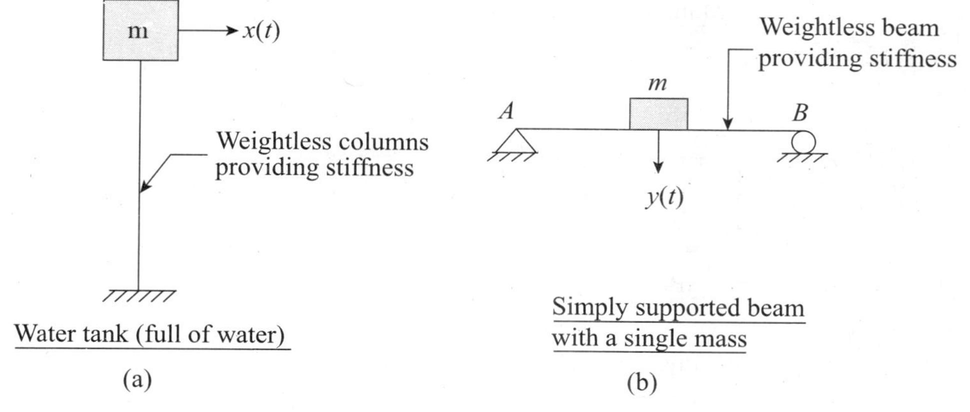

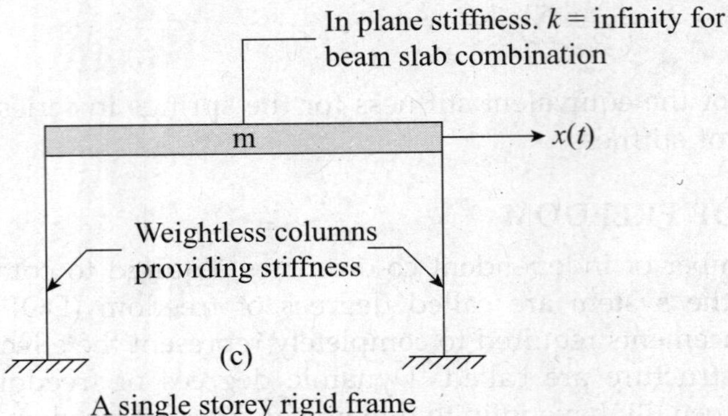

FIG. 9.13 Idealisation of SDF systems

Multi-degree freedom system: More than one co-ordinate is required to describe the dynamic motion of the system. Typical examples of MDF system are shown in Fig. 9.14.

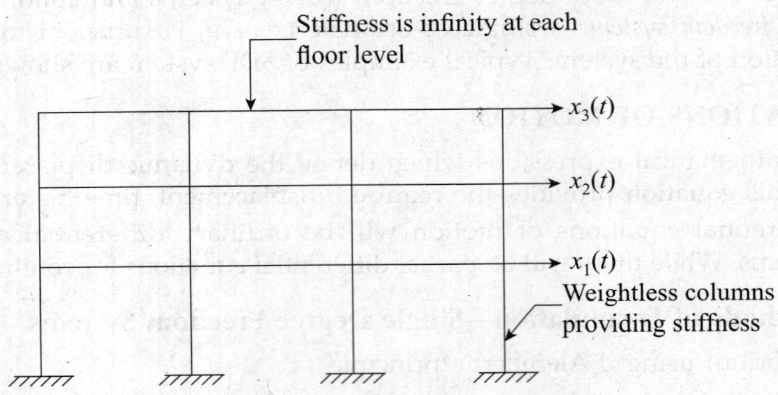

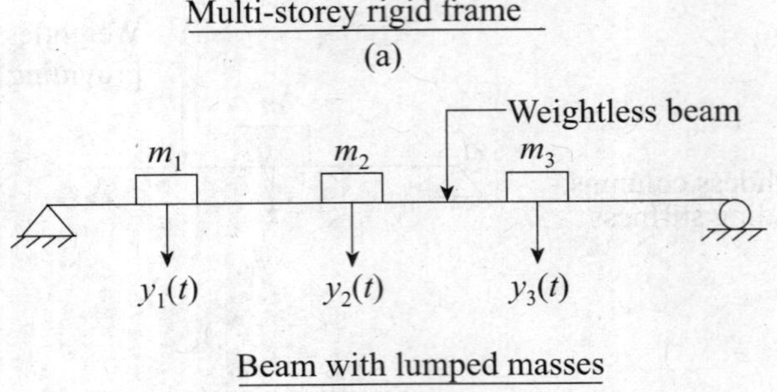

FIG. 9.14 Idealisation of MDF systems

Continuous system: These systems are analyzed with infinite degrees of freedom with the governing equations being partial differential equations. These systems become extremely complex for practical systems and solutions are available for standard cases. These systems can be converted to discrete finite MDF systems. A typical continuous system is shown in Fig. 9.15.

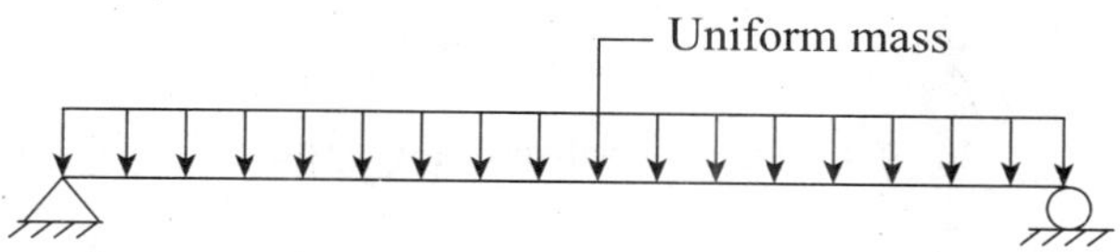

FIG. 9.15 Idealisation of continuous mass system

Statement:

The rate of change of momentum of a mass m is equal to the force acting on the mass. Mathematically, this is expressed as

$$P(t) = \frac{d}{dt}(mdv) \tag{11}$$

$$= \frac{d}{dt}\left(m\frac{dx}{dt}\right) \tag{12}$$

$$= m\frac{d^2x}{dt^2} \tag{13}$$

$$= m\ddot{x} \tag{14}$$

In the above expression, x represents the displacement and dv represents the change in velocity. It will be assumed that the mass does not vary with time.

$$P(t) = m\ddot{x} \rightarrow \text{Newton's 2nd law} \tag{15}$$

$$P(t) = m\ddot{x} = 0 \rightarrow \text{d'Alembert's principle} \tag{16}$$

The concept that a mass develops the inertia force proportional to its acceleration and opposing motion is known as d'Alembert's principle.

The equations of motions are expressed as the equations of dynamic equilibrium. $P(t)$ represents the net force and it includes the elastic spring force, damping force and the external excitation force.

Consider the mathematical model of SDF systems

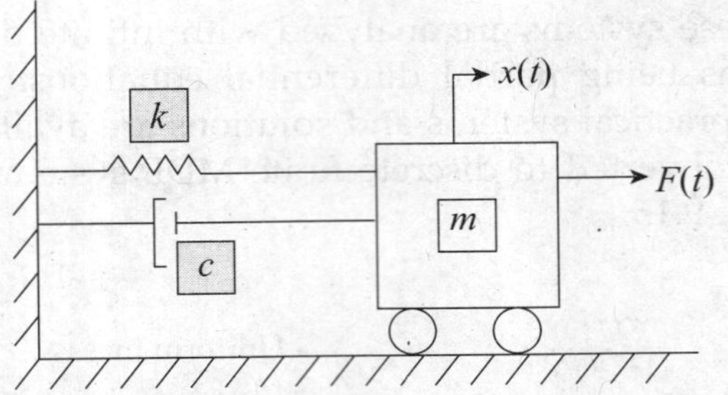

FIG. 9.16 Idealised SDF system

FIG. 9.17 Free body diagram

In the above free body diagram,

$$F_I = \text{Inertia force} = m\ddot{x}$$

$$F_D = \text{Damping force} = c\dot{x}$$

$$F_S = \text{Spring force} = kx$$

From d'Alembert's principle

$$F(t) - (F_I + F_D + F_S) = 0 \tag{17}$$

$$\boxed{m\ddot{x} + c\dot{x} + kx = F(t)} \tag{18}$$

The equations of equilibrium can also be derived from (i) virtual work method (ii) Hamiltaon's principle and (iii) energy concepts.

9.10.2 Undamped Free Vibration Analysis of SDF Systems

In structural systems, many times damping is small and it can be neglected. Also if the vibrations are initiated by the initial disturbance such as initial displacement and initial velocity without any dynamic force, we get the condition of free undamped vibrations. The mathematical model representing free undamped vibrations is shown in Fig. 9.18.

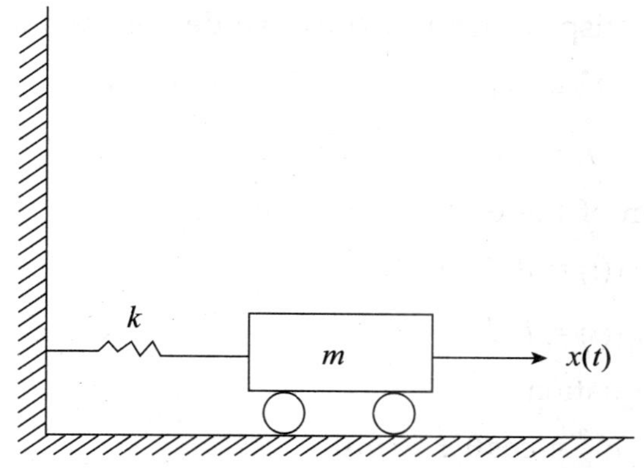

FIG. 9.18 Idealisation of undamped system

The free body diagram for the mathematical model gives

$$F_I + F_S = 0 \tag{19}$$

$$m\ddot{x} + kx = 0 \tag{20}$$

The above equation represents the ordinary differential equation for free undamped vibrations. The equation is linear with constant coefficients.

Solution of the differential equation of the motion.

The solution of the above differential equation can be expressed as

$$x = Be^{\lambda t} \tag{21}$$

where B and λ are constants. Differentiating twice;

$$\ddot{x} = B\lambda^2 e^{\lambda t} \tag{22}$$

Substituting x and $\ddot{x}$ in the differential equation,

$$mB\lambda^2 e^{\lambda t} + kBe^{\lambda t} = 0 \tag{23}$$

i.e.

$$Be^{\lambda t}\,(m\lambda^2 + k) = 0 \tag{24}$$

As $\quad Be^{\lambda t} \neq 0$; then

$$m\lambda^2 + k = 0 \tag{25}$$

$$\lambda^2 = \frac{-k}{m} = \omega_n^2 \tag{26}$$

where ω is a quantity which denotes the physical characteristics

$$\omega = \sqrt{\frac{k}{m}} \tag{27}$$

The expressions for displacements $x(t)$ can be derived as follows;

Consider $$\lambda^2 = -\omega_n^2 \tag{28}$$

$$\lambda = \pm i\omega_n \tag{29}$$

The general solution of the undamped system is

$$x(t) = B_1 e^{\lambda_1 t} + B_2 e^{\lambda_2 t} \tag{30}$$

$$x(t) = B_1 e^{i\omega_n t} + B_2 e^{-i\omega_n t} \tag{31}$$

Using the Euler's equation;

$$e^{\pm i\theta} = \cos\theta \pm i\sin\theta \tag{32}$$

and hence

$$x(t) = B_1(\cos\omega_n t + i\sin\omega_n t) + B_2(\cos\omega_n t - i\sin\omega_n t) \tag{33}$$

where $$\theta = \omega_n t \tag{34}$$

i.e. $$\boxed{x(t) = c_1\cos\omega_n t + c_2\sin\omega_n t} \tag{35}$$

The expression for velocity $\dot{x}(t)$ is obtained by differentiating Eq. (35) and hence

$$\dot{x}(t) = -c_1\omega_n\sin\omega_n t + c_2\omega_n\cos\omega_n t \tag{36}$$

The constants c_1 and c_2 are found from known values of $x_0(t)$ and $\upsilon_0(t)$ which form the initial condition. Thus, the above problem is aptly called initial value problem. Using the conditions,

For $$t = 0; \quad x(t) = x_0 \tag{37}$$

$$t = 0; \quad \dot{x}(t) = \upsilon_0 \tag{38}$$

Substituting the values of Eqs. (37) and (38) in Eqs. (35) and (36), we obtain the expression for displacement as

$$x(t) = x_0\cos\omega_n t + \frac{\upsilon_0}{\omega_n}\sin\omega_n t \tag{39}$$

Frequency and period of undamped single degree of freedom system.

In Eq. (39), the factor sin*ωnt* and cos*ωnt* are periodic functions and hence the resulting response will also be periodic. The motion repeats itself after a particular time interval. This time interval is known as the natural period of undamped field vibration T_n.

The natural period of undamped field vibration can be defined as the time required to go through one cycle of undamped field vibration and is expressed in seconds;

$$T_n = \frac{2\pi}{\omega_n} \tag{40}$$

(or)

$$\omega_n = \frac{2\pi}{T_n} \tag{41}$$

The number of times the motion repeats itself in one second is called natural cyclic frequency of undamped field vibration, denoted by f_n.

$$f_n = \frac{1}{T_n} = \frac{1}{2\pi/\omega_n} \tag{42}$$

$$\therefore \quad \omega_n = 2\pi f_n \tag{43}$$

In field vibration analysis without damping ω_n, T_n & f_n are called the basic vibration characteristics and depends on the stiffness and mass of the system. The units for ω_n in rad/sec, T_n is in seconds and f_n in Hertz or cycles per second; k is in kN/m or N/m and m is N sec/m or kN sec/m.

Amplitude of motion and phase angle.

Consider the response

$$x(t) = \dot{x}_0 \cos\omega_n t + \frac{\dot{x}_0}{\omega_n} \sin\omega_n t \tag{44}$$

i.e.

$$x(t) = c_1 \cos\omega_n t + c_2 \sin\omega_n t \tag{45}$$

The constants c_1 and c_2 can be represented by means of vector triangle.

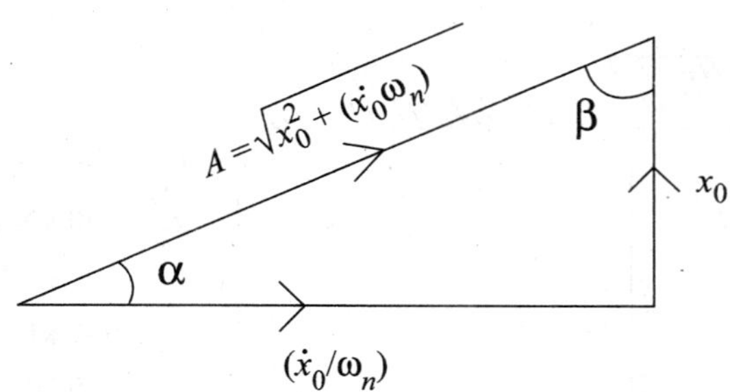

FIG. 9.19 Definition of angle α

In the above vector representation,

A = amplitude of motion

α, β = phase angles

β = phase angle with which the resulting motion with respect to initial displacement

Thus, from the above triangular representation

$$A = \sqrt{x_0^2 + \left(\frac{\dot{x}_0}{\omega_n}\right)^2} \tag{46}$$

and

$$\beta = \tan^{-1}\left(\frac{\dot{x}_0}{\omega_n x_0}\right) \tag{47}$$

The resulting response of free undamped vibrations is shown below.

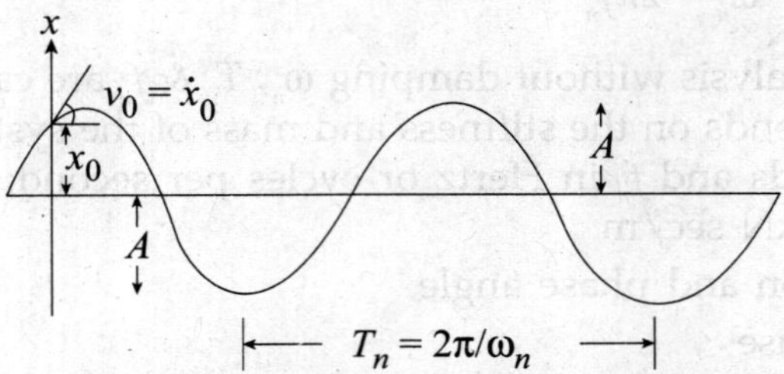

FIG. 9.20 Undamped free vibration response

Example 9.1 A mass is connected to the foundation through a network of springs as shown in Fig. 9.21. Determine the free vibration characteristics of the system. The system is undamped.

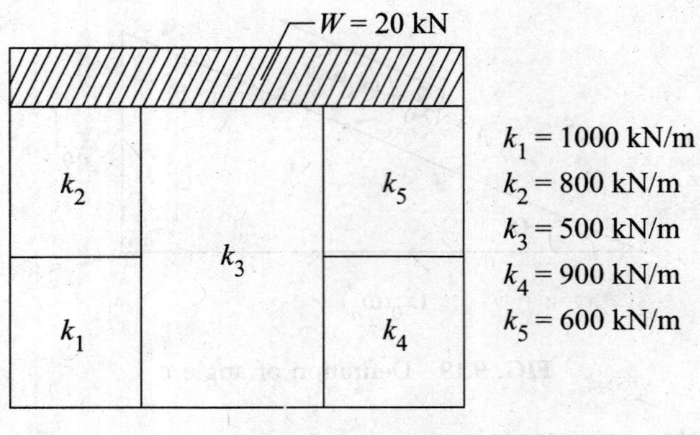

FIG. 9.21

Solution: The springs k_1 and k_2 can be considered springs in series and similarly k_4 and k_5 can be considered springs in series. The equivalent system with springs can be represented as

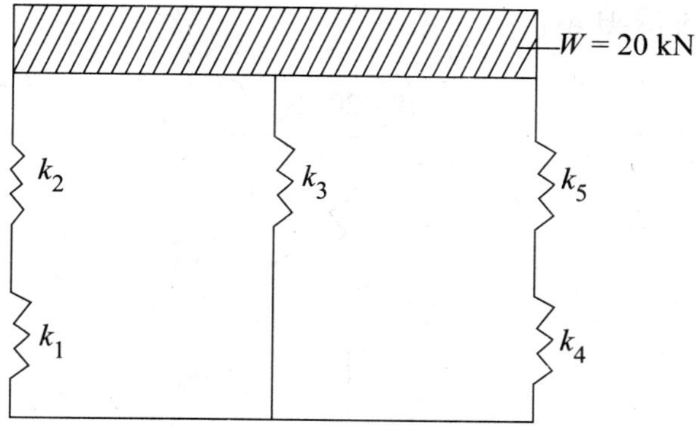

FIG. 9.22 Equivalent system of springs

$$\frac{1}{k_{eq})_{1,2}} = \frac{1}{k_1} + \frac{1}{k_2}$$

$$= \frac{1}{1000} + \frac{1}{800} = 2.25(10)^{-3}$$

$$k_{eq})_{1,2} = 444.44 \text{ kN/m}$$

$$\frac{1}{k_{eq})_{4,5}} = \frac{1}{k_4} + \frac{1}{k_5}$$

$$= \frac{1}{900} + \frac{1}{600} = 2.778(10)^{-3}$$

$$\frac{1}{k_{eq})_{4,5}} = 360 \text{ kN/m.}$$

The problem is simplified as

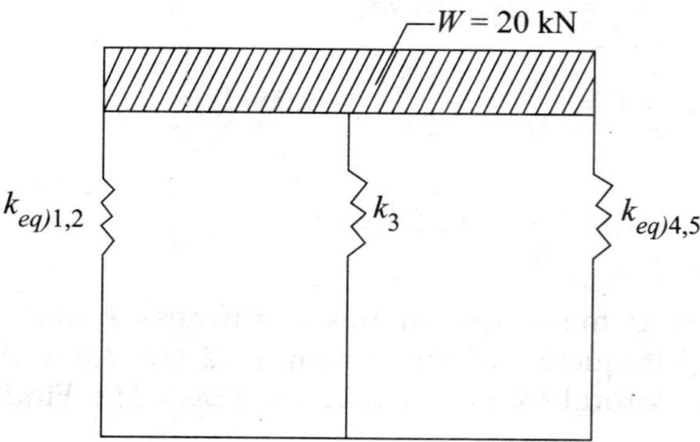

FIG. 9.23 Reduced equivalent system

Finally, this is idealised as

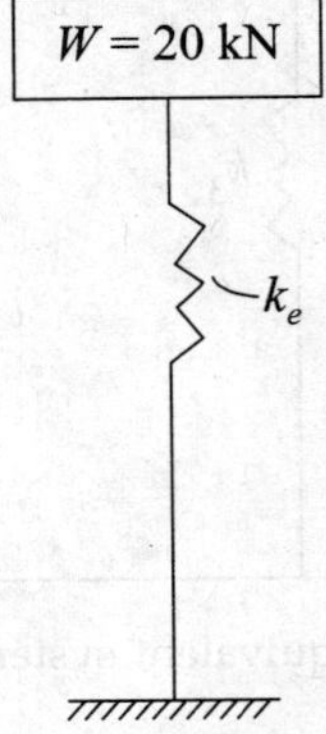

FIG. 9.24

$$k_e = k_{eq}\big)_{1,2} + k_3 + k_{eq}\big)_{4,5}$$

$$k = 444.44 + 500 + 360$$

$$k = 1304.44 \text{ kN/m}$$

Note: The above problem can be considered an elevated water tank supported on columns and connected through a network of horizontal bracings. Each column can be treated as springs in series and the equivalent analysis can be considered:

$$\omega_n = \sqrt{\frac{k_e}{m}}$$

$$\omega_n = \sqrt{\frac{1304.44}{(20/9.81)}}$$

$$\omega_n = 25.29 \text{ rad/sec}$$

$$f_n = \frac{\omega_n}{2\pi} = \frac{25.29}{2\pi} = 4.03 \text{ H}_z$$

$$T_n = \frac{1}{f_n} = 0.25 \text{ secs}$$

Example 9.2 A spring mass system has a stiffness k and the mass is m. The corresponding natural frequency of the system is 12 Hz. An additional 20 kg mass is coupled to m and the natural frequency decreases by 2 Hz. Find k and m.

Solution: The given data is pictorially represented as

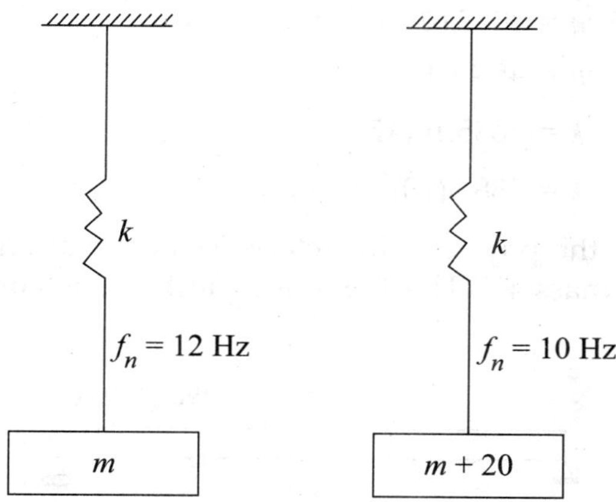

FIG. 9.25

Then

$$f_n = \frac{\omega_n}{2\pi}$$

i.e.

$$\omega_n = 2\pi f_n$$

$$\omega_n = 2\pi(12) = 75.40 \text{ rad/s}$$

$$\omega_n = 2\pi(10) = 62.83 \text{ rad/s}$$

$$75.40 = \sqrt{\frac{k}{m}}$$

$$5685.16 = \frac{k}{m}$$

$$\boxed{k = 5685.16 \text{ m}}$$

Similarly,

$$62.83 = \sqrt{\frac{k}{(m + 20)}}$$

$$3947.61 = \frac{k}{(m + 20)}$$

$$\boxed{k = 3947.61(m + 20)}$$

As the value of k is same in both the systems

$$5685.16\, m = 3947.61(m + 20)$$

$$m = 45.44 \text{ kg}$$

and

$$k = 5685.16(45.44)$$

$$k = 258.3(10)^3 \text{ N/m}$$

Example 9.3 Calculate the physical characteristics of vibration of a simply supported beam having a central mass W. The flexural rigidity EI is constant. Neglect the self weight of the member.

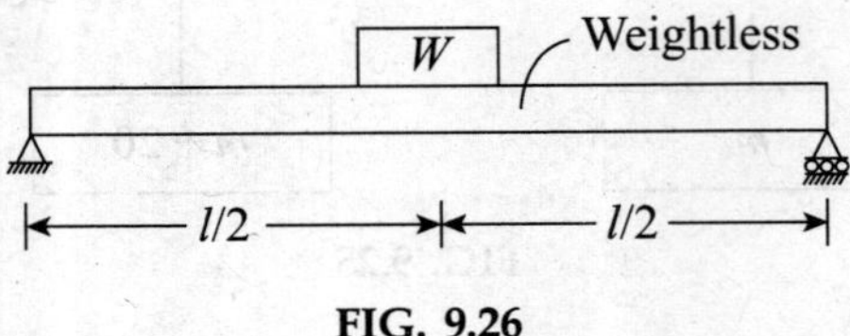

FIG. 9.26

Solution: We know

$$P = k\delta$$

and k is the force required for unit displacement.

and

$$k = \frac{P}{\delta}$$

$$k = \frac{P}{Pl^3/48EI}$$

$$\boxed{k = \frac{48EI}{l^3}}$$

$$\omega_n = \sqrt{\frac{k}{m}}$$

$$\omega_n = \sqrt{\frac{48EIg}{l^3W}} = 21.67\sqrt{\frac{EI}{l^3}} \text{ rad/sec}$$

$$f_n = \frac{\omega_n}{2\pi} = 21.67\sqrt{\frac{EI}{Wl^3/2\pi}}$$

$$T_n = \frac{1}{f_n} = 0.2896\sqrt{\frac{Wl^3}{EI}}$$

Example 9.4 A simply supported beam supports a weight as shown in the Fig. 9.27. The modulus of elasticity of the material is 200 Gpa and the second moment of the area is $8(10)^6$ mm^4. Calculate the free vibration of the system.

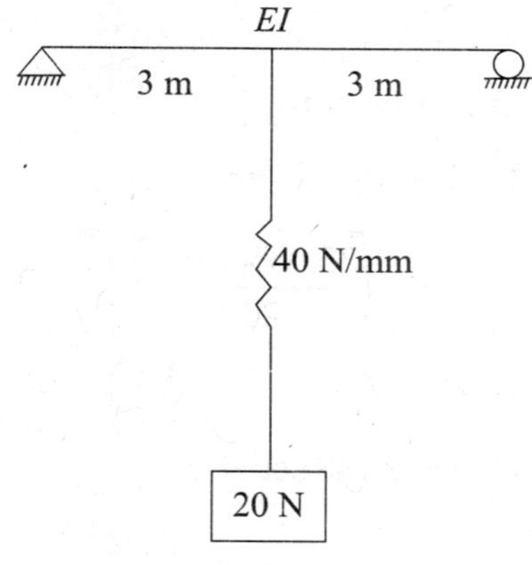

FIG. 9.27

Solution: The load is pulling both the spring and the beam. Force in the spring and beam are same but the displacements are different. Hence, this can be considered the spring in series.

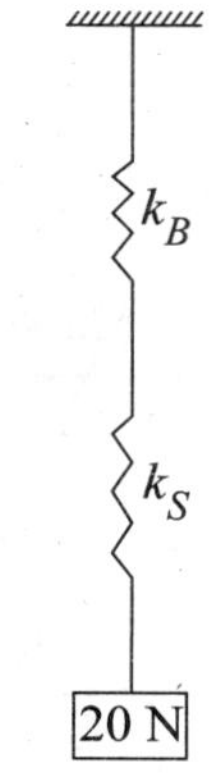

FIG. 9.28

$$k_B = \text{stiffness of the beam} = \frac{48EI}{l^3}$$

$$k_B = \frac{48(200 \times 10^3) \times 8(10)^6}{6000^3}$$

$$k_B = 355.55 \text{ N/mm}$$

and

$$k_s = \text{stiffness of the spring} = 40 \text{ N/mm}$$

$$\frac{1}{k_e} = \frac{1}{k_B} + \frac{1}{k_S}$$

$$k_e = 35.955 \text{ N/mm}$$

i.e.

$$k_e = 35.955(10)^3 \text{ N/mm}$$

$$\omega_n = \sqrt{\frac{k_e}{m}} = \sqrt{\frac{35.955(10)^3}{20}}$$

$$\omega_n = 42.4 \text{ rad/sec}$$

$$T_n = \frac{2\pi}{\omega_n} = 0.148 \text{ sec}$$

$$f_n = \frac{1}{T_n} = 6.757$$

Example 9.5 Determine the natural period for the system shown in Fig. 9.29. Assume that the beams and springs supporting the weight *W* are massless.

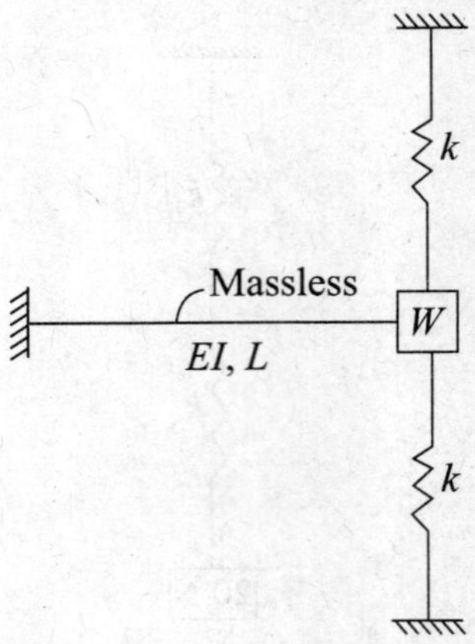

FIG. 9.29

Solution: The deflection of the beam at the free end and the deflection of the springs will have the same deflection. from the compatibility condition.

Therefore, the system can be considered springs in parallel. The equivalent stiffness of the system can be calculated as

$$k_e = k_{\text{beam}} + k_{\text{springs}}$$

$$k_e = \left(\frac{3EI}{l^3} + 2k\right)$$

$$k_e = \frac{3EI + 2kl^3}{l^3}$$

$$\omega_n = \sqrt{\frac{k_e}{m}}$$

$$\omega_n = \sqrt{\frac{3EI + 2kl^3}{l^3 m}}$$

$$\omega_n = \sqrt{\frac{(3EI + 2kl^3)}{l^3 (W/g)}}$$

$$\omega_n = \sqrt{\frac{g(3EI + 2kl^3)}{Wl^3}}$$

$$T_n = \frac{2\pi}{\omega_n}$$

$$T_n = 2\pi \sqrt{\frac{Wl^3}{g(3EI + 2kl^3)}}$$

i.e.

$$T_n = 2\pi l \sqrt{\frac{W}{g}\left(\frac{l}{3EI + 2kl^3}\right)}$$

Example 9.6 A man weighing 70 kg stands at the end of a cantilever diving board of span 1.2 m. The man oscillates at a frequency of 3 Hz. What is the flexural rigidity of the diving board?

Solution

Frequency $f_n = 3$ Hz

$$\omega_n = 2\pi f_n$$

$$\omega_n = 6\pi \text{ rad/s}$$

As it is a cantilever, the stiffness is obtained from Table 9.1 as

$$k = 3EI/l^3$$

and hence

$$\omega_n = \sqrt{\frac{k}{m}}$$

$$6\pi = \sqrt{\frac{3EI}{1.2^3 \times 70}}$$

$$EI = 2387.65 \text{ Nm}^2$$

Example 9.7 Determine the free vibration characteristics of a rigid frame shown in Fig. 9.30. Assuming an initial displacement of 25 mm and initial velocity of 25 mm/sec. Calculate the amplitude with respect to displacement, velocity and acceleration. Also calculate the responses at t = 1 sec. Sketch the response of the system.

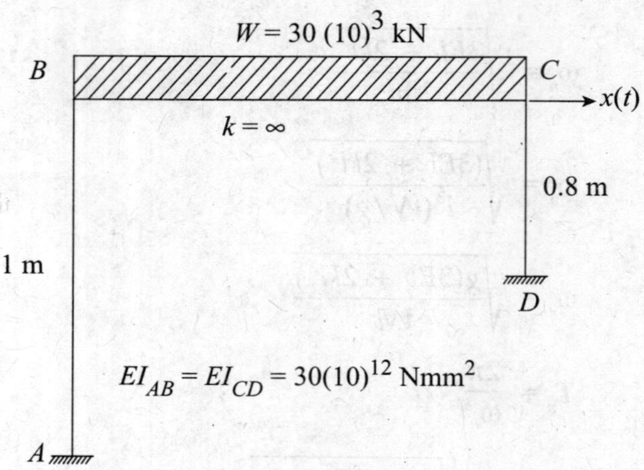

FIG. 9.30

Solution: The columns can be treated as springs in parallel. Assuming that the beam-slab combination makes the girder very rigid at the top.

The lateral stiffness of such system with the column fixed at the foundation level can be derived as follows.

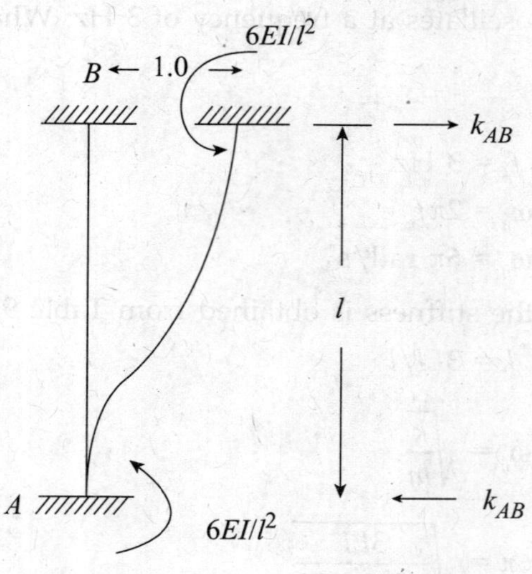

FIG. 9.31

The value of k_{AB} is obtained by taking moments about A

$$\Sigma M_A = 0;$$

$$k_{AB}(l) - \left(\frac{6EI}{l^2}\right) - \left(\frac{6EI}{l^2}\right) = 0$$

$$\boxed{k_{AB} = \frac{12EI}{l^3}}$$

On similar lines, we can find the translational stiffness of the column hinged at the base as $(3EI/l^3)$. For the given problem, the equivalent stiffness

$$k_e = \left(\frac{12EI}{l^3}\right)_{AB} + \left(\frac{12EI}{l^3}\right)_{CD}$$

$$k_e = \frac{12 \times 30 \times 10^{12}}{1000^3} + \frac{12 \times 30 \times 10^{12}}{800^3}$$

$$k_e = 1063.13(10)^3 \text{ N/mm}$$

$$\omega_n = \sqrt{\frac{k_e}{m}} = \sqrt{\frac{1063.13(10)^3}{(30 \times 10^3/9.81)}}$$

$$\omega_n = 18.65 \text{ rads/s}$$

$$T_n = \frac{2\pi}{\omega_n} = 0.33 \text{ sec}$$

$$f_n = \frac{1}{T_n} = 3.03$$

Maximum displacement amplitude

$$A = \sqrt{x_0^2 + \left(\frac{x_0}{\omega_n}\right)^2}$$

$$A = \sqrt{25^2 + \left(\frac{25}{18.65}\right)^2}$$

$$A = 25.03 \text{ mm}$$

Maximum velocity amplitude $= A\omega_n$

$$= 25.03 \times 18.65$$

$$= 466.8 \text{ mm/sec}$$

Maximum acceleration amp $= A\omega_n^2$

$$= 25.03 \times (18.65)^2$$

$$= 8705.99 \text{ mm/s}^2$$

$$\beta = \tan^{-1}\left(\frac{\dot{x}_0}{\omega_n x_0}\right)$$

$$\beta = 0.05 \text{ rad}$$

Responses

$$x(1) = 25.03 \cos (18.65 \times 1 - 0.05)$$

$$x(1) = 23.72 \text{ mm}$$

$$\dot{x}(1) = -25.03(18.65) \sin (18.65 \times 1 - 0.05)$$

$$\dot{x}(1) = -148.89 \text{ mm/s}$$

$$\ddot{x}(1) = -25.03(18.65)^2 \cos (18.65 \times 1 - 0.05)$$

$$\ddot{x}(1) = -8251.26 \text{ mm/s}^2$$

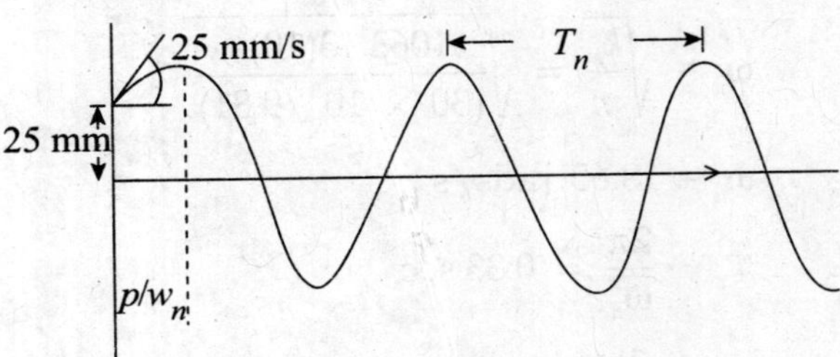

FIG. 9.32

Example 9.8 Determine the natural frequency of the system shown in Fig. 9.33. (VTU, Jan 2009)

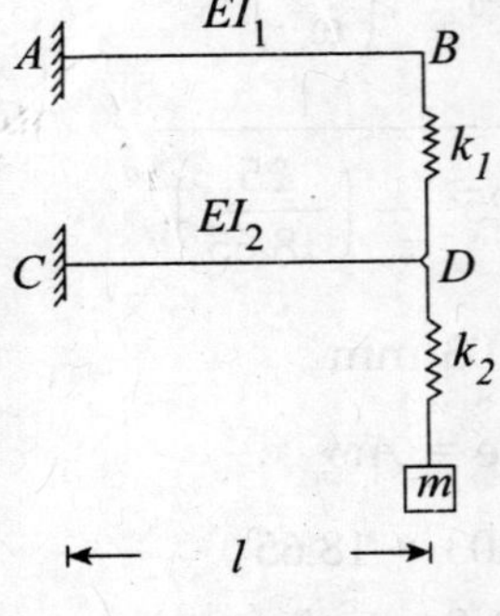

FIG. 9.33

Solution The points *B* and *D* of the beam experience the base force along with springs. Therefore, the beams and the springs are in series.

$$\frac{1}{k_e} = \frac{1}{k_{\text{beam}})_1} + \frac{1}{k_{\text{beam}})_2} + \frac{1}{k_1} + \frac{1}{k_2}$$

$$= \frac{1}{3EI_1/L^3} + \frac{1}{3EI_2/L^3} + \frac{1}{k_1} + \frac{1}{k_2}$$

$$= \frac{L^3}{3E}\left[\frac{1}{I_1} + \frac{1}{I_2} + \frac{1}{k_1} + \frac{1}{k_2}\right]$$

$$\frac{1}{k_e} = \frac{L^3}{3E}\left[\frac{k_1k_2(I_1 + I_2) + I_1 + I_2(k_1k_2)}{I_1I_2k_1k_2}\right]$$

$$k_e = \frac{-3EI_1I_2k_1k_2}{L^3[k_1k_2(I_1 + I_2) + I_1I_2(k_1 + k_2)]}$$

$$\omega_n = \sqrt{\frac{k_e}{m}}$$

Example 9.9 A mass of 0.01 kN is attached to a spring of stiffness of 0.7 kN/mm. Determine the natural frequency and coefficient of critical damping. (VTU, 2010)

Solution

$$m = 0.01 \text{ kN} = 1 \text{ kg}$$

and

$$\omega_n = \sqrt{\frac{0.7}{1}} = 0.837 \text{ rad/sec}$$

$$C_{cr} = 2m\omega_n = 1.4$$

Example 9.10 Determine the natural angular frequency, cyclic frequency and period of oscillation for spring mass system with mass 10 kg and stiffness 1000 N/m. If the system is given an initial displacement of 0.1 m and an initial velocity of 0.2 m/sec; obtain the equation of motion. Also find the displacement, velocity and acceleration after 0.2 sec. (VTU, Dec 2010)

Solution Natural angular frequency,

$$\omega_n = \sqrt{\frac{k}{m}}$$

$$\omega_n = \sqrt{\frac{1000}{10}} = 10 \text{ rad/sec}$$

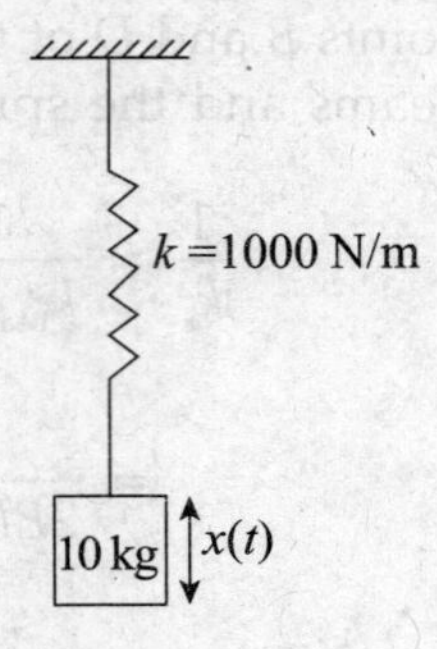

FIG. 9.34

$$\text{Cyclic frequency } f_n = \frac{\omega_n}{2\pi} = \frac{10}{2\pi} = 1.59 \text{ Hz (cycles)}$$

$$\text{Period of oscillation } T_n = \frac{1}{f_n} = \frac{2\pi}{\omega_n} = 0.628 \text{ sec}$$

The equation of motion is given by

$$x(t) = A \cos(\omega_n t - \beta)$$

$$A = \sqrt{x_0^2 + \left(\frac{\dot{x}_0}{\omega_n}\right)^2}$$

$$A = \sqrt{0.1^2 + \left(\frac{0.2}{10}\right)^2}$$

$$\boxed{A = 0.102}$$

$$\beta = \tan^{-1} \frac{\dot{x}_0}{\omega_n x_0} = \tan^{-1}\left(\frac{0.2}{10 \times 0.1}\right)$$

$$\boxed{\beta = 0.2 \text{ rads}}$$

$$x(t) = 0.102 \cos(10t - 0.2) \text{ m}$$

$$\dot{x}(t) = -0.102 \times 10 \sin(10t - 0.2) \text{ m/s}$$

$$\ddot{x}(t) = -0.102(10)10 \cos(10t - 0.2) \text{ m/s}^2$$

$$x(0.2) = -0.023 \text{ m}$$

$$\dot{x}(0.2) = -0.993 \text{ m/s}$$

and $$\ddot{x}(0.2) = 2.317 \text{ m/s}^2$$

Example 9.11 A mass of 45 kg is dropped on to the end of a cantilever beam, with a velocity of 0.5 m/sec. The mass sticks to the beam and vibrates with the beam. The I beam made of steel (E = 210(10)^9 N/m^2) is 3.5 m long, has a depth of 20 cm.

I = 3(10)$^{-5}$ m^4. Find the maximum stresses developed in the beam, as it vibrates.

(VTU, 2011).

Solution: This is an initial value problem $x_0 = 0$ and $\dot{x}_0 = 0.5$ m/sec. The stiffness of the cantilever beam (Table 9.1) is computed and used as follows.

45 kg

3.5 m

FIG. 9.35

$$k = 3EI/l^3$$

$$k = \frac{3(210)10^3 \times 3 \times 10^{-5}}{3.5^3}$$

$$k = 440.82(10)^3 \text{ N/m}$$

Natural angular frequency $= \omega_n = \sqrt{\dfrac{k}{m}}$

$$\omega_n = \sqrt{\frac{440.82(10)^3}{45}}$$

$$\omega_n = 98.97 \text{ rad/sec}$$

Maximum displacement,

$$A = \sqrt{x_0^2 + \left(\frac{\dot{x}_0}{\omega_n}\right)^2}$$

$$A = \sqrt{0 + \left(\frac{0.5}{98.97}\right)^2}$$

$$A = 5.052 \times 10^{-3} \text{ m}$$

Maximum force

$$P_{max} = kA$$

$$= 440.82(10)^3 \times 5.052 \times 10^{-3}$$

$$P_{max} = 2227 \text{ N}$$

$$\text{Maximum bending moment} = 2227 \times 3500 = 7.8(10)^6 \text{ N/mm}$$

$$\text{Maximum stress} = \frac{7.8\,(10)^6 \times 100}{3 \times 10^{-5} \times 10^{12}}$$

$$\text{Maximum stress} = 25.98 \text{ N/mm}^2$$

Example 9.12 Determine the natural frequency and period of the system as shown in Fig. 9.36. Take $E = 2.1(10)^5$ N/mm^2 and $I = 13(10)^6$ mm^4. (VTU, 2012)

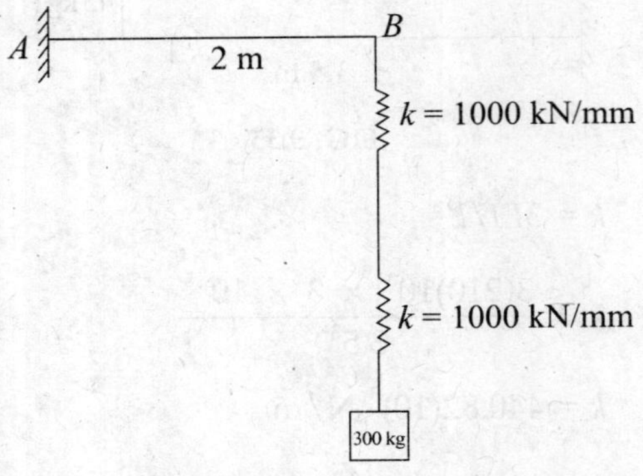

FIG. 9.36

Solution: Point B experiences the same force along with the springs. Therefore, the beam AB and the two springs are in series. The beam can be replaced by a spring k_1 of stiffness $48EI/l^3$ where EI is the flexural rigidity of the beam and l is the length. Thus, the three springs (beam + 2 springs) are in series and are replaced by an equivalent spring having stiffness k_{eq}.

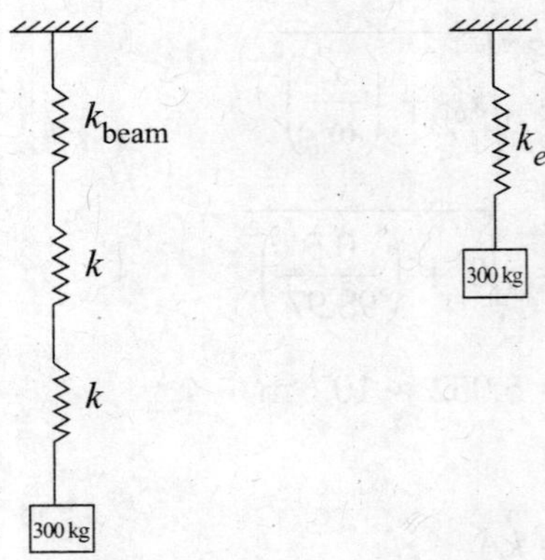

FIG. 9.37

$$k_{beam} = \frac{3EI}{l^3} = \frac{3(2.1)10^5 \times 13 \times 10^6}{2000^3} = 1023.75 \text{ N/mm}$$

i.e.
$$k_{beam} = 1023.75(10)^3 \text{ N/m}$$

$$\frac{1}{k_e} = \frac{1}{k_{beam}} + \frac{1}{k} + \frac{1}{k}$$

$$\frac{1}{k_e} = \frac{1}{1023.75(10)^3} + \frac{1}{1000(10)^6} + \frac{1}{1000(10)^6} \text{ N/m}$$

$$k_e = 1021.66(10)^3 \text{ N/m}$$

$$\omega_n = \sqrt{\frac{k_e}{m}} = \sqrt{\frac{1021.66(10^3)}{300}} = 58.36 \text{ rads/s}$$

$$T_n = \frac{2\pi}{\omega_n} = 0.108 \text{ secs}$$

9.10.3 Moment Distribution Procedure for Calculating the Lateral Translational Stiffness

- In some numerical problems, if the expression for deflection cannot be remembered then the stiffness can be calculated using the moment distribution method.
- In this method, a unit displacement is given at a point where the entire mass is assumed to be concentrated.
- The fixed end moments are calculated and the distribution factors are evaluated based on the boundary conditions.
- Moment distribution is carried out and the final moments are obtained.
- Using the free body diagram of the element, the translational stiffness is evaluated.

Example 9.13 Compute the translational stiffness of a simply supported beam having a central mass, *W*. The flexural rigidity *EI* is constant. Neglect the self weight of the member and use the moment distribution method.

Solution: Consider a fixed beam subjected to a unit deflection at midspan as shown in Fig. 9.38.

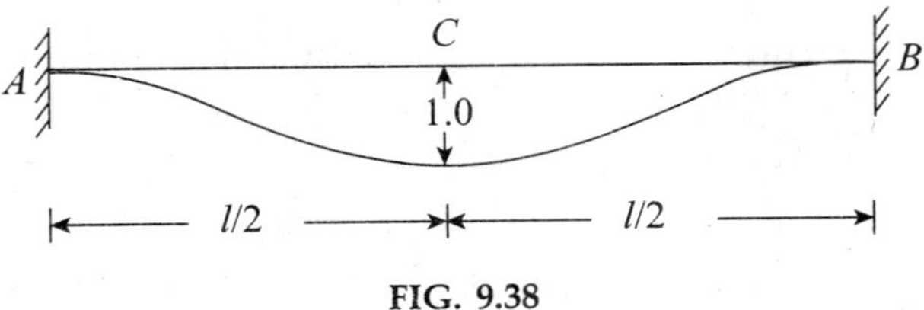

FIG. 9.38

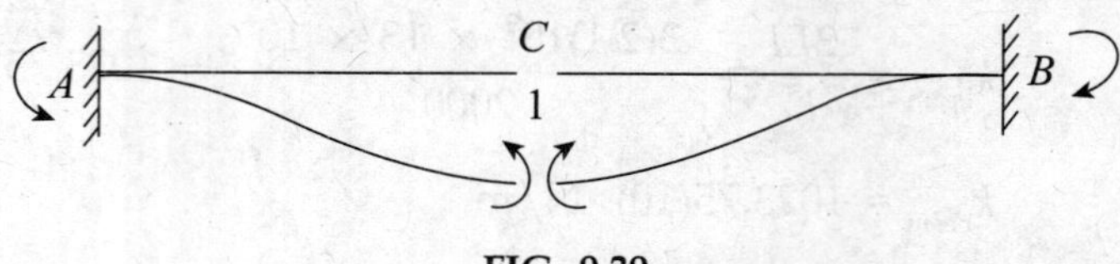

FIG. 9.39

The moments induced due to unit settlement (deflection) is computed as

$$M_{FAC} = M_{FCA} = \frac{-6EI(1)}{(l/2)^2} = \frac{-24EI}{l^2}$$

$$M_{FCB} = M_{FBC} = \frac{+6EI(1)}{(l/2)^2} = \frac{+24EI}{l^2}$$

(**Note:** The clockwise moments are taken as positive).

As the given beam is supported, as shown below. The moments are zero at the ends. Hence, the moment distribution is done as follows:

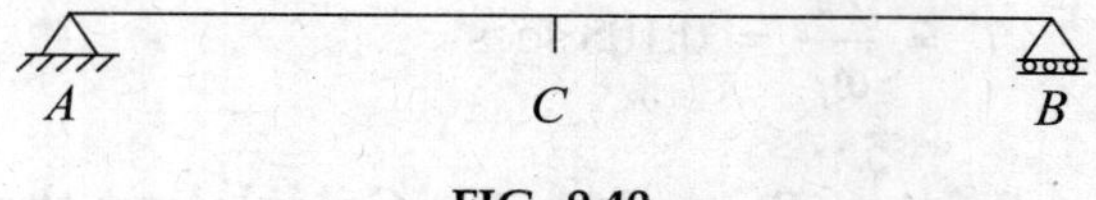

FIG. 9.40

−24	−24	+24	+24	(EI/l^2)
+24			−24	Balance
	+12	−12		Co
0	−12	+12	0	Final

Drawing the free body diagram;

FIG. 9.41

and the values of k_{CA} and k_{CB} are calculated by taking moments about A and B respectively.

$$\Sigma M_A = 0;$$

$$k_{CA}(l/2) - (12EI/l^2) = 0$$

$$k_{CA} = 24EI/l^3$$

$\Sigma M_B = 0;$

$$(12EI/l^2) - k_{CB}(l/2) = 0$$

$$k_{CB} = 24EI/l^3$$

Hence,
$$k = k_{CA} + k_{CB}$$

$$\boxed{k = 48EI/l^3}$$

Example 9.14 Determine the natural frequency of a fixed beam carrying a concentrated weight W at its centre as shown below. Neglect the mass of the beam.

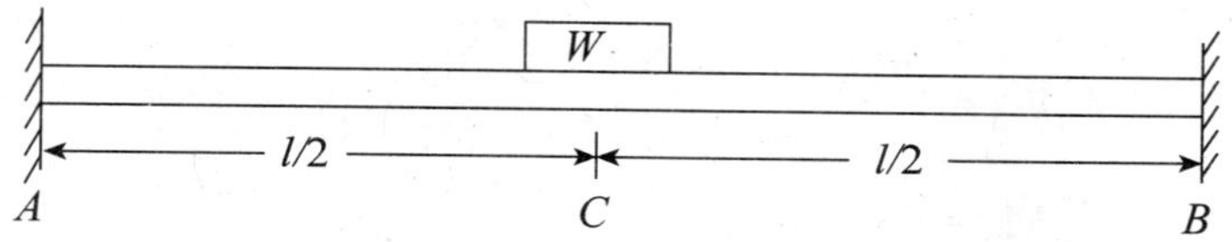

FIG. 9.42

Solution: Let the centre of the fixed beam settle by unity

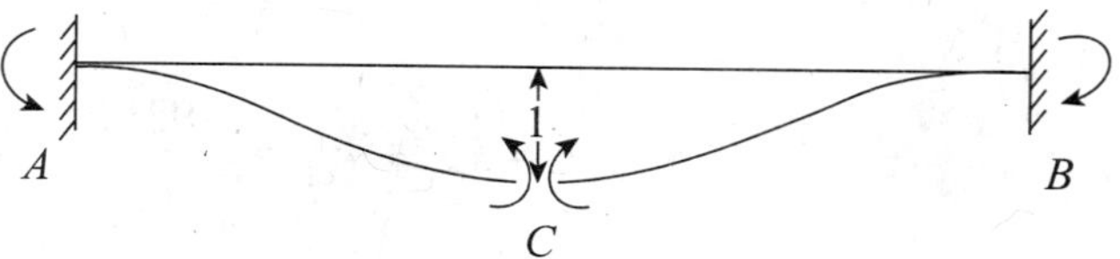

FIG. 9.43

The fixed end moments are obtained as

$$M_{FAC} = \frac{-6EI(1)}{(l/2^2)} = \frac{-24EI}{l^2}$$

$$M_{FCA} = -24EI/l^2$$

$$M_{FCB} = \frac{+6EI(1)}{(l/2)^2} = \frac{+24EI}{l^2}$$

$$M_{FBC} = +24EI/l^2$$

Drawing the free body diagram the translational stiffness is obtained as follows.

FIG. 9.44 Free body diagram

$$\Sigma M_A = 0$$

$$\frac{-24EI}{l^2} - \frac{-24EI}{l^2} + k_{CA}\left(\frac{l}{2}\right) = 0$$

$$k_{CA} = \frac{96EI}{l^3}$$

$$\Sigma M_B = 0$$

$$\frac{24EI}{l^2} + \frac{24EI}{l^2} - k_{CB}\left(\frac{l}{2}\right) = 0$$

$$k_{CB} = \frac{96EI}{l^3}$$

$$\therefore \quad k = k_{CA} + k_{CB} = \frac{192EI}{l^3}$$

$$\omega_n = \sqrt{\frac{k}{m}} = \sqrt{\frac{192EI}{l^3}\left(\frac{g}{W}\right)}$$

$$\omega_n = 8\sqrt{\frac{3EIg}{l^3W}}$$

$$f = \frac{\omega_n}{2\pi} = \frac{4}{\pi}\sqrt{\frac{3EIg}{Wl^3}}$$

Example 9.15 Obtain expressions for free vibration characteristics of a propped cantilever subjected to a central mass. The flexural rigidity EI is uniform throughout.

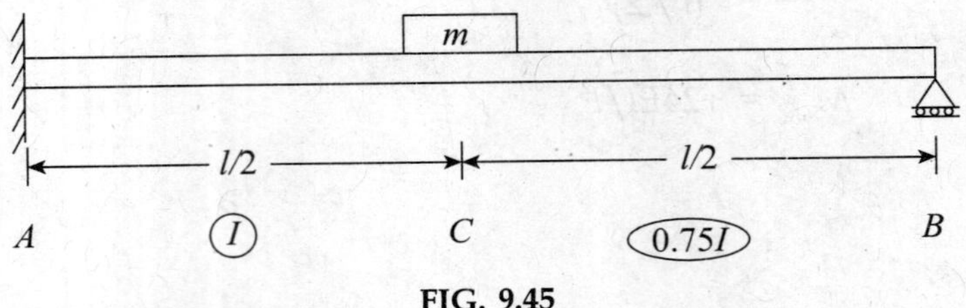

FIG. 9.45

Solution: The propped cantilever beam is non-prismatic as moment of inertia of *AC* is *I* and that of *CB* is 0.75*I* and that moment distribution can be used to determine the translational can be used to determine the translational stiffness. Consider the corresponding fixed beam and apply a unit displacement at *C*.

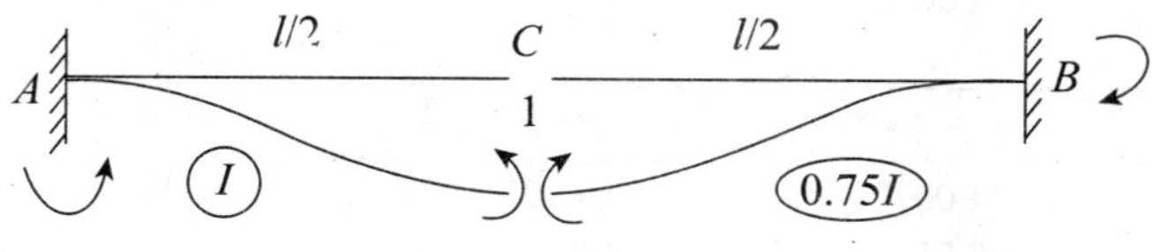

FIG. 9.46

The moments developed due to unit vertical displacements are

$$M_{FAC} = \frac{-6EI}{(l/2)^2} = \frac{-24EI}{l^2}$$

$$M_{FCA} = \frac{-6EI}{(l/2)^2} = \frac{-24EI}{l^2}$$

$$M_{FCB} = \frac{+6E(0.75I)}{(l/2)^2} = \frac{+18EI}{l^2}$$

$$M_{FBC} = \frac{+6E(0.75I)}{(l/2)^2} = \frac{+18EI}{l^2}$$

Distribution factors

$$S_{CA} = \frac{I}{(l/2)} = \frac{2I}{l}$$

$$S_{CB} = \frac{0.75I}{(l/2)} = \frac{1.5I}{l}$$

$$d_{CA} = \frac{S_{CA}}{S_{CA} + S_{CB}} = 0.57$$

$$d_{CB} = \frac{S_{CB}}{S_{CA} + S_{CB}} = 0.43$$

Moment distribution table

AB	*CA*	*CB*	*BC*	*Members*
0	0.57	0.43	1	Distn Factor
−24.00	−24.00	+18.00	+18.00	FEMS
			−18.00	Bal
	–	−9.00	0.00	Co
−24.00	−24.00	+9.00	0.00	Total
	+8.55	+6.45		Bal
+4.56				Co

Free body diagram

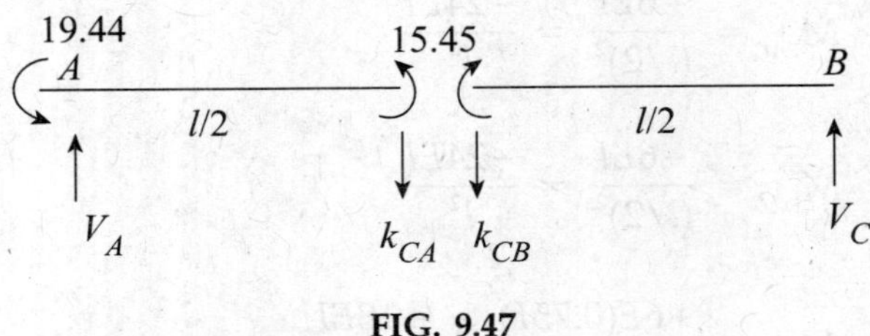

FIG. 9.47

Referring to the free body diagram and taking moments about *A*, the value of k_{CA} is obtained. Similarly, by taking moment about *B*, we obtain the value of k_{CB}. Thus, the translational stiffness is obtained by the summation of stiffnesses of k_{CA} and k_{CB} respectively.

$$\Sigma M_A = 0$$

$$(-19.44 - 15.45)\frac{EI}{l^2} + k_{CA}\left(\frac{l}{2}\right) = 0$$

$$k_{CA} = 69.78\frac{EI}{l^3}$$

$$\Sigma M_B = 0$$

$$+15.45EI/l^2 - k_{CB}(l/2) = 0$$

$$k_{CB} = 30.9EI/l^3$$

Thus, $$k = k_{CA} + k_{CB} = 100.68EI/l^3$$

$$\omega_n = \sqrt{\frac{k}{m}} = \sqrt{\frac{100.68EIg}{Wl^3}}$$

$$T_n = \frac{2\pi}{\omega_n} = 2\pi\sqrt{\frac{Wl^3}{100.68EIg}}$$

$$f_n = \frac{1}{2\pi}\sqrt{\frac{100.68EIg}{Wl^3}}$$

Table 9.1 Stiffness of the springs

S. No.	Type	Details	Stiffness
1.	k_1, k_2	Springs in series	$k = \dfrac{k_1k_2}{k_1 + k_2}$
2.	k_1, k_2	Springs in parallel	$k = k_1 + k_2$
3.	l	Axial stiffness	$k = AE/l$
4.	l, k	Cantilever with unit deflection at free end	$k = 3EI/l^3$
5.	$l/2$, k, $l/2$	Hinge and roller support with central load	$k = 48EI/l^3$
6.	a, k, b	Simply supported beam with eccentric load	$k = 3EIl/a^2b^2$
7.	1	Fixed beam with unit settlement at ends	$k = 12EI/l^3$
8.	1	Fixed end settles by unity in a propped cantilever	$k = 3EI/l^3$
9.	$l/2$, k, $l/2$	Fixed beam with central deflection of unity	$k = 192EI/l^3$

Example 9.16 A steel beam of length 1 metre carries a mass of 50 kg at its free end as shown below. Find the free vibration characteristics of the system. It can be assumed that the weight of the member is negligible. When compared to the mass.

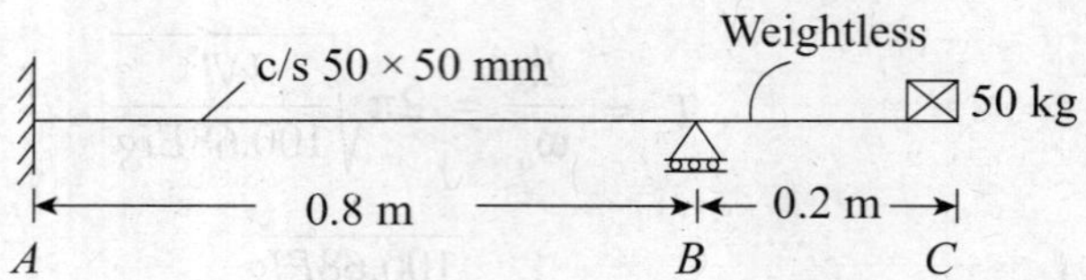

FIG. 9.48

Solution: The stiffness of the beam at the free end is calculated based on moment distribution method. As the end is free end, the value of M_c is zero. Hence, provide a roller support at C and moment distribution is carried out by satisfying the above condition that M_c is zero. The moments due to unit vertical displacement are found and moment distribution is done.

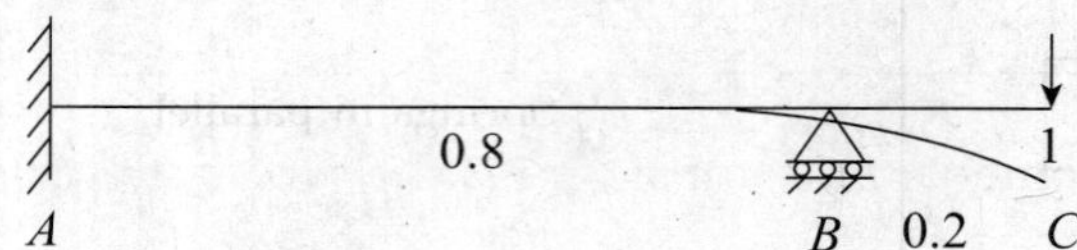

FIG. 9.49

Fixed end moments

Due to unit displacement at C;

$$M_{FBC} = \frac{-6EI(1)}{0.2^2} = -150EI$$

$$M_{FCB} = \frac{-6EI(1)}{0.2^2} = -150EI$$

Distribution factor

Assuming an imaginary roller support at C

$$S_{BA} = I/0.8 = I/0.8 = 1$$

$$S_{BC} = \left(\frac{3}{4}\right)\left(\frac{I}{0.2}\right) = \frac{3I}{0.8} = 3$$

$$k_{BA} = \frac{S_{BA}}{S_{BA} + S_{BC}} = 0.25$$

$$k_{BC} = \frac{S_{BC}}{S_{BA} + S_{BC}} = 0.75$$

Moment distribution table

AB	*BA*	*BC*	*CB*	*Members*
0.00	0.25	0.75	1.00	DF
		−150.00	−150.00	FEM
			+150.00	Bal
		+75.00		Co
		−75.00		Total
	+18.75	+56.25		Bal
+9.38				Co
+9.38	+18.75	−18.75	0.00	Final

Using the above moments and considering the free body diagram of *BC*

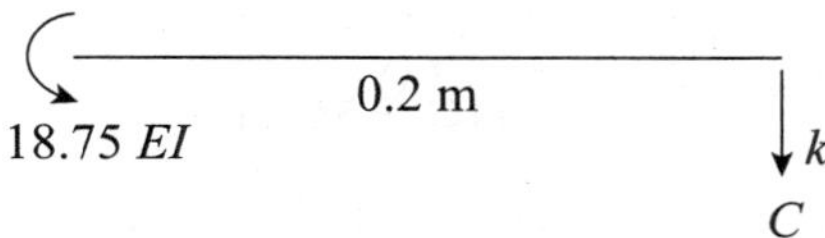

FIG. 9.50

$$\Sigma M_B = 0$$

$$k(0.2) - 18.75EI = 0$$

$$k = 93.75EI$$

$$k = 93.75 \times 10^6 (2 \times 10^5) \times \frac{50 \times 50^3}{12 \times 10^{12}} = 9.765 \times 10^3 \text{ N/m}$$

$$\omega_n = \sqrt{\frac{9.765(10^3)}{50 \times 10^3/9.81}} = 1.38$$

$$f_n = \frac{1.38}{2\pi} = 0.22$$

Example 9.17 A vibration test was conducted on the model of an tank. A cable is attached to the tank induces a force of 16.4 N horizontally and pulls the tank by 2 mm. The cable is cut and the resulting free vibration is recorded. At the end of four complete cycles, the time elapsed is 2 seconds and the amplitude is 0.5 mm. Based on the above data, compute the damping ratio, natural period of vibration (undamped), stiffness, weight of the system, damping coefficient. Number of cycles required for the displacement amplitude to decrease to $\left(\frac{1}{10}\right)^{th}$ of the initial amplitude and the amplitude after 20 cycles.

Solution: The mathematical model and the response representing a problem is indicated in Fig. 9.51.

$$x_0 = 2 \text{ mm} \quad ; \quad x_4 = 0.5 \text{ mm}$$

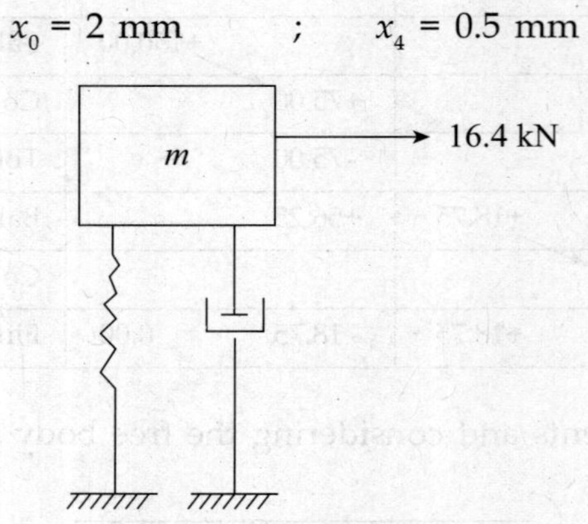

FIG. 9.51

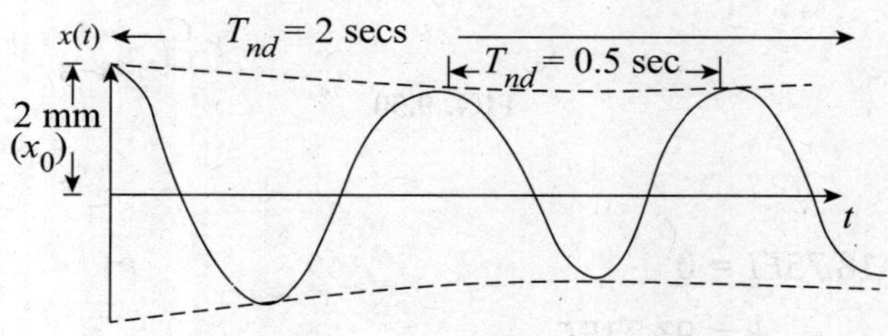

FIG. 9.52

Stiffness $$k = \frac{16.4}{2} = 8.2 \text{ N/mm} = 8.2 \text{ kN/m}$$

$$T_{nd} = \frac{2}{4} = 0.5 \text{ sec}$$

Logarithmic decrement

$$\delta = \frac{1}{N} \log_e \left(\frac{x_0}{x_N}\right)$$

$$\delta = \frac{1}{4} \log_e \left(\frac{2}{0.5}\right)$$

$$\delta = 0.346$$

$$\xi = \frac{\delta}{\sqrt{\delta^2 + 4\pi^2}} = \frac{0.346}{\sqrt{0.346^2 + 4\pi^2}}$$

$$\boxed{\xi = 0.0551(5.51\%)}$$

$$\omega_{nD} = \frac{2\pi}{T_{nD}} = \frac{2\pi}{0.5} = 12.57 \text{ rad/sec}$$

$$\omega_{nD} = \omega_n\sqrt{1 - \xi^2}$$

$$\boxed{\omega_n = 12.58 \text{ rad/sec}}$$

$$T_n = \frac{2\pi}{\omega_n} = \frac{2\pi}{12.58}$$

$$\boxed{T_n = 0.499}$$

$$\omega_n^2 = \frac{k}{m}$$

$$m = \frac{8.2}{(12.58^2)}$$

$$\boxed{m = 51.8 \text{ kg}}$$

$$W = mg$$

$$W = 51.8 \times 9.81$$

$$\boxed{W = 508.15 \text{ N}}$$

$$\xi = \frac{c}{c_{cr}}$$

$$\therefore \quad c = \xi \times 2 \times \sqrt{km}$$

$$c = 0.0551 \times 2 \times \sqrt{8.2 \times 508.15 \times 10^3}$$

$$\boxed{c = 71.8 \text{ NS/m}}$$

Number of cycles required

$$N = \frac{1}{8}\log_e\left(\frac{x_0}{x_N}\right)$$

$$= \frac{1}{0.346} \log_e \left(\frac{2}{0.5} \right)$$

$$\boxed{N = 4 \text{ cycles}}$$

Amplitude for 20 cycles

$$\frac{x_0}{x_N} = e^{N\delta}$$

$$\frac{2}{x_N} = e^{20 \times 0.346}$$

$$\boxed{x_N = 1.97(10)^{-3} \text{ mm}}$$

Example 9.18 A machine has a mass of 200 kg. It is placed on an isolator and the free vibration record is shown below. Determine the damping characteristics and the undamped natural frequency.

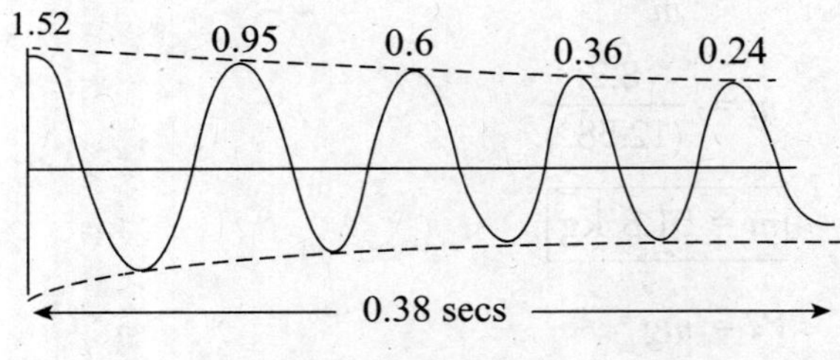

FIG. 9.53

Solution

$$T_{nD} = \frac{0.38}{4} = 0.095$$

$$\omega_{nD} = \frac{2\pi}{T_{nD}} = 66.13$$

$$\omega_{nD} = \omega_n \sqrt{1 - \xi^2}$$

$$\omega_{nD} = \frac{66.13}{\sqrt{1 - \xi^2}}$$

$$\delta = \frac{1}{N} \log_e \left(\frac{x_0}{x_N} \right)$$

$$\delta = \frac{1}{4}\log_e\left(\frac{1.52}{0.24}\right)$$

$$\delta = 0.4614$$

$$\xi = \frac{\delta}{\sqrt{\delta^2 + 4\pi^2}} = 0.0745$$

$$\omega_n = \frac{66.13}{\sqrt{1 - 0.0745^2}}$$

$$\boxed{\omega_n = 66.312 \text{ rad/s}}$$

9.11 FREE VIBRATION OF DAMPED STRUCTURAL DYNAMIC FORCE SYSTEM

Practical structural systems always possess a small amount of energy dissipating capacity. This energy dissipating capacity is represented by viscous damping. Mathematical model, free body diagram and the differential equation of motion is indicated.

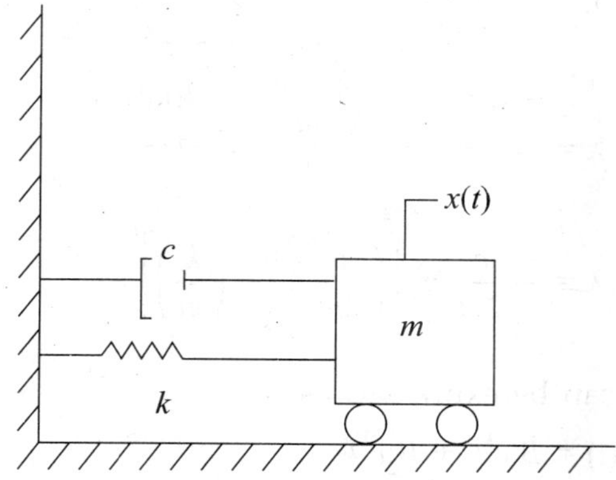

FIG. 9.54 Idealisation of damped system

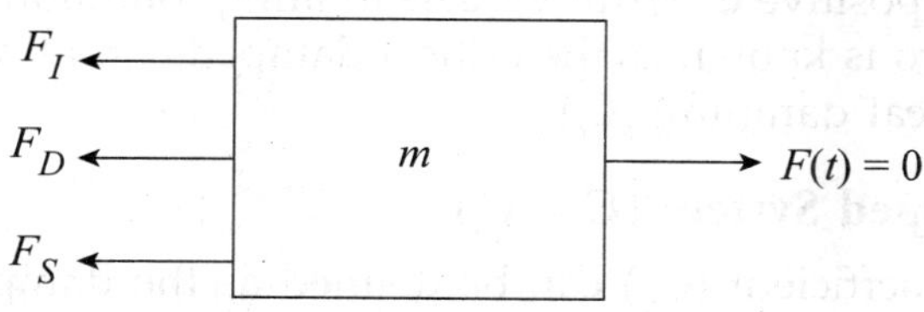

FIG. 9.55

$$F_I + F_D + F_S = 0 \qquad (1)$$

$$m\ddot{x} + c\dot{x} + kx = 0 \qquad (2)$$

The above equation is a linear ordinary differential equation with constant coefficient and order of two. The solution for the above differential equation can be expressed as

$$x\ (t) = Be^{\lambda t} \tag{3}$$

$$\dot{x}(t) = B\lambda e^{\lambda t} \tag{4}$$

$$\ddot{x}(t) = \lambda^2 Be^{\lambda t} \tag{5}$$

Substituting Eq. (2) in Eqs. (3–5);

$$m(B\lambda^2 e^{\lambda t}) + c(B\lambda e^{\lambda t}) + k(Be^{\lambda t}) = 0 \tag{6}$$

$$Be^{\lambda t}\ (m\lambda^2 + c\lambda + k) = 0 \tag{7}$$

As $Be^{\lambda t} \neq 0$

$$m\lambda^2 + c\lambda + k = 0 \tag{8}$$

The above equation is known as characteristic equation and is quadratic in λ. Dividing by m

$$\lambda^2 + \left(\frac{c}{m}\right)\lambda + \frac{k}{m} = 0 \tag{9}$$

$$\lambda = \frac{-(c/m) \pm \sqrt{\left(\frac{c}{m}\right)^2 - 4(k/m)}}{2} \tag{10}$$

i.e.

$$\lambda = -\frac{c}{2m} \pm \left[\left(\frac{c}{2m}\right)^2 - \left(\frac{k}{m}\right)\right]^{1/2} \tag{11}$$

The general solution can be expressed as

$$x(t) = B_1 e^{\lambda_1 t} + B_2 e^{\lambda_2 t} \tag{12}$$

The actual solution depends on the sign of the expression under the radical (square root). This can be zero, positive or negative. The limiting condition in which the quantity under the radical is zero is known as the critical damped condition. The corresponding damping is called critical damping (C_r).

9.11.1 Critical Damped System ($C = C_r$)

The critical damping coefficient (C_{cr}) can be defined as the damping coefficient which makes the expression under radical as zero

$$\left(\frac{C_{cr}}{2m}\right)^2 - \frac{k}{m} = 0 \tag{13}$$

$$C_{cr}^2 = 4mk \tag{13a}$$

$$C_{cr} = 2\sqrt{km} \tag{14}$$

Unit for C_{cr}

$$\sqrt{km} = \sqrt{\frac{\text{kN}}{\text{m}} \times \frac{\text{kN}}{\text{m/sec}^2}} = \text{kNsec/m} = \text{Ns/m}$$

The critical damping coefficient can also be expressed in terms of natural frequency

$$\omega_n^2 = \frac{k}{m} \tag{15}$$

$$k = m\,\omega_n^2 \tag{15a}$$

Substituting Eq. (14);

$$C_{cr} = 2\sqrt{m\omega_n^2 m} \tag{16}$$

$$\boxed{C_{cr} = 2m\omega_n} \tag{17}$$

In a critically damped system, expression in the radical is zero.
In critically damped system, root of the characteristic equation is zero.

$$\lambda_1 = \lambda_2 = -\frac{C_{cr}}{2m} = \frac{2m\omega_n}{2m} = \omega_n \tag{18}$$

$$\lambda_1 = \lambda_2 = -\omega_n \tag{19}$$

For a case of the repeated roots, the solution of the differential equation can be expressed as

$$x(t) = A_1 e^{-\omega_n t} + A_2 t e^{-\omega_n t} \tag{20}$$

$$x(t) = e^{-\omega_n t}(A_1 + A_2 t) \tag{21}$$

The above solution does not show any oscillation. Therefore, a critically damped system does not vibrate. The response decreases as the time increases and it becomes zero as $t \to \infty$. The displacement becomes to zero with an exponential dacay.

A critical damping system is a condition having minimum amount of damping so that no vibration occurs in the free response. It represents the dividing line between oscillatory and non-oscillatory motions.

The constants A_1 and A_2 can be evaluated if required based on the initial condition at $t = 0$

At $\quad t = 0; \quad x(0) = x_0 \quad (22)$

$$\dot{x}(0) = \dot{x}_0 \tag{23}$$

The solution can be expressed after finding the constants

$$A_1 = x_0 \tag{24}$$

$$A_2 = \dot{x}_0 + \omega_n x_0 \tag{25}$$

and

$$x(t) = [x_0(1 + \omega_n)t + x_0 t]e^{-\omega_n t} \tag{26}$$

The natural frequency term in the above expression is only hypothetical.

Note: Critical damped systems are useful in the design of instruments to measure dead weights and recoil springs in guns.

9.11.2 Overdamped Systems

In an overdamped systems, the damping present will be greater than the critical damping ($C > C_{cr}$). The expression under the radical is positive and the two roots of the characteristic equation are real and distinct.

A dimensionless factor called damping ratio (or damping factor) can be defined as the ratio of the damping coefficient C to C_{cr} denoted by the symbol ξ (*Zeta*)

$$\xi = \frac{C}{C_r} = \frac{C}{2m\omega_n} \tag{27}$$

$$C = 2m\omega_n\xi \tag{28}$$

Substituting for C in the expression for the general solution, the constants λ_1 and λ_2

$$\lambda_1, \lambda_2 = \frac{-2m\omega_n\xi}{2m} \pm \sqrt{\left(\frac{2m\omega_n\xi}{2m}\right)^2 - \left(\frac{k}{m}\right)} \tag{29}$$

$$\lambda_1, \lambda_2 = -\omega_n\xi \pm \omega_n\sqrt{\xi^2 - 1} \tag{30}$$

For overdamped system, $\xi > 1$ expressing

$$\omega_n\sqrt{\xi^2 - 1} = \omega_{0D} \tag{31}$$

where ω_{0D} is the overdamped natural frequency.

From Eqs. (30) and (31), we obtain

$$\lambda_1, \lambda_2 = -\omega_n\xi \pm \omega_{0D} \tag{32}$$

Substituting for λ_1 and λ_2 in the general solution in terms of ω_{0D}, we get

$$x(t) = B_1 e^{\lambda_1 t} + B_2 e^{\lambda_2 t} \tag{33}$$

$$x(t) = B_1 e(-\omega_n\xi + \omega_{0D})^t + B_2 e^{-\omega_{0D} t} \tag{34}$$

i.e.

$$x(t) = e^{-\omega_n \xi} [B_1 e^{\omega_{0D} t} + B_2 e^{-\omega_{0D} t}] \tag{35}$$

The constants B_1 and B_2 are evaluated based on the initial condition;

At $t = 0$; $x(0) = x_0$ (36)

Using the above condition

$$B_1 + B_2 = x_0 \tag{37}$$

At $t = 0$; $\dot{x}(0) = \dot{x}_0$ (38)

Using the above condition;

$$\dot{x}_0 = e^{-\xi\omega_n t} [B_1 e^{\omega_{0D} t} (\omega_{0D} - \xi\omega_n) - B_2 e^{-\omega_{0D} t} (\omega_{0D} + \xi\omega_n)] \tag{39}$$

In the above expression $t = 0$

$$\dot{x}_0 = B_1(\omega_{0D} - \xi\omega_n) - B_2(\omega_{0D} - \xi\omega_n) \tag{40}$$

Solving Eqs. (37) and (40) the constants B_1 and B_2 can be evaluated. The response decreases with increase in time t and becomes zero as $t \to \infty$. As in critically damped system, the response of an overdamped system does not show any oscillations. The displacement $x(t)$ will not change the sign above time. With initial conditions, a overdamped system will show the displacement returning to zero with an exponential dacay. Since, damping is greater than critical damping, the return to the equilibrium position will delay. Greater the damping, greater is the resistance to motion and more time will be required to return to equilibrium position. Overdamping system finds applications in automatic closing of doors. Typical response of C_{cr} and overdamping system are shown in Fig. 9.56.

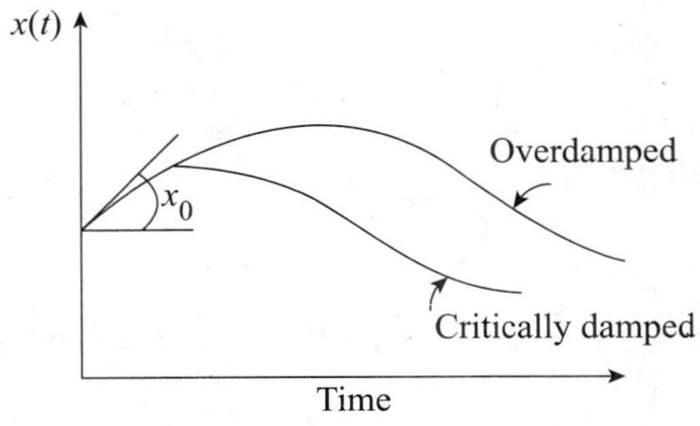

FIG. 9.56

9.11.3 Underdamped Systems

This is the normal case that exists in the physical systems. The damping present in an underdamped system is less than C_{cr}($C < C_{cr}$). Most of the physical systems have damping less than C_{cr}. Therefore, under normal situations, if a system is said to be damped it is considered undamped system. The expression under the radical is negative. The roots λ_1 & λ_2 can be expressed as

$$\lambda_1, \lambda_2 = -\xi\omega_n \pm \sqrt{\xi^2 - 1}\,\omega_n$$

For an undamped system, $\xi < 1$

λ_1, λ_2 is expressed as

$$\lambda_1, \lambda_2 = -\xi\omega_n \pm i\omega_n \sqrt{1 - \xi^2}\,(\xi < 1)$$

Expressing

$$\omega_{nD} = \omega_n \sqrt{1 - \xi^2}$$

where, ω_{nD} = Damped natural frequency

$$\lambda_1, \lambda_2 = -\xi\omega_n \pm i\omega_{nD}$$

Damped natural frequency will always be less than undamped natural frequency. Since, the damping ratios in the structural system are usually small, the variations between the two natural frequencies will be negligible. For example, if ξ = 20% then

$$\omega_{nD} = \omega_n \sqrt{1 - 0.2^2} = 0.98\omega_n$$

A plot of $\left(\frac{\omega_{nD}}{\omega_n}\right)$ and damping ratio is shown in Fig. 9.57.

$$\left(\frac{\omega_{nD}}{\omega_n}\right) = \sqrt{1 - \xi^2}$$

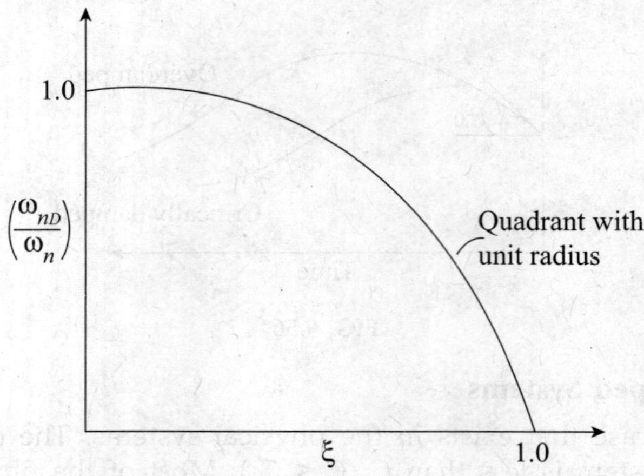

FIG. 9.57

Note: The decrease in ω_n with ξ is small initially and it becomes steep as the value of ξ increases for lightly damped systems as in the case of RCC and steel structures

$$\frac{\omega_{nD}}{\omega_n} \simeq 1$$

Hence,
$$\omega_{nD} \simeq \omega_n$$

Substituting for λ_1 and λ_2 in the general expression of response, we get

$$x(t) = B_1 e^{(-\xi\omega_n t + i\omega_{nD} t)} + B_2 e^{(-\xi\omega_n t - i\omega_{nD} t)}$$

$$x(t) = e^{-\xi\omega_n t}\,[B_1 e^{i\omega_{nD} t} + B_2 e^{-i\omega_{nD} t}]$$

The above equation is simplified using Euler's equation

$$e^{\pm i\theta} = \cos\theta \pm i\sin\theta$$

Thus,

$$x(t) = e^{-\xi\omega_n t}\,[C_1 \cos\omega_{nD} t + C_2 \sin\omega_{nD} t]$$

where C_1 and C_2 are the new constants to be obtained from the initial condition.

The quantity inside the bracket represent "simple harmonic motion" with natural angular while $e^{-\xi\omega_{nD}}$ represents a damper to the amplitude of motion, that is, the amplitude decreases as 't' increases to evaluate the constants C_1 and C_2 based on the initial conditions.

$$C_1 = x_0$$

and
$$C_2 = \frac{x_0\xi\omega_n + \dot{x}_0}{\omega_{nD}}$$

$$x(t) = e^{-\xi\omega_n t}\,[C_1 \cos\omega_{nD} t + C_2 \sin\omega_{nD} t].$$

At $t = 0$; $x(0) = x_0$ and hence

$$\boxed{C_1 = x_0}$$

$$\dot{x}(0) = \dot{x}_0$$

and using the above condition

$$\boxed{C_2 = \frac{x_0\xi\omega_n + \dot{x}_0}{\omega_{nD}}}$$

Substituting the values of C_1 and C_2:

$$x(t) = e^{-\xi\omega_n t}\left[x_0 \cos\omega_{nD} t + \left(\frac{x_0\xi\omega_n + x_0}{\omega_{nD}}\right)\sin\omega_{nD} t\right]$$

Amplitude of free-damped vibrations.

Consider the vector representation of initial conditions and the amplitude as shown in Fig. 9.58.

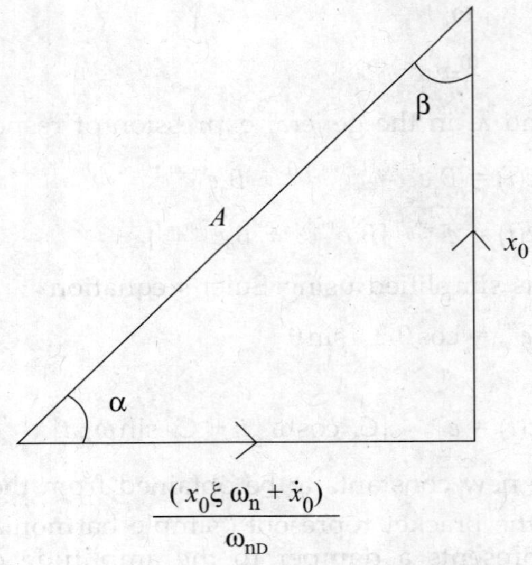

FIG. 9.58

Substituting the above values in the response expression,

we get

$$x(t) = e^{-\xi\omega_n t} A[\cos\omega_{nD}t\cos\beta + \sin\omega_{nD}t\sin\beta]$$

Using the trigonometric relationship;

$$\cos A\cos B + \sin A\sin B = \cos(A - B)$$

$$x(t) = Ae^{-\xi\omega_n t}[\cos(\omega_{nD}t - \beta)]$$

In the above expression A represents the amplitude which always be diminishing with respect to time and β represents the phase angle by which the resultant portion lags behind the initial displacement component.

From the trigonometric transformations, we note that

$$A^2\sin^2\beta + A^2\cos^2\beta = (x_0)^2 + \left(\frac{x_0\xi\omega_n + \dot{x}_0}{\omega_{nD}}\right)^2$$

$$A = \sqrt{x_0^2\left(\frac{x_0\xi\omega_n + \dot{x}_0}{\omega_{nD}}\right)^2}$$

and $$\tan\beta = \frac{x_0 \xi \omega_n + \dot{x}}{\omega_{nD} x_0}$$

A plot of free vibration response for an undamped system is indicated as shown in Fig. 9.59.

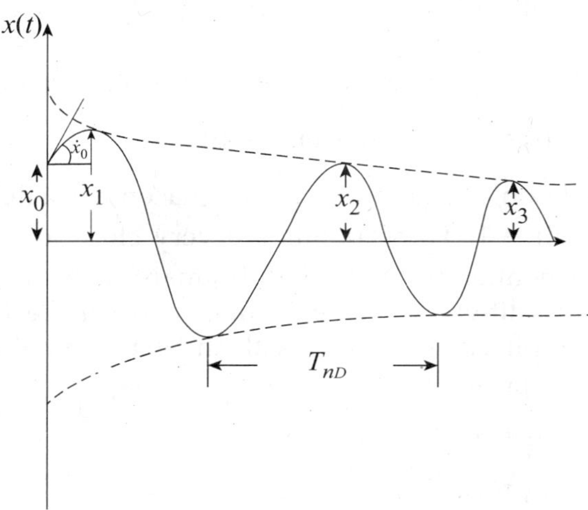

FIG. 9.59

From the above response curve, we note the mass vibrates with a constant angular frequency ω_{nD} about a station equilibrium position, due motion is oscillatory but not periodic. The amplitude of vibration goes on decreasing in an exponential manner and the reduction in the magnitude of the amplitude for each successive cycle given a measure of damping in the system. However, the reduction in magnitude of the amplitude for each successive cycle gives a measure of damping in the system. However, the reduction in magnitude of the amplitude occurs at equal intervals of time and this is known as damped natural period of vibration T_{nD}. Mathematically, T_{nD} can be expressed as

$$T_{nD} = \frac{2\pi}{\omega_{nD}}$$

i.e. $$T_{nD} = \frac{2\pi}{\omega_n \sqrt{1 - \xi^2}}$$

The damped natural period will be marginally greater than the undamped natural period for small values of damping ratio. For many of the practical system where

$$\xi \leq 20\%, \qquad T_{nD} \simeq T_n$$

9.11.4 Logarithmic Decrement (δ)

It is a concept used to measure the damping ratio of a single degree freedom (SDF) system. It is a practical method of finding damping in the system by initiating free vibrations and obtain the rate of dacay of amplitude of motion. It is a measure of the decay of two successive amplitudes constituting one cycle.

Logarithmic decrement is defined as the natural logarithm of the two successive amplitudes constituting one cycle denoted by δ

Mathematically, $$\delta = \log_e \frac{x_1}{x_2}$$

In the above expression, x_1 and x_2 represent two successive peak amplitudes

Consider $$x(t) = Ae^{-\xi\omega_{nD}t}\,[\cos\omega_{nD}t - \beta]$$

From the above equation, we note that the displacement will be on the exponential curve when $(\cos\omega_{nD}t - \beta) = 1$. These points are very close to the peak displacement. The points on the exponential curve is slightly/marginally to the right of the peak amplitude. However, the difference is very small and can be ignored. It is usual to assume that the peak amplitude coincides with the exponential curve.

Using the above concept, we have

$$x_1 = Ae^{-\xi\omega_n t_1}$$

$$x_2 = Ae^{-\xi\omega_n (t_1 + T_{nD})}$$

As $x_1 > x_2$;

$$\frac{x_1}{x_2} = e^{\xi\omega_n T_{nD}}$$

Taking natural logarithm on both sides

$$\log_e\left(\frac{x_1}{x_2}\right) = \log_e(e^{\xi\omega_n T_{nD}})$$

$$\delta = \log_e\left(\frac{x_1}{x_2}\right) = \xi\omega_n T_{nD}$$

$$\delta = \frac{\xi\omega_n 2\pi}{\omega_n\sqrt{1-\xi^2}}$$

$$\boxed{\delta = \frac{2\pi\xi}{\sqrt{1-\xi^2}}}$$

From the above expression, we note that the logarithmic decrement representing the ratio of two successive amplitudes constituting one cycle in an underdamped system is constant and is a function of damping ratio only.

For low damping,

$$\sqrt{1 - \xi^2} \simeq 1$$

$$\delta = 2\pi\xi \text{ (or)}$$

$$\xi = \frac{\delta}{2\pi}$$

Consider $$\frac{x_1}{x_2} = e^{\xi\omega_n T_{nD}}$$

$$\frac{x_1}{x_2} = e^{\delta}$$

i.e. $$\delta = \xi\omega_n T_{nD}$$

If a system has 5% of critical damping, then the logarithmic decrement, δ will be using the expression,

$$\delta = 0.1\pi$$

Which indicates the successive peaks,

$$\frac{x_1}{x_2} = e^{\delta} = e^{0.1\pi}$$

$$\frac{x_1}{x_2} = 1.369 \simeq 1.37$$

It indicates that the successive peak is 1.37 times the preceding one; in other words, each peak will have a value of 0.73 times the preceding one.

Therefore, the effect of damping can easily be visualised through the logarithmic decrement. To find the exact expression for the damping ratio,

Consider

$$\delta = \log_e \frac{x_1}{x_2}$$

$$\delta = \frac{2\pi\xi}{\sqrt{1 - \xi^2}}$$

Squaring on both sides and simplifying

$$\xi = \frac{\delta}{\sqrt{\delta^2 + 4\pi^2}}$$

For lightly damped systems, better accuracy can be obtained in finding the damping ratio. By considering the response peaks which are n cycles apart.

If x_0 represents the amplitude at the particular maxima (usually the initial displacement given) and x_N represents the amplitude after N cycles;

$$\frac{x_0}{x_N} = \left(\frac{x_0}{x_1}\right)\left(\frac{x_1}{x_2}\right)\left(\frac{x_2}{x_3}\right) \ldots\ldots \left(\frac{x_{n-1}}{x_n}\right)$$

$$\frac{x_0}{x_N} = e^{\delta} \times e^{\delta} \times e^{\delta} \times \ldots \times e^{\delta}$$

$$\frac{x_0}{x_N} = e^{N\delta}$$

Taking the natural logarithm on both sides, we have

$$\log_e\left(\frac{x_0}{x_N}\right) = N\delta$$

$$\boxed{\delta = \frac{1}{N}\log_e\left(\frac{x_0}{x_N}\right)}$$

Example 9.19 A mass of 10 kg is resting on a spring of stiffness 10 N/m and a damper with damping coefficient 0.135 Ns/mm. If a velocity of 10 cm/sec is applied to the mass at rest position, what will be the position at the end of 1 sec.

Solution: Assuming vibrations in vertical direction, the initial conditions are

$$y_0 = y(0) = 0$$

$$\dot{y}_0 = \dot{y}(0) = 0.1 \text{ m/s}$$

and required is

$$y(t) = y(1) = ?$$

We know

$$\xi = \frac{C}{C_{cr}} = \frac{135}{2\sqrt{km}} = \frac{135}{2\sqrt{10 \times 10}} = 6.75$$

Here $\xi > 1$, i.e., overdamped system

The response of a overdamped system is given by

$$y(t) = e^{-\xi\omega_n t}\left[B_1 e^{\omega_{OD} t} + B_2 e^{-\omega_{OD} t}\right]$$

$$\omega_{OD} = \omega_n\sqrt{\xi^2 - 1}$$

and

$$\omega_n = \sqrt{\frac{k}{m}} = 1 \text{ rad/sec}$$

$$\omega_{0D} = 1\sqrt{6.75^2 - 1}$$

$$\omega_{0D} = 6.675 \text{ rad/sec}$$

Substituting initial conditions in the response equation
We obtain

Put $\qquad t = 0; \qquad y(0) = 0$

i.e. $\qquad B_1 + B_2 = 0 \qquad (1)$

$$\boxed{B_1 = -B_2}$$

$$\dot{y}_0(t) = e^{-\xi\omega_n t}\,[B_1 e^{\omega_{0D} t}(\omega_{0D} - \xi\omega_n) - B_2 e^{-\omega_{0D} t}(\omega_{0D} + \xi\omega_n)] \qquad (2)$$

Substituting $B_1 = B_2$ and putting $t = 0$;

$$B_2 = 7.4(10)^{-3}$$

$$B_1 = -7.4(10)^{-3}$$

Substituting all the values

$$y(t) = e^{-6.75(1)1}\,[-7.4(10)^{-3}\,e^{6.675\times 1} + 7.4(10)^{-3}\,e^{-6.675\times 1}]$$

$$y(t) = 6.85 \text{ mm}$$

The response of the problem is indicated in Fig. 9.60.

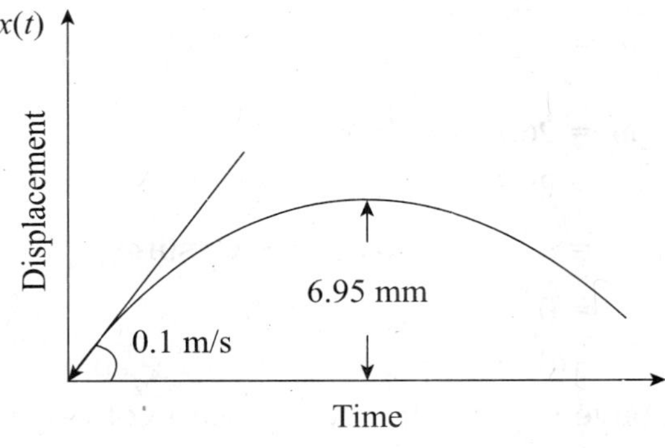

FIG. 9.60

Example 9.20 A mass of 70 kg is mounted on a spring for which the total stiffness is 50 kN/m and the total damping is 1.2 kNs/m. Find an expression for the motion using the following conditions $x(0) = 0$ and $\dot{x}_0(0) = 100$ mm/sec.

Solution: The given mass-spring-dashpot system is shown below.

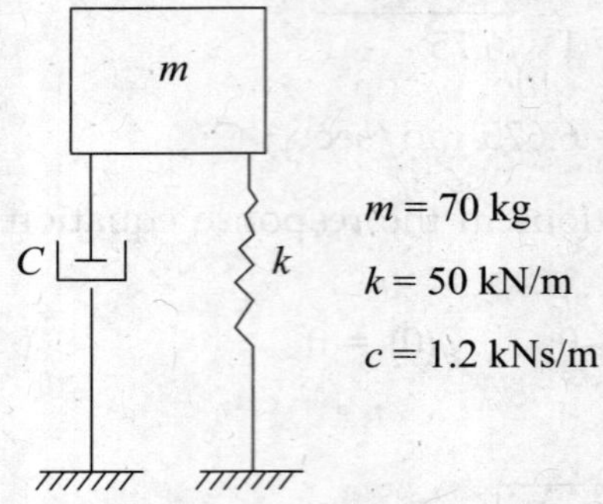

FIG. 9.61

$$\xi = \frac{C}{C_{cr}} = \frac{1.2}{2\sqrt{km}} = \frac{1.2(10)^3}{2\sqrt{50 \times 70 \times 10^3}} = 0.32 < 1.0$$

For an underdamped system

$$\omega_n = \left(\sqrt{1 - \xi^2}\right)\omega_n$$

$$= \sqrt{1 - 0.33^2 \omega_n}$$

$$\omega_{nD} = 0.947\ \omega_n$$

$$\omega_n = \sqrt{\frac{k}{m}}$$

$$\omega_n = \sqrt{\frac{50(10)^3}{70}}$$

$$\omega_n = 26.73 \text{ rad/sec}$$

$$\omega_{nD} = 25.3 \text{ rad/sec}$$

$$x(t) = e^{-\xi\omega_n t}\,[C_1 \cos\omega_{nD} t + C_2 \sin\omega_{nD} t]$$

At $t = 0; \quad x(0) = 0$

$$x(0) = C_1 = 0$$

Differentiating the above expression and the value of constant is obtained as

$$C_2 = \frac{x_0\xi\omega_n + \dot{x}_0}{\omega_{nD}}$$

$$\therefore \quad C_2 = \frac{0 + 0.1}{25.3}$$

$$C_2 = 3.95(10)^{-3}$$

$$\boxed{y(t) = e^{-8.906t} \times 3.95 \times 10^{-3} \sin(25.3)t}$$

The presence of damping reduces the response quickly as seen in the above problem. In practical systems, if external dampers are provided, the response decreases very quickly. For the same problem, the response can also be expressed in terms of amplitude and phase angle.

$$A = \sqrt{C_1^2 + C_2^2}$$

$$A = C_2 = 3.95(10)^{-3} \text{ m}$$

$$\boxed{C_2 = 3.95 \text{ mm}}$$

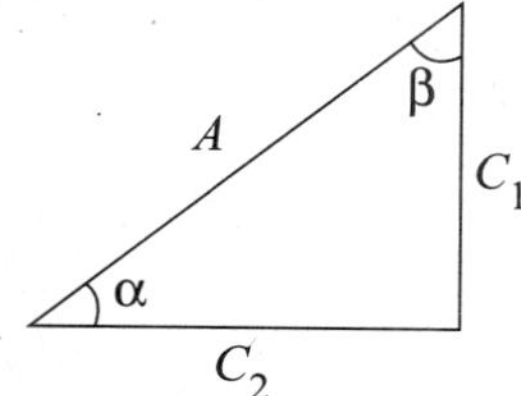

FIG. 9.62

$$\cos\beta = \frac{C_1}{A} = 0$$

$$\beta = 90°$$

A and C_2 are coinciding, it is obvious from the vector diagram that $\alpha = 0$ and $\beta = 90°$ clearly stating that the amplitude coincides with C_2.

$$y(t) = Ae^{-\xi\omega_n t} (\cos\omega_n t - \beta)$$

$$y(t) = 3.95(10)^{-3}\, e^{-8.517t} \sin 25.3t$$

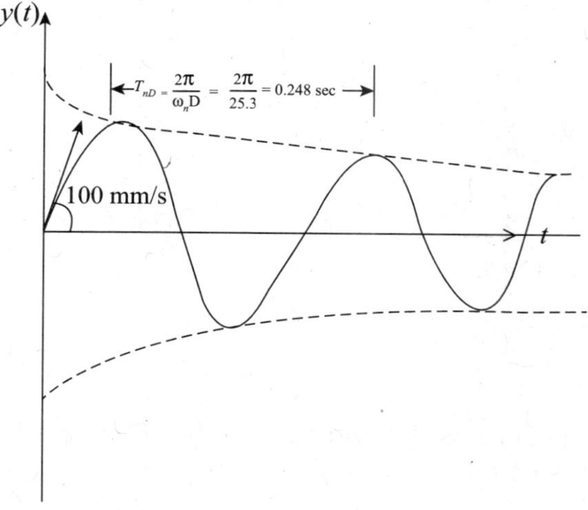

FIG. 9.63

Example 9.21 A steel portal frame shown in Fig. 9.64 is subjected to free vibration while giving an initial displacement without velocity. Taking damping as 4% of critical, find the characteristics of motion.

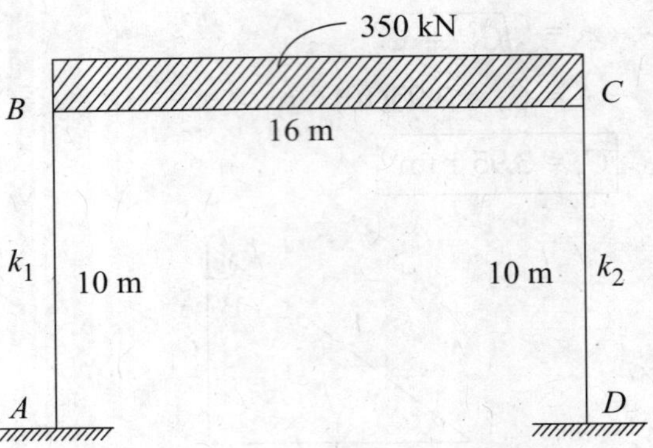

FIG. 9.64

Solution: Let the horizontal deflection be δ at the top of the member fixed at both ends. The moments and forces developed at the supports is shown below. For such a member, the stiffness is $(12EI/l^3)$ for a unit displacement at the top support. The above frame consists of two vertical columns and a rigid beam.

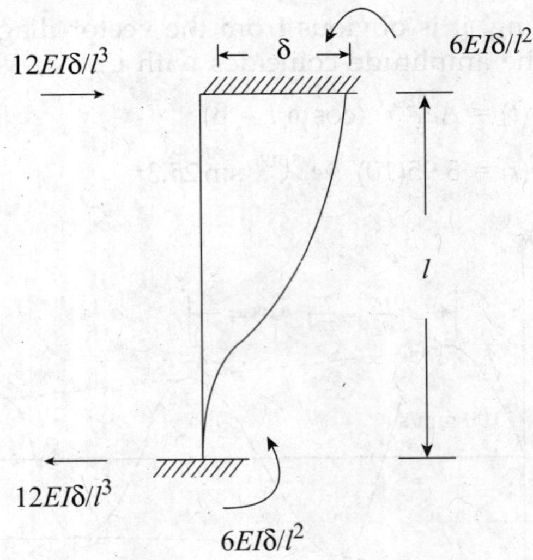

FIG. 9.65

The stiffness of two columns AB and CD can be considered two springs connected in parallel. The stiffness of the columns is

$$k_{AB} = k_{CD} = \frac{12EI}{l^3}.$$

and hence the equivalent stiffness in the columns is

$$k_{eq} = \frac{12(EI)_{AB}}{L_{AB}^3} + \frac{12(EI)_{CD}}{L_{CD}^3}$$

where L_{AB} and L_{CD} are the lengths of the members AB and CD respectively.

$$k_{eq} = \frac{24EI}{l^3}$$

$$k_e = \frac{24(2 \times 10^5)I}{10000^3}$$

$$\boxed{k_e = 4.8(10)^{-6}I}$$

$$\omega_n = \sqrt{\frac{k_e}{m}}$$

$$\omega_n = \sqrt{\frac{4.8(10)^{-6}\, I}{(350/9.81)}}$$

$$\omega_n = 3.66(10)^{-4}\sqrt{I}$$

Assume the value of I for each column = $9.28(10)^8$ mm^4.

$$\omega_{nD} = 11.07 \text{ rads/sec}$$

$$\omega_n = 11.1 \text{ rads/sec}$$

We know

$$x(t) = e^{-\xi\omega_n t}\,[C_1 \cos\omega_{nD}t + C_2 \sin\omega_{nD}t]$$

At $\qquad t = 0; \qquad x(0) = x_0 \Rightarrow C_1 = x_o$

At $\qquad t = 0; \qquad \dot{x}(0) = 0 \Rightarrow C_2 = 0.04x_0$

$$x(t) = e^{-0.04(11.1)t}\,[x_0 \cos(11.07)t + 0.04x_0 \sin(11.07)t]$$

$$\boxed{x(t) = e^{-0.445t}\, x_0\,[\cos(11.07)t + 0.04 \sin(11.07)t]}$$

9.12 FORCED VIBRATIONS

Free vibration analysis is carried out to determine the natural frequency and period of the given structural system. Different types of dynamic loads acting on the systems will result in response which have to be evaluated. The response is calculated depending on the type of dynamic load acting on the system.

The governing differential equation for forced vibration analysis is expressed as

$$m\ddot{x} + c\dot{x} + kx = F(t)$$

The type of analysis depends on the external dynamic force which can be harmonic, period or general loading. Harmonic excitations are analyzed by calculating the particular integral by assuming them as sine or cosine function. The periodic loadings are analyzed using the concept of Fourier series, while the general dynamic loading is analyzed using Duhamel's integral.

If closed form solutions are not possible, numerical methods are adopted. Large systems are analyzed based on matrix analysis.

REVIEW QUESTIONS

Remembrance

9.1. Define vibration.
9.2. What is free vibration?
9.3. What is forced vibration?
9.4. Mention the elementary parts of a vibration system.
9.5. What is degree of freedom?
9.6. What is natural frequency?
9.7. What is stiffness?
9.8. What is linear vibration?
9.9. What is non-linear vibration?
9.10. What is resonance?
9.11. What is periodic motion?
9.12. Distinguish between periodic and non-periodic vibration?
9.13. What do you mean by single degree freedom system?
9.14. How many degrees of freedom a continuous structure has?
9.15. Explain period and frequency?
9.16. What is damping?
9.17. What is the range of coefficient of viscous damping in a single degree freedom system?
9.18. What is damped natural frequency?

Understanding

9.1. What is initial value problem?
9.2. Name the diagram isolated from all other bodies showing the external forces in the system including the initial force?
9.3. The value of damping coefficient for real structure is less than critical damping. Is it true?

9.4. Dynamic response indicates the displacement time history. Is it true?

9.5. If external force is zero, the resulting motion is called free vibration? Is it true?

9.6. All structures can be reduced to systems consisting of springs, masses and dashpot system. Is it true?

9.7. Give examples of periodic loading.

9.8. What is the use of logarithmic decrement?

EXERCISE PROBLEMS

9.1. Determine the natural frequencies of the following systems shown below. Assume $E = 2(10)^5$ MPa.

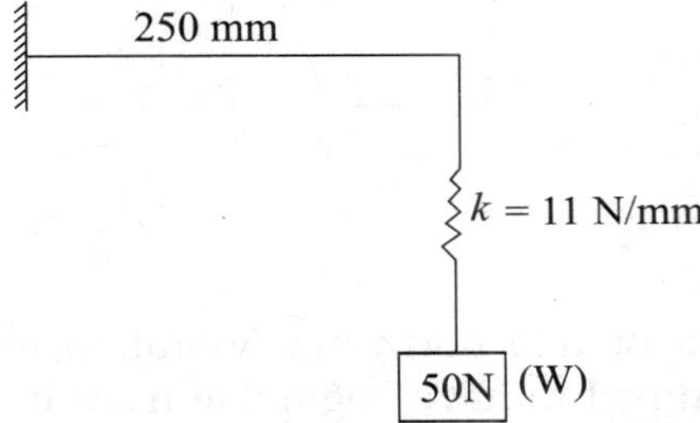

The cross section of the beam is 30 mm × 10 mm.
(**Ans:** k_e = 9870 N/m; ω_n = 44 rad/sec; T_n = 0.143; secs).

9.2. Determine the natural frequency of the following systems shown below. Assume $E = 2(10)^5$ MPa.

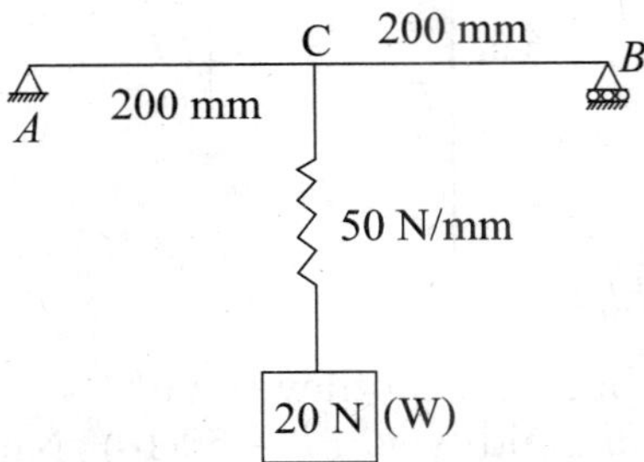

The cross section of the beam is 400 × 150 mm.
(**Ans:** $k_e = 50(10)^3$ N/m; ω_n = 156.6 rad/sec; T_n = 0.04; sec).

9.3. A diver weighing 60 kg stands at the end of a cantilever diving board of span 1.0 m. He oscillates at a frequency of 2 Hz. If $E = 1(10)^5$ N/mm^2 and the width is 450 mm, calculate the thickness of the diving board?
(**Ans:** 4.32 mm).

9.4. A single degree freedom system has an undamped natural frequency of 10 rads/s and a damping factor of 10%. The initial conditions are $x_0 = 0$ and $\dot{x}_0 = 0.7$ m/s. Determine the damped natural frequency and the equation of motion.
(**Ans:** 9.95 rad/s; $x(t) = 0.07 \quad e^{-t} \cos(0.95t - \pi/2)$)

9.5. For a single degree freedom, if $m = 10$ kg, $k = 10$ N/m and C = 4 Ns/m, find the damping factor, the logarithmic decrement and the ratio of successive amplitudes.
(**Ans:** 20 Ns/m, 1.28, 3.61)

9.6. Determine the effective stiffness of the springs of the figure shown below.

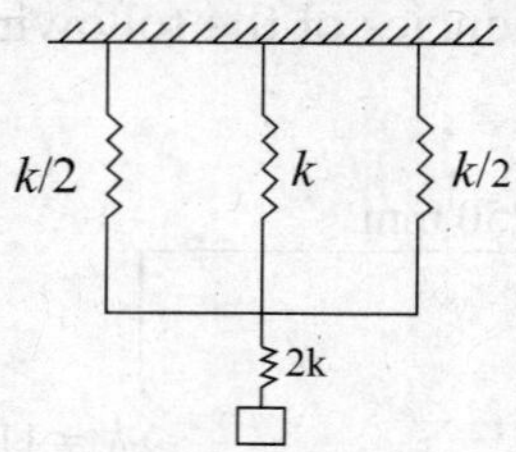

(**Ans:** k)

9.7. Determine the period of free horizontal vibration of the frame shown below. The columns are assumed to have negligible mass in comparison with the total mass of the girder.

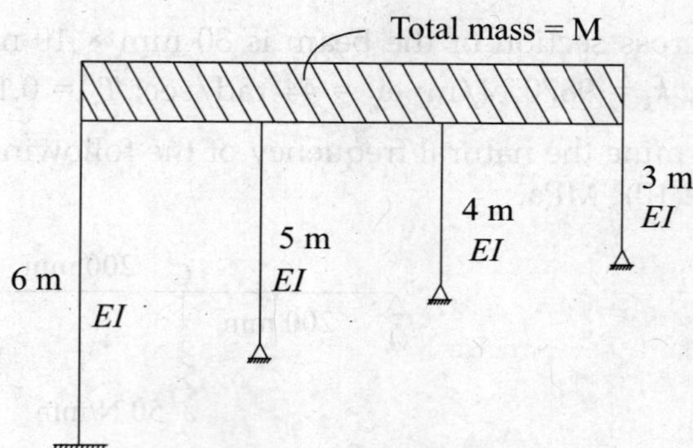

(**Ans:** $T = 12.89 \sqrt{M/EI}$)

9.8. A water tower is modelled as a cantilever with $W = 500$ kN. The height of tower is 45 m, and its flexural rigidity is $EI = 50(10)^9$ Nm^2. Find the natural angular frequency and natural period of vibration. If the motion is initiated in the horizontal direction by 2 mm, what is the displacement at one eighth of natural period of vibration?
(**Ans:** $\omega = 0.18$ rad/sec, $T = 34.9$ sec, $x = 1.41$ mm)

9.9. An oscillator with single degree freedom is set to vibrate. The initial displacement $x(0) = 1$. The ratio of initial displacement to the succeeding displacement is 1.20. $m = 25$ kg and $k_w = 20$ N/mm.
(**Ans:** $\omega = 0.894$ rad/sec, $\delta = 0.182$, $\xi = 0.029$, $c = 1.29$ Nsec/mm, $\omega_d = 0.893$)

References

1. Peter Marti, Theory of Structures, Wiley, Ernst & Sohn, 2013.
2. Seetharamulu Kaveti, Advanced Structural Analysis, Tata McGraw-Hill Education Pvt. Ltd., 2013.
3. James C. Anderson and Fazad Naeim, Basic Structural Dynamics, John Wiley and Sons, 2012.
4. Muthu KU, Azmi Ibrahim, Vijayanand M and Maganti Janardhana, Basic Structural Analysis, I.K. International Publishing House Pvt. Ltd., 2011.
5. Ghali A, Nevile AM and Brown TG, Structural Analysis, A Unified Classical and Matrix Approach, Spon, London, 2009.
6. Pandit GS and Gupta SP, Structural Analysis: A Matrix Approach, Tata McGraw-Hill Education Pvt. Ltd., 2008.
7. Devadass Menon, Structural Analysis, Narosa Publishing House, New Delhi, 2008.
8. Ashok K. Jain, Advanced Structural Analysis, Nemchand Bros., Roorkee, 2006.
9. Thandavanmoorthy, Analysis of Structures, Oxford University Press, 2005.
10. Krishna Raju N and Muthu KU, Numerical Methods in Engineering Problems, McMillan India Ltd., New Delhi, 2003.
11. Sarwar A Raz, Analytical Methods in Structural Engineering, New Age International Publishers, 2001.
12. Madhujit Mukhopadhyay, Vibrations, Dynamics and Structural Systems, Balkema/ Rotterdam 2000.
13. Bhavikatti SS, Structural Analysis Vol. II, Vikas Publishing House Pvt. Ltd., 1999.
14. Negi LS and Jangid RS, Structural Analysis, Tata McGraw-Hill Pvt. Ltd., New Delhi, 1997.
15. Ramamurutham and Narayan R, Theory of Structures, Dhanapat Rai Publishing Company, New Delhi, 1996.
16. Prakash Rao DS, Structural Analysis, University Press (India) Pvt. Ltd., 1996.
17. Norris CH, Wilbur JB and Utku S, Elementary Structural Analysis, McGraw Hill Book Company, Singapore, 1991.
18. Wang CK, Intermediate Structural Analysis, McGraw Hill International Students Edition, 1989.
19. Behr RA, Goodspeed CH and Henry RM, Potential errors in approximate methods of structural analysis, J. *Struct. Engg*, ASCE, 115(4), 1989, 1002 - 1005.
20. Epstein HI, Approximate location of inflection points, J. *Struct. Engg*, ASCE, 114(6), 1988, 1002 - 1005.
21. Beaufeit FW, Basic Concepts in Structural Analysis, Prentice Hall, Englewood Cliffs, NJ, 1987.

22. Mariopaz, Structural Dynamics, Van Nastrand Ranihold Company, 1985.
23. Jindal RL, Inderminate Structures, S Chand and Company Ltd., 1984.
24. John T. Dewolf, Preliminary member sizes: Rigid frames, J. *Struct. Div*, ASCE, 107(5), 1981, 987 – 992.
25. Moshe F Rubinstein, Matrix Computer Analysis of Structures, Prentice Hall Inc, Engle Wood Cliffs, NJ, 1980
26. Laursen, Structural Analysis, McGraw Hill, New York, 1978.
27. Brain J Bell, Advanced Theory of Structures, ELPS, MacDonald and Erans, 1978.
28. Brohn DM and Cowan J, Teaching towards an improved understanding of structural behaviour, *The Structural Engineer*, 55(1), 1977, 9–17.
29. Mohan Kalani, Basic Concepts and Conventional Methods of Structural Analysis, IITB, 1973.
30. Vazirani and Rutwani, Analysis of Structures, Khanna Publishers, 1972.
31. Darkov and Kuznetsov, Structural Mechanics, Mir Publishers Moscow, 1970.
32. Matheson JAL, Hyperstatic Structures, ELBS Butterworth & Co, London, 1966.
33. Nayagam KPS, Theoretical Problems for Structural Designers, Batsford Ltd., London, 1962.
34. Junnarkar, Mechanics of Structures, Vivek Publications, Bombay, 1961.
35. Borg SF and Gennaro JJ, Advanced Structural Analysis, D Van Nostrand Company, Princeton, NJ, 1960.
36. Pippard AHS and Baker JF, The Analysis of Engineering Structures, Edward Arnold & Co, London, 1953.

Index